MW01140089

Woody-Plant Seed Manual

PREPARED BY THE FOREST SERVICE

U. S. DEPARTMENT OF AGRICULTURE

Miscellaneous Publication No. 654   Issued June 1948

Reprint of First Edition, Published 1948 by the U.S. Department of Agriculture

Woody-Plant Seed Manual:
Miscellaneous Publication No. 654, U.S. Department of Agriculture

ISBN: 1-930665-63-6

Library of Congress Card Number: 2002094157

THE BLACKBURN PRESS
P. O. Box 287
Caldwell, New Jersey 07006
U.S.A.
973-228-7077
www.BlackburnPress.com

ACKNOWLEDGMENT

THIS MANUAL brings together information on all phases of seed handling and presents the results of more than 20 years of studies. Although extensive literature in this field was reviewed critically and authentic information from nearly 400 published articles drawn upon, the manual is in large measure based on unpublished information.

Forest Service field personnel at several experiment stations and regional offices furnished a backlog of source material for treatments of individual genera. Special mention should be made of the Lake States Forest Experiment Station which supplied important data on seed extraction, storage, and germination behavior for some 260 species, and the California Forest and Range Experiment Station for 85 species, of the 444 species and varieties finally included. The Southern Forest Experiment Station also made a considerable contribution to this information. Acknowledgment is made to the Soil Conservation Service, the Bureau of Plant Industry, Soils, and Agricultural Engineering, and to some State agencies, chiefly the Wisconsin and New York Conservation Departments, and Louisiana State University, for their additions to the hitherto unpublished information.

The Lake States Forest Experiment Station directed the preparation of the manual. The individual sections were prepared by personnel of the various forest and range experiment stations as follows: Seed and Its Development, Eugene I. Roe, Lake States; N. T. Mirov, California; and Arthur G. Chapman, Central States. Seed Production and Dispersal, William E. McQuilkin, Northeastern. Source of Seed, Paul O. Rudolf, Lake States. Collection, Extraction, and Storage, Philip C. Wakeley, Southern; and Paul O. Rudolf. Treatment of Seed Prior to Sowing, Philip C. Wakeley and Paul O. Rudolf. Seed Testing, Eugene I. Roe and Philip C. Wakeley. The seed-testing technique described was based on work of C. G. Bates, Lake States Forest Experiment Station, and P. C. Wakeley. The authors of part 1 with assistance on some species by R. K. LeBarron, Lake States Forest Experiment Station, collaborated in the preparation of the species descriptions which make up part 2. H. L. Shirley, formerly of the Lake States Forest Experiment Station and now assistant dean of the New York State College of Forestry, outlined the scope and began the compilation of material. Henry I. Baldwin of the New Hampshire Forestry Department was a valued collaborator during the early stages of this publication. Raphael Zon, former director, Lake States Forest Experiment Station, made many helpful suggestions on presentation.

The line illustrations were drawn by Leta Hughey of the Washington Office, United States Forest Service, W. H. Lindemann and A. Kuban of the Lake States Forest Experiment Station, and N. T. Mirov and Helen M. Dille of the California Forest and Range Experiment Station. Many others have also contributed to the preparation of the manual. It was reviewed in its final form, in whole or in part, by H. L. Shirley and by Paul J. Kramer, professor of botany, Duke University. Suggestions and criticisms concerning the botanical material have been offered by W. A. Dayton, Elbert L. Little, Jr., and Doris W. Hayes of Dendrology and Range Forage Investigations, Washington Office, United States Forest Service. Dr. H. P. Brown of the New York State College of Forestry reviewed all seed and seedling drawings in this manual.

The manuscript of this manual was completed for publication at the beginning of World War II. In the main, the bibliography includes nothing published since 1941.

CONTENTS

PART 1

PART 2

PART 1
INTRODUCTION

FROM 1929 to 1939 revegetation of cut-over and burned lands and eroded soils for timber production, watershed protection, wildlife food and cover, and shelterbelt planting was undertaken on a scale never before approached in the United States. In 1929, for example, about 111,000 acres were reforested. In 1939 about 484,000 acres were planted, and the total acreage planted at the close of that year was about 4,700,000 acres.

Planting at the 1939 rate requires annually about 145 tons of forest seed, valued at about $500,000. With more than 45,000,000 acres in need of planting, assuming that the job can be done in the course of 25 years, this would call for the annual planting of some 1,800,000 acres and require the use of about 600 tons of seed per year. Whereas reforestation was drastically curtailed during World War II, forest planting of large areas doubtless will figure prominently as a postwar measure.

Revegetation has been practiced for more than a century in the Old World, and for a few decades in the New. Hence a large fund of knowledge has been built up, chiefly on an empirical basis. Extensive as it is, however, this information is adequate only for about a dozen European species and perhaps an equal number of North American species, most of them conifers. If, in the past, reforestation was confined to comparatively few species, today, because of the multiplicity of purposes for which planting is being done, hundreds of species are in use.

All this has created a need for reliable information—information based on field practices and laboratory tests as to the time of seed ripening and therefore the proper time for its collection, behavior of the seed after planting (whether it comes up immediately or must go through a period of dormancy), viability, storage of seed in case of species which do not bear crops frequently, and pretreatment to induce prompt germination. Insufficient knowledge of any part of this chain may nullify the best of practices in others.

The need for such information became particularly acute after the planting of shelter belts was begun on a large scale in the Prairie-Plains region, large-scale revegetation was undertaken by the Soil Conservation Service in erosion control, and planting of shrubs and trees became widespread in connection with the growing activities of wildlife management. The necessary use of several hundred species of trees and shrubs about the seed handling of which little was known has resulted in many cases in failures and disappointments.

In the purchases of agricultural seed, much of the uncertainty has been eliminated through certification; in some cases, such as that of alfalfa, the law requires that seed of foreign origin must be distinctively stained before it may be imported into this country.

Trade in forest seed, however, is on a very uncertain basis. It is difficult to obtain seed of known origin or true botanical identity, or of known age, purity, and viability, except from a few well-known dealers. Buying seed on the market is consequently somewhat of a gamble.

To meet the need for more reliable information, the United States Forest Service has prepared this seed manual. The best information on all phases of seed handling has been collected as a guide to the most effective and economical seed and reforestation practices.

The manual consists of two main parts. Part 1 formulates general principles on the various phases of seed handling from formation of the seed to sowing. Part 2, which forms the larger part of the manual, provides relatively detailed but concise information for 444 species and varieties of trees and shrubs; this includes data on distribution and use, discussions of seeding habits, methods of seed collection, extraction and storage, seed germination, and nursery and field practice. Although many of these species or varieties are not yet in wide use, all are of potential value for conservation planting.

This manual is, in a sense, the first step toward organization of the whole tree-seed enterprise on a sound and scientific basis, and for this reason it should be of interest not only to public agencies but also to seed dealers, horticulturists, and commercial nurserymen.

SEED AND ITS DEVELOPMENT

MOST vegetation used by man, whether natural or planted, is propagated from seed. An understanding of the development and behavior of the seed, therefore, is essential to any activity directed toward revegetating the land.

Seed plants (spermatophytes) are divided into two main groups. Seedsmen deal extensively with seeds of both. The conifers (pines, firs, spruces, hemlocks, junipers, yews, etc.) have naked seeds borne usually in cones and are classed as gymnosperms. The true flowering plants (oaks, elms, willows, poplars, maples, viburnums, etc.) have the seeds enclosed in a fruit and are called angiosperms. Thus, seeds may develop either from cones or from true flowers. In this manual the young cones of gymnosperms are considered as flowers under the usage of foresters, and both types of seed development are discussed together.

FLOWERING

A general knowledge of the characteristics and development of flowers is of considerable value to the seed collector in making estimates of seed crops.

Typically, a flower (fig. 1) consists of the following parts: flower stalk (peduncle), the receptacle (the enlarged end of the peduncle, to which the other parts are attached), the calyx (composed of sepals), corolla (composed of petals), stamens (composed of anthers and filaments), and one or more pistils or carpels (composed of the stigma, style, and ovary).

FIGURE 1.—Structure of a complete flower: *A*, Face view showing the calyx of 5 sepals, the corolla of 5 petals, the 10 stamens, and the pistil. *B*, Longitudinal section showing the relation between the parts: *a*, Receptacle; *b*, sepal; *c*, petal; *d*, stamen; *e*, pistil, with ovary cut lengthwise, exposing the ovules. (After Sinnott, *Botany: Principles and Problems.*)

In the flowers of some trees and shrubs, one or more parts are lacking, such as the corolla (in sweetgum and walnut), calyx and corolla (in willow), and stamens in some flowers and pistils in others (as in alder, birch, and hickory). In still other species all the floral parts are present, but instead of being separate they are more or less united.[1]

Flowers of trees and shrubs vary considerably in size, ranging from the minute, inconspicuous ones of the willows to the showy ones of the magnolias, which are several inches in diameter. There is also considerable variation in color. Many species have attractive blossoms, but in others the flowers are clearly distinguishable from the foliage only at close view.

From the standpoint of the man dealing with seed, the only floral parts that warrant much consideration are the stamens and pistils. These produce, respectively, the sperm or male nuclei and the egg or female nuclei which, when united, lead to seed formation. The primary function of the calyx and corolla, both of which are modified leaves, is to protect the rather delicate stamens and pistils until they are mature. The calyx and corolla in many species also attract by their brilliant coloring, odor, and nectar supplies, the bees, flies, moths, and other insects which play a large part in pollination—which is the transfer of pollen from the stamens to the pistils of the same or different flowers. Trees and shrubs which lack a corolla, or both calyx and corolla, are usually wind-pollinated (poplar, walnut).

The flowers of gymnosperms, represented in this manual by conifers, more properly are cones, because the floral parts, such as calyx, corolla, stamens, and pistil, are lacking. Instead, the male and female elements are borne in separate cones, consisting of crowded scales on an axis, both kinds of cones usually on the same plant. The male flowers, or staminate cones, of conifers are small, yellowish, numerous, and short-lived. They produce great quantities of yellow pollen grains which may be carried long distances by the wind. The female flowers, or ovulate cones, fewer in number and usually small, have scales bearing naked ovules, which develop into seeds.

Flower Arrangement

The fruit is borne upon the parent plant in much the same manner as the flowers. Therefore, a knowledge of floral arrangement is useful in helping the collector form an estimate of potential seed crops and working out collection and extraction techniques. Generally speaking, flowers are arranged in one of two ways—as solitary individuals on a single peduncle, usually occurring within the axils of leaves (magnolia, pawpaw, yellow-poplar),

[1] Sepals (buttonbush, viburnum), petals (catalpa, honeysuckle), stamens (baccharis, lupine), pistils (grape, rhododendron).

or in clusters. The cluster is the more common and occurs in a variety of forms, of which the simpler, shown in figure 2, are raceme (black cherry, currant, striped maple), catkin (alder, birch, poplar), spike (amorpha), head (baccharis, buttonbush), cyme (American elder, viburnum), [2] umbel (wild-sarsaparilla), and panicle (buckeye, creeper, madrone).

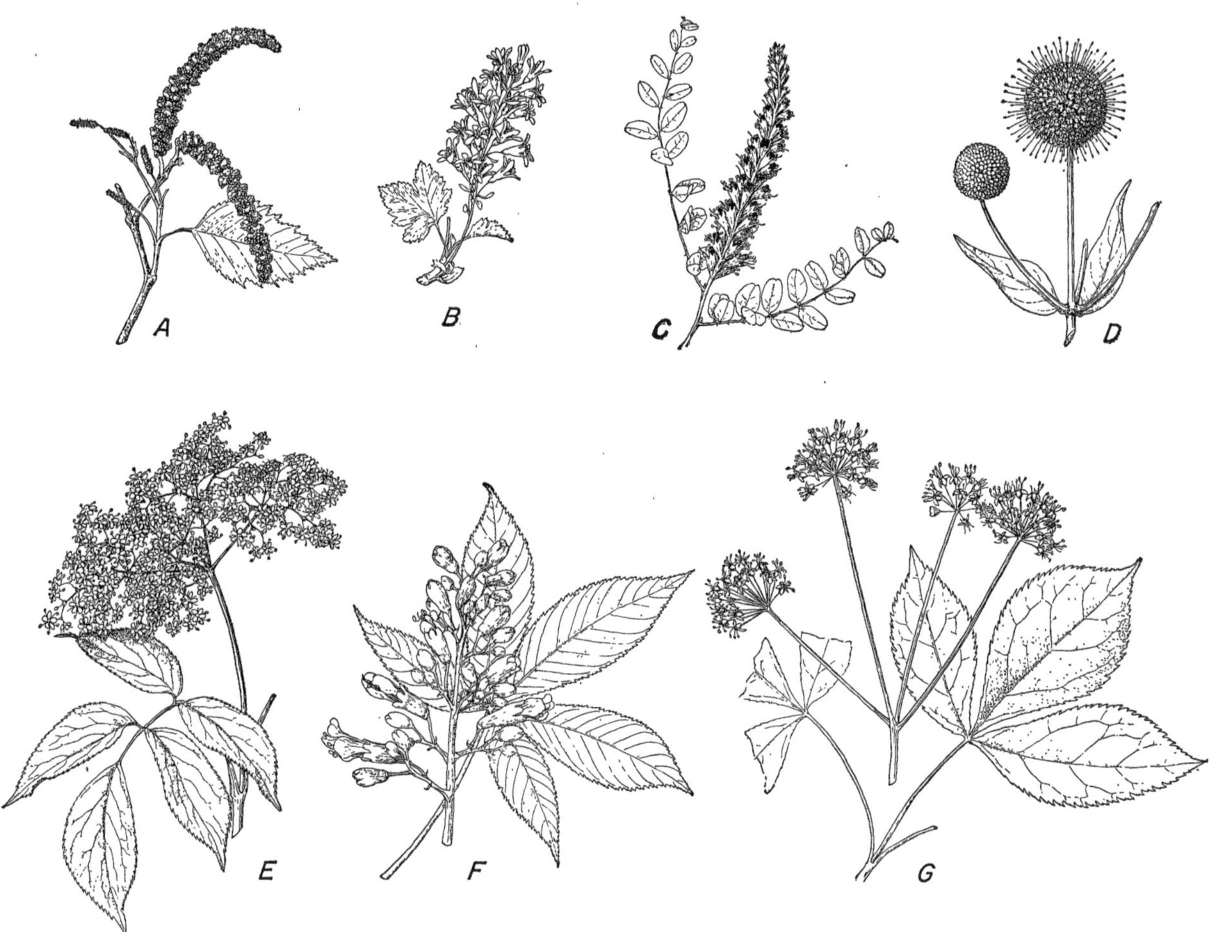

FIGURE 2.—Common types of flower clusters: *A,* Catkin (*thinleaf alder*); *B,* raceme (*winter currant*); *C,* spike (*California amorpha*); *D,* head (*common buttonbush*); *E,* cyme (*American elder*); *F,* panicle (*woolly buckeye*); *G,* umbel (*wild-sarsaparilla*).

To avoid overestimating the size of a potential seed crop during the blossoming period, a knowledge of the flower-seed relationships described below is of prime importance to the seed collector.

Typically, a flower bears both stamens and pistils, and hence is bisexual or perfect. Plants with perfect flowers can produce seeds. Many species, however, particularly trees, bear unisexual or imperfect flowers. When staminate (male) flowers and pistillate (female) flowers, or in conifers the staminate cones and ovulate cones, are borne on the same plant as in alder, birch, striped maple, and most conifers, including Douglas-fir, fir, larch, and pine,[3] the species is described as monoecious. When the staminate flowers and the pistillate flowers (or staminate and ovulate cones) occur on separate plants (as in persimmon, poplar, willow, and most species of juniper), the species is termed dioecious. Of course, the plants which produce only staminate flowers will never bear seed. In some species (for example, buckeye) perfect flowers as well as both staminate flowers and pistillate flowers are found on the same plant; these are termed polygamo-monoecious. Still other species (for example, bittersweet, common buckthorn, Dahurian buckthorn, and soapberry), bear perfect flowers and, in addition, either staminate flowers or pistillate flowers on the same plant; these are described as polygamo-dioecious. Both polygamo-monoecious and polygamo-dioecious plants can produce seed.

In some monoecious species (Norway maple) the

[2] A corymb resembles a cyme, but the order of blooming is from the outside of the cluster toward the center instead of the opposite.

[3] The two kinds of flowers may be borne on different parts of the tree (in firs, the staminate cones are confined to the upper half of the crown, the ovulate cones to the tip of the crown), or in different parts of the same cluster (in mountain maple the staminate flowers occur at the apex of a many-flowered compound raceme, the pistillate toward its base).

staminate flowers mature before the pistillate on some individuals, while in others the pistillate mature first. Such an arrangement, which brings about cross-pollination and tends to produce progeny superior in vigor to that from self-pollinated plants, is also found in many species with perfect flowers (amorpha, hawthorn, pawpaw). In the latter case the stamens may shed pollen before the pistil is ready to receive it, or the pistil may mature before the pollen can be supplied by the stamens.

POLLINATION AND SEED FORMATION

Seed formation involves pollination and fertilization and the production of a new, though rudimentary, plant (embryo) with stored food and a protective covering. In the angiosperms the seeds are borne within an enclosed ovary, the enlarged lower part of the pistil (see fig. 1). When a pollen grain is carried by wind, insects, or gravity from an anther (pollen-bearing part of a stamen) to a sticky stigma (apex of a pistil) in the process of pollination, it germinates, producing a long, microscopic tube that grows down through the style into the ovary. Here the pollen tube containing two male nuclei or sperms penetrates an ovule, or rudimentary seed, and double fertilization occurs. Each ovule contains within its embryo sac eight nuclei, among them an egg nucleus and two polar nuclei. In the process of fertilization one male nucleus unites with the egg nucleus, the fertilized egg developing into the embryo, or young plant, of the seed. The other male nucleus unites with the two polar nuclei, and the fused nucleus develops into the endosperm, a food-storage tissue for the growing embryo or the young seedling which arises from it. One or two seed coats (integuments), usually thickened and hard, are formed on the outside of the ovule, but a small pore, the micropyle, remains. The mature ovule with all its parts now is a seed. The ripened ovary, containing the seeds, composed of the usually thickened ovary wall (pericarp) and any other closely associated parts, is known as a fruit. Sometimes other floral parts, such as the calyx (wintergreen) or receptacle (apple and pear), are adherent to the ovary and indistinguishable from the pericarp and are considered part of the fruit.

In the gymnosperms the seeds are not enclosed in a fruit but are borne loosely between the flattened scales of the ovulate cones. During pollination the wind-borne pollen grains come in direct contact with the exposed ovules. Though the minute details differ from those in angiosperms, the egg nucleus of the ovule is fertilized by a male nucleus from the pollen, the fertilized egg growing into the embryo plant. The endosperm of gymnosperms is formed directly from the ovule and not as a result of a second fertilization. The female flower enlarges, often greatly, and becomes the familiar, hard cone (as in pines, firs, spruces), bearing the exposed or naked seeds. However, in some groups, such as junipers, the cone scales grow together to form a berrylike structure around the seeds.

In most species of trees and shrubs, fertilization occurs in the spring, and the seeds ripen relatively soon thereafter, usually in 3 to 6 months. In others, such as most of the pines, the egg is not fertilized until a year after pollination, the seed ripening in the fall of the second year. In three species, Chihuahua pine, Italian stone pine, and Torrey pine, the seeds ripen in the fall of the third year. The oaks also vary in the length of time required for seed ripening, white oaks mostly maturing their seeds in 1 year, and black oaks in 2 years.

SEED STRUCTURE

As already shown, a seed is a mature ovule. The fertilized egg has developed into an embryo, which is usually embedded within an endorsperm (in some cases the endosperm is very thin, or even absent); and the whole is enclosed in one or two hardened integuments or seed coats (fig. 3).

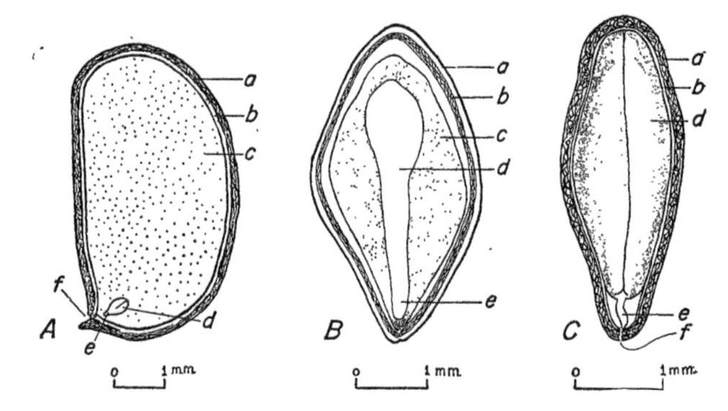

FIGURE 3.—Longitudinal view of typical seeds. *A, Wild-sarsaparilla,* a seed with large endosperm and minute embryo; *B, eastern hemlock,* a seed with large embryo surrounded by endosperm; and *C, serviceberry,* a seed with no endosperm, the embryo practically filling the seed cavity. *a,* Outer seed coat, or testa; *b,* inner seed coat; *c,* endosperm; *d,* cotyledon; *e,* radicle; *f,* micropyle.

The Embryo or Rudimentary Plant

The embryo, sometimes called the germ, is composed of seed leaves (cotyledons), bud (plumule), stem (hypocotyl), and rudimentary root (radicle); it is a plant in miniature. These parts can often be readily distinguished by dissection.

Most species discussed in this manual have 2 cotyledons and belong to the class of angiosperms designated accordingly dicotyledons. The other class of angiosperms, the monocotyledons, which is not represented in this manual, includes plants with 1 cotyledon, such as palms, grasses, and lilies. In the conifers the number of cotyledons ranges from 2 (arborvitae, juniper, yew family) to 15 (Douglas-fir, 6–12; fir, 4–10; hemlock, 3–6; pine, 4–15).

Cotyledons may manufacture food for the developing seedling until true leaves are formed, or they may contain stored food which is used to nourish the developing seedling for a large part of its early life. In the former case, the cotyledons are thin, of medium to large size, and resemble true leaves, because of the presence of chlorophyll, which enables them to synthesize food. Such cotyledons always emerge above ground and become green; germination of this type is called epigeous (dogwood, red maple).

Cotyledons which serve as food-storage organs are thick and fleshy and do not resemble true leaves. Germination of seeds with this type of cotyledons may be epigeous (beech, locust), in which case the cotyledons after emergence tend to become greenish in color and manufacture some food; or it may be hypogeous, the thick, often hemispherical cotyledons remaining below the ground for weeks or months (buckeye, oak, silver maple, walnut).

The plumule or embryonic bud, readily visible in some species, but microscopic in others, is the source of stem elongation.

The hypocotyl or stem of the embryo points toward the micropyle, and arches above the ground in species with epigeous germination. At the lower end of the hypocotyl is the radicle, or rudimentary root.

Endosperm

As already noted, the cotyledons in seeds of some species serve as food-storage organs. In many others, however, part of the reserves of carbohydrates, fats, and proteins which furnish the energy for germination and early growth of the seedling are stored in a separate tissue—the endosperm. The amount of endosperm relative to the size of the embryo varies widely among species. At one extreme are those seeds in which endosperm is lacking (locust, oak), or present in so thin a layer as to be readily overlooked. Such seeds have comparatively large embryos in which the parts are readily distinguishable, with most or all of the food reserves carried in the cotyledons. At the other extreme are those seeds in which the endosperm contains most or all the food reserves and forms a large part (barberry, pine) or the bulk of the kernel (currant and gooseberry, yew); the embryo in such seeds typically is especially small or minute.

Seed Coats

The chief function of the seed coats is to protect the embryo from injury by crushing, drying out, or attack by insects, fungi, and bacteria until it can begin active growth and develop into a new plant capable of survival. The outer coat (usually called the testa) is typically hard and crustaceous; the inner coat is generally thin and membranous. Often the latter cannot be distinguished from the former except with a microscope.

Variations from the general type of seed coat are common. In some species, such as fir, poplar, and willow, the testa is relatively soft, in fact so soft in fir as to make mechanical removal of the seed wings injurious. In other species, the testa is hard and bony (hawthorn, holly, rose), or contains cutinized layers (locust), which interfere with germination. In still others, both seed coats may be membranous (elm), the outer membranous and the inner fleshy (maple), or the outer part of the testa soft and fleshy and the inner part hard and somewhat bony (magnolia).

The seed coat in gymnosperms is simpler in structure than in angiosperms. The coat may be rather hard (pine),[4] soft (fir), or leathery (cypress). Some conifers have resin vesicles on the surface or within the seed coats (fir, hemlock, incense-cedars); this resin tends to make the seed sticky and somewhat hard to handle.

Sometimes the seed coat is expanded into a wing (catalpa, trumpetcreeper) or bears a tuft or coma of long soft hairs (poplar, willow). The wing in conifer seeds is considered by some plant anatomists to have been derived from the upper part of the cone scale on which the seed is borne. All such appendages are of considerable aid in seed dispersal.

TYPES OF SEED

From the standpoint of collection and extraction, seeds fall into three groups:

1. *True seeds,* readily extracted from dry fruits or cones. This group includes most conifers (fir, hemlock, larch, pine) and species bearing dehiscent fruits such as pods (honeylocust, locust, yellowwood) and capsules (fremontia, poplar, willow). Extraction of the seeds from such fruits usually involves drying by solar or artificial heat, followed by threshing or shaking. Commercial seed is almost invariably the true seed.

2. *Dry fruits,* with seed surrounded by a tightly adhering pericarp. These are of three main types: achenes, free of appendages (eriogonum), or retaining the feathery styles (clematis, cliffrose); nuts (chestnut, filbert, oak); and samaras, or key fruits (ash, elm, maple). Seeds of this group are rarely extracted from the fruit. For practical purposes the entire fruit is the seed, although sometimes the styles of some achenes and the wings of samaras are removed to reduce bulk and facilitate handling.

3. *Fleshy fruits,* such as accessory fruits (buffaloberry, wintergreen); aggregate fruits (raspberry); berries (barberry, currant, honeysuckle); drupes (cherry and plum, dogwood, walnut); multiple or collective fruits (mulberry, osage-orange); and pomes (apple, pear). In all cases, the seeds [5] can be easily extracted by macerating the fruit in water and allowing the fleshy pericarp to float away, or, as in the walnut, by removing the pericarp or husk in a corn sheller. Sometimes fleshy fruits are dried and sown without extracting the seeds. Seeds also may be removed by maceration from the berry-like cones of juniper and the fleshy arils of yew.

[4] Seed coats range from relatively thin in jack pine to very hard, thick, and woody in Digger pine.

[5] In the drupaceous fruits, such as cherry and walnut, what is commonly called the seed consists of the stony inner part of the pericarp (endocarp) as well as the true seed.

SEED RIPENING

As the seeds ripen on the plant a series of physical and chemical changes occur. Although the chemical changes are of paramount importance, they require complicated laboratory technique for determination, and thus cannot be used by the seed collector to indicate degree of maturity. Fortunately, however, chemical changes are often accompanied by easily recognized changes in color, taste, odor, and texture. Thus, fleshy fruits (berries, drupes, and the like) which depend on animal agencies for seed dispersal during ripening turn from green to such colors as red (barberry), orange (common persimmon), blue (alternate-leaf dogwood), or purple (serviceberry). The flesh, originally green, dry, sour, bitter, or astringent, becomes yellow or reddish, juicy, and often edible.

Other fruits, such as those which are wind-dispersed, usually change from green to straw color or brown (elm, maple, yellow-poplar). The seed coats usually darken (from yellowish to brown or black), and become harder and firmer as water is lost from the tissues. The embryo and the endosperm, when present, also lose water and change from milky and soft to white or cream color and firm. At this stage, the embryo is generally fully developed, except in such species as European ash and Swiss stone pine, where development apparently takes place after the seed has fallen.

All these physical changes are accompanied by alterations in chemical composition. Soluble organic compounds such as simple sugars, fatty acids, and amino acids are gradually converted into more complex carbohydrates, fats and oils, and proteins. Part of the soluble inorganic substances from the soil are incorporated into some of the organic compounds; other parts may remain inorganic, become insoluble, and help make up the ash content. Often minute quantities of some of the following organic substances are also present in the seed or fruit: organic acids, alkaloids, glucosides, tannins, hydrocarbons, pigments (anthocyanins, xanthophyll, carotene, chlorophyll), vitamins, hormones, enzymes, and essential oils.

Chemical Composition of Ripe Seed

The chemical composition of seeds of typical forest trees is presented in table 1. Note that there is considerable variation between species in all constituents. This is even true within genera. The pines are noteworthy for their high fat content, with some species high in protein as well. Acorns, on the other hand, are high in carbohydrates and relatively low in protein. Acorns of the two black oaks (northern red and scarlet oaks) have a much higher fat content than those of white oaks.

Knowledge of the chemical composition and nutritive value of seeds should be a great help in selecting plants to use as food for wildlife and forage for livestock.

Moisture Content

Table 1 shows the striking difference in moisture content between tree seeds. Pine seeds, with a low moisture content, can be stored for a relatively long time; but acorns, which have a high moisture content, cannot. The moisture content of most conifer seeds can be reduced even more without decreasing viability; in fact, pine seed with a low moisture content keeps much better, particularly at high temperatures, than seed which has not been well dried. Barton[6] has found that seed of grand fir, reduced from 12.4 percent to 5.8 percent moisture, still had about half its original viability of 98 percent after 11 years of sealed storage at 41° F.

On the other hand, seeds of many angiosperms, such as some species of buckeye, maple, and oak, show complete loss of viability if subjected to appreciable drying. White oak acorns (*30*) lose their germinability when the moisture content is 25 to

[6] Letter from L. V. Barton, Boyce Thompson Institute, Yonkers, N. Y., to Lake States Forest Experiment Station, St. Paul 1, Minn. February 10, 1941. [On file at station.]

TABLE 1.—*Chemical composition of seeds of various forest trees, based on dry weight*

Species	Ash	Protein	Fat	Carbohydrates		Total	Other substances[1]	Moisture content
				Soluble	Insoluble			
	Percent	*Percent*	*Percent*	*Percent*	*Percent*	*Percent*	*Percent*	*Percent*
Acorns 1 month old, without pericarp (*30*)[2]:								
Chestnut oak	2.26	8.50	4.57	14.83	32.20	47.03	37.64	91.2
Northern red oak	2.62	7.16	22.50	10.58	23.89	34.47	33.25	49.0
Scarlet oak	2.06	7.75	30.83	9.41	24.26	33.67	25.69	35.0
White oak	2.56	7.42	6.81	10.47	47.93	58.40	24.81	65.4
Without seed coat:								
Digger pine[3]	5.0	29.6	56.6	---	---	8 8	---	5.1
Swiss stone pine (*50*)	3.05	17.24	50.25	16.84	7.43	24.27	5.19	---
Pinyon[3]	2.9	15.1	64.1	---	---	17.9	---	3.4
Norway spruce (*44*)	4.74	19.12	35.31	5.43	7.00	12.43	28.40	---
With seed coat:								
Eastern white pine	5.58	30.21	35.44	1.99	2.82	4.81	23.96	7.53
Longleaf pine	5.39	35.24	31.66	2.38	2.15	4.53	23.18	6.87

[1] Include crude fiber, tannin, etc.
[2] Italic numbers in parentheses refer to Literature Cited, p. 51.
[3] Basis for original values (*56*) included water; present values adapted therefrom.

50 percent of the dry weight of the embryo. Freshly collected seeds of silver maple contain about 58 percent moisture; viability is completely lost when this falls below 30 to 34 percent (*28*).

It is readily apparent that moisture content may mean the difference between retention and complete loss of viability. Hence considerable effort must be made to dry seeds of some species before they are put into storage and to prevent others from becoming too dry while stored. Once the proper moisture content is reached, it can usually be maintained by storing the seed in moistureproof containers.

GERMINATION

The emergence of a seedling or embryo plant and its development to a point where it can maintain itself is called germination. Before germination can take place, two conditions are necessary: (1) the seeds must be ready to germinate; and (2) external factors must be favorable.

Some seeds, as soon as mature, will germinate readily and completely under favorable conditions,[7] but a substantial majority of the seeds of trees and shrubs, at least in the cooler regions, show little or no immediate response to these conditions. Such seeds are said to be dormant. The difficulty may be caused by seed coat impermeability (amorpha, honeylocust, locust); conditions within the seed (beech, birch, maple, walnut); or a combination of these (some species of serviceberry, ceanothus, basswood, rose, and sumac). The causes of dormancy and methods of controlling it are discussed under Treatment of Seed Prior to Sowing, page 31.

The following environmental conditions are required for ready germination: an abundance of water, favorable temperature, and sufficient oxygen. Sometimes light is necessary.

Moisture Requirements

The cells of the germinating seed cannot carry on the vital processes of absorption, digestion, food transfer, assimilation, respiration, and growth without an abundance of water. By softening the seed coats, water renders them more permeable to oxygen and carbon dioxide.

Water is absorbed by seeds in relatively large amounts. For example, it was found that eastern white pine seeds, with an initial moisture content of about 7 percent (dry-weight basis), did not begin to germinate until their moisture content reached about 45 percent. Absorption of water then increased rapidly. The newly germinated seedlings showed a moisture content of 172 percent of that of the dry seed, or 23 times as much.[8]

Water absorption may vary considerably within a genus. For example, after 155 hours of soaking, the moisture content of chestnut oak acorns rose from 91 to 145 percent (of dry weight) but of northern red oak acorns only from 49 to 63 percent (*30*).

The rate of water absorption is largely dependent upon the degree of seed-coat permeability. Some seeds absorb the water required for germination in a very short period; others take much longer. For instance, Scotch pine seeds will absorb their water requirements, 35 to 37 percent, in 48 hours, but yew seeds require as many as 18 days to reach this level (*31*).

Temperature

Seeds of many species germinate well, i.e., completely and rapidly, under a wide range of temperatures; others germinate poorly or not at all except at a specific and rather narrow temperature range. For instance, ponderosa pine seeds germinate well at temperatures fluctuating from 57° to 78° F., or kept constant at 70° and 80° (*6*). Seeds of Scotch pine and Norway spruce germinate most rapidly at 77°, although total germination is no higher than at lower temperatures (*23*).

Tests at the Lake States Forest Experiment Station have shown that seeds of some angiosperms germinate well only within a specific temperature range (*41*). For example, seeds of boxelder, hitherto a poor germinator in the laboratory, after having been stratified for 90 days at 41° F. to break internal dormancy, showed 67 percent germination at 50° to 77° (temperatures alternated diurnally), but only 12 percent germination at temperatures fluctuating between 68° and 86°, and 7 percent at a constant temperature of 50°. On the other hand, stratified seeds of American plum from northern Minnesota showed best germination at 50°; Norway maple seeds respond best at 41° to 50°; and American bittersweet seeds at 50° to 77°. Although definite proof is lacking, it is believed that these species germinate well under natural conditions only when temperatures prevail similar to those of early spring.

Some seeds germinate at relatively high temperatures. Haasis (*24*) found that pitch pine seeds germinate at temperatures up to 135° F. Such instances are undoubtedly rare, for extremely high temperatures may change the chemical composition of seeds, causing abnormal germination and plant development. For example, germination of longleaf pine becomes abnormal, reduced, or inhibited at temperatures above 80°.

Oxygen

When seeds begin to germinate, considerable energy is required for assimilation. The necessary energy is supplied by a process called respiration, which consists of the oxidation of sugars, derived in some cases from the digestion of other substances such as fats and proteins. Respiration is essentially the same in animals and plants. Seeds will not germinate if the supply of oxygen is inadequate, as for instance if the seed coat is not permeable to

[7] This is true of some species of alder, catalpa, a few legumes (peashrub, tesota), poplar, willow, and many pines. Many readily germinable seeds (poplar, willow) are extremely short-lived unless special provision is made for their storage.

[8] Based on unpublished data on seeding habits, seed handling, germination, and nursery practice for various trees and shrubs. On file at Lake States Forest Experiment Station, St. Paul 1, Minn.

gases, the germination medium is too wet, or the seeds are planted too deep. In the latter case the available oxygen in the soil is probably used up by bacteria and fungi living nearer the surface, and none is available to the seeds.

Light

It has long been supposed that germination, at least in some species, is influenced by light. Recent studies at the Southern Forest and Range Experiment Station, New Orleans, La., have indicated that the seeds of southern pines germinate more rapidly in diffused light than in total darkness. On the other hand, for many species, such as jack pine, germination in complete darkness may be 100 percent. In nurseries the influence of light is kept at a minimum and often completely eliminated if the seeds are covered with soil or sand, but may be appreciable when the beds are mulched with cloth, as is done in most southern pine nurseries. The possible influence of light should, therefore, always be kept in mind when difficulties are encountered in the germination of unfamiliar seeds.

BIOCHEMICAL CHANGES DURING GERMINATION

During germination the biochemical processes are generally the reverse of those occurring during seed ripening. The stored products are digested into simpler and more easily transported substances. Thus, soluble sugars are formed from starch and fats; insoluble proteins are digested into soluble amino acids from which new proteins are built up in the growing embryo. Sugars, chiefly glucose, are used partly in building new cells and are partly oxidized in respiration. In fact, so much sugar is used in the respiratory process that the seedlings in early stages of development actually weigh less than the original seeds. None of these biochemical changes occurs, of course, unless the environment is favorable.

TYPES OF GERMINATION

The early stages of germination are similar in all seed plants consisting of general swelling of the seed, followed by emergence of the radicle and development of a primary root. From this stage on germination is either epigeous (cotyledons appearing above the ground) or hypogeous (cotyledons remaining below the surface) (fig. 4).

In epigeous germination, a typical hypocotyl arch may be seen emerging from the soil; the rapidly elongating hypocotyl straightens out, bringing the cotyledons, usually with seed coat still attached (pine), above ground. Within a few days the cotyledons, nourished by the endosperm, force their way out of the seed coat, which then falls to the ground along with the shriveled endosperm. This is usually soon followed by the growth of the plumule. Sometimes the seed coat is sloughed off before the

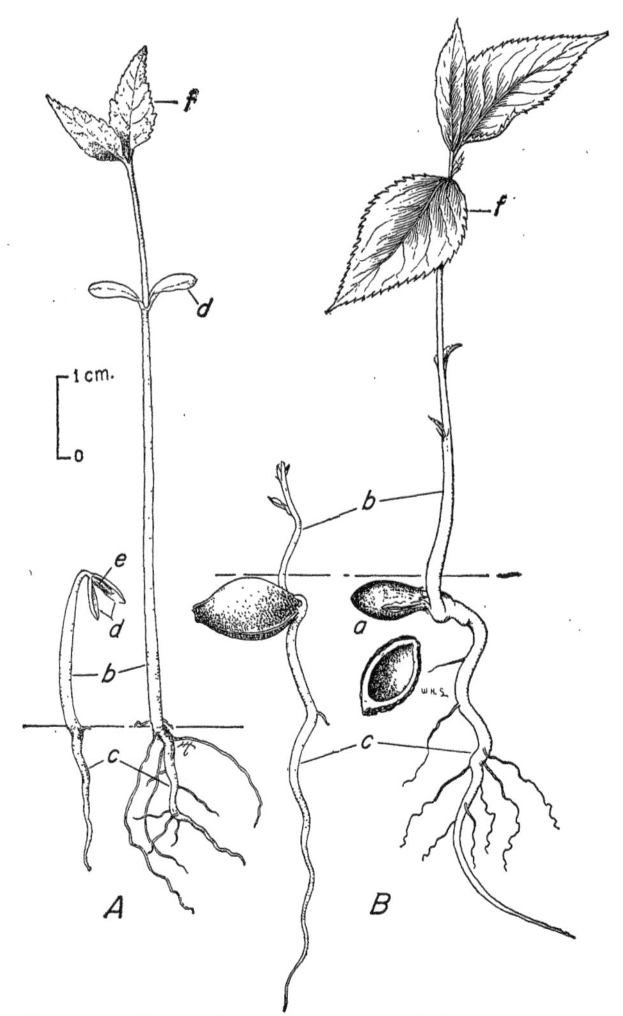

FIGURE 4.—Types of seed germination. *A*, Epigeous germination as shown by pin cherry seedlings at 1 and 10 days; and *B*, hypogeous germination of Allegheny plum seedlings at 1 and 9 days. *a*, Seed; *b*, hypocotyl; *c*, primary root; *d*, cotyledons; *e*, plumule; *f*, primary leaves.

hypocotyl emerges, particularly in the case of seeds with little or no endosperm (locust, mesquite).

In hypogeous germination, root growth is accompanied by rapid development and elongation of the plumule, the part of the seedling above the cotyledons. The hypocotyl does not lengthen; hence the cotyledons stay below the soil surface, where they remain attached to the seedling for weeks or months. Hypogeous germination (buckeye, filbert, oak, walnut) makes for more rapid growth than epigeous germination because there is not only a larger reserve of food in the seed, but the earlier development of green leaves adds to the food supply. As might be anticipated, seeds that germinate hypogeously are usually rather large, although not all large seeds germinate in this way. Nor are the types of germination constant within a genus. Germination of common chokecherry seed is epigeous; that of black cherry, similar in size, hypogeous. Some buckthorns have epigeous germination, but in at least one species germination is of the opposite type.

SEED PRODUCTION AND DISPERSAL

IN ORDER to collect seed successfully, it is necessary to know when seed are produced and dispersed by different species, and also how factors such as age, size, vigor, and soundness of desirable seed trees affect production and dispersal. The ability to estimate seed crops will come with an understanding of the effects of weather on such crops, the intervals at which various species mature their seed, which insects reduce seed crops, and which species have unisexual plants and therefore produce seed only on trees bearing female or pistillate flowers. Successful seed collecting also depends on a knowledge of the manner of seed dispersal, how long after ripening one may forego collection, and which species produce seed that are relished by animals and consequently hoarded where they may later be collected.

PRODUCTION

The fundamental biological processes of flowering, fertilization, seed development, and ripening are profoundly influenced by (1) internal factors—those associated with the biological characteristics of the plant—and (2) external factors—both physical and biological, which are part of the environment and modify the effects of internal factors.

Internal Factors

AGE, SIZE, AND VIGOR

The time of seed bearing is determined by age rather than size of the plant. Each species tends to begin seed production within a definite, relatively narrow age range. This range varies widely between species, even within the same genus. The Digger pine of California and jack pine of the Lake States, for example, may bear cones as early as the third year; and lodgepole and pitch pine at 5 or 6 years; while eastern white pine, longleaf and shortleaf pine, and others, rarely bear cones until after 10 years. Most tree species do not produce seed in significant quantity in the first several decades of life, in which the trees have the greatest vegetative vigor and attain most of their height growth. Most trees produce their best seed crops during middle age. During this period—which may last from several decades to a century or more—the comparatively dense, luxuriant foliage of the earlier decades gives way to more open types of crown, and height growth declines to a moderate rate. Almost all of a tree's energy during the first decades is utilized in vegetative growth, whereas in middle age and later, a greater part of it is diverted to reproduction.

Seed production tends to decline as the physiological and pathological symptoms of overmaturity appear. An exception is the so-called distress crops, which usually are induced by injury, insect attack, or abnormal weather. Trees suppressed by larger individuals or subjected to other adverse conditions may never produce appreciable quantities of seed. There appears to be little difference in seedlings grown from seeds collected from parent trees of different ages, except from very young or very old trees. This matter is discussed in greater detail on page 18.

Large trees normally produce more seeds than smaller trees of the same age. However, size of tree affects quantity of seeds produced mainly during middle age and later. A middle-aged tree which, because of poor site or for other reasons, has remained relatively small, may bear far more seeds than a vigorous young tree of the same or greater size. Within the age range of prolific seed bearing, seed production is much more directly associated with size of crown than height or trunk diameter, i. e., the number of fruits is directly related to the number of branchlets.

Vigor may be an expression of youth, of favorable site and climatic conditions, or of inherent tendencies, but in the sense of thrifty growth it is essential up to a certain point for good seed production. An individual tree or stand which, because of poor site, competition, or other reasons is in a static or deteriorating condition, will not produce large quantities of seeds. In even-aged stands the more vigorous, dominant individuals receive the light and attain the crown size necessary for prolific seed production. Without such expression of dominance, both vegetative and reproductive processes practically cease. Exceptional rates of growth and luxuriance of foliage that are decidedly above average for the species, however, are usually associated with low seed production.

In general, a greater percentage of high-quality seeds is produced in years of heavy rather than light seed crops, except perhaps for distress crops. Many of the seeds of some species produced during the early years of rapid vegetative growth are empty, or if viable, result in relatively weak seedlings. Exceptional vegetative vigor probably is associated with relatively poor quality of seed because most of the food materials are diverted to the development of vegetative parts. Bates (7) says that "the strength of Norway pine (*Pinus resinosa* Ait.) seeds and the immediate vigor and hardiness of seedlings developing therefrom vary inversely as the vegetative vigor of the parent trees.'' This may also be true for some other species. The common belief that overmature and decadent trees produce lower quality seeds than middle-aged trees probably is wrong. Bates considers this belief fallacious not only in the case of *Pinus resinosa* but probably for forest trees in general.

POLLINATION AND FERTILIZATION

Cross-fertilization (i.e., fertilization by pollen from another plant) is regarded as biologically superior to self-fertilization. The former results in stronger progeny in many herbaceous plants, and in some species there exists a definite incompatibility between pollen and pistil of the same plant. Less is known about the importance of cross-pollination in forest trees. However, Dengler (*13*) reports that inbreeding through self-pollination of Scotch pine and Swiss mountain pine resulted in numerous instances of degeneration, evidenced by low germinative capacity, high seedling mortality, and stunted or deformed growth. Hence, it seems reasonable to assume that seeds produced by isolated trees (self-fertilized for the most part) will generally yield weaker progeny than seeds produced in stands where cross-pollination prevails.

Furthermore, because of inadequate pollination, an isolated tree may set a compartively small number of sound seeds even though displaying an abundance of fruit. Hence, it is usually recommended both from the standpoint of the genetic properties of seeds and quantity of sound seeds per unit measure of fruit, that when possible, seeds should be collected from groups or stands, rather than from isolated trees.

Many species of plants exhibit characteristics which favor cross-pollination but do not preclude self-pollination. In the case of forest trees, chief among these characteristics are (1) segregation of pistils and stamens in separate flowers, often with the pistillate ones limited to the topmost portions of the crown where they are least likely to receive pollen from the mother tree, and (2) different dates of maturation by the pollen and stigmas of the same tree.

In a dioecious species the staminate plants of course bear no seeds, and the pistillate ones bear only when growing in reasonably close proximity to staminate plants. Recognition of dioecism is of paramount importance in attempts to regenerate forests by means of seed trees.

MATURATION OF SEED

Most plant species complete the processes from flowering and pollination to seed ripening in one season, but some tree species require 2 or even 3 years. In most of the latter species the fruit remains relatively small during the first year after flowering. In the pines, fertilization takes place about a year after pollination. In a few species, such as certain junipers, the fruit attains practically full size the first year, though it does not ripen until the second. This habit must be recognized in order to prevent the collection of much immature and worthless seed. In general, the period required for maturation is not important, except as it affects the length of time the developing fruit is exposed to unfavorable weather, insects, and other hazards.

CYCLE OF SEED PRODUCTION

Many forest trees and shrubs produce good seed crops at regular intervals. Some bear good crops every year, some every second or third year, and some at longer intervals—up to 10 or more years. Between the good years, seed crops are much lighter —sometimes total failures. The shorter the period between good crops, the more regular is the production cycle. In some species this pattern is distinct enough to indicate an inherent rhythmic tendency independent of site or climatic factors. Why this is so is not fully known.

A partial explanation, widely accepted, is based on Hartig's work with European beech—that food reserves stored in the stems are exhausted by production of a heavy seed crop, and a number of years is required to replenish them. However, the apparent cycle frequently is disturbed, probably because of variations in weather. To the seed collector it is important to recognize that there are definite seed years and off years for many species. He can then collect several years' supply during the good years, and store them carefully to tide him over the lean years.

INHERITED VARIABILITY

In addition to periodicity, there is some evidence also of inherited variability of seed production within species of forest trees. Recognition of such variability is important in the selection and establishment of tree-seed farms, and in possible future work in genetics.

TREE-SEED FARMS

Trees of good form which produce abundant seeds of high quality should be selected by the seed collector and some of them preserved for seed production. In many cases their value for this purpose will be greater than for wood products. This is true not only of such species as red pine, of which only a few virgin stands are left over most of its range, but also, for western pines, of which many virgin stands remain. In the latter case, there is a wide latitude for selecting the most desirable stands, well dispersed over its botanical range. Later on there may develop plantations of known good seed origin that can be set aside as tree-seed farms.

External Factors

CLIMATE

Good seed crops are produced under the same moisture and temperature conditions that favor vegetative growth. Seeds, however, are often borne in good quantity in years considerably below the optimum for vegetative growth.

According to some European reports, the formation of flower buds is greatest in dry, sunny summers, possibly as a result of a higher carbon-nitrogen ratio built up by high photosynthetic activity and a somewhat reduced nutrient absorption. Species that mature seeds in one season, therefore, might be expected to bear heavily the year following a dry season; species that require 2 years would fruit heavily the second year after a dry season, etc. It is impossible to make definite generalizations about these relationships. Obviously when climatic factors are added to the inherited tendency of a given species toward periodic seed production, and due allowance is made for weather vagaries in the seasons between flower-bud formation and fruit ripening, forecasts of seed crops cannot be made with accuracy before the fruits actually swell on the tree.

Probably the most direct relationship between weather and seed production occurs at the time of flowering. Many forest trees are wind pollinated; hence, a prolonged rainy period can so reduce the flight of pollen that fertilization will be much below normal. Insect pollinators may also be curtailed by bad weather.

After fertilization, reasonably favorable weather is necessary for satisfactory ripening of fruit. A certain amount of failure and abscission from unknown causes commonly occurs between fertilization and ripening in many species. Excessive premature drop of fruits or failure to ripen may occur, however, when the weather is too wet, too cold, or too dry. Such difficulties are most commonly encountered with species planted outside their natural range. Locally violent storms, particularly wind and hail, sometimes take a heavy toll of immature fruits.

SOIL CONDITIONS

Seed production is usually greatest when the supply of available soil nutrients as well as the physical conditions are most favorable for vegetative growth. The production of seeds requires large amounts of mineral nutrients—potassium, phosphorus, nitrogen, and others. Obviously, an ample supply of these elements in the soil helps to build up the food reserves necessary for heavy seed bearing. Similarly, soil texture and structure conducive to aeration, water absorption, and the activities of soil organisms, also favor food synthesis and seed production. For some species, soil acidity may be important.

The soil factors must be in proper balance. An excess of available nitrogen may result in vigorous vegetative growth and comparatively low reproductive activity. Flowering and fruiting are governed to some extent by the carbon-nitrogen (C:N) ratio within the plant. For maximum seed production, absorption of nutrients must be in proper balance with the processes of food synthesis. High carbohydrate production, hence a high C : N ratio, favors flowering and fruiting; a low C : N ratio, resulting from growth in weak light, is unfavorable.

Good seed crops are not limited to the most fertile soils. On the whole, soil fertility controls seed production probably less than it does vegetative growth. If the necessary food materials can be mobilized, plants sometimes fruit at the expense of vegetative growth, especially when the light is adequate and the C : N ratio high. However, the basic fact is that seeds can be borne only to the extent that the soil supplies the requisite nutrients; consequently, the closer the soil approaches the optimum for a species, the heavier the seed crop, other things being equal.

COMPETITION

Competition for light in the forest results in certain trees attaining a favorable position, while the remainder suffer more or less from suppression. Trees or other plants growing in open or uncrowded situations usually are the most prolific seeders because of the greater spread of tops or crowns and greater density of fruits per unit of crown space. At the other extreme, stands may be so dense that both vegetative and reproductive processes are practically at a standstill. These variations are due to competition for soil nutrients, moisture, or light, or all of them to some degree. Thus, intense competition for nutrients may create a condition essentially comparable to very infertile soil; similarly, intense competition for moisture may create conditions comparable to drought. In all these cases, seed production is reduced roughly in proportion to the deficiency of critical elements.

Except under the most adverse conditions, competition for soil nutrients and moisture usually relaxes enough during parts of the year, and occasionally for a whole year, so that some seeds are borne, if there is adequate light. Many understory trees of light-demanding species, however, exist for years without bearing seeds, despite ample nutrients and moisture. Thus, it has been reported (*57*) that in a year of moderately good seed crop, 98.8 percent of the seeds of western white pines in a given area were produced by the two upper crown classes (dominants and codominants), 1.2 percent by the intermediates, and none by the more suppressed trees in the stand. The failure of trees to fruit in weak light may be due to a lack of food reserves, or in part to low C : N ratios brought about by low photosynthetic activity.

INSECTS

Insects influence seed production in the following ways: (1) As pollinators for certain species;

891183°—50——2

(2) as destructive agents reducing vigor or causing death of the plants; and (3) by consuming seeds before they mature. Though the majority of commercial forest tree species are wind-pollinated, insect pollination is the rule in several groups. Among insect-pollinated trees are willows, basswoods, buckeyes, catalpas, some of the maples, and leguminous trees such as black locust. Insect pollination is also the more common method among the undershrubs whose habit of growth is much less favorable for wind pollination than that of tall trees. Trees or shrubs growing within their natural range usually do not suffer from lack of insect pollination. Occasionally, however, insect pollinators become inactive during inclement weather, and the tree or shrub later fails to set seed.

Many species, after suffering severe or fatal injury from insect attacks (such as bark beetles on pines) or physiological disturbances, tend to produce a usually heavy, final distress crop of seeds. Some species will even flower far beyond the normal season. Seeds produced in distress crops are usually below average in size and viability.

Insects feeding on the fruit of trees and shrubs destroy much seed. Damage of this kind is especially common in some pines and in nut-bearing species such as oaks, hickories, and walnuts. Few tree species are free from such depredations. Most insects of this type do their damage during the larval stages, the eggs being deposited in the flower or developing fruit. Infestations of from 20 to 50 percent of the fruit are not unusual among certain nut-bearing trees; in some years, the seed crop may be almost a total loss. The scattered red pine stands of lower Michigan, for example, usually yield poor cone crops largely because of chronic heavy insect infestation.

DISEASE

Diseases caused by fungi, bacteria, and viruses are relatively unimportant in seed production of forest trees. The major forest pathogens attack some part of the vegetative structure, and insofar as they reduce growth or vigor, or cause death, they affect seed production. Heart rot, although damaging to timber, does not seem to impair seed crops. Diseases of this kind are not usually regarded as causes of distress crops.

Blights are perhaps most conspicuous on fleshy-fruited species, such as plum pockets on wild plum, caused by *Taphrina pruni*. There are, however, some diseases which reduce seed production in conifers, such as the cone rust caused by *Cronartium strobilinum* on longleaf and slash pines (*25*), that caused by *Diplodia pinea* on the cones and seeds of Austrian pine and sometimes the twig blight of Scotch and Austrian pine. Doubtless many blights on fruits of forest trees and shrubs curtail seed production, but not seriously enough to warrant much study by pathologists.

There is considerable controversy but little scientific evidence on the question of whether or not it is advisable to collect seeds from diseased trees. Collecting seeds from overmature trees affected by heart rots, or other diseases of old age, is probably safe enough. Trees deformed by disease should be avoided. Studies have shown that, for Douglas-fir, Scotch pine, loblolly pine, and other species, some genetic strains are more resistant to certain diseases than to others, and individual trees in a stand probably display similar differences. Where there is a choice between sound and diseased trees, therefore, it is good practice to collect seeds from the former only.

DISPERSAL

From the standpoint of seed collection, the time of dispersal is of far greater significance than the method. The major agencies of seed dispersal are wind and animals, including birds.

Wind Dispersal

Seeds dependent on wind dispersal have a variety of structural modifications which assist flight, and form part of the seed proper (pine, spruce, fir, poplars, willows, catalpa, and the exotic *Paulownia* or princess tree) or of the fruit (birch, elm, ash, maple, yellow-poplar). In a few instances the instruments of flight are bracts attached to, or surrounding the fruit (basswood, hophornbeam).

Wings are the most common flight structure, associated with comparatively light seed weight. They are generally of the single-wing marginal type, as in birch and redwood seed, or the terminal type, as in ash, yellow-poplar, and pine seed. The latter type is by far the most effective in slowing down the rate of fall and thereby increasing the distance of flight. A few trees, such as some species of silverbell have four-winged fruits, which do not travel very far. In addition to wings, the hairy or cottony structures found on poplar and willow seeds are efficient for dispersal purposes. Catalpa seeds have hairy wings. A few shrubs of the family Compositae, such as *Baccharis* spp., produce achenes with a hairy pappus. The bladdernuts, whose fruits are inflated, buoyant capsules, exemplify a rare type of adaptation to wind dispersal among forest plants.

Some tree and shrub seeds are wind dispersed without special structural adaptations. Members of the Ericaceae, such as rhododendron and sourwood, produce extremely minute powdery seeds that are shaken out of dehiscent capsules and wafted away almost like dust. Some of the leguminous trees like redbud and locusts hold their fruits until they are torn loose by strong winds which carry the pods considerable distances. Other relatively large, lightweight fruits or fruit clusters, such as those of sweetgum and sycamore, are wind-distributed to some extent.

The distances to which fruits or seeds are carried by wind vary from a few hundred feet up to several miles. As a rule, seeds are not blown more than 100 to 500 feet from the parent stand in numbers sufficient to produce a full stand. The outstanding exceptions are willows and poplars, whose voluminous cottony seeds may be wafted in quantity over much greater distances.[9]

Dispersal by Mammals and Birds

The seeds of forest trees and shrubs do not display so many structural adaptations as are found among seeds of herbaceous species for distribution by animals. Bur fruits and barbed or hooked appendages are rare, and even the burs on the seed of such species as chestnut and chinquapin are not of much aid in distribution. Sticky seeds or fruits which adhere to mammals or birds are also uncommon, except those of the parasitic mistletoes, apparently carried by birds.

The most important means of distribution by animals in the forest are (1) the eating of fleshy fruits with hard seed which pass intact through their digestive tracts, and (2) hoarding. Birds eat and distribute most of the small-seeded berries and berrylike fruits though certain mammals such as bears and skunks also help.

Aldous (*1*) reports that *Vaccinium* seed have been germinated from bear dung and rose seed from grouse droppings, and that seeds of *Chiogenes, Rubus,* and *Vaccinium* passed through chipmunks apparently intact, though unable to germinate. Tests at the Lake States Forest Experiment Station showed that seed of the following species gave reasonably good germination after passage through birds: Blackcap raspberry, blackberry, Missouri gooseberry, American elder, Tatarian honeysuckle, black cherry, poison sumac, and meadow rose. There is some evidence that passage of juniper seed through birds is an aid to germination. Presumably, the grinding action in the bird's crop, or enzyme action in the digestive tract, renders the seed coats permeable. Birds often carry fruits such as cherries and plums to a convenient perch, consume them and discard the seed.

Dispersal of seed by hoarding is effected primarily by squirrels, and to some extent by mice and birds. Squirrels hoard mostly cones and nut fruits; mice collect conifer and other comparatively small seed. In Europe, nutcrackers (jaylike birds) are reported to hoard the seed of *Pinus cembra* (*43*) and various nut fruits. Usually some of the hoarded seed is untouched during the winter and germinates the following spring. Nuts require moist conditions for survival during winter, and burial by squirrels under litter is an excellent storage and planting method.

The hoarding habits of animals are vital to the forest succession of nut trees. Most of these species are somewhat tolerant; they tend to follow and replace pioneer trees, such as certain pines, aspens, and pin cherry that typically spring up after fire or on old fields. However, the invasion of pioneer forest communities by oaks, hickories, and similar species depends almost entirely on the hoarding activities of animals—primarily squirrels. Without this hoarding, plant succession in many forest communities would be slowed down, and might actually be different.

Other Methods of Dispersal

Gravity is important in seed dispersal on steep slopes, especially with heavy, globose fruits such as acorns, walnuts, apples, and persimmons. Running water may be a factor occasionally, but is important in the forest only for flood-plain species like willows, alders, red birch, sycamore, etc., and the seeds of even these species are adapted to other methods of dispersal. Landslides or snowslides, rain and ocean currents occasionally help to distribute seeds. The forceful ejection of seeds from the fruit is rare in trees, most common in herbaceous species, and occasionally found in shrubs, of which the witchhazel is a classic example. The Para rubber tree (*Hevea brasiliensis*) is one of the few forest species bearing explosive fruits.

[9] The seed of one species, *Populus tremula,* has been observed to travel 1,700 feet. Slope may markedly affect the distance of dispersal. Theoretically, a seed floating down at a 45° angle to level ground will travel a distance equal to the height of the tree. On a 30° slope, however, a seed falling at a 45° angle will travel 2.69 times as far down slope as on level ground before coming to earth, but only 0.73 as far up slope.

SOURCE OF SEED

IMPORTANCE OF A KNOWN SEED SOURCE

SEED source is second in importance only to choice of species in reforestation practice. For this reason the collector and vendor of forest seeds must have accurate and adequate information as to the origin of each seed lot, and the nurseryman and planter should use only seeds from sources suitable for his locality.

In the early days of forestry it was believed that any seeds of suitable species and good viability were usable, and that not to use the cheapest seeds was foolish. However, as large-scale plantations developed in Europe, observers noticed that trees grown from imported seeds were generally inferior to those of local origin. In some cases results were disastrous, as in the enforced eradication of several thousand acres of Scotch pine plantations in Germany during the early part of the present century. One eminent German forester estimated that the use of improper seeds had cost the country several million marks (*29*).

Most countries in northern and central Europe have enacted rather stringent laws covering the importation and use of forest tree seeds. Sweden placed a duty on imported tree seeds as long ago as 1888 (*29*), and the large German seed dealers submitted to voluntary regulation of imports about 1906. Later German laws required certification by appointed forest officers of stands and even individual trees before seeds could be collected; adequate labeling of seed as to origin; restriction on the movement of seeds within the country by definite zones; and heavy import duties on foreign seeds, which had to be dyed red for ready identification. Other European countries enacted similar laws.

The importance of seed origin has been largely overlooked in the United States until recently. So far there have not been the unsatisfactory results from planting trees of improper seed origin observed in Europe. Experimental evidence shows, however, that American species are not different from their foreign counterparts in this respect. The safest procedure is to use seeds of local origin wherever possible, and to plant other seeds only if they come from areas similar to the planting site in climatic and soil characteristics, or if they have been proved by scientific evidence to be better than local seeds.

CERTIFICATION OF SEED

Present knowledge of seed source and racial variation of forest trees and shrubs is sufficient to point out certain precautions necessary to make reforestation work effective. The sad experience in Europe through failure to recognize the importance of proper seed origin should be a warning to American tree planters.

There is no Federal legislation and few State laws to enforce the use of forest tree and shrub seeds of known origins. Some dealers who sold seed in quantity to European buyers were forced by pressure from their clients to list data as to origin. In 1939 the United States Department of Agriculture adopted the following forest seed policy:

Recognizing that trees and shrubs, in common with other food and fiber plants, vary in branch habit, rate of growth, strength and stiffness of wood, resistance to cold, drought, insect attack, and disease, and in other attributes which influence their usefulness and local adaptation for forest, shelterbelt, and erosion-control use, and that such differences are largely of a genetic nature, it shall be the policy of the United States Department of Agriculture insofar as practicable to require for all forest, shelterbelt, and erosion-control plantings, stocks propagated from segregated strains or individual clones of proven superiority for the particular locality or objective concerned.

Furthermore, since the above attributes are associated in part with the climate and to some extent with other factors of environment of the locality of origin, it shall be the policy of the United States Department of Agriculture:

1. To use only seed of known locality of origin and nursery stock grown from such seed.
2. To require from the vendor adequate evidence verifying place and year of origin for all lots of seed or nursery stock purchased, such as bills of lading, receipts for payments to collectors, or other evidence indicating that the seed or stock offered is of the source represented. When purchases are made from farmers or other collectors known to operate only locally, a statement capable of verification will be required as needed for proof of origin.
3. To require an accurate record of the origin of all lots of seed and nursery stock used in forest, shelterbelt, and erosion-control plantings, such records to include the following minimum standard requirements to be furnished with each shipment:
   (1) Lot number
   (2) Year of seed crop
   (3) Species
   (4) Seed origin:
      State
      County
      Locality
      Range of elevation
   (5) Proof of origin
4. To use local seed from natural stands whenever available unless it has been demonstrated that seed from another specific source produces desirable plants for the locality and uses involved. Local seed means seed from an area subject to similar climatic influences and may usually be considered as that collected within 100 miles of the planting site and differing from it in elevation by less than 1,000 feet.
5. When local seed is not available, to use seed from a region having as nearly as possible the same length of growing season, the same mean temperature of the growing season, the same frequencies of summer droughts, with other similar environment so far as possible, and the same latitude.

6. To continue experimentation with indigenous and exotic species, races, and clones to determine their possible usefulness, and to delimit as early as practicable climatic zones within which seed or planting stock of species and their strains may be safely used for forest, shelterbelt, and erosion control.

7. To urge that States, counties, cities, corporations, other organizations, and individuals producing and planting trees for forest, shelterbelt, and erosion-control purposes, the expense of which is borne wholly or in part by the Federal Government, adhere to the policy herein outlined.[10]

This policy, if followed by tree-planting agencies, would go far toward promoting the proper use of seeds of known origin. About the same time, the Minnesota State Highway Department set up rules delimiting the region from which they would purchase stock for roadside planting. Some States, such as New York and Georgia, have enacted seed laws which cover the handling of tree seeds, but such action should be Nation-wide to be effective.

The traffic in forest seed should meet two main standards established by voluntary action or regulation:

1. Seed collectors should label their seeds accurately as to species, time and place of collection (showing both geographic location and altitude, and preferably information as to stand and soil conditions).

2. Dealers should purchase only properly labeled seeds, and from collectors whose reliability has been established by reputation or some system of licensing or examination.

The users of seeds or nursery stock should demand adequate information as to seed origin, and buy only stock of local origin or of proven adaptability to local conditions.

DEVELOPMENT OF RACES

Well known is the fact that there are numerous varieties of wheat and most other cultivated plants, some of which are much better adapted than others to a particular locality. Likewise, a single species of wild plants may include a diversity of unique forms, strains, varieties, or races. Hence, it is important that seed of the most suitable origin be used in any given locality.

The importance of seed origin of forest trees was first recognized in Europe. Scotch pine, the leading European timber species, became the first subject of investigation. English and French shipbuilders had long used the straight tall, full-boled timbers from the Scotch highlands for masts. In the early 1800's the accessible supply of Scotch timbers had declined seriously, and other sources of suitable mast timbers were sought. At the same time, some efforts were made to develop supplies of mast timbers through reforestation, frequently with stock grown from foreign seed.

Foresters at that time gave opposite reasons for the differences in Scotch pine growing in different localities. Some ascribed these differences entirely to environmental conditions and treatment, others to heredity. To settle this question, Scotch pines from seeds obtained in many different localities in Europe were planted side by side in many experiments beginning in 1821 (*29*) and extended to an international scale in 1907 and 1908.

Although racial differences soon became evident, little was published about these tests until after World War I. Since then a mass of evidence has appeared showing the existence of several climatic races in Scotch pine and other European forest species (Norway spruce, European larch, English oak, European beech, and European white birch).

In the United States experiments modeled after those made in Europe were begun as early as 1911 with several western species. Only ponderosa pine and Douglas-fir tests have yielded definite results.[11] Tests under way long enough to show definite results agree with those made in Europe in showing the existence of climatic races.

Tree races develop not only in different latitudes, but also at different altitudes within mountainous regions. Since climate changes markedly with altitude as well as latitude, both kinds of development are included in the term climatic races. In addition, soil or site races may develop in areas similar climatically but characterized by different soil or site conditions.

There are still many important species for which we have neither reliable observations nor experimental evidence to indicate the presence of climatic races. Even for those species given most attention there is no definite understanding as to how many distinct races actually exist. Much investigation is still needed, first to determine the presence of races in species not yet studied, and second to learn the range or distribution of individual races.

What is Known About Forest Tree Races

Following are the major findings determined by experiments and careful field observations:

1. Several important forest species have developed races which must be taken into account in reforestation.

2. It is likely that most species which grow under a variety of climatic and soil conditions have developed races.

3. Different species do not react in the same way to environmental conditions in producing races; hence it is unsafe to assume that results for one species can be applied to another, even though their ranges are similar.

4. As the climate of the place of seed origin increases in warmth and mildness: (*a*) Susceptibility, especially of young shoots, to frost and snow damage increases. (*b*) Resistance to certain fungi increases. (*c*) Rate of growth increases.

[10] U. S. Department of Agriculture, Bureau of Plant Industry. Memorandum for the Secretary, dated May 16, 1939, signed by M. A. McCall, chairman, Seed Policy Committee. 3 pp. [Mimeographed.]

[11] Later, tests were also made with lodgepole pine, Engelmann spruce, red pine, white spruce, green ash, and slash pine.

(*d*) Also, for conifers, buds open later in the spring; shoot growth and leaf formation begin later; growth of these organs ceases later; new shoots harden off later; and winter buds are formed later. For hardwoods, the reverse appears to be true. (*e*) There is a tendency to form thicker stems and longer, heavier, more persistent, irregular, and numerous branches, culminating in strongly developed wide crowns. (*f*) Fruit and seed size increases.

When trees are planted in foreign localities they suffer increasingly in direct relation to the degree that the planting site varies in climate from the seed source (fig. 5). In Sweden, therefore, it is recommended that Scotch pine seeds be restricted to localities in which the mean summer temperature does not differ more than 1°C. from the point of origin. In general, results are best when local seeds are used. However, there appear to be superior strains of some species, such as the East Prussian or Baltic race of Scotch pine or the Burmese race of teak, which may produce better trees in foreign localities than would seed from the local sources.

F-330382

FIGURE 5.—View of Scotch pines of different seed origins 5 years after planting in northeastern Minnesota. The tree on the left, of central European origin, has inherited rapid growth but relatively poor cold resistance, and poor and ultimately unprofitable growth habit. The tree on the right, from seed collected in Norway, has grown more slowly but has better cold resistance and good form, and should develop into a good forest tree.

Our present knowledge of the occurrence of tree races is summarized by species in table 2. Despite the fact that there is some knowledge of racial development for 32 American and 35 foreign species, the study of this problem has hardly begun. Present knowledge seems only to indicate that there is racial variation. How many races there are and their approximate delimitations, have been worked out partially and somewhat tentatively for 2 American species (ponderosa pine and Douglas-fir) and 3 European species (Scotch pine, Norway spruce, and European larch). For about one-third of the 67 species listed, there is no adequate experimental background to prove the existence of races, although observations definitely indicate such development. Finally, the number of species listed includes only a small percentage of all those used or useful in reforestation.

TABLE 2.—*Racial variation in American and foreign species of forest trees and shrubs*

AMERICAN SPECIES

Species	Number of races known[1]	Remarks
Red alder	2	Differ in frost hardiness with altitude.
Green ash	2	Differ in drought resistance and foliage color.
Quaking aspen[2]	3	------------------------
Douglas-fir	3	Three main forms, green or Pacific Coast, gray or inland, and blue or Rocky-Mountain.
Grand fir	5	Two main forms, green or Pacific Coast, and gray or inland.
Eastern hemlock	2	Differ in seed size.
Western hemlock	2	Differ in frost hardiness.
Black locust[2]	2	Differ in form, growth rate, frost hardiness, and wood quality.
Bur oak[2]	2	Differ in leaf and fruit shape.
Northern red oak[2]	2	------------------------
Pawpaw[2]	( )	Differ in fruit characteristics.
Eastern white pine	2	Differ in seed size.
Jack pine	2	Differ in seed size, leaf coloration.
Jeffrey pine	2	Differ in seed color, size, and weight.
Loblolly pine	4	Differ in growth rate and disease resistance.
Lodgepole pine	9	Differ in seed size, growth rate, form, and disease resistance.
Longleaf pine	2	Differ in root habit, growth, and needle color and character.
Pitch pine[2]	2	Differ in form.
Ponderosa pine	4	Two main forms, Rocky Mountain and West Coast; differ in several characteristics.
Red pine	3	Differ in seed size, frost hardiness, and growth.
Shortleaf pine	2	Differ in growth rate and period.
Slash pine	2	Differ in frost hardiness.
Torrey pine[2]	2	Differ in seed shape.
Western white pine	2	Differ in seed size.
Pinyon	2	do.
American plum	2	Differ in germination characteristics.
Red raspberry	2	Differ in hardiness.
Fourwing saltbush	([3])	Differ in growth rate.
Engelmann spruce	2	Differ in frost hardiness.
Red spruce	2	Differ in seed size.
Sitka spruce	5	Differ in frost hardiness and growth rate.
White spruce	2	do.

[1] There are probably more races than the number listed for many of the species.

[2] Not based on experimental evidence. [3] Several.

TABLE 2.—*Racial variation in American and foreign species of forest trees and shrubs*—Con.

FOREIGN SPECIES

Species	Number of races known[1]	Remarks
Catechu acacia	2	Differ in foliage color.
European alder (*9*)	2	Differ in growth habit.
Guamachil apes-earring.	2	Differ in frost hardiness.
Apple[2]	([3])	Differ in hardiness, fruit form, and characteristics.
Apricot[2]	([3])	Differ in hardiness.
European ash	2	Differ in leaf size, drought resistance, and height growth.
European beech	([3])	Differ in form, rate of growth, and phenology.
European white birch[2].	([3])	------------------------
Cedar-of-Lebanon[2]	2	Differ in frost hardiness.
Siberian elm[2]	([3])	Differ in size, growth rate, and hardiness.
Silver fir	2	------------------------
Common jujube[2]	([3])	Differ in size and form.
Dahurian larch	2	Differ in frost hardiness, rate of growth, and insect resistance.
European larch	3	Differ in seed size and viability, growth rate and period, form, and disease and insect resistance.
Japanese larch[2]	3	Differ in bark characteristics and insect resistance.
Siberian larch	2	Differ in frost hardiness.
Sycamore maple	2	Differ in growth rate and phenology.
European mountain-ash.	2	Differ in seed viability.
White mulberry[2]	2	Differ in drought resistance and genetic composition.
Cork oak[2]	2	Western variety ripens fruit in 2 years as compared to 1 for typical form.
English oak	([3])	Differ in rate and time of growth, frost, drought and disease resistance, etc.
Peach[2]	([3])	Recognized by horticulturists.
Common pear[2]	([3])	Differ in hardiness.
Austrian pine[2]	([3])	Differ in form, growth rate, etc.
Balkan pine[2]	2	------------------------
Cluster pine	3	Differ in frost hardiness, growth characteristics, etc.
Japanese black pine[2]	2	Differ in form.
Merkus pine	2	Differ in growth rate.
Scotch pine	5	Differ in growth rate and habit, seed size, frost resistance, form, etc.
Swiss mountain pine	3	Differ in growth habit.
Swiss stone pine[2]	2	------------------------
Malabar simaltree	2	Differ in foliage characteristics.
Norway spruce	2	Differ in seed size, cone color, crown shape, needle color, growth rate, phenology, disease and insect resistance, and root development.
Common teak	3	Differ in seed size; germinative capacity, seedling development, and foliage characteristics.
Persian walnut[2]	3	Differ in frost hardiness and fruit characteristics.

[1] There are probably more races than the number listed for many of the species.
[2] Not based on experimental evidence. [3] Several.

Difficulties of Distinguishing Between Races

For those species which have been studied most intensively, particularly Scotch pine, Norway spruce, ponderosa pine, and Douglas-fir, races can be distinguished with a fair degree of accuracy on the basis of morphological and physiological characteristics.

The important question for the purchaser or inspector of seed or nursery stock, however, is whether or not races can be distinguished by characteristics of seeds or stock. Studies on Scotch pine have disclosed some racial differences in weight, optimum germination temperature (*29*), protein-serological content, and fermentation activity (*47, 48*) of the seed; phototropic responses of young seedlings (*48*); and needle length and color, rate of growth, form, cold resistance, sugar content, and dry weight of nursery stock (*29*). None of these differences, however, is very precise.

So far, then, there is no very exact way of determining racial origin on the basis of seeds or young seedlings for most plants used in reforestation. The purchaser of seed or stock must depend chiefly on the certification of the collector or dealer.

INDIVIDUAL VARIATIONS

Not only do trees develop races under varying conditions of climate and soil, but neighboring trees of the same species and race may vary in seed size and seedling development, stem form and structure, and phenological and physiological characteristics. Accordingly, seed collectors should gather seeds from the individual plants or stands which have the most desirable characteristics.

Seed Size and Seedling Development

Trees of the same species and of similar age and development, growing side by side, may show considerable differences in seed size and weight, and in fact even the seeds from an individual tree may vary a good deal. This is in addition to the fluctuation in seed size and quality which occurs in good and poor crop years. A number of studies of the effect of seed size on the resulting plants have been made in Europe, Asia, and North America. In general, the results, summarized in table 3, indicate that the larger the seed the larger the seedling. The advantage of the larger seeds probably persists longest in those species, like oaks, whose seeds contain considerable amounts of stored food.

There is also apparently some connection between size of seed and size of fruit. For instance, it has been found that large cones of jack pine contain more and heavier seeds than small cones,[12] and that cone size provides a good index of seed

[12] JACK PINE CONE STUDY. 1935. [Unpublished data on file at Lake States Forest Experiment Station.]

TABLE 3.—*Relation of seed size to seedling development*

Species	Relation between size of seed and resulting seedlings	Duration of effect (*years*)
European chestnut	Direct in one case, reverse in another.	
Silver fir	do	2–4
European larch	Direct	1–4
Norway maple	Heaviest seeds produce largest seedlings.	3+
Russian mulberry	Large seeds produce taller and better seedlings than smaller or medium seeds.	
Durmast oak	Direct	2+
English oak	do	1+
Northern red oak	do	1+
Austrian pine	do	2–7
Chir pine	Smallest seeds produce smaller plants than medium or large seeds.	
Scotch pine	Direct	2–6
Sal shorea	do	—
Norway spruce	do	2–7
Teak	Large seeds produce largest seedlings in one case; not in another.	

yield of red pine;[13] the production of viable seeds increases with cone size.

The practical application of these studies is as follows: (1) Seeds of several species can be graded approximately by size and quality classes by sorting cones into size or volume groups. (2) By using the larger seeds, larger 1–0 seedlings can be grown in the nursery. In some cases this might permit field planting with younger stock or might reduce the number of weedings or cleanings required to establish the plantations. Grading should be done with seeds of known origin, otherwise the segregation of the seeds by size may be the equivalent of selecting certain races.

Mother Tree and Seedling Development

Trees of different ages and vigor classes of a given species, even though growing under the same conditions, produce varying quantities of seeds. This fact early led to speculation as to whether there might not be differences in the vigor and development of seedlings grown from mother trees of different ages and crown classes. Studies of this matter have been made on Scotch pine, Austrian pine, Norway spruce, silver fir, ponderosa pine, Douglas-fir, jack pine, deodar cedar, pindrow fir, Himalayan spruce, sal shorea, teak, and sissoo.

Analysis of the results leads to the following conclusions: (1) Age, size, and dominance of mother tree in themselves do not have any measurable effect on the progeny. (2) Such differences in vigor and development of seedlings as have been noted are related directly to seed size. These findings indicate that the seed collector should gather seeds only from those trees which bear fruits or seeds of average or larger size.

[13] LAKE STATES FOREST EXPERIMENT STATION. ANNUAL INVESTIGATIVE REPORT. 19 pp. February 1941. [Manuscript.]

Differences in Form and Structure

Trees of a species growing in the same stand may differ greatly in shape of crown, stem form, branching habit, bark characteristics, grain and figure of the wood, color and form of foliage and fruit, and proportion of heartwood. To some extent, all of these characteristics are not heritable but are affected by age, density of stand, or injuries. This is especially true of crown shape and some types of stem deformation. Yet, even these have become part of the racial characteristics in some species. Scotch and Austrian pine in Austria and Norway spruce in Europe indicate a tendency to inherit narrow or spreading crowns (*29*). Also, some types of stem form are inherited, as shown by studies on European larch in Austria, Germany, and Switzerland; Norway spruce and silver fir in Germany; European beech in Denmark; and English oak in Austria and Russia (*29*). Inherited differences in bark have been found in English oak in Germany (*29*) and in babul acacia and sal shorea in India (*10*). These differences, which probably occur in many other species, may be connected with differences in resistance to fire or other damage.

The grain or figure in wood is of considerable economic importance. Spiral grain, which occurs in many woody plants, generally detracts from their commercial value, but such patterns as bird's-eye maple, curly birch and flowered teak bring much higher prices than the plain wood of the same species. A comprehensive study on chir pine made in India (*10*) led to the conclusion that the tendency to produce spiral grain (left-handed twist) was inherited, and that certain environmental conditions help to develop and accentuate such tendencies. Less reliable tests and observations have indicated the heritability of spiral grain in Jeffery pine, ponderosa pine, and sugar pine (*10*). There is no adequate information to show that bird's-eye or other figured grain is inherited.

Likewise, there is no experimental evidence that differences in the proportion of heartwood to sapwood—a matter of considerable commercial importance—are heritable (*10*).

Variations in foliage and fruit forms between localities have been noted for deodar in India (*10*), Norway spruce in Europe (*29*), and Douglas-fir and blue spruce in America. These variations, which exclude horticultural varieties, are of importance chiefly in ornamental use. How far they are due to hybridization or genetic characteristics is not known.

Phenological and Physiological Differences

Within a given locality, trees of the same species frequently show differences in resistance to various types of injury. Norway spruce, for instance, often has both late and early budding forms, with consequent differences in frost damage (*29*). This may, however, be a result of mixing strains in reforestation practice. A somewhat similar condition occurs in the case of English oak (*29*). Ordinarily the least resistant strains will be eliminated by natural selection.

Differences within races have been noted for Scotch pine and Norway spruce in resistance to disease (*29*), and for ponderosa pine with regard to mistletoe (*10*). Somewhat similar evidence is at hand concerning insect attacks. Nun moths have shown preference for certain Norway spruce trees (*29*), and lac insects for certain types of Malay lactree and Bengal kino (*10*). Deer have been reported to show preferences in browsing certain ponderosa pine trees (*49*). The seed collector should be alert to gather seed from the resistant trees.

Strains of Scotch pine, Austrian pine, and Norway spruce, growing under the same conditions (*29*), may exhibit markedly different rates of growth. The most conclusive evidence of such variations, however, has been found in the case of ponderosa pine (*49*). For many purposes the seed collector should prefer the fast-growing trees, but for some uses slower growth may be preferable. Certain trees, particularly of black walnut and chestnut, regularly produce more abundant crops of nuts than do their neighbors.

The vast hereditary variation in latex yield of the Para rubbertree is well known. Likewise resin yields vary greatly from tree to tree in longleaf and slash pines in the United States, cluster pine in France, and chir pine in India (*10*). So far, little has been done to select and breed high-yielding strains of these turpentine trees.

HYBRIDS AND TREE BREEDING

Many natural tree hybrids are known and can be utilized. A recent Canadian publication (*27*) lists 405 hybrids of known parentage in "28 genera of forest trees represented in North America by indigenous species of considerable economic importance," as follows:

Genus:	*Number of hybrids*		*Number of hybrids*
Fir *(Abies)*	3	Walnut *(Juglans)*	16
Maple *(Acer)*	9	Larch *(Larix)*	12
Buckeye *(Aesculus)*	11	Magnolia *(Magnolia)*	5
Alder *(Alnus)*	8	Spruce *(Picea)*	6
Madrone *(Arbutus)*	1	Pine *(Pinus)*	20
Birch *(Betula)*	13	Sycamore *(Platanus)*	1
Hickory *(Carya)*	7	Poplar *(Populus)*	121
Chestnut *(Castanea)*	15	Oak *(Quercus)*	77
Catalpa *(Catalpa)*	2	Locust *(Robinia)*	4
Dogwood *(Cornus)*	5	Willow *(Salix)*	42
Hawthorn *(Crataegus)*	5	Yew *(Taxus)*	2
Cypress *(Cupressus)*	1	Basswood *(Tilia)*	8
Honeylocust *(Gleditsia)*	1	*Hemlock (Tsuga)*	1
Holly *(Ilex)*	3	Elm *(Ulmus)*	6

In addition to these, there are many forest tree hybrids of foreign genera or parents which are not definitely known.

Not only should the seed collector be able to recognize races and select the most desirable trees and stands, but he must also be on the lookout for natural hybrids.

Because of his intimate knowledge as to the location of natural hybrids and stands or individual trees of noteworthy development the experienced seed collector can be a valuable ally of the tree breeder. The collector can aid greatly in discovering the plants which are needed for breeding special varieties or artificial hybrids.

Artificial tree breeding is a relatively new field, and, because of the long time required for trees to mature, is only on the threshold of producing plants which might enter commerce. Tree breeding began as early as 1845, when Klotsch in Germany crossbred two species each of pine, alder, oak, and elm, and considerable work has since been done in other foreign countries (Austria, Canada, Denmark, England, Russia, and Switzerland). The United States, although it started late, has more than held its own in this field.

Some of the major developments are: (1) Oak hybrids were produced in Texas (*49*) in 1909. (2) The same year, the Division of Forest Pathology, United States Department of Agriculture, began work in breeding blight-resistant chestnuts which has been carried on for many years (*12*). (3) In 1924 a paper company in Maine began an intensive study of poplar hybridization. (4) In 1925 the Institute of Forest Genetics (then called the Eddy Tree Breeding Institute),[14] the first organization devoted solely to tree breeding, began to work with pines. (5) In 1930 the Brooklyn Botanical Gardens undertook the breeding of chestnuts combining blight resistance and good timber form. (6) The Forestry Division of the Tennessee Valley Authority is endeavoring to breed trees on a large scale which will combine good timber qualities with the annual production of fruit or nut crops of high quality and quantity. (7) Some of the forest experiment stations of the United States Forest Service have carried on tree breeding studies for several years (*49*). (8) A project for selecting and propagating strains of eastern white pine resistant to blister rust is under way at the University of Wisconsin. (9) The Division of Forest Pathology, United States Department of Agriculture, is hybridizing elms to produce trees resistant to Dutch elm disease and phloem necrosis.

The products of the tree breeder probably will be of little importance in the seed trade in the near future, but as improved varieties and promising hybrids are developed their seed probably will be as widely handled by dealers as are comparable agricultural seeds today.

[14] Since 1935 part of the California Forest and Range Experiment Station of the U. S. Forest Service.

COLLECTION, EXTRACTION, AND STORAGE

OBTAINING and maintaining supplies of good seeds depend on an understanding of the best techniques for collecting, extracting, and storing seeds and the natural factors which influence their viability.

COLLECTION

The seed collector must know (1) where sufficient seeds can be found on plants of desirable form and development, (2) when seeds are ripe enough to gather and over how long a period collection can be made safely, and (3) how to collect seeds, whether from the plants, from the ground, or from animal hoards.

Where to Collect Seeds

The collector usually travels through the country locating suitable collecting grounds, checking the quality of the seeds, and estimating the quantities available and the difficulties of collection. For large-scale collection accurate estimates some time in advance of seed ripening are necessary as a basis for hiring labor, contracting for equipment, and making competitive bids.

There are several points to be observed:

1. *The parent plants should be of desirable form and development.* Desirable form, best judged on mature plants, may be tall, straight, clean stems for forest trees to be grown primarily for timber production, a low spreading growth habit for erosion control planting, or thicket formation and abundant fruit production for wildlife purposes.

2. *The position of the tree in a stand largely governs its ability to produce seeds.* Ordinarily the bulk of the crop is produced on the dominant trees, those whose crowns receive light from above and the sides. Open-grown trees often produce abundant seeds but their quality is often low because of self-pollination. This is discussed more fully under Seed Production and Dispersal, page 9.

3. *There is a relationship between flowering habits and seed production.* Among such trees as the poplars and willows, in which each tree bears only one kind of flower, only the "female" trees can produce seeds. (See p. 3.) Such "female" trees may make up only one-third of the stand, as in European aspen (*42*). Flowering habit also determines the part of the crown in which fruits are borne, as described on page 10. For example, in black spruce the cones are clustered near the tip of the tree, while in red pine they occur near the ends of branchlets over much of the crown.

4. *Estimates should be based on good sampling.* Most collectors, especially those lacking long experience, should base their estimates on actual counts of fruits on representative trees or small sample plots well distributed over the collecting area (*51*). Sampling should take into account variations in age, density, and composition of stand.

5. *Mature stands of particularly good development or plantations of known good seed source which produce seeds in reasonable abundance should be set aside as "tree-seed farms."* Such stands will provide desirable local collecting areas. (See p. 10).

6. *Collectors in some regions can obtain information on the general likelihood of good local crops within a broad territory from local seed-crop reporting services.*[15] Such reports are based on systematic, region-wide surveys made shortly before the fruits mature. Seed-crop surveys provide a desirable basis for planning current collection operations, and will build up the information needed to locate areas of good and poor seed production. Nation-wide coverage would be very convenient, but is not feasible at this time because of the scarcity of qualified observers.

7. *The soundness of seeds in individual localities or even on individual plants should be tested.* This can be done in the course of sampling the stand by cutting open a portion of the fruits and determining the percent of filled seeds.

8. *Some idea of next year's potential crop can be obtained* for such trees as the pines, the black oaks, and others which require 2 years to mature their fruits, by estimating the yield of first-year fruits. Such estimates are only approximate, however, since weather, insects, and other factors can reduce the crop seriously in any one year.

When to Collect Seeds

Ripeness of the seeds and the length of time they may remain on the plant or on the ground without deterioration or injury determine the time of collection.

Ability to recognize maturity of the seeds is of first importance. In general this seems simple enough. The color or general appearance of the fruit usually changes at the time the seeds ripen, that is, when they are no longer dependent upon the parent plant for food and moisture. For making small-scale collections it may be sufficient to know what color the fruits of various species are when ripe, but this is not enough when large-scale collections must be made in the limited time usually available.

[15] Such surveys were made up to 1932 by the Lake States Forest Experiment Station and have been conducted during recent years by the Northeastern and Southern Forest Experiment Stations. Other local surveys are made on many of the national forests and some of the State forests.

Unfortunately when fruits are fully ripe they may have shed much of their best seed; they may be eaten by birds and other animals, or they may fall to the ground and make economical collection difficult. It is necessary, therefore, to recognize the earliest stage of ripening at which a good yield of viable seeds can be obtained. Experienced collectors usually determine the proper degree of ripeness of fruits by their general appearance, color, the degree of "milkiness" of the seeds or hardness of the seed coats, their attractiveness to squirrels or other animals, or some combination of these factors. Inexperienced collectors, however, find considerable difficulty in determining fruit ripeness, since no one of the criteria already mentioned appears to be an infallible guide by itself.

The general season in which to make seed collections is known for a great many species, some of which are as follows:

Season:	*Species*[1]
Spring	River birch, cottonwoods, elms (except Chinese), red maple, silver maple, poplars, willows, Berlandier ash.
Summer	Cherries, Douglas-firs, elders, alpine larch, magnolias, red maple, mulberries, Siberian peashrub, plums, serviceberries, California sycamore.
Fall	Ashes (except Berlandier), beeches, birches (except river birch), boxelder, catalpas, cherries, Douglas-firs, Chinese elm, firs, hickories, junipers, larches (except alpine), magnolias, maples (except red and silver), oleasters, Osage-orange, pecan, most pines, plums, spruces, sycamores, walnuts.
Winter	Ashes (except Berlandier), yellow birch, boxelders, catalpas, Osage-orange, black spruce, Norway spruce, sycamores, walnuts.
Any season	Aleppo pine, bishop pine, jack pine, lodgepole pine, Monterey pine, pond pine, sand pine.

[1] More detailed information for a large number of species can be found in part 2 of this manual.

The exact time for starting seed gathering, however, must be determined for each species in each locality each year, and reliable means for readily determining seed ripeness are needed.

Accurate guides to cone ripeness, based on the specific gravity of freshly picked sound cones are available for a few pines, and even the inexperienced collector can use them easily. Usually 2 to 5 cones per tree are dropped into a suitable test liquid, and when more than half of the cones from the majority of the trees float, collection can begin. For loblolly, longleaf, ponderosa, red, and slash pines (and apparently also shortleaf pine), the cones are ripe enough to collect when their specific gravity in place on the tree has dropped to 0.88-0.86, that is, when they will float in lubricating oil S.A.E. 20 (specific gravity 0.88) or in a half-and-half mixture of kerosene and linseed oil (specific gravity 0.86). Eastern white pine cones are ready to pick at a slightly higher specific gravity (0.97-0.92), or when they will float in linseed oil (specific gravity about 0.93) (*45*).

In many of the legumes, maturity is marked by the shriveling of the cord which connects the seed to the pod.

Although color is not considered a reliable index of cone ripening for ponderosa pine (*34*), it appears to be a fairly good criterion for Austrian pine, eastern white pine, jack pine, Japanese red pine, and red pine. However, color is neither so easily defined nor so accurately measured as is specific gravity.

The unevenness of fruit ripening common to many species adds to the difficulties of collection. Ripening on individual plants in some species, such as the pines, is fairly uniform, but neighboring plants may differ considerably. In red pine, there may be a difference of 10 or more days in the time that cones on neighboring trees are ripe enough to collect. Fruits on individual plants of the plums and cherries and the mulberries ripen very unevenly. To collect most of the ripe fruits on Russian mulberry, for example, it is necessary to go over the plants every 2 to 3 days for a 2-to 3-week period.

The best quality seeds are generally obtained from well-ripened fruits, but there are exceptions to this rule. Some species yield good seeds when the fruits are somewhat green. As a matter of fact, the ordinarily dormant seeds of some species show high germination the first spring if the fruits are collected slightly green in late summer and the seed sown immediately. Among such species are alternate-leaf and gray dogwoods, Peking cotoneaster, American hornbeam, eastern redbud, witch-hazel, prairie crab, eastern hophornbeam, black jetbead, and arrowwood viburnum (*46*). On the other hand, seeds of northern catalpa and American sycamore give much better germination if they are collected late in the winter rather than in the fall.

It is essential to know how long a collecting period may be expected for various plants. The poplars and willows shed their seeds within a few days of ripening, and the fruits should be gathered just before they are ready to open. Eastern white pine sheds most of its seeds within a few weeks of ripening, but Scotch pine retains a large part of its seeds until spring, and jack, lodgepole, and sand pines retain many of their seeds in closed cones for several years. Some of the fleshy fruits, such as the plums, usually drop promptly upon ripening; others, such as the barberries, may remain on the plants over winter; many of both types, however, may be eaten by birds and animals if not picked promptly. The characteristics of each species should be known.

How to Collect Seeds

Seeds are collected from standing or felled trees, from the ground, from rodent caches, or occasionally from the surface of still water or from drifts along the shore.

The most common method is to collect the seeds from standing plants. Tall trees usually must be climbed and the fruits or seeds detached by hand picking; by cutting them off with hooks, clippers, or rakes; or by knocking them off with poles, flails, or clubs. The most useful tools for detaching cones and other fruits beyond reach of the hands are poles with hooked heads that will cut or catch on the thrust as well as on the pull (fig. 6).

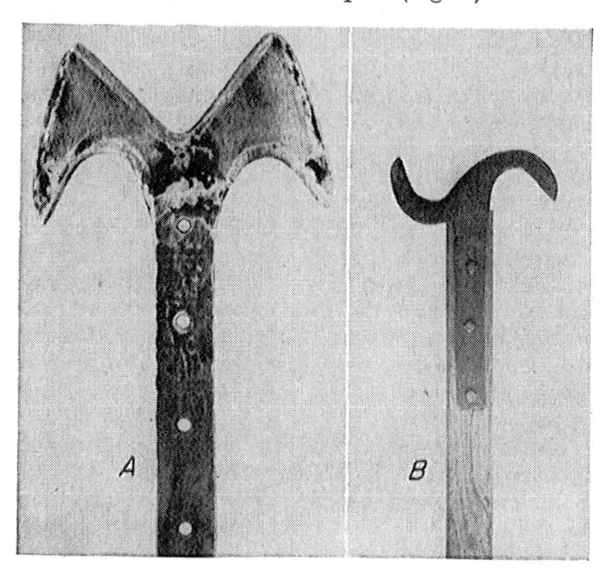

FIGURE 6.—Two types of seed collecting or cone hooks for cutting fruit, cones, or small limbs from trees: *A*, A double-hook detacher made from two discarded mower blades welded to a flat piece of steel 8½ inches long. The hooks are mounted on bamboo or other light poles about 10 feet long for use when working on tree crowns, or on much longer poles for use from the ground; *B*, A "tumbled-S" hook hammered out of scrap iron.

For climbing, ladders, life lines, safety belts, or climbing irons may be required. In hand picking, the seeds or fruits are usually placed in buckets, baskets, or bags. If the fruits are cut or knocked off, they are usually caught in burlap or light canvas sheets spread below. Seed collection from low trees or shrubs is done mostly without climbing, by hand picking or flailing. The choice between collection methods depends upon a number of factors besides the type and abundance of fruits. On ground clear enough to spread a cloth, small fruits such as those of chokecherry, redcedar, or even black locust can be collected best by stripping, whipping, or shaking. A seed cloth is unnecessary for fruits large enough to be readily seen on rough ground. On calm days elm, ash, and similar winged seeds can be collected best in a large square of light fabric tied at the corners to bamboo poles and raised close under the branches as the seeds are whipped or shaken off.

Seeds of most forest trees can be collected much more cheaply from felled than from standing trees. Arrangements can often be made to collect during logging operations. Care must be taken to collect only from trees logged after the fruits have matured and which still retain their seeds.

Collection of cones from squirrel hoards is not widely applicable because of the limited number of species to which it is adapted, the desirability of knowing the character of the parent trees, and the experience and skill required to locate squirrel hoards. It is true, however, that in the Lake States and in many parts of the West, squirrel hoards frequently yield sound seeds (*37*) at a low cost and may be rifled even after seeds have been released from cones still on the trees.

Gathering seeds or fruits from water surfaces or from drifts along the shore is practicable for only a few species used in conservation planting such as baldcypress and some of the willows.

After seeds or fruits are removed from the plants, care must be taken to prevent spoilage or reduction in viability. Moist or fleshy fruits may begin to heat [16] or ferment after a few hours if they are kept in piles or containers. Elm seeds have been known to begin heating within an hour after collection. Fruits should be transported to the extraction point with a minimum of delay, and in some cases they should be spread out and dried partially before shipment. Fleshy fruits should not be crushed, nor dried too long.

Arrangements Prior to Collection

Seed collectors should make certain arrangements prior to collection, in addition to locating collecting grounds, assembling equipment, and hiring labor: (1) They should obtain written permission to gather seeds on property other than their own. (2) They should consult their State Entomologist to make sure they will violate no quarantine regulation. (3) They should have a written agreement with purchasers covering such points as the species, place, and date of collection, standards of quality, and treatment before delivery; the unit of measurement (including the moisture-content percent at which the seeds shall be weighed, if weight units are used); price per unit and point of delivery; frequency of shipment by collector or of pick-ups by buyer; the largest quantity the buyer will accept at the contract price; and the date when payment is to be made. (4) *They should understand their liability in the case of injury to workmen and train their crews in safety matters.*

EXTRACTION AND CLEANING

To prevent spoilage, conserve space and weight in shipment and storage, and facilitate handling and sowing, seeds of many species must be separated from the fruits and cleaned of fruit parts and debris.

Methods of Extraction

Seeds are separated from the fruits by drying; threshing; depulping; or fanning, sieving, or other cleaning procedures. The methods of seed extraction

[16] When moist organic matter is confined so that aeration is poor, it usually begins to decompose, and heat is generated in the process. This is commonly called heating.

commonly used for a number of species are as follows:

Extraction method:	*Species*
Air drying or kilns...	Arborvitaes, baldcypress, ceanothuses, chamaecyparises, chestnut, chinquapins, chokeberries, cypresses, Douglas-firs, elders, elms, eucalyptus, firs, hemlocks, hollies, California incense-cedar, larches, manzanitas, mountain-ashes, pines, poplars, common pricklyash, redwood, Russian-olive, spruces, sweetgum, viburnums, willows.
Kilns	Aleppo pine, bishop pine, jack pine, lodgepole pine, Monterey pine, pond pine, sand pine.
Threshing or screening..	Acacias, alders, baccharises, beeches, catalpas, Kentucky coffeetree, filberts, fremontias, hickories, honeylocusts, American hornbeam, common lilac, locusts, Siberian peashrub, eastern redbud, rhododendrons, silktree, sourwood, sumacs, walnuts, witch-hazel.
Depulping	Apples, aralias, barberries, blackberries, buffaloberries, lilac chastetree, cotoneasters, creepers, elders, grapes, hollies, honeysuckles, huckleberry, common jujube, junipers, red mahonia, manzanitas, mountain-ashes, mulberries, Osage-orange, common pear, common persimmon, European privet, raspberries, meadow rose, sassafras, common seabuckthorn, serviceberries, silverberry, snowberries, western soapberry, common spicebush, tupelos, viburnums, yews.
Cleaning	Apacheplume, ashes, birches, antelope bitterbrush, elms, hackberries, eastern hophornbeam, common hoptree, lindens, mountain-mahoganies, oaks, Carolina silverbell, tanoak, common winterfat, yellow-poplar.

DRYING FRUITS

Fruits may be spread out to dry and release their seeds at ordinary air temperatures or in heated kilns. After drying, fruits such as cones usually are tumbled to loosen the seeds.

At Air Temperatures

The simplest method of air drying is to spread the fruits in shallow layers on cloths or platforms outdoors. In dry weather, this method is quite satisfactory for many species. Outdoor drying permits only a minimum of control, however, and there may be a considerable loss of seeds to birds and rodents and a large outlay for burlap or canvas.

If the air is moist or the quantities of fruit so great as to make the cost of frequent covering prohibitive, drying is better done under a roof. The fruits are spread out on the floor, in bins, or on a series of shelves or racks of trays. Indoor extraction is usually somewhat slower than outdoor extraction, but is easier to control.

In air-drying there must be free circulation of air across and around each fruit. This requires (1) that any building used for drying provide good ventilation—a suitable type is shown in figure 7 [17] —and (2) that the fruits be spread one layer deep. Small fruits require more space than equal volumes of large fruits (longleaf pine cones require about 8 square feet per bushel when spread in a single layer, whereas the smaller shortleaf pine cones, require from 16 to 20 square feet per bushel). Fruits exposed to air on one side only should be turned or stirred frequently. If wire-mesh trays or shelves are used, a mesh large enough to let the seeds fall through provides better ventilation; but if the seeds on the separate trays and shelves are from different lots, the mesh should be small enough to retain the seeds to prevent mixing.

F-310759

FIGURE 7.—Cone-curing shed, 80 feet long, used by the Southern Region of the U. S. Forest Service.

Air drying is used not only for extracting seeds from cones, but also for curing certain small fleshy fruits such as those of the buckthorn, the small cherries, or the mountain-ashes. These dried fruits are used as commercial seed.

In Heated Kilns

Many cones will open in from a few hours to 2 days under artificial heat as compared to 1 to 3 weeks in air-drying. Cones of jack pine and lodgepole pine, which do not open well at air temperature, open readily at higher temperatures. With properly designed kilns, temperatures and humidities can be closely controlled and the time required for extraction reduced without harm to the seeds.

For satisfactory results cones of several species must be precured by air-drying for 2 to several weeks before they are placed in kilns at high temperatures. The moist longleaf pine cones collected in the early part of the season must be precured,

[17] The U. S. Forest Service and several State Forest Services use this type of cone-drying shed. It is 80 feet long, ventilated by hinging alternate boards in the side, and equipped with 5 or 6 shelves 16 by 80 feet on each side of a central 6-foot aisle. As the side shelves are loaded, the aisle between is bridged with removable shelving, giving a total shelf capacity (in a shed 6 shelves high) of about 8,000 square feet. Such a shed, without overloading, will hold 1,000 to 1,500 bushels of longleaf pine cones for air extraction or twice that many for precuring prior to kiln extraction. Specifications for this shed can be obtained from the U. S. Forest Service, Atlanta, Ga.

whereas those collected late in the season seldom need precuring. Red pine cones require 2 to 3 weeks precuring and can be held in cone sheds advantageously for 4 to 5 months before extraction. In precuring, the cones can be spread in layers four or five cones deep. Species like jack pine need no precuring.

Kilns should reduce cone moisture as quickly and efficiently as possible. They must provide heat for evaporating moisture, air circulation for circulating heat, and control of both temperature and atmospheric humidity. Two general types of kilns are used for extracting seeds from cones; the simple convection type and the forced-air type. The former has been used long and widely; the latter has been developed since 1934.

Simple Convection Kilns.—Convection kilns may be anything from a stove-heated room to fairly elaborate special structures. Ordinarily they consist of a heating unit above which is an extraction chamber, preferably made of masonry or other fireproof material. Heat enters the chamber through one or more flues and dries the cones as it passes up through them arranged in tiers of trays. Since the air becomes cooler and moister as it passes up through the cones, the efficiency of extraction is not uniform in all parts of the kiln. Hence the trays must often be shifted to different levels or rerun. A convenient arrangement is a two-story building on a sidehill, with the furnace, cone-tumbler, and seed-cleaning apparatus on the lower floor, and the tiers of trays on the second floor above the furnace.

Unopened cones from a curing shed or from trucks can be delivered to the second floor by way of a level gangway from the sidehill, and after opening can be dropped through a chute into the cone-tumbler below. Convection kilns are relatively cheap, they may be installed by ordinary labor, and they require no special technical ability to operate. However, they are relatively inefficient and are even less efficient if overloaded with cones; the fire risk is often high and temperature control is seldom adequate; and the output in pounds of seeds per day is often relatively low.

A modification and improvement of the simple convection type kiln is the extraction drum type. This has been developed and used chiefly in Ontario.[18] This differs from the typical convection kiln in two respects:
(1) Instead of being placed in trays, the cones are put in hexagon-shaped, wooden framework drums, covered with heavy steel screening. The drums, placed in two tiers, are turned several times at half-hour intervals by means of a crank projecting outside the heating chamber. (2) The warm air does not rise into a dead chamber, but is kept in continual circulation by means of a cold-air shaft which opens on the heat chamber wall below the bottom drums and drops to the floor below (*38*).

Forced-air Kilns.—A forced-air kiln operated on the same principles as those applied successfully in lumber drying has been developed by the Forest Products Laboratory and is in use by the Forest Service.[19] This kiln consists of a chamber, preferably constructed from moistureproof, fireproof, and fairly good insulating materials, and a separate furnace room, on the same or a lower floor level. The chamber accommodates 2 skid loads of metal-screen cone trays (16 to 33 per skid load, depending on spacing between trays). At the base of the chamber is a fresh-air intake; in the upper part are two 24-inch disk fans and subdivided steam heating coils; and at the top is a vent blower. The fans provide forced circulation of warm air through the trays of moist cones. Heat and humidification are supplied by steam, and low relative humidity is maintained through ventilation. An electrically operated recorder-controller keeps the temperature and relative humidity at the proper levels and also automatically controls the amount of air vented by the exhaust blower. Further details concerning this kiln may be found in United States Department of Agriculture Technical Bulletin 773 (*39*).

The advantages of the forced-air kiln are: Positive air circulation, close automatic control of temperature and relative humidity, and low fire risk. As a result, cones can be opened quickly and efficiently. The disadvantages are: Expensive construction (the extraction of at least 5,000 bushels of cones per year is necessary to justify such a kiln (*39*)), and skilled labor for its installation and operation.

In operating any kind of kiln, temperatures must be maintained below the point at which the seeds will be injured. At the same time, since extraction is usually more rapid at higher temperatures, the highest safe temperature should be used. Too much desiccation must also be avoided. Species vary greatly in their ability to withstand high temperature and drying. Under certain conditions of air circulation and relative humidity, red pine and jack pine can be extracted without injury at 150° to 170° F., eastern white pine at 140° F., and longleaf pine at 115° F. Longleaf pine showed a significant loss in viability when extracted under the same conditions at 120° F. The time required for extraction by different methods of drying is given in table 4 for several pines.

Tumbling of Cones

After cones have been run through a kiln or exposed to air-drying, the seeds usually have to be shaken out by tumbling or raking the cones. Generally the cones are run through tumblers, revolving boxes, or drums with screened sides. On small operations the intermittent type, which is filled and emptied through a door in the side, is commonly used. On large-scale operations, how-

[18] Specifications may be obtained from Forestry Branch, Ontario Department of Lands and Forests, Toronto, Ontario, Canada.

[19] Specifications for this kiln can be obtained from the U. S. Forest Service, Atlanta, Ga., or Milwaukee, Wis.

TABLE 4.—*Comparison of drying methods for extracting pine seeds*

Species	Air drying, time required	Simple-convection kilns		Forced-air kilns		
		Temperature	Time required	Temperature	Relative humidity	Time required
	Hours	*°F.*	*Hours*	*°F.*	*Percent*	*Hours*
Eastern white pine	24–72	120	8–12	140	17	8
Jack pine	----------	145–150	2–7	170	20	6
Loblolly pine	300–1,000	120–140	6–48	----------	----------	----------
Longleaf pine	300–1,000	115	12–72	115	20–30	8–16
Ponderosa pine	100–150	120 –	3	----------	----------	----------
Red pine	----------	130–140	24–72	170	21	5
Shortleaf pine	400–1,200	130–140	6–48	----------	----------	----------
Slash pine	300–+1,000	120	6–48	----------	----------	----------
Scotch pine	----------	130	5–24	130	----------	4–8

ever, a continuous or progressive tumbler is preferred. This is a 4-sided wire cage slightly larger at one end than at the other, and turning on a horizontal axis. Cones are fed in at the small end and move gradually down to and out of the larger end as the tumbler revolves. The seeds fall out through the wire sides into a container below. A satisfactory progressive tumbler for longleaf pine cones is shown in figure 8. To prevent their reclosing, cones should be tumbled promptly upon removal from the kiln; air-dried cones should be tumbled on a moderately dry day.

F-322906

FIGURE 8.—Early model of progressive cone tumbler. Tumbler tapers toward end near camera.

THRESHING DRY FRUITS

The seeds of many species must be separated from the pods or capsules enclosing them; the fruits of others must be broken from the bunches in which they grow. Beating with a flail or treading under foot are quite effective on a small scale with such species as the locusts, catalpas, and ashes. Sometimes agricultural machinery can be used. Black walnuts, for instance, can be hulled in a corn sheller. Frequently, however, special apparatus is necessary for best results. A macerator, developed by the Forest Service for use on the Prairie States Forestry Project, and a hammer mill have been widely useful.

The macerator (fig. 9) is made of metal, and the parts through which the seeds pass are watertight to permit the use of running water while macerating fleshy fruits. Fruits or seeds pass down through a hopper, between concave blades and a revolving cylinder like those of a threshing machine, and out through an opening the size of which is ad-

F-342190

FIGURE 9.—Extracting Osage-orange seed with Forest Service macerator.

justed by means of a slide. Speed is controlled by graduated pulleys. In a single day such a machine has produced 400 to 600 pounds of clean honeylocust seeds from previously unbroken pods and greater quantities for some other fruits (*16*).

The hammer mill, which was developed for grinding feed materials is also adaptable for seed extraction. It consists essentially of a hooded inlet or hopper, a central chamber containing a series of hammers which rotate about a central shaft, and removable outlet screens of different sizes. When used with dry-fruited species such as black locust, northern catalpa, eastern redbud, eastern hophornbeam, American hornbeam, and honeylocust an outlet screen with openings large enough to pass the seed is employed. Fruits can be fed into the mill continuously during extraction, and it is reported 500 pounds can be cleaned per hour (*53*). Care must be taken to run the hammer shaft at comparatively low speeds (about 400 to 800 r.p.m.) to avoid injury to the seeds.

DEPULPING FLESHY FRUITS

Some small fleshy fruits can be dried and used in that form, but the seeds of most fleshy or pulpy fruits must be removed promptly to prevent spoilage. Nearly all the different methods of depulping require the use of running water. Small lots can be cleaned by hand, by treading in tubs, or by rubbing through screens with hand brushes and water from a hose. Food choppers, concrete mixers, feed grinders, cider mills, wine presses, and restaurant potato peelers have also been used for extracting seeds from fleshy fruits, but none of these is widely applicable.

Probably the most useful apparatus for depulping is the Forest Service macerator. In one day this machine can turn out, for example, about 1,000 pounds of clean plum seeds or 4,000 pounds of Russian-olive (*16*).

The hammer mill has also been used successfully for depulping fleshy fruits. An outlet screen usually is employed with openings small enough to hold back the seeds but let the pulp pass through. While the machine is running, water is forced through it with hoses. The machine is filled with fruits (about 25 pounds) and run a few minutes until they are clean. It is reported that by this method over 500 pounds of fruit per hour can be handled (*53*).

Some fleshy fruits, such as mulberry, chokecherry, and Osage-orange require crushing and soaking in water before being washed through the macerator or through screens. Such fruits have a tendency to ferment during soaking. Although fermentation has been advocated sometimes as an aid in depulping and even in improving germination (probably by overcoming seed coat dormancy), it is generally advisable to avoid fermentation, especially if the seeds are to be stored. Seeds from fruit which has fermented until acetic acid is formed may be badly injured.

EXTRACTION BY CLEANING METHODS

Seeds of several species require no extraction other than removal of chaff or trash. Elm and maple seeds, for instance, are usually first separated from leaves and twigs, and acorns often are merely screened to remove cups or floated to cull those which are weeviled.

WHEN THOROUGH EXTRACTION PAYS

Often not all the seeds can be separated from the fruits by a single process. Whether or not the additional cost of more thorough extraction is justified by the extra yield of seeds can readily be determined by means of a simple formula $X=\frac{C+R+S}{W_1+W_2}$, in which X = final cost per pound, C = cost of total quantity of fruits, R = total cost of first extraction process, S = total cost of additional process, W_1 = total weight of seed extracted by first process, and W_2 = total additional seed extracted by extra process.[20]

Methods of Cleaning

To facilitate storage and handling, seeds of many species must be cleaned after separation from the fruits. Chaff, trash, adhering fruit parts, and empty seeds usually are removed by one or more of the following four general methods of cleaning: Dewinging, screening, fanning, and flotation. Sometimes cleaning is combined with the extraction process, and often a combination of methods is required to clean the seeds. For instance, seeds of most conifers require both dewinging and fanning.

DEWINGING

Seeds of most conifers should be dewinged after extraction. Small lots of pine,[21] spruce, or similar seeds can be dewinged quite satisfactorily by rubbing them between moistened hands. Other simple methods are: (1) Tying the seeds loosely in a sack and then beating with flails, trampling, or banging the sack against trees or walls; (2) spreading the seeds in layers ½ to 1 inch deep,

[20] For example, a nurseryman had extracted 495 pounds of uncleaned seeds from 356 bushels of cones by raking them on tarpaulins after air-drying. The cones had cost \$352.44 delivered at the nursery and the raking had cost \$37.38. By hauling the cones to a nearby building and running them through a tumbler, at an additional cost of \$36.47, he obtained an extra 163 pounds of cleaned seeds. Using the formula, the cost per pound of the seeds without shaking was $\frac{\$352.44+\$37.38+0}{495+0}=\$0.79$, and the cost per pound with shaking was $\frac{\$352.44+\$37.38+\$36.47}{495+163}=\$0.65$. The extra process in this case not only added 33 percent to the original supply of seeds but decreased the cost per pound about 18 percent. (From article by J. B. Ely, Jr., NURSERYMAN STUDIES SEED COSTS. U. S. Forest Serv. Planting Quart. 9(1):11. 1940 [Mimeographed.])

[21] Longleaf pine seeds, unlike those of most other pines, have a firmly attached wing which cannot be removed readily. Ordinarily the seeds are used with the wings attached or with only part of the wing broken off.

moistening them, and raking at frequent intervals until dry.[22] Hand-rubbing is one of the least injurious methods of dewinging seeds, beating or trampling in sacks causes some injury, and moistening the seeds in layers may bring about heating or molding and preclude long-time storage. All these methods are too slow and expensive for large-scale use.

Dewinging machines which can be effective and safe are better suited to large-scale operations. Such machines usually consist of sets of fiber or stiff hair brushes mounted on arms projecting from a horizontal shaft, and revolving inside horizontal cylinders lined with wire or made of corrugated and perforated sheet metal (fig. 10).[23] Part of the debris drops out through the wire or perforations; the remainder is largely removed by fanning. Such dewinging machines, like cone tumblers, may be either intermittent, with the seeds being inserted and removed through a door in the top; or progressive, with the seeds being inserted at the small end and discharged at the large end of a tapered chamber. The Forest Service macerator can also be used for dewinging seeds (see p. 25).

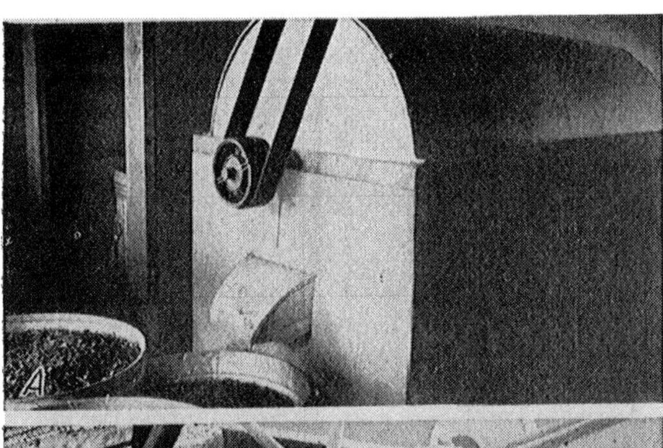

F-301102, 301105

FIGURE 10.—*A*, Exterior view of continuous type dewinger with truncate conical chamber 66 inches long; *B*, interior of continuous dewinger, showing brushes mounted at one-third intervals around circumference of corrugated perforated chamber. Looking toward seed inlet.

From 10 to 30 percent of the seeds may be injured by cracking or heat if the dewinging machine is not properly used and adjusted. In one case, injury to red pine seeds ceased when the speed of the machine was reduced from 120 to 60 r.p.m. (*15*). A study made at the Lake States station with jack pine seeds showed a difference of 25 to 30 percent in real germination between the following methods, ranging from best to worst: (1) Rubbing over a screen by hand or by power-driven brushes, (2) hand rubbing, (3) beating in a bag, (4) leaving wings undisturbed, (5) trampling in a sack and then putting through small fanning mill.

SCREENING

In many cases seeds can be cleaned satisfactorily by running them through screens, either dry or with running water. Screens must be of proper size to separate the seeds from the chaff or debris. Often two screens are used in a series, one with a mesh large enough to pass the seeds but hold back larger objects, and a second with a mesh small enough to hold the seeds but to pass smaller material. After screening, seeds often are further cleaned by fanning.

FANNING

Fanning is used to remove wings or light chaff, and is useful also in winnowing empty seeds. This method is not practicable with birch, poplar, or other very light seeds.

There are several simple methods of fanning. Dewinged conifer seeds can be placed in a thin layer in a wire-bottomed tray, the tray suddenly lowered 2 to 2½ feet and moved aside to avoid the wings which come down more slowly than the seeds. Seeds may be winnowed by passing them back and forth from one container to another outdoors on a windy day or in front of a fan, the

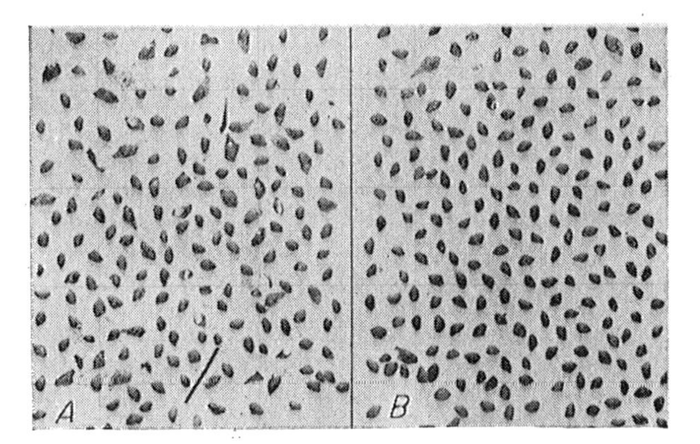

F-322884

FIGURE 11.—*A*, Slash pine seed dewinged by hand and winnowed in the wind; *B*, seed of the same original lot dewinged by wetting and stirring and then fanned in an agricultural seed mill. In addition to visible impurities, the wind-winnowed sample includes appreciably more empty seed than the mill-cleaned lot.

[22] Nurseryman R. C. Allen of the Mississippi State Forest Service has very successfully combined this method with several other operations. He first stratifies loblolly pine seeds with the wings on in moist sand and peat to stimulate germination; the wings are then loosened by the wetting and stirring operation. Peat, empty seeds, and loosened wings are removed in a single fanning operation immediately before sowing.

[23] Specifications for this machine can be obtained from the U. S. Forest Service at Atlanta, Ga.

chaff being blown aside. Such methods are not well adapted to large-scale use and often do not give very uniform results.

Standard agricultural seed fanning or cleaning mills usually give better and more uniform results (fig. 11). Most of these mills combine a fan with two screens placed at a fixed or adjustable angle. The screens are oscillated by an eccentric from a power-driven shaft to move the seeds along at a suitable rate of speed. The upper screen is coarse enough to stop and divert foreign material larger than the seeds, and the lower screen stops all seeds but passes trash and chaff which is then discharged. At some point in the system the seeds pass through a transverse or nearly vertical blast from the fan, which blows out any chaff or material of the same size as the seeds but lighter (such as empty seeds). Unless fanning is done skillfully the seeds will contain too many impurities, or too many good seeds will be blown out.

The dust from fanning mills, dewingers, and tumblers is a fire and explosion hazard and possibly detrimental to the health of workmen. These hazards should be controlled by adequate safety measures and training.

FLOTATION

Seeds of most pulpy or fleshy fruits can be cleaned easily and effectively by flotation in water. Sound seeds usually sink; poor seeds, skins, and pulp either float or sink more slowly than sound seeds. A simple technique for flotation is as follows (*16*): (1) Place a quantity of macerated fruit in a large washtub which is tilted slightly; (2) direct a stream of water from an ordinary garden hose into the tub at such an angle as to create a rotary swirl, and a lifting effect upon the material; (3) stir the material in the bottom of the tub slightly. The pulp and light seeds will be brought to the surface and carried over the edge of the tub by the overflow water.

Some dry fruits are also cleaned by water flotation. Freshly gathered acorns often are separated from the cups and weeviled fruits by this means. For loblolly pine in which the thick seed coat makes empty seeds fairly heavy, flotation is often more effective than fanning.

For species which have seeds light enough to float in water even when sound, flotation in ether, studied at the Lake States station, or other liquids (*3*) has shown promise in laboratory trials.

THE EXTRACTION FACTOR

Knowledge of the amount of cleaned seeds that will be produced by a given quantity of fresh fruits, the extraction factor, is a necessary basis for planning the amount of fruit to gather to supply specific sowing or market requirements. Ordinarily the extraction factor is expressed as pounds or ounces of cleaned seed produced per bushel of fruits or as pounds of cleaned seeds per 100 pounds of fresh fruits. It varies with the character of the collecting season, the dryness of the fruit when collected, the thoroughness of extraction, and other factors, but for each species there is an average value or range of values, which gives a good guide for collecting operations. The amount of cleaned seeds produced per 100 pounds of fruit as it is usually collected ranges from 30 to 50 pounds for many species,[24] and may range from about 1 pound for common lilac to nearly 100 pounds for some oaks.

Relation Between Seed and Seedling Costs

There is a definite relation between the cost of seeds and the ultimate cost of seedlings produced. The number of usable seedlings produced per pound of seeds sown is the chief measure of the true value of the seeds. This factor is dependent upon the number of seeds per pound, the purity of the seeds, and their viability, as well as the survival of the seedlings in the nursery. Nature largely controls the number of seeds per pound, but man, by the methods of collection, extraction, cleaning, and storage he uses, can exercise considerable control over seed purity and viability. As the number per pound, purity, and viability of seeds increases, the number of seedlings per dollar's worth of seeds also increases. The cost of seeds required to produce 1,000 seedlings can be determined readily from a simple formula:

$$S = 1{,}000\ [D \div (C \times P \times G \times L)],$$

in which S = cost of seed alone per 1,000 seedlings produced; D = cost of seed per pound; C = number of seeds per pound; P = purity percent, expressed as a decimal; G = germination percent, expressed as a decimal; and L = percent of germinating seedlings surviving until lifting time, also expressed as a decimal.

STORAGE

Why Storage is Needed

Frequently seeds cannot or should not be sown immediately after extraction and cleaning, but must be stored until sowing time. Most commonly, seeds are extracted in the fall and held until the following spring. Many species, however, produce good seed crops only at intervals of several years, and the seed collected in the good years must be stored for several years to provide for sowings in the lean years. The aim of storage practices is to maintain the highest possible seed viability. For some species this is a simple matter; for others it is difficult; and for many, suitable storage practices have not been determined.

Storage Methods

Seed characteristics and length of storage largely determine the methods that should be used. Generally, seeds keep best at low moisture

[24] Values for many species are given in part 2 of this manual.

content and low temperatures. However, there are many exceptions to this general rule. Some seeds keep well at ordinary temperatures, and some are injured if their moisture content is reduced slightly or at all. The storage methods recommended for some commonly used species are as follows:

Method:	*Species*
Dry, cold storage in sealed containers.	Apples, arbovitaes, ashes, barberries, birches, antelope bitterbrush, blackberries, silver buffaloberry, ceanothuses, lilac chastetree, cypresses, Douglas-firs, elders, elms, firs, riverbank grape, hackberries, hemlocks, honeylocusts, common hoptree, black huckleberry, junipers, larches, black locust, maples (other than silver), mountain-ashes, oleasters, Osage-orange, pines, poplars, common pricklyash, raspberries, eastern redbud, redwood, sassafras, giant sequoia, snowberries, spruces, sumacs, sweetgum, sycamores, witch-hazel, yellow-poplar.
Moist, cold storage..	Beeches, buckeyes, chestnut, chinquapins, filberts, hickories, silver maple, oaks, tanoak, tupelos, walnuts, yews.
At room temperature.	Acacias, Kentucky coffeetree, eucalyptus, fremontias, common lilac, lindens, common pear, Siberian peashrub, European privet, meadow rose, fourwing saltbush, common seabuckthorn, common winterfat.
Under partial vacuum.	Poplars.

COLD STORAGE

Widely employed and generally effective is storage at low temperatures, usually just a little above freezing. The low temperatures slow down the life processes within the seed and yet maintain its viability. Low temperatures are particularly necessary to maintain the viability of seeds which have a high fat or oil content.

The best temperatures for storing seeds of most species lie between 33° and 50° F. The best seed moisture content is below 10 percent of oven-dry weight (*36*) for many conifers; but it must be kept above 35 percent for oaks, hickories, and silver maple; and seeds of southern magnolia apparently should not be permitted to dry at all.

A refrigerator or cold room is usually required for proper cold storage. Control should be sufficiently good to hold the temperatures constant or at least within a range of a few degrees.

Seeds which have been extracted properly usually are ready for immediate storage. Kiln-dried seeds or seeds air-dried in dry climates, as in the Rocky Mountain region, ordinarily have a low moisture content suitable for storage. In humid climates further drying often is necessary prior to storage. This can be done by treating the seeds in these ways: Sunning in shallow layers, placing them loosely in sacks of open weave and frequently stirring or shaking, spreading out under fans, rerunning through a kiln, or using a rotary drier (fig. 12). Drying should be rapid enough to prevent heating and molding, but not so rapid or at such high temperatures as to injure the seeds. To maintain the proper moisture content in storage the seeds must either be kept in sealed containers or the entire storage chamber must be kept at the proper relative humidity.

F-342191

FIGURE 12.—Rotary seed drier turned by an electric motor. The cylinder is 32 inches in diameter and 32 inches long.

STORAGE BY STRATIFICATION

In moist storage the seeds usually are mixed with one to three times their volume of moist peat moss, sand, chopped sphagnum, or other porous or granular substance. This provides the seeds with moisture and oxygen, while keeping them more or less separate from each other. They are then placed in the ground under a straw or leaf mulch (in cool climates) or stored in refrigerators. In other words, they are placed in cold stratification just as is done with many species to overcome dormancy. Outdoor storage is suitable only over winter, and indoor stratification often is used for only a few months. Many of the nuts are usually stored moist, but sometimes pine seeds, usually stored dry, may also be held over winter in stratification.

STORAGE AT ORDINARY AIR TEMPERATURES

The simplest and oldest method of storage is to hold the seeds at ordinary air temperatures. Seeds of some species, especially in dry climates, can be held in this manner for many years. They are sometimes stored in sacks but most commonly in sealed containers. Seeds of many legumes are stored in this way. Seeds kept in open containers must be protected from rodents.

Seeds are stored sometimes in cool cellars, or frequently in special storage sheds—well-insulated buildings in which temperatures, while not controlled, show a gradual rise and decline with the seasons.[25] Seeds of many species can be kept satisfactorily for one or more years in such sheds; for longer storage they must be held at low temperatures.

[25] Specifications for such seed storage buildings may be obtained from the U. S. Forest Service, Milwaukee, Wis.

STORAGE UNDER PARTIAL VACUUM

The life processes of seeds can be slowed down by reducing the oxygen supply as well as by lowering temperatures. Ordinarily short-lived seeds of poplars can be kept fairly well for several months in sealed containers from which much of the air has been exhausted (down to 1 mm. pressure) by suction pumps. Black jetbead, however, does not retain its viability any better under partial vacuum than at normal air pressure (*19*).

Vacuum storage has been attempted on a laboratory scale only. While it seems promising, particularly in combination with low temperatures for use with short-lived seeds, its application on a large scale appears somewhat doubtful.

Effectiveness of Storage

The best storage methods have been worked out for only a few species, but reasonably good practices are known for many.[26] Seed storage for 3 to 5 years will be adequate for most purposes, with 10 years about the maximum storage necessary for species having good seed crops at long intervals.

The effectiveness of various storage methods can best be illustrated by citing some examples of actual storage tests:

1. Red pine seeds at 2 moisture contents, stored under 4 temperature conditions, showed no significant loss of viability in 10 years at low temperatures and moisture contents (*40*) (table 5).

[26] Such methods are given for many species in part 2 of this manual.

TABLE 5.—*Germination of red pine seeds after storage in sealed containers at different temperatures and moisture conditions*

Storage temperature range (°F.)	Seed moisture content at beginning of storage	Germinative capacity	
		After 6 years' storage	After 10 years' storage
	Percent	*Percent*	*Percent*
32–39	4.7	95.4	[1]97.5
	6.5	94.8	[1]97.0
41–50	4.7	90.8	[1]93.2
	6.5	90.6	57.4
32–68	4.7	94.4	58.9
	6.5	89.6	33.7
0–100	4.7	66.8	46.2
	6.5	56.8	26.4
Average, all temperatures	4.7	86.9	76.6
	6.5	83.1	53.7

[1] Increases probably due to sampling error.

2. Longleaf pine seeds, ordinarily difficult to store, were dried to moisture contents of 9 and 22 percent and some of each stored for 2½ months in both sealed and open containers at 38° F. and 75° F. At the low temperature all lots of seeds kept well. At the higher temperature the dry seeds kept perfectly in sealed containers, but the moister seeds lost all viability. In unsealed containers both dry and moist seeds dropped to intermediate viability.

3. Under open, dry storage at room temperatures, seeds of plume albizzia have kept 50 years with fair viability (33 percent), and those of candlenuttree still gave fairly high germination (74 percent) after 79 years of such storage (*17*).

4. Black poplar seeds maintained their original viability for 1 to 2 months when stored at room temperatures under partial vacuum (1 mm. pressure), for 4 months when stored in sealed containers at 36° F., and for more than 10 months when stored under partial vacuum at 36° F. (*42*).

Storing Unfamiliar Seeds

Nature provides some clues to the methods required for storing unfamiliar seeds. For example, seeds of many acacias, which grow in very dry climates, usually maintain viability for a long time in dry storage at ordinary temperatures; seeds of the oaks, which fall to the ground as soon as they ripen and either germinate immediately or lie over winter on moist ground or in the litter, usually require moist cold storage; seeds of the ashes which usually hang on the plants in the fruits well into winter usually stand dry storage well; seeds of most spruces which usually lie on the ground over winter generally require cold storage.

These are clues, but they do not provide a sufficient basis for successful storage practices. There are too many exceptions to the general rules, and too many species require specific moisture or temperature conditions during storage. For some species there is a range of only a few degrees in temperature or a few percent in moisture content at which storage is satisfactory. Accordingly, exploratory tests should be made with unfamiliar species, or in using new methods with familiar seeds. These will point to the best methods for large-scale storage.

TREATMENT OF SEED PRIOR TO SOWING

PERFECTLY sound and uninjured seeds of many trees and shrubs often fail to germinate even when conditions of temperature, moisture, oxygen, and light are suitable. Such seeds are called dormant, and special treatment to induce germination is required before or in connection with sowing.

Seed dormancy is common and frequently is an important factor in the survival of the species; it may prevent germination during the first intermittent warm periods in the spring, and permit it to take place only after the weather becomes favorable for seedling survival. To the nurseryman, however, dormancy often is a vexatious matter. Without pretreatment, dormant seeds often germinate irregularly over 2 or 3 years; make spotty seedbeds containing stock of different ages and sizes; and tie up the nursery area for an extra year or two, increasing costs of production. Although two-thirds of some 440 species studied have some form of seed dormancy, the species most commonly used in reforestation (chiefly pines and spruces) fortunately have seeds that show either no dormancy or dormancy of a type that may be overcome by generally known simple treatments (often by fall-sowing). However, many of the new species coming into more general use for shelter-belt, wildlife, and erosion planting are unfamiliar to most nurserymen and often have dormant seeds.

The best known methods for pretreating seeds of many of these plants are given by species in the second part of this manual. Exploratory tests should be made of any species not discussed if large-scale sowings are to be undertaken. The knowledge of the causes of dormancy and how to overcome it is therefore a very important basis for nursery work and also for obtaining natural regeneration in the forest.

WHAT CAUSES DORMANCY?

Dormancy may arise from the hereditary characteristics of the seed itself, or from conditions induced by extraction or storage. There are two main causes of seed dormancy: (1) An impermeable or hard seed coat which prevents water and oxygen from reaching the embryo or in some cases prevents the embryo from breaking through the seed coat[27] even though water has been able to pass in; and (2) conditions of the embryo or stored food within the seed which prevent germination. Frequently only one type of dormancy is present, but in many species there exists a double dormancy caused by a combination of seed coat and internal conditions.

The biochemical and physical changes taking place during germination are the reverse of those occurring during ripening. Ripening[28] involves an inflow of relatively simple soluble compounds (simple sugars, amino acids, etc.) from other parts of the plant, their transformation into more complex and less soluble substances, and, usually, more or less dehydration of the living tissues of the seed and hardening of the outer seed coat. In germination the seed coats are softened or split, and the complex stored food products are reconverted into simpler substances easily transported and used for the development of the embryo into a seedling. Any influences which interfere with this reconversion of stored food or with the development of the embryo, despite exposure of the seed to warmth, water, oxygen, and possibly light, may induce dormancy.

The extent of dormancy may vary from a barely perceptible sluggishness in development to a complete and absolute failure to germinate even after 2 or 3 years' exposure to suitable conditions. In the latter case the seed, unless dormancy is replaced by decay, does not spoil; it simply remains alive without commencing growth.

CAN DORMANCY BE PREDICTED?

Some simple and reliable guides for predicting the occurrence of seed dormancy would be very convenient for the nurseryman and seed technician. It is generally known that certain species always have dormant seed, others have it sometimes, and still others never do, but unfortunately, seed dormancy shows no particular relationship to taxonomy, distribution, or other bases commonly used in classifying plants. In short, there is no reliable way of predicting the occurrence of dormancy, and the nurseryman must either know through experience or study when to expect it or resort to tests of his own to determine whether or not seeds are dormant.

OVERCOMING SEED DORMANCY

The nature of the treatments required to break dormancy differs considerably from species to species, within species, and even within individual lots of seed. In some species dormancy is easily broken by any of several methods; in others, it responds only to some very exactly controlled treatment, or, in extreme cases, seems incapable of being broken. The special treatment may coincide with part of the storage period, or it may

[27] In the following discussion seed coat is construed as the outer covering of commercial seed. It may be the true seed coat or the pericarp or other fruit parts outside of the seed coat proper.

[28] See page 6 for more complete discussion.

follow or even precede the storage period and differ entirely from the storage treatment. In many instances seeds probably pass into and out of dormancy between extraction and sowing without the nurseryman's being aware of it. This probably accounts for some of the discrepancies between laboratory and nursery germination.

The seed technician or nurseryman must understand the basic principles involved in breaking dormancy, because of the uncertainty or dearth of information concerning some species, and also because of the danger of injuring or killing the seeds by treating them incorrectly. Although failure to pretreat has often nullified the best of nursery practice, attempts to stimulate germination have occasionally resulted in total loss of the treated seeds.

A single type of pretreatment may be sufficient to overcome either seed coat or internal dormancy, but generally the best treatments for the two are somewhat different.

Seed Coat Dormancy

Seed coat dormancy is usually revealed by the appearance of the seeds or through cutting or soaking tests, and is comparatively simple to overcome. Out of 444 species studied, 7 percent had impermeable seed coats. These included most of the legumes, and also representatives of several other families such as the bumelias (Sapotaceae), huckleberries (Ericaceae), buckthorns (Rhamnaceae), sumacs (Anacardiaceae), and soapberries (Sapindaceae).

ACID TREATMENT

One of the commonest methods of pretreating seeds with impervious coats, such as that of black locust, is to soak them in concentrated sulfuric acid. The treatment is highly effective with many species; in extreme cases it has increased germination from less than 10 percent to more than 90 percent.

The only special materials and equipment required are:

(1) A supply of acid sufficient to cover the seeds; (2) containers (wood or earthenware preferred) not too readily acted upon by the acid; (3) wire containers and screens for handling, draining, and washing the seeds; (4) a supply of running water; (5) some place to drain away the dilute acid resulting from rinsing the seeds and (6) facilities for drying the seeds after rinsing. **Another requirement should be that workmen understand and carry out precautions needed to handle sulfuric acid safely.**

The steps in the acid treatment are as follows:

1. Thoroughly mix all the seeds to be treated as one lot. This is very important, since dormancy differs because of individual tree characteristics, and because there are variations in extraction treatment and moisture content in various containers during storage; consequently, there may be considerable differences in the optimum period of exposure to the acid.

2. Determine the optimum period of immersion in acid (see following paragraph) by first treating several small samples for different periods and either germinating them, soaking them for 1 to 15 days (depending on species) in water at room temperature and observing the percentage of swelled seeds, or more approximately, by examination of the seed coats. Insufficient treatment leaves the seed coats of some species glossy; overtreatment pits them and may even expose the endosperm; correctly treated seeds are dull, but not deeply pitted. This is of prime importance because the correct period of exposure to the acid differs not only from species to species, but also from lot to lot within species.

3. Cover the dry seeds with concentrated sulfuric acid—the commercial grade (specific gravity 1.84, 95 percent pure) is satisfactory—and allow them to soak for the required period (usually 15 to 60 minutes). Careful stirring of the seeds reduces the length of treatment necessary. Violent stirring, however, must be avoided lest it injure the seeds. Treatment should be carried out at a temperature range of 60° to 80° F. If lower temperatures are used the seeds must be soaked longer; if higher temperatures are used, they must be soaked less.

4. Remove the seeds from the acid and wash them over wire screen in cool, running water for about 5 to 10 minutes to remove any residue of acid. Water should be applied copiously at the start and the seeds stirred carefully during rinsing.

5. The seeds should be dried carefully before using them, unless wet sowing is preferred.

Small lots can be treated by placing them in a suitable vessel, and following the steps just described.

Fifty-pound lots can be treated more easily by placing them in screen-wire cylinders reinforced with heavier wire, lowering them into sufficient acid, stirring as already described, draining them in the cylinder, and washing them for 10 minutes in a wire-bottomed box, first with several buckets of water and then with a hose (figs. 13 and 14). They should then be dried on screens or tarpaulins, or in a rotary drier. In large-scale treatments special care must be taken to see that temperature does not get out of control, with resultant injury to the seeds.[29] Costs depend upon wages and the price of acid, but have been reported as about 4 cents per pound in large operations.

A variation of the method just described is sometimes used. The seeds are spread out in a rather thick layer, enough acid poured on to moisten them thoroughly without permitting an excess to run off,

[29] More complete details, and thoroughgoing comparisons of acid treatments, mechanical scarification, and hot-water treatments of black locust seed, with untreated checks, based on both laboratory and field sowings, are given by Meginnis (*35*).

and the seeds stirred until every one is thoroughly coated with acid. At the end of the desired period of treatment, the seeds are washed and dried in the usual manner. There is no recovery of acid by this method, but it involves less danger of injuring the seeds by heating, especially in large lots (*16*).

F-291624

FIGURE 13.—Apparatus for acid treatment of black locust seeds.

The advantages of the acid treatment are: (1) It is highly effective for many species; (2) it requires little or no special equipment; (3) it can be done at a reasonable cost; (4) most of the acid can be recovered and reused several times at a considerable saving in cost; (5) the treated seeds can be held from a week to a month or more before sowing, without appreciable deterioration; (6) since the process leaves the seeds dry, firm, and unswollen, they can be sown with mechanical seeders as well as by hand; and (7) the resulting seedlings may be less subject to attack by pathogenic soil organisms than those from seeds treated by other methods (*35*), although in some instances scarified seeds have excelled acid-treated seeds in this respect (*11*).

F-291626

FIGURE 14.—Apparatus for rinsing acid-treated black locust seeds. Bottom of box consists of 16-mesh screen supported by poultry netting.

The disadvantages and limitations of the method are chiefly these:

1. The length of treatment must be worked out carefully and the temperature controlled, especially in large lots, to prevent serious injury to the seeds.

2. Acid, seeds, and containers must be handled with great care to avoid destruction of clothing and injury to workmen; in particular, care must be taken not to splash water into acid, as the resulting violent reaction may splatter acid into workmen's eyes.

MECHANICAL SCARIFICATION

Seeds of the legumes, eastern redcedar, and a few other species (*11, 16, 26, 35, 52, 54*) may be made permeable to water and their germination greatly improved by scratching the seed coats with abrasives. This may be done in a number of ways—by hand, by tumbling or churning the seeds in drums lined with sandpaper or in drums or concrete mixers containing coarse sand or sharp gravel, by abrading them between a revolving sandpaper disk or a stone wheel and a stationary rubber disk, and by other means. The technique is widely used with seeds of leguminous forage plants, and a number of devices for scarification are on the market. The coarseness of abrasive, speed of operation, and duration of treatment will vary from species to species and even from lot to lot. Sand or gravel used in a revolving drum must be of such a size as to be separated easily from the seeds by screening; for black locust, gravel passing a ¾-inch but stopped by a ½-inch sieve has worked well.

The treatment may be controlled by germination or soaking tests, as described for treatment by acid, or often by examination with a hand lens and comparison of treated with untreated seeds. Saving time and labor is an important factor in choosing the equipment, but the paramount consideration, of course, must be the effectiveness and uniformity of the results and the ability to stop short of any appreciable injury to the seeds.

A scarifier for forest-tree seeds which seems particularly flexible, efficient, and easily controlled in its action consists of 6 vertical, sandpaper-covered disks, 0.25 inch thick, 10.5 inches in diameter, and mounted on a 0.75-inch steel shaft turning at 500 to 900 r.p.m. inside a horizontal cylinder also lined with sandpaper (*54*). The cylinder is 24 inches long and 11 inches in diameter inside, is made of 28-gage galvanized iron, with ends of 1-inch pine, and is divided lengthwise into halves, which are fastened together with hinges to permit filling and emptying, and the replacement of worn abrasive (preferably No. 2½, grade 30E, silicon carbide paper, fastened on with casein glue). A narrow, screened slit running the length of the bottom half lets dust escape. Removal of a pin permits the opened cylinder to turn about the shaft so that seed can be poured out (fig. 15).

F-327823-327822

FIGURE 15.—Disk scarifier developed at Lake States Forest Experiment Station: *A,* Exterior view; *B,* interior view.

In operation the lower half of the cylinder is filled with seeds. An 11 by 24-inch cylinder holds about 10 pounds of tree seeds of any size. Enough movement and mixing takes place to insure very uniform scarification. Only a few minutes' treatment is required for black locust and similar seeds, but harder-coated species require appreciably longer treatment. The seeds must be essentially free of resin or soft pulp to prevent clogging the abrasive surfaces. For large-scale production, larger cylinders are suggested, possibly with silicon carbide wheels instead of sand-covered disks. The rate of revolutions and length of treatment must be worked out by actual test for each size of scarifier and species of seeds (*16*).

The chief advantages of scarification are: (1) It is highly effective for many species; (2) it requires no temperature controls; (3) it involves little or no danger of injury to workmen; (4) the seeds keep well for a short time following treatment, although with some species they endure long-time storage less well than unscarified seeds; and (5) the seeds may immediately be sown dry, thus making possible the use of drill seeders.

The major disadvantages of scarification are: (1) It requires special equipment, except for very small lots; (2) the seeds must be largely free of resin or soft pulp; and (3) scarified seeds often are susceptible to injury from pathogenic organisms.

SOAKING

Soaking in hot water is frequently recommended for seeds of legumes, particularly the larger seeded species. Ordinarily the seeds are placed in four to five times their volume of water which has been brought to a boil (temperature usually varies from 170° to 212° F. at the time the seeds are immersed) and allowed to soak in the gradually cooling water for about 12 hours. Occasionally the seeds are actually boiled in the water for several minutes, but this is injurious to most species.

The favorable aspects of the hot water treatment are: (1) It is simple to apply and inexpensive; (2) no special material or equipment is needed; (3) it is fairly effective with some species.

Hot water treatment has these unfavorable aspects: (1) With many species it leaves the seeds soft, often somewhat swollen, and frequently adhering to each other, so that mechanical sowing is difficult and the likelihood of injury to the seeds is great; and (2) results have been erratic. The erratic results very likely have been due to failure to specify the temperature of the water and especially to allow for different rates of cooling if, as is usually directed, the seeds are placed in hot water and allowed to cool. Water near boiling is usually recommended, but it will stay near boiling much longer in a warm room than in a cold shed. The temperature of the seeds and the ratio of their weight to the weight of the water will also affect rate of cooling, as will the total mass of water and seeds once they have come into equilibrium with each other. A spoonful of seed placed in a cupful of water at 200° F. and put aside to cool for 12 hours will be subjected to conditions very different from those surrounding 100 pounds of seed placed in 20 gallons of water initially at the same temperature and set aside to cool for the same period. All these factors require standardization if the method is to become very dependable.

A modification of the hot water treatment involves soaking or boiling the seeds in hot water to which ordinary baking soda has been added. In one test made by the Lake States Forest Experiment Station boiling for 3 minutes in a 1-percent solution of baking soda proved injurious to black locust seed. It softened the seed coats, however, and the treatment may be useful with other species or if applied under different conditions.

Seeds approaching the border line between dormancy and merely sluggish germination respond to soaking in cold water. If they require soaking for more than 24 hours, it is advisable to change the water once or twice daily, or use running water to prevent oxygen depletion and "souring."

A water-soluble substance in the seed balls of sugar beets (*55*) has been found to inhibit germination unless removed by soaking, preferably in several waters. A similar condition is reported for guayule (*Parthenium argentatum* Gray) seeds, and may be true of the seeds of some forest trees. The improvement in germination caused by soaking seeds of several species in cold water is, however, ordinarily attributed to softening of the seed coat or the insuring of complete absorption by the living tissues. Seeds of green ash, for example, will germinate in a relatively short time if sown soon after collection from the tree, but after 3 to 4 years' dry storage it requires as much as 2 weeks' soaking to make the seed coat permeable.

VARIATION IN SEED COAT DORMANCY AND PRETREATMENT REQUIRED

Oftentimes freshly collected seeds do not have inpermeable seed coats, whereas after drying seed coats of the same species are impermeable. Three examples may be cited to illustrate this variation: (1) Freshly collected seeds of Indian-currant coralberry sown in a Wisconsin nursery gave satisfactory germination without pretreatment, while seeds that were dried before sowing were dormant; (2) fresh American elder seeds are ground up in the gizzards of ring-necked pheasants, while dried seeds largely pass through intact; (3) freshly collected hawthorn seeds are injured by sulfuric acid which penetrates and destroys the embryo, but after several weeks of drying at room temperature such treatment can be used safely (*20*).

The most effective treatment for breaking seed coat dormancy varies from species to species, even within the same genus. The information in table 6 illustrates this point. Figure 16 shows the effectiveness of several pretreatments on overcoming dormancy of one sample of black locust seed.

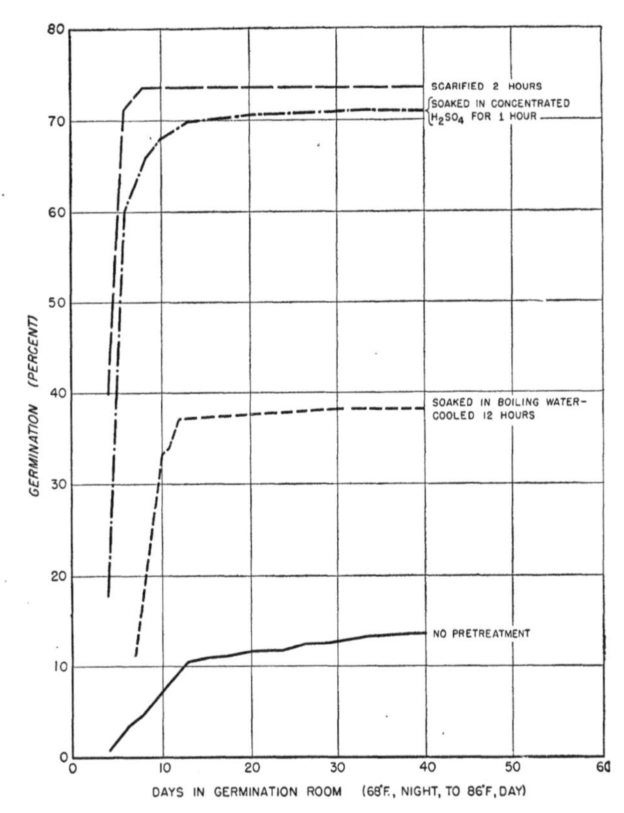

FIGURE 16.—Effect of several pretreatments to overcome seed coat dormancy in black locust (*Robinia pseudoacacia*).

Internal Dormancy

The commonest type of seed dormancy is caused by internal conditions of the stored food or of the embryo itself; 43 percent of 444 species studied showed this type of dormancy. In most cases certain chemical changes in the stored food or embryo must take place before germination can begin, but in others the embryo is unusually small at the time of seed dispersal, and it must undergo some growth before germination is possible. The latter is true of European ash (*32*) and Swiss stone pine (*43*). Dormancy is localized in a certain part of the embryo of some species. In several viburnums it is the plumule that is dormant (*22*). Unfortunately there is no way to distinguish seeds with internal dormancy by their appearance or by simple, quick tests. With unfamiliar seed, the safest procedure is to make some small exploratory germination tests involving a few simple pretreatments and a control.

TABLE 6.—*Variation in effective pretreatments to overcome seed coat dormancy in Amorpha and Rhus*

Species	Occurrence of dormancy	Most effective pretreatment known
Dwarf-indigo amorpha	General; high proportion of dormant seed in each lot.	Soak in concentrated sulfuric acid for 5–8 minutes.
Leadplant amorpha	do	Scarify seed coats lightly.
Laurel sumac	General	Expose seed 5 minutes to temperature of 200–240° F.
Lemonade sumac	do	Soak in concentrated sulfuric acid for 4 or more hours; longer treatments for older seed.
Smooth sumac, staghorn sumac	do	Soak in concentrated sulfuric acid for 60–80 minutes.
Sugar sumac	do	Soak in concentrated sulfuric acid for 1–6 hours.

If the untreated control germinates satisfactorily, no special treatments are necessary. If not, dormancy is indicated, and the most promising type of pretreatment should be followed.

Seeds having internal dormancy usually are dispersed in the autumn, lie on the ground, often partially covered with litter over winter, and germinate in the spring. Some of them may lie over until the second or third spring before they germinate; occasionally some lie dormant for several years. In the latter cases seed coat dormancy probably is also involved. Treatments to overcome internal dormancy must induce the same changes in the seeds as occur in nature.

COLD STRATIFICATION

The most widely used method of breaking internal dormancy is to expose the seeds to abundant moisture and adequate oxygen, at a temperature usually between 32° and 41° F. for a period of 1 to 4 months. This approximates conditions to which the seeds are exposed in overwintering on the ground out-of-doors, and is commonly known to horticulturists, nurserymen, and seed technicians as cold stratification.[30] It has sometimes resulted in germination of 80 to 90 percent in 15 to 20 days, as contrasted with 10 to 20 percent in 60 to 90 days for untreated seed.

To apply the cold stratification treatment the following materials and equipment are necessary or desirable:

1. A suitable moisture-retaining medium, such as sand,[31] acid granular peat, or chopped sphagnum moss, to separate and aerate the seeds. Well-weathered pine sawdust has been used without mishap, but sawdust in general, especially when fresh, should be tested carefully on a small scale to make sure it has no toxic or harmful heating effect.

2. Suitable containers, such as trays, boxes, tanks, cans, or barrels. The container should protect the seeds, etc., from drying but be well drained from the bottom. Boxes, tanks, drums, or barrels must have either perforated or false bottoms several inches above the true bottom, to keep water from accumulating and shutting off the supply of oxygen from part of the seeds. Shallow trays may work all right without perforations.

3. A refrigerator, cold room (see fig. 17), cold storage plant, or ice house. Temperatures controlled within a range of 32° to 50° F. are preferred. Closer control is necessary for best results with some species.

F-433112

FIGURE 17.—Cold room used for stratifying and storing seed. A constant temperature of 41° F. is maintained in the room, and small compartments can be kept at either 34° or 50° F.

In carrying out cold stratification:

1. The seeds should either be mixed uniformly with about 1 to 3 times their volume of the medium, or layers of seeds ½ to 3 inches thick should be alternated with similar layers of the medium. Thin layers of seeds may be kept separate from, yet in intimate contact with, the medium by separators or sacks of cheesecloth, or other light fabric. This method is particularly useful where many small samples must be stratified and danger of mixing must be avoided.

2. The containers of seeds and moist medium should be placed in the cold room or its equivalent and held at the desired low temperature (commonly 41° F.) for the pretreatment period (usually 30 to 90 days). The moisture content of the stratification medium apparently should be as uniform as possible within the container and evidently not so high as to prevent aeration, but otherwise may be varied within wide limits.[32] Peat, sphagnum, or sawdust usually are moistened enough at the start so that a little free water can be squeezed out easily by hand, and ample drainage is provided below to take off any excess water that may be added during treatment.

[30] Some technicians dislike this term because the seeds often are mixed freely with the moist medium rather than placed in strata or layers. There is also some quarrel with other terms used to denote this treatment, especially "storage" and "after-ripening." The former is in disrepute because the object of the process is not to maintain the seeds in a constant, stable state as in true storage; the latter because no food substances are added to the seeds nor, so far as is known, are any complicated food substances built up. Such terms as "incipient germination" and "pregermination" have been suggested as preferable. However, because the term is widely used and generally understood, "cold stratification" is used in this bulletin to refer to this method of treatment.

[31] Sand, because it reduces the tendency to heat, is reported to be superior to any organic materials for stratifying seeds that retain an appreciable coating of pulp from the fruit. Ordinarily "sterile" sand obtained from deep wells, etc., is preferred over surface sand, and washed over unwashed sand. It is always preferable to use the safer of any two media recommended; where accessibility or cost dictate the use of the less preferred, a small-scale "contrast treatment" should be set up with the other medium.

[32] MacKinney and McQuilkin (*33*) stratified loblolly pine seeds in both sand and peat, each at moisture contents of 25, 50, 75, and 100 percent of water-holding capacity. They found both media and all four moisture contents equally effective in stimulating germination.

Stratification has been tried at temperatures from slightly below 32° to about 68° F. (*2, 4, 5, 33*), and a number of optima have been worked out. In general, freezing is of doubtful benefit or distinctly harmful; the higher temperatures are often less effective and more likely to result in premature sprouting or injurious heating of the seed; and the best range of temperatures appears to be from 34° to 41°, although 50° seems to be optimum for some species. If evaporation can take place freely from moist peat at 38°, the peat and seeds may freeze as a result of the further cooling induced by evaporation.

Recommended periods of treatment differ widely. Some species require 3 or 4 months or more. Others respond well in a month, and a few after only 15 days of stratification. In cases of doubt, 30 to 45 days should be tried, with a "contrast treatment" of double that period tested or sown simultaneously with the main lot. A striking illustration of the effects of different stratification periods is shown in figure 18, in which low and sluggish germination of untreated balsam fir seed is contrasted with the increasingly rapid and more complete germination with greater stratification periods.

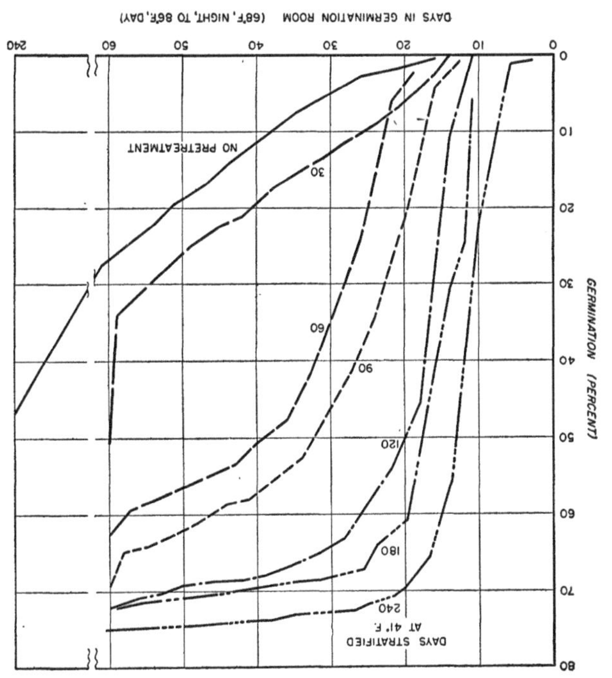

FIGURE 18.—Effect of stratification period on germination of balsam fir (*Abies balsamea*) seeds.

3. Remove the containers from the cold room and separate the seeds from the stratification medium. If the seeds have been kept in sacks or separated by cloth, this is a simple matter. If the seeds are intermixed with the stratification medium, they may be separated out by flotation (as described on page 44) by washing in a device like a placer-miner's cradle, or by drying seeds and medium just enough to permit cleaning by screems or in a fanning mill. If screens are used, a medium such as sand or sawdust should be screened in advance to a size that will permit easy separation from the seeds when stratification is complete. (In some cases it may, of course, be feasible to sow the medium with the seeds.) Cleanings between pretreatment and sowing must be made carefully to avoid injuring the now highly sensitive seeds by bruising or too much drying. They also tend to float or blow out a certain percent of either empty or sound seeds, thus altering the germination percent and consequently the quantity of seeds to be sown on each bed. If the sowing rate can be controlled, however, this final removal of empties may be advantageous.[33]

Seeds should be sown soon after removal from stratification. If allowed to dry out, they may be injured or go into secondary dormancy, which often is much harder to break than the original dormancy.

The idea of cold stratification was arrived at largely through consideration of both the life history of seed under natural conditions, and of the empirical successes attained by fall sowing in the nursery (*4*). Stratification may give good results even when neither temperature nor period of treatment are closely controlled. Hardwood seeds stored in moist sand in a pit or pile outdoors under cover enough to prevent freezing, or pine seeds sown in the winter in the Gulf States, or fall-sown and heavily mulched in the North or West, must go through about the same physiological processes as those pretreated under more precise control in a refrigerator. Certainly all give good results.

For testing purposes seeds may be sown in flats and the entire flat held at low temperatures. At the end of the stratification period the flat may be removed and placed in the germination room without disturbing the seeds. Where tests are made in Jacobsen or other standard germinators, a variation of stratification known as prechilling frequently is used. The seeds are placed on moist blotting paper, or comparable medium, in the germinator and the whole apparatus held at low temperatures (usually 41° F.) for the desired length of time.

Large-scale stratification has often been done in the following manner: Pine seeds, for example, are weighed out in 20-pound lots, and each lot tied very loosely in a cotton sack, like an old flour sack. Seeds and sacks are soaked overnight—no longer—in cold water. The next morning five such sacks are packed in a wooden barrel or metal drum the bottom of which contains several holes to permit drainage. The seeds in each sack are spread out to make a layer about 3 inches deep, and these cloth-enclosed layers are alternated with layers of wet peat or sphagnum moss. Each barrel, when full, contains six layers of peat alternated with five of seeds, representing an equivalent of 100 pounds of seeds on a dry-weight basis. The barrels are then

[33] One southern nurseryman, as described in the footnote on page 27, uses the cold-stratification process to dewing pine seeds, and the cleaning process between pretreatment and sowing to fan out not only the peat moss, but also the wings and empty seeds. This eliminates practically the entire cost of the usual cleaning after extraction.

refrigerated for the desired period. They are inspected weekly and, if need be, rewatered and repacked to prevent drying or heating. When treatment is complete and the seeds ready to sow, the sacks of seed may be separated from the moss or other medium without loss, and the known dry-weight of each sack makes it easy to apportion accurately the amount of seeds to be sown per unit of seedbed.

The advantages of stratification are:

1. It not only hastens but also frequently improves germination for many species. The combination of cold and moisture has a marked and favorable effect on transformation of complex food substances into simpler substances readily utilized for germination, and in some cases softens the seed coat and brings about other beneficial changes. As a result germination not only is better but usually is much more uniform.

2. Both dry-stored and moist-stored seeds may be treated safely. With dry-stored seeds, as of the pines, stratification restores their moisture content to the level necessary for active growth. With moist-stored seeds, such as acorns, it would be difficult to say where storage ended and cold stratification began, but germination of stratified samples usually is superior to that of freshly collected ones.

3. It does not require specialized equipment or material if applied as outdoor stratification.

4. There is little danger of injury to the seeds through reasonable overtreatment or of injuries to workmen in applying the treatment.

Some disadvantages of cold stratification are: (1) The treatment often requires a long time; (2) as usually practiced, it requires some special equipment such as cold rooms and special containers; (3) difficulties of separating the seeds from the stratifying medium may increase the cost unduly; (4) calculation of sowing rates is made more difficult by increases in seed weight through absorption of moisture; (5) seeds must be sown soon after removal from low-temperature conditions or heavy losses may ensue; (6) if the seeds have been injured in dewinging or cleaning, cold stratification may intensify the injury, and (7) in large-scale treatments the seeds may heat severely, despite the cold, and cause serious losses.

WARM FOLLOWED BY COLD STRATIFICATION

The seeds of some species normally are dispersed in late summer or early fall, and usually are exposed to a period of warm weather and then the cold of winter before they germinate in the spring. Seeds of some such species do not germinate normally until a succession of similar conditions has been experienced. Seeds of some viburnums subjected to ordinary germination temperatures produce roots but no shoots. However, when stratified at these warmer temperatures and then at low temperatures they germinate normally (*22*). Warm followed by cold stratification has been found effective for a considerable number of species, and it is likely that further tests will show it to be useful for many more.

To carry out this practice, the seeds are placed in moist sand, peat, or other media just as in cold stratification. Then they are held at germination temperatures, commonly 68° (night) to 86° F. (day), for 1 to several months, after which they are shifted to cold chambers or refrigerators and held at low temperatures, usually 41°, for 1 or more months longer. In practice, these conditions can often be met by sowing the seeds soon after collection in the late summer or early fall. Such treatment has proved highly effective with such species as American hornbeam, prairie crab, and eastern hophornbeam (*46*). Where summer sowing is impossible and stratification facilities are not available, the seeds might just as well be held for sowing until the following summer, since seedling production will be delayed until the second spring anyhow.

CHEMICAL TREATMENT

Some work has also been done on breaking of dormancy by means of chemicals. Thus Deuber (*14*), following leads obtained by other workers in breaking the dormancy of potato tubers, considerably increased the rapidity and extent of germination of red oak and black oak acorns both by soaking them in a 3-percent solution of thiourea for 15 minutes, and by exposing them for 24 to 96 hours to the fumes of 4 milliliters of ethylene chlorhydrin per 100 acorns and per liter of atmosphere in sealed bottles. A few such methods are noted in part 2 of this manual, but none appears to have been applied to any great extent.

VARIATION IN INTERNAL DORMANCY AND PRETREATMENT REQUIRED

Some of the various treatments required to break internal dormancy in different species, even those within the same genus, are shown in table 7.

Double Dormancy

Out of the 444 species described in this manual, 17 percent have combined seed coat and internal dormancy. In some of these species, both kinds of dormancy are severe; in others, both are mild; while in many, one kind may be severe and the other mild. In a good many species some lots of seeds may have only one kind of dormancy (usually internal), while others have double dormancy.

To overcome double dormancy it is necessary to treat the seeds so as to make the seed coat permeable and to induce in the embryo or stored food the changes essential for germination. Sometimes cold stratification is sufficient, but more often hot-water soaking, acid treatment, or scarification followed by cold stratification is necessary. In many cases warm followed by cold stratification gives good results. In a test made at the Lake States Forest Experiment Station, pin cherry seeds germ-

TABLE 7.—*Pretreatments to overcome internal dormancy*

Species	Occurrence of dormancy	Most effective pretreatment known
Firs	Variable; in some, but not all lots for all species so far tested.	Stratify in moist sand or peat at 41° F. for 30–90 days.
Fourwing saltbush	General	Hold in dry storage for 12–18 months.
Black spruce, red spruce	do	Stratify in moist sand at 41° for 60–90 days.
Blue spruce, Brewer spruce	Variable; some seed in each lot.	Stratify in moist sand at 41° for 30–90 days.
Sitka spruce	Variable; some lots only	Stratify in moist sand at 41–50° for 60–90 days.
White spruce	General	Stratify in moist sand at 41° for 60–90 days.
Blackhaw viburnum	do	Stratify in moist sand or peat at 68° (night) to 86° (day) for 150–200 days; then at 41° for 30–45 days.
European cranberrybush viburnum.	do	Stratify in moist sand or peat at 68° (night) to 86° (day) for 60–90 days; then at 41° for 30–60 days.
Mapleleaf viburnum	do	Stratify in moist soil or peat at 68° (night) to 86° (day) for 180–520 days; then at 41° for 60–120 days.
Witherod viburnum	do	Stratify in moist sand at 68° (night) to 86° (day) for 60 days; then at 50° for 90 days.

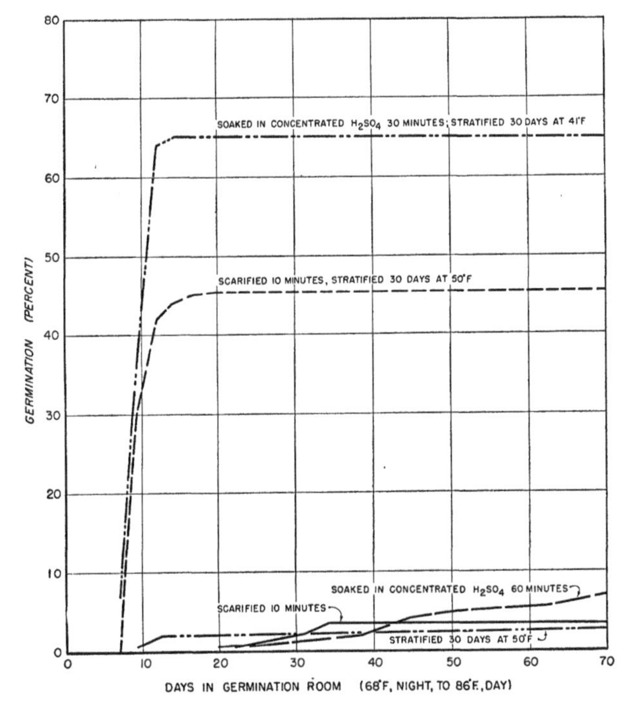

FIGURE 19.—Effect of several pretreatments to overcome double dormancy in eastern redbud (*Cercis canadensis*).

inated 60 percent in 11 days after being stratified for 60 days at 68° (night) to 86° F. (day) and an additional 90 days at 41°; seeds stratified 90 days at 41° gave no germination in 60 days, and untreated seeds gave none in 200 days. The variability in dormancy and the kinds of pretreatment required are shown in table 8. For the effect of several pretreatments to overcome double dormancy in eastern redbud see figure 19.

Possible Injuries in Large-Scale Pretreatment

There is a pronounced tendency for pretreatment practices worked out on an experimental scale to develop flaws in large-scale application. Scarifiers of efficient size for large lots of seeds do not work at the same rate as small laboratory models. Temperature during treatment with acid gets out of control more easily with large masses of seeds than with small, with resultant injury to the seeds.

Probably the most difficult pretreatment hazard to combat, however, is that of heating of seeds in the refrigerator. Seeds placed in commercial refrigeration in barrels for pretreatment have occasionally heated so severely, despite the cold, that the hand could not be thrust down into the barrel, and

TABLE 8.—*Pretreatment to overcome double dormancy*

Species	Occurrence of dormancy	Most effective pretreatment known
Blueblossom ceanothus, buckbrush ceanothus, coast whitethorn ceanothus, deerbrush ceanothus, hoaryleaf ceanothus, Jerseytea ceanothus, snowbrush ceanothus, squawcarpet ceanothus, trailing ceanothus.	General	Place in hot water at 170°-212° F. for 12 hours (overnight), then stratify in moist sand at 41° for 60–90 days.
Jimbrush ceanothus	do	Boil in water for 5 minutes, then stratify in moist sand at 41° for 90 days.
European birdcherry, Appalachian sand cherry, pin cherry.	do	Stratify in moist sand at 68° (night) to 86° (day) for 60–90 days, then at 41° for 60–120 days.
Black cherry	do	Soak in concentrated sulfuric acid for 30 minutes, then stratify in moist sand at 41° for 120 days.
Amur maple	In some lots; others appear to have embryo dormancy only.	Scarify lightly, then stratify in moist sand at 41° for 60–90 or more days.

have been ruined by this extreme temperature.[34] In cases reported with pine seeds, the heating has usually, if not always, occurred after the fiftieth day of treatment. Soaking the seeds in cold water may offset this tendency. Some nurserymen mix a considerable quantity of cracked ice with the seeds and medium when they are packed in the barrels or trays, partly to maintain correct moisture by melting, but largely to insure thorough initial chilling of the mass. Frequent inspection and, if need be, repacking of the seeds, may prevent heating, but the best safeguard is to avoid long periods of treatment. Experience in large-scale practice with pines indicates that safe and effective periods for cold stratification of many species may be 30 to 45 days, even though laboratory tests may have given better results after fully 2 to 3 months.

VARIATIONS IN OCCURRENCE OF SEED DORMANCY

That seed dormancy varies among closely related species not only in occurrence but also in type and intensity has already been mentioned. To illustrate this variation, a few examples are cited. Most maples so far studied have internal dormancy. However, in California boxelder some lots are nondormant while others have internal dormancy; in Amur maple all lots have internal dormancy, but some have seed coat dormancy in addition; while in silver maple there is no seed dormancy. Feltleaf ceanothus, hairy ceanothus, and Monterey ceanothus have general seed coat dormancy, but Jerseytea ceanothus, hoaryleaf ceanothus, and all other species studied have combined seed coat and internal dormancy. Flowering dogwood and possibly silky dogwood have general internal dormancy, whereas alternate-leaf dogwood, roughleaf dogwood, bunchberry dogwood, and gray dogwood have combined seed coat and internal dormancy, pacific dogwood has double dormancy in older seeds but none in fresh seeds, and redosier dogwood has internal dormancy alone in some lots and double dormancy in others. White spruce, black spruce, and red spruce have general internal dormancy; in Brewer spruce and blue spruce, some seed in each lot have internal dormancy, whereas others are nondormant; in Sitka spruce, some lots have internal dormancy and others none; and in Engelmann spruce and western white spruce, seeds are nondormant.

Among the pines there is great variability. Such species as Coulter pine, shortleaf pine, eastern white pine, loblolly pine, and Torrey pine have general internal dormancy as a rule. Others, such as Japanese red pine, border limber pine, spruce pine, sugar pine, Austrian pine, pitch pine, pond pine, and Japanese black pine, commonly have internal dormancy in some lots but none in others. In still others, such as bristlecone pine, jack pine, slash pine, lodgepole pine, Jeffrey pine, longleaf pine, ponderosa pine, and Scotch pine, internal dormancy may occur rarely or occasionally. A few, such as whitebark pine, limber pine, singleleaf pinyon, western white pine, and Digger pine, may have both seed coat and internal dormancy but both types are not always present. One species, Swiss stone pine, appears to have double dormancy as a general rule. Many other examples could be given.

[34] It has been suggested that in such cases the masses of seeds and peat acted as insulators to retain heat, and that either the respiration of the seeds in the center offset the effect of the refrigerator from the start, or that after stimulation by the treatment, the seeds began to respire faster during defrosting of the refrigerator and to be warmed up more and more by the heat of their own respiration, so that they could not be cooled down again.

UNSOLVED PROBLEMS IN SEED DORMANCY

Much has been learned about seed dormancy, and workable pretreatments have been developed for many species. However, there are a considerable number of species of present or potential value for conservation planting for which some pretreatments have been worked out that either give inadequate germination or are too complicated or time-consuming for practical purposes. Among the species in this category are Saskatoon serviceberry, wild-sarsaparilla, American hornbeam, American chestnut, stiff bushpoppy, the fremontias, Fremont silktassel, witch-hazel, oldfield common juniper, Sierra juniper, eastern hophornbeam, Swiss stone pine, American black currant, blueberry elder, baldcypress, common prickly-ash, and common jujube.

In addition, there are several species—even whole genera—for which no consistently successful pretreatments have been discovered. These include some plants long grown by horticulturists through vegetative propagation. This method of regeneration, however, is relatively expensive and does not lend itself well to the large-scale practice required in reforestation. Accordingly, workable means of artificial reproduction by seed should still be sought. Among the species in this group are such as: Sweetfern, the euonymuses, the hollies (other than common winterberry), Alpine larch, American fly honeysuckle, limber honeysuckle, Tatarian honeysuckle, mountain-holly, mountain-ash (American species), the yews, American basswood, and California torreya. It is quite possible that some combinations of chemical and mechanical treatments along with stratification will suffice for many of these species. Perhaps new chemicals are needed, or different alternations of warm and cold stratification.

In certain cases some new type of pretreatment not yet developed may be necessary. For some species which have seeds that no known treatments seem to stimulate into germinating, it may be that natural evolution has tended to foster suckering or other vegetative methods of reproduction at the expense of seedling regeneration. It is likely that little can be done to solve the problems of seed dormancy for such plants.

SEED TESTING

PURPOSE

SEED testing is essential in determining the amount of seeds to be sown per unit of area and the number of healthy seedlings that can be obtained. Seeds are generally tested for genuineness, purity, number per pound, moisture content, and viability. The tests are applied to seeds available in the trade or collected for more or less immediate sowing.

The rate of sowing in nursery beds is determined largely by the quality of seeds used, which, in turn, depends on the source from which they were obtained, the manner in which extracted, age, and the conditions under which they have been kept. In large quantities of seeds, these factors are largely obscured, but their influence on the seeds can be determined by certain measures and tests. Seed testing is relatively simple and can be done by the nurseryman, if he has the facilities and time, or by some agency which furnishes such service.

In most European countries forest tree seeds cannot be sown until their quality has been established at government-owned laboratories, but in this country no certification of forest seeds is required except in New York and Georgia. Since few American seed dealers even certify seeds as to their origin and age or guarantee their viability, it behooves purchasers to make some provision for testing in order to avoid losses.

Whether the nurseryman or the seed laboratory does the testing, certain basic principles must be followed in order that the test samples may give a reliable evaluation of the seed lots as a whole.

EXAMINATION BEFORE GERMINATION

Selection of Samples

Since all tests, except those for number of seeds per pound and purity, usually destroy the seeds tested, and since the counting, drying, germinating, and cutting involved in tests make large demands upon time and equipment, samples from seed lots are necessarily small. Care in their selection is necessary, for samples are drawn from lots as large as several thousand pounds. An accurate estimate of the quality of the seeds cannot be obtained unless the sample tested is truly representative of the entire lot.

Representative sampling is attained either by very thorough mixing of the seeds before sampling or by drawing a number of small subsamples of equal size, from different parts of the lot, with care to make the number from each part porportionate to the quantity of seed in that part. The reason for this apportionment of subsamples is that variations are likely to exist between and within the parts of the seed lot. Seeds in one container may deteriorate more rapidly than those in another, and if the poor container contains twice as many seeds as the good one, it should contribute twice as many seeds to the sample representing the entire lot. In like manner seeds in any one container, unless thoroughly mixed immediately before sampling, should be sampled by drawing subsamples from the top, side, center, and bottom, because of the tendency of empty seeds to work to the top, seeds next to the side to fluctuate over a wider range of temperature during shipping, and so on. Representative sampling may be accomplished as follows:

1. Preferably sample every container. However, where both seed and storage conditions are quite uniform and there are more than 30 containers in the lot, it may be permissible to sample only alternate containers.

2. Draw the subsamples from the different containers in such a manner that until the sample is taken, any one seed has just as good a chance of being included as any other seed. Small seed may be sampled by means of probes (fig. 20) thrust in closed position to the bottom of the container, opened and allowed to fill with seeds and then closed before removal. Obtain several such samples to insure that the sides and center are represented as well as the top and bottom of the container. If a probe is unavailable, or large seeds are to be sampled, the lot may be poured from the container and several samples drawn from the top, middle, and bottom.

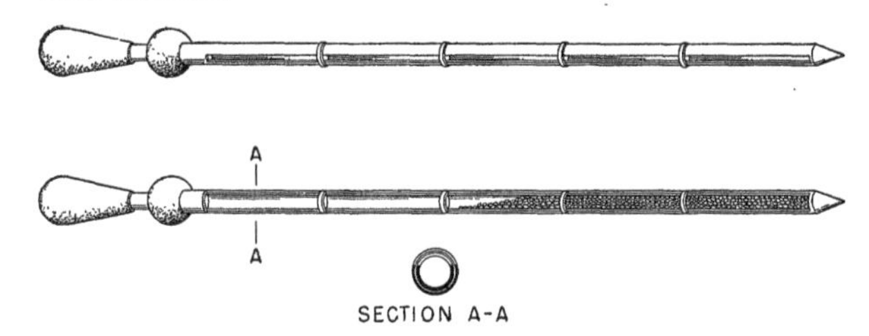

FIGURE 20.—Probe used to draw samples of small seed: *A*, Partly open; *B*, fully open and ready to discharge seed.

3. Mix the subsamples together thoroughly and reduce the whole to a convenient working size, which will depend upon the purpose of the test. One of the best methods is to heap the seeds in a cone-shaped pile, pouring them through a rigidly supported funnel. The seeds are then scooped continually from one side of the cone and piled in a second cone in the same way. This process should be repeated at least four times to insure good mixing. The seed pile may then be quartered by divid-

ing it vertically and one of the quarters treated in a like manner until the desired quantity of seeds for testing is obtained. Mixing and sampling in the manner described may be practiced easily by using two 1-pound lots of seeds, one of which has been dyed with ink, mercurochrome, or any other convenient stain. These operations should be repeated until consistent 50–50 counts are obtained from small lots. This practice will be even more instructive if one of the lots is made up of blind or light seeds.

For exploratory studies, samples of 50 or 100 seeds may suffice. Such studies may be used to see whether seeds of an unfamiliar species germinate promptly or are dormant, or for a rough check (cutting test) on whether seeds have been fanned or have been injured in kiln extraction or storage. A homogeneous seed lot of 100 pounds or less requires about 400 seeds, tested separately in four equal parts, for a germination test sensitive enough to determine rate of sowing in the nursery. A lot larger than 100 pounds, or lacking homogeneity, or both, may require 800 to 1,000 seeds for the same purpose.

Theoretically, fewer seeds are required for tests of lots with germination above 80 percent or below 20 percent than for those with germination of about 50 percent. In practice the number required at a germination of 50 percent should always be used, since the attainment of germination above 80 or below 20 percent cannot be assured until the test is about complete. This applies to routine "service" germination tests for the computation of sowing rate. Various research germination tests may require from 100 to 1,000 or 2,000 seeds of each treatment, depending on the design of the individual experiment. A more complete discussion is given under Number to be Tested, page 47.

The sizes of samples used in determining purity, moisture content, and number of seeds per pound are governed in part by the uniformity and other characteristics of the material (the cleaner and more uniform the seeds, and the smaller their size, the smaller the weight of the sample), but to a large extent also by the sensitivity of the balances available.

Sampling should be done by a trained man, since despite all the mechanical devices suggested for obtaining a uniform, representative sample, there is need for constant alertness and common sense in recognizing sources of variation in seeds and in adapting sampling to include these variations.

Genuineness

The botanical identity of seeds is determined by comparing a representative sample of the lot in question with samples whose identity is definitely known. In most cases an examination made with a hand lens of external characteristics such as color—solid or spotted—form, size, texture, and appearance of the micropyle, will be sufficient. In other cases, it may be necessary to examine the seeds in cross section so that such characteristics as color and texture of the inner seed coat and endosperm and the size, form, and location of the embryo will be visible. Some species have typical odors,[35] which may prove useful in their identification. Another characteristic of possible value in seed identification is the chemical nature of water-soluble extractives. For instance, some samples of white ash seeds soon after sowing stain the surface of the soil covering a dark brown, whereas green ash seeds, which are often difficult to distinguish from those of white ash, do not appear to do this. Certain species of buckthorn and viburnum also produce such stains.

Since most nurserymen do not have access to a collection of authentic seed samples, they are ordinarily unable to check genuineness. The nurseryman can, however, keep in close contact with the source of his seed supply, buy seeds only from collectors and dealers who will guarantee their genuineness, or send samples to a local State seed laboratory for checking.

Purity

The percent by weight of clean whole seeds true to species in a sample containing seeds and mixed impurities is the commonly accepted expression of seed purity. Written as a formula,

$$\text{Purity} = \frac{\text{Weight of clean seeds}}{\text{Weight of seeds plus impurities}}$$

For instance, a sample of eastern white pine seed as received for testing weighed 27.70 grams, and the 1,303 clean seeds it contained weighed 24.49 grams. By substitution, its purity $= \frac{24.49}{27.70}$ or 88.4 percent.

Purity is determined by individual inspection of the test sample, the pure or clean seeds being all those which appear sufficiently well formed, whole, normal, and uninjured as to give promise of germination. This practice differs from that used with agricultural seeds, in which any fragment larger than one-half a seed is considered a pure seed. On the basis of the above definition, any seed outwardly normal in appearance, even though it may lack internal development, is counted as a pure seed since inspection alone does not reveal the defect. On the other hand, a seed which shows checks or fissures in the seed coat, although an apparently sound kernel may be present, is classed with the impurities, as is also a seed with worm holes.

Seeds of larger-seeded species generally are of high purity, and those of small-seeded species, such as alder and birch, usually are relatively impure because of cone scales, leaves, and other debris difficult to remove in the cleaning process.[36]

[35] Seeds of certain species of ash, notably such members of the subgenus *Bumelioides* as black ash, blue ash, and Manchurian ash, have characteristic aromatic odors not known to occur in those of other species outside this group.

[36] Certain large seeds, notably those of baldcypress, however, may also be low in purity because the cone scales and seeds are quite similar in size and specific gravity and hence are separable only by hand cleaning.

For species in which cleaned seeds ordinarily are not sown, the purity percentage is determined on the basis of the material commonly used, whether it be entire fruits, fruit fragments, or seeds only partially dewinged. Examples are the seeds of longleaf pine, the corky wing bases of which it is virtually impossible to remove; or seeds from drupelike fruits such as those of hackberry, which usually are sown with the dried pulp still adhering to the stone. For all practical purposes, purity determinations to the nearest 1 percent are sufficiently accurate.

Number Per Pound

The number of seeds per pound, a highly important figure in determining the rate of sowing, is affected by their size, moisture content, and the percent of sound or filled seeds in the lot. Since the seeds used by the nurseryman are rarely 100 percent pure, this value is usually expressed in two ways: Number of clean seeds per pound of unclean seeds (clean seeds plus impurities); and the number of clean seeds per pound of pure seed. For instance, the sample of eastern white pine seed already mentioned, including seeds and impurities, had 21,337 clean seeds per pound, but after being freed of its impurities had 24,134 clean seeds per pound. The first expression commonly is used in reporting tests on seed samples submitted by the nurseryman, for it represents the number of clean seeds there are per pound in the seed lot as he intends to sow it. The latter is used as a standard for evaluating the relative size of seeds. These values can be obtained at the same time that purity is determined by weighing out a sample of 1,000 or more seeds, depending on the size and the balance available.[37] The seeds are then counted out and all impurities, such as needles, twigs, wing fragments, resin grains, and split or otherwise deformed seeds, removed. This waste is weighed. The two values may then be computed as follows:

(1) Number of clean seeds per pound of unclean seeds=

$$\frac{\text{Total number of clean seeds} \times \text{number of units of weight in 1 pound}}{\text{Weight of sample of unclean seeds.}}$$

Similarly, (2) number of clean seeds per pound of pure seeds=

$$\frac{\text{Total number of clean seeds} \times \text{number of units of weight in 1 pound}}{\text{Weight of sample of pure seeds.}}$$

Example: A sample of white spruce seeds as submitted for testing weighed 5.95 grams; after cleaning, 5.55 grams; and contained 2,624 seeds.

1. Number of clean seeds per pound of unclean seeds= $\frac{2{,}624 \times 453.6}{5.95}$ or 200,041.
2. Number of seeds per pound of pure seeds= $\frac{2{,}624 \times 453.6}{5.55}$ or 214,459.

[37] For very small seeds, an analytical balance should be used, and weights obtained to the nearest milligram.

Moisture Content

Moisture content is usually expressed as a percent of the oven-dry weight of commercially cleaned seeds, not on a "pure seed" basis. Usually the seeds are dried to constant weight in an oven at just over the boiling point of water (101° to 105° C. or 214° to 221° F.). Since such high temperatures undoubtedly drive off some of the more volatile hydrocarbons, especially in the case of conifer seeds, it is believed that the use of lower temperatures with adequate humidity control, although requiring a longer time, is more satisfactory.

Moisture-content determinations are useful primarily as a guide to conditioning seeds for storage, but may also provide an index of their quality. Freshly collected seeds of certain maples, the white oaks, and buckeye have a high moisture content, 50 to 100 percent. If this is lowered very much, the seed quality is impaired. Drying silver maple seeds to less than 30-percent moisture content causes 100-percent loss in viability. On the other hand, a high moisture content in some conifer seeds, particularly if they are not fresh, should also be viewed with suspicion.

Excessive drying of seeds may be shown externally by wrinkled or checked seed coats, or internally by wrinkled, shriveled, or loose kernels.

TESTS FOR VIABILITY

Viability, sometimes called seed soundness, is the percent of seeds capable of germinating when exposed to the most favorable conditions. This may be determined either directly by germination tests or indirectly by such physical measures as cutting tests, the growth of excised embryos, flotation, size, color, and such biochemical methods as staining of embryos and the measurement of enzyme activity. The direct germination test is, of course, the most satisfactory method, but indirect methods can often be used in cases where the necessary time or the facilities for making germination tests are unavailable. This is particularly true in the case of some of the maples, for which no satisfactory germination technique has as yet been worked out, or the cucumbertree and certain shrubs such as elder and snowberry which require considerable periods of time for pretreatment before germination tests can be made.

Physical Measures

CUTTING TESTS

Cutting tests are useful for determining the percent of filled seeds—which, along with seed size, modifies the number of seeds per pound[38] —and to provide a quick approximate measure of the viability of seeds slow to germinate. They are not

[38] In addition to the variation in the number of seeds per pound caused by differences in seed size, there is, of course, considerable variation due to differences in the percent of blind or empty seeds present.

891183°—50——4

completely reliable because of the difficulty of distinguishing good seeds from rancid seeds and seeds injured during handling. They should be used as an accessory to all germination tests so that the results of the latter may be based upon the number of filled seeds sown (real germination) as well as the total number of seeds sown (germinative capacity). Cutting tests are often used as an approximate check on the development of cones and other fruits on prospective collection areas and as a measure of the degree to which empty seeds have been removed by fanning and other cleaning operations.

To make such a test the seeds are cut open[39] or broken with a hammer, the cross sections examined with a hand lens or reading glass, and the percent of viable seeds determined visually. Seeds which in cross section appear firm, plump, and of good color are considered viable. Color alone, however, is not enough; sugar maple seeds stored for 10 years had embryos as green as when they ripened, but the seeds were dead when tested for germination. Most viable seeds have whitish or cream-colored kernels. Those of immature seeds are frequently milky and do not have the firm texture of mature seeds. Classed as nonviable are those seeds which are moldy, decayed, shriveled, rancid smelling, blind (those lacking complete embryo and endosperm), abortive, filled with woody material (yellow-poplar) or with resin (fir), or wormy.[40]

GROWTH OF EXCISED EMBRYOS

A more exact estimate of viability can be obtained by observing the growth of excised embryos. Embryos dissected from the seeds are placed on moist, sterile cotton or in petri dishes and kept at about 70° F. for from several days to 3 weeks. During this period, viable embryos show various types and degrees of development, such as enlargement of the cotyledons, development of green color, and even emergence of the radicle, but those which are not viable turn yellow or brown and soon become moldy and decay (*21*).

The technique used for the dissection depends on the seed type. In some seeds with hard outer coats, such as plum and chokecherry, this is accomplished by removing outer coats by cracking in a small vise or in some such device as a pair of pliers equipped with an adjustable stop screw. (Properly, it is not the seed coat but the endocarp which is cracked in these cases.) In others, notably buffaloberry, hawthorn, eastern redbud, and snowberry, the outer coats are removed by soaking in concentrated sulfuric acid for a period sufficient to soften them. Still others, primarily those with relatively soft coats, such as apple, Douglas-firs, pear, and some pines, need only a short period in moist peat, or overnight soaking in water at room temperature to permit ready excision.

Once the outer seed coat has been removed, the seeds are soaked in water overnight and the inner coat slit by making an incision with a scalpel at the end opposite the embryo. By skillful manipulation (gentle prying and squeezing) the embryo can be made to slip out.

The chief advantage of this method lies in the fact that it permits the determination within a few days or weeks of the viability of seeds requiring several months for germination in soil or other media. In all but a few instances, however, it is believed that this advantage is outweighed by the difficulty of removing viable embryos without injuring them. This method, of course, can hardly be used for such small seeds as birch, tamarack, and northern white-cedar.

FLOTATION

Since the empty seeds of many species float in water while sound ones sink, this method is sometimes used to measure seed quality, particularly in heavy-seeded species such as oaks and beech. However, if the seeds are not freshly collected and have become rather dry, many sound ones may also float.

For seeds not as heavy as water, light liquids, as ethyl ether, alcohol, and xylene, have been used to separate viable from empty seeds. Unfortunately these liquids often are injurious to seeds with thin coats, although they may probably be used safely for seeds with hard coats.

SIZE

The number of sound seeds per pound is occasionally used as an index of seed quality. Large seeds from a given climatic zone usually are preferable to small seeds. The former tend to produce larger and stronger seedlings. However, the best drought-hardiness, frost-hardiness, and similar qualities are not necessarily associated with seed size within a race.

COLOR

Seed color furnishes a rough index of quality for some species. For instance, light-colored seeds of Scotch pine, Austrian pine, and Jerseytea ceanothus are almost invariably blind.

Good seed, at least in conifers, should as a general rule have bright, clean coats. Stained or moldy coats are an indication that the seed may have been subjected to heating or otherwise improper storage. Occasionally seed stored at a low temperature and high humidity will develop spots of mold on the surface of the coats, but unless such seed have been so stored for a considerable period, the moldy appearance probably has little significance.

Biochemical Methods

Another group of indirect methods for determining seed viability depends upon various chemical measures to detect the presence of life in seeds.

[39] Small seeds may often be cut and examined more easily if they are soaked in water for several hours prior to cutting.

[40] This refers to seeds which show no external signs of being wormy and as a result are not considered as impurities at the time of the test. In some seeds such as acorns, weevils do not always prevent germination.

These fall into two categories: (1) the staining of embryos, and (2) the measurement of enzyme activity. Both require training and facilities ordinarily found only in a seed laboratory.

STAINING SEED EMBRYOS

This method is based on the principle that certain dyes are retained only by living, others only by dead material. Indigo carmine, for instance, will stain dead or injured tissue but not uninjured living cells. Salts of selenium, on the other hand, will stain living embryos red.[41]

MEASURING ENZYME ACTIVITY

The method of determining seed viability by measuring the enzyme activity assumes that oxidizing and reducing enzymes are associated with living tissue alone. Thus any activity shown by these substances in seeds is considered an indication of their viability.[42]

Direct Germination Tests

The most reliable method for determining the true viability of a lot of seed is to germinate a representative sample. Germination tests upon which to base large-scale sowing or storage or for research purposes are best conducted by qualified technicians in a well-equipped laboratory. Tests requiring less precision can be carried on by anyone who has the time and a few facilities and who is willing to follow a few basic principles.

MEDIA

Two main types of media are used for germination tests—natural and artificial.

Fine, nonalkaline sand, acid peat, or sifted sphagnum moss are the preferred natural media. Nonalkaline sand contains only a few micro-organisms and hence can easily be sterilized, and peat and sphagnum moss are practically free from the fungi that cause damping-off. Garden soil often contains large amounts of organic matter and is likely to be teeming with micro-organisms such as those causing damping-off. This disease is apt to be troublesome when soils containing considerable amounts of silt and clay are used for tests. The use of alkaline sands can also cause heavy damping-off loss. Tests using these media are run in flats, clay pots, greenhouse benches, or in compressed peat mats.

At the Lake States Forest Experiment Station germination tests are run in redwood flats, 12 by 12 by 3½ inches, with yellow dune sand as a medium (fig. 21). After a thorough washing in a mortar box with large amounts of water at 150° F., the sand has a moisture equivalent of 1.81 percent. Little trouble is experienced with damping-off because the washing, in addition to carrying off most of the silt and clay, probably has somewhat of a pasteurizing effect. (If desired, sand may be sterilized by means of live steam or, if low electric rates are available, by heating. These methods may

M-31273F, 268197

FIGURE 21.—Natural media are preferred for testing seed germination. *A*, Sand flats used at the Lake States Forest Experiment Station. Note the division of flats into quadrats. *B*, Peat mat used at the Southern Forest Experiment Station. The seed is sown in shallow grooves and the dish then covered with a piece of glass to prevent drying.

[41] Staining seed embryos involves their dissection from the seed. Although results may be obtained after exposure of the embryos to these compounds for a few hours to a few days, the dissection of enough sound, uninjured embryos to give reliable results is probably even more laborious than making simple germination tests. As in the case of the method based on the growth of excised embryos, it is possible to determine in a short time the viability of seeds requiring lengthy periods of stratification, but even here the staining tests do not reveal how the seeds will respond to any specific treatment or set of germination conditions. And this response, after all, is of utmost concern to the nurseryman.

[42] Several investigators have found an apparent correlation between the activity of some of these enzymes and seed viability; others report contradictory results. This is particularly true of catalase, which has been studied perhaps more than any other enzyme and which has been found present in dead as well as viable seeds. In view of the conflicting evidence concerning the reliability of enzyme activity as an indication of seed quality and the fact that its determination often requires special equipment, this method of determining seed viability appears for the present to be of research interest only.

also be used with garden soil.) When tests are completed, the sand down to the seed level is discarded and the remainder saved and rewashed prior to use a second time. The flats are also washed each time they are used. Peat has also been used in flats, but on the basis of a few tests appears to have no advantages over the sand generally used. Very small seeds are usually germinated in petri dishes with either sand or peat as a medium.

Germination tests at the Southern Forest Experiment Station are carried on in peat mats which are placed in glass baking dishes (fig. 21) and covered with heavy glass. The seeds are sown uncovered in a series of 10 shallow grooves which are molded into each mat. Enough water is poured into the dish so that after the peat has absorbed all it will take up, a shallow layer will remain free in the bottom. This medium keeps the seeds well supplied with water, permits free access of light, and discourages molds more than do other media on which the seeds are exposed directly to the air. The peat appears to stimulate the germination of seeds of certain species, particularly slash pine. Exposing the seeds to low temperatures as a preliminary to testing is particularly easy in connection with the peat-mat test; all that is necessary is to place the mat, with the seeds in it, in a suitable refrigerator. The peat mat requires much less space than the standard sand flat, weighs less, and does not contain any grit.

Artificial media include germinators of the Jacobsen, Geneva, and Stainer types, and petri dishes. The seed is germinated on porous plates, blotters, or filter paper, or on agar, and not on a soil substratum. Artificial germinators are commonly used by many of the European seed-testing agencies, but they are less common in this country.

The chief advantage of using natural substrata like sand, peat, or soil is that they give results close to those obtained in the nursery. Such media insure a better balance between moisture and aeration, discourage the growth of mold on the surface of the seed, and largely eliminate germinable but weak seeds which would have no chance in the nursery seedbeds.

The advantages of using porous plates, blotters, or other artificial media are: less room and time are required for germination; because of the smaller bulk, temperature and moisture conditions are more easily controlled.

DEPTH OF COVER

In sand or soil, seeds should be sown at a depth at least equal to their diameter measured at midpoint. (At the Lake States Forest Experiment Station, seeds are generally sown at twice their diameters.) When peat mats are used, the seeds are not usually covered. In the case of the extremely small seeds of certain of the heath family, good results may be obtained by sprinkling them on the surface of peat and then covering them lightly with dry sand from a large salt shaker.

MOISTURE SUPPLY

In germination tests the moisture content of the medium should be kept at a moderate and fairly constant level. If moisture is too abundant, damping-off may become troublesome; also, aeration will not be sufficient. The construction of standard germinators tends to keep the moisture content relatively constant. Moisture regulation is also a rather simple matter with peat mats. More careful attention is required in the case of sand [43] and soil flats. After having taken a few moisture samples as checks, one soon learns by the color of the substratum if there is too much or too little moisture. If too much water is supplied to sand, it will soon percolate to the bottom of the flat and gradually leak out. Therefore, flats should never have watertight bottoms.

Clay pots may be set in pans of water so that they are irrigated from below by capillary action. However, the water must be kept at a relatively low level to prevent saturation.

TEMPERATURE CONTROL

A daily alternation of temperatures over a rather narrow range is more favorable for germination of most species than are constant temperatures. Since most seeds germinating in nature are probably exposed to daily fluctuations in temperature, except possibly those which germinate very early in the spring underneath leaf litter, this fact is not surprising. A temperature range comparable to that of the summer months—a daily alternation of from about 68° to 86° F. (20° to 30° C.)—has given good results for most species tested at the Lake States Forest Experiment Station. The following species, however, germinate better at lower temperatures:

At 50° (night) to 77° F. (day): Barberry, American bittersweet, boxelder, Appalachian sand cherry, pin cherry, common chokecherry, eastern hophornbeam, common hoptree, red maple, fourwing saltbush.

At 50° (night) to 77°F. (day) or warmer:[44] Jerseytea ceanothus, black cherry, scarlet elder, American elm, swamp fly honeysuckle, Tatarian honeysuckle, alligator juniper, common prickly-ash, red raspberry, eastern redcedar, and coast rhododendron.

At 50° F.: Wild sweet crab, black huckleberry, Amur maple, common pear, Allegheny plum, Amer-

[43] The sand used as a testing medium at the Lake States Forest Experiment Station gives excellent results if kept at a moisture content (oven-dry weight basis) of from 8 to 12 percent. This is easily maintained by watering the flats twice a day with distilled water during the colder months (when humidity is low as a result of artificial heating) and less often in the summer. Distilled water is used because it is much less likely than is tap water to contain spores of the fungi which cause damping-off, and also because the alkalinity of the tap water favors the development of this disease.

[44] Daily alternations of 68° to 86° F. are equally effective.

ican plum (seeds from northern Minnesota).[45]

Between 40° and 50° F.: Norway maple, American mountain-ash.

At 41° F.: Meadow rose, serviceberry.[46]

Whenever possible, the temperatures used for germination tests of some of the more refractory species should be made to coincide as closely as possible with those prevailing outdoors at the time nursery germination occurs.

Although germination tests are often run in greenhouses, this is not very satisfactory because temperatures can be controlled accurately only at night and on cloudy days and these, largely during the winter. If steam is available the year round, it is relatively simple (except in the South and parts of California), with the aid of a thermostat, a unit heater, and one or two large electric fans, to devise a germination room in which the temperature can be controlled on all save a very few days during the warmest summer months.

A more elaborate example is provided by the main germination room of the Lake States Forest Experiment Station, which handles 800 to 1,100 flats a year at a temperature varying diurnally from 68° to 86° F. Diurnal temperatures are maintained by batteries of three 150-watt incandescent bulbs with reflectors (one battery over each of four tables), the position of which is changed morning and afternoon so as to distribute the heat more equitably, supplemented by heat from a circulating heater connected to a high-pressure steam line through a reducing valve. Heat is thus available at all times of the year. The heater is controlled by means of a day-night thermostat; the temperature cycle is raised and lowered by a time switch; the lights are turned on and off by another time switch.

Nocturnal temperatures are lowered to 68° F. by means of a fan with a capacity of 1,200 cubic feet per minute which brings in cold air from outdoors to the floor of the room, the warm air escaping through a window open at the top. This fan is connected to a day-night thermostat and also to the time switch that operates the light cycle. Circulation in all parts of the room is assured by two or three electric fans which operate day and night.

If for any reason the temperature in the room gets below 68° F., as during some winter nights, the heater thermostat cuts in, and heat is circulated until the temperature reaches the desired point. Similarly, if the temperature gets above 86° during the day, the cold-air fan operates until the temperature is lowered.

LIGHT

Light does not appear to be necessary for the germination of forest seeds of most species. However, recent investigations at the Southern Forest Experiment Station indicate that germination of the seeds of some of the southern pines sown uncovered on peat mats is decidedly better in a well-lighted room than in dim light or in the dark. With these species, also, a ¼-inch covering of sand or soil in sand flats or nursery beds has generally given slower and poorer germination than a ⅛-inch covering. However, with seeds sown in flats and covered with ¼ to 1 inch of sand, it does not seem that light can play a very important role in germination.

NUMBER TO BE TESTED

The sample to be tested should be just large enough to insure results of sufficient precision to serve the purposes of the test. As a general rule, not less than 400, and preferably 800 to 1,000, seeds should be used for a test. With large-seeded species such as hickories, oaks, and walnuts, it is often impractical to use more than 200 seeds per test; but not less than 100 should ever be used. "Feeler" tests of an unfamiliar species may be run with as few as 50 or 100 seeds.

REPLICATION AND TECHNIQUE

To insure more reliable results all test samples to be sown should be divided into four or more subsamples (*18*), each of which is sown, recorded, and analyzed separately. Samples are usually designated by serial numbers and subsamples by letters. Wherever possible each subsample should be sown in a separate flat or germinator to reduce the possibility of losing all subsamples should some-

F-433113

FIGURE 22.—Accessories used in making germination tests in sand flats. The 4/64 round-hole zinc grain screen, used in washing out seeds for examination after the test is completed, holds the following (bottom to top): Block, used to help distribute seed within a quadrat at time of sowing; set of gages (⅛ to 1 inch), used to insure correct sowing depth; cross, used to form an isolation strip; and tamper, used when flats are being filled with sand.

[45] American plum seeds from Nebraska germinate best at 70° to 80° F.

[46] Daily alternations of 68° to 86° F. also effective.

thing like damping-off cause trouble and to disclose any lack of uniformity that may exist between different flats or germinators. Ideally, subsamples should be scattered at random in the testing chamber, but any advantages so gained are usually more than offset by the difficulties thus created in keeping records of germination.

The work of counting seeds, controlling depth of sowing and covering, and keeping subsamples separated can be done more quickly and easily by the use of certain accessories (fig. 22). For example, at the Lake States Forest Experiment Station, the flats, after being filled level full, are cleared of sand to the sowing depth desired by means of hardwood depth gages so made as to remove all sand to a level of from 1/8 to 1 inch below the top. Subsamples are kept well separated from each other by means of a wooden cross set in the flat on its narrow edge (1 inch wide) so as to form four quadrats of even size. An isolation strip is thus afforded which will prevent possible contamination of any subsample by its neighbors. After the seeds have been covered with sifted sand, the latter is lightly pressed into position and then leveled off by running the upper edge of the depth gage across the flat and flush with the two sides.

Brown, Toole, and Goss (*8*) have described a vacuum counter and setter which works well with absolutely clean, smooth seeds of regular shape and moderate size. It should materially lighten the burden of setting up tests in sand flats, on blotters, or on peat mats without grooves. A metal head perforated with 50 or 100 holes picks up a corresponding number of seeds from a shallow tray when suction is applied, and deposits them at uniform spacing on the germination medium when the suction is cut off again.

DURATION

The information desired determines the length of time that germination tests are allowed to run. Tests comparing the effects of various pretreatments may be continued for several weeks or even months. Routine tests for germinative capacity (total germination) of well-known species are often run for only 30 or 40 days. Ordinarily, if a test has not shown full germination by the end of 60 days, it should be discontinued and an attempt made to determine the reasons for the incomplete germination. Seeds of many species reach the peak of germination (germinative energy) at or before 30 days if handled properly, and it should be possible to reach the same goal with many other species once the optimum pretreatment and germination conditions have been worked out for them. Germination prolonged over a period of much more than 30 days produces stands of uneven height in the nursery.

GERMINATION COUNTS AND RECORDS

In some European seed-testing laboratories germination is counted at 5, 10, 20, 30, 60, and 100 days. In this country counts usually are made at least every 2 or 3 days until the peak of the germination curve has been reached and about once a week after this time.

In making germination counts, only seedlings normal in every respect, with the exception of those which have been killed by damping-off fungi,[47] are recorded as having germinated. Seedlings which emerge from the seed coat, cotyledons first, are considered abnormal since they usually fail to develop normally. Albinos are also discarded for the same reason. Seedlings from a polyembryonic seed are counted as one seedling only, even though more than one of these may develop into normal seedlings.

Records should be kept of all abnormal and diseased seedlings to help determine the causes of possible variability when the test is completed. A very high degree of accuracy in counting seeds and computing percents is necessary if tests are to be reliable.

Seeds which have not germinated are often examined at the conclusion of the test by making a cutting test of all, or a representative sample, of them to determine how many are still sound and potentially germinable. Viable seeds can usually be distinguished quite easily from those which have spoiled. Unless all seeds are cut, however, it is sometimes difficult to determine accurately the potential germination (total germination plus sound ungerminated seeds), because of empty coats of germinated seeds left in the medium after the seedlings have emerged. Where many seeds are to be examined, a number of grain testing screens (4/64 round-hole zinc and 30-mesh wire screens) will prove very useful in separating the seeds from sand but are not necessary where peat mats, blotters, or standard germinators have been used. Such "post mortems" are extremely useful in evaluating the results of various pregermination treatments of the seeds of little-known species.

INTERPRETATION OF SEED TESTS

Germination in the seed laboratory takes place under favorable and closely controlled conditions of temperature, moisture, medium, and depth of cover. Consequently, laboratory germination is almost always higher than that in the nursery, [48] and the

[47] If discovered soon enough, this disease may often be controlled by soaking the medium with a solution of aluminum sulfate, cupric oxide, semesan, and similar products. Infected flats should be isolated to minimize the chances of spreading the disease to other flats. Such flats should be watered sparingly and, when the test is over, the sand or soil discarded and the flat washed with hot water.

[48] In a few cases, however, nursery germination may be higher because temperature conditions in the laboratory are too dissimilar from those in nature. For example, boxelder seeds, which have long shown good germination in the nursery, gave poor results in the laboratory until the proper germination temperature was worked out.

differences between the two may be considerable, as shown in the tabulation that follows.

Species:	*Nursery germination (percent) considering laboratory germination = 100 percent*[1]
Western hemlock	50
Jeffrey pine	78
Lodgepole pine	60
Ponderosa pine	75
Red pine	75
Scotch pine	70
White pine	80
Pinyon	80–85
Port-Orford-cedar	50
Yellow-poplar	50

[1] These data are illustrative, being based on experience at several different laboratories and nurseries.

As a rule, nursery germination closely approaches the laboratory figure only in the case of rapidly germinating seeds of high quality. The poorer and more sluggish the seed, the farther field germination falls below that shown by the laboratory test. The testing methods, species, local nursery conditions, season, and many other factors contribute to the variation between laboratory and nursery germination. For this reason seed-testing results cannot be applied directly to the nursery. Each nurseryman should prepare a set of "experience tables" for each species, based on several years' comparison of laboratory [49] with seedbed germination, to be used as a correction factor.

The nurseryman must know not only how to relate laboratory germination to that expected in his nursery, but also how to predict tree or plant percent—the number of usable seedlings which can be produced from a certain amount of viable seeds. This calls for the building up of another set of tables based on experience in growing seedlings over a several-year period. Tree percent, which is the result of the usual and unavoidable oversummer loss, naturally shows considerable fluctuation. Fairly reliable averages are tabulated below for 22 trees and shrubs described in part 2 of this manual. Unpredictable losses such as those caused by severe storms, the depredations of birds, epidemics of insects, and disease are guarded against, not by increasing the sowing rate per bed but by sowing extra beds.

Species:	*Average percent of usable seedlings produced from viable seed sown*
Baldcypress	40–50
European white birch	15
Dahurian buckthorn	25
Silver buffaloberry	12
Lilac chastetree	16
Siberian crab	15
Desertwillow	15
Elms	12
Hackberries	30
Tatarian honeysuckle	15
Common jujube	33
European larch	10
Japanese larch	20
Siberian larch	30
Common lilac	12
Black locust	25
Russian mulberry	12
Nannyberry	25
Siberian peashrub	33
Torrey pine	60
Redwood	20
Spruces	50

[49] A complete set of tables should be made for each laboratory involved if seeds are tested by several different agencies.

APPLICATION OF TEST RESULTS

Seed testing is sometimes used for research purposes or as a check on legally prescribed seed standards, [50] but its main objective is to determine the correct rate of nursery sowing needed to produce stands of a given density. Such a value is based on two sets of data—one worked out in the laboratory and the other obtained in the nursery. The testing laboratory supplies the number per pound, purity, and germination percent of the seed to be sown. Based on his experience, the nurseryman must supply a correction to relate laboratory to nursery germination and an adjustment for the normal oversummer loss in the seedbeds.

Although the rate of sowing can be computed in several different ways, the following formula is believed to take into account more fully than do others the factors affecting seedling establishment. Thus, if A = area of seedbed in square feet; S = number of living seedlings desired per square foot at end of growing season; C = average number of seeds per pound at the moisture content at which the seed is sown; P = average purity percent of seeds (expressed as a decimal); G = average effective germination percent in the laboratory (expressed as a decimal);[51] L=average percent of germinating seeds that will be living trees at the end of the season (expressed as a decimal); and W = the weight of seeds to sow per bed in pounds; then

$$W = \frac{A \times S}{C \times P \times G \times L}$$ [52]

Example: If seedbed area (A) = 400 square feet; desired density (S) is 30 seedlings per square foot;

[50] Research tests are often made for the purpose of learning the effects of various treatments on seed germination; check tests, which in this country are seldom applied to forest tree seeds, usually are made to see that purity percent, germination percent, and content of noxious weed seeds meet certain legal standards. A detailed discussion of these two types of tests, however, is not appropriate here.

[51] This value will have to be adjusted if the experience tables show a need for it.

[52] This formula, transformed to read $S = \frac{C \times P \times G \times (W \times L)}{A}$, can be used, when proper statistical data from seed tests are available, to predict the range within which the actual average stand of seedlings may be expected to be found within certain limits, for instance, 19 times out of 20. Such calculations will show strikingly how much less error is involved if such values as C, P, and G are based on 4 observations instead of 2, for example. Ordinarily, however, the nurseryman need not go so far in his calculations.

number of seeds per pound (C)[53] is 4,410; purity (P) is 90 percent; germination (G) is 65 percent, and experience tables indicate that 88 percent (L) of all seeds germinating will be living trees at the end of the summer, then

$$W = \frac{(400)\ (30)}{(4410)\ (.90)\ (.65)\ (.88)} =$$

$$\frac{12000}{2270.268} = 5.29 \text{ pounds per bed.}$$

To obtain the full benefits of seed testing in terms of uniform and correct density in the seedbeds, the seeds must be thoroughly mixed before sowing. Seeds from all containers in the lot to be sown should be emptied in thin layers on top of one another on a canvas or smooth floor and then stirred with rakes or shovels. The importance of thorough mixing can be illustrated by the following example.

Assume a sowing lot of 2 cans, one of seed germinating 100 percent, the other, 0 percent. The nurseryman knows the viability of the entire lot, but not of the individual cans. Correct sampling and careful laboratory testing have indicated an average germination, for the lot, of approximately 50 percent, and a weight of seeds to produce a stand of 35 trees per square foot is computed on this basis. But, unless the contents of the 2 cans are thoroughly mixed together, half the beds will have a stand of 70 trees per square foot, and half will have no stand at all.

Seed testing provides the information for sowing seeds in the right amount to produce nursery stock most efficiently and economically. Used by skilled and experienced nurserymen, it becomes the cornerstone of successful nursery practice.

[53] Fluctuations in C between the time of determination and the time of sowing are ordinarily unimportant, but may sometimes become very important, as when longleaf pine seed, extracted at a seed moisture content of 35 percent, is counted, and then stored in an atmosphere that reduces it to 12 percent, or when any seeds are stratified in moist sand or peat after counting. In such cases C must be redetermined immediately before sowing or else corrected by calculations.

LITERATURE CITED

(1) ALDOUS, S. E.
1941. FOOD HABITS OF CHIPMUNKS. Jour. Mammal. 22:18-24.

(2) BALDWIN, H. I.
1930. THE EFFECT OF AFTER-RIPENING TREATMENT ON THE GERMINATION OF EASTERN HEMLOCK SEED. Jour. Forestry 28: 853-857, illus.

(3) ———
1932. ALCOHOL SEPARATION OF EMPTY SEED, AND ITS EFFECT ON THE GERMINATION OF RED SPRUCE. Amer. Jour. Bot. 19: 1-11, illus.

(4) BARTON, L. V.
1928. HASTENING THE GERMINATION OF SOUTHERN PINE SEEDS. Jour. Forestry 26: 774-785, illus.

(5) ———
1935. STORAGE OF SOME CONIFEROUS SEEDS. Boyce Thompson Inst. Contrib. 7: 379-404, illus.

(6) BATES, C. G.
1930. THE PRODUCTION, EXTRACTION, AND GERMINATION OF LODGEPOLE PINE SEED. U. S. Dept. Agr. Tech. Bul. 191, 92 pp., illus.

(7) ———
1931. A NEW PRINCIPLE IN SEED COLLECTING FOR NORWAY PINE. Jour. Forestry 29: 661-678, illus.

(8) BROWN, E., TOOLE, E. H., AND GOSS, W. L.
1928. A SEED COUNTER. U. S. Dept. Agr. Cir. 53, 4 pp., illus.

(9) BÜSGEN, M.
1929. BAU UND LEBEN UNSERER WALDBÄUME. Ed. 3, 436 pp., illus. London.

(10) CHAMPION, H. G.
1933. THE IMPORTANCE OF THE ORIGIN OF SEED USED IN FORESTRY. Indian Forest Rec. 17: 76 pp., illus.

(11) CHAPMAN, A. G.
1936. SCARIFICATION OF BLACK LOCUST SEED TO INCREASE AND HASTEN GERMINATION. Jour. Forestry 34: 66-74, illus.

(12) CLAPPER, R. B.
1943. NEW CHESTNUTS FOR OUR FORESTS? Amer. Forests 49: 331-333, 365, illus.

(13) DENGLER, A.
1939. UEBER DIE ENTWICKLUNG KÜNSTLICHER KIEFERNKREUZUNGEN. Ztschr. f. Forst u. Jagdw. 71: 457-485, illus.

(14) DEUBER, C. G.
1932 CHEMICAL TREATMENTS TO SHORTEN THE REST PERIOD OF RED AND BLACK OAK ACORNS. Jour. Forestry 30: 674-679, illus.

(15) ELIASON, E. J., AND HEIT, C. E.
1940. THE RESULTS OF LABORATORY TESTS AS APPLIED TO LARGE SCALE EXTRACTION OF RED PINE SEED. Jour. Forestry 38: 426-429, illus.

(16) ENGSTROM, H. E., AND STOECKLER, J. H.
1941. NURSERY PRACTICE FOR TREES AND SHRUBS SUITABLE FOR PLANTING ON THE PRAIRIE-PLAINS. U. S. Dept. Agr. Misc. Pub. 434, 159 pp., illus.

(17) EWART, A. J.
1908. ON THE LONGEVITY OF SEEDS. Roy. Soc. Victoria, Proc. 21. 210 pp.

(18) FISHER, R. A.
1936. STATISTICAL METHODS FOR RESEARCH WORKERS. Ed. 6, 339 pp. London.

(19) FLEMION, F.
1933. PHYSIOLOGICAL AND CHEMICAL STUDIES OF AFTER-RIPENING OF RHODOTYPOS KERRIOIDES SEEDS. Boyce Thompson Inst. Contrib. 5: 143-159, illus.

(20) ———
1938. BREAKING THE DORMANCY OF SEEDS OF CRATAEGUS SPECIES. Boyce Thompson Inst. Contrib. 9: 409-423, illus.

(21) ———
1941. FURTHER STUDIES ON THE RAPID DETERMINATION OF THE GERMINATIVE CAPACITY OF SEEDS. Boyce Thompson Inst. Contrib. 11: 455-464.

(22) GIERSBACH, J.
1937. GERMINATION AND SEEDLING PRODUCTION OF SPECIES OF VIBURNUM. Boyce Thompson Inst. Contrib. 9: 79-90, illus.

(23) HAACK.
1912. DIE PRÜFUNG DES KIEFERNSAMENS. Ztschr. Forst u. Jagdw. 44: 193-222, 273-308.

(24) HAASIS, F. W.
1928. GERMINATIVE ENERGY OF LOTS OF CONIFEROUS-TREE SEED, AS RELATED TO INCUBATION TEMPERATURE AND TO DURATION OF INCUBATION. Plant Physiol. 3: 365-412, illus.

(25) HEDGECOCK, G. G., HAHN, G. G., AND HUNT, N. R.
1922. TWO IMPORTANT PINE CONE RUSTS AND THEIR NEW CRONARTIAL STAGES. Phytopathology 12: 109-122, illus.

(26) HURST, W. M., HUMPHRIES, W. R., AND MCKEE, R.
1934. THE BARREL SEED SCARIFIER. U. S. Dept. Agr. Leaflet 107, 5 pp., illus.

(27) JOHNSON, L. P. V.
1939. A DESCRIPTIVE LIST OF NATURAL AND ARTIFICIAL INTERSPECIFIC HYBRIDS IN NORTH AMERICAN FOREST-TREE GENERA. Canadian Jour. Res. Sect. C, Bot. Sci. 17: 411-444.

(28) JONES, H. A.
1920. PHYSIOLOGICAL STUDY OF MAPLE SEEDS. Bot. Gaz. 69: 127-152, illus.

(29) KALELA, A.
1937. ZUR SYNTHESE DER EXPERIMENTELLEN UNTERSUCHUNGEN ÜBER KLIMARASSEN DER HOLZARTEN. Inst. Forest Fenniae Commun. 26, 445 pp. Helsinki.

(30) KORSTIAN, C. F.
1927. FACTORS CONTROLLING GERMINATION AND EARLY SURVIVAL IN OAKS. Yale Forest School Bul. 19, 115 pp., illus.

(31) LAKON, G.
1911. BEITRÄGE ZUR FORSTLICHEN SAMENKUNDE. I. DER KEIMVERZUG BEI DEN KONIFEREN-UND HARTSCHALIGEN LEGUMINOSENSAMEN. Naturw. Ztschr. Forst u. Landw. 9: 226-237, illus.

(32) ———
1911. BEITRÄGE ZUR FORSTLICHEN SAMENKUNDE. II. ZUR ANATOMIE UND KEIMUNGSPHYSIOLOGIE DER ESCHENSAMEN. Naturw. Ztschr. Forst u. Landw. 9: 285-298.

(33) MACKINNEY, A. L., AND MCQUILKIN, W. E.
1938. METHODS OF STRATIFICATION FOR LOBLOLLY PINE SEEDS. Jour. Forestry 36: 1123-1127.

(34) MAKI, T. E.
1940. SIGNIFICANCE AND APPLICABILITY OF SEED MATURITY INDICES FOR PONDEROSA PINE. Jour. Forestry 38: 55-60, illus.

(35) MEGINNIS, H. G.
1937. SULPHURIC ACID TREATMENT TO INCREASE GERMINATION OF BLACK LOCUST SEED. U. S. Dept. Agr. Cir. 453, 35 pp., illus.

(36) MILLER, E. C.
1938. PLANT PHYSIOLOGY, WITH REFERENCE TO THE GREEN PLANT. Ed. 2, 1201 pp., illus. New York and London.

(37) MOORE, A. W.
1940. WILD ANIMAL DAMAGE TO SEED AND SEEDLINGS ON CUT-OVER DOUGLAS FIR LANDS OF OREGON AND WASHINGTON. U. S. Dept. Agr. Tech. Bul. 706, 28 pp., illus.

(38) RICHARDSON, A. H.
1925. GATHERING AND EXTRACTING RED PINE SEED. Jour. Forestry 23: 304-310.

(39) RIETZ, R. C.
1939. KILN DESIGN AND DEVELOPMENT OF SCHEDULES FOR EXTRACTING SEED FROM CONES. U. S. Dept. Agr. Tech. Bul. 773, 70 pp., illus.

(40) ROE, E. I.
1940. LONGEVITY OF RED PINE SEED. Minn. Acad. Sci. Proc. 8: 28-30, illus.

(41) ———
1941. EFFECT OF TEMPERATURE ON SEED GERMINATION. Jour. Forestry 39: 413-414.

(42) ROHMEDER, E.
1941. DIE VERMEHRUNG DER PAPPELN DURCH SAMEN. Forstarchiv 17: 73-80, illus.

(43) ———, AND LOEBEL, M.
1940. KEIMVERSUCHE MIT ZIRBELKIEFER. Forstwiss. Centbl. 62: 25-36, illus.

(44) RONGGER, N.
1899. ÜBER DIE BESTANDTEILE DER SAMEN VON PICEA EXCELSA (LINK) UND ÜBER DIE SPALTUNGSPRODUKTE DER AUS DIESEN SAMEN DARSTELLBAREN PROTEINSTOFFE. Landw. Vers. Sta. 51: 89-116.

(45) Rudolf, P. O.
1940. WHEN ARE PINE CONES RIPE? Minn. Acad. Sci. Proc. 8: 31-38, illus.

(46) Sandahl, P. L.
1941. SEED GERMINATION. Parks & Recreation 24: 508.

(47) SCHMIDT, W.
1937. NEUE WEGE DER RASSENFORSCHUNG UND KIEFERNANERKENNUNG. DIE PHYSIOLOGISCHE KIEFERN-RASSENDIAGNOSE BEI DER SAATGUTANERKENNUNG. Jahrb. Gruppe Preussen-Schliesien (Schlesischer Forstver.) Deut. Forstver. 1936: 31-57, illus.

(48) ———
1938. DIE KLIMA-RASSENDIAGNOSE BEI PINUS SILVESTRIS. Internatl. Seed Testing Assoc. Proc. 10: 256-258.

(49) SCHREINER, E. J.
1937. IMPROVEMENT OF FOREST TREES. U. S. Dept. Agr. Yearbook 1937: 1242-1279, illus.

(50) SCHULZE, E., AND RONGGER, N.
1899. UBER DIE BESTANDTEILE DER SAMEN VON PINUS CEMBRA (ZIRBELKIEFER ODER ARVE). Landw. Vers. Sta. 51: 189-204.

(51) SCHUMACHER, F. X., AND CHAPMAN, R. A.
1942. SAMPLING METHODS IN FORESTRY AND RANGE MANAGEMENT. Duke Univ. Forestry Bul. 7, 213 pp., illus.

(52) SPAETH, J. N.
1934. A PHYSIOLOGICAL STUDY OF DORMANCY IN TILIA SEED. N. Y. (Cornell) Agr. Expt. Sta. Mem. 169, 78 pp., illus.

(53) STEAVENSON, H. A.
1940. THE HAMMER MILL AS AN IMPORTANT NURSERY IMPLEMENT. Jour. Forestry 38: 356-361, illus.

(54) STOECKELER, J. H., AND BASKIN, L. C.
1937. THE DENBIGH DISC SCARIFIER, A NEW METHOD OF SEED TREATMENT. Jour. Forestry 35: 396-398, illus.

(55) TOLMAN, B., AND STOUT, M.
1940. TOXIC EFFECT ON GERMINATING SUGARBEET SEED OF WATER-SOLUBLE SUBSTANCES IN THE SEED BALL. Jour. Agr. Res. 61: 817-830, illus.

(56) WOODS, C. D., AND MERRILL, L. H.
1900. NUTS AS FOOD. Maine Agr. Expt. Sta. Ann. Rpt. 1899: 71-92, illus.

(57) ZON, R.
1915. SEED PRODUCTION OF WESTERN WHITE PINE. U. S. Dept. Agr. Bul. 210, 15 pp.

PART 2

INTRODUCTION

A KNOWLEDGE of the general biological and technical information on seed, given in part 1, will make possible more effective use of the detailed material for individual species and varieties of trees and shrubs that follows. Originally, more than 680 species and varieties were listed for inclusion in this publication, but more than 200 had to be omitted for lack of authentic information. It is believed that the 444 species and varieties, both native and introduced, for which it was possible to obtain enough reliable data include those in present use and those that are likely to be used for planting in the near future.

Large as this task was, it involved only a small part of the 3,000-odd species of trees and shrubs in our abundant native flora. In preparing this manual authentic information from nearly 400 articles published prior to 1942 was drawn upon. However, several genera and a good many species within other genera are written up wholly or largely on the basis of unpublished information.

Synonomy of scientific names, and data on natural ranges and dates of introduction were taken largely from Rehder's Manual of Cultivated Trees and Shrubs, Edition 2, 1940.

Complete data were not available for many of the species; only those were included for which there was enough information to provide a reliable clue to their handling. The information given for each species covers the following points:

1. Scientific and common names, and synonyms.
2. Growth habit, natural range, chief uses, and how long in cultivation.
3. Kind and time of flowering, fruit ripening, seed dispersal, seed description, and occurrence of climatic races.
4. Time and methods of seed collection, extraction, and cleaning; yield of cleaned seeds per bushel or 100 pounds of fresh fruits, number of cleaned seed per pound (low, average, and high, and number of samples on which based); purity, soundness, and cost of commercial seed (as of 1940); effectiveness of known storage methods.
5. Time, place, and type of natural germination; kind of dormancy and methods of overcoming it; number of seed, medium, period, and temperature required for germination tests; amount and period of germinative energy; and germinative capacity (low, average, and high, and number of tests on which based).
6. Time, method, depth, density and rate of sowing; shading requirements, diseases and insect pests and their control; nursery germination and plant percent; kind of stock for field planting; preferred planting sites; and other feasible methods of propagation.

The information is presented by genera, arranged alphabetically by scientific name. Nearly 70 genera are represented by one species only; others include from 2 to 48 species. Within those genera for which more than one species is included, material pertaining to individual species is arranged in tabular form and alphabetically by scientific name, while material pertaining to the genus as a whole, or that which is insufficient to afford comparison by species, is given in the text.

Seedlings shown in the drawings are sketches of plants grown at the Lake States Forest Experiment Station and California Forest and Range Experiment Station seed laboratories. It should be remembered, therefore, that they may be more spindly than seedlings grown in the open.

ABIES Mill. Fir

(Pine family—Pinaceae)

Distribution and Use.—The firs consist of about 40 species of large evergreen trees native to the temperate parts of the Northern Hemisphere and on the higher mountains south to Guatemala, northern Africa and the Himalayas. They produce soft perishable wood which is valuable for pulp, and in some species for lumber. The prominent resin vesicles of the bark are a source of the balsam used in making slides for microscopic work. While young, some species have decided ornamental value as Christmas trees or for landscape planting, but they are likely to become thin and unattractive as they grow older. The seeds are eaten by rodents and the branches of some species are browsed by large game animals. Nine indigenous species with one additional variety and one species native to Europe have been used or are potentially valuable for conservation planting in the United States. The distribution and chief uses of these 10 species and 1 variety are given in table 1.

Only five of the species listed in table 1 are used in reforestation work in the United States, and these to a very small extent. They are *Abies alba, A. balsamea, A. concolor, A. grandis, and A. lasiocarpa* var. *arizonica.* The first species, however, has not flourished in this country. Considerably greater use has been made of the firs in ornamental planting than for reforestation. *A. concolor, A. magnifica,* and *A. procera* are often planted as park trees in western and central Europe and also in the eastern United States, the first and last species doing particularly well as far north as eastern Massachusetts. *A. venusta* has been successfully planted in California and the Mediterranean region of Europe. *A. amabilis, A. balsamea, A. fraseri,* and *A. lasiocarpa* have been used occasionally in ornamental planting in the eastern United States and in Europe but have shown little value.

Seeding Habits.—The flowers of *Abies* are unisexual, the two kinds being borne in small conelike clusters in the early spring on branchlets of the previous season's growth and in different parts of the same tree. The male cones are oval or cylindrical and hang singly from the lower side of the branches of the upper half of the crown; the female flowers are globose or ovoid and stand erect and singly on the uppermost part of the crown; this arrangement favors cross-fertilization. The fruit, an erect cone with thin closely overlapping scales—and in some species conspicuous bracts—ripens the same fall. When mature, the scales become separated from the central spikelike axis and fall away with their large-winged seeds, two of which are borne at the base of each scale, only the central spike remaining. No fertile seeds are borne at the ends of the cone.

The mature seed is usually ovoid or oblong and consists of a thin, rather soft seed coat with several resin vesicles on its surface, and an embryo with 4 to 10 cotyledons surrounded by a fleshy endosperm (figs. 1, 2, and 3). Seed dispersal is effected by wind and to some extent by rodents. Details of the time of flowering, cone ripening, and frequency of seed production are given in table 2.

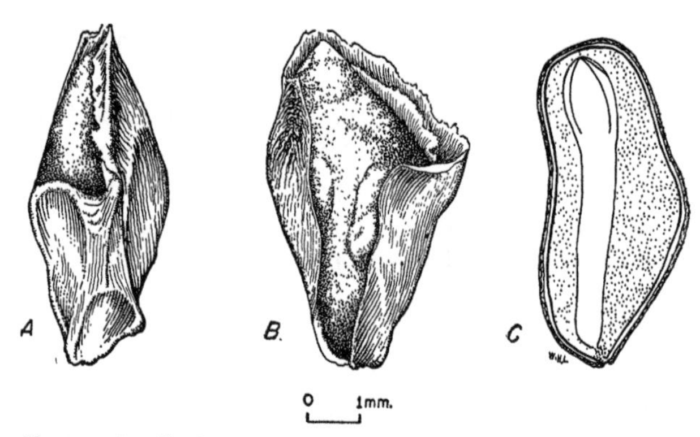

Figure 1.—Seed views of *Abies balsamea: A* and *B*, Exterior views from two angles; *C*, longitudinal section.

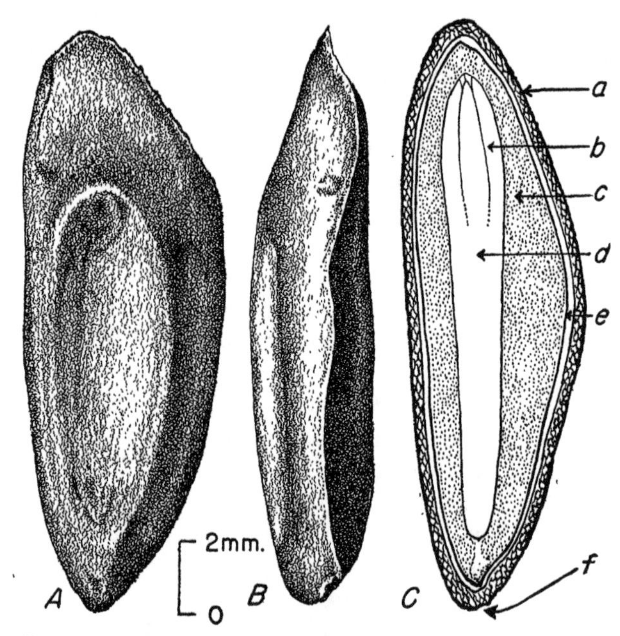

Figure 2.—Seed views of *Abies magnifica* var. *shastensis. A* and *B*, Exterior views from two angles. *C*, Longitudinal section of seed: *a*, Seed coat; *b*, cotyledons; *c*, endosperm; *d*, embryo; *e*, nucellus; *f*, micropyle.

TABLE 1.—*Abies: Growth habit, distribution, and uses*

Accepted name	Synonyms	Growth habit	Natural range	Chief uses	Date of earliest cultivation
A. alba Mill.[1] (silver fir).	*A. pectinata* DC., *A. picea* Lindl. not Mill.	Large forest tree.	Mountains of central and southern Europe.	Lumber, pulpwood; source of Strasbourg turpentine.	(2)
A. amabilis (Dougl.) Forb. (Pacific silver fir).	*Picea amabilis* Dougl. (silver fir, red fir, lovely fir).	-----do------	Lower slopes of canyons and on flats, southern Alaska, Coast and Cascade Ranges, British Columbia to Oregon.	Pulpwood, lumber, landscape planting.	1830
A. balsamea (L.) Mill.[3] (balsam fir).	balsam ---------------	Tree of medium size.	Moist, well-drained soils and swamps, Labrador to northwestern Canada and south to West Virginia, Iowa, and Alberta.	Pulpwood, boxboards, Christmas trees, browse for game animals, Canada balsam, medicinal bark.	1698
A. concolor (Gord. and Glend.) Hoopes[4] (white fir).	concolor fir, Colorado fir.	Tree of medium to large size.	Moist, well-drained soils, Rocky Mountains and Great Basin from western Wyoming and northern Utah to northern Mexico; in Sierra Nevadas from southern Oregon to Lower California.	Lumber, pulpwood, landscape planting.	1851
A. fraseri (Pursh) Poir. (Fraser fir).	southern balsam fir----	Tree of medium size.	Mountains of Virginia, Tennessee, and North Carolina, above 4,000 feet.	Pulpwood, protection forest, Christmas trees.	1811
A. grandis (Dougl.) Lindl.[5] (grand fir).	*A. gordoniana* Carr. (lowland white fir, giant fir).	Largest of American firs.	Moist soils, British Columbia to northern California and east to western Montana.	Lumber, pulpwood, occasionally in ornamental planting.	1831
A. lasiocarpa (Hook.) Nutt. (alpine fir).	*A. subalpina* Engelm. (Rocky Mountain fir).	Tree of medium size.	Mountain slopes and summits, Alaska south to Oregon, Utah, and northern New Mexico.	Protection forest, game food, pulpwood, fuel.	1863
A. lasiocarpa var. *arizonica* (Merriam) Lemm. (corkbark fir).	*A. arizonica* Merriam (cork fir, Arizona fir).	-----do------	Gravelly and rocky soils, mountains of northern Arizona, New Mexico.	Protection forest; potential source of cork.	1901
A. magnifica A. Murr.[6] (California red fir).	*A. nobilis* var. *magnifica* (A. Murr.) Kellogg (red fir).	Large forest tree.	Moist soil of mountain slopes, southern Oregon and California.	Lumber, pulpwood, fuel.	1851
A. procera Rehd. (noble fir).	*A. nobilis* (Dougl.) Lindl. (larch).	-----do------	Deep soils of lower slopes, Coast and Cascade Ranges, Washington to northern California.	Interior finish and specialty products, ornamental planting; potentially valuable for aircraft and pulpwood.	1830
A. venusta (Dougl.) K. Koch (bristlecone fir).	*A. bracteata* (D. Don) Nutt. (Santa Lucia fir).	-----do------	Santa Lucia Mountains of California.	Protection forest, ornamental planting.	1853

[1] Includes several horticultural varieties.
[2] Long cultivated.
[3] Includes var. *macrocarpa* Kent. Valuable information on the collection, extraction, cleaning, and weight of fruits and seeds of *A. balsamea* and 19 other species was obtained from a manuscript prepared by A. H. Richardson of the Ontario Department of Lands and Forests, Toronto, Ontario, Canada.
[4] Includes var. *lowiana* (A. Murr.) Lemmon and several horticultural varieties.
[5] Unpublished information on this species obtained from L. V. Barton, Boyce Thompson Institute, Yonkers, N. Y. Correspondence on file at Lake States Forest Experiment Station, St. Paul, Minn. Feb. 10, 1941.
[6] Wood more durable than most firs. Includes var. *shastensis* Lemmon.

A. amabilis

A. concolor

A. fraseri

A. grandis

A. lasiocarpa

A. procera

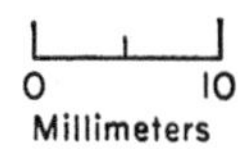

FIGURE 3.—Seed views of six species of *Abies.*

Data on the occurrence of geographic strains in *Abies* are rather limited. In *A. grandis,* five climatic races have been distinguished differing in color of needles, bark thickness, and properties of the wood, i.e., specific gravity, shrinkage, warping, and splitting. In Norway, plants of *A. amabilis* that originated from about 3,000 feet above sea level in western Washington were somewhat susceptible to frost damage but nevertheless perfectly hardy in years which were not too unfavorable. Considerable difference was found in plants of *A. procera;* stock from about 1,500 feet suffered considerable winter injury, but stock from 3,000 feet was entirely hardy. Trees of *A. concolor* of California origin are said to grow better in England than those from Rocky Mountain seed; in the eastern United States, the opposite is true. Further study would probably show that similar races exist in some of the other species, particularly among those which occur over a wide range of climatic conditions, such as *A. balsamea.*

COLLECTION, EXTRACTION, AND STORAGE.—Cones are usually collected by hand picking from standing trees[1] or from those recently felled in logging operations, as soon as the seed is ripe but before the cone scales begin to separate. Squirrel caches can sometimes be used as a source of supply for *Abies lasiocarpa* and probably other species if the seed appear to be firm and fully developed. Because freshly collected cones contain considerable moisture, they should be spread out to dry in shallow layers in the sun, in well ventilated sheds, or indoors at room temperature. In *Abies* the cone curing and seed extraction processes usually coincide, the cones breaking up within a few weeks.

Seeds may be separated from the cone scales and bracts by screening or running through a fanning mill. Because the seed coat in most species is soft and affords little protection to the seed, the more common types of dewinging are likely to be very injurious. Although commercial seed is usually sold clean, dewinging cannot be recommended as a general practice until less drastic methods of cleaning are worked out.[2] After a short period of drying the seed is ready for storage or for use. Data on the yield of cones and cost of seed are given in table 3.

Although seeds of some firs, such as *Abies concolor,* are rather perishable and cannot be stored even at low temperatures in airtight containers for more than 3 or 4 years, that of others (*A. balsamea, A. procera*) will retain viability for as long as 5 years if stored in sealed containers and at temperatures of 36° to 39° F. or lower. Seed of *A. grandis* with an initial moisture content of about 6 percent, and stored 11 years at about 40°, still showed approximately 50 percent of its original germination.

Under ordinary storage conditions seeds of most species will show little or no germination after 1 year. Viability of *Abies alba* seed, however, is not completely lost until the end of the third year.

GERMINATION.—In nature, *Abies* seeds usually germinate in the spring after lying on the ground over winter. Moist humus or mineral soil under partial to heavy shade apparently form equally good natural seedbeds. Germination, which is of the epigeous type (fig. 4), is seldom above 50 percent because of dormancy of the embryos, a typically high percent of seed injured during dewinging, insect infestation, and the perishable nature of the seed.

[1] This involves considerable hazard since the cones are invariably in the top of the trees and the wood of most species is very brittle; safety belts are essential.

[2] Seed dewinged by hand rubbing has given excellent germination.

TABLE 2.—*Abies: Time of flowering and cone ripening, and frequency of seed crops*

Species	Time of—			Commercial seed-bearing age		Seed year frequency	
	Flowering	Cone ripening	Seed dispersal	Minimum	Optimum	Good crops	Light crops
				Years	*Years*		
A. alba	May to mid-June	Mid-Sept. to mid-Oct.	Mid-Sept. to mid-Oct.	65–70	(1)	Every 2–3 years	Intervening years.
A. amabilis	Spring	September	October			do.	do.
A. balsamea	May	Late Aug. to early Sept.	Sept.–Nov.	20	30+	Every 2–4 years	do.
A. concolor	May–June	Sept.–Oct.	Sept.–Oct.	40	50–100	do.	do.
A. fraseri	Mid-May to early June	Sept. to mid-Oct.	Oct. to early Nov.				
A. grandis		September		20	(2)	Every 2–3 years	Intervening years.
A. lasiocarpa				20		Every 3 years	Intervening years; failures occasionally.
A. lasiocarpa var. *arizonica*.	Late June	Mid-Sept. to early Oct.	Late Sept. to mid-Oct.	50	150–200	Every 2–3 years	Intervening years.
A. magnifica	June	August	Sept.–Oct.		(3)	do.	do.
A. procera		Early Sept.	October	50–60	(4)	Infrequent	Some seed every year.
A. venusta		Late August	September			Every 3–5 years	

[1] Maximum commercial seed-bearing age for *A. alba* is 400+ years.
[2] Ability of *A. grandis* to bear commercial seed increases to old age.
[3] In *A. magnifica*, commercial seed are borne by moderately old trees.
[4] Ability of *A. procera* to bear commercial seed increases to old age.

TABLE 3.—*Abies: Yield of cleaned seed, and purity, soundness, and cost of commercial seed*

Species	Cleaned seed					Commercial seed		
	Per bushel	Per pound			Basis, samples	Purity[1]	Soundness[1]	Cost per pound
		Low	Average	High				
	Ounces	*Number*	*Number*	*Number*	*Number*	*Percent*	*Percent*	*Dollars*
A. alba		8,200	10,400	18,600	48	88 (57–100)	46 (30–81)	1.50–2.50
A. amabilis	48	8,200	11,300	14,900	8	91	[2]45	2.75–3.50
A. balsamea	[3]37–42	30,000	59,800	94,500	58	90 (42–100)	51	1.75–2.25
A. concolor	48–82	[4]8,200	15,100	27,200	97	91 (58–100)	60	1.25–3.00
A. fraseri	35	44,000	56,000	69,000	3		[5]30	3.00–5.00
A. grandis		12,600	23,200	44,300	31	83 (40–97)	48	1.75–2.50
A. lasiocarpa		23,900	37,500	51,300	16	94	[2]53	2.50–3.50
A. lasiocarpa var. *arizonica*	16–24	17,600	22,300	25,800	9	66 (42–88)	45	6.00
A. magnifica		4,000	6,600	11,000	36	80 (42–99)	[5]44	1.75–2.50
A. procera	40	11,200	14,600	19,300	9	92	[2]48	2.50–3.50
A. venusta								5.00

[1] Since part of the seed wing is persistent, it is impossible to remove all empty and abortive seed during the fanning process; hence, soundness for the species listed is relatively low. First figure in column is average percent; figures in parentheses indicate range from lowest to highest.
[2] Two samples.
[3] One bushel of cones weighs 35 pounds.
[4] Seed of *A. concolor* var. *lowiana* is apparently larger than that of the type; 13 samples showed: low, 8,200; average, 10,900; and high, 14,000 clean seed per pound.
[5] One sample.

TABLE 4.—*Abies: Dormancy and method of seed pretreatment*

Species	Dormancy: Kind	Dormancy: Occurrence	Stratification: Medium	Stratification: Temperature	Stratification: Duration
				°F.	*Days*
A. alba	Embryo	Some but not all lots	Sand or peat	41	30-60
A. amabilis	do	do[1]	do	41	60-90
A. balsamea	do	do	Sand	41	[2]90
A. concolor	do	do	do	40	[3]60-90
A. fraseri	do	do	Peat	[4]41	40+
A. grandis	([5])	([5])	do	41	30
A. lasiocarpa	Embryo	Some but not all lots	Sand	41	60
A. lasiocarpa var. *arizonica*	do	do	Peat	41	[6]60
A. magnifica	do	do	Sand	41	60
A. procera	do	do	Sand or peat	[4]41	
A. venusta	do	([5])	do	41	90

[1] All lots show some dormant seed; in some, however, the percent of dormant seed is much greater than in others.
[2] Stratification for periods as long as 180–240 days greatly increases rate of germination in some lots.
[3] Untreated seed germinates fairly well.
[4] Treatment suggested; experimental data not complete.
[5] Not definitely established; probably much as in other species.
[6] Stratification 30 days at 32° F. can also be used.

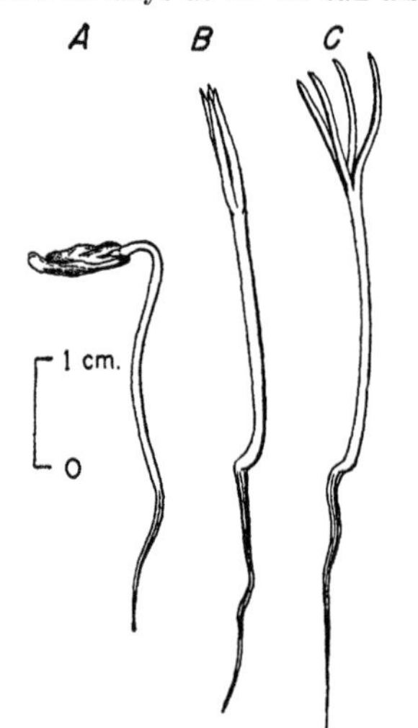

FIGURE 4.—Seedling views of *Abies balsamea: A,* At 2 days; *B,* at 5 days; *C,* at 7 days.

In addition to reductions made in the natural seed supply by rodents and birds, the seed of some species is sometimes so heavily infested with insects as to make it worthless. Among the more common insect pests of this type are the chalcid flies (*Megastigmus strobilobius* Ratz.) which attack *Abies alba* and *A. amabilis,* and *M. pinus* Parf. which infests *A. amabilis, A. concolor,* and other species. Seeds of *A. magnifica* and *A. procera* are also often heavily infested with insects.

Since most lots of *Abies* seed contain seeds with dormant embryos, germination is usually hastened and increased by stratification in sand or peat for 30 to 90 days at 41° F. In the case of *A. balsamea,* a much longer period is apparently necessary, at least in some lots, to complete germination within a short time. (See fig. 19, pt. 1, p. 39.) Such information as is available on dormancy and the treatment suggested to overcome it is given in table 4.

TABLE 5.—*Abies: Recommended germination test duration, and summary of germination data*

Species	Test duration recommended: Nondormant lots	Test duration recommended: Dormant lots untreated	Test duration recommended: Dormant lots stratified	Germinative energy		Germinative capacity: Low	Germinative capacity: Average	Germinative capacity: High	Germinative capacity: Basis, tests
	Days	*Days*	*Days*	*Percent*	*Days*	*Percent*	*Percent*	*Percent*	*Number*
A. alba		50-60		2-63	20-30	5	23	80	213
A. amabilis	30			55	30	1	22	55	6
A. balsamea	30	[1]210	[2]60-120	40-70	20	1	22	[3]74	52
A. concolor	30-60		30	57-67	20-30	1	34	94	71
A. fraseri	35		30-45	25	30	10	42	85	4+
A. grandis	30			18-60	20	1	28	98	30
A. lasiocarpa	30		30	8-52	20	7	38	54	15
A. lasiocarpa var. *arizonica*	30		20-30	11-70	20	12	31	70	10
A. magnifica		100+		14-53	30	5	25	59	20
A. procera	30	100				2	24	[4]72	7
A. venusta			40				51		1

[1] Germination much lower than that obtained from stratified seed.
[2] Depending on length of stratification period.
[3] Seed of Danish sources (21 samples): low, 9 percent; average, 33 percent; high, 68 percent.
[4] Seed of European sources (25 samples): low, 5 percent; average, 28 percent; high, 70 percent.

Germination may be tested in sand flats, peat mats, or standard germinators, using 800 to 1,000 seeds per test, these to be stratified in the case of lots known or suspected of being dormant. Optimum temperatures have not been worked out; diurnal alternations from 70° to 85° F. are believed to be satisfactory for most species; however, lower temperatures are sometimes used. Tests in British Columbia indicate that seed of *Abies grandis* germinate much better in the dark than under full light, temperature conditions being equal. Other details of technique and results that may be expected from germination tests are given in table 5.

NURSERY AND FIELD PRACTICE.—Fir seed should be sown in the fall, or they can be spring sown if stratified seed are used or the seed lots are definitely known to be nondormant. A well-drained sandy loam to loam forms the best seedbed and the seed may be sown broadcast or in drills at a rate to produce 60 to 80 seedlings per square foot. Seed should be covered with 1/8 to 3/8 of an inch of nursery soil and protected from birds and rodents by means of screen. Fall-sown beds should be mulched over winter, preferably with burlap. Since seedlings of most firs are highly susceptible to damping-off, the seedbeds should be treated with sulfuric acid, aluminum sulfate, or ferrous sulfate, depending on the acidity of the soil and the need for weed control, prior to sowing. Germination of untreated, dormant seed begins 10 to 35 days after spring sowing and is prolonged over a considerable period; fall-sown or stratified seed will germinate more promptly. Nursery germination varies from 15 to 35 percent.

Seedlings are extremely subject to heat injury and should be given light to half shade for at least the first season. Growth in most species is rather slow; therefore 2-1 or 2-2 transplant stock is generally used for field planting. Fir seedlings in the nursery may sometimes be attacked by the Pales weevil, *Hylobius pales*. In the Northeastern States nursery stock of *Abies balsamea* which is snow-covered a large part of the winter is sometimes subject to snow-mold caused by species of the fungus *Phacidium*. This defoliating disease can be controlled by spraying the beds in the late fall with dormant-strength lime-sulfur.

891183°—50——5

ACACIA Mill. Acacia

(Legume family—Leguminosae)

Distribution and Use.—The acacias include about 500 species of deciduous, or sometimes evergreen, trees and shrubs widely distributed in the tropics and warmer temperate areas. Nearly 300 species are found in Australia; there are about 70 American species. Some 75 species are of known economic value, and about 50 of these are generally cultivated. Acacias are valuable for many purposes: collectively they yield lumber, furniture wood, fuel wood, tannin, and such products as gum arabic, resins, medicine, fibers, perfumes, and dyes; some are useful for reclamation of sand dunes, shelter belts, ornamental purposes and street trees, forage, and as a host for the valuable lac insect. The 2 species with 1 additional variety listed in table 6 have been planted rather extensively in the warmer parts of the United States.

Seeding Habits.—Acacia flowers are perfect or polygamous, many of them are yellow, and they appear in the spring. The fruit is a two-valved or indehiscent pod or legume which ripens in the late summer. One or more kidney-shaped seeds occur per fruit, and usually are released by the splitting of the pod. They contain no endosperm (fig. 5). Details of the seeding habits of two acacias are as follows:

	A. decurrens var. *mollis*	*A. melanoxylon*
Time of—		
Flowering	June and later	February-March
Fruit ripening	June-October	July-November
Seed dispersal	June-October	July-December, or later.
Good seed crops borne	Almost annually	

Seeding habits of *A. decurrens* var. *normalis* are said to be similar to those of *A. decurrens* var. *mollis*.

Information is not available as to the development of geographic strains or races.

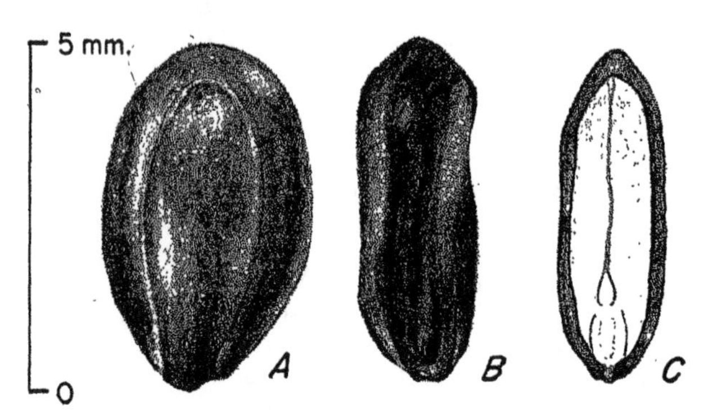

Figure 5.—Seed views of *Acacia melanoxylon*: *A* and *B*, Exterior views from two angles; *C*, longitudinal section.

Collection, Extraction, and Storage.—Ripe acacia pods should be picked from standing trees, or fallen seed can be swept up underneath the trees (this may include seed 1 to 6 years old). Seed can be extracted by trampling or flailing the pods and then using a blower to remove pod fragments and debris; they can also be extracted by running the pods through a threshing machine. Information as to seed characteristics is meager: One sample of *Acacia melanoxylon* seed averaged 33,000 per pound, 96 percent in soundness, and 89 percent in purity. (Three samples of *A. decurrens* averaged 36,000 seed per pound and it is likely that vars. *normalis* and *mollis* are similar in yield.) The cost per pound of *A. decurrens* var. *mollis* seed runs from $0.25 to $1.00; that of *A. melanoxylon* seed is about $4.25.

Acacia seeds are among the most durable of forest seeds and need not be kept in sealed containers. Stored in the open, they germinate after many years. (Some species retain a little viability up to

Table 6.—*Acacia: Growth habit, distribution, and uses*

Accepted name	Synonyms	Growth habit	Natural range	Chief uses	Date of earliest cultivation
A. decurrens var. *mollis* Lindl. (blackgreen-wattle acacia).	*A. mollissima* Willd.	Moderate-sized evergreen tree.	Australia; grows well in California.	Street tree, tanbark, sand dune reclamation, timber, posts, mine props, fuel wood.	Before 1840.
A. decurrens var. *normalis* Benth. (queenwattle acacia).	Sydney black wattle.	do	Restricted range along coast of Australia.	do	1850 in California.
A. melanoxylon R. Br. (blackwood acacia).	Tasmanian blackwood.	Large evergreen tree.	Australia; grows well in California.	Street tree, furniture and cabinet wood, fuel wood.	

TABLE 7.—*Acacia: Methods of seed pretreatment for dormancy*

Species	Pretreatment recommended	Other methods of pretreatment	Remarks
A. decurrens var. *mollis*	Scarification	1. Immerse in boiling water and allow to cool, 12 to 24 hours. 2. Boil in water 5 minutes. 3. Soak in concentrated sulfuric acid for 2 hours.	The boiling-water and sulfuric acid methods give results only slightly poorer than scarification, but germination is less rapid.
A. decurrens var. *normalis*	do	do	do.
A. melanoxylon	Soak in concentrated sulfuric acid 20 minutes.	1. Immerse in boiling water and allow to cook, 12 to 24 hours. 2. Place in water, bring to a boil, and allow to soak 24 hours.	The boiling-water methods give results only a little poorer than the acid treatment, but germination is less rapid.

68 years.) Seed of *A. decurrens* var. *mollis* retains its viability for several years in cool dry storage. It is probable that seed of *A. decurrens* var. *normalis* has similar properties. *A. decurrens* seed, after soaking 2 hours in acid, germinated 63 percent after 17 years open storage. Seed of *A. melanoxylon,* which was first air-dried to a constant weight and then stored in sealed containers, retained its viability unimpaired for at least 3 months; seed stored in the open still retained 12 percent viability after 51 years. It is reported that in Australia the seed of some species remain dormant in the soil for several years.

GERMINATION.—Natural germination of *Acacia decurrens* var. *mollis* seems particularly good in a mixture of ashes and mineral soil following forest fires; information for other species is not available. Germination is epigeous in all species for which data have been collected. Weevils sometimes ruin the seed by laying their eggs in the flower buds and larvae appear later in the pods. The seeds of most species have hard coats which cause poor germination unless they are first scarified, treated with hot water, or soaked in sulfuric acid. Some species also appear to require 2 to 4 months "after ripening" in dry storage before good germination may be obtained. In Australia it is sometimes recommended that the seeds be prepared for germination by placing them in the dying embers of a fire. Further information on pretreatment of the seed of three species is given in table 7.

Germination tests should be made in flats with soil or sand, or in standard germinators using 400 properly pretreated seeds per test. Results of tests for three species of acacia are given in table 8. Recommended procedures and results vary with the species.

NURSERY AND FIELD PRACTICE.—Properly pretreated seed of *Acacia decurrens* var. *mollis* should be sown in the spring and covered with about one-half inch of soil. One ounce of pretreated seed will produce about 600 first-year seedlings. Seed of *A. decurrens* var. *normalis* can probably be handled in the same manner. *A. melanoxylon* seed should be sown in flats in the greenhouse and later transplanted to seedbeds. Acacias may also be propagated by cuttings of half-ripened wood taken with a heel. Seedlings of *A. decurrens* var. *mollis* are sometimes attacked by a bagworm, but those of *A. decurrens* var. *normalis* are less susceptible to such damage.

TABLE 8.—*Acacia: Recommended conditions for germination tests, and summary of germination data*

Species	Test conditions recommended			Germination data from various sources					
	Apparatus	Temperature	Duration of test	Germinative energy		Germinative capacity			Basis, tests
						Low	Average	High	
		°F.	*Days*	*Percent*	*Days*	*Percent*	*Percent*	*Percent*	*Number*
A. decurrens var. *mollis*	Soil flats	60	14			50	72	92	14+
A. decurrens var. *normalis*	Germinator					48	74	98	4+
A. melanoxylon	Soil flats		15	32–68	6–13	40	60	85	5+

ACER L. Maple

(Maple Family—Aceraceae)

Distribution and Use.—Maples are deciduous, rarely evergreen, trees or shrubs; they include about 115 species in North America, Asia, Europe, and northern Africa. Some produce valuable lumber, others are used for production of maple sugar; many species have considerable ornamental value, either because of their handsome foliage, their flowers, or their fruit. They are, as a result, widely used for street and park planting. Others provide food and cover for wildlife. Of the more important species described in this bulletin, 9 are native to the United States and 3 to the Old World. Their distribution and chief uses are given in table 9.

With the possible exception of *Acer saccharum* which is sometimes planted in the Northeastern and Lake States, none of the maples are used to any great extent for reforestation in the United States. *A. ginnala* has been successfully planted for ornamental and shelter-belt purposes in the Intermountain States, and in South Dakota, and Minnesota. *A. macrophyllum,* in addition to being planted within its native range, is occasionally cultivated in the Eastern States. *A. negundo* has been extensively used for shelter belts in the Prairie-Plains region and farther west. The two European maples, *A. platanoides* and *A. pseudoplatanus,* and their several varieties have been widely planted in the Eastern States in yards and on streets. This is also true of *A. saccharinum* and *A. saccharum. A. pensylvanicum, A. rubrum,* and *A. spicatum* have been used to a limited extent in the Lake States for game food and cover planting. *A. pensylvanicum* has been also occasionally planted in Europe.

Seeding Habits.—The flowers in *Acer* are regular, usually polygamo-dioecious or polygamo-monoecious (rarely all perfect or dioecious) and are borne in clusters produced from separate lateral or terminal buds, appearing in early spring before, with, or after the leaves. The fruit is composed of 2 samaras, which eventually separate from a small persistent pedicel; each is compressed laterally and prolonged on the back into a large papery wing thickened on the outer edge and contains typically a single seed without endosperm (figs. 6 and 7). Ripening occurs in late summer or fall—in some species in spring—and is indicated by the light brown to straw color of the samaras. Commercial seed consists of the ripe samaras from which the wings are sometimes removed. The seeding habits of 11 species and 1 variety are detailed in table 10.

No data are available on the age for commercial seed-bearing of any of the 12 maples except *A. saccharinum* which begins to bear at 35 to 40 years. Seed is largely dispersed by wind; that of a few species such as *Acer negundo, A. rubrum,* and *A. saccharinum* may also be disseminated by water.

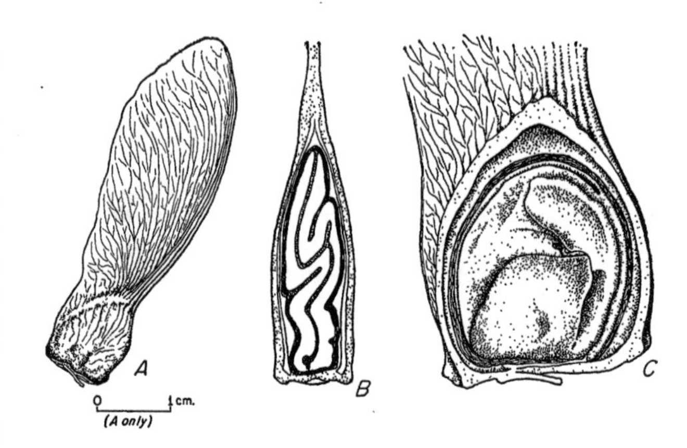

Figure 6.—Seed views of *Acer platanoides*: *A,* Exterior view; *B* and *C,* longitudinal sections in two planes, enlarged.

Very little reliable information exists as to the development of geographic strains in the maples. In the case of *Acer pseudoplatanus,* studies in Germany and Switzerland have shown that stock grown from seed of sources with colder climatic conditions grow more slowly, have a shorter growing season, and shed leaves earlier than that from warmer localities. Because the remaining species all have a wide range and several varieties have been described for some of them, it is more than likely that they also have developed geographic strains.

Collection, Extraction, and Storage.—Maple seed is usually collected from standing trees, but sometimes from trees soon after they are felled in logging operations. In the former case, the seed is shaken or whipped from trees with light poles and collected on canvas spread on the ground. That of some species like *Acer negundo, A. rubrum,* and *A. saccharinum* can also be gathered from lawns, pavements, etc., or from the surface of water in pools or streams. After collection, leaves and other debris can be removed by hand, screening, or fanning.

No special treatment of the seed of fall-ripening species is necessary other than to make sure that it is superficially dry if storage is contemplated. Seed of *Acer saccharinum,* however, has a moisture content at the time of dispersal of about 60 percent and must be protected against excessive drying; if the moisture content falls below 30 to 34 percent, the seed shows complete loss of viability.

The wings of the large-seeded species are sometimes removed by threshing or in a coarse dewinger to facilitate handling. For instance dewinging the seed of *Acer saccharum* probably reduces its weight by 20 percent and its bulk by 50 percent. Other than this the seed is not extracted from the samaras. Data on size, purity, soundness, and cost of seed for 11 species and 1 variety are given in table 11.

Although storage of *Acer* seed for considerable periods is seldom attempted, the seed ordinarily being sown soon after collection, it is believed that most species can be kept for at least a year or two with little loss in viability if stored at low temperatures and protected against excessive drying.

TABLE 9.—*Acer: Growth habit, distribution, and uses*

Accepted name	Synonyms	Growth habit	Natural range	Chief uses	Date of earliest cultivation
A. ginnala Maxim. (Amur maple).	*A. tataricum* var. *ginnala* Maxim. (Siberian maple).	Shrub to small tree.	Central and northern China, Manchuria, and Japan.	Ornamental and shelter-belt planting.	1860
A. glabrum Torr. (Rocky Mountain maple).	dwarf maple	do	Rocky Mountain and far Western States	Erosion control; game food.	1882
A. macrophyllum Pursh (bigleaf maple).	Oregon maple	Tree of small to medium size.	Pacific coast from Alaska to California.	Furniture and cabinet making; flooring; street planting; sometimes for sugar.	1812
A. negundo L. (boxelder)	*Negundo fraxinifolium* Nutt., *N. aceroides* Moench (Manitoba maple, ashleaf maple).	do	New England to Minnesota, Florida, and Texas.	Shelter-belt planting; game food.	1688
A. negundo var. *californicum* (Torr. & Gray) Sarg. (California boxelder).	*A. californicum* (Torr. & Gray) Dietr., *N. californicum* Torr. & Gray	do	Valleys of California.	Shelter-belt and ornamental planting.	1865
A. pensylvanicum L. (striped maple).	*A. striatum* Lam. (moosewood).	do	Quebec to Wisconsin, south to Georgia.	Game food; ornamental planting.	1755
A. platanoides L. (Norway maple).		Tree of medium size.	Europe and Caucasus.	Ornamental planting	[1]
A. pseudoplatanus L. (sycamore maple).	planetree maple, SPN.	do	Europe and western Asia.	do	[1]
A. rubrum L.[2] (red maple)	scarlet maple, swamp maple.	Large forest tree	Newfoundland to Florida, Minnesota, and Texas.	Lumber; game food; ornamental planting.	1656
A. saccharinum L. (silver maple).	*A. dasycarpum* Ehrh. (soft maple, river maple, water maple).	do	New Brunswick to Florida, South Dakota, and Oklahoma.	Ornamental and shelter-belt planting; stream-bank protection; sometimes for lumber and furniture.	1725
A. saccharum Marsh. [syn. *A. saccharophorum* K. Koch.] (sugar maple).	*A. barbatum* Michx. (hard maple, rock maple, sugar-tree).	do	Eastern Canada to Georgia, Minnesota, and Texas.	Lumber, furniture, and specialty products; cooperage and handle stock; maple sirup; ornamental planting.	1753
A. spicatum Lam. (mountain maple).	*A. montanum* Ait. (moose maple).	Tall shrub to small bushy tree.	Newfoundland to Saskatchewan, Lake States, and south in Appalachian Mountains.	Browse and game cover	1750

[1] Long cultivated.
[2] Unpublished information on this species obtained from R. M. Fisher, Illinois Department of Conservation, Jonesboro, Ill. Apr. 8, 1942. Correspondence on file at Lake States Forest Experiment Station, St. Paul, Minn.

ACER

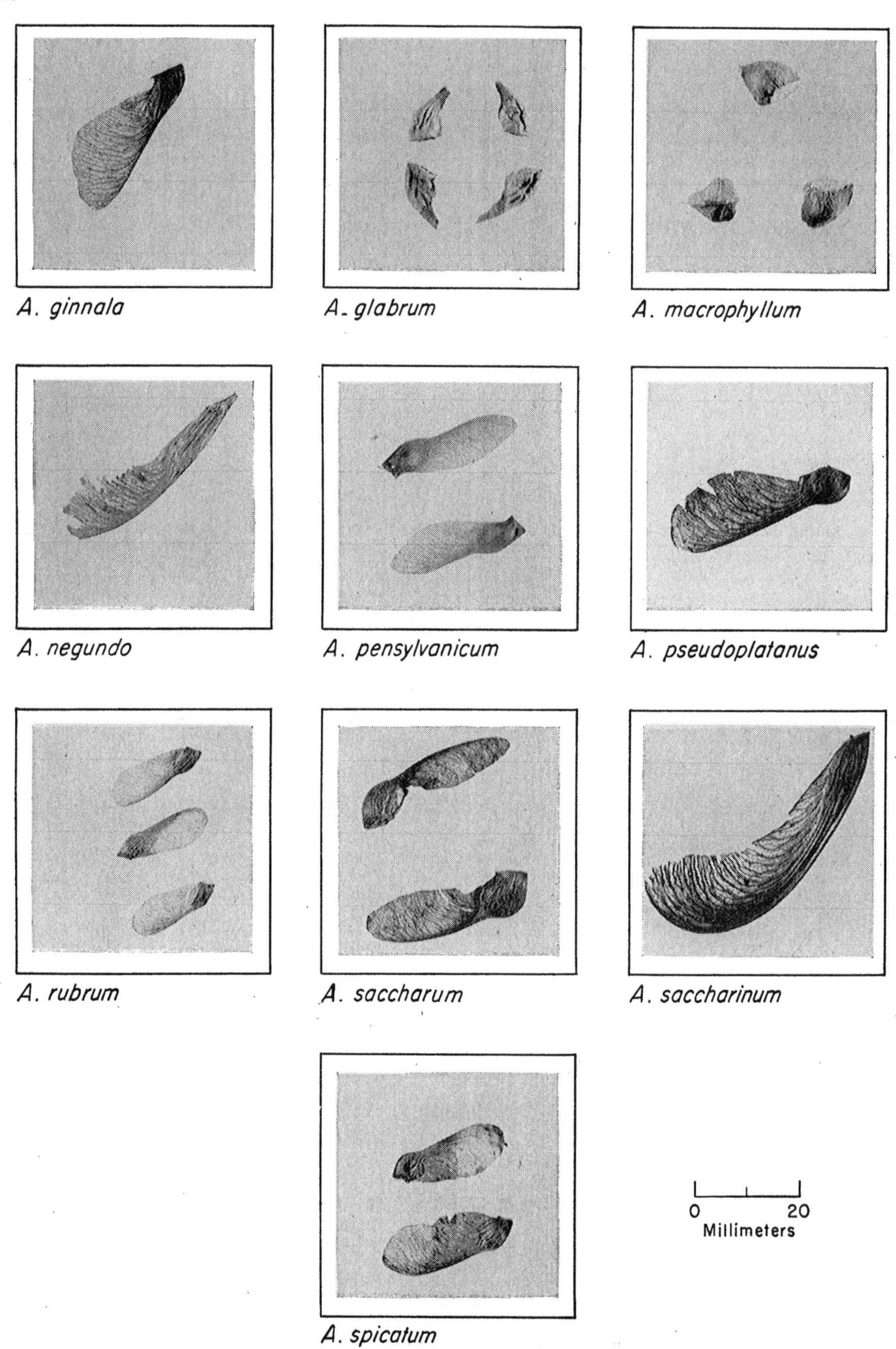

FIGURE 7.—Seed views of 10 species of *Acer*.

TABLE 10.—*Acer: Time of flowering and fruit ripening, and frequency of seed crops*

Species	Time of—			Seed year frequency	
	Flowering	Fruit ripening	Seed dispersal	Good crops	Light crops
A. ginnala	May–June	August–September	September to early winter		
A. glabrum	May	Mid-August to September.[1]	September–October		
A. macrophyllum	April–May	September–October	Late fall to early winter		Every year.
A. negundo	March–May 15	August–October[1]	September to early spring	Nearly every year.	Intervening years.
A. negundo var. *californicum.*		September–October[1]	Fall to early winter	do	
A. pensylvanicum	May–June 15	September–October 15	October–November		
A. platanoides	April–June	do	do		
A. pseudoplatanus	April–May	August–October	Fall		
A. rubrum	February–May	March to late June[1]	April–July	Nearly every year.	
A. saccharinum	February–April	April–June 15[1]	April to mid-June	do	Intervening years.
A. saccharum	March–May	Mid-September to October	October–December	Every 3 to 7 years.	do.
A. spicatum	May–June	Mid-September to mid-October.	do		

[1] Some trees are sterile due to the fact that they bear only staminate flowers, the pistillate flowers being produced by different individuals.

Seeds of the spring-ripening species, *A. rubrum* and *A. saccharinum,* ordinarily considered of transient vitality, have been successfully stored for longer periods than usually considered possible; the former for 21 months in sealed containers at 41° F. with only about 15 percent loss. Even that of *A. saccharinum* will retain its original viability for at least a year if stored cold and moist at about 32° to 50°. Seed of *A. ginnala* has been kept for 2½ years in sealed containers at 41° with no loss in germinability; that of *A. negundo* and *A. pensylvanicum* for 18 months under these conditions with little to no loss. Seed of *A. platanoides* stored under the same conditions lost only about 20 percent of its initial germination during one year.

Since seed of *A. pseudoplatanus* is reported to keep for 1 to 2 years under ordinary storage conditions, its longevity also can likely be extended

TABLE 11.—*Acer: Yield of cleaned seed, and purity, soundness, and cost of commercial seed*

Species	Cleaned seed				Commercial seed		
	Per pound			Basis, samples	Purity[1]	Soundness[1]	Cost per pound
	Low	Average	High				
	Number	*Number*	*Number*	*Number*	*Percent*	*Percent*	*Dollars*
A. ginnala	10,400	15,200	19,500	6	90	88 (60–99)	1.50–1.70
A. glabrum	13,200		20,300	2		[2]80	.65–1.50
A. macrophyllum	2,800	3,100	3,400	3		[2]87	1.25–1.40
A. negundo	8,200	11,800	15,000	14	92 (75–98)	62 (0–94)	.25–0.75
A. negundo var. *californicum*		10,000		1+	1+		1.25
A. pensylvanicum	9,700	11,500	15,600	4	92	89 (79–95)	2.00–3.50
A. platanoides	1,300	2,600	4,600	11	92 (78–100)	93	.75–1.20
A. pseudoplatanus	5,100	5,900	7,200	7	82 (73–100)	[2]94	1.00–1.80
A. rubrum	12,700	22,800	38,200	12	92 (80–100)	87 (76–96)	1.50–2.50
A. saccharinum	900	1,400	1,900	6	97	97	.60–1.00
A. saccharum	3,200	6,100	9,100	15	94	[3]52 (38–67)	.60–1.65
A. spicatum	15,300	22,200	27,800	5	93	73 (57–84)	3.25–5.00

[1] First figure is average percent; figures in parentheses indicate range from lowest to highest.
[2] Based on 1 sample.
[3] This low value is due to the fact that the samaras occur in pairs, one of which is invariably empty. Soundness of more than 50 percent is probably due to removal of some of the empty seeds by fanning; those less than 50 percent, to the picking up of empty seeds from the ground.

considerably by use of lower temperatures and airtight containers. *A. saccharum* will remain viable for at least a year if stored in open or sealed containers at 36° to 40° F. The only maple thus far known whose seed cannot be stored either at room or low temperatures for even a short time is *A. macrophyllum.* Seed of *A. spicatum* may also have a more transient vitality than that of some of the other species, but even this has shown at least one-fourth its original germination at the end of 2 years in sealed cold storage.

GERMINATION.—Natural germination in most of the fall-ripening species is believed to take place in the early spring following dispersal. Seed of the summer-ripening *Acer saccharinum* germinates immediately following its fall, and that of some lots of *A. rubrum* germinates soon after falling in early summer, while others probably do not germinate until the next spring. Germination is epigeous in most species (figs. 8 to 12), and hypogeous in at least one *(A. saccharum).* Moist mineral soil with considerable organic matter is probably the most favorable seedbed, although for some species hardwood leaf litter appears to be equally good.

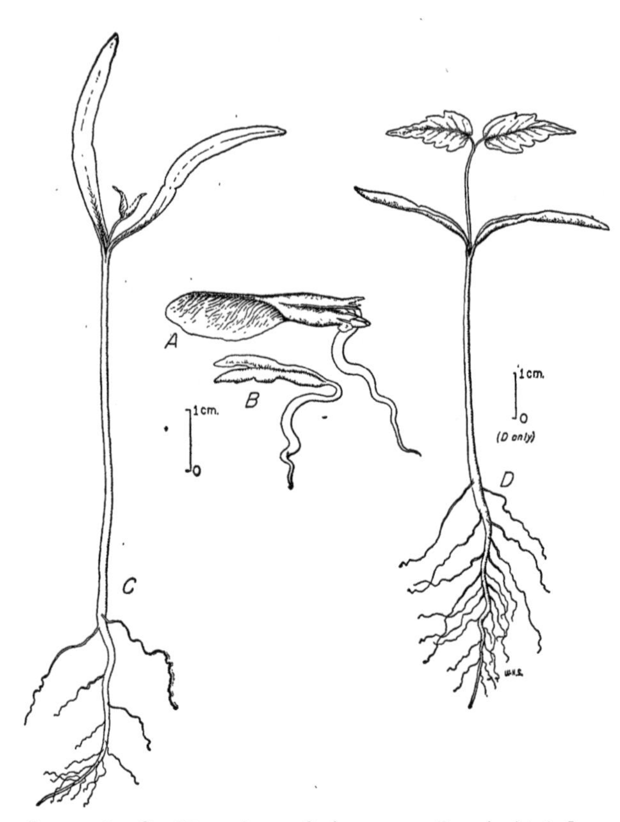

FIGURE 9.—Seedling views of *Acer negundo: A,* At 1 day; *B,* at 2 days; *C,* at 7 days; *D,* at 12 days.

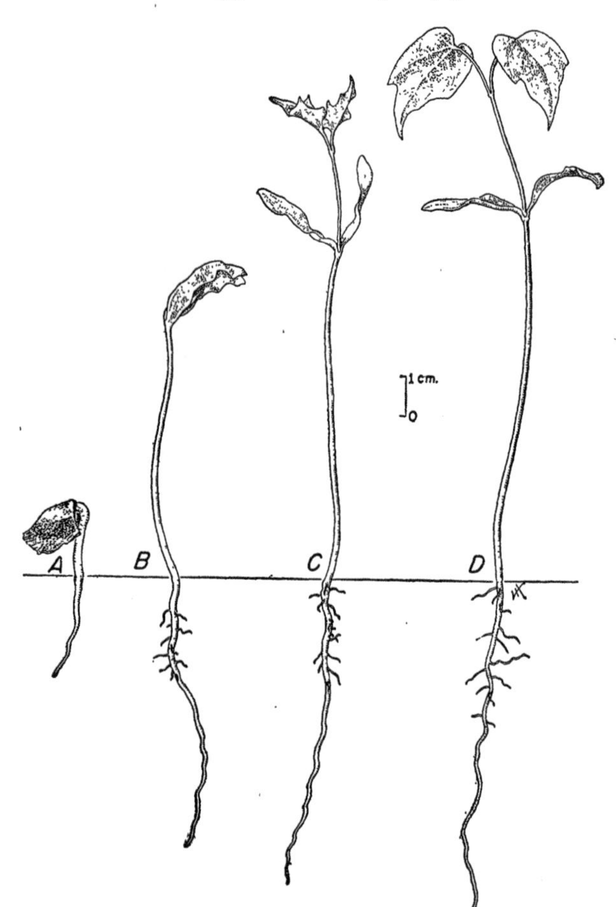

FIGURE 8.—Seedling views of *Acer platanoides: A,* At 1 day; *B,* at 3 days; *C,* at 7 days; *D,* at 19 days.

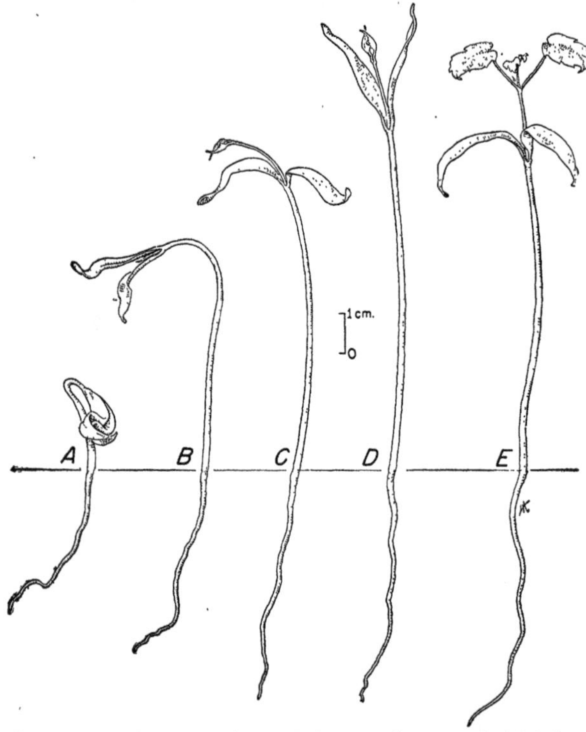

FIGURE 10.—Seedling views of *Acer saccharum: A,* At 1 day; *B,* at 4 days; *C,* at 7 days; *D,* at 14 days; *E,* at 43 days.

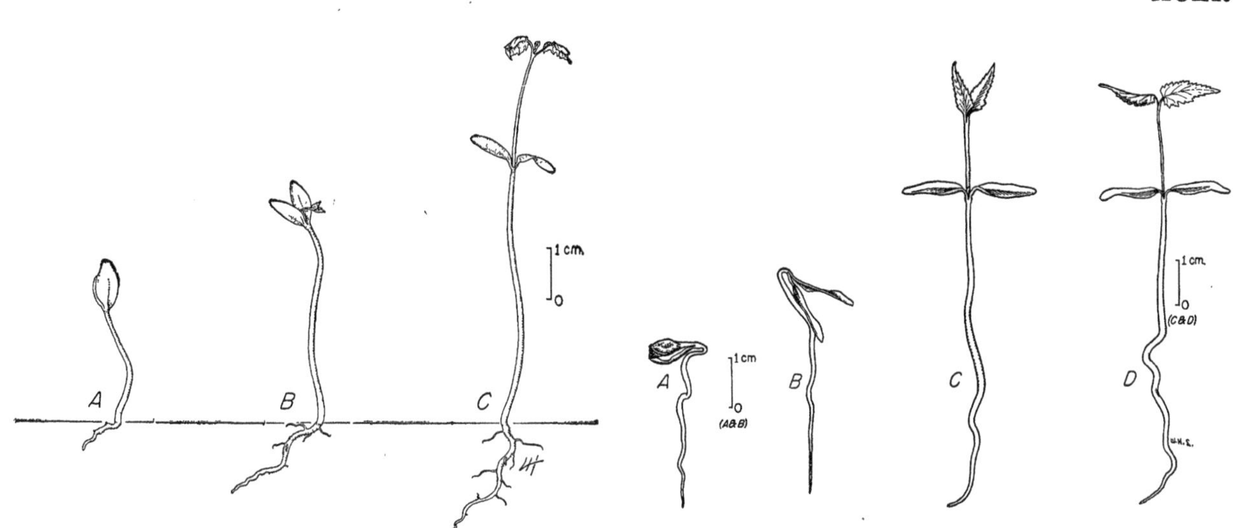

FIGURE 11.—Seedling views of *Acer ginnala: A,* At 1 day; *B,* at 2 days; *C,* at 15 days.

FIGURE 12.—Seedling views of *Acer rubrum: A,* At 1 day; *B,* at 2 days; *C,* at 9 days; *D,* at 11 days.

All 11 species herein discussed, with the exception of *Acer saccharinum,* show delayed germination due to dormant embryos. In some of these, notably *A. ginnala* and possibly also in some lots of *A. spicatum* and *A. pensylvanicum,* germination is retarded in addition by the pericarp and possibly by the seed coat proper. Dormancy may be overcome in the former species by simple stratification at low temperatures in moist sand, peat, or a mixture of these media. For seed lots of those species which appear to show seed coat impermeability as well as embryo dormancy, some form of scarification is suggested in combination with stratification. More detailed information is given in table 12.

Germination tests are best made in flats, using 200 to 400 or more seeds per test, these to be properly pretreated in the case of species showing dormancy. For *Acer saccharinum* which is usually of high soundness, 100 to 200 seeds may be sufficient. As might be expected in a genus of wide distribution, procedures and results show considerable variation between species (table 13).

TABLE 12.—*Acer: Dormancy and methods of pretreatment*

Species	Dormancy		Stratification			Other methods of pretreatment
	Kind	Occurrence	Medium	Temperature	Duration	
				°F.	*Days*	
A. ginnala	Impermeable seed coat and dormant embryo.	Some lots[1]	Sand	41	150+	Light scarification and stratification for 90 days at 41° F.[2]
A. glabrum[3]	Embryo		do	41	90	
A. macrophyllum	do		do	41	60	
A. negundo	do	All lots	do	41	90	Soak for 2 weeks in cold running water.
A. negundo var. *californicum.*	do	Some lots	do	41	90	
A. pensylvanicum[3]	do[4]	All lots	do	41	90–120	
A. platanoides	Embryo	do	do	41	90–120	
A. pseudoplatanus[3]	do		do	41	90+	
A. rubrum	do	Some lots; others not dormant.	do	41	60– 75	Soak in cold running water for about 5 days.
A. saccharinum	None					
A. saccharum	Embryo		Sand or peat	36–41	60– 90	
A. spicatum[3]	do[4]		Sand	41	90–120	(5)

[1] Others appear to have embryo dormancy only.
[2] Suggested additional treatment for lots showing seed coat impermeability and embryo dormancy. Cutting through pericarp and seed coat at hilum partly successful. Stratification for those showing embryo dormancy only.
[3] Treatment suggested; experimental data not complete.
[4] Some lots may have seed coat impermeability also.
[5] Best results (experimental scale) obtained from seed which has been extracted from pericarps and seed coats scratched in process.

ACER

TABLE 13.—*Acer: Recommended conditions for germination tests, and summary of germination data*

Species	Test conditions recommended			Germination data from various sources							Remarks
	Temperature		Duration[1]	Germinative energy		Germinative capacity			Potential germination	Basis, tests	
	Night	Day				Low	Average	High			
	°F.	*°F.*	*Days*	*Percent*	*Days*	*Percent*	*Percent*	*Percent*	*Percent*	*No.*	
A. ginnala	50	50	50–60	24–28	17–20	1	24	52	86–90	4	Temperature recommendations only tentative; alternations of 68°–86° F. also fairly effective.
A. glabrum[2]	68	86									
A. macrophyllum[2]	68	86				32		90			
A. negundo	50	77	50–60	19–62	8–30	1	33	80		15	Temperature recommendations based on northern seed.
A. negundo var. *californicum*.			10–20					60		1	
A. pensylvanicum						0.8	1.4	1.9	70–82	3	
A. platanoides	40–50	40–50					30	94		9	Alternations of 68°–86° and 50°–77° F. are less effective.
A. pseudoplatanus				[3]37	20	0	12	50	1–91	8	
A. rubrum	50	77	30–40	39–70	9–22	1	46	74	75–84	9	Alternations of 68°–86° F. are almost as effective.
A. saccharinum[2]	75	85	20–30	73–99	8–14	34	76	99		7	
A. saccharum	68	86	30	16–60	4–24	16	39	68		10	Alternations of 50°–77° F. are practically as effective.
A. spicatum				[3]32	31	0	8	34	12–100	5	

[1] Dormant seed which have not been pretreated require from 90 to 300 days and then germinate very poorly.
[2] Temperatures for tests suggested; experimental data not complete.
[3] 1 sample.

NURSERY AND FIELD PRACTICE.—The seed of most species of *Acer* with a few exceptions should be sown in the fall in mulched beds or if stratified seed is used, in the spring. In the case of *A. saccharum* and probably other species, the seed must not be stratified too long or germination will occur during the stratification period. The seed is usually sown from ¼ to 1 inch deep, broadcast or in drills spaced about a foot apart and at the rate of about 20 viable seeds per foot. Germination is generally completed within a few weeks in the spring. However, seed lots of *A. ginnala* which have impermeable coats will probably not germinate until the second spring unless the seed is sown early enough in the fall to be subjected to a month or two of weather warm enough to permit fungal and bacterial action on the seed coats. Some seed lots of *A. pensylvanicum* and *A. spicatum* are also probably in this class, but because of a characteristic later ripening they cannot be sown soon enough to obtain germination the following spring.

Seed of *Acer saccharinum* and of most lots of *A. rubrum* should be sown in late spring as soon after collection as possible. Seed of other lots of *A. rubrum,* possibly due to a slight lowering of its moisture content, apparently will not germinate in any great quantity until the next spring. Since such lots cannot be distinguished in advance, it is safer to sow *A. rubrum* seed in the fall or stratify it prior to spring sowing. Seedbeds used for these two species must be kept constantly moist.

Nursery germination: *A. rubrum,* 35 to 80 percent; *A. pensylvanicum,* 41 percent; *A. platanoides,* 20 percent; *A. saccharinum,* 18 percent; *A. saccharum,* 15 percent.

Seedlings of most maples, with the possible exception of *Acer rubrum* and *A. saccharinum,* should be given some shade while becoming established in the beds. Insect pests or fungous diseases cause little damage to maple seedlings in the nursery; probably the most serious disease is *Verticillium* which results in wilting of the tops of small trees of *A. platanoides* and *A. saccharum.* One-year-old seedlings are usually large enough (*A. rubrum* 6 to 12 inches tall) for planting in the field. Most species do best if planted on fertile soil, such as loams, with ample to abundant moisture. In addition to propagation from seed some species, such as *A. negundo,* can also be propagated from hardwood cuttings or by division.

AESCULUS L. Buckeye, horsechestnut

(Horsechestnut family—Hippocastanaceae)

DISTRIBUTION AND USE.—The buckeyes, occurring in North America, southeastern Europe, and eastern and southeastern Asia, include about 25 species of deciduous trees or shrubs. They are cultivated for their dense shade or for their ornamental flowers, and the wood of some species is occasionally used for lumber and paper pulp. The shoots and seeds of some buckeyes are poisonous to stock. Of the 4 species described in table 14, 3 are native to the United States, and the remaining one, horsechestnut, was introduced into this country from southern Europe.

None of the four species are much used in reforestation but all are used for ornamental planting. This is particularly true of *Aesculus hippocastanum* which has been widely planted as a shade tree in Europe and also in the eastern United States, where it sometimes escapes from cultivation. *A. glabra* and *A. octandra* are sometimes planted in Europe and eastern United States, the former species having been successfully introduced into Minnesota, western Kansas, and eastern Massachusetts. *A. californica* is also planted occasionally in Europe and to a somewhat greater extent in the Pacific Coast States.

SEEDING HABITS.—The flowers in *Aesculus* are irregular, white, red or pale yellow in color, and are borne in showy clusters which appear after the leaves. Only those flowers near the base of the branches of the cluster are perfect and fertile; the others are staminate. The fruit is a somewhat spiny or smooth, leathery, round or pear-shaped capsule, with three cells, each of which may bear a single seed. Sometimes only one cell develops, the remnants of the abortive cells and seeds being plainly visible at maturity. When only one cell is developed, the large seed (fig. 13) is round to flat in shape, dark chocolate to chestnut-brown in color, smooth and shining, and has a large, light-colored

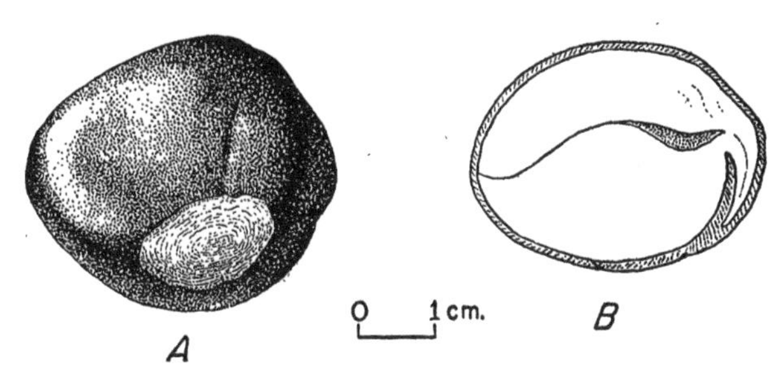

FIGURE 13.—Seed views of *Aesculus octandra: A*, Exterior view; *B*, longitudinal section.

TABLE 14.—*Aesculus: Growth habit, distribution, and uses*

Accepted name	Synonyms	Growth habit	Natural range	Chief uses	Date of earliest cultivation
A. *californica* (Spach) Nutt. (California buckeye).	----------------	Low, much branched tree.[1]	Dry gravelly soils, lower slopes of Coast Range and Sierra Nevada, California.	Erosion control, ornamental planting, fuel; seeds roasted and used by Indians as food.	1855
A. *glabra* Willd.[2] (Ohio buckeye).	A. *ohioensis* DC. (fetid buckeye, American horsechestnut).	Forest tree.	Moist rich soils, Pennsylvania, south in Allegheny Mountains to Alabama, west to southeastern Nebraska and eastern Oklahoma.	Artificial limbs, woodenware, wood pulp, lumber, ornamental purposes, game food.	1809
A. *hippocastanum* L.[2] (horsechestnut).	chestnut, bongay	Large tree..	Balkan Peninsula-----------	Street and landscape planting, occasionally in cabinetry, bark used in medicine.	1576
A. *octandra* Marsh.[2] (yellow buckeye).	A. *flava* Ait., A. *lutea* Wangenh. (sweet buckeye).	Large forest tree.	Moist rich soils, Pennsylvania, south in Allegheny Mountains to north Alabama, west to Missouri, Oklahoma, and northwestern Texas.	Same as A. *glabra*; fruit sometimes eaten by cattle.	1764

[1] More often a shrub, forming extensive dense thickets.
[2] Includes several varieties.

AESCULUS

TABLE 15.—*Aesculus: Time of flowering and seed dispersal*

Species	Time of—		
	Flowering	Fruit ripening	Seed dispersal
A. californica	April–August	September–October	November.
A. glabra	March–May	September to mid-October	Early September to late October.
A. hippocastanum	Late April to early June	Mid-September to early October	Mid-September to mid-October.
A. octandra	April–June	September	September.

hilum resembling the pupil of an eye. It contains no endosperm, the cotyledons being very thick and fleshy. Ripening occurs in the fall, at which time the capsules split and release the seed. Detailed data on the seeding habits of the four species of *Aesculus* discussed here are given in table 15.

No data are available on the age for commercial seed bearing, the periodicity of seed crops, or the existence of geographic strains for any of the four species discussed here. Seed dispersal is by gravity and animals, and sometimes by water.

COLLECTION, EXTRACTION, AND STORAGE.—The seed may be collected by picking or shaking from the trees as soon as the capsules turn yellowish and begin to split open, or they may be gathered from the ground soon after falling. A short period of drying at room temperature may be used to free the seeds from any part of the capsules that may still adhere to them, but great care should be taken that the drying is not so excessive that the seed coats become wrinkled and lose their waxy appearance. The seed should be sown at once in a nursery or stratified and sown in the spring. No data on the optimum storage conditions for *Aesculus* seeds are available, although seed of *A. hippocastanum* stored in cloth sacks at 40° to 50° F. and at low humidity dropped from 90 to 29 percent viability in 5 months. It is believed that much better retention of viability will be obtained if the seeds are kept in sealed containers to prevent drying and at temperatures at or slightly below 32°.

Such data as are available on seed size and cost are given in detail in table 16. Purity and soundness are invariably high, generally being close to 100 percent.

GERMINATION.—Natural germination in *Aesculus* normally takes place in the early spring after the seeds have lain on the ground over winter; that of *A. californica,* however, takes place just after the winter rains have begun, usually in November. Rich, moist soil near streams forms the most favorable seedbed. Germination is hypogeous (fig. 14).

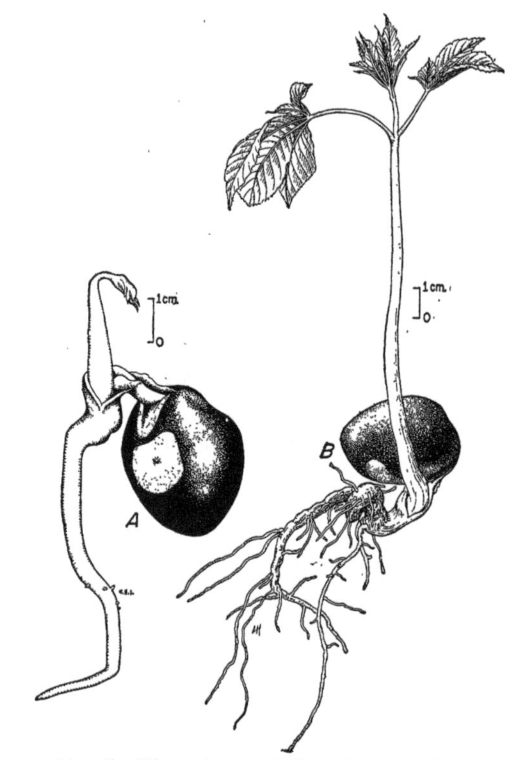

FIGURE 14.—Seedling views of *Aesculus octandra: A,* At 2 days; *B,* at 4 days.

TABLE 16.—*Aesculus: Yield of cleaned seed, and cost of commercial seed*

Species	Cleaned seed				Commercial seed, cost per pound
	Per pound[1]			Basis, samples	
	Low	Average	High		
	Number	*Number*	*Number*	*Number*	*Dollars*
A. californica		10		1	1.20–2.50
A. glabra	48		67	2	.30–1.35
A. hippocastanum		109		1	.04– .15
A. octandra	27		30	2	.35– .70

[1]This value varies, of course, not only with seed size but also with moisture content, which, in *Aesculus* seeds, is initially rather high. One sample of *A. octandra* showed a moisture content of 95 percent after it had been kept at room temperature for 36 days after collection.

TABLE 17.—*Aesculus: Recommended germination test duration, and summary of germination data*

Species	Test duration recommended		Germination data from various sources				
			Germinative capacity			Potential germination	Basis, tests
	Stratified seed	Untreated seed	Low	Average	High		
	Days	*Days*	*Percent*	*Percent*	*Percent*	*Percent*	*Number*
A. californica		20	36		75		2
A. glabra		230+		28		100	1
A. hippocastanum		230+		18		100	1
A. octandra	20	160+	49	70	84	93	3

All of the species described here except *Aesculus californica* show delayed germination due to embryo dormancy. This may be overcome by stratification in moist sand at 41° F. for about 120 days.[3] Germination tests are most easily and successfully made in flats, using 100 to 200 stratified seeds per test. Temperatures alternating from 68° (night) to 86° F. (day) appear satisfactory but may not be optimum. Results that may be expected under these conditions are given in table 17.

NURSERY AND FIELD PRACTICE.—*Aesculus* seed is practically always sown in the fall as soon after collection as possible to prevent its drying out. If desired, however, the seed of those species showing dormancy can be stratified and sown in the spring. The seed is sown in nursery rows; germination is nearly complete. Ordinarily, 1–0 stock is large enough for field planting. With the exception of *A. californica* which occurs naturally on dry slopes, *Aesculus* seedlings should not be planted in dry situations. *Aesculus* can also be propagated by grafting or budding and the shrubby species by layering.

[3] The optimum periods for the various species have not been worked out. However, limited experience shows that 60 days is not sufficient to give prompt and complete germination of *A. glabra* and *A. hippocastanum* nor 90 days quite enough for *A. octandra.*

AILANTHUS ALTISSIMA (Mill.) Swingle Ailanthus

(Ailanthus family—Simaroubaceae)

Also called tree of heaven, tree-of-heaven ailanthus, SPN; Chinese sumac. Botanical syns.: *A. glandulosa* Desf., *A. cacodendron* Schinz and Thell., *A. japonica* Hort. Includes vars. *erythrocarpa* (Carr.) Rehd., *pendulifolia* (Carr.) Rehd., and *sutchuenensis* (Dode) Rehd. and Wils.

DISTRIBUTION AND USE.—Native to China, this short to medium-tall deciduous tree is of value chiefly for shade and ornamental purposes, particularly in cities where soils are poor and the atmosphere smoky. It is also planted for shelter belts, for game food and cover, and occasionally for timber in New Zealand. *Ailanthus* was introduced into cultivation in 1751 and brought to America in 1784. It has become naturalized in many parts of the eastern United States and Canada, and in Kansas; in some localities it is so well at home that it appears to be a part of the native flora.

SEEDING HABITS.—Commercial "seed" consists of the one-celled, one-seeded, oblong, thin, spirally twisted samaras. These are 1 to 1½ inches long, light reddish brown in color, and bear the seed at about the middle (fig. 15). Flowers open in mid-April to July; the large crowded clusters of bright red seed ripen in September to October of the same season; seed dispersal occurs from October until the following spring; prolific seeder. Trees 12 to 20 years old bear seed in considerable quantities.

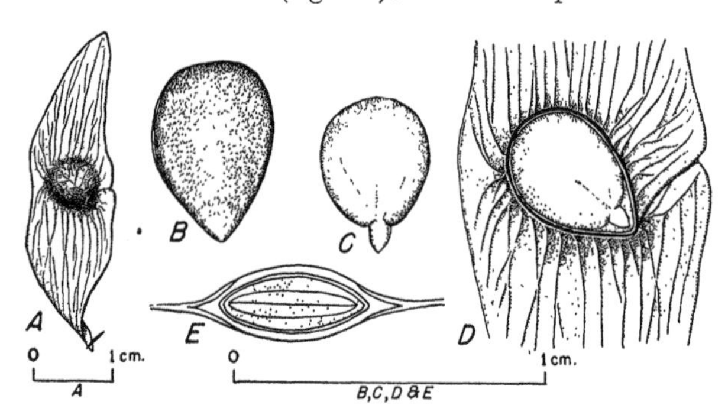

FIGURE 15.—Seed views of *Ailanthus altissima: A*, Exterior view of samara; *B*, exterior view of seed; *C* and *D*, exterior view of seed with coat removed; *E*, longitudinal section.

COLLECTION, EXTRACTION, AND STORAGE.—*Ailanthus* fruits are picked from standing trees by hand or flailed or stripped onto canvas at any time during the late fall and early winter. After collection the seed should be spread out to dry to get rid of superficial moisture; it may then be run through a macerator and fanned to remove impurities, or it may be flailed or trampled in a grain bag and run through a fanning mill. One hundred pounds of fruits yield from 30 to 90 pounds of cleaned seed. Number per pound (with wings, 6 samples): low, 12,700; average, 14,600; high, 16,500; average without wings, about 17,000. Commercial seed averages in purity about 88 percent; in soundness about 88 percent; in cost per pound, $0.35 to $1.40. Seed probably should be stored dry in sealed containers at a temperature of about 40° F. One lot stored in sacks for somewhat over a year at temperatures ranging from 20° to 105° F. showed germination of 75 percent.

GERMINATION.—It is not thoroughly established whether or not *Ailanthus* seed is dormant. A limited number of tests made by the Lake States Forest Experiment Station, however, indicate that it has a dormant embryo, and that germination is greatly benefited by stratification in moist sand for 60 days at 41° F. The Russians recommend this. Soaking in water for 10 days seems to give poorer results than no treatment. Test methods recommended: sand flats; temperature, alternating diurnally from 60° to 80° F.; seeds required, 1,000; duration, 30 days for stratified seeds, 60 to 80 days for unstratified seed. Under these test conditions *Ailanthus* in 2 tests showed a germinative energy for stratified seed of 8 to 52 percent in 9 to 26 days; and in 3 tests of stratified seed, a germinative capacity of—low, 14 percent; average, 48 percent; high, 75 percent.

NURSERY AND FIELD PRACTICE.—Seed should be stratified over winter and sown in spring in drills or broadcast, covering it with one-half inch of soil. About 15 to 25 percent of the viable seed sown produce usable 1–0 seedlings. This species reproduces from sprouts as well as by seed. It grows well on a variety of soils but does best on light, rather moist soils.

ALBIZIA JULIBRISSIN Durazz. Silktree

(Legume family—Leguminosae)

Also called mimosa, silky acacia, silktree albizzia, SPN. Botanical syns.: *Acacia* j. (Durazz.) Willd., *A. nemu* Willd. Includes var. *rosea* (Carr.) Mouillef.

Distribution and Use.—Native to Asia from Persia to China, this handsome leguminous tree was introduced into cultivation in 1745 and has been planted widely for ornamental purposes in the southern United States where it has become established locally from Virginia to Louisiana; it is also hardy in California. In addition to its ornamental value, the tree is useful for wildlife cover and browse.

Seeding Habits.—The flat, light brown to greenish-brown seeds are oval in shape and about one-half inch in length (fig. 16). They are borne in flat, linear, straw-colored 8- to 12-seeded pods about 6 inches long which occur in rather large clusters. Light pink flowers are borne in June to August; the papery pods ripen from September to November of the same year and begin to disintegrate on the trees soon thereafter, although they may remain attached to the tree for some time.

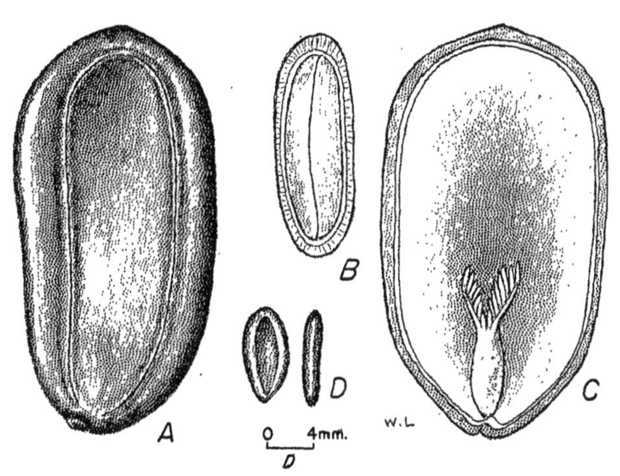

Figure 16.—Seed views of *Albizia julibrissin*: *A*, Exterior view of seed; *B*, cross section; *C*, longitudinal section, showing large plumule and cotyledons; *D*, exterior views from two planes.

Collection, Extraction, and Storage.—Silktree seed collection should begin as soon as the seed coats become hard and flinty. The pods may be picked or shaken from the trees and collected on canvas. Seed are extracted readily from the pods by flailing or by threshing, and a hammer mill could probably be used to advantage for this process. The resulting debris is easily removed from the seed by running it through a fanning mill. Three samples of clean seed averaged 11,100 (ranging from 10,600 to 11,700) to the pound. Commercial seed is practically 100 percent pure, 90 to 100 percent sound, and costs $1 to $1.40 per pound. Practically no data are available on proper storage methods. A small sample of seed which had been kept in a loosely corked bottle in a laboratory for almost 5 years showed 90 percent germination. Seed of the related *Albizia lophantha* still gave 33 percent germination after 50 years dry, open storage.

Germination.—Silktree seed is dormant because of an impermeable seed oat. This condition can be overcome by treatment in a mechanical scarifier until breaks begin to appear in the seed coat. Brief sulfuric acid treatment might be of benefit. Germinative capacity in three tests of scarified seed was as follows: low, 80 percent; average, 87 percent; high, 100 percent in 3 days. Germination is epigeous (fig. 17).

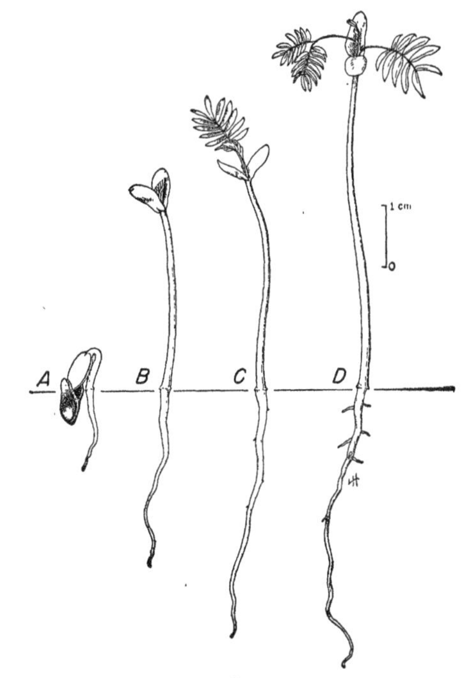

Figure 17.—Seedling views of *Albizia julibrissin*: *A*, At 1 day; *B*, at 3 days; *C*, at 5 days; *D*, at 8 days.

ALNUS Mill. Alder

(Birch family—Betulaceae)

DISTRIBUTION AND USE.—Native to swamps, stream bottoms, or high mountains, the alders consist of about 30 species of deciduous trees and shrubs distributed chiefly in the Northern Hemisphere and in the Andes Mountains to Peru and Bolivia. Some species are valuable timber trees; others are useful for their ornamental qualities, as a source of food and shelter for game, or for erosion control. Because the roots of several species foster nitrogen-fixing organisms, they increase soil fertility. Of the 5 species described here, 3 are confined to the United States, 1 occurs only in the Eastern Hemisphere, and 1 is found in both this country and Europe. More specific information regarding their distribution and chief uses is given in table 18.

None of the alders is much used in reforestation in this country but in Europe both *Alnus glutinosa* and *A. incana* are planted fairly extensively for this purpose. The former species has long been grown in the eastern United States where it has escaped from cultivation and become established locally in New Jersey, southern New York, and eastern Massachusetts. In Europe it frequently hybridizes with *A. incana.*

SEEDING HABITS.—The flowers in *Alnus* are usually borne in the spring. They are monoecious and are produced in catkins which are formed in the preceding season and expand before or with the leaves. The staminate catkins occur in small clusters. They are naked and erect during winter, becoming much elongated and pendulous at maturity. The pistillate aments, also erect and usually naked during winter, occur in short racemes or elongated panicles, enlarge but slightly at flowering, and develop into ovoid strobiles or conelets. The seed—strictly speaking, a fruit—consists of a small, compressed, narrowly winged or practically wingless nutlet without endosperm (figs. 18 and 19) borne in pairs at the base of each "cone" scale. It ripens in autumn at which time the strobiles have become thick and woody, and dark brown in color. Dispersal, which is effected usually by wind and often to a considerable distance, takes place from shortly after ripening until early in the spring. Details of seeding habits for five species are given in table 19.

Good crops of seed of *Alnus rubra* are borne about every fourth year with light crops in intervening years. Optimum seed-bearing age is 25

TABLE 18.—*Alnus: Growth habit, distribution, and uses*

Accepted name	Synonyms	Growth habit	Natural Range	Chief uses	Date of earliest cultivation
A. crispa (Ait.) Pursh (American green alder).	*A. mitchelliana* M. A. Curt., *A. viridis* Am. auth. (green alder, mountain alder).	Handsome aromatic shrub.	Newfoundland to British Columbia, south to North Carolina, Michigan, and Minnesota.	Game food and cover; potentially valuable for ornamental planting.	1782
A. glutinosa (L.) Gaertn.[1] (European alder).	*A. rotundifolia* Mill., *A. vulgaris* Hill, *A. communis* Desf. (black alder, European black alder).	Tree of medium size.	Europe, western and northern Asia, northern Africa.	Woodenware, cooperage, charcoal, bark for dye and tanning.	(2)
A. incana (L.) Moench[1] (speckled alder).	hoary alder, gray alder.	Tall shrub to small tree; thicket forming.	Europe, Caucasus, in North America from Newfoundland to Saskatchewan, south to Pennsylvania and Nebraska.	Game food and cover; stream-bank protection.	(2)
A. rubra Bong. (red alder).	*A. oregona* Nutt. (Oregon alder).	Medium to large tree, occuring in groves.	Pacific coast from southern Alaska to southern California.	Furniture and cabinet work, fuel, game food, wood used for smoking salmon.	1884
A. tenuifolia Nutt. (thinleaf alder).	*A. incana* var. *virescens* S. Watson, *A. occidentalis* Dipp. (mountain alder).	Tall shrub to small tree; thicket forming.	British Columbia to lower California, east to Saskatchewan, Colorado, and New Mexico.	Erosion control, fuel, game food and cover; of secondary importance as forage plant.	1880

[1] Includes several varieties.
[2] Long cultivated.

TABLE 19.—*Alnus: Time of flowering and seed dispersal*

Species	Time of—		
	Flowering	Seed ripening	Seed dispersal
A. crispa	June	Late August to mid-October	Soon after ripening.
A. glutinosa	Early spring	Fall	Late fall to early spring.
A. incana	April–May	September–October	Mid-September to early winter.
A. rubra	Early spring	November	Late November–December.
A. tenuifolia	do	Fall	

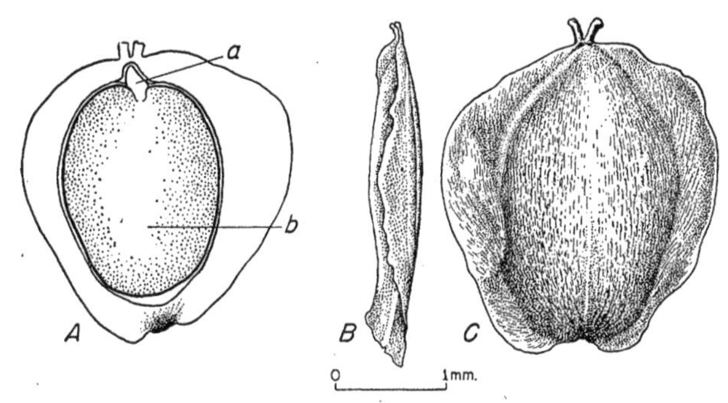

FIGURE 18.—Seed views of *Alnus rubra. A*, Longitudinal section: *a*, Radicle; *b*, cotyledon. *B* and *C*, Exterior views in two planes.

years with maximum seed production occurring at about 100 years. Details are lacking for the other species. The existence of geographic strains in *Alnus* has not been demonstrated, but in view of the wide geographic range of the several species discussed and the recognition of numerous varieties in *A. glutinosa* and *A. incana* it is more than likely that such strains do occur.

COLLECTION, EXTRACTION, AND STORAGE.—The fruit is usually collected from standing trees or trees recently felled in logging operations as soon as the scales show signs of separating. They are then spread out to dry on canvas in a warm room or in a kiln operated at a rather low temperature (probably between 80° and 100° F.). Most of the seed will fall out during the drying process; the remainder can be extracted by tumbling the fruits in a drum or similar device so constructed as to allow the seed to fall through. An alternative method used with *Alnus glutinosa* in Europe is to cut off small cone-bearing branches and hang them up in bundles, shaking these to extract the seed when the leaves have dried. The seed is then spread in layers to dry. Cleaning may be accomplished by the use of a fanning mill or by screening. Soundness is usually low as it is difficult to fan out empty seeds. Detailed data on cleaned and commercial seed for five species are given in table 20.

FIGURE 19.—Seed views of two species of *Alnus.*

TABLE 20.—*Alnus: Yield of cleaned seed, and purity, soundness, and cost of commercial seed*

Species	Cleaned seed					Commercial seed		
	Yield per bushel of fruits	Per pound			Basis, samples	Purity[1]	Sound-ness[2]	Cost per pound
		Low	Average	High				
	Pounds	*Number*	*Number*	*Number*	*Number*	*Percent*	*Percent*	*Dollars*
A. crispa		696,000		1,864,000	2	[3]61	42–93	
A. glutinosa		289,000	352,000	639,000	86	59 (24–90)	39	0.60–1.25
A. incana		473,000	666,000	890,000	123	[4]41 (12–89)	51	1.50–2.15
A. rubra	[5]1.12	363,000	666,000	1,087,000	4		70	[6]2.06
A. tenuifolia			675,000		1			

[1]First figure is average percent; figures in parentheses indicate range from lowest to highest.
[2]Based on two samples each for *A. crispa, A. glutinosa,* and *A. incana.*
[3]2 samples.
[4]European seed is often adulterated with that of the cheaper *A. glutinosa.*
[5]Uncleaned; average number of seeds per fruit, 100.
[6]Five-year average cost of U. S. Forest Service collection.

891183°—50——6

ALNUS

TABLE 21.—*Alnus: Dormancy and method of seed pretreatment*

Species[1]	Dormancy		Stratification		
	Kind	Occurrence	Medium	Temperature	Duration
				°F.	*Days*
A. crispa	Embryo	At least some lots	Sand	41	60
A. glutinosa	None[2]				
A. incana[3]	Embryo	Some lots	Sand	41	90+

[1]Definite information on *A. rubra* and *A. tenuifolia* is lacking. Fresh seed germinates poorly, possibly because of dormancy or possibly because of a high proportion of empty seed.
[2]Germination of some lots appears to be benefited slightly by stratification.
[3]European seed appears to germinate well without pretreatment.

Little is known concerning the optimum storage conditions for *Alnus* seed. Since, however, one sample of *A. glutinosa* dropped only 8 percent in germination (from 36 to 28 percent) after 16 months' storage in a sealed jar at 41° F., and *A. rubra* seed is known to lose its viability rather quickly when stored in open containers, it seems likely that dry storage in sealed containers at a low temperature would prove beneficial to most species.

GERMINATION.—Natural germination in most species is believed to take place in the spring following dispersal. Seed of some species show embryo dormancy; those of others appear to be variable—some seeds in a given lot being dormant, the rest not dormant, while entire lots may not show any dormant seeds. Germination is epigeous (fig. 20) and in most species occurs best on moist mineral soil. More detailed information on the germination of three species is given in table 21.

Germination tests can be made in sand flats or standard germinators using 1,000 seeds per test, these to be pretreated in the case of those lots showing dormancy. Temperatures alternating diurnally from 68° to 86° F. have been used successfully for *Alnus crispa* and *A. glutinosa* and probably can be used for other alders. Other conditions for testing and results that may be expected are given in table 22.

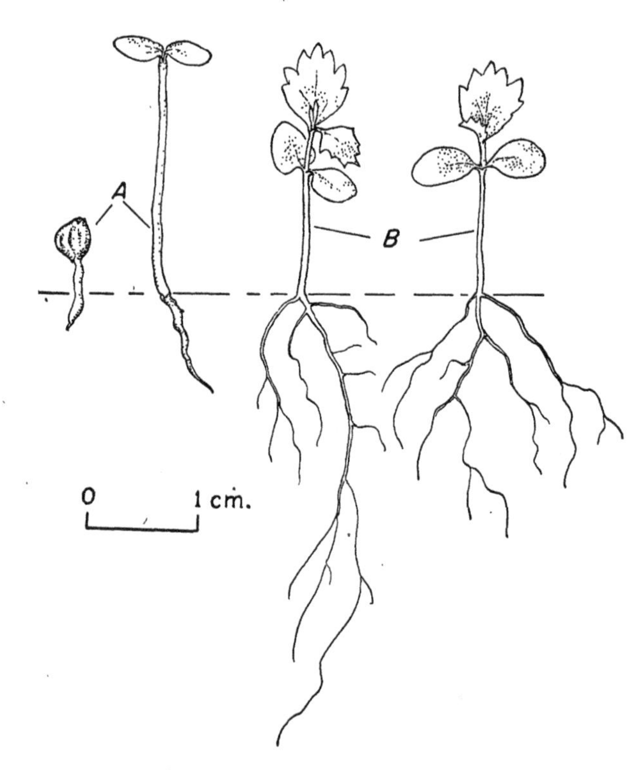

FIGURE 20.—Seedling views: *A*, *Alnus glutinosa* at 1 day and 7 days; *B*, two older seedlings of *A. tenuifolia*.

TABLE 22.—*Alnus: Recommended germination test duration, and summary of germination data*

Species	Test duration recommended	Germination data from various sources						
		Germinative energy		Germinative capacity			Potential germination	Basis, tests
				Low	Average	High		
	Days	*Percent*	*Days*	*Percent*	*Percent*	*Percent*	*Percent*	*Number*
A. crispa	30–40	28	12	0	23	40	30–40	3
A. glutinosa	30	1–70	10	1	27	82		492
A. incana	30	1–76	10	2	32	88		337
A. rubra				14		40		2
A. tenuifolia						16		1

Nursery and Field Practice.—Seeds of those lots which show no or only slight dormancy may be sown in the spring; otherwise sowing is done in the fall or the seed may be stratified for 60 to 90 days prior to spring sowing. The seed is sown broadcast or in drills and is covered lightly with soil. A light covering of coarse, well-washed sand or of sand mixed with hardwood humus has been found superior to nursery soil or leaf litter for seeds of *Alnus incana*. The beds should be kept moist and shaded until late summer of the first season. Germination usually begins in 35 days (*A. incana*) to 45 days (*A. rubra*) after sowing; that of *A. tenuifolia* begins in 8 to 10 days and is usually complete in 30 days. About 20 percent of the filled seeds of *A. rubra* germinate in the nursery. In field planting, 2–0 stock of *A. glutinosa* should be satisfactory but 1–2 is sometimes recommended. This species grows well on damp soil; it may also be propagated from hardwood cuttings or from layers, as can the shrubby species.

AMELANCHIER Med. Serviceberry

(Rose family—Rosaceae)

Distribution and Use.—Native to North America, Europe, northern Africa, southwestern and eastern Asia, the serviceberries consist of about 25 species of deciduous unarmed trees or shrubs which are valuable chiefly for their ornamental qualities and their edible fruit. The hard, heavy and strong wood of the arborescent species is sometimes used in turnery. Most species are valuable sources of food for wildlife. Three species for which detailed information is available are of present or potential use in conservation planting and are described here; their distribution and chief uses are given in table 23.

Relatively little use is made of any of the serviceberries in conservation planting at the present time. *Amelanchier alnifolia* is sometimes used in shelterbelt planting and erosion control in the northern part of the Prairie-Plains region; *A. arborea* has been planted occasionally for wildlife food and cover and also for erosion control. Most species have possibilities for landscape planting and deserve considerably more usage for this purpose.

Seeding Habits.—The perfect, white, regular flowers, in some species very attractive, appear in terminal clusters in the spring either before or after the leaves. The fruit, which in many species is sweet and juicy, is bluish black to reddish purple or sometimes orange in color and ripens in early to late summer. It is a berrylike pome crowned by the usually reflexed calyx lobes and contains 4 to 10 small seed, some of which are usually abortive. Fertile seeds are dark brown in color with a leathery seed coat and with the embryo filling the seed cavity (figs. 21 and 22). Commercial seed generally consists of the dried fruit although cleaned seed is also available. Seeding habits are given in detail for 3 species in table 24. Seed dispersal is almost entirely by animals and birds and usually takes place about as soon as the fruit ripens.

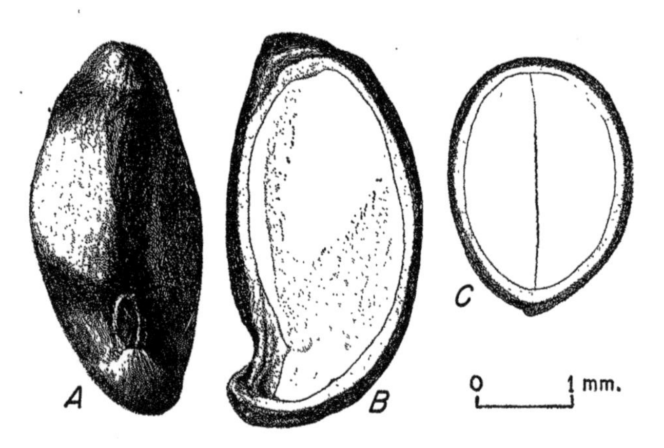

Figure 21.—Seed views of *Amelanchier sanguinea: A*, Exterior view; *B*, longitudinal view; *C*, cross section.

Table 23.—*Amelanchier: Growth habit, distribution, and uses*

Accepted name	Synonyms	Growth habit	Natural range	Chief uses	Date of earliest cultivation
A. alnifolia Nutt.[1] (saskatoon serviceberry).	*A. canadensis* var. *alnifolia* (Nutt.) Torr. & Gray (common serviceberry, alderleaf serviceberry, Pacific serviceberry).	Low shrub to small tree; thicket forming.	Upper Michigan west and north to Saskatchewan and Alaska, and south to Nebraska, New Mexico, and California.	Fruit used as food by man and wildlife; valuable browse for livestock and deer; erosion control.	1826
A. arborea (Michx. f.) Fern. (downy serviceberry).	*A. canadensis* auth. not (L.) Medic., *A. c. tomentula* Sarg. (Juneberry, shadbush, shadblow serviceberry).	Tree to 60 feet.	Maine to Minnesota, south to Georgia, Louisiana, and Oklahoma.	Fruit eaten by wildlife, ornamental planting; wood sometimes used for handles.	1623
A. sanguinea (Pursh) DC. (roundleaf serviceberry).	*A. rotundifolia* Roem. not (Lam.) Dum.-Cours.; *A. spicata* Robins. & Fern., not (Lam.) K. Koch.	Slender, straggling, shrub.	Quebec to Minnesota, south to New York, Indiana, and Nebraska.	Fruit used as food by man and animals.	1824

[1] Includes also the closely related species *A. florida* Lindl., which is sometimes considered a variety of *A. alnifolia.*

TABLE 24.—*Amelanchier: Time of flowering and seed production*

Species	Time of— Flowering	Time of— Fruit ripening	Color of ripe fruit	Seed production
A. alnifolia	May–June	July–August	Dark purple	Good crops borne nearly every year.
A. arborea	Late March–May	Late June to early July	Reddish purple[1]	
A. sanguinea	May	Late July–August	Black	Some borne every year.[2]

[1]Fruit dry and scarcely edible.
[2]Begins at about 8 years, is at optimum from 10 to 15 years.

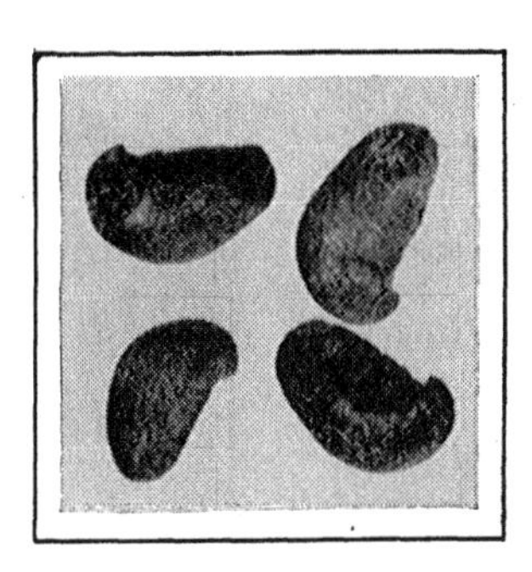

FIGURE 22.—Seed view of *Amelanchier alnifolia.*

Nothing is known about the existence of geographic strains in *Amelanchier.* Such strains probably occur in some of the species such as *A. alnifolia* and *A. arborea,* which are of wide distribution, but this is a question which will have to await a satisfactory revision of the group. Possibly some of the closely related species, which are often separated from each other with extreme difficulty, will be found to be geographic strains of a single polymorphic species.

COLLECTION, EXTRACTION, AND STORAGE.—The fruit is collected[4] by picking from the trees or shrubs as soon as possible after ripening to prevent its consumption by birds. Unless the seeds are to be extracted immediately, the berries should be spread out to dry in shallow layers; otherwise heating will occur. Extraction may be easily accomplished by macerating the fruit in water. This method causes most of the pulp to be floated away. After drying and rubbing, the remainder may be winnowed or run through a fanning mill to remove debris and small abortive seeds which are extremely numerous. Seed is probably best removed from the dried fruit of commerce by rubbing followed by fanning. Such information as is available on cleaned and commercial seed is given in table 25.

[4] Collection cost of *A. alnifolia,* 20 cents per pound of fruit.

Little is known concerning the proper storage conditions for *Amelanchier* seeds. Since an unidentified species showed high germination after 1 year's storage in the fruit in a sealed container at 41° F., it is recommended that *Amelanchier* seed be kept at low temperatures in sealed containers.

GERMINATION.—Although definite information is lacking, it is believed that natural germination of *Amelanchier* seed occurs mostly during the spring after the seed is dispersed. Germination is epigeous (fig. 23).

The seed is dormant probably because of embryo conditions. In some species, the seed coat may also retard germination—immersion of the seed in concentrated sulfuric acid for 30 minutes, while possibly of no advantage, at least had no harmful effects. Dormancy can be overcome, at least in part, by low-temperature stratification. Scarification or

TABLE 25.—*Amelanchier: Yield of cleaned seed, and purity, soundness, and cost of commercial seed*

Species	Cleaned seed: Yield per 100 pounds of fruit	Cleaned seed: Per pound Low	Cleaned seed: Per pound Average	Cleaned seed: Per pound High	Cleaned seed: Basis, samples	Commercial seed: Purity	Commercial seed: Soundness	Commercial seed: Cost per pound
	Pounds	*Number*	*Number*	*Number*	*Number*	*Percent*	*Percent*	*Dollars*
A. alnifolia	2	[1]51,000	82,000	113,000	5		74	[2]0.75–2.50
A. arborea	1	50,000		81,000	5	93	[3]82 (54–95)	[2]4.50 [4]12.00
A. sanguinea			84,000		1	95	83	

[1]Dried berries per pound, 5,100.
[2]Dried berries.
[3]Average percent; figures in parentheses indicate range from lowest to highest.
[4]Clean seed.

AMELANCHIER

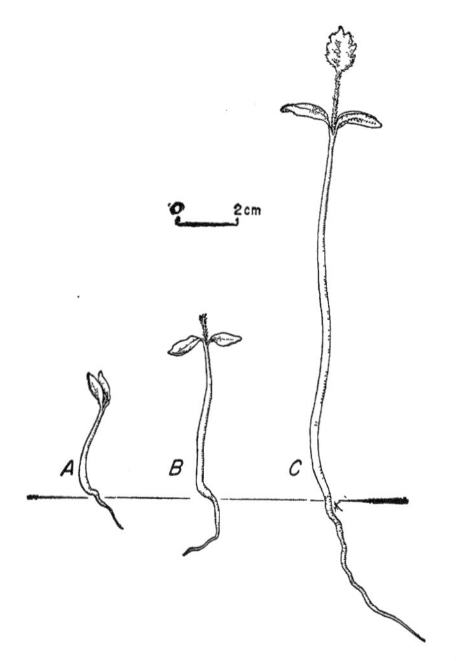

FIGURE 23.—Seedling views of *Amelanchier* spp: *A*, At 3 days; *B*, at 5 days; *C*, at 7 days.

acid treatment in combination with stratification is believed worthy of trial. Such details of pretreatment as are known are given below for two species of *Amelanchier*.

	A. alnifolia	*A. arborea*
Dormancy	Embryo	Embryo
Stratification:		
Medium	Peat	Peat
Temperature °F.	35–37	33–41
Duration days	180+	90–120

Further study is needed for both species; the stratification period for *A. alnifolia* can probably be shortened.

Germination tests can be made in flats containing sand, or sand and peat in mixture, and require 400 to 800 properly pretreated seeds per test. Optimum temperature is not known; daily alternations of 68° to 86° F. and a constant temperature of 41° F. have proved equally successful for some species. *Amelanchier alnifolia* will germinate at 35° to 37° F. when dormancy is broken. Results that may be expected from germination tests are as follows:

	A. alnifolia	*A. arborea*
Basis, tests	2	2
Test duration days	30	--
Germinative capacity:		
Low percent	50	38
High do	95	70

NURSERY AND FIELD PRACTICE.—*Amelanchier* seed is usually either fall sown or stratified and sown in the spring. Results thus obtained are only fair, many seedlings not germinating until the second spring. It is suggested that the seed be sown as soon as possible after collection and the beds kept mulched until the following spring when good germination should occur.

The seed should be sown in drills at the rate of 25 viable seeds per linear foot and covered with one-fourth inch of nursery soil. Nursery germination of *Amelanchier arborea* is 40 percent. The beds (at least for *A. alnifolia*) should be given half shade the first year. Growth of most species is slow, the seedlings being transplanted after 1 year in the seedbed and then field planted 2 to 3 years later. *Amelanchiers* can also be propagated from cuttings taken in the fall or spring, or from suckers.

AMORPHA L. Amorpha, false indigo

(Legume family—Leguminosae)

Distribution and Use.—Consisting of about 15 closely related species of deciduous shrubs or subshrubs[5] native to North America, the false indigoes are valuable chiefly for ornamental purposes because of their handsome flowers and foliage. Some species furnish food and cover for game; others are useful in erosion control. The distribution and chief uses of the three most important species of *Amorpha* are given in table 26. Little use is made of the various species of *Amorpha* at the present time. *A. fruticosa* is grown to some extent for game food planting, and all three species listed in table 26 are planted to some extent for erosion control. *A. fruticosa* is also planted in Russia.

Seeding Habits.—The irregular perfect flowers of false indigo are blue to violet purple in color and are borne in the spring or summer in dense terminal spikes; those of *Amorpha nana* are fragrant. The fruit, a short indehiscent, somewhat curved, and often gland-dotted pod containing one (sometimes two) small glossy seeds (figs. 24 and 25) ripens from midsummer to late summer. Ripening is indicated by the light-brown color of the pods. Commercial seed usually consists of the dried pods. Details of flowering and seed production for three species of *Amorpha* are as follows:

Time of—	*A. canescens*	*A. fruticosa*	*A. nana*
Flowering	June–July	May–June	May–June.
Fruit ripening	Aug.–Sept.	Aug.	July.

No data are available on the age at which the shrubs begin to bear seed or on the frequency of

[5] Some species die back every year almost to the ground.

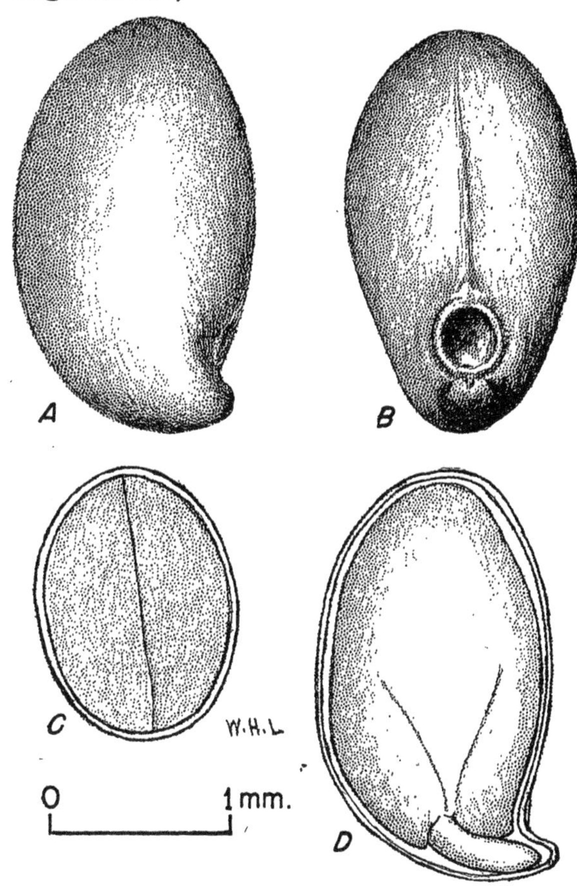

Figure 24.—Seed views of *Amorpha canescens*: *A* and *B*, Exterior views in two planes; *C*, cross section; *D*, longitudinal section.

Table 26.—*Amorpha: Growth habit, distribution, and uses*

Accepted name	Synonyms	Growth habit	Natural range	Chief uses	Date of earliest cultivation
A. canescens Pursh[1] (leadplant amorpha).	shoestrings	Bushy shrub	Dry hills and prairies Manitoba, south to Indiana, Louisiana, and New Mexico.	Game food and cover, landscape planting, potentially useful for soil cover.	1812
A. fruticosa L.[2] (indigobush amorpha).	false indigo	Tall shrub	Along streams and lakes Connecticut to Minnesota, south to Florida and New Mexico.	Landscape planting, game food, erosion control.	1724
A. nana Nutt. (dwarf-indigo amorpha).	*A. microphylla* Pursh (fragrant false indigo, fragrant dwarf indigo).	Low bushy shrub.	Prairies Manitoba and Saskatchewan, south to Iowa and New Mexico.	Valuable chiefly for game food and erosion control.	1811

[1] Very drought resistant. Includes one or more varieties. Valuable information on this and 10 other species obtained from W. H. Brener, Wisconsin Conservation Department, Wisconsin Rapids, Wis. Correspondence on file at Lake States Forest Experiment Station, St. Paul, Minn. May 28, 1941.
[2] Includes 1 or more varieties.

AMORPHA

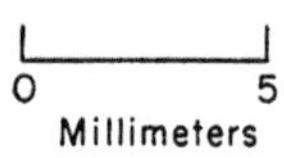

FIGURE 25.—View of fruit and seed of *Amorpha fruticosa.*

seed crops. It is likely, however, that good crops are borne rather frequently. Dispersal probably occurs during the fall months and is believed to be accomplished mostly by animals.

The only evidence available as to the existence of geographic strains is the fact that at least three geographic varieties have been recognized in *Amorpha fruticosa.*

COLLECTION, EXTRACTION, AND STORAGE.—*Amorpha* seed can be collected by stripping the ripe pods from the branches. The pods should be spread out for a few days to permit superficial drying. Extraction is probably unnecessary, as the entire pods may be sown;[6] if desired, however, the seeds may be extracted by flailing or beating. Extraction of *A. canescens* appears to be very difficult. Such data as are available on seed size, purity, soundness, and cost are given in table 27.

Nothing definite is known regarding the optimum storage conditions for *Amorpha* seed. However, since seed of *A. canescens* stored in a sealed bottle for 22 months at 41° F. followed by 16 months storage at room temperature had dropped in germination only from 86 to 78 percent, it is likely that sealed storage at continuous low temperatures would prolong vitality for many years. Seed of *A. fruticosa* is said to retain its viability 3 to 5 years under ordinary storage conditions.

[6] These are usually one-seeded, thin, and soft enough that germination is not inhibited.

GERMINATION.—The time of natural germination of *Amorpha* seed is not definitely known but is believed to take place the spring after dispersal. Germination is epigeous (fig. 26). Seed lots of *A. fruticosa* and *A. nana* show a high percent of dormant seeds, due to an impermeable seed coat.[7] This may be overcome in *A. fruticosa* by light scarification and in *A. nana* by soaking for 5 to 8 minutes in concentrated sulfuric acid. The latter method can probably be applied to *A. fruticosa* as well. (See caution pp. 32–33.) Soaking in hot water is also of benefit.

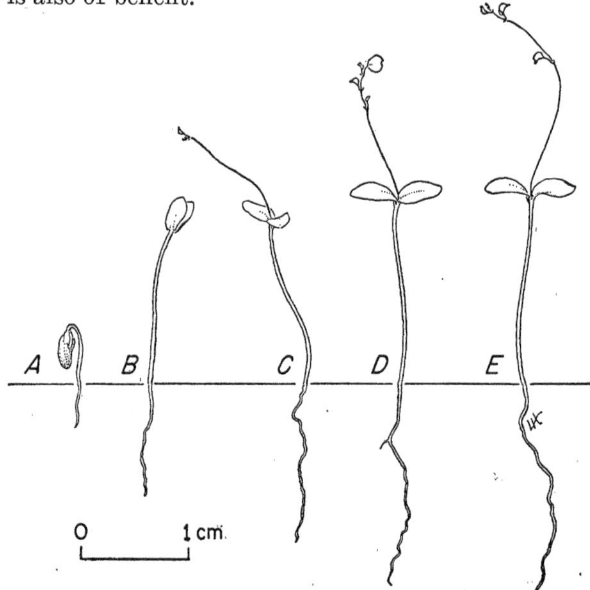

FIGURE 26.—Seedling views of *Amorpha canescens: A,* At 1 day; *B,* at 2 days; *C,* at 8 days; *D,* at 20 days; *E,* at 52 days.

Germination tests can be made in sand flats or standard germinators, using 400 to 1,000 pretreated seeds per test. The procedures to be followed and the results that can be expected are given in table 28.

[7] *A. canescens* may also show this. The only sample tested, however, gave excellent germination without treatment although the drastic method employed to extract the seeds from the pods may have rendered the seed coats permeable to moisture.

TABLE 27.—*Amorpha: Yield of ripe fruit and cleaned seed, and purity, soundness, and cost of commercial seed*

Species	Ripe fruit				Cleaned seed				Commercial seed		
	Per pound			Basis, samples	Per pound			Basis, samples	Purity	Soundness	Cost per pound
	Low	Average	High		Low	Average	High				
	Number	*Number*	*Number*	*Number*	*Number*	*Number*	*Number*	*Number*	*Percent*	*Percent*	*Dollars*
A. canescens	88,000	96,000	106,000	3	--------	296,000	--------	1	--------	99	2.10
A. fruticosa[1]	37,000	52,000	93,000	5+	72,000	--------	82,000	2	97	[2]73 (40-92)	0.85-2.00
A. nana	--------	60,000	--------	1	--------	--------	--------	--------	--------	86	.55-2.00

[1] 100 pounds of dried fruit will produce about 60 pounds of clean seed.
[2] 10 samples. First figure is average percent; figures in parentheses indicate range from lowest to highest.

TABLE 28.—*Amorpha: Recommended conditions for germination tests, and summary of germination data*

Species	Test conditions recommended			Germination data from various sources					
	Temperature		Duration	Germinative energy		Germinative capacity			Basis, tests
	Night	Day				Low	Average	High	
	°F.	*°F.*	*Days*	*Percent*	*Days*	*Percent*	*Percent*	*Percent*	*Number*
A. canescens	68	86	40	[1]81	13	30	64	86	3
A. fruticosa	68	86	15–20	----------	----------	42	67	85	8+
A. nana	[2]68	[2]86	30–40	----------	----------	----------	70	----------	1

[1]1 test.
[2]Suggested; experimental data not complete.

NURSERY AND FIELD PRACTICE.—The seed is usually sown in the nursery in the fall although sometimes it is stratified prior to spring sowing. Spring sowing of seed treated to overcome seed-coat impermeability would probably result in heavier stands. The seed may be sown in the pods in rows about $\frac{3}{16}$ to $\frac{1}{4}$ inch in depth; germination of *Amorpha fruticosa* is 45 to 50 percent. One pound of commercial seed of *A. canescens* will produce about 22,000 usable plants; of *A. fruticosa,* 1,000 to 5,600 plants. The various species of *Amorpha* can also be propagated from cuttings, layers, or suckers.

APLOPAPPUS PARISHII (Greene) Blake — Parish goldenweed

(Composite family—Compositae)

Botanical syns.: *Bigelovia parishii* Greene, *Ericameria parishii* (Greene) Hall.

Parish goldenweed is an erect shrub 3 to 8 feet high that grows in the lower portion of the chaparral belt of California. Its primary value is for erosion control of dry slopes. Members of this genus are a potential source of rubber similar to that produced by guayule. The yellow flowers bloom in summer and autumn. Clusters of fruiting heads are collected in sacks during October and November, and seed (fig. 27) is separated from bristles by rubbing and fanning. One pound contains about 1,633,000 cleaned seeds—achenes—(1 sample). Freshly collected seed occasionally germinates up to 95 percent, but ordinarily germination is much lower (about 20 percent) on account of a large percentage of defective seeds (2+ samples). Parish goldenweed may also be propagated by cuttings.

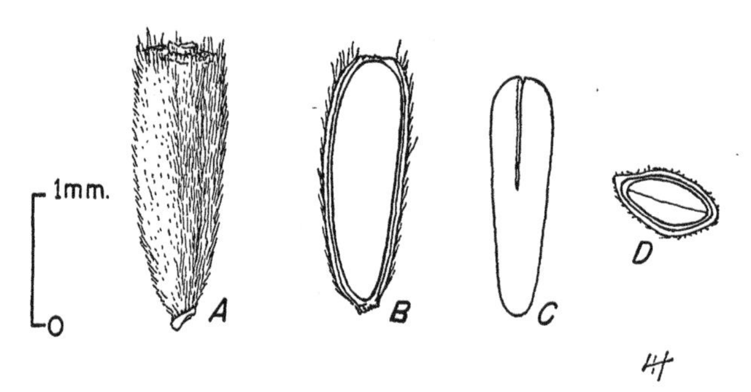

FIGURE 27.—Seed views of *Aplopappus parishii*: *A*, External view; *B*, longitudinal section; *C*, embryo; *D*, cross section.

ARALIA L. Aralia

(Ginseng family—Araliaceae)

DISTRIBUTION AND USE.—*Aralia* includes about 20 species of deciduous trees, shrubs, or herbs found in North America, Asia, Malaya, and Australia. The tree and shrub species are spiny and the herb species are either spiny or smooth-stemmed. They are useful chiefly for ornamental purposes and for wildlife food; the bark and roots have medicinal value. None has been planted very widely. Three American species, which are of potential value for planting and for which reliable information is available, are described in table 29. *A. spinosa* has been introduced successfully into the Pacific Northwest, eastern Massachusetts, and western Europe.

SEEDING HABITS. — The small, greenish-white flowers which bloom in the late spring or summer are perfect in some cases and in others the male and female flowers are borne separately on the same or different plants. They are borne in clusters. The fruit, which matures in the late summer or fall, is a small, berrylike drupe containing two to five compressed, crustaceous light reddish-brown stones or nutlets, which are round, oblong, or egg-shaped. Each nutlet contains one compressed thin-coated, light-brown seed (figs. 28 and 29), the coat of which is united with the thin, fleshy endosperm. The nutlet is the seed of commerce. Comparative seeding habits of three species are given in table 30. Scientific information is not available concerning racial variation among the aralias.

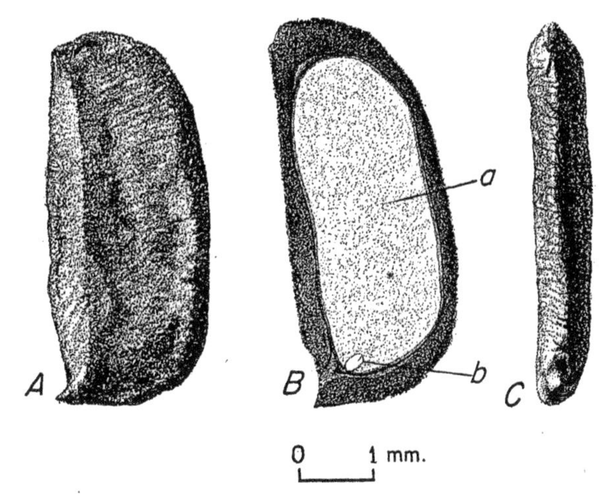

FIGURE 28.—Seed views of *Aralia nudicaulis*. *A* and *C*, Exterior views in two planes. *B*, Longitudinal section showing: *a*, large endosperm; *b*, minute embryo.

FIGURE 29.—Seed views of *Aralia spinosa*.

TABLE 29.—*Aralia: Growth habit, distribution, and uses*

Accepted name	Synonyms	Growth habit	Natural range	Chief uses	Date of earliest cultivation
A. hispida Vent. (bristly aralia).	wild elder, bristly sarsaparilla.	Low shrub	Newfoundland to North Carolina and west to Minnesota and Indiana.	Wildlife food, bark of medicinal value.	1788
A. nudicaulis L. (wild-sarsaparilla).	small spikenard	Low perennial herb.	Newfoundland to North Carolina and west to Manitoba and Missouri.	Wildlife food, roots of medicinal value.	1731
A. spinosa L. (devils-walkingstick).	Hercules-club, angelica-tree.	Short tree	Pennsylvania to Florida westward to southwestern Iowa and eastern Texas.	Wildlife food, ornamental, bark for medicinal purposes.	1688

ARALIA

TABLE 30.—*Aralia: Time of flowering and fruit ripening*

Species	Time of—		Ripe fruit	
	Flowering	Fruit ripening and seed dispersal	Color	Diameter
				Inches
A. hispida	June–July	August–September	Purple black	1/4–5/16
A. nudicaulis	May–June	August	do	
A. spinosa	July–August	September–October	Black	1/4

COLLECTION, EXTRACTION, AND STORAGE.—*Aralia* fruits are collected when they begin to fall from the plants in the autumn. The seeds are ripe when their coats become hard and brittle, and this ripening may occur somewhat later than the ripening of the pulp. To prevent fermentation, immediately after picking, the fruits should be run through a macerator with water and the pulp and empty seed floated off or screened out. Small samples can be pulped by rubbing in water between No. 16 mesh screens. Yield, size, purity, soundness, and cost of commercial cleaned seed vary by species as is shown in table 31. Information as to the longevity and behavior of *Aralia* seed in storage is not available.

GERMINATION.—*Aralia* seed have dormant embryos and those of some species also appear to have impermeable seed coats. Germination, which is epigeous, is slow and poor without pretreatment. *A. spinosa* is reported to have mild embryo dormancy which can be overcome satisfactorily by stratification at low temperatures. Satisfactory methods of pretreating seed of *A. nudicaulis* and *A. hispida* have not yet been worked out. Warm plus cold stratification has shown some result with the former species, but essentially none with the latter (best germination 0.5 percent—potential germination 84 to 90 percent). With *A. nudicaulis* best germination (24 percent—potential germination 66 to 92 percent) occurred after the seed had been stratified for 60 days at 68° F. (night) to 86° (day), plus 60 days at 41°, plus 60 days at 68° to 86°, plus 60 days at 41°. Obviously a simpler pretreatment should be found. Perhaps soaking in concentrated sulfuric acid or scarification to soften the seed coat can be substituted for part of the stratification period, or a longer period (90 to 120 days) in cold stratification, after the initial 60 days' warm stratification, may suffice.

Germination tests may be made in sand flats or peat mats, using 400 to 1,000 pretreated seeds per test. Since the best results for *Aralia nudicaulis* and *A. hispida* have already been mentioned and no information is available for *A. spinosa,* further details are not given here.

NURSERY AND FIELD PRACTICE.—Little detailed information is available. However, seed of this genus should be sown in the fall, or if pretreated, in the spring. The temperature of the soil in seed flats should be kept higher than that of the air above the flats. The aralias may also be propagated by root cuttings.

TABLE 31.—*Aralia: Yield of cleaned seed, and purity, soundness, and cost of commercial seed*

Species	Cleaned seed					Commercial seed		
	Yield per 100 pounds of fruit	Per pound			Basis, samples	Purity	Soundness	Cost per pound
		Low	Average	High				
	Pounds	*Number*	*Number*	*Number*	*Number*	*Percent*	*Percent*	*Dollars*
A. hispida		94,000		99,000	2	99	99	3.00
A. nudicaulis		84,000	99,000	111,000	3	98	96	
A. spinosa	11	105,000		157,000	2		86	1.25–3.40

ARBUTUS MENZIESII Pursh Pacific madrone

(Heath family—Ericaceae)

Also called madrone, madroña.

DISTRIBUTION AND USE.—Native to foothills and mountains, Coast Ranges, Sierra Nevadas, California and north to British Columbia, this small to medium tall, handsome evergreen tree is valuable for wood products, as browse and game food, and is often cultivated in parks. It was introduced into cultivation in 1827. Occasionally it is planted in southern and western Europe.

SEEDING HABITS.—The perfect, white flowers open from March to June, and semifleshy, orange-red, berrylike drupes containing above five seeds (fig. 30) are ripe from October to December.

COLLECTION, EXTRACTION, AND STORAGE.—Berries are collected from standing trees, and seeds are either separated from the pulp and thoroughly dried, or dried berries can be stored at room temperature for 1 or 2 years. Both cleaned seed and dried berries should be stored in airtight containers at 35° to 40° F. Number of dry fruit per pound, 2,000 (1 sample). Cost of commercial seed ranges from $1.65 to $3 per pound.

GERMINATION.—In nature berries are devoured by birds early in the winter and seeds germinate in the spring. Cleaned seeds should be used in germination tests and planting. Seeds need after-ripening, and germination can be induced by stratification in light soil at 35° to 40° F. for 3 months. Germination ranged from 2 to 55 percent in 30 days, excluding stratification period, 2+ tests.

NURSERY PRACTICE.—Madrone seed should be sown early in the spring or in the fall. In California it is raised in flats and transplanted to pots for landscape planting. The madrone grows best in well-drained soils not exposed to dry winds. Propagation by cuttings, grafting, or layers is sometimes used.

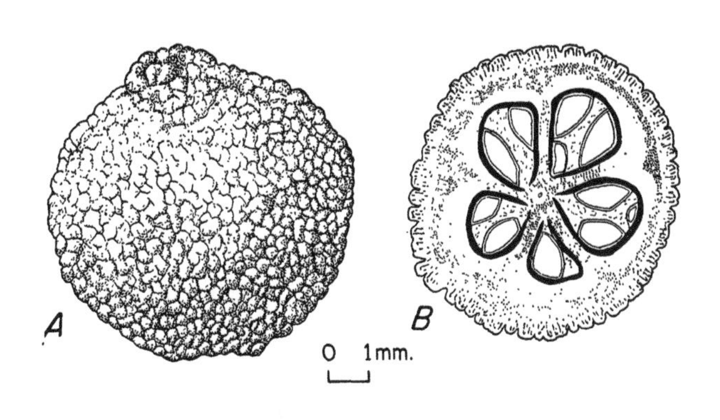

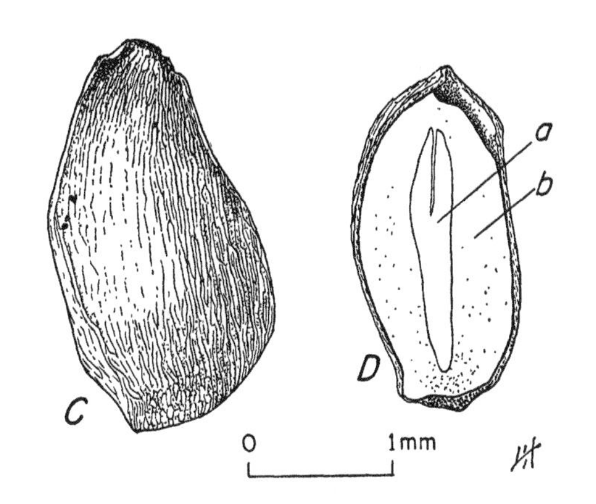

FIGURE 30.—Seed views of *Arbutus menziesii*. *A*, Exterior view of fruit. *B*, Transverse section of fruit showing five irregular cells. *C*, Exterior view of seed. *D*, Longitudinal section of seed showing: *a*, Embryo; *b*, endosperm.

ARCTOSTAPHYLOS Adans. Bearberry, manzanita

(Heath family—Ericaceae)

DISTRIBUTION AND USE.—*Arctostaphylos* contains about 50 species of evergreen shrubs, rarely small trees, native to North America, south to Central America and 1 species circumpolar. They are useful as wildlife food, for erosion control, and as honey plants; the fruits of some are used to make preserves and fruit drinks, and the leaves have medicinal value. Two species, valuable for conservation planting although little used so far, and for which reliable information is available, are described in table 32.

SEEDING HABITS.—Small white or pink, perfect flowers bloom in the early winter to spring. The fruit is a dry, fleshy, or mealy drupe which is red when ripe and resembles a very small apple (manzanita in Spanish); it ripens in summer to fall. Each fruit contains 4 to 10 bony, 1-seeded nutlets which are more or less firmly consolidated (figs. 31 and 32). The seeds are dispersed chiefly by birds and animals. Details of the seeding habits of 2 species of *Arctostaphylos* are as follows:

	A. patula	*A. uva-ursi*
Time of—		
Flowering	--------	April to July
Fruit ripening	August to October	July to October
Seed dispersal	--------	July to March

No scientific information is available as to racial variation in *Arctostaphylos* species. It is probable that such development has taken place in *A. uva-ursi* with its widespread distribution.

COLLECTION, EXTRACTION, AND STORAGE.—The ripe fruits should be picked from the bushes by hand and either dried and used in that form, or they may be run through a macerator with water and the pulp floated off or screened out. Commercial seed may consist of dried fruits, or a mixture of whole stones, stone pieces, or individual nutlets in varying amounts, depending on the species. Fruits of *Arctostaphylos patula* usually are collected from August to October and those of *A. uva-ursi* from July to the following spring.

The characteristics of *Arctostaphylos* seed vary according to species as shown in the following tabulation:

	A. patula	*A. uva-ursi*
Cleaned seed:		
Per 100 pounds of fruit ....lbs..	--	12
Per pound:		
Low ....number..	--	27,000
Average ....do....	[1]18,000	--
High ....do....	--	[2]58,000
Basis, samples ....do....	1	2
Commercial seed:		
Purity ....percent..	--	99
Soundness ....do....	--	76
Cost per pound ....dollars..	--	1–2

[1] Dried fruits, 1,450 to 1,700 in 2 samples.
[2] Individual nutlets.

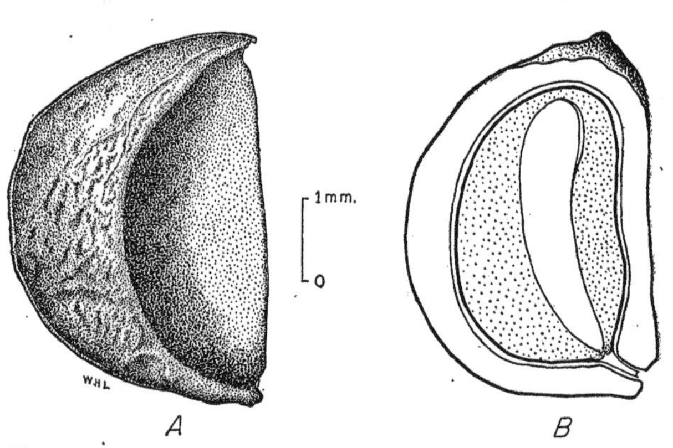

FIGURE 31.—Seed views of *Arctostaphylos uva-ursi*: *A*, Exterior view of seed; *B*, longitudinal section.

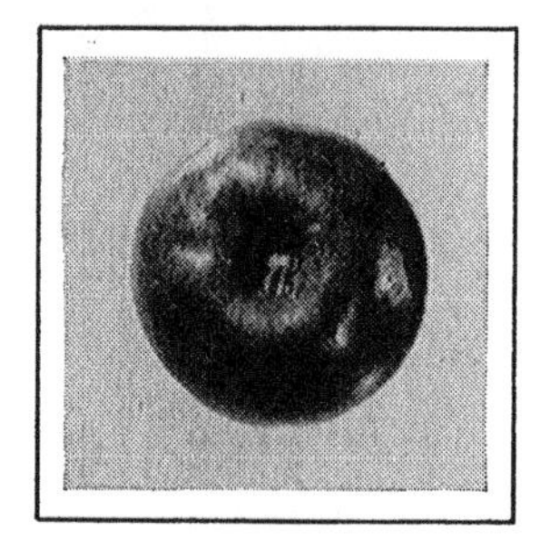

FIGURE 32.—Views of fruit and seed of *Arctostaphylos patula*.

TABLE 32.—*Arctostaphylos: Growth habit, distribution, and uses*

Accepted name	Synonyms	Growth habit	Natural range	Chief uses	Date of earliest culti-vation
A. patula Greene (green-leaf manzanita).	--------------	Widely branching, low shrub.	Mountains of California and adjacent Oregon.	Erosion control; wildlife food.	------
A. uva-ursi (L.) Spreng. (bearberry).	*Uva-ursi uva-ursi* (L.) Britt.	Prostrate shrub.	Labrador to Alaska, south to Virginia, Illinois, Nebraska, and in the mountains to New Mexico and northern California; northern Europe, Asia.	Wildlife food; erosion control.	1800

No information is available as to the behavior of *Arctostaphylos* seed in storage.

GERMINATION.—*Arctostaphylos* seed has epigeous germination (fig. 33). It is slow to germinate because of a hard, impermeable seed coat in combination with a dormant embryo. Seeds of species in which the stones break up into individual nutlets germinate most readily, usually the first year in the nursery. Seeds of those species having the nutlets cohering as a single stone do not germinate for 2 or 3 years, and then poorly. The most promising treatment for inducing reasonably prompt and satisfactory germination consists of (1) soaking the seeds in concentrated sulfuric acid, then (2) stratifying in moist sand at 68° to 86° F., followed by (3) a short period of stratification at 41° F. Details of the pretreatment recommended for 2 species of *Arctostphylos* are shown below:

	A. patula	*A. uva-ursi*
Soak in concentrated sulfuric acid.	--------	3–6+ hours.
Stratify in moist sand, soil, or peat.	40° F., 90 days	60°–86° F., 60 days; plus 40°, 60 days.

Results for *A. patula* could probably be improved by using warm plus cold stratification following acid treatment; and for *A. uva-ursi*, a constant temperature of 77° F. can be used for warm stratification instead of the 68° to 86° F. listed above.

Germination tests for *Arctostaphylos* should be run in sand flats or peat mats at 68° (night) to 86° F. (day) for 30 days, using 500 properly pretreated seeds per test. Average results for the 2 species are as follows:

		A. patula	*A. uva-ursi*
Basis, tests		1+	7
Germinative capacity:			
Low	percent	--	12
Average	do	--	24
High	do	16	49
Potential germination	do	--	60–65

NURSERY AND FIELD PRACTICE.—Good germination (60 to 75 percent) can be obtained in the nursery by soaking the seed in concentrated sulfuric acid—3 to 5 hours for entire stones and stone pieces and 2 to 3 hours for individual nutlets—and then sowing the seeds in drills during the early summer months, covering them with 1/4 to 3/8 of an inch of soil. Germination will occur the following spring. The seedbeds should be mulched over winter. *Arctostaphylos* species can be propagated quite readily by means of cuttings.

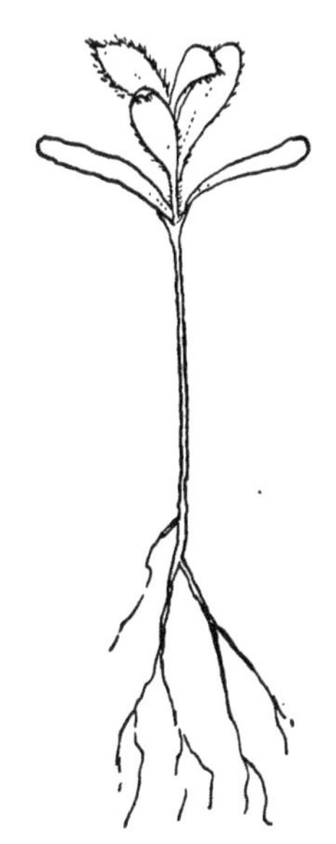

FIGURE 33.—Seedling of *Arctostaphylos patula* at 1 month, actual size.

ARONIA Med. Chokeberry

(Rose family—Rosaceae)

DISTRIBUTION AND USE.—The chokeberries consist of three closely related species of deciduous shrubs native to North America which are of value for fall and winter wildlife food and, because of their handsome leaves, attractive flowers, and decorative fruits, for ornamental purposes. None of the chokeberries has yet been cultivated extensively. Since all three species may be of value for planting and reliable information for them is available, they are described in table 33.

SEEDING HABITS.—The attractive, white perfect flowers bloom in the spring. The fruits, which ripen in late summer, are rather dry, berrylike pomes, each containing one to five seeds (fig. 34), some of which may be abortive. Natural seed dispersal is chiefly by animals. The fruits of *Aronia melanocarpa* shrivel soon after ripening and most of them drop; those of *A. prunifolia* shrivel at the beginning of winter, while those of *A. arbutifolia* remain plump and bright far into the winter. Good seed crops are produced almost every year by *A. arbutifolia,* and about every second year by *A. melanocarpa.* The seeding habits of the three species are given in table 34.

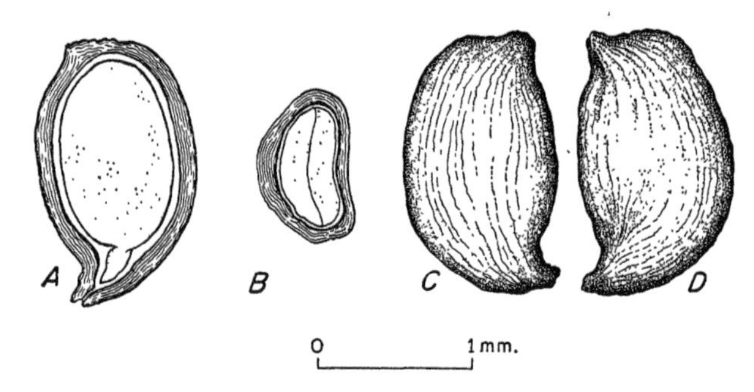

FIGURE 34.—Seed views of *Aronia melanocarpa: A,* Longitudinal section; *B,* cross section; *C* and *D,* exterior views in two planes.

TABLE 33.—*Aronia: Growth habit, distribution, and uses*

Accepted name	Synonyms	Growth habit	Natural range	Chief uses	Date of earliest cultivation
A. arbutifolia (L.) Ell. (red chokeberry).	*Pyrus arbutifolia* (L.) L.f., *Sorbus arbutifolia* (L.) Heynh.	Medium to tall shrub.	Massachusetts to Florida, west to Ohio, Arkansas, and Texas.	Wildlife food, ornamental.	About 1700.
A. melanocarpa (Michx.) Ell. (black chokeberry).	*Aronia nigra* Dipp., *Pyrus melanocarpa* (Michx.) Willd., *Sorbus melanocarpa* (Michx.) Heynh.	Usually low shrub.	Nova Scotia to Florida, west to Ontario and Minnesota.	do	do.
A. prunifolia (Marsh.) Rehd. (purplefruit chokeberry).	*A. atropurpurea* Britt., *A. floribunda* Spach, *Sorbus arbutifolia* var. *atropurpurea* (Britt.) Schneid.	Medium to tall shrub.	Nova Scotia to Florida, west to Indiana.	do	1800 (?).

TABLE 34.—*Aronia: Time of flowering, and fruit color and size*

Species	Time of— Flowering	Time of— Fruit ripening	Time of— Seed dispersal	Fruit Color	Fruit Average diameter
					Inches
A. arbutifolia	March–May	September–October	September–winter	Bright or dull red	3/16–1/4
A. melanocarpa	April–May	August–October	August–winter	Lustrous black or purple black.	1/4–5/16
A. prunifolia	do	August–September	August–September	Lustrous purplish black	5/16–3/8

Scientific information as to racial variation among the chokeberries is lacking. However, such development is quite likely since all three species have rather extensive ranges and several varieties are recognized within *Aronia arbutifolia* and *A. melanocarpa*. One of the former is confined within definite geographic boundaries and may be a climatic race.

COLLECTION, EXTRACTION, AND STORAGE.—Fruits of *Aronia* should be picked from the bushes by hand as soon as they ripen to prevent their consumption by birds. Since cleaning is not very practical, the fruits usually are dried and used in that form. Accordingly, commercial seed usually consists of the dried "berries." The seed can be extracted from the fruits, however, by rubbing the latter over screens. Yield, size, purity, soundness, and cost of commercial cleaned seed vary considerably by species as shown in table 35. The longevity and behavior of chokeberry seed under storage conditions are not known.

GERMINATION.—Chokeberry seed has a dormant embryo which can be stimulated into germination by stratification at temperatures a little above freezing. Germination is epigeous. Methods of pretreatment recommended for chokeberry seed are as follows:

Stratify in:	*A. arbutifolia*	*A. melanocarpa*	*A. prunifolia*
Moist peat	33°–41° F., 90 days	------	33°–50° F., 60 days.
Moist sand	------	41° F., 90+ days	------

Stratification longer than 90 days for *A. melanocarpa* would probably give better results, but stratification longer than 60 days gives poorer results for *A. prunifolia.*

Germination tests for *Aronia* should be run for 30 days at temperatures of 68° F. (night) to 86° (day) or at constant temperatures of about 70°, using 400 to 1,000 properly pretreated seeds per test. Methods recommended and average results for the 3 species are given in table 36.

NURSERY AND FIELD PRACTICE.—In nursery practice the dried fruits usually are soaked in water for a few days, mashed, and then the whole mass stratified until spring, when it is sown. Germination takes place within a few days after sowing. The chokeberries do best on moist sites, but *Aronia melanocarpa* also grows well on drier and rocky soils.

TABLE 35.—*Aronia: Yield of cleaned seed, and purity, soundness, and cost of commercial seed*

Species	Cleaned seed					Commercial seed		
	Yield per 100 pounds of fruit	Per pound			Basis, samples	Purity	Soundness	Cost per pound[1]
		Low	Average	High				
	Pounds	*Number*	*Number*	*Number*	*Number*	*Percent*	*Percent*	*Dollars*
A. arbutifolia	4–32	----------	[2]256,000	----------	1	----------	70	0.75–1.25
A. melanocarpa	8–24	219,000	276,000	356,000	4	82	95	.75–1.00
A. prunifolia	----------	----------	----------	----------	----------	----------	----------	.75–1.25

[1]Dry "berries."
[2]Or 7,355 dried "berries."

TABLE 36.—*Aronia: Recommended conditions for germination tests, and summary of germination data*

Species	Test conditions recommended				Germination data from various sources						
	Temperature		Duration		Germinative energy		Germinative capacity			Potential germination	Basis, tests
	Night	Day	Pretreated seed	Untreated seed	Amount	Period	Low	Average	High		
	°F.	*°F.*	*Days*	*Days*	*Percent*	*Days*	*Percent*	*Percent*	*Percent*	*Percent*	*Number*
A. arbutifolia	68	70	30	150+	--------	--------	92	94	100	--------	4
A. melanocarpa	68	86	30	620+	18–22	11–14	19	22	26	45–58	4
A. prunifolia	68	70	30	150+	--------	--------	92	--------	100	--------	2

891183°—50——7

ASIMINA TRILOBA (L.) Dunal Pawpaw

(Annona, or custard-apple family—Annonaceae)

Distribution and Use.—Native to rich moist soils from western New York and western New Jersey west to southeastern Nebraska and south to Florida and eastern Texas, this small deciduous tree is valued chiefly for its fruits which are eaten widely by man and by game. It also forms thickets which furnish cover for birds and animals. Due to its ornamental character, it is sometimes cultivated in the eastern United States where it is hardy as far north as eastern Massachusetts. It was introduced into cultivation in 1736.

Seeding Habits.—Pawpaw fruit consists of a dark brown to black fleshy berry 3 to 6 inches long and 1 to 1½ inches in diameter which contains several dark brown shiny seeds embedded horizontally. The latter are oblong in shape, flattened, and about 1 inch in length (fig. 35). Flowering occurs from March to May or June; fruit ripens in September to October of the same season, and occasionally earlier or later. Trees begin to produce fruit at 6 to 8 years and continue to bear until death. The existence of geographic strains has not been proved. There are, however, two distinct types of fruit; one with white barely edible flesh, the other much larger in size with orange-colored succulent pulp. Trees from which these two types arise do not appear to differ botanically, at least outwardly.

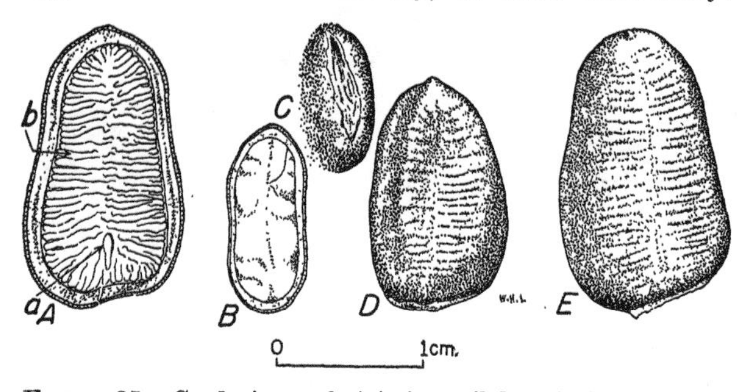

Figure 35.—Seed views of *Asimina triloba*. *A*, Longitudinal section: *a*, Minute embryo; *b*, large endosperm. *B*, Cross section; *C*, exterior view of end showing hilum and micropyle; *D* and *E*, exterior views from two planes.

Collection and Extraction.—The fruit should be collected as soon as the flesh becomes soft. It may then be shaken from the tree onto canvas. Extraction, which is accomplished by macerating the fruit in water and then floating off pulp, may or may not be necessary. Sometimes the entire fruit is sown without pulping; in fact, it is the opinion of some that the fermentation attending such sowing is necessary to obtain germination. In one case 100 pounds of fruit yielded 11 pounds of clean seed. Cleaned seed per pound (3 tests): low 294; average, 697; high, 1,200. Purity, 100 percent; soundness, 96 percent. Cost: $0.75 to $1.90 per pound.

Germination.—Relatively little is known about the germination of pawpaw seed except that it is hypogeous (fig. 36) and has a dormant embryo. Although stratification for 100 days at 50° F. is recommended, this may not always prove effective, and in some tests 41° appeared better than 50°. Germination of seeds that are sown in the nursery in the fall or stratified and then sown in spring is slow and irregular. The seeds do not usually germinate before July or August and some may lie dormant until the following spring. This suggests the existence of an impermeable seed coat in addition to a dormant embryo. Germinative capacity of seed stratified 60 days at 41° (3 tests): low, 50 percent; average, 62 percent; high, 82 percent.

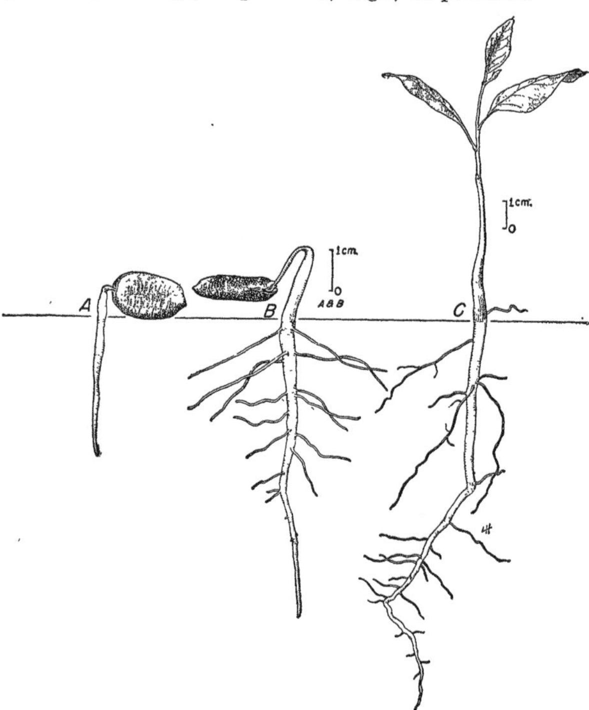

Figure 36.—Seedling views of *Asimina triloba: A*, At 2 days; *B*, at 9 days (seed drops off 9 to 18 days); *C*, soon after seed has dropped off.

Nursery and Field Practice.—Seed should either be fall sown or stratified and then sown in the spring. They should be covered to a depth of three-fourths of an inch, and some shade appears to be necessary. One recommendation is to place the seed in sand-filled pots before they dry and keep them in a cool cellar. The sprouting seeds are then picked out and planted in the spring. Seedlings grow slowly the first few years, and they grow best on moist, rich soils. Pawpaw may be propagated also by layers or by root cuttings.

ATRIPLEX CANESCENS James Fourwing saltbush

(Goosefoot family—Chenopodiaceae)

Also called chamiza, bushy atriplex, shad-scale. Botanical syn.: *A. occidentalis* D. Dietrich

Distribution and Use.—Fourwing saltbush[8] is a medium-sized, half-evergreen shrub native to dry plains from South Dakota and Oregon to Texas and Lower California. It furnishes valuable browse for stock, food and cover for game, and is useful for erosion control. Seed of this shrub is relished by cattle, sheep, and goats. It has been cultivated since 1870, and planted quite extensively in southwestern United States.

Seeding Habits.—*Atriplex* "seed" as gathered is, in fact, a dry fruit marginated by four rounded, very conspicuous wings (fig. 37). The dioecious flowers bloom in June to August. Fruits ripen in late August and September, at which time the wings are dry and yellow. Dispersal is by wind and gravity. Plants 2 years old have been known to bear seed; good seed crops are borne almost annually. Seed from different localities produce plants of different size.

[8] Valuable unpublished information on this and six other species of southwestern plants was furnished by L. A. Prichard, U. S. Forest Service, Superior, Ariz. On file at Lake States Forest Experiment Station, St. Paul, Minn. 1938.

Collection, Extraction, and Storage.—The "seed" is rather persistent; it can be collected from September to December, or before the spring winds begin, by stripping it from dense clusters. One man can collect a 3-gallon bucket of seeds in 3 minutes. Where produced abundantly, the cost of gathering the seed is only a cent or two a pound. If much of the seed is to be gathered, stock should be kept away from the area where the bushes are growing. No cleaning is needed. The seed should be stored in a dry place. Cleaned "seed" per pound (5+ samples): low, 10,500; average, 22,500; high, 40,000. Dewinged "seeds" about double these values. Cleaned "seeds" average 83 percent in purity, 50± percent in soundness; they do not ordinarily appear on the market. Apparently the germinating qualities of fourwing saltbush seed do not deteriorate appreciably, at least not until the sixth

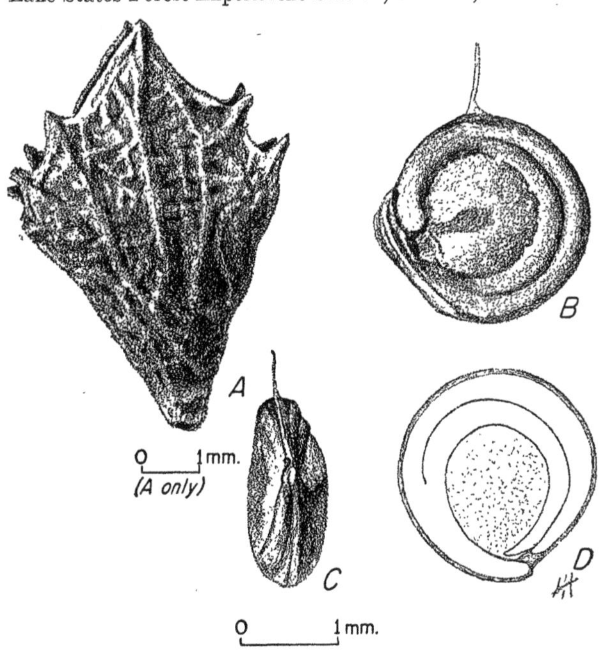

Figure 37.—Seed views of *Atriplex semibaccata: A,* Exterior view of fruit; *B* and *C,* exterior view of seed in two planes; *D,* longitudinal section showing embryo coiled around endosperm.

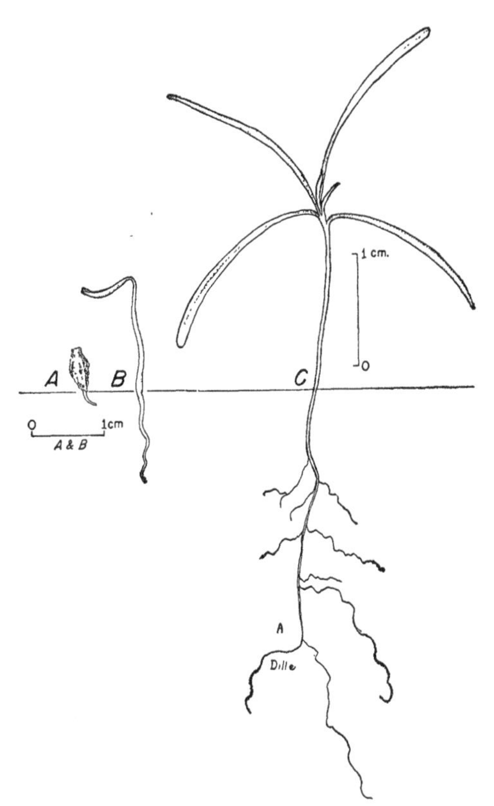

Figure 38.—Seedling views of *Atriplex canescens: A,* At 1 day; *B,* at 2 days; *C,* older seedling.

or seventh year after gathering. One lot germinated 19 percent after 9 years in dry open storage.

Germination.—The best natural seedbed is a fine alkaline alluvial soil. Natural reproduction is often abundant. Germination is epigeous (fig. 38). It was found that *Atriplex* seed did not germinate well if tested within a few weeks after maturity and at the temperatures generally used for the germination of seeds of ordinary farm crops. There appears to be variable dormancy; some lots germinate promptly without pretreatment whereas others in the nursery sometimes show germination straggling over 1 to 2 years after sowing. In one series of tests, however, stratification in moist sand for 30 days at 41° F. did not improve or speed-up germination. Plump green seed have given prompt germination, whereas fully ripened seed appear to germinate better if held in dry storage for 12 to 18 months. Tests should be run in sand flats at 50° (night) to 77° (day) for 20 to 30 days, using 400 seeds per test. Germinative capacity (12 tests): low, 4 percent; average, 18 percent; high, 47 percent. The highest value was given by seed which had been dewinged. This treatment may have had a beneficial influence on seed-coat conditions.

Nursery and Field Practice.—Good stands of fourwing saltbush have been obtained by rolling the seed after broadcast sowing it fairly thickly and covering with one-eighth inch of soil and one-fourth inch of sifted sand. Seeding may be done at any time of the year, but the best results will probably be obtained by seeding in late spring or early summer. Good stands are often obtained by seeding as late as the first of January or even the first of February. Although half-shading is sometimes used, the beds should be left unshaded to avoid soft growth and consequent sun injury on exposure. The seedlings are very susceptible to damping-off during the first 2 weeks of their life. Seedbeds should be protected against birds and rodents which relish the seed.

Direct seeding in the field under scattered brush has been successful and favorable results have also followed broadcasting the seed on unplowed ground at the rate of 8 or 10 pounds to the acre and then single disking. Land that does not give a reaction for calcium carbonate seldom gives satisfactory results with fourwing saltbush. Planting should be done on soils of sandy loam or finer texture, although good growth sometimes occurs on sand dunes.

BACCHARIS L. Baccharis

(Composite family—Compositae)

DISTRIBUTION AND USE.—The genus *Baccharis* consists of some 250 species of deciduous or evergreen shrubs or herbs native to North America and, especially, South America. Some species are used as ornamentals; some for erosion-control planting; some for medicinal purposes; and in certain localities they provide homemade brooms. They are of poor forage value and some are poisonous to livestock. Of the 23 species native to the United States, 16 are found in the far West. Three species, which have been cultivated to some extent for conservation purposes, are described in table 37.

SEEDING HABITS.—The white or yellowish male and female flowers, borne separately on different plants, are in heads which occur in clusters. The female flowers develop into compressed, usually 10-ribbed achenes (the seed of commerce) which are tipped by a pappus of bristly hairs one-half inch long or less (fig. 39). The seeds are dispersed chiefly by the wind. The comparative seeding habits of 3 species of *Baccharis* are as follows:

	B. pilularis	*B. sarothroides*	*B. viminea*
Time of—			
Flowering	Aug.–Oct.	Early fall	Spring–summer.
Fruit ripening	Sept.–Nov.	Oct.–Nov.	do.
Seed dispersal	Fall	Nov.–Dec.	May–June.

Seed crops are borne annually as a rule. No information is available concerning the development of climatic races in these species.

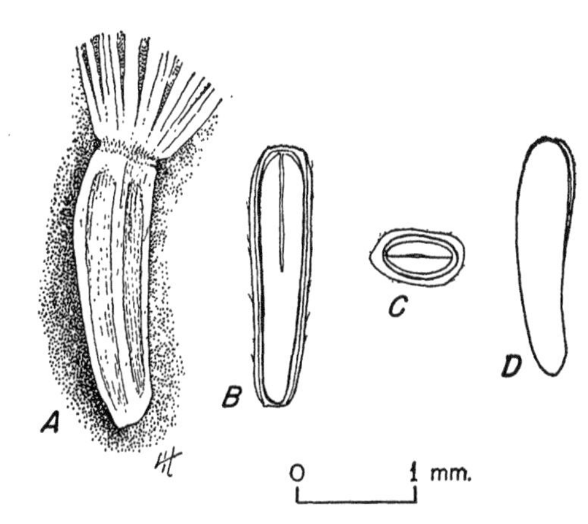

FIGURE 39.—Seed views of *Baccharis viminea: A,* Exterior view of fruit; *B,* longitudinal section; *C,* cross section; *D,* dissected embryo.

COLLECTION, EXTRACTION, AND STORAGE.—The ripe fruits of *Baccharis* are collected by hand or by brushing them onto canvas spread beneath the shrubs. They should be spread out to dry in a warm well ventilated room or in the sun, protected from the wind. When dried, the fruits may be rubbed between the hands or treated en masse to remove the pappus. Sometimes the entire fruits are used without removing the pappus. The size of the fruit varies per pound according to the species as follows: *B. pilularis,* 1 sample, 82,000 fruits; *B. viminea,* 1

TABLE 37.—*Baccharis: Growth habit, distribution, and uses*

Accepted name	Synonyms	Growth habit	Natural range	Chief uses	Date of earliest cultivation
B. pilularis DC. (kidneywort baccharis).	*B. consanguinea* DC.	Small ascending evergreen shrub, usually thicket forming.	Oregon to southern California	Erosion control on sand dunes; wildlife food.	Before 1910.
B. sarothroides A. Gray (broom baccharis).	greasewood, rosinbrush.	Green, broomlike shrub.	Arizona, New Mexico, and adjacent Mexico.	Erosion control	------
B. viminea DC. (mulefat baccharis).		Large, loosely branched shrub.	Flood beds of streams in the Sierra Nevada foothills and the San Joaquin Valley south to coastal southern California.	do	------

sample, 50,000 fruits. Comparable information for *B. sarothroides* is lacking. Seed of these species have not entered commercial trade so prices are not available.

Cleaned seed of *Baccharis sarothroides* may be stored dry over winter satisfactorily; possibly longer storage is feasible but information is lacking. No information is available as to longevity under storage for *B. pilularis* and *B. viminea.*

GERMINATION.—Natural germination (fig. 40) usually takes place in the spring following seed dispersal. Disturbed, moist mineral soil is a favorite seedbed. *Baccharis sarothroides* often seeds in on new road fills. Germination of the three species listed here seems to take place quite readily without any special treatment of the seed.

Germination tests for *Baccharis* can be run in soil or sand flats at temperatures of 68° (night) to 86° F. (day) for 15 to 30 days, using 400 to 1,000 seeds or fruits per test. Average results for the 3 species under these conditions are as follows:

	B. pilularis	*B. sarothroides*	*B. viminea*
Basis, tests	1+	1	2
Germinative capacity:			
Low .... percent	--	--	75
Average .... do	--	64	--
High .... do	92	--	82

NURSERY AND FIELD PRACTICE.—*Baccharis sarothroides* seed can be sown broadcast in the nursery in March or April and should be covered to a depth of one-eight inch. Seedlings usually appear within 4 days. Although information is lacking, similar practices probably would prove successful for the other two species discussed here. *Baccharis* plants may also be propagated by stem cuttings; *B. viminea* is particularly easily propagated by this means.

FIGURE 40.—Two-month-old seedling of *Baccharis pilularis,* twice actual size.

BERBERIS L. Barberry

(Barberry family—Berberidaceae)

Distribution and Use.—The barberries include about 175 species of evergreen or deciduous spiny shrubs (rarely small trees) found most abundantly in eastern and central Asia and South America, with a few species in North America (about 12), Europe, and northern Africa. Most species are desirable ornamentals because of their handsome foliage, bright flowers in spring, and colored fruits in the fall, which often persist into winter. They also have some value as wildlife food. Two species that are widely planted in the United States and for which reliable information is available are described in table 38. *Berberis vulgaris* has been actively eradicated in wheat-producing areas because it is an alternate host of the fungus which causes black stem rust of wheat. *B. thunbergii,* on the other hand, is not an alternate host of this organism and can be planted safely in wheat-producing areas.

Seeding Habits.—Barberry has perfect yellow flowers which are borne in the spring in various shaped clusters, in spikes, or individually, depending on the species. By fall they develop into one- to several-seeded red to black berries (fig. 41). Natural seed dispersal is chiefly by birds. Fruit crops are borne annually. The comparative seeding habits of the two species discussed are as follows:

	B. thunbergii	*B. vulgaris*
Time of—		
Flowering	May–June	April–June.
Fruit ripening	Late summer	Late summer–early fall.
Seed dispersal	Late summer–over winter	Late summer–over winter.
Color, ripe fruit	Bright red	Scarlet to purple.
Seeds per fruit	1–2	2–3.

No scientific information is available as to the development of climatic races in these two barberry species. Both, however, have numerous recognized varieties and may have developed climatic races.

Collection, Extraction, and Storage.—Ripe barberry fruits should be picked by hand (heavy gloves are advisable) from the bushes in the fall

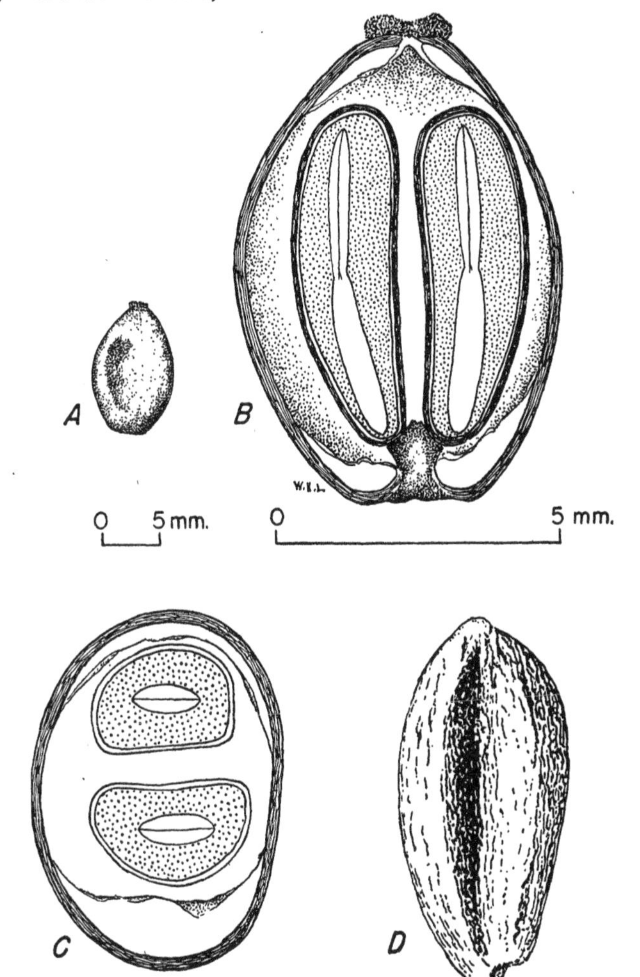

Figure 41.—Seed views of *Berberis thunbergii: A,* Exterior view of fruit; *B,* longitudinal section of fruit; *C,* cross section of fruit; *D,* exterior view of seed.

Table 38.—*Berberis: Growth habit, distribution, and uses*

Accepted name	Synonyms	Growth habit	Natural range	Chief uses	Date of earliest cultivation
B. thunbergii DC. (Japanese barberry).	*B. sinensis* K. Koch not Desf., *B. japonica* Hort.	Deciduous, much-branched shrub.	Japan	Ornamental; food for birds.	About 1864.
B. vulgaris L. (European barberry).		Deciduous, medium-sized shrub.	Europe; frequently naturalized in eastern North America.	Ornamental; food for birds; occasionally used for jelly making.	[1]

[1]Long cultivated.

or winter. The whole berries may be sown immediately, or the seed may be extracted by running the fruits through a macerator with water, and screening out or floating off the pulp. After extraction the seed should be dried and then sown, or stored in sealed containers at temperatures slightly above freezing. Information as to longevity under storage is lacking. The characteristics of commercial cleaned seed vary by species as shown in the following tabulation:

	B. thunbergii	*B. vulgaris*
Cleaned seed:		
Per 100 pounds fruit.......pounds..	16–25	------
Per pound:		
Low....................number..	25,000	34,000
Average....................do....	27,000	------
High.......................do....	28,000	41,000
Basis, samples.............do....	3	2
Commercial seed:		
Purity..................percent..	93	------
Soundness..................do....	95	------
Cost per pound cleaned seed..dollars..	2.25–3.00	0.70–1.00
Cost per pound dried fruits....do....	1.50	------

Germination.—Natural germination which is epigeous (fig. 42) takes place the spring following dispersal and occurs quite readily on bare soil and in pastures. Most barberry seed have embryo dormancy which can be overcome by stratification in moist sand or peat for 15 to 40 days at 32° to 41° F.

Germination of barberry seed may be tested in sand flats, peat mats, or standard germinators at 50° to 60° F. (night) to 70° to 80° (day) for 40 days using 400 to 500 stratified seeds per test. The average germinative capacity of *Berberis thunbergii* in 2+ tests was 80 percent (low) and 99 percent (high); in 2+ tests of *B. vulgaris* it was 86 percent (low) and 96 percent (high).

Nursery and Field Practice.—Whole berries or cleaned seed may be fall sown, or stratified seed may be used in the spring. Injury from molds is more likely if berries are used. Seedbeds should be treated to prevent damping-off. The barberries may be propagated by cuttings, layering, or grafting as well as by seeding. For best results, they should be planted in a moist, but well-drained, light loam, although the two species discussed here will grow on fairly dry soils.

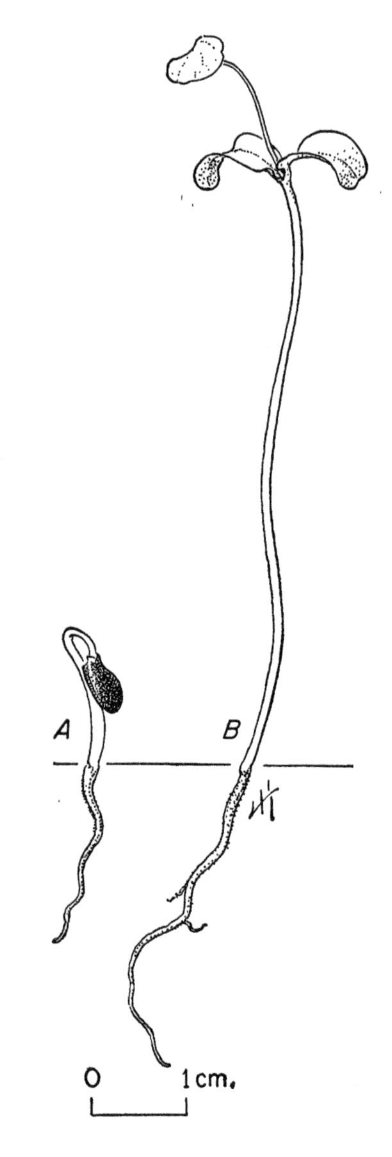

Figure 42.—Seedling views of *Berberis thunbergii*: *A*, At 1 day; *B*, at 16 days.

BETULA L. Birch

(Birch family—Betulaceae)

DISTRIBUTION AND USE.—The birches consist of about 40 species of deciduous trees and shrubs which occur in the cooler parts of the Northern Hemisphere. Several species produce valuable lumber; others, because of their graceful habit, handsome foliage and bark, are useful for ornamental planting; still others, although short-lived, are important in that they are pioneer species and quickly establish cover on cut-over and burned lands. Of the more important species 7 are native to the United States; 2 are indigenous to Europe and Asia. Their distribution and chief uses are given in table 39.

TABLE 39.—*Betula: Growth habit, distribution, and uses*

Accepted name	Synonyms	Growth habit	Natural range	Chief uses	Date of earliest cultivation
B. glandulifera (Reg.) Butler.	*B. pumila* var. *glandulifera* Reg. (dwarf birch, bog birch, swamp birch, low birch).	Thicket-forming shrub.	Bogs, Ontario to Saskatchewan, south to Indiana and Minnesota.	Browse and cover for game animals.	------
B. glandulosa Michx. (bog birch).	resin birch, scrub birch, dwarf birch.	Shrub	Bogs, Newfoundland to Alaska, south to New York, Michigan, Colorado, and Oregon.	Browse and cover for game animals, forage for livestock.	1880
B. lenta L. (sweet birch).	*B. carpinifolia* Ehrh. (cherry birch, black birch).	Large tree	Rich woods, northeast to Ohio and south to Alabama and Georgia.	Lumber, furniture, woodenware, fuel, oil of wintergreen, twigs used for tea.	1759
B. lutea Michx. f. (yellow birch).	*B. excelsa* Pursh not Ait., *B. alleghaniensis* Britt. (gray birch, silver birch).	do	Rich woods, Newfoundland to Manitoba, south to Georgia, Ohio, and Indiana.	Furniture, flooring, interior finish, veneer, specialty products, wood distillation, fuel.	1800
B. nigra L. (river birch).	*B. rubra* Michx. (red birch, water birch).	Small tree	River and stream banks, Massachusetts to Florida east of Allegheny Mountains, Gulf States to Texas and in Mississippi basin, north to Wisconsin and Minnesota.	Used to some extent for wood products; erosion control planting; potentially valuable for landscape planting.	1736
B. papyrifera Marsh.[1] (paper birch).	*B. alba* var. *papyrifera* (Marsh.) Spach, *B. papyracea* Ait. (canoe birch, white birch).	Large tree	Labrador to British Columbia, south to North Carolina, Indiana, Iowa, Nebraska, Montana, Oregon.	Spools, toothpicks, and other specialty products; the bark, for canoes and Indian ornaments; tree a pioneer on burned-over land; landscape planting.	1750
B. pendula Roth.[1] (European white birch).	*B. verrucosa* Ehrh., *B. alba* L. in part.	do	Dry soils, Europe to Japan.	Lumber, ornamental purposes, game food and cover.	(2)
B. populifolia Marsh.[1] (gray birch).	*B. alba* var. *populifolia* (Marsh.) Spach (oldfield birch, white birch).	Small tree	Dry soils, northeastern United States and adjacent Canada.	Specialty products—spools, toothpicks, shoe pegs, hoops—pulpwood, fuel, sometimes as a source of tannin; good nurse tree for more desirable species.	1750
B. pubescens Ehrh.[1]	*B. odorata* Bechst., *B. alba* L. in part (European white birch).	Large tree	Moist soils, northern and central Europe to eastern Siberia.	Lumber, ornamental purposes; game food and cover.[3]	1789

[1] Includes several recognized varieties.
[2] Long cultivated.
[3] Hardy in northeastern United States.

BETULA

Of the nine species of *Betula* listed, only *B. lutea* is used in reforestation in this country, and this to a very limited extent; considerable amounts of *B. pendula* and *B. pubescens* are planted in Scandinavian countries and the U.S.S.R.; *B papyrifera, B. pendula,* and *B. pubescens* or their several varieties are often used for landscape planting both in this country and Europe.

SEEDING HABITS.—The apetalous flowers are monoecious and are borne in catkins. The pendulous staminate catkins are formed in late summer or autumn, remain naked during winter, and open after considerable elongation in early spring; the less conspicuous pistillate catkins—terminal on short spurlike lateral branchlets, their scales closely overlapping—appear with the leaves and become brown and woody when ripe, at which time they are erect or pendulous conelets or strobiles. The latter usually ripen in autumn, each scale bearing a single small, winged nutlet or seed (figs. 43 and 44) which is oval in outline and bears at its apex the two persistent stigmas. The conelets slowly disintegrate on the trees after ripening much as in the manner of *Abies* cones, their axes persisting on the branches. Seed dispersal is usually by wind, sometimes by water. Details of seeding habits for the seven species discussed are given in table 40.

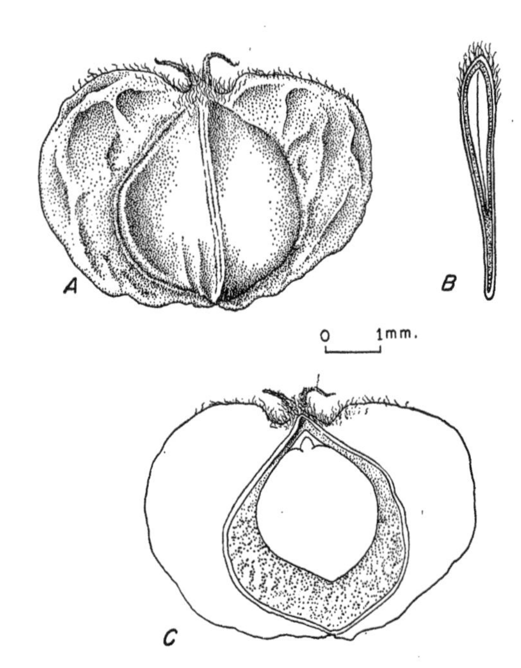

FIGURE 43.—Seed views of *Betula nigra: A,* Exterior view of fruit; *B* and *C,* longitudinal section of fruit in two planes.

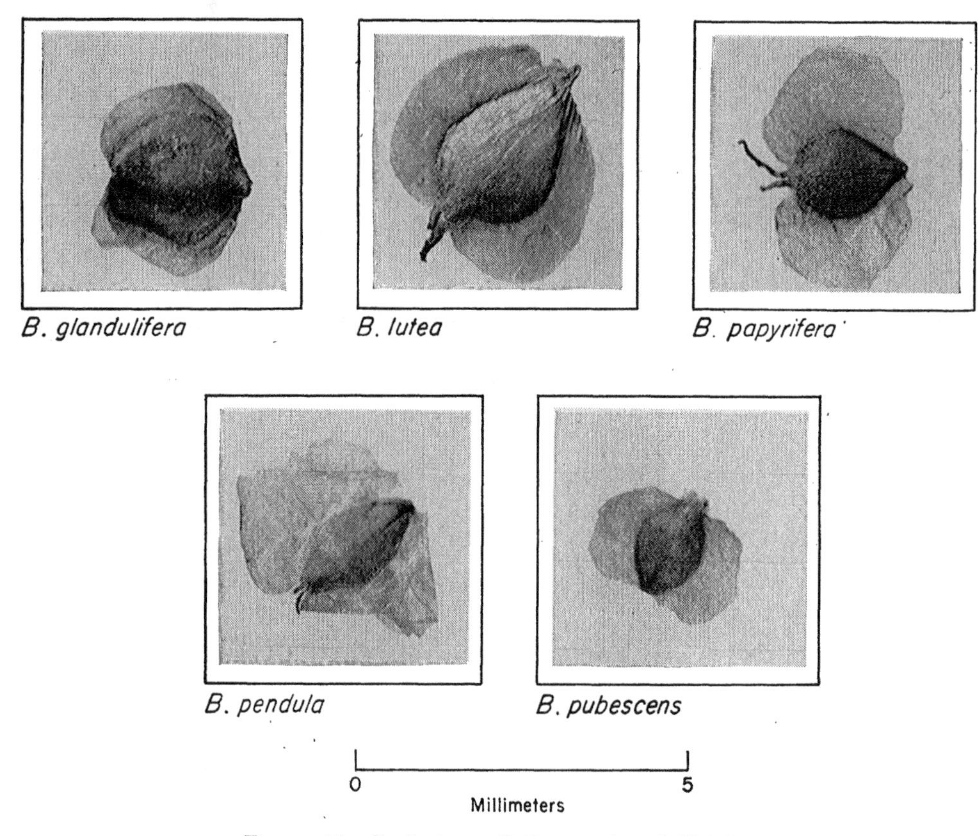

FIGURE 44.—Seed views of five species of *Betula.*

TABLE 40.—*Betula: Time of flowering and seed ripening, and frequency of seed crops*

Species	Time of— Flowering	Seed ripening	Seed dispersal	Seed year frequency Good crops	Light crops
B. glandulifera	Spring	September	September–June		
B. lenta	April to mid-May	Mid-August to mid-September.	September–November	Every 1 to 2 years.	Intervening years.[1]
B. lutea	Mid-April to late May	September to mid-October.	November–February	do.	do.
B. nigra		April–June	April–June		
B. papyrifera	Mid-April to early June	Early August to mid-September.	Sept. 1–Apr. 1	Almost every year.	Intervening years.[2]
B. pendula	April to mid-May	Mid-July to late August	July–September	Periodically	Every year.
B. populifolia	Late April–late May	Early September to mid-October.	October to mid-winter	do.	do.[3]

[1]Seed production begins at about 40 years.
[2]Commercial seed-bearing begins at 15 years, is optimum at 40 to 70 years.
[3]Commercial seed-bearing begins at 8 years, is optimum at 30 years; maximum age is 50 years.

No direct evidence is known of the existence of geographic strains among the various species of *Betula*. However, in view of the fact that most of these occur over a wide geographic range and that a number of varieties are recognized by horticulturists in several of the species, it seems likely that such strains have developed.

COLLECTION, EXTRACTION, AND STORAGE.—Birch seed is collected by picking or stripping the conelets while they are still green enough to hold together, from standing trees or shrubs or from trees recently felled on logging operations.[9] The fruits are generally gathered into collecting bags, rather than being allowed to fall on canvas, to prevent their breaking up and the attendant loss of seed. The freshly collected fruits are rather green and thus subject to heating; they should be spread out to dry for several weeks during which time they will disintegrate to a more or less degree.

[9] The bog species are collected most easily early in the winter after the bogs have frozen.

Most commercial seed consists of a mixture of seed and cone scales and partially disintegrated strobiles, and is consequently low in purity. Purity, however, can be greatly increased if the fruits are broken up by light flailing or shaking and the seed separated from most of the scales and other debris by screening[10] and fanning. Although the seed may be cleaned further by removal of the wings, this operation is not essential.[11] Birch seed, in general, is light in weight, the size varying in different seed lots and by species. Data on this factor and on purity, soundness, and cost are given in table 41.

[10] Round-hole grain screens of the following sizes have proved satisfactory for this purpose: *Betula glandulifera*, 6/64-inch; *B. lutea*, 8/64-inch; *B. nigra*, 10/64-inch; *B. papyrifera*, 8/64-inch; *B. pendula* and *B. pubescens*, 6½/64-inch. Remaining scales can be removed by fanning.

[11] The Forestry Branch, Ontario Department of Lands and Forests, removes much of the wing from yellow birch and paper birch seeds by running the fruits and seed through a winging machine for about 2 minutes, followed by sifting in 1/10- and 5/64-inch screens, respectively.

TABLE 41.—*Betula: Yield of cleaned seed, and purity, soundness, and cost of commercial seed*

Species	Cleaned seed per pound Low	Average	High	Basis, samples	Commercial seed Purity[1]	Soundness[1]	Cost per pound
	Number	*Number*	*Number*	*Number*	*Percent*	*Percent*	*Dollars*
B. glandulifera	1,396,000	2,422,000	3,470,000	4	[2]64 (22–92)	[2]38 (14–66)	
B. lenta	493,000	646,000	933,000	13	5–66	[3]72	3.00
B. lutea[4]	278,000	447,000	907,000	23	63 (54–89)	56 (27–74)	1.50–2.50
B. nigra	287,000	375,000	548,000	13	58 (25–96)	[3]42	2.00–2.50
B. papyrifera[4]	610,000	[5]1,380,000	4,120,000	28	59 (21–97)	70 (58–89)	1.50–2.00
B. pendula	1,510,000	2,400,000	5,040,000	144	27 (12–43)	(0–77)	1.50
B. populifolia	3,581,000		4,930,000	2	88	(59–82)	2.00
B. pubescens	750,000	1,720,000	4,500,000	45	28 (13–76)	29 (0–84)	.80–1.50

[1]First figure is average percent; figures in parentheses indicate range from lowest to highest.
[2]Experimental samples.
[3]1 sample.
[4]Yield of dewinged seed per bushel of fruit: *B. lutea*, 1.0 to 3.5 pounds; *B. papyrifera*, 2.0 to 3.4 pounds.
[5]1 sample dewinged seed, 1,393,000 per pound.

BETULA

Seed of the few species of birch on which storage studies have been made will keep for several years without appreciable loss in germinability. That of *Betula lenta* will retain most of its initial viability for at least 18 months when stored at room temperature providing the moisture content is kept at 8 percent or lower; if the moisture content is no higher than 12 percent, the seed will also keep well, but only if stored at a low temperature (45° F.).

Seed of *Betula lutea* stored in a tightly closed bottle at about 40° F. for 12 years has shown a germination of 44 percent (original germination not known); when stored dry in open bottles at room temperature, practically all its viability is lost in 18 months. Seed of *B. nigra* stored with a moisture content of 16.6 percent dropped from 22 percent to no germination after 2 years at 41° F. That of *B. papyrifera* appears to keep well for at least 18 months at room temperature if its moisture content is maintained at about 1 percent,[12] but at a temperature of about 45° the moisture content may be kept at about 12 percent without any serious loss in vitality.

Seed of *Betula pendula* can be stored up to 2 years if it is kept in a sealed container at 32° to 50° F. That of *B. populifolia* can be stored at room temperature for 18 months with little loss only if the moisture content is maintained at about 5 percent; with a moisture content higher or lower than this (12 and 1 percent) the seeds keep well only at low temperatures (45° F.).

GERMINATION.—Natural germination of the seed of fall-ripening species of *Betula* occurs in the spring following dispersal, although some may lie over until the second spring; that of *B. nigra* germinates soon after it falls. Moist mineral soil and rotten logs form the best seedbed for upland species. Seed of bog species probably germinate best on moist peat. Germination is epigeous (fig. 45).

[12] This low moisture content may be attained by suspending air-dry seed in a closed container, the bottom of which is covered with quicklime.

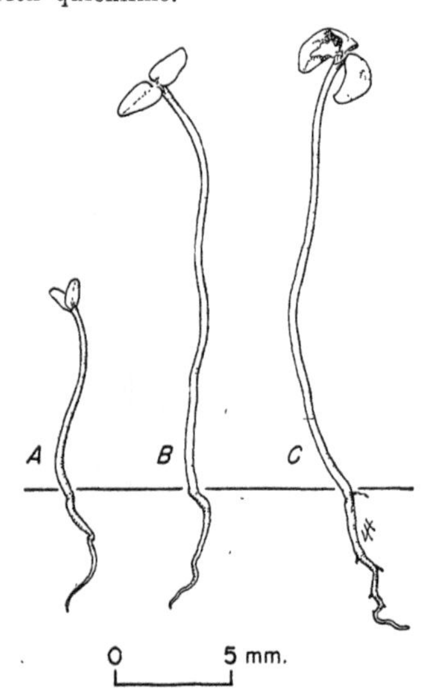

FIGURE 45.—Seedling views of *Betula populifolia: A,* At 1 day; *B,* at 10 days; *C,* at 40 days.

Birch seed is relatively free from attack by insects and other pests. Seed of *Betula nigra* is sometimes attacked by the fungus *Sclerotinia betulae,* but the effect is not known to be serious.

Seed lots of all fall-ripening species show delayed germination to a more or less degree, due apparently to embryo dormancy.[13] This can be easily overcome by stratification at a low temperature in moist sand or peat as indicated in table 42.

TABLE 42.—*Betula: Method of seed pretreatment for dormancy*

Species	Stratification		
	Medium	Temperature	Duration
		°F.	*Days*
B. glandulifera[1]	Sand	41	60
B. glandulosa[2]	Soil	--	83
B. lenta[3]	Sand or peat	32–41	42–70
B. lutea	--do	41	30–60
B. papyrifera	Sand or peat	41	60–75
B. pendula	--do	32–50	30–60
B. populifolia[3]	--do	32–50	60–90+

[1] Tentative recommendation; experimental data not complete.
[2] In cold frame during winter.
[3] Untreated seed germinates well at high temperatures.

Germination tests may be made in sand flats, peat mats, or Jacobsen germinators, using 1,000 seeds per test, these to be pretreated in the case of those species showing dormancy. Testing procedures that appear to be fairly successful, and results that may be expected, are given in table 43.

NURSERY AND FIELD PRACTICE.—Seed of the fall-ripening birches may be sown in the fall or stratified from 4 to 8 weeks in moist peat or sand prior to spring sowing. Sowing on top of the snow in early winter is also practiced, particularly in nurseries well protected against blowing. Seed of *Betula nigra* is sown in early summer without treatment. Birch seed is sown broadcast and should be covered with $\frac{1}{16}$ to $\frac{3}{16}$ of an inch of light nursery soil; if the surface can be kept moist, the seed may be sown without covering. Germination is usually complete in 4 to 6 weeks after spring sowing. Birch seedlings are delicate and require light shade for 2 to 3 months during the first summer. Tree percent is rather low; for example, Russian experience shows that only about 15 percent of *B. pendula* seed will produce 1–0 seedlings. Stock generally used for field planting is 1–0 or 2–0 seedlings. Shrubby species may also be propagated by layering.

[13] Reports of European investigators, notably Rafn, indicate that most birch seeds will complete their germination in 30 days, a preliminary period of stratification being unnecessary. American workers on the other hand have generally found seed of the fall-ripening species to be markedly benefited by stratification.

TABLE 43.—*Betula: Recommended conditions for germination tests, and summary of germination data*

Species	Test conditions recommended: Temperature, Night	Test conditions recommended: Temperature, Day	Test conditions recommended: Duration	Germinative energy		Germinative capacity: Low	Germinative capacity: Ave.	Germinative capacity: High	Potential germination	Basis, tests	Remarks
	°*F.*	°*F.*	*Days*	*Percent*	*Days*	*Percent*	*Percent*	*Percent*	*Percent*	*Number*	
B. glandulifera	50	77	20–40	-------	-------	0	2	5	13	3	Alternations of 68°—86° F. equally as good.
B. glandulosa	(1)	(1)	30	-------	-------	-------	-------	24	-------	1	-------
B. lenta	59	90	30	4–60	20	5	43	87	-------	13	Constant temperatures of 59°, 68°, and 77° F. also good.[2]
B. lutea	59	90	30–40	6–48	20	6	27	48	21–52	22	do.[2]
B. nigra	68	86	30	1–80	10–14	1	34	80	-------	13	-------
B. papyrifera	59	90	30–40	11–87	10–20	0	34	92	-------	28	Constant temperatures of 59°, 68°, and 77° F. also good.[2]
B. pendula	68	86	30–40	1–84	10	2	36	88	-------	143	-------
B. populifolia	68	86	40	-------	-------	38	64	82	-------	3	(2)
B. pubescens	68	86	30	10–71	10	5	40	72	-------	44	-------

[1] Greenhouse.
[2] Unstratified seed germinates best at about 90° F.

BUMELIA LANUGINOSA (Michx.) Pers. Gum bumelia

(Sapote family—Sapotaceae)

Also called gum elastic, woolly bumelia, woolly buckthorn, chittimwood.

Distribution and Use.—Native from Virginia southward and westward in the Coastal Plain to eastern Texas, and northward in the Mississippi basin to Kansas and southern Illinois, gum bumelia is a deciduous[14] spiny shrub or small tree, valuable for game food, ornamental purposes, as a honey plant, and to some extent for shelter-belt planting. The shrub is extremely drought-resistant and is hardy as far north as eastern Massachusetts. It was introduced into cultivation in 1806.

Seeding Habits.—The light- to dark-brown, shiny seed (fig. 46) is about one-fourth inch long, oval in outline, and is borne singly within a fleshy drupe about one-half inch long. The mostly perfect white flowers open from June to July; the fruit, which is produced in abundance, ripens from September to October and remains on the shrubs until winter.

[14] Evergreen in the South.

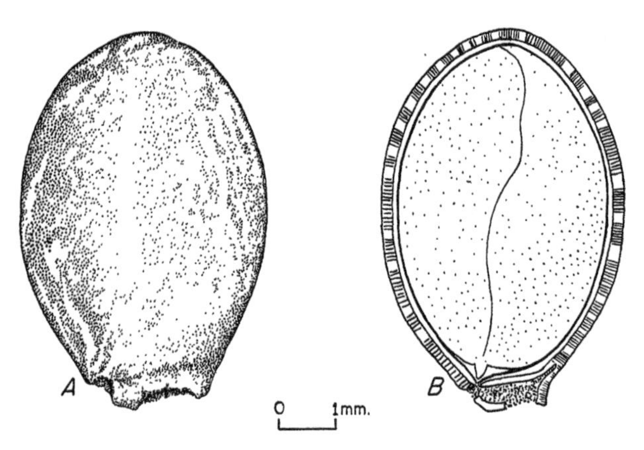

Figure 46.—Seed views of *Bumelia lanuginosa: A,* Exterior view; *B,* longitudinal section.

Collection, Extraction, and Storage.—The fruit should be collected by picking from the bushes as soon as it turns purplish black in color to avoid its dispersal by birds. Approximate cost of collection is $0.35 per pound. It may be cleaned by careful maceration in water so as not to injure the rather brittle seed coat. Cleaned seed: yield per 100 pounds of fresh fruit, 20 to 25 pounds; number per pound (4 samples): low, 4,890; average, 5,700; high, 7,180; purity, 94 percent; soundness, 88 percent.

Germination.—Gum bumelia seed germinates rather slowly, possibly because of a somewhat impermeable condition of the seed coat and possibly because of a dormant embryo. Germination is of the hypogeous type and, although no supporting data are available, has been reported to be hastened if the seed is soaked in sulfuric acid for 20 minutes. Since stratification for 60 days at 41° F. has also proved to be of some benefit, more studies of this seed are needed. Tests may be run in sand flats and require about 400 seeds. Germinative capacity, 44 to 58 percent (2 tests).

Nursery and Field Practice.—Twenty-five viable seeds should be sown per linear foot in the nursery. The species is taprooted and will probably have to be grown for 2 years in the nursery before it can be field planted.

CAMPSIS RADICANS (L.) Seem. Common trumpetcreeper

(Bignonia family—Bignoniaceae)

Also called trumpet-honeysuckle, trumpet-vine. Botanical syns.: *Bignonia radicans* L., *Tecoma radicans* (L.) Juss.

Distribution and Use.—Native to eastern United States from Pennsylvania to Missouri and south to Florida and Texas, trumpetcreeper[15] is a deciduous woody climber valuable for erosion control and game food; it has been introduced into New England. Date of earliest cultivation is 1640.

Seeding Habits.—The large, orange to scarlet perfect flowers appear from mid-June throughout the summer to late September, and the fruit is a two-celled capsule maturing in September to October. Seeds, many in each fruit, are flat with two membranaceous wings (fig. 47), dispersed chiefly by wind. Seed crops are borne almost every year.

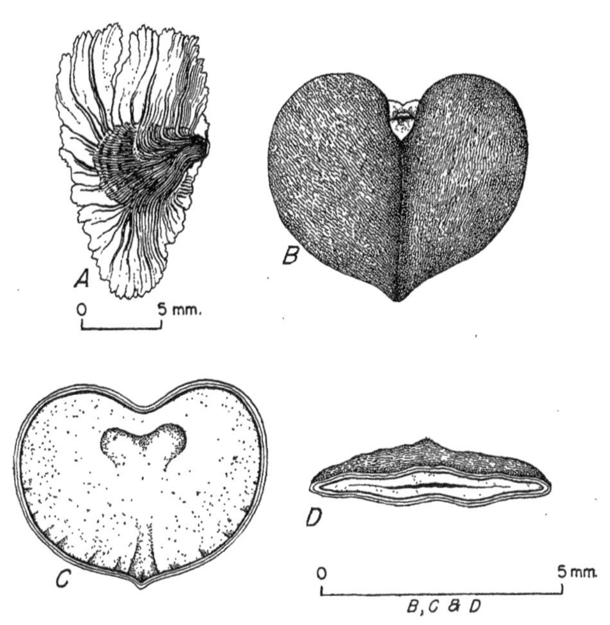

Figure 47.—Seed views of *Campsis radicans: A,* Exterior view of seed with appendages; *B,* exterior view of seed; *C,* longitudinal section; *D,* cross section.

Collection, Extraction, and Storage.—Ripe fruits of trumpetcreeper should be gathered by hand when they turn brown in the fall and before they split open. Cleaned seeds per pound, 136,000 (1 sample); purity, 98 percent; soundness, 52 percent. Cost of commercial seed, $1.75 to $3.20 per pound. Information on storage practices and results is lacking.

[15] Valuable unpublished information on this species was provided by the U. S. Bureau of Plant Industry, Soils, and Agricultural Engineering, Washington, D. C.

Germination.—Moist sandy localities provide the best natural seedbed. The vine is vigorous and spreads rapidly once it has become established. Germination is epigeous (fig. 48). The seeds exhibit embryo dormancy, which appears to vary somewhat from lot to lot. Stratification in moist sand for 60 days at 41° to 50° F. speeds up but does not increase germination. Tests should be run for 30 days in sand flats at 68° (night) to 86° (day) using 1,000 stratified seeds per test. Germinative energy of stratified seed, about 51 percent in 19 days; germinative capacity of stratified seed (4 tests): low, 54 percent; average, 66 percent; high, 86 percent.

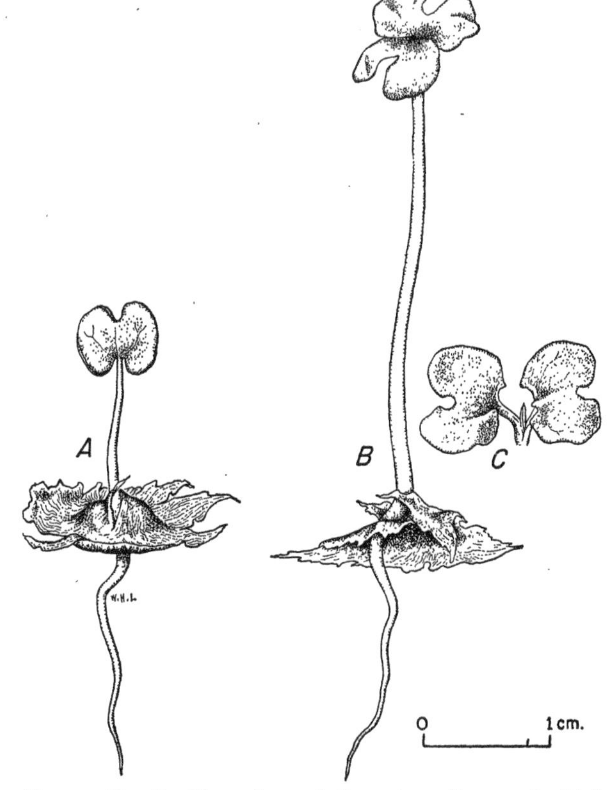

Figure 48.—Seedling views of *Campsis radicans: A,* At 1 day; *B,* at 9 days; *C,* details of cotyledons showing plumule at 9 days.

Nursery and Field Practice.—It is suggested that the seed be sown in the fall or stratified seed used in the spring. In horticultural practice propagation usually is by means of cuttings.

CARAGANA ARBORESCENS Lam. Siberian peashrub

(Legume family—Leguminosae)

Also called pea-tree, caragana. Botanical syn.: *C. caragana* Karst. Includes f. *lorbérgii* Koehne, var. *nana* Jaeg., and var. *pendula* Carr.

Distribution and Use.—Native to Siberia and Manchuria, the Siberian peashrub is one of the most cold- and drought-resistant deciduous shrubs or small trees in cultivation and is the best known of some 50 species in this genus. Its chief use is for the establishment of shelter belts but it is also of value for wildlife food and cover, erosion control, and ornamental purposes. Introduced into cultivation in 1752, it has been widely planted in the prairie regions of North America.

Seeding Habits.—The fruit is a sessile, linear pod or legume, 1 to 2 inches long and sharply pointed, and brown and glabrous when ripe. Each pod contains several, usually six or more, oblong or nearly spherical reddish-brown seeds (fig. 49). Showy, yellow flowers appear in May and June, and the pods ripen in July, soon splitting open and dispersing the seed. The seed are completely dispersed by mid-August on the average. Although no experimental evidence is available, it appears likely that geographic strains may have developed in view of the rather wide range of the species and the recognition of several varieties.

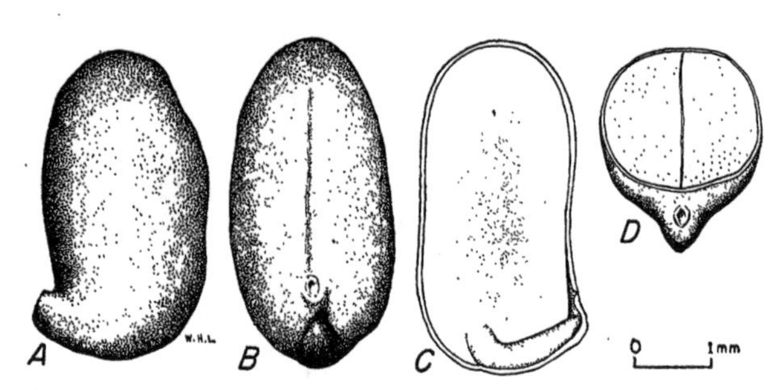

Figure 49.—Seed views of *Caragana arborescens*: *A* and *B*, Exterior view, two planes; *C*, longitudinal section; *D*, cross section.

Collection, Extraction, and Storage.—The collecting season for Siberian peashrub is seldom longer than 10 days. Pods should be collected from the shrubs by hand just as they begin to open, usually in July, and spread out to dry in a cone shed or granary, leaving them until they have popped open. After the pods are thoroughly dry the seed may be removed by light sifting, beating, and fanning. Yield of cleaned seed per 100 pounds of fruits averages 13 to 20 pounds, depending on dryness of pods. Cleaned seed per pound (30+ samples): low, 13,000; average, 17,000; high, 22,000. Commercial seed averages in purity 98 percent; soundness, 98 percent; and its cost ranges from $0.70 to $1.50 per pound. Seed should be stored dry in sacks over winter. Longer storage probably demands sealed containers and low temperatures. One test with seed stored in sacks at relative humidity of 60 to 75 percent and temperature of 41° F. showed complete loss of viability at the end of 5 years. Russian experience, however, indicates satisfactory viability up to 5 years of ordinary dry storage.

Germination.—Natural germination takes place the spring following seed dispersal. Exposed mineral soil is a favorite seedbed. Contrary to some reports, there appears to be no dormancy. Neither sulfuric acid treatments nor stratification in moist sand at 41° F. for 60 days has improved germination. Tests should be run in sand flats or Jacobsen germinators for 30 to 60 days at 68° to 86° using 1,000 seeds per test. Germinative energy under these test conditions averaged 45 to 72 percent in 25 to 41 days, and germinative capacity (33 tests) was: low, 55 percent; average, 76 percent; and high, 100 percent.

Nursery and Field Practice.—Seed of this species may be sown in drills or broadcast in late summer or spring. Dry-stored seed sown in the spring will germinate more promptly if soaked in lukewarm water for 10 to 12 hours before sowing. A sowing depth of one-half inch is recommended. About 33 percent of the viable seed sown produce usable seedlings. Germination in the nursery is largely completed in 20 days, but it may continue for 60 days. Late summer sowings produce a 2- to 3-inch growth before freeze-up. In such cases, after the leaves are dropped, cover the trees with soil; remove the soil in the spring and mow the trees just above the ground line to produce well-branched stock. Nursery stock should be sprayed to prevent damage by aphids, blister beetles, and other leaf-eating insects.

CARPENTERIA CALIFORNICA Torr. Carpenteria, bush-anemone

(Saxifrage family—Saxifragaceae)

Restricted in its native range to a few localities in the foothills of the Sierra Nevada in Fresno County, Calif., carpenteria is a medium-sized evergreen shrub which is valuable as an ornamental. It was introduced into cultivation in 1880.

The large, white, perfect flowers bloom in June to August; the fruit is a conical, leathery capsule containing numerous oblong tiny seeds (fig. 50); the time of seed collection is September and October. Cleaned seed per pound (2 samples): 15,010,000 to 21,460,000. Purity of one sample was 25 percent.

Germination without treatment of the seeds is excellent. The maximum recorded: 80 percent in 71 days (1+tests). The first seedlings appeared 17 days after sowing. Careful watering of the flats is essential to prevent loss of the fine seed. The seedbed should be in a sunny, sheltered position on a well-drained soil. Carpenteria is also propagated by cuttings and from suckers.

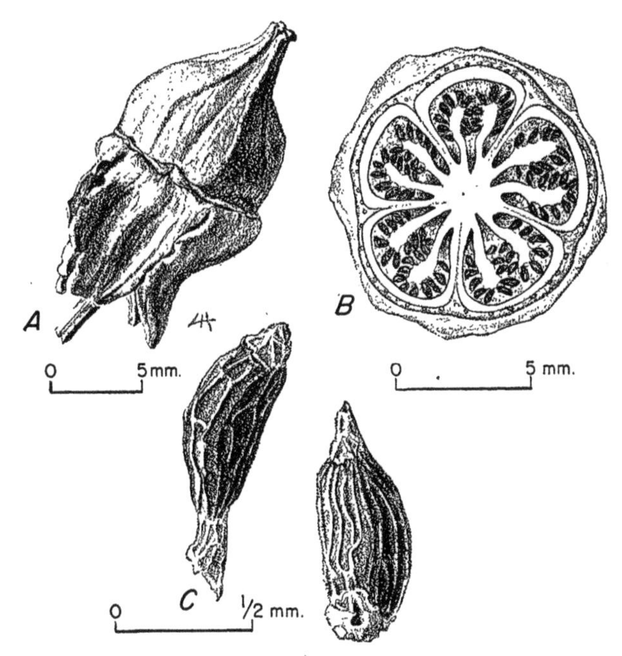

FIGURE 50.—Seed views of *Carpenteria californica*: *A*, Exterior view of fruit; *B*, cross section of fruit; *C*, exterior view of seeds in two planes.

891183°—50——8

CARPINUS CAROLINIANA Walt. American hornbeam

(Birch family—Betulaceae)

Also called hornbeam, ironwood, water beech, blue beech. Botanical syns.: *C. americana* Michx., *C. virginiana* Michx. not Mill.

DISTRIBUTION AND USE.—Found from Nova Scotia south to Florida, westward to southwestern Ontario, northern Minnesota, eastern Nebraska, Kansas, Oklahoma, and Texas, and occurring on mountains of southern Mexico and Central America, American hornbeam is a small, deciduous tree which attains its largest size and most common occurrence on west slopes of southern Allegheny Mountains and in southern Arkansas and eastern Texas. It occurs mainly as an understory tree in rich, moist soils on bottom lands, coves, and lower protected slopes. It is commercially valuable for its hard, tough wood for handles, mallets, golf-club heads, and less valuable for its limited bird food. It was introduced into cultivation in 1812.

SEEDING HABITS.—The male and female flowers, occurring separately on different trees, appear in April to early June; fruits are ribbed nutlets with bracts borne in catkins (fig. 51) ripening in August to October. Seeds are blown only a short distance and dispersal is mainly by birds. Trees reach the commercial seed-bearing stage in about 15 years—optimum, 25 to 50 years; maximum, 75 years. Good seed crops are produced from 3 to 5 years, with light intervening crops. Botanists recognize a northern form and a southern form which differ in bark, leaf, and bract characteristics and which may be geographic races.

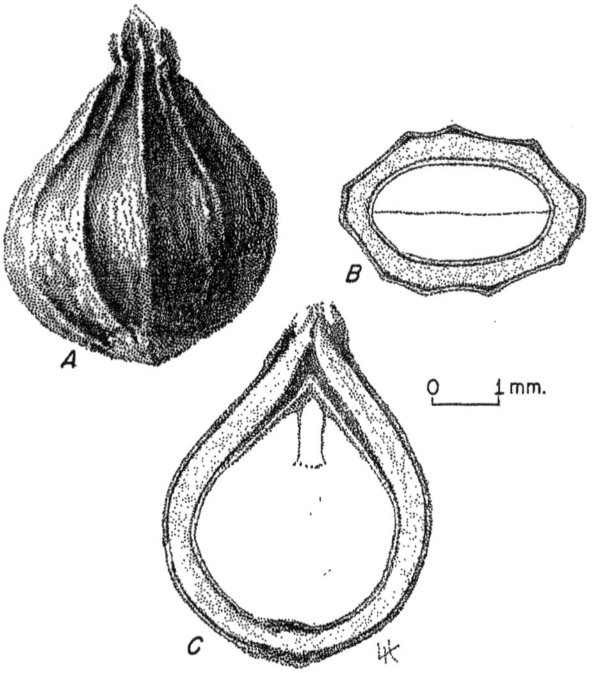

FIGURE 51.—Seed views of *Carpinus caroliniana*: *A*, Exterior view, showing part of adhering pedicel; *B*, cross section; *C*, longitudinal section.

COLLECTION, EXTRACTION, AND STORAGE.—Fruits should be collected by hand-picking from the tree in late summer or fall before they become dry. Pale greenish-brown color of bracts or involucres and light greenish-gray color of seeds indicate ripeness. After collection, the fruits should be spread out to dry and subsequently placed in a dewinging machine or beaten in bags to separate seed from the involucre. Seed can be separated from the chaff by screening and fanning. Cleaned seed per pound (7 samples): low, 15,000; average, 30,000; high, 45,000. Commercial seed averages in purity about 96 percent, ranges in soundness from 30 to 95 percent, and costs per pound from $1.00 to $2.50. The seed should be stratified in a mixture of sand and peat for over-winter storage; for longer storage it should be sealed in containers soon after collection and stored at 35° to 45° F.

GERMINATION.—The optimum natural seedbed is continuously moist, rich, loamy soil protected from extreme atmospheric changes. Germination is epigeous and occurs from April to June in spring succeeding seed maturity. Dormancy, apparently caused by conditions in embryo and endosperm, may be broken by stratification for 100 to 120 days at 35° to 45° F. in sand or peat, preferably in mixture of the two materials. It is possible that seed collected slightly green will germinate promptly. Effective tests may be made on 400 pretreated seeds in flats of sand or of sand and peat in 45 days at 60° to 80°. Germinative capacity is usually low, 1 to 5 percent in 2 tests; potential germination ranges up to 70 percent.

NURSERY PRACTICE.—Sowing may be in fall or spring. Fall seeding should be done in well-prepared beds soon after seed are collected, the seed being covered approximately one-fourth inch deep with firmed soil. Beds should be mulched with burlap, straw or other material, and held in place until after last frost in spring. Drills 8 to 12 inches apart are preferred to broadcast sowing because this facilitates subsequent weeding and cultivation. Spring seeding should employ stratified seed. In one case seed collected slightly green in August and sowed immediately germinated 100 percent the following spring. Surface soil should be kept moist until after germination. Partial shade until after seedling establishment is recommended.

CARYA Nutt. Hickory

(Walnut family—Juglandaceae)

DISTRIBUTION AND USE.—The hickories, including about 20 species, are all native to the temperate region of eastern North America and to the highlands of northern Mexico, except for 2 species which occur in eastern Asia. All develop into sizable trees, most being highly valuable for wood products or for production of food for man and animals, some being valuable for both. The seven species described in table 44 are considered to have the greatest value for use in the United States. *Carya illinoensis* is cultivated for nuts in large plantations in the lower South and west to eastern Texas and in northern Mexico. Many horticultural varieties and hybrids are propagated for their superior nut qualities. *C. ovata* and *C. laciniosa* are planted for purposes of both wood and nut production. *C. glabra* is planted only for its high-quality wood.

SEEDING HABITS.—Staminate and pistillate flowers appear in spring on the same tree (monoecious). Staminate flowers are in catkins, developing from axils of leaves of the previous season or from the inner scales of the terminal buds, located at the base of current year's growth. Pistillate flowers, from 2 to 10, appear in short spikes on peduncles terminating current year's shoots. Fruits, a type of dry drupe or the seed an achene type, ripen in the fall of the same year. Fruits are ovoid, globose, or pear-shaped with thin or thick husks hardening at maturity and splitting away from the nut into 4 valves along sutures. The hardened endocarp or nut shell is 4-celled at the base and 2-celled at the apex. The nut is the seed of commerce (figs. 52 and 53). Flowering and seeding descriptions are summarized for 7 species in table 45.

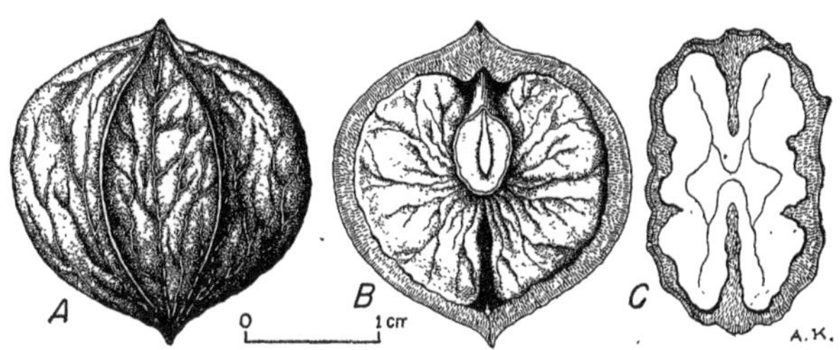

FIGURE 52.—Seed views of *Carya ovata: A,* Exterior view; *B,* longitudinal section; *C,* cross section.

TABLE 44.—*Carya: Distribution and uses*

Accepted name	Synonyms	Natural range	Chief uses	Date of earliest cultivation
C. aquatica (Michx. f.) Nutt.[1] (water hickory).	*Hicoria aquatica* (Michx. f.) Britt. (bitter pecan).	River swamps, Virginia to Florida, west to Texas, Mississippi Valley north to Illinois.	Fuel	1800
C. cordiformis (Wangenh.) K. Koch (bitternut hickory).	*H. cordiformis* (Wangenh.) Britt., *C. amara* Nutt., *H. minima* (Marsh.) Britt. (bitter nut, swamp hickory).	Rich, moist woods, bottom lands Maine to Florida, Minnesota, southwest to Texas.	Wildlife food, wheel stock, tool handles, fuel.	1689
C. glabra (Mill.) Sweet (pignut hickory).	*H. glabra* (Mill.) Britt., *C. porcina* (Michx. f.) Nutt. (broom hickory, black hickory).	Dry woods, Maine to Florida, west to southern Ontario, southwest to Texas.	Wildlife food, tool handles, wheel stock, sporting goods, fuel.	1750
C. illinoensis (Wangenh.) K. Koch (pecan).	*H. pecan* (Marsh.) Britt., *C. pecan* Engl. & Graebn., *C. olivaeformis* Nutt.	Rich bottom lands of Mississippi River from Iowa to Gulf of Mexico and of lower tributaries, and mountains of Mexico.	Human and animal food, fuel, lumber, and other wood products.	1766
C. laciniosa (Michx. f.) Loud. (shellbark hickory).	*H. laciniosa* (Michx. f.) Sarg., *C. sulcata* (Willd.) Nutt. (big shellbark, king nut, bottom shellbark, bigleaf shagbark hickory).	Rich bottom lands, New York to North Carolina, west to Iowa and Oklahoma.	Human and animal food, wagon and car stock, tool handles.	1800
C. ovata (Mill.) K. Koch (shagbark hickory).	*H. ovata* (Mill.) Britt., *C. alba* Nutt. not (L.) K. Koch (shellbark hickory).	Rich, moist soils from Maine to Florida, west to Nebraska and Texas.	Human and animal food, wheel stock, tool handles, ladders, sporting goods, fuel.	1911
C. tomentosa Nutt. (mockernut hickory).	*C. alba* (L.) K. Koch not Nutt., *H. alba* (L.) Britt.	Dry slopes, ridges, well-drained lowlands Maine to Florida, west to Iowa and eastern Texas.	Human and wildlife food, tool handles, wheel stock, fuel.	1766

[1] Valuable unpublished information on this and 14 other southern species was furnished in two progress reports by J. R. Dilworth. On file at Louisiana State University, Department of Forestry, Baton Rouge. June 24 and 26, 1941.

CARYA

TABLE 45.—*Carya: Time of flowering and seed dispersal, commercial seed-bearing age, and frequency of seed crops*

Species	Time of—			Commercial seed-bearing age			Frequency of good crops[1]
	Flowering	Fruit ripening	Seed dispersal	Minimum	Optimum	Maximum	
				Years	*Years*	*Years*	*Years*
C. aquatica	March–May	Sept.–Nov.	Sept.–Dec.	20	40– 75	125	1–2
C. cordiformis	April–May	Sept.–Oct.	do	30	50–125	175	3–5
C. glabra	do	do	do	30	75–200	300	1–2
C. illinoensis	March–May	do	do	20	75–225	300	1–2
C. laciniosa	April–June	Sept.–Nov.	do	40	75–200	350	1–2
C. ovata	do	Sept.–Oct.	do	40	60–200	300	1–3
C. tomentosa	April–May	do	do	25	40–125	200	2–3

[1] Light seed crops are borne in intervening years.

C. cordiformis

C. illinoensis

0 20 Millimeters

FIGURE 53.—Seed views of two species of *Carya*.

No information on the existence of geographic races is available. However, as in the case of other species with a wide geographic range, it is likely that such races may have developed also in some of the hickories.

COLLECTION, EXTRACTION, AND STORAGE.—Nuts are collected in the fall from the ground soon after natural seed-fall or after shaking from trees. When ripe, husks turn brown and split along sutures. Those of *Carya tomentosa, C. ovata, C. laciniosa,* and *C. illinoensis* split to the base releasing the nuts, those of the latter species often persisting on the tree throughout winter. Husks of *C. glabra, C. cordiformis,* and *C. aquatica* split only to middle or slightly beyond and generally cling to the nuts. Persisting husks may be removed by hand, by trampling, or by running the fruits through a corn sheller. Thrifty *C. ovata* and *C. laciniosa* trees often produce, respectively, 1½ to 2 and 2 to 3 bushels of nuts each. Collection costs range from $0.75 to $1.50 per bushel. The size, purity, soundness, and cost of commercial clean seed are indicated in table 46.

Rather uniform storage treatments may be applied to the several species of hickories. For prolonged storage, 3 to 5 years' duration, seed should be placed in closed containers at 41° F. soon after collection to prevent excessive drying, maintaining a relative humidity of at least 90 percent. For indoor storage over winter, nuts should be stratified in trays in sand, in a mixture of sand and peat, or in sandy loam soil at 35° to 45°. In outdoor storage, nuts can be placed in a moist, well-drained sandy loam soil or other medium and protected from freezing and sudden temperature changes with about 2 feet of compost, leaves, soil, or other material.

TABLE 46.—*Carya: Yield of cleaned seed, and purity, soundness, and cost of commercial seed*

Species	Cleaned seed					Commercial seed		
	Yield per 100 pounds fruit	Per pound			Basis, samples	Purity	Soundness	Cost per bushel[1]
		Low	Average	High				
	Pounds	*Number*	*Number*	*Number*	*Number*	*Percent*	*Percent*	*Dollars*
C. aquatica			199		1			
C. cordiformis	65–85	125	156	185	3	100	95	1.25–2.00
C. glabra	65–85	175	200	225	3+	100	95	1.25–2.00
C. illinoensis	50–75	55	100	160	5+	100	95	3.00–6.00
C. laciniosa	15–25	25	30	35	3+	100	95	1.25–2.50
C. ovata	[2]25–38	80	100	150	7+	100	95	1.00–2.50
C. tomentosa	50–80	34	90	113	3			1.00–1.75

[1]If purchased in small quantities from dealers, per pound costs range from $0.20 to $0.65.
[2]Clean seed: weight per bushel, 37.7 lbs.; number per bushel, 6,200.

TABLE 47.—*Carya: Dormancy and method of seed pretreatment*

Species	Stratification			Remarks
	Medium	Temperature	Duration	
		°F.	*Days*	
C. aquatica	Sand, peat, sand and peat, sandy loam soil.	30–35	90–150	30 to 60 days' stratification often sufficient for seed which have been dry-stored for approximately a year.
C. cordiformis	do.	32–45	90–120	do.
C. glabra	do.	32–45	90–120	do.
C. illinoensis	do.	35–45	30– 90	Dormancy in cultivated varieties can sometimes be broken by various acids and hydroxides.
C. laciniosa	do.	32–45	120–150	30 to 60 days' stratification often sufficient for seed which have been dry-stored for approximately a year.
C. ovata	do.	35–45	90–150	do.
C. tomentosa	do.	30–35	90–150	do.

GERMINATION.—Moist, fertile woodland soils offer the best natural seedbed. Germination, hypogeous, under these conditions occurs from late April to early June after seed-fall in the autumn. Under natural conditions, nuts seldom remain viable until the second spring. Dormancy, caused largely by conditions in the embryo, may be broken by stratifying seed at low temperatures. Details of dormancy and pretreatment are given in table 47. Adequate germination tests may be made on 100 to 200 pretreated seeds in flats of sand, peat, sand and peat, or sandy loam at temperatures alternating diurnally from about 68° to 86° F. Details for methods of testing and expected results are given in table 48.

NURSERY PRACTICE.—Drilling of nuts may be done in either fall or spring. Fall seeding usually gives better results if beds are protected against rodents and excessive soil washing. Six to eight nuts should be drilled per linear foot in rows 8 to 12 inches apart, and then covered with ¾ to 1½ inches of firmed soil. Beds should be mulched with straw or leaves and screened until germination time in spring. Only stratified seed, which should not be allowed to dry after removal from stratification, are used in spring beds. Beds should be mulched until germination is complete. Generally no shading is required for seedlings. *Carya laciniosa,* probably the most tolerant of all the species, may be profitably shaded. Vegetative propagation is best done by budding. *C. cordiformis* stock is probably used most in general budding or grafting work with the exception of *C. illinoensis.*

TABLE 48.—*Carya: Recommended germination test duration, and summary of germination data*

Species	Test duration recommended		Germination data from various sources					
	Stratified seed	Untreated seed	Germinative energy		Germinative capacity			Basis, tests
					Low	Average	High	
	Days	*Days*	*Percent*	*Days*	*Percent*	*Percent*	*Percent*	*Number*
C. aquatica	----	----	----	----	----	----	81	1
C. cordiformis	30–45	250–300	40	30	20	55	80	3
C. glabra	30–45	250–300	----	----	----	85	90	2
C. illinoensis	45–60	200–300	36–90	20–25	9	50	100	9
C. laciniosa[1]	45–60	300–350	----	----	----	----	----	----
C. ovata	45–60	300–350	75	40	62	80	88	6
C. tomentosa	----	----	----	----	64	66	69	3

[1]Suggested treatment; experimental data not complete.

CASTANEA DENTATA (Marsh.) Borkh. American chestnut

(Beech family—Fagaceae)

Distribution and Use.—Native to the Appalachian Mountains and adjacent highlands west to Arkansas and north to Michigan, Ontario, and Maine, American chestnut is a medium-sized, deciduous tree which formerly ranked as one of our most important and valuable timber species; in addition, the nuts provided game food and were extensively marketed for human consumption, the wood was an important source of tannic acid, and the bark had medicinal value. During the past 35 years, however, it has been almost completely destroyed as a commercial species by the chestnut blight. From some scattered trees—mostly infected —which still persist in the southern Appalachians, and from sprouts, which often escape the blight for a few years, a limited amount of seed can still be obtained. Propagation of the species is of course futile anywhere within the natural range of the genus, except for experimental or other special purposes. It has been cultivated since 1800.

Seeding Habits.—The fruiting structure is a very spiny, globose involucre, 2 to 2½ inches in diameter, which encloses one to three more or less flattened nuts ½ to 1 inch wide and about 1 inch long (fig. 54). The nuts, which are the true fruits but for practical purposes are treated as seeds, are bright brown in color, with a broad scar at the base and finely hairy toward the apex. Each nut normally contains one fully developed true seed, the bulk of which consists of the two sweet, fleshy cotyledons of the embryo. The male and female flowers, occurring separately on different trees, appear in June or July; the fruit ripens in September and October, and the burs open and release the nuts soon after the onset of killing frosts.

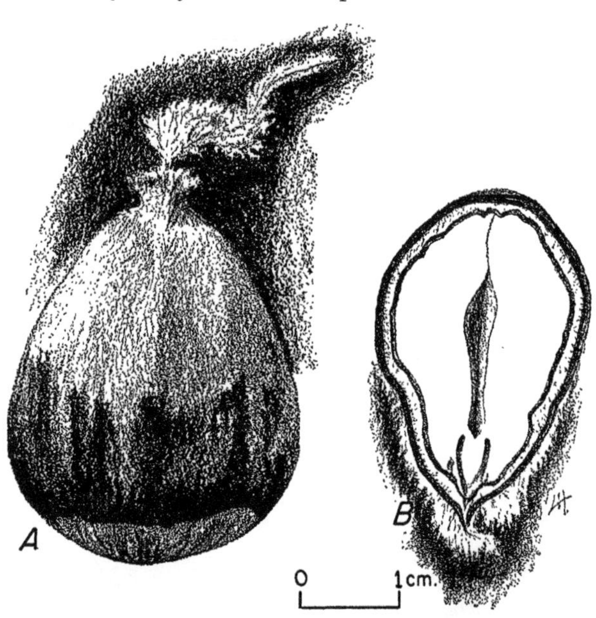

Figure 54.—Seed views of *Castanea dentata: A,* Exterior view; *B,* longitudinal section.

Collection, Extraction, and Storage.—The burs or nuts may be collected from the trees or from the ground after they have fallen naturally, or after vigorously shaking the trees or branches. Unopened burs should be spread to dry and open, preferably in a drying shed out of direct sunlight. The nuts must always be guarded against excessive drying. As soon as all the burs have opened, the nuts should be separated from the empty husks and, if not planted at once, stratified in moist sand, sawdust, or peat at 32° to 40° F. If cold storage facilities are not available, the nuts can be kept fairly satisfactorily over winter by stratification in an outdoor pit at least 2 feet deep and well covered with earth and straw. In either place, the seed should be examined occasionally after February for sprouting. When germination starts, the nuts should be either planted or subjected to temperatures slightly below 32° in order to inhibit growth to a minimum. Storage longer than over winter is difficult and has been little investigated. There are 100 to 160 nuts per pound (4+ samples). Prices are quoted at $0.50 to $1.50 per pound.

Germination.—American chestnuts apparently require a period of after-ripening before germination. In nature the nuts after-ripen in the litter over winter and are ready to begin growth as soon as spring temperatures permit. Germination is hypogeous. When planted in the fall or stratified over winter, after-ripening proceeds as in nature. Germination may even begin in the stratification medium if the temperatures are above 32° F. Germinative capacity is 65 to 80 percent (4+ tests).

Nursery and Field Practice.—Where rodent depredations are not serious, direct seeding in the fall immediately after the nuts are collected is regarded as preferable to other methods of propagation. Nuts stratified over winter may be direct seeded in the spring also, as soon as the frost is out of the ground. Depth of planting should be about 1 inch. In the nursery, seeding is done in rows at the rate of 10 to 12 nuts per linear foot. The seedlings may be transplanted to the field after either 1 or 2 years.

CASTANOPSIS (D. Don) Spach — Chinquapin, evergreen-chinkapin

(Beech family—Fagaceae)

Distribution and Use.—Chinquapins consist of about 30 species of evergreen trees or shrubs native chiefly to the tropical and subtropical mountains of southern and eastern Asia, with 2 species in North America, both in California. Several species are cultivated for ornamental purposes, some produce lumber, some are useful for erosion control, and most of them provide nuts which are food for wildlife. The two species native to the United States are described in table 49.

Seeding Habits.—The male and female flowers are borne separately on the same tree—the male flowers in upright spikes and the female flowers at the base of the staminate spikes or in separate catkins. By the end of the second season the latter develop into ripe fruits which consist of a spiny bur enclosing one to three ovoid or globose nuts which are more or less angular and usually one-seeded (figs. 55 and 56). The comparative seeding habits of two species of *Castanopsis* are as follows:

	C. chrysophylla	*C. sempervirens*
Time of—		
Flowering	Irregularly June–Feb.	July–Aug.
Fruit ripening	Aug.–Sept.	Sept.–Oct.
Seed dispersal	Fall	Fall.
Color of—		
Nut	Light yellow brown	------------
Seed	Dark purple red	------------

No data are available concerning the age for commercial seed bearing, the periodicity of seed crops, nor the development of climatic races. It is known that the shrubby form of *Castanopsis chrysophylla* (sometimes designated var. *minor*) is hardier than the typical form.

Collection, Extraction, and Storage.—The ripe burs should be collected in late summer or early fall, before they open, and spread out to dry in the sun or in a warm room. They may then be run through a fruit disintegrator or shaker to

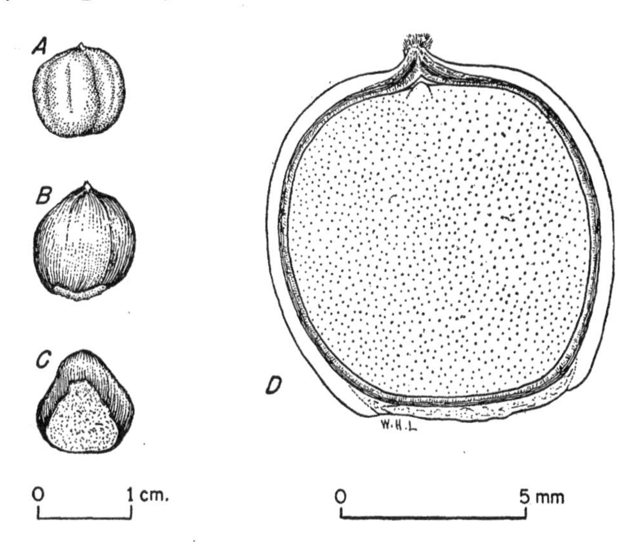

Figure 55.—Seed views of *Castanopsis chrysophylla*: *A*, Embryo; *B* and *C*, exterior view in two planes; *D*, longitudinal section.

Figure 56.—Seed views of *Castanopsis sempervirens*.

Table 49.—*Castanopsis*:[1] *Growth habit, distribution, and uses*

Accepted name	Synonyms	Growth habit	Natural range	Chief uses	Date of earliest cultivation
C. chrysophylla (Dougl.) A. DC. (golden chinquapin).	giant evergreen-chinkapin.	Medium to tall tree; sometimes shrubby.	Mountain slopes from southwestern Washington to California and Nevada.	Furniture; wildlife food.	1845
C. sempervirens (Kell.) Dudley (Sierra evergreen-chinkapin).	brush chinquapin	Spreading shrub	Mountain slopes from southern Oregon to southern California.	Erosion control	------

[1]*Castanopsis chrysophylla* is grown more commonly than *C. sempervirens* in temperate Europe, although neither species has been cultivated very widely.

separate the nuts from the burs. Nuts may also be picked up from the ground after dispersal when crops are plentiful. The nuts are then ready for sowing or storage. The size, purity, and soundness of *Castanopsis* seed varies by species as shown in the following tabulation:

	C. chrysophylla	*C. sempervirens*
Nuts per fruit.......number....	1–2	1–3
Cleaned seed per pound:		
Low................do......	700	--
Average............do......	900	1,200
High...............do......	1,100	--
Basis, samples.......do......	3	2
Commercial seed:		
Purity............percent....	100	86
Soundness..........do......	95	--

Castanopsis seeds retain their viability well for at least 2 years, and probably longer, when stored in sealed containers at 41° F. One sample of *C. chrysophylla* seed stored in this manner lost only 6 percent of its viability in 5 years. It dropped from 50 to 44 percent.

Germination.—Germination is hypogeous (fig. 57). *Castanopsis* seed does not show very high germination as a rule, but since stratification does not improve results something besides seed dormancy must be involved. Germination of *Castanopsis* seed may be tested in sand flats or peat mats for 70 to 80 days, using 100 to 200 seeds per test. The germinative capacity for *C. chrysophylla* in 3 tests was as follows: low, 14 percent; average, 33 percent; and high, 53 percent. *C. sempervirens* in 1+ tests showed a high of 30 percent in 25 days.

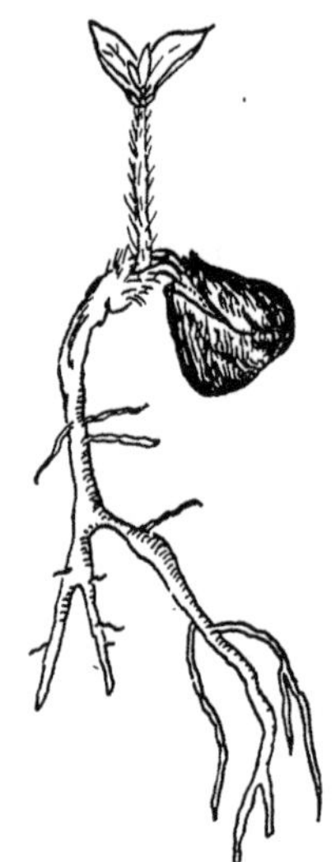

Figure 57.—Seedling of *Castanopsis sempervirens* at 1 month, actual size.

Nursery and Field Practice.—The seeds should be sown in drills in the fall, or after stratification or cold storage in the spring, covering them with about 2 inches of soil. Fall sown seed must be specially protected against rodents. Propagation by layering, grafting, or budding is also feasible.

CATALPA Scop. Catalpa

(Bignonia family—Bignoniaceae)

Distribution and Use.—The catalpas include about 10 species of deciduous, or rarely evergreen, trees native to North America, the West Indies, and eastern Asia. Some species have been cultivated for ornamental purposes or for the production of fence posts, poles, and railroad ties, for which their durable wood is well suited. The 2 species native to the United States, *C. bignonioides* and *C. speciosa*, are described in detail in table 50. Both catalpas have been planted quite widely outside their native range; however, *C. speciosa* has been planted somewhat more extensively because of its greater hardiness and more durable wood. Both species have been grown to some extent in Europe.

Seeding Habits.—The attractive, perfect flowers are borne in the spring in heads or clusters at the ends of branchlets. By fall they develop into long, round pods (the fruits), brownish in color and 6 to 20 inches long, which split into halves. Each pod contains numerous oblong, thin and papery, winged seeds (1 to 2 inches long and 1/4 inch wide in the American species), which bear tufts of long, white hairs at each end (fig. 58). Comparative seeding habits of the 2 species of *Catalpa* discussed here are as follows:

Time of—	*C. bignonioides*	*C. speciosa*
Flowering	May–July	May–June
Fruit ripening	October	October
Seed dispersal	October–spring	October–March

The minimum commercial seed-bearing age for *Catalpa speciosa* is 20 years; it bears good seed crops every 2+ years and light seed crops in intervening years. No scientific information is available as to the development of climatic races among the catalpas, and such development is apt to be further obscured by the tendency of catalpa species to hybridize with one another. Several varieties of *C. bignonioides* are recognized.

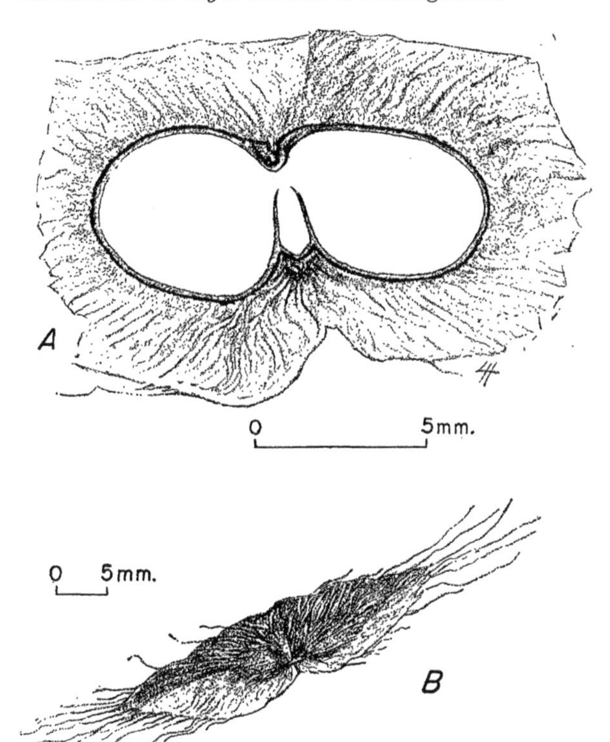

Figure 58.—Seeds views of *Catalpa speciosa: A*, Exterior view; *B*, longitudinal section.

Table 50.—*Catalpa: Distribution and uses*

Accepted name	Synonyms	Natural range	Chief uses	Date of earliest cultivation
C. bignonioides Walt. (southern catalpa).	*C. catalpa* (L.) Karst., *C. syringaefolia* Sims, *C. cordifolia* Moench (common catalpa, Indian bean).	Georgia to Florida and Mississippi; naturalized as far north as New York.	Ornamental; fence posts and rails; shelter belts.	1726
C. speciosa Warder (northern catalpa).	*C. cordifolia* Jaume St.-Hil. not Moench (western catalpa, hardy catalpa).	Southern Illinois and Indiana to western Tennessee and northern Arkansas; naturalized elsewhere.	Ornamental; fence posts and rails, poles; occasionally furniture and interior trim; shelter belts.	1754

CATALPA

Collection, Extraction, and Storage.—Pods should be collected only after they have become brown; they can be picked by hand from standing or newly felled trees in late fall to early spring, and should be dried for a time. The seed can then be separated by light beating and shaking. There is some evidence that pods of *Catalpa speciosa* collected in late winter (February and March) yield seed of higher and more lasting viability than those collected in the fall. The yield, size, purity, soundness, and cost of commercial seed vary according to species as shown in the following tabulation:

	C. bignonioides	*C. speciosa*
Cleaned seed:		
Per 100 pounds of fruit..pounds	35	25–35
Per pound:		
Low..number	14,000	16,000
Average..do	20,000	21,000
High..do	37,000	30,000
Basis, samples..do	7+	24
Commercial seed:		
Purity..percent	60	90
Soundness..do	96	90
Cost per pound..dollars	1.00±	0.60–1.00

The seed should be stored dry, and it keeps satisfactorily over winter in ordinary storage at 50° F. It is reported that seed of *Catalpa bignonioides* retains its viability for at least 2 years in dry storage. One lot of late-collected *C. speciosa* seed, which was stored in burlap sacks in a seed warehouse, germinated 76 percent the second spring after collection as compared to 94 percent the previous spring. Another lot stored under the same conditions—collected green in the fall—germinated 60 percent the following spring and was dead 3 years later.

Germination.—Catalpa seed germinates well without pretreatment and appears to receive no benefit from moist, cold stratification. Germination is epigeous (fig. 59). Germination of catalpa seed may be tested in sand flats or standard germinators for 30 to 60 days at 68° (night) to 86° F. (day), using 400 seeds per test. Average results for the 2 species discussed are as follows:

	C. bignonioides	*C. speciosa*
Basis, tests	6+	25
Test duration..days	30	60
Germinative energy:		
Period..do	12	9–67
Amount..percent	72	60
Germinative capacity:		
Low..do	41	17
Average..do	79	75
High..do	98	96

Two samples of *C. speciosa* seed collected in February and March in 2 different localities germinated 96 and 94 percent in 30 days.

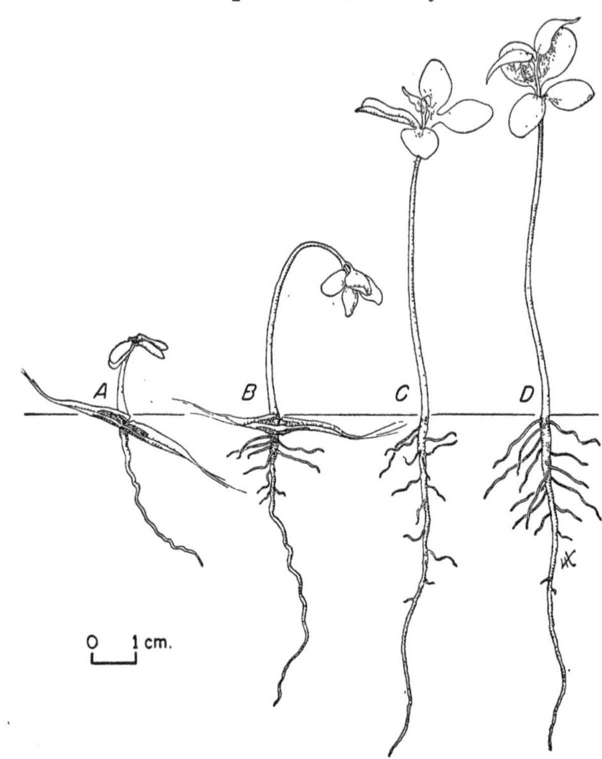

Figure 59.—Seedling views of *Catalpa bignonioides*: *A*, At 1 day; *B*, at 5 days; *C*, at 8 days; *D*, at 20 days.

Nursery and Field Practice.—Catalpa seed should be sown in the late spring—after the soil has warmed up—in drills at the rate of about 30 per linear foot, and covered with about one-eighth inch of soil. A pine needle mulch is recommended for *C. bignonioides*. In Louisiana this species starts germination about 12 days after March sowing and germinates about 80 percent. Catalpas should be field planted as 1-0 stock, and they do best on a deep, rich, fairly moist soil. In the nursery they are troubled occasionally with parasitic diseases; and in the South and East are frequently infested with nematodes. Powdery mildews often occur on the leaves, but appear to do little harm. In many localities older trees are seriously threatened by the catalpa sphinx and other insects.

CEANOTHUS L. Ceanothus, redroot

(Buckthorn family—Rhamnaceae)

DISTRIBUTION AND USE.—The genus *Ceanothus* comprises some 60 species[16] native to North America, chiefly in the Pacific coast region. They are mostly deciduous or evergreen shrubs and but rarely small trees. Although seldom used in reforestation, the ceanothi are important sources of food and shelter for wildlife; because of their ability to grow on dry soils many could be used for erosion control; and several species are cultivated as ornamentals. The 14 species described in detail in table 51 are of present or potential usefulness.

[16] Because several species hybridize freely in nature and in cultivation, ready identification is sometimes difficult.

TABLE 51.—*Ceanothus: Growth habit, distribution, and uses*

Accepted name	Synonyms	Growth habit	Natural range	Chief uses	Date of earliest cultivation
C. americanus L. (Jerseytea ceanothus).	redroot, New Jersey tea.	Low bushy, erect shrub.	Dry soils from eastern Canada to Manitoba and South Carolina, Texas.	Wildlife food, ornamental.	1713
C. arboreus Greene (feltleaf ceanothus).	island lilac	Shrub or small tree	Channel Islands off southern California.		
C. crassifolius Torr. (hoaryleaf ceanothus).		Spreading shrub	Hills and mountains of California.	Wildlife food, erosion control [1]	
C. cuneatus (Hook.) Nutt. (buckbrush ceanothus).	buckbrush, wedgeleaf ceanothus.	Tall evergreen shrub	California and southern Oregon	Erosion control,[1] wildlife food.	1848
C. diversifolius Kell. (trailing ceanothus).		Trailing or prostrate shrub.	Sierra Nevada and northern California Coast Ranges.		
C. impressus Trel. (Santa Barbara ceanothus).	*C. dentatus* T. & G. var. *impressus* Trel.	Shrub	Restricted area near coast in southern San Luis Obispo County and northern Santa Barbara County, Calif.	Ornamental	
C. integerrimus Hook. & Arn. (deerbrush ceanothus).	deer brush, sweet birch.	Tall, erect shrub	Mountains of California	Wildlife browse, erosion control.[1]	
C. oliganthus Nutt. (hairy ceanothus).		Shrub	California Coast Ranges of San Luis Obispo and Santa Barbara Counties, and the San Gabriel Mountains.	Erosion control,[1] wildlife browse.	
C. prostratus Benth. (squaw-carpet ceanothus).	mahalamats	Prostrate shrub	Forests of central California north to Washington.	Wildlife food, nurse crop for conifer reproduction.	
C. rigidus Nutt. (Monterey ceanothus).		Low shrub	Sand flats and hills around Monterey Bay, Calif.	Ornamental	
C. sanguineus Pursh (redstem ceanothus).	red soap-bloom	Tall shrub	British Columbia to northern California.	Erosion control, wildlife food and cover.	1853
C. sorediatus Hook. & Arn. (jimbrush ceanothus).		Thicket-forming shrub.	California Coast Range		
C. thyrsiflorus Eschsch. (blueblossom ceanothus).	blueblossom, blue myrtle.	Evergreen shrub or small tree.	Coast of central California to Oregon.	Ornamental, erosion control.	1837
C. velutinus Dougl. (snowbrush ceanothus).	tobacco brush, mountain balm, sticky laurel.	Tall evergreen, thicket-forming shrub.	Mountains of British Columbia south to California and Colorado.	Wildlife food and cover, erosion control.[1]	1853

[1] Potentially useful; not yet in general use.

CEANOTHUS

TABLE 52.—*Ceanothus: Time of flowering and fruit ripening, and flower color*

Species	Time of—		Flower color
	Flowering	Fruit ripening and seed dispersal	
C. americanus	May–July	August–early October	White.
C. arboreus	July–August		Pale blue to white.
C. crassifolius		May–June	
C. cuneatus	March–May	April–June	Dull white, or bluish.
C. diversifolius	May–June	June–July	
C. impressus	February–April	June	
C. integerrimus	April–July	June–August	Blue, sometimes white.
C. oliganthus		May–June	
C. prostratus	April–May	July	Blue.
C. rigidus	March–April	May–June	do.
C. sanguineus	May–June	June–July	White.
C. sorediatus		May–July	
C. thyrsiflorus	March–June	April–July	Blue, rarely white.
C. velutinus	May–July	July–August	White.

SEEDING HABITS.—The small but showy perfect flowers are borne in clusters, and the fruit is a 3-lobed, dry capsule which at maturity splits lengthwise forcefully ejecting 3 nutlets or seeds (figs. 60 and 61). The seeds are flattened and albuminous with a thin, crustaceous seed coat; the fleshy embryo has flat cotyledons. Details of the seedling habits of 14 species discussed are shown in table 52. Seed crops are borne annually in *Ceanothus americanus;* information for other species is lacking.

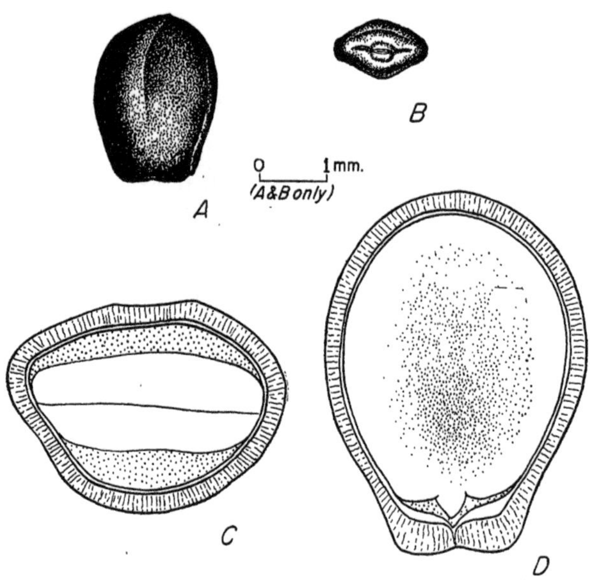

FIGURE 60.—Seed views of *Ceanothus americanus: A* and *B*, Exterior views in two planes; *C*, cross section; *D*, longitudinal section.

Scientific information is not available as to the development of races or geographical strains among the *Ceanothi.* However, it is likely that strains or races have developed among species of wide distribution such as *C. americanus, C. sanguineus,* and *C. velutinus* and perhaps even those of more restricted geographic range but growing at a spread of altitude in the mountains.

COLLECTION, EXTRACTION, AND STORAGE.—Ceanothus fruits ripen quickly and open suddenly; in a few days green and succulent capsules may become dry and empty. Accordingly, ceanothus fruits must be watched closely and picked by hand just before they are ready to open. Immediately after collection the capsules should be spread out in a warm, dry, well-ventilated place and covered loosely with paper or cloth to prevent the loss of seeds when the capsules burst. The seeds should then be separated from the empty capsules by means of a suitable mesh sieve (fine debris may be removed by winnowing) and stored in air-dry containers at about 40° F. until they are to be used. The characteristics of ceanothus seeds vary considerably by species as shown in table 53.

Commercial seed of many ceanothus species is not available and only small quantities of others appear on the market with attendant high prices.

Information on seed-storage practices is scarce for the *Ceanothi.* Field observations have indicated that seeds of *C. integerrimus* and *C. velutinus* are long-lived under natural conditions. Seed of *C. americanus* stored dry in sealed containers at 41° F. showed no appreciable loss in viability in 2 years. It is probable that other species could be stored satisfactorily under similar conditions.

GERMINATION.—The seeds of most ceanothus species have impermeable seed coats and some of them also have dormant embryos, particularly those growing high in the mountains or under continental conditions. Accordingly, the seeds must be soaked in hot water[17] and in addition some of them must be stratified in moist sand at low temperatures for 2 to 3 months. The beneficial

[17] Water at a temperature of 170°–180° F. should be poured over the seeds and they should be allowed to soak in the gradually cooling water for several hours. Boiling water can be poured over most ceanothus seeds without harm.

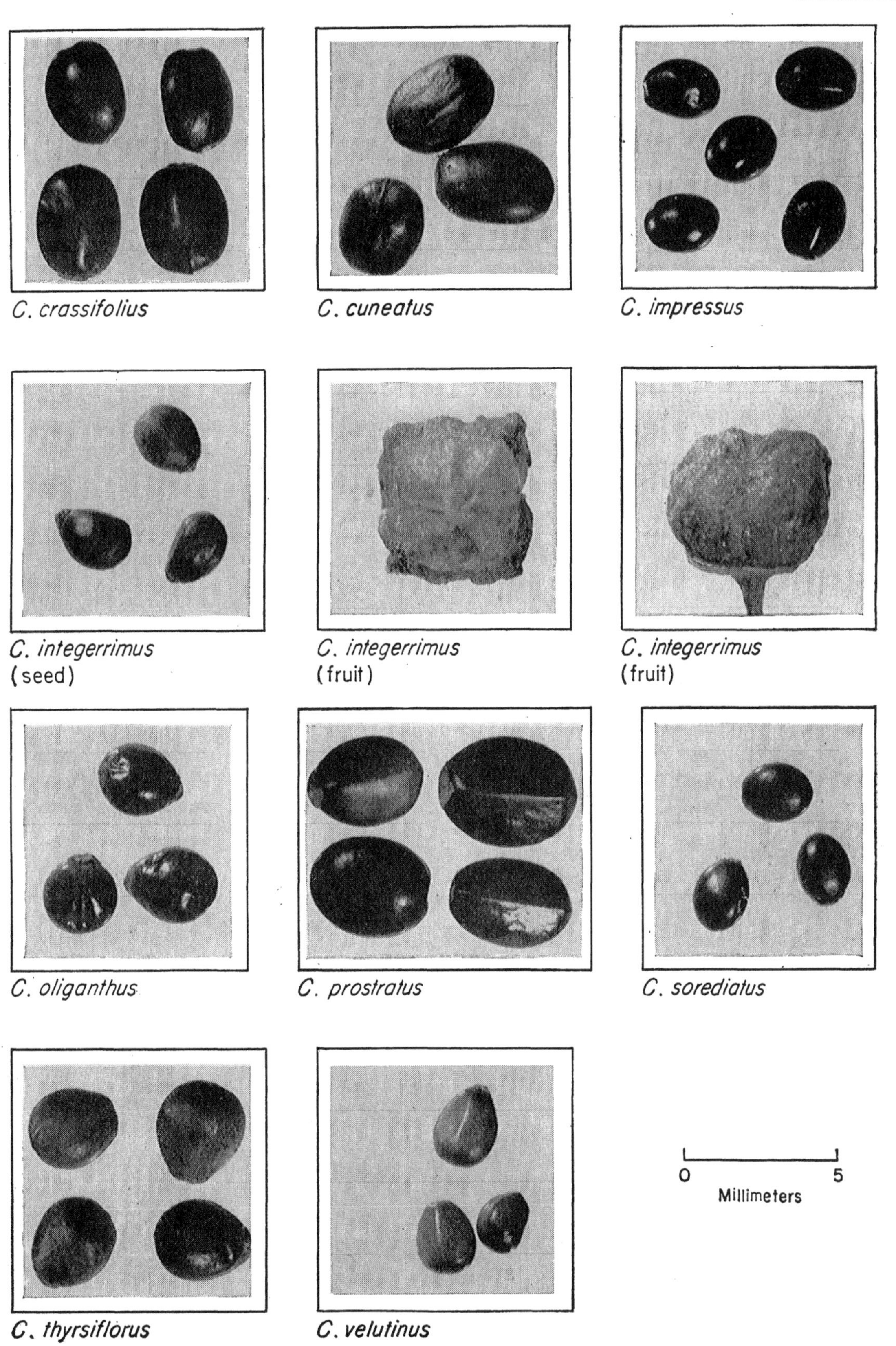

FIGURE 61.—Seed views of nine species of *Ceanothus.*

CEANOTHUS

TABLE 53.—*Ceanothus: Yield of cleaned seed, and purity, soundness, and cost of commercial seed*

Species	Cleaned seed per pound				Commercial seed		
	Low	Average	High	Basis, samples	Purity	Soundness	Cost per pound
	Number	*Number*	*Number*	*Number*	*Percent*	*Percent*	*Dollars*
C. americanus[1]	96,000	112,000	132,000	5	98	[2]61	1.25–2.20
C. arboreus	48,000	----------	50,000	2	----------	----------	20.00
C. crassifolius	33,000	53,000	65,000	3	92	92	----------
C. cuneatus	36,000	49,000	56,000	3	97	96	----------
C. diversifolius	----------	84,000	----------	1	----------	----------	----------
C. impressus	----------	111,000	----------	1	99	82	----------
C. integerrimus	58,000	----------	81,000	2	----------	----------	----------
C. oliganthus	62,000	----------	73,000	2	94	92	----------
C. prostratus	37,000	----------	42,000	2	85	76	----------
C. rigidus	----------	72,000	----------	1	----------	----------	----------
C. sanguineus	----------	128,000	----------	1	----------	----------	----------
C. sorediatus	121,000	----------	122,000	2	97	98	----------
C. thyrsiflorus	48,000	----------	151,000	2	94	88	4.00
C. velutinus	71,000	----------	152,000	2	96	----------	[3].45

[1]One hundred pounds of *C. americanus* fruits yield 30 pounds of cleaned seed on the average. Comparable information for other species is not available.
[2]Light-colored seed seem to be abortive in most cases.
[3]Per packet.

TABLE 54.—*Ceanothus: Dormancy and methods of seed pretreatment*

Species	Seed dormancy		Pretreatment			Remarks
	Kind	Occurrence	Medium	Temperature	Duration	
				°F.	*Days*	
C. americanus	Embryo	General	Moist sand	41	60	Recommend soaking in hot water over night, then stratifying 60 days.
	Seed coat	Variable within lots	Hot water	170–212	0.50	
C. arboreus	Seed coat	General	do	175–195	0.50	Stratification does not increase germination.
C. crassifolius	Embryo	Probably general	Moist sand	41	90	Recommend soaking in hot water over night, then stratifying 90 days.
	Seed coat	do	Hot water	160	0.50	
C. cuneatus	Embryo	Probably variable	Moist sand	41	90	Hot water treatment alone gives good results; with stratification results are improved moderately.
	Seed coat	General	Hot water	160	0.50	
C. diversifolius	Embryo	General	Moist sand	41	60	With 90 days stratification, some seeds germinate during stratification.
	Seed coat	do	Hot water	170–212	0.50	
C. impressus	Embryo	do	Moist sand	41	60	Hot water alone or stratification alone gave only about half as good germination as the combination.
	Seed coat	do	Hot water	170–212	0.50	
C. integerrimus	Embryo	Probably general	Moist sand	41	90	Soak in hot water first; boiling 1 minute alone gives fair results.
	Seed coat	do	Hot water	175	0.50	
C. oliganthus	Seed coat	General	do	175	0.50	----------
C. prostratus	Embryo	Probably general	Moist sand	41	90	Soak in hot water first, then stratify.
	Seed coat	do	Hot water	170–212	0.50	
C. rigidus	Seed coat	Probably general	do	160	0.50	Stratification does not improve germination. Boiling 1 minute gives fair results.
C. sorediatus	Embryo	Probably general	Moist sand	41	90	Boil seeds in water 5 minutes, then stratify 90 days.
	Seed coat	do	Hot water	212	0.003	
C. thyrsiflorus	Embryo	Probably variable	Moist sand	36	90	Stratification after hot water treatment increases germination slightly. Soaking H_2SO_4 for 1 hour is not as good as hot water treatment.
	Seed coat	Probably general	Hot water	160	0.50	
C. velutinus[1]	Embryo	General	Moist sand	41	90	Soak in hot water, then stratify 90 days. Boiling in water for 5 minutes plus stratification is nearly as good.
	Seed coat	do	Hot water	175	0.50	

[1] No information is available for *C. sanguineus* but it is probably safe to assume that hot water treatment plus stratification in moist sand for 90 days at 41° F. will induce satisfactory germination.

TABLE 55.—*Ceanothus: Recommended conditions for germination tests and summary of germination data*

Species	Test conditions recommended			Germination data from various sources				
	Temperature		Duration	Germinative capacity			Potential germination	Basis, tests
	Night	Day		Low	Average	High		
	°F.	*°F.*	*Days*	*Percent*	*Percent*	*Percent*	*Percent*	*Number*
C. americanus	68	86	30		32		84	1
C. arboreus			40	61		90		3+
C. crassifolius			90			[1]76		1+
C. cuneatus			90			[2]92		1+
C. diversifolius			[3]60			61		1+
C. impressus			30			[4]73		1+
C. integerrimus			20			85		1+
C. oliganthus			70			62		1+
C. prostratus			30			71		1+
C. rigidus			60	77		92		2+
C. sorediatus			30			[5]100		1+
C. thyrsiflorus			60			[6]83		1+
C. velutinus			30	68		73		2+

[1]Germination of seed treated with hot water alone was 48 percent.
[2]Germination of seed treated with hot water alone was 60–81 percent.
[3]Suggested period; experimental data not complete.
[4]Germination of seed treated with hot water alone was 40 percent; that of seed stratified only was 32 percent.
[5]Germination of seed treated with hot water alone was 38 percent.
[6]Germination of seed treated with hot water alone was 73 percent; that of seed soaked for 1 hour in sulfuric acid was 46 percent.

effects of high temperatures in inducing germination are shown in nature by the abundant seedlings following fires in such species as *C. cuneatus* and *C. velutinus*. Germination is epigeous in *C. americanus* (fig. 62); information for other species is lacking. Details as to germination characteristics of 13 species of *Ceanothus* are given in table 54.

Germination tests should be made in sand flats using 400 to 1,000 properly pretreated seeds per test. Seedlings are subject to damping-off which can be controlled by sprinkling cupric oxalate on the seed flats just after the seeds have been sown. Recommended procedure and results vary with the species as shown in table 55.

NURSERY AND FIELD PRACTICE.—Little information on nursery practice for the ceanothus species is available. Spring sowing of properly pretreated seed is probably advisable. Horticultural varieties are also propagated by cuttings of mature wood in the autumn (placed in coldframe or greenhouse), softwood cuttings taken from forced plants in early spring, layering, and by grafting onto roots of *C. americanus*. Ceanothus species will grow in most soils but do best in those that are light and well-drained. Most California species prefer sunny locations.

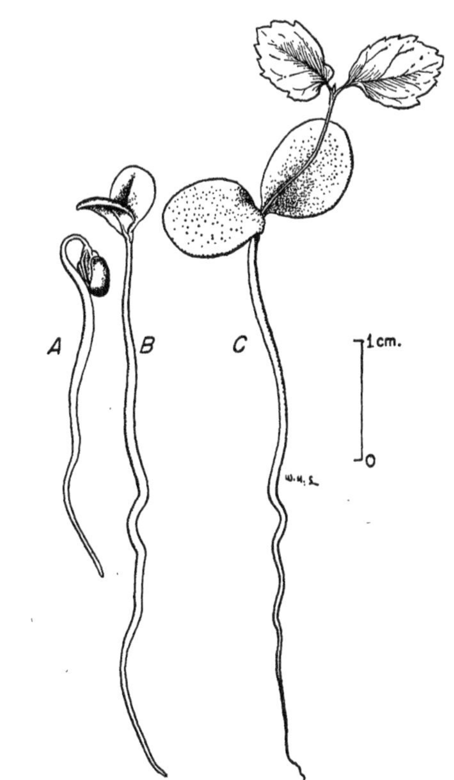

FIGURE 62.—Seedling views of *Ceanothus americanus: A,* At 1 day; *B,* at 5 days; *C,* at 15 days.

CEDRUS Trew True cedars

(Pine family—Pinaceae)

DISTRIBUTION AND USE.—The true cedars, consisting of three or four closely related species of large evergreen trees, are native to the Syrian Mountains, the Himalayas, the Atlas Mountains, and Cyprus. The oily, sweet-scented wood is very durable, and is an important source of timber in the Himalayas and North Africa. In addition an oil is obtained from the distillation of the wood. Three species of *Cedrus,* all of which are planted to some extent in the United States, are described in table 56.

Hardy races of *Cedrus libani* can be grown as far north as Massachusetts; this species has been known for its durable wood since Biblical times. *C. atlantica* can be grown as far north as New York in sheltered positions, while *C. deodara* can be grown safely only in California and the Southern States.

SEEDING HABITS.—The male and female flowers of the true cedars are borne separately on the same or on separate trees. The male flowers, which bear pollen only, are in upright, cylindrical catkins about 2 inches long. The female flowers, which develop into cones, are small upright, ovoid bodies, greenish or purplish in color, and about one-half inch long and one-fourth inch in diameter. Although pollination takes place in the fall, the cones do not begin to grow until the following spring and do not attain full development until the second or sometimes the third year. The mature, barrel-shaped cones are brown, large (2 to 5 inches long), erect, and resinous; are borne on short, stout stalks, and are characterized by numerous closely appressed, very broad scales, each containing two seeds. The cones break up from fall to spring following maturity, leaving the central axis on the tree as in *Abies.* The mature seed is rather soft and oily and is irregularly triangular in shape with a membranous, broad wing several times larger than the seed (fig. 63). Available information as to the comparative seeding habits of the three species discussed is as follows:

Time of—	*C. atlantica*	*C. deodara*	*C. libani*
Flowering	Summer-early fall	Sept.-Oct.	Summer-early fall
Cone ripening	----------	Sept.-Nov.	----------
Seed dispersal	Fall-spring	Sept.-Dec.	Fall-spring

The commercial seed-bearing age for *C. deodara* begins from 30 to 45 years; good seed crops are borne every 3 years with light crops in the intervening years.

There is little information available as to the development of climatic races in *Cedrus.* It is

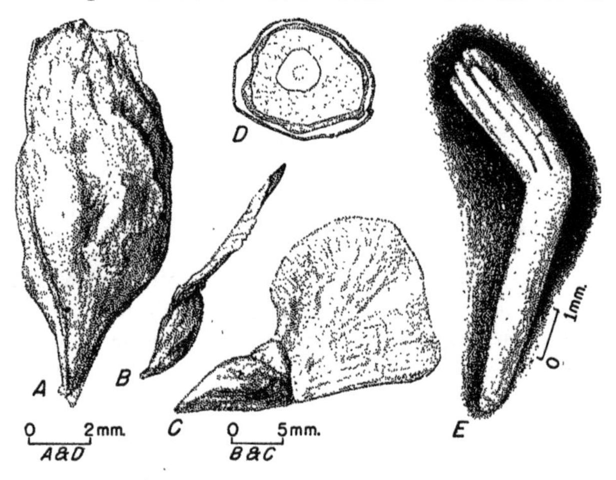

FIGURE 63.—Seed views of *Cedrus libani: A,* Exterior view; *B* and *C,* exterior views from two planes, seed with wings; *D,* cross section; *E,* embryo.

TABLE 56.—*Cedrus: Distribution and uses*

Accepted name	Synonyms	Natural range	Chief uses	Date of earliest cultivation
C. atlantica Manetti (Atlas cedar).	*C. libani* var. *atlantica* Hook. f. (Algerian cedar).	Atlas Mountains in Algeria and Morocco.	Timber, ornamental purposes.	Before 1840
C. deodara (Roxb.) Loud. (deodar cedar).	*C. libani* var. *deodara* Hook. f. (Himalayan cedar).	Western Himalayas from Afghanistan to Garhwal in temperate zone at moderate to high elevations.	Timber, railroad ties, furniture, oil, ornamental purposes.	1831
C. libani Loud. (Cedar-of-Lebanon).	*C. libanotica* Link, *C. cedrus* Huth.	Asia Minor to northern Africa.	Timber, ornamental purposes.	1638

TABLE 57.—*Cedrus: Yield of cleaned seed, and purity and cost of commercial seed*

Species	Cleaned seed per pound				Commercial seed	
	Low	Average	High	Basis, samples	Purity	Cost per pound
	Number	*Number*	*Number*	*Number*	*Percent*	*Dollars*
C. atlantica	3,400	5,600	7,800	17	89	2.50–4.50
C. deodara	2,300	3,600	4,300	170	85	2.50–4.00
C. libani	2,500	4,900	6,800	9	87	2.50–4.00

known that stock of *C. libani* grown from seed collected at the highest elevations where the species occurs in Asia Minor has proved hardy in Massachusetts whereas other sources tried must be grown farther south. It is possible that similar racial development might be found for *C. deodara* which grows under a considerable range of climatic conditions.

COLLECTION, EXTRACTION, AND STORAGE.—Cones can be gathered from standing or felled trees in the fall just before they are completely ripe. Small collections of seed may also be made by raking up fallen cone scales beneath the trees. The cones may be opened by soaking them for 48 hours in warm water (in *Cedrus atlantica* and *C. libani*) or by drying them in the sun (*C. deodara*). After the scales are dry they can be put through a screen-covered shaker to remove the seed. The size and purity of commercial cleaned seed vary according to the species as is shown in table 57.

Cedrus seeds are oily, and under ordinary storage conditions do not keep well. It is reported that the entire cones of *C. atlantica* can be stored dry over winter with satisfactory results. No information is available as to the keeping qualities of *Cedrus* seed in cold storage. Ordinarily the seed should be sown not later than the spring following ripening.

GERMINATION.—Natural germination of *Cedrus* seed usually takes place the first spring following seed dispersal and is usually best on mineral soil or in ashes following a fire. Germination is epigeous (fig. 64). The natural seed supply is considerably reduced by the activities of cone weevils (*Euzophera cedrella* in *C. deodara*) and birds. The seedlings need full light but are highly susceptible to drought damage. The seeds do not exhibit dormancy and consequently need no pretreatment to induce germination.

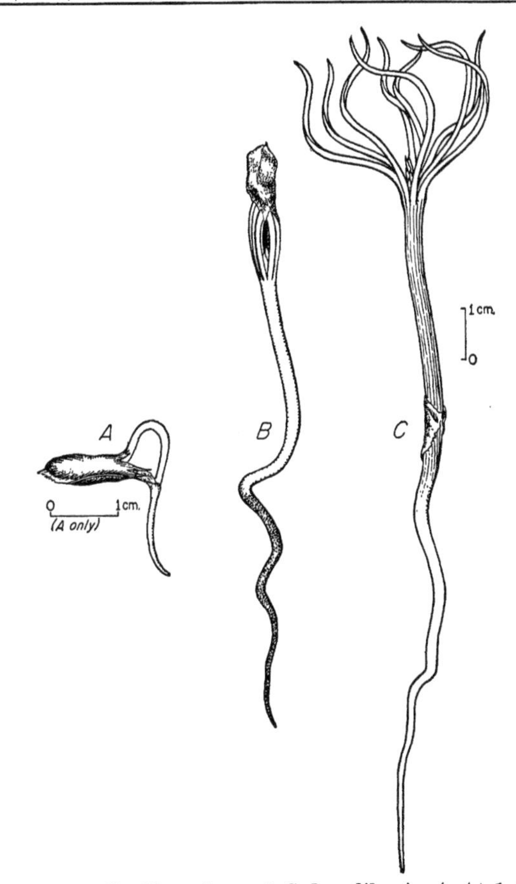

FIGURE 64.—Seedling views of *Cedrus libani*: *A*, At 1 day; *B*, at 4 days; *C*, at 8 days.

The germination of *Cedrus* seed should be tested in sand flats or standard germinators using 500 to 1,000 seeds per test. Methods recommended and average results for the 3 species are given in table 58.

TABLE 58.—*Cedrus: Recommended conditions for germination tests, and summary of germination data*

Species	Test conditions recommended[1]			Germination data from various sources					
	Temperature		Duration	Germinative energy		Germinative capacity			Basis, tests
	Night	Day		Amount	Period	Low	Average	High	
	°F.	*°F.*	*Days*	*Percent*	*Days*	*Percent*	*Percent*	*Percent*	*Number*
C. atlantica	68	86	30–40	27	20	9	43	74	17
C. deodara	68	86	30–40	43	20	13	79	98	89
C. libani	68	86	30	23	20	13	42	92	9

[1] Suggested treatment; experimental data not complete.

891183°—50——9

CEDRUS

Nursery and Field Practice.—*Cedrus* seed should be sown in the fall as soon after collection as possible in drills 4 to 6 inches apart and covered with about one-half inch of soil. It has been recommended that the seed be soaked for 2 to 3 hours in water before sowing. In the nursery the seedlings are often attacked by white grubs and cutworms. Precaution should also be taken against damping-off. *Cedrus*[18] species may also be propagated by veneer grafting or by cuttings of adventitious shoots.

[18] For *Cedrus deodara* it is recommended in India that ½–1½ stock be used for field planting and that planting be confined to cool, damp situations. Direct seeding using 20 to 25 pounds of seed per acre is also practiced in India with this species.

CELASTRUS SCANDENS L. American bittersweet

(Staff-tree family—Celastraceae)

Also called climbing bittersweet, false bittersweet, shrubby bittersweet, waxwork.

DISTRIBUTION AND USE.—American bittersweet[18a] is valuable for ornamental purposes, game food and cover, Christmas decorations, and the bark for medicinal uses. This deciduous woody vine grows in woods and along fence rows from Quebec to North Carolina and westward to Manitoba, Kansas, Oklahoma, and New Mexico. It was introduced into cultivation in 1736.

SEEDING HABITS.—The light to reddish seed are about one-fourth inch long and are borne in fleshy arils, two of which are usually found in each of the two to four cells composing the fruit, a dehiscent capsule (fig. 65.) The polygamo-dioecious or dioecious flowers open from May to June, and the yellow to orange capsules ripen from late August to October. They split open soon thereafter, exposing showy red arils, which may persist throughout much of the winter. Good seed crops are borne almost annually.

[18a] Valuable unpublished information on this species was furnished by W. C. Secrist of the Pennsylvania Department of Forests and Waters, Mont Alto, Pa. 1939 [Letter on file at the Lake States Forest Experiment Station.]

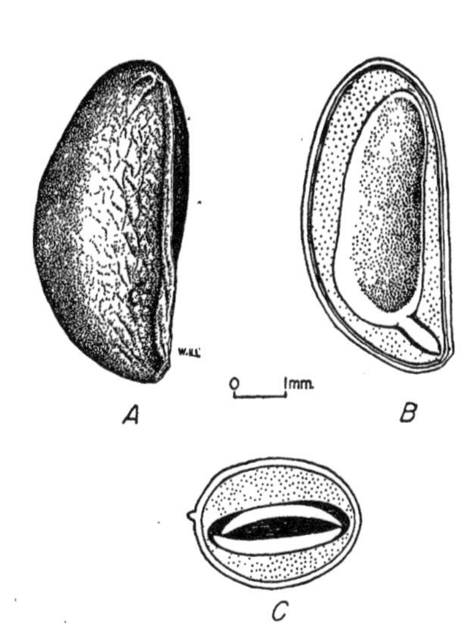

FIGURE 65.—Seed views of *Celastrus scandens: A,* Exterior view; *B,* longitudinal section; *C,* cross section.

COLLECTION, EXTRACTION, AND STORAGE.—The ripe fruit should be collected as soon as the capsules separate, exposing the arils, or from about mid-September as long as they hang on the vines. They should be spread out in shallow layers and allowed to dry for 2 or 3 weeks after harvesting. The scales of the capsules are rather difficult to remove; it is likely they can be broken up by flailing. Running the open fruit with water through a hammer mill might also prove practicable. Except for removing the scales, it is not essential to extract the seed from the fleshy arils which are often dried and then sown. Removing the pulp costs from $0.75 to $1 per pound based on labor at 40 cents per hour and no special equipment. Cleaned seed: number per fruit, 4 to 8;

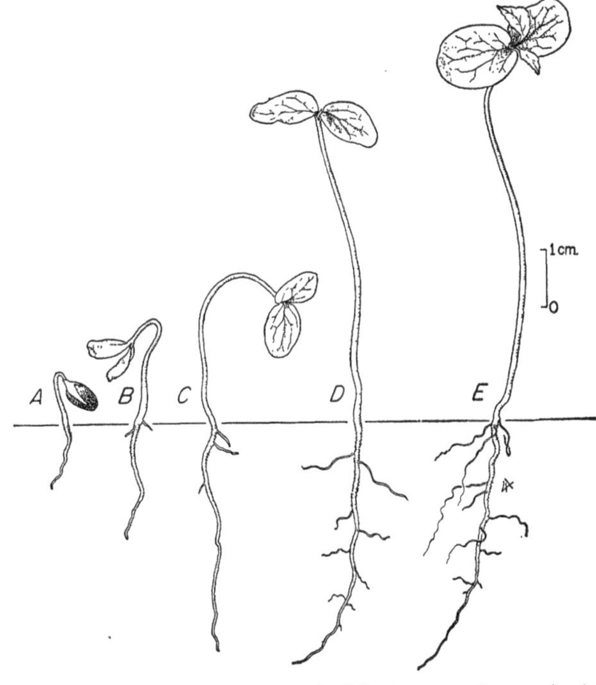

FIGURE 66.—Seedling views of *Celastrus scandens: A,* At 1 day; *B,* at 2 days; *C,* at 5 days; *D,* at 10 days; *E,* at 39 days.

number per pound (10 samples): low, 12,000; average, 26,000; high, 40,000. Average purity, 98 percent; average soundness, 85 percent. Cost of commercial seed: clean, $2 to $3 per pound; dried fruit, about $1 per pound. Few data are available on storage. However, a test made by the Lake States Forest Experiment Station on cleaned seed which had been stored for 20 months in a sealed jar at 41° F. showed 9 percent germination in the spring of 1940 but only 0.2 percent in August of the same year, and none the following year.

GERMINATION.—Seed of American bittersweet have a dormant embryo and thus require after-ripening for germination. Although there is some evidence that the seed coat may have an inhibiting effect on germination, good results may be obtained simply by fall sowing or by stratification in moist sand or peat for 3 months at 41° F. It seems to make little difference whether cleaned seed or dried fruit is sown; however, it appears that both cleaned seed and fruit should be dried at room temperature for 2 or 3 weeks after harvesting before they are sown. Germination is epigeous (fig. 66). Test methods recommended: Sand flats; temperature alternating diurnally from 50° to 77° F.; seeds required, 400; duration, 30 days. Germinative capacity, stratified seed (6 tests): low, 9 percent; average, 47 percent; high, 80 percent. Potential germination, 9 to 93 percent.

NURSERY PRACTICE.—Seed should be sown either clean or in the dried pulp in the fall, or stratified in January and sown in the early spring. Young seedlings are somewhat susceptible to damping-off. American bittersweet grows well on a variety of soils, either in the sun or in the shade. Propagation by root cuttings, layers, or stem cuttings is also sometimes practiced.

CELTIS L. Hackberry

(Elm family—Ulmaceae)

Distribution and Use.—The hackberries include about 70 species of trees or rarely shrubs of the Northern Hemisphere which are deciduous except for a few occurring in tropical and subtropical regions. They are useful chiefly for ornamental purposes when not too disfigured by broomlike, insect galls; for shelter-belt planting and to some extent for wildlife food. Only a very few species have received attention by landscape gardeners and forest planters. Two species are described in table 59. The species of the genus as a whole are relatively unimportant economically.

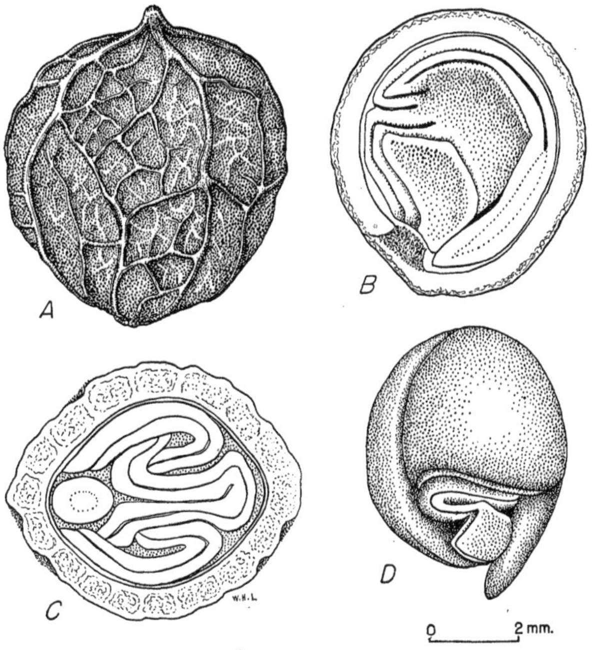

Seeding Habits.—Flowers of the hackberries, sometimes perfect and sometimes unisexual on the same tree, appear with the leaves in the spring. Fruit ripens in the fall of the current year. Commercial seed consists of the fruit, a spherical drupe with thin pulp containing a single bony smooth or pitted nutlet in the American species (figs. 67 and 68). Most species are prolific seeders, the fruits often remaining on the trees until midwinter. *Celtis laevigata* and *C. occidentalis* are in flower from April to May; fruit ripening occurs from September to October; and the seed is dispersed from fall to winter. These two species bear good seed crops most years and light seed crops in intervening years.

Figure 67.—Seed views of *Celtis occidentalis: A,* Exterior view. *B,* longitudinal section; *C,* cross section; *D,* embryo.

Figure 68.—Seed views of *Celtis laevigata.*

Table 59.—*Celtis: Distribution and uses*

Accepted name	Synonyms	Natural range	Chief uses	Date of earliest cultivation
C. laevigata[1] Willd. (sugarberry).	*C. mississippiensis* Bosc (sugar hackberry, SPN).	Virginia to Florida, west to Mexico and New Mexico, north in Mississippi Valley to Indiana, Missouri, and Kansas.	Ornamental and shelter-belt planting, bird food.	1811
C. occidentalis[2] L. (hackberry).	*C. crassifolia* Lam. (common hackberry, SPN; sugarberry, nettletree, false elm).	Quebec to Manitoba, south to North Carolina, Alabama, and Oklahoma.	Shelter-belt planting, fuel, bird food; formerly important Indian food.	1656

[1] Includes var. *texana* (Scheele) Sarg. and var. *smallii* (Beadle) Sarg.
[2] Includes var. *canina* (Raf.) Sarg. and var. *crassifolia* (Lam.) Gray.

CELTIS

COLLECTION, EXTRACTION, AND STORAGE.—The fruit may be picked from the trees by hand or stripped onto canvas as soon as it has turned orange red in color in case of *Celtis laevigata* and dark purple or black in case of *C. occidentalis*. It may be collected as late as midwinter. Collection is more efficient after leaves have fallen. Average cost per pound for gathering seed is 25 to 50 cents. Fruits may or may not require drying, depending upon time of collection. Although no cleaning except to remove twigs and other debris by screening or fanning is really necessary, the pulp may be removed by running the fruit with water through a hammer mill, a process costing approximatly 2½ cents per pound of seed. The number of cleaned fruits per pound varies: *C. laevigata,* 2,000 to 2,400 (2 samples); *C. occidentalis* (22 samples): low, 1,580; average, 2,050; and high, 2,380. Yield, size, purity, and soundness for these 2 species are as follows:

	C. laevigata	*C. occidentalis*
Cleaned seed:		
Per 100 pounds fruit...pounds..	50–75	40–75
Per pound:		
Low...number..	3,700	3,500
Average...do....	4,400	4,300
High...do....	5,600	5,400
Basis, samples...do....	5	12+
Commercial seed:		
Purity...percent..	98	99
Soundness...do....	94	95
Cost per pound...dollars..	0.40–1.50	0.40–1.50

Seed may be stored equally well as dried fruit or as cleaned seed. It is better to store seed in sealed containers, although dried fruits of *Celtis laevigata* in cloth sacks at 40° F. dropped in germinative capacity only from 50 to 40 percent in 3 years. Dried fruits of *C. occidentalis* have been stored at 41° in sealed containers for 5½ years without loss of viability. It is recommended that either dried fruits or clean seed be stored in sealed containers at 35° to 40°. Fruits collected in late fall or winter usually do not need further drying before storage. For short periods, 12 to 16 months, seed may be placed in moist sand at low temperatures without serious loss of viability.

GERMINATION.—Hackberry seed germinates early in the spring after lying on the ground over winter. Germination is epigeous (fig. 69). Best natural conditions for germination are a rich, moist loamy soil. The seed has a dormant embryo and possibly an impermeable seed coat, although the evidence for the latter is not conclusive. Dormancy may be broken by stratification in moist sand at 41° F. for 60 to 90 days. Stratification hastens, increases, and makes germination for *Celtis laevigata* and *C. occidentalis* more uniform; fruits with macerated pulp respond more quickly to stratification than do fruits with their outer rather firm skins intact.

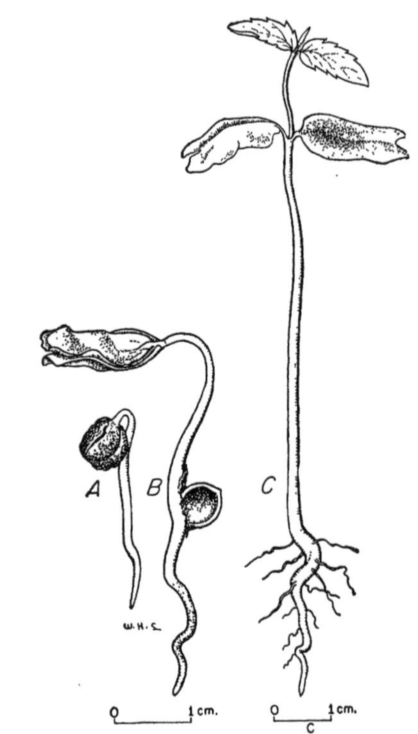

FIGURE 69.—Seedling views of *Celtis laevigata: A,* At 1 day; *B,* at 2 days; *C,* at 5 days.

Germination tests may be made on 100 to 200 seeds in flats of sand, sand and peat, or sandy loam soil. Test conditions recommended for *Celtis laevigata* and *C. occidentalis* are temperatures of 68° (night) and 86° F. (day) for 60 days in the case of stratified seed and 60 to 90 days if untreated. Results that may be expected are:

	C. laevigata	*C. occidentalis*
Basis, tests...number....	6+	18
Germinative energy:		
Amount...percent....	30–50	19–75
Period...days....	25–30	13–41
Germinative capacity:		
Low...percent....	35	3
Average...do......	55	41
High...do......	80	77

NURSERY AND FIELD PRACTICE.—Hackberry seed may be sown in the nursery in the fall or, if stratified, in the spring. Beds should be well prepared before seed are sown in rows 8 to 10 inches apart and covered with about one-half inch of firmed soil. Density of sowing in row should be approximately three times the desired density of seedlings. Seeded beds should be mulched with straw or leaves which should be held in place with bird screens until time for germination in spring. Surface soil must be kept moist during period of germination. *Celtis laevigata* and *C. occidentalis* may be propagated by cuttings.

CEPHALANTHUS OCCIDENTALIS L. Common buttonbush

(Madder family—Rubiaceae)

Also called button willow, button tree.

Distribution and Use.—The common buttonbush grows on wet lands from New Brunswick to Florida, westward to southern Minnesota, Nebraska, Oklahoma, southern New Mexico, California, and Mexico. It occurs also in Cuba and eastern Asia. Over most of this wide range it is a shrub, but in the Southern States and the Southwest it becomes a small tree. It is useful as food for birds, for ornamental purposes, and as a honey plant. The date of earliest cultivation is 1735.

Seeding Habits.—Commercial "seed" are dry, podlike, somewhat conical-shaped fruits about one-eight inch long (fig. 70), which separate eventually from the base into two, sometimes three or four, one-seeded nutlets. The fruits are borne in globular heads or balls about three-fourths inch long. Perfect cream-white fragrant flowers open from June to September and the brown, densely packed heads of fruit ripen from September to October. Sometimes they are only partly developed and gradually fall to pieces in the late autumn and winter months; some fruit may persist until spring. Common buttonbush is a prolific seeder.

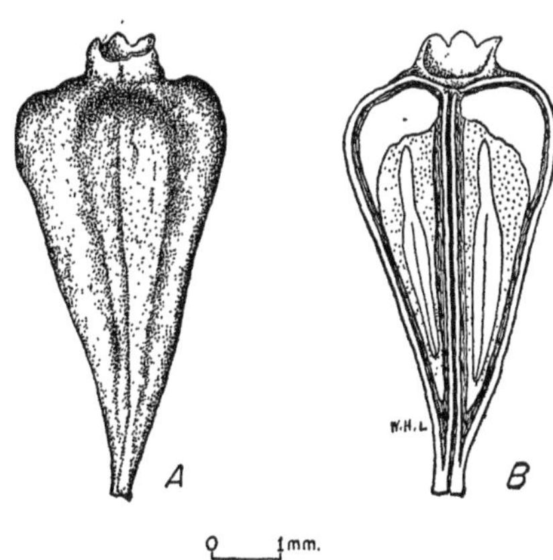

Figure 70.—Seed views of *Cephalanthus occidentalis: A,* Exterior view of fruit; *B,* longitudinal section showing two nutlets.

Collection, Extraction, and Storage.—The fruit balls may be collected as long as they hang on the branches, and collection should begin as soon as they turn brown and show signs of disintegrating. If perfectly dry, light flailing will break them up into the component fruits. Four samples showed 118,000 to 160,000 (average 134,-000) commercial "seeds" per pound, or about twice that number of actual seeds. Purity was 96 percent; soundness is rather difficult to evaluate (95 percent in 1 sample). Cost of commercial "seed," $0.85 to $1.90 per pound. Very little is known about proper storage conditions for this seed. Seed stored in wax-sealed jars at 41° F. showed 6 percent germination at the end of 4 years. Since the original germination was not known, it cannot be stated whether those conditions are optimum, but since the germination was one of the best values so far obtained, it can be concluded that storage probably does not reduce the viability of the seed.

Germination.—Buttonbush seed germinates without special treatment, but germination which is epigeous (fig. 71) is very low. The seed coats appear to be readily permeable to water. Tests may be conducted in sand flats at a temperature fluctuating diurnally from 70° to 85° F. Seeds required: at least 1,000 (about 500 fruits); test duration, 30 to 40 days. Germinative capacity (4 samples): low, 1 percent; high, 7 percent.

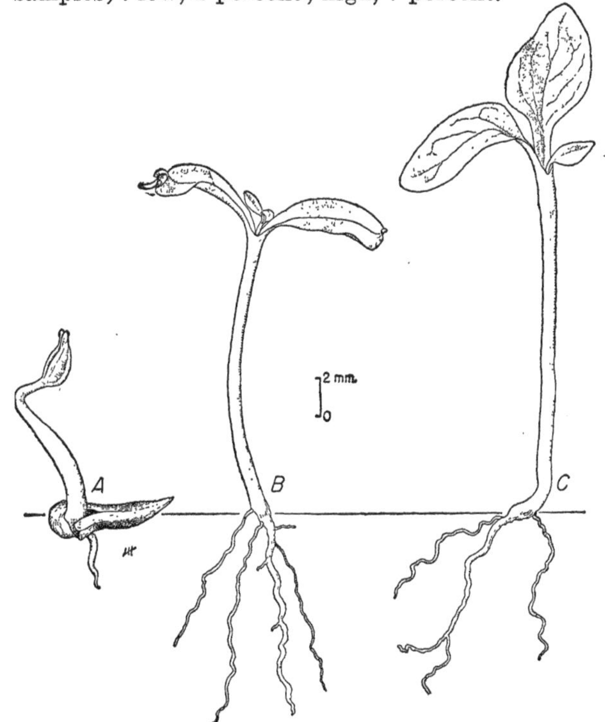

Figure 71.—Seedling views of *Cephalanthus occidentalis: A,* at 1 day; *B,* at 23 days; *C,* at 49 days.

Nursery and Field Practice.—Buttonbush can be propagated in the nursery by means of seeds or by cuttings. It grows best on sandy, rather moist soils, but thrives on any good garden soil.

CERCIS CANADENSIS L. Eastern redbud

(Legume family—Leguminosae)

Also called redbud, Judas-tree.

Distribution and Use.—Occurring from New York southward to Florida, westward to the valley of the Brazos River in Texas, northward through eastern Oklahoma, Kansas, and Nebraska to southern Ontario, eastern redbud is a small deciduous tree which grows in rich soils along stream borders and bottom lands, in forest understory, and on cut-over lands. The best development occurs in southwestern Arkansas, southeastern Oklahoma, and eastern Texas. Eastern redbud is valuable for game food, for landscaping because of its showy pink or rose-colored flowers and its foliage, and as a honey plant. It has been cultivated since 1641.

Seeding Habits.—Showy pink, perfect flowers occur from March 1 to May 15. The fruits, which mature in September and October of the first year, are flat, thin pods about 3 inches long, each containing several flattened seeds (fig. 72). These pods often remain on the tree until midwinter. Although some open on the tree, others fall to the ground and usually open after weathering or disturbance by birds and animals. Seeds are dispersed short distances by wind, but mainly by birds and animals. Age for commercial seed production: Minimum, 5 years; optimum, 20 to 40 years; maximum, 50 to 75 years. Average good seed crops occur in alternate years with intervening light crops.

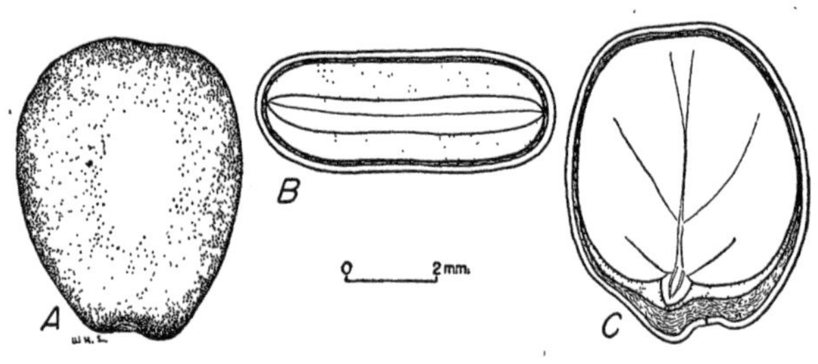

Figure 72.—Seed views of *Cercis canadensis*: *A*, Exterior view; *B*, cross section; *C*, longitudinal section.

Collection, Extraction, and Storage.—Seed collection may begin when pods and seed turn brown, and continue through September, October, and November. Most efficient methods of collection are hand picking from standing trees or shaking or flailing the pods onto canvas, after which they are placed in loosely woven sacks or spread

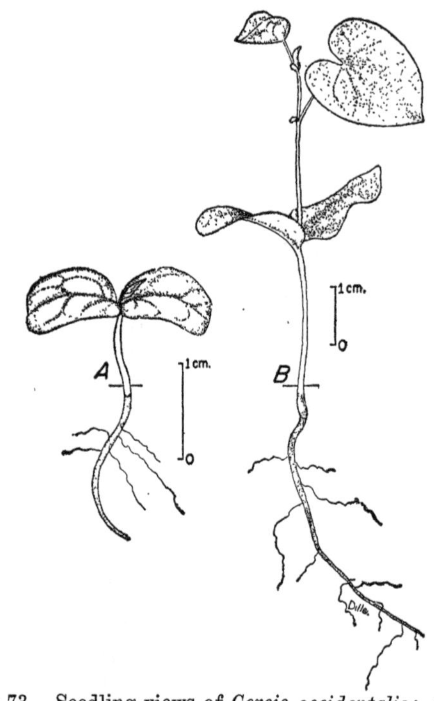

Figure 73.—Seedling views of *Cercis occidentalis*: *A*, Young seedling; *B*, seedling about 1 month old.

out to air-dry. Seeds should be threshed from the dried pods and separated from the chaff by screening and fanning. Approximately 20 to 35 pounds of seed are extracted from 100 pounds of fruits. Seed per pound (9 samples): low, 14,000; average, 18,000; high, 25,000. Commercial seed averages in purity about 90 percent; in soundness, about 85 percent; and in cost, from \$1 to \$2 per pound. Soon after cleaning, seed should be stored dry in sealed containers; storage at 41° F. prolongs viability.

Germination.—Rich, well-drained, mineral loamy soils, protected from drying winds, form optimum natural seedbeds in which germination takes place during April and May. Germination is epigeous (fig. 73) and may occur during the first or second spring following seed maturity. Seed dormancy is due to an impermeable coat and in at least some lots also to conditions within the embryo. Lots showing seed-coat impermeability only may be sacrified, soaked in concentrated sulfuric acid, or in hot water; lots with double dormancy may also be given these treatments prior to cold stratification. Best results have been obtained by 30 minutes' soaking in sulfuric acid followed by 60 days' stratification in moist sand at 41° F.

Adequate tests of eastern redbud may be made using 400 pretreated seeds in sand flats at temperatures of 68° (night) to 86° F. (day) for 30 days. Using the recommended pretreatment, germinative energy is 70 to 80 percent in 8 to 14 days, and germinative capacity is 76 to 85 percent (2 tests). Other methods give lower results.

Nursery and Field Practice.—Properly pretreated seed should be sown, preferably drilled, in well-prepared seedbeds during the last of April or early May. The covering of firmed soil should not exceed one-fourth inch and bed surfaces should be kept moist during the germination period. Acceptable results may be obtained from fall sowings of untreated seed. In one case seed collected slightly green in early September and sown immediately germinated 90 percent the following spring. Mulching fall-sown beds with burlap, straw, or other effective loose mulch is necessary, but it should be removed in spring when germination starts. Eastern redbud grows best in rich, sandy, rather moist, loams. Propagation by layering and cuttings is sometimes practiced.

CERCOCARPUS H. B. K. Mountain-mahogany

(Rose family—Rosaceae)

DISTRIBUTION AND USE.—The mountain-mahoganies include about 20 species of evergreen or half-evergreen shrubs or small trees native to the dry interior or mountainous regions of western North America from Oregon and Montana south to southern Mexico. Locally the heavy, close-grained wood is manufactured into small articles or is used for fuel or producing charcoal. Some species are used to a limited extent as ornamentals, and many of them are useful for erosion control on dry mountain slopes. Three species that are of potential value for reforestation, and for which reliable information is available, are described in table 60.

SEEDING HABITS.—The small, greenish white or reddish perfect flowers bloom in the spring and by fall develop into one-seeded achenes—the fruit—distinguished by a feathery style 2 to 3 inches long (fig. 74). The seed has a conspicuous hilum, membranous testa, and an embryo that fills the seed cavity. Seed dispersal is chiefly by wind, occasionally by animals. The comparative seeding habits of the three species of *Cercocarpus* discussed are as follows:

Time of—	*C. betuloides*	*C. ledifolius*	*C. montanus*
Flowering	May	May–June	May–June.
Fruit ripening	June–Sept.	May–Aug.	Aug.–Sept.
Seed dispersal	Late summer	Summer	Summer.

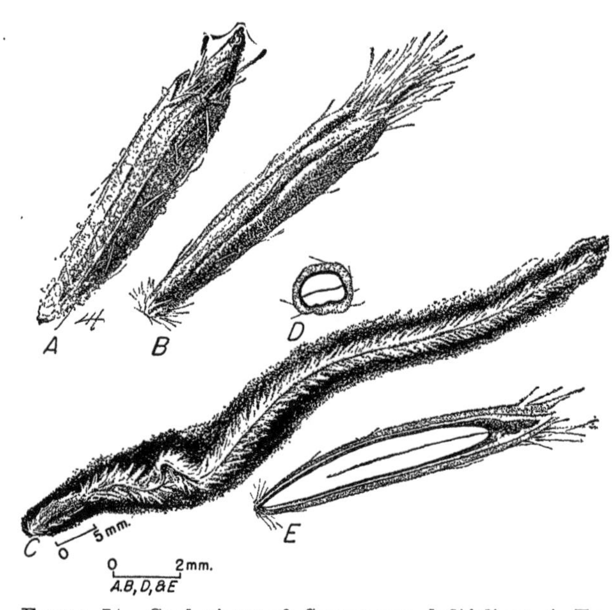

FIGURE 74.—Seed views of *Cercocarpus ledifolius: A*, Exterior view of seed in calyx, style broken off; *B*, exterior view of seed out of calyx, style broken off; *C*, exterior view of complete seed; *D*, cross section; *E*, longitudinal section.

TABLE 60.—*Cercocarpus: Growth habit, distribution, and uses*

Accepted name	Synonyms	Growth habit	Natural range	Chief uses	Date of earliest cultivation
C. betuloides Nutt. (birchleaf mountain-mahogany).	*C. parvifolius* var. *betuloides* (Nutt.) Sarg. (hard tack).	Shrub or small tree.	Dry slopes in mountains of California	Erosion control	[1877].
C. ledifolius Nutt. (curlleaf mountain-mahogany).		Small tree	Dry mountain slopes from Washington and Montana south to California and Arizona.	Erosion control, deer browse.	About 1879.
C. montanus Raf. (true mountain-mahogany).	*C. parvifolius* Nutt.	Shrub up to 6 feet tall.	Dry mountain slopes from Montana and South Dakota, south to Utah, New Mexico, and Kansas.	Possible erosion control, ornamental.	1913.

TABLE 61.—*Cercocarpus: Yield of cleaned seed, and purity and soundness of commercial seed*

Species	Cleaned seed per pound				Commercial seed	
	Low	Average	High	Basis, samples	Purity	Soundness
	Number	*Number*	*Number*	*Number*	*Percent*	*Percent*
C. betuloides		30,000		1		
C. ledifolius	40,000	44,000	50,000	3	49	
C. montanus	40,000		42,000	3	65	80

COLLECTION, EXTRACTION, AND STORAGE.—In the late summer or fall[19] the ripe fruits should be shaken off the shrubs or trees onto canvas. Since the fruits usually are dry at the time of collection no further drying is necessary, but they should be separated from leaves and debris by winnowing. Sometimes the entire fruits are used in nursery sowing; at other times the hairy styles are removed by rubbing over a screen and then fanning.

The size, purity, and soundness of commercial cleaned seed vary according to species as shown in table 61. One unidentified lot (probably *Cercocarpus montanus*) of fruits yielded 22 pounds of cleaned seed per 100 pounds of fruit. One lot of *C. montanus* seed still retained high viability after dry storage for 5 years in burlap bags in a warehouse. Comparable information for the other species is lacking.

GERMINATION.—Germination is epigeous (fig. 75). There appears to be some variability in seed dormancy among the various *Cercocarpus* species. *C. betuloides* seed germinates promptly and well without pretreatment; *C. ledifolius* germinates very slowly without pretreatment; and *C. montanus* indicates variability between lots. Dormancy probably can be overcome by stratifying the seed in moist sand or peat for 30 to 90 days at 41° F.

Germination tests of *Cercocarpus* seed should be made in sand or soil flats for 20 to 60 days (untreated seeds of some species require 250 to 300 days), using 1,000 nondormant or pretreated seeds per test. Temperatures of 68° (night) and 86° F. (day) probably will be suitable. Average results for the 3 species discussed, using untreated seed, are as shown in table 62.

NURSERY AND FIELD PRACTICE.—*Cercocarpus* seed may be sown in the fall, or nondormant seed in the spring, and covered with one-fourth inch of screened sand. If the entire fruits are sown, they should first be soaked in water for 30 minutes; otherwise the long feathery tail of the seed will uncurl as it becomes moist after sowing, and expose the seed. The seedbeds should be kept well moistened from the time of sowing until germination begins. With *C. montanus* seed germination begins from 7 to 30 days after spring sowing, and may be as high as 90 percent. On alkaline soils, damping-off may cause heavy losses; under such conditions only about 2 percent of the seed sown produce usable 2-0 seedlings. For field planting, 2-0 stock is probably most suitable. Ordinarily, *Cercocarpus* species should be planted on well-drained soils in sunny locations. *C. ledifolius* will thrive on locations too dry for most other shrubs. *Cercocarpus* species may also be propagated by cuttings of half-ripened wood under glass.

[19] *Cercocarpus montanus* seed has been collected as late as November in New Mexico.

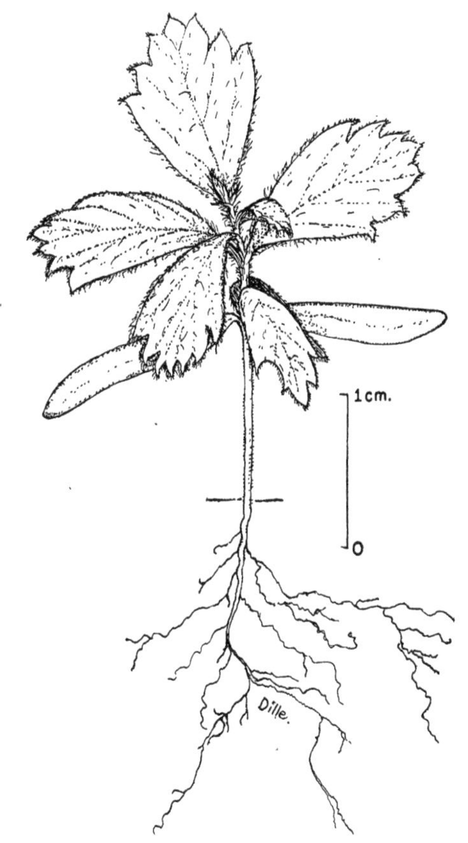

FIGURE 75.—Older seedling of *Cercocarpus betuloides*.

TABLE 62.—*Cercocarpus: Germinative capacity of untreated seed*

Species	Duration of test	Germinative capacity			Basis, tests
		Low	Average	High	
	Days	*Percent*	*Percent*	*Percent*	*Number*
C. betuloides	20	34	--	89	2
C. ledifolius	263	1	29	44	3
C. montanus	30+	3	34	62	4

CHAMAECYPARIS Spach White-cedar

(Pine family—Pinaceae)

Distribution and Use.—*Chamaecyparis* includes six species of medium to large evergreen trees native to North America, Japan, and Formosa. The very durable, nonresinous wood is used for lumber, poles, and posts and sometimes furniture, but the white-cedars are cultivated chiefly for ornamental purposes because of their somber beauty and variety of form. Some species are used for hedges and windbreaks. The three species native to the United States, all of which are planted to some degree, are described in table 63. *C. lawsoniana* is planted in California, the Middle Atlantic States, and the temperate parts of Europe. *C. nootkatensis* is also planted in Europe, and occasionally *C. thyoides.*

Seeding Habits.—The yellow (red in *Chamaecyparis lawsoniana*) male flowers bearing pollen only, and the female flowers which produce cones, are borne on different branches of the same tree; both are minute and inconspicuous. The very small, spherical cones, which stand erect on the branchlets, are one-fourth to one-half inch across, and become reddish brown (bluish purple in *C. thyoides*) and bloomy. They mature in the late summer or early fall of the first year. However, those of *C. nootkatensis* are reported to mature at the end of the second year in some cases. The cones open slowly to release the seed and sometimes remain on the tree for another year. Each cone has from 6 to 12 scales; each scale bears from 1 to 5 minute, slightly compressed seeds which have broad, gauzy wings. The seed coat consists of 2 layers: the outer thin and membranous, and the inner thick and hard (figs. 76 and 77). Comparative seeding habits of the 3 species discussed are given in table 64. In *C. thyoides,* seed are produced prolifically and good crops are borne almost annually; mature stands may release 8 or 9 million seed per acre in a season. *C. lawsoniana* bears good seed crops every 4 to 5 years; good seed crops of *C. nootkatensis* are only occasional.

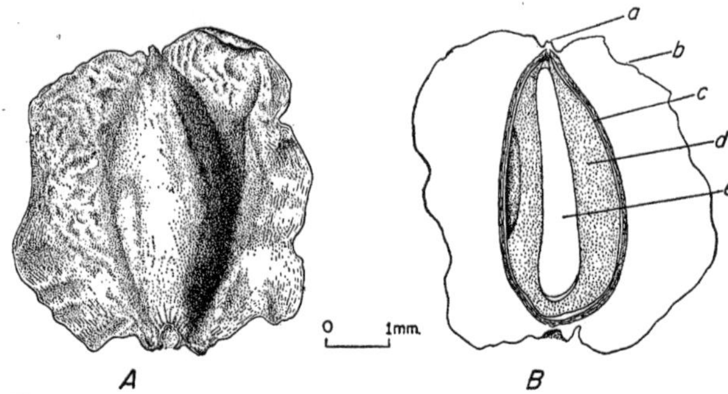

Figure 76.—Seed views of *Chamaecyparis lawsoniana*. *A,* Exterior view. *B,* Longitudinal section: *a,* Micropyle; *b,* seed coat; *c,* nucellus; *d,* endosperm; *e,* cotyledons.

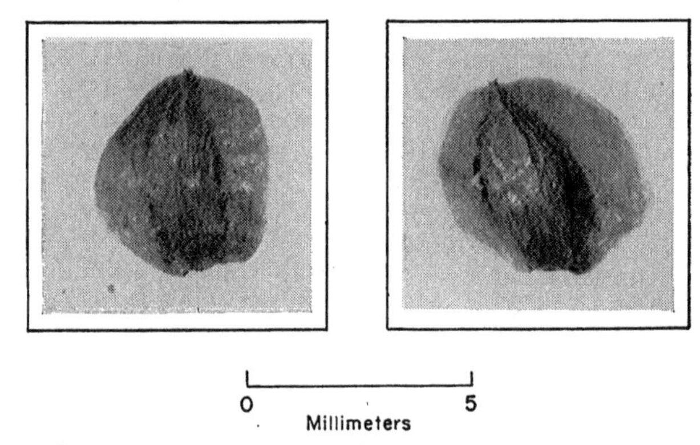

Figure 77.—Seed views of *Chamaecyparis nootkatensis.*

Table 63.—*Chamaecyparis: Distribution and uses*

Accepted name	Synonyms	Natural range	Chief uses	Date of earliest cultivation
C. lawsoniana (A. Murr.) Parl. (Port-Orford-cedar).	*Cupressus lawsoniana* A. Murr. (Port Orford cedar, Lawson cypress).	Southwestern Oregon and northwestern California.	Lumber, flooring, boatbuilding, ties, furniture, posts; windbreaks; ornamental.	1854
C. nootkatensis (D. Don) Spach (Alaska-cedar).	*Chamaecyparis nutkatensis* Spach (Nootka cypress, yellow cypress).	Pacific coast from Alaska to northern Oregon.	Boatbuilding, general carpentry, cabinet work; ornamental; wildlife food and cover; erosion control.	1853
C. thyoides (L.) B. S. P. (Atlantic white-cedar).	*Cupressus thyoides* L., *Chamaecyparis sphaeroidea* Spach (white cedar, southern white cedar).	Fresh water bogs and swamps of Atlantic Coastal Plain from Maine to Mississippi.	Poles, posts, shingles, lumber, mine timber, ornamental.[1]	1727

[1] *C. thyoides* is the hardiest but least ornamental species in the genus.

TABLE 64.—*Chamaecyparis: Time of flowering and cone ripening, and commercial seed-bearing age*

Species	Time of—			Seeds per cone scale	Cone diameters	Commercial seed-bearing age		
	Flowering	Cone ripening	Seed dispersal			Minimum	Optimum	Maximum
				Number	*Inches*	*Years*	*Years*	*Years*
C. lawsoniana	Spring	Sept.–Oct.	October	2–4	¼	8	100	
C. nootkatensis	Early spring	Fall–spring		2–4	½			
C. thyoides	March–April	Fall	Oct.–late fall	1–2	¼	4	20+	(1)

1 From maturity until death.

No scientific information is available as to the development of climatic races in *Chamaecyparis.* That there is tremendous variation within the individual species is shown by the fact that there are recognized some 70 varieties of *C. lawsoniana,* 15 varieties of *C. nootkatensis,* and 12 varieties of *C. thyoides.* It is quite likely that climatic races have developed, and care should be taken to obtain seed from trees of the form desired.

COLLECTION, EXTRACTION, AND STORAGE. — The ripe cones should be collected by hand from standing or felled trees in the fall before seed dispersal has advanced far, and spread out to dry in the sun, in a warm room, or in a kiln. A temperature of 90° to 110° F. is suitable for *Chamaecyparis lawsoniana. Chamaecyparis* seed are easily injured and should not be dewinged. The yield, size, purity, soundness, and cost of cleaned commercial seed vary considerably by species as shown in table 65.

No scientific data are at hand concerning longevity of *Chamaecyparis* seed in storage. It is known that seed of *C. lawsoniana* spoil quickly under ordinary storage conditions. Seeds of *C. thyoides* stored naturally in the peaty forest floor have been known to retain their viability for at least 1 year. If *Chamaecyparis* seed must be stored, best results will probably be given by dry, sealed storage, at temperatures near freezing.

GERMINATION.—Natural germination is best on moist surfaces—mineral soil, peat, muck, or rotten logs—in partial shade. *Chamaecyparis lawsoniana* and *C. thyoides* reproduction come in well after logging and the former also comes in after fires. Although *Chamaecyparis* seed are usually tested without pretreatment, germination characteristically is low. This is due in part to poor seed quality, but it also appears to be due partly to embryo dormancy for at least some species. Stratification in moist sand at temperatures just above freezing should overcome such dormancy. It is recommended that such stratification be carried out for 60 to 90 days at 41° F. with *C. nootkatensis* seed; shorter periods will probably suffice for the other species.

Recommendations are to test the germination of *Chamaecyparis* seed in sand flats or standard germinators at 68° F. (night) to 86° F. (day) for 60 days, using 1,000 pretreated seeds per test. Average results with untreated seeds of the 3 species discussed are given in table 66.

NURSERY AND FIELD PRACTICE.—Recommendations are available for *Chamaecyparis lawsoniana* only. Seed of this species usually is sown broadcast in the spring[20] and should be covered with one-fourth inch of soil and given half shade to midseason of the first year. Germination usually takes place 6 to 8 weeks after sowing and is about 50 percent of the laboratory germination. The seedlings are moderately susceptible to damping-off; they are also attacked occasionally by cedar blight caused by *Phomopsis juniperovora* which can be controlled partially by frequent spraying with 4-4-50 bordeaux, colloidal sulfur, or normal semesan. *Chamaecyparis* species can be propagated quite readily by cuttings.

20 Fall sowing or the use of stratified seed in the spring might be better, especially for the other species.

TABLE 65.—*Chamaecyparis: Yield of cleaned seed, and purity, soundness, and cost of commercial seed*

Species	Cleaned seed					Commercial seed		
	Per 100 pounds fruit	Per pound			Basis, samples	Purity	Soundness	Cost per pound
		Low	Average	High				
	Pounds	*Number*	*Number*	*Number*	*Number*	*Percent*	*Percent*	*Dollars*
C. lawsoniana	20	80,000	210,000	600,000	63	84	46	1.60–2.65
C. nootkatensis		66,000	108,000	180,000	8	84		6.75–7.00
C. thyoides	10	420,000	460,000	500,000	11			

CHAMAECYPARIS

TABLE 66.—*Chamaecyparis: Summary of germination data from various sources*

Species	Germinative energy		Germinative capacity			Potential germination	Basis, tests
	Amount	Period	Low	Average	High		
	Percent	*Days*	*Percent*	*Percent*	*Percent*	*Percent*	*Number*
C. lawsoniana	20–28	30–38	0	52	87		60
C. nootkatensis			0		2	22–57	7
C. thyoides			61	80–88	96		11+

CHILOPSIS LINEARIS (Cav.) Sweet   Desertwillow

(Bignonia family—Bignoniaceae)

Also called flowering willow, catalpa willow.

Distribution and Use.—Occurring from western Texas westward through southern Utah to southern California and southward into Mexico, desertwillow is a deciduous shrub or low tree valuable for shelter belts, game cover, and erosion control in arid regions. The durable wood is used locally for fence posts but is of little commercial importance. It is occasionally cultivated in the southern United States and Mexico.

Seeding Habits.—The light-brown oval seeds (fig. 78), about one-fourth inch long exclusive of the long fringe of thin, soft, white hairs present at each end, are borne in two-celled capsules or pods. The latter are from 7 to 12 inches in length, taper from the middle to each end, and contain numerous seed. Handsome, perfect flowers appear in April and continue to be produced until about August; the capsules ripen in the fall and persist on the branches during the winter. Desertwillow is a prolific seeder.

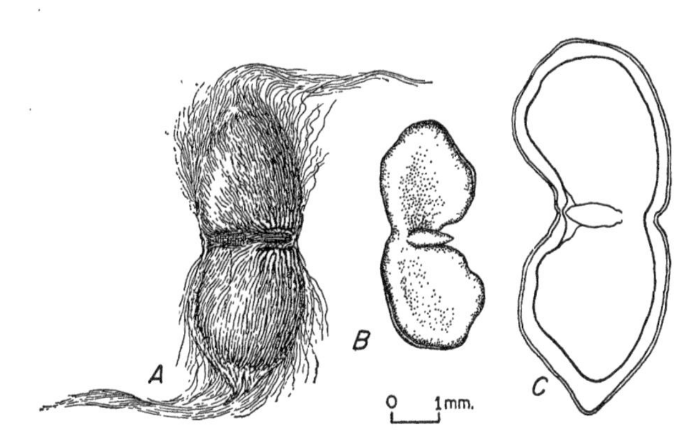

Figure 78.—Seed views of *Chilopsis linearis: A,* Exterior view; *B,* embryo; *C,* longitudinal section.

Collection, Extraction, and Storage.—The pods can be picked into sacks from the branches from late September, when the seeds begin to ripen, until late in the winter months. Because of the long flowering season, the seeds ripen unevenly; hence care must be taken to avoid picking immature pods. Collection costs 25 to 50 cents per pound of pods. If the pods are not thoroughly dry, they should be spread out in thin layers. Seeds may be extracted from the pods by light beating followed by shaking and sifting to separate the seeds from the capsule fragments and other foreign material. No further cleaning is practicable. Cleaned seed: yield per 100 pounds dried pods, 30 to 50 pounds; number per pound (11 samples): low, 40,000; average, 86,000; high, 128,000. Commercial seed averages 92 percent in purity; 87 percent in soundness. The seed should be stored in a dry cold place until spring. No data on longevity are available.

Germination.—Desertwillow seed is not dormant and germinates readily. The seed will germinate faster if kept in wet sand for several days; this probably helps it to absorb water. Test methods recommended are sand flats or water with temperatures alternating diurnally from 68° to 86° F. There is some indication that temperatures above 80° are preferable. Duration of tests is 30 to 60 days, using 1,000 seeds. Average results: germinative energy, 30 to 60 percent in 9 to 30 days; germinative capacity (11 samples): low, 26 percent; average, 55 percent; high, 100 percent.

Nursery and Field Practice.—Desertwillow seed should be sown in the nursery in the spring after the soil has warmed up, otherwise the seed may decay. The seed is sown in rows at a depth of one-fourth inch and at a ratio of seven times as much viable seed as the desired number of usable seedlings. The tree may also be progagated from cuttings.

CHIONANTHUS VIRGINICUS L. Fringetree

(Olive family—Oleaceae)

Also called white fringetree, SPN; old mans beard.

Distribution and Use.—Fringetree occurs on rich well-drained soils of stream banks, coves, and lower protected slopes in partial shade and openings from southern Pennsylvania southward to Tampa Bay, Fla., westward through the Gulf States to the Brazos River, Tex., and northward to southern Missouri. It is a large deciduous shrub or small tree, useful for ornamental planting within and beyond its range in eastern United States. The date of earliest cultivation is 1736.

Seeding Habits.—Male and female flowers of fringetree usually occur on separate trees and open from May to June; the fruits are one-seeded, dark blue, oval drupes about three-fourths inch long (fig. 79), and ripen in September and October. Seed dispersal beyond the immediate vicinity of the tree is by birds and rodents. Plants produce seed at 5 to 8 years.

Collection, Extraction, and Storage.—The fruits should be hand-picked from branches in September and October after they turn purple, and the pulp macerated by rubbing over hardware cloth mesh fine enough to retain seed; pulp may then be washed away. One pound of fruits yields about 5.5 ounces of seed. Seeds per pound (3 samples): low, 1,700; average, 1,800; high, 2,000. Commercial seed averages in purity about 99 percent, in soundness about 95 percent, and it costs from $1.50 to $2.60 per pound. Viability is retained in stratified seed for 1 or 2 years.

Germination.—A natural seedbed of moist, rich soil on stream banks, coves, or lower protected slopes is best for fringetree. Germination is hypogeous (fig. 80) and occurs in spring—often in

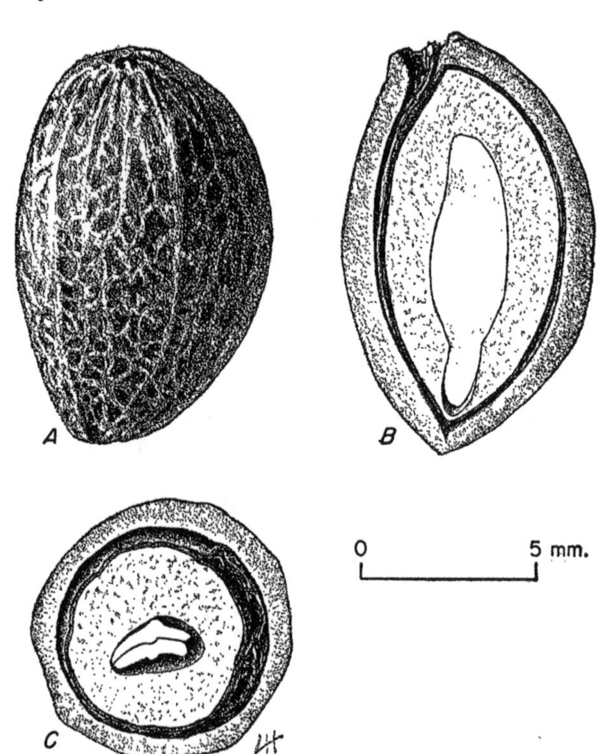

Figure 79.—Seed views of *Chionanthus virginicus*: *A*, Exterior view; *B*, longitudinal section; *C*, cross section.

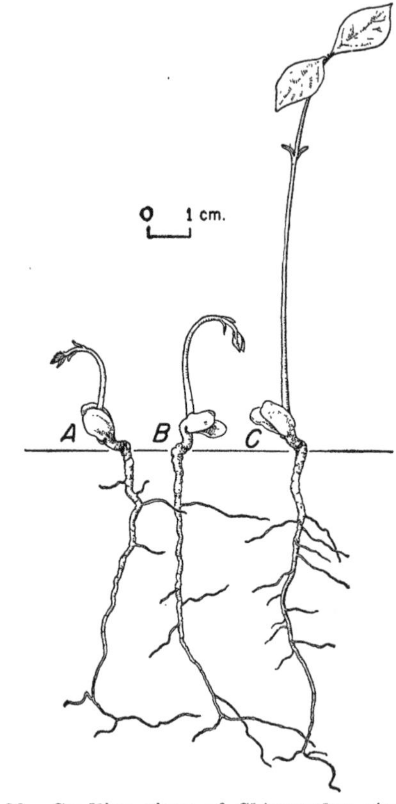

Figuure 80.—Seedling views of *Chionanthus virginicus*: *A*, At 1 day; *B*, at 4 days; *C*, at 7 days.

the second spring after seed fall. Dormancy in seed is due to conditions in the embryo and endosperm, and may be broken at least partly by stratification over winter in mixture of sand and peat at 41° F. Adequate tests are difficult to make because of slow germination even of pretreated seeds. Using 200 seeds which have been stratified first at 68° for 1 or more months followed by 1 or more months stratification at 41°, tests should be run in sand, sandy loam, or mixture of sand and peat flats at 68° (night) to 86° (day) for approximately 1 year. Roots appear at the higher temperature and shoots at the lower temperature. Germinative capacity is about 40 percent; potential germination, 50 to 60 percent (2 tests). Tests can be made quickly (20 to 30 days) and with high results when excised embryos are used.

Nursery Practice.—Fringetree seed may be fall- or spring-sown. Soon after the seed are cleaned, fall sowing should be done in well-prepared beds in rows 8 to 12 inches apart and the seed covered one-fourth to one-half inch with firmed soil. Burlap, straw, or leaf mulch should be held in place with shade or bird screens until after the last frost in spring. Spring sowing involves use of seed stratified for 1 or 2 years after collection. Surface soil of the beds should be kept moist until after the germination period. Propagation by layering, grafting, or budding onto ash seedlings is sometimes practiced.

891183°—50——10

CHRYSOTHAMNUS NAUSEOSUS (Pursh) Britton — Rubber rabbitbrush

(Composite family—Compositae)

Native from the Dakotas and Saskatchewan to Texas, northern Mexico, California, Washington, and British Columbia, this low shrub has many varieties, 11 being found in California alone. It can be used for erosion control and is a valuable honey plant. Latex from this species has been found to yield rubber similar to that produced by guayule.

The perfect, yellow flowers of rubber rabbitbrush occur in heads and appear mostly from August to October, although some varieties begin to bloom in June. Fruiting heads can be collected in October and November and cleaned by rubbing in sacks; chaff is removed by fanning. One sample contained approximately 335,000 cleaned seed (fig. 81). The percentage of empty seed is high. Commercial seed costs $1.50 per ounce. Seed germinates without treatment a few days after sowing. Highest germination obtained: 36 percent in 24 days (1 sample).

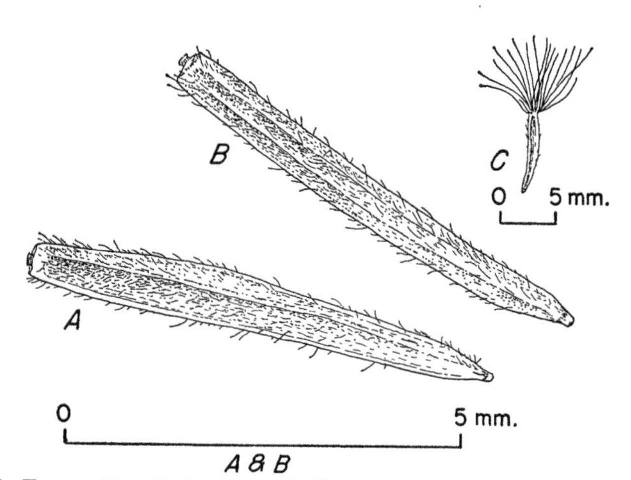

FIGURE 81.—Seed views of *Chrysothamnus nauseosus*: *A* and B, Exterior views of commercial seed in two planes; *C*, exterior view of fruit.

CLADRASTIS LUTEA (Michx. f.) K. Koch Yellowwood

(Legume family—Leguminosae)

Also called American yellowwood, SPN; virgilia.

DISTRIBUTION AND USE.—The native range of yellowwood extends from North Carolina into eastern and central Tennessee, northern Alabama, and central Kentucky, the glades country of southwestern Missouri, and into central and northern Arkansas. This species was introduced into cultivation in 1812. Locally, it grows on limestone cliffs in rich soils, and the greatest abundance is in Missouri and in the vicinity of Nashville, Tenn. The wood is hard, close-grained, and bright yellow in color, turning to light brown on exposure; commercially, it is a substitute for walnut in gunstocks and a source of clear yellow dye. Yellowwood is hardy as far north as New England and is often planted for its ornamental value.

SEEDING HABITS.—The fragrant, perfect, white, showy flowers bloom in June—usually in alternate years—and the fruit ripens in late August or September of the same year. Fruits fall and split open soon after maturing, and the seed are dispersed by birds and rodents. Each fruit contains four to six short, oblong, compressed seed with thin, dark-brown seed coats and without endosperm (fig. 82). Good seed crops are produced generally in alternate years.

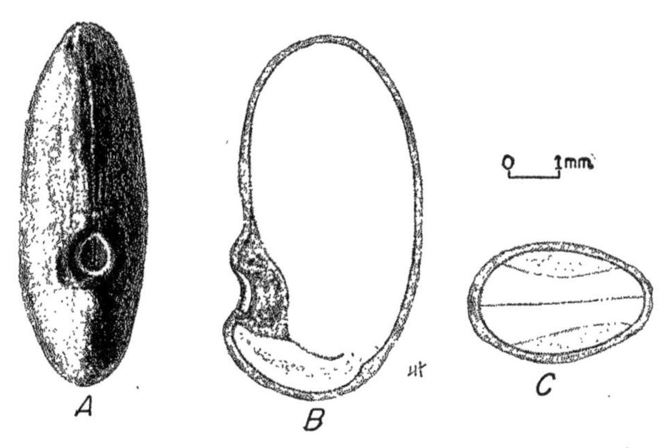

FIGURE 82.—Seed views of *Cladrastis lutea: A,* Exterior view; *B,* longitudinal section; *C,* cross section.

COLLECTION, EXTRACTION, AND STORAGE.—Fruits should be collected soon after maturity by handpicking from the tree or by shaking or whipping them onto outspread canvas. Pods turn brown and split open easily at maturity. They should be allowed to dry soon after gathering and then opened by beating in sacks or threshing in a bean huller. The seed may be separated from the pod remnants by screening and fanning. Seed average about 11,300 to 14,600 per pound (2 samples). Average purity and soundness of commercial seed are, respectively, 82 percent and 67 percent. Costs range from $2.50 to $3 per pound. Seed should be stored soon after extraction in sealed containers at 41° F. for prolonging viability. For over-winter storage, stratification in sand or sand and peat is recommended.

GERMINATION.—Optimum natural seedbed conditions for yellowwood are in rich, moist limestone soils protected by partial vegetation or litter cover. Germination, epigeous (fig. 83), takes place in the

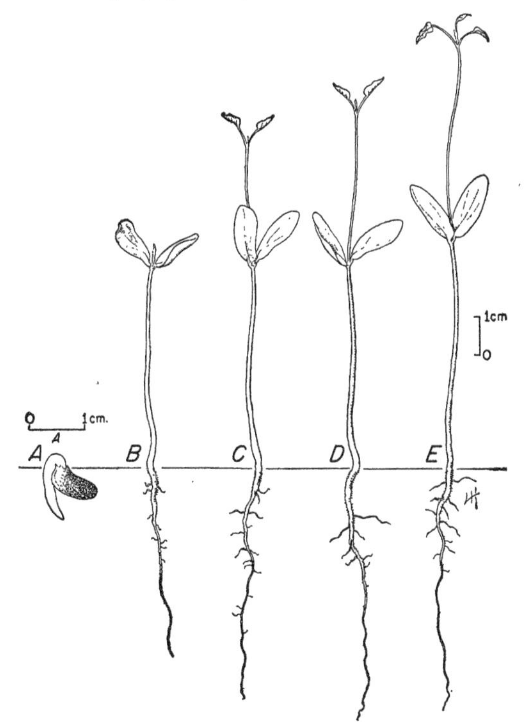

FIGURE 83.—Seedling views of *Cladrastis lutea: A,* At 1 day; *B,* at 6 days; *C,* at 10 days; *D,* at 16 days; *E,* at 20 days.

spring following seed fall. Dormancy, caused chiefly by an impermeable seed coat and to a lesser degree by conditions in the embryo, may be broken by stratification for 90 days at 41° F. in sand or sand and peat, by scarification and storage for about 30 days, or by acid treatment for 30 to 60 minutes. Adequate tests may be made on 200 treated seeds in sand or sand and peat flats in 30 days at 68° to 86°. Average germinative energy and germinative capacity for acid-treated seed are, respectively, 51 to 66 percent in 8 to 11 days and 56 to 67 percent (2 tests).

Nursery Practice.—Seeding may be done in the fall or spring. Beds should be well prepared and drilled with rows 8 to 12 inches apart, to facilitate weeding and cultivation, and the seed covered with about one-fourth inch of firmed soil. Untreated seed are sown in fall beds which should be mulched with straw or leaves and protected with bird or shade screens until after late frosts in spring. Erecting side boards simplifies mulching and screening. Stratified seed or dry-stored seed which have been scarified or acid treated are used in spring beds. Shading of seedlings is unnecessary.

CLEMATIS L. Clematis

(Buttercup family—Ranunculaceae)

DISTRIBUTION AND USE.—The clematises include over 200 species of climbing vines, or erect or ascending perennial herbs (sometimes woody) widely distributed through the temperate regions, chiefly in the Northern Hemisphere. They are grown primarily for ornamental purposes, but some of them are useful also for erosion control and wildlife food or forage. Three species that can be grown in the southern United States are described in table 67.

SEEDING HABITS.—The perfect flowers bloom in the summer and the fruits ripen soon after the cessation of flowering. Clematis fruits consist of a cluster of one-seeded achenes with persistent feathery styles, which are the seeds of commerce (fig. 84). The seeds are dispersed by the wind in late summer or fall, and seed crops are borne annually. Comparative seeding habits of the three species discussed here are given in table 68.

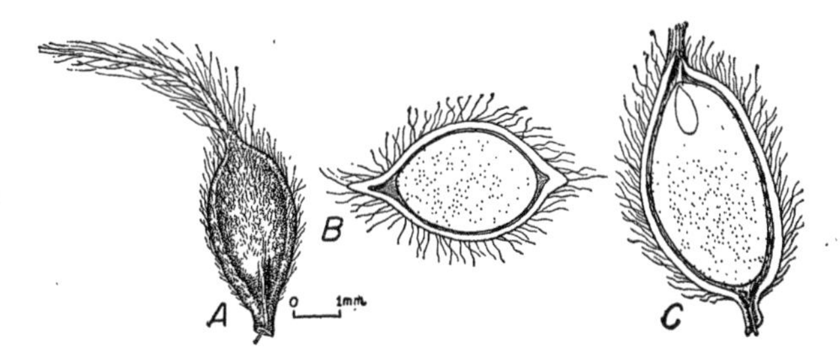

FIGURE 84.—Seed views of *Clematis virginiana*: *A*, Exterior view; *B*, cross section; *C*, longitudinal section.

Scientific information is not available as to the development of climatic races among the clematises. Both *C. flammula* and *C. viticella* include one or more cultivated varieties, but it is not known if any of these are actually climatic races.

COLLECTION, EXTRACTION, AND STORAGE.—The ripe fruits should be collected by hand from the

TABLE 67.—*Clematis: Growth habit, distribution, and uses*

Accepted name	Synonyms	Growth habit	Natural range	Chief uses	Date of earliest cultivation
C. flammula L. (plume clematis).	*C. pallasii* J. F. Gmelin.	Woody climber	Mediterranean region, extending into Persia.	Ornamental, could be used for wildlife food and cover, and erosion control.	1590
C. vitalba L. (travelersjoy).	old mans beard	Handsome climber	Southern Europe, northern Africa, and the Caucasus region.	do	(1)
C. viticella L. (Italian clematis).	vine bower	Slender, woody climber.	Southern Europe and western Asia.	do	1597

[1] Long cultivated.

TABLE 68.—*Clematis: Time of flowering, fruit ripening, and seed dispersal*

Species	Time of—		Flower color
	Flowering	Fruit ripening and seed dispersal	
C. flammula	August–October	August–October	White.[1]
C. vitalba	July–September	July–September	do.[1]
C. viticella	June–August	June–August	Purplish.

[1]Fragrant flowers.

plants and the seeds extracted, presumably by drying and shaking the fruits. Size, purity, and cost of commercial cleaned seed vary considerably by species, as shown in table 69. Information is not available as to the longevity of clematis seed under storage conditions.

Germination.—Germination tests for the 3 clematises studied were made in Jacobsen germinators and showed at best 1 percent germination. There appears to be embryo dormancy, so it is recommended that the seed be stratified in moist sand or peat at 32° to 50° F. for 30 to 90 days to stimulate germination. It is also recommended that germination tests be run in sand flats for 40 to 60 days at 68° F. (night) to 86° (day) using 400 to 1,000 properly stratified seeds per test.

Nursery and Field Practice.—Although no information as to nursery practice is available, it is recommended that seed be sown in the fall or that stratified seed be sown in the spring. Clematises grow best on rich, well-drained, light loams. A light admixture of lime is usually beneficial. The cultivated clematises usually are propagated by grafting (generally on roots of *C. flammula* and *C. viticella*), cutting, or layering. They are subject to root injuries caused by a nematode.

Table 69.—*Clematis: Yield of cleaned seed, and purity and cost of commercial seed*

Species	Cleaned seed per pound				Commercial seed	
	Low	Average	High	Basis, samples	Purity	Cost per pound
	Number	*Number*	*Number*	*Number*	*Percent*	*Dollars*
C. flammula	------------	25,000	------------	1	95	2.00
C. vitalba	------------	320,000	------------	1	92	1.25
C. viticella	18,000	27,000	38,000	3	95	2.50

COMPTONIA PEREGRINA (L.) Coult. Sweetfern

(Waxmyrtle family—Myricaceae)

Syns.: *C. asplenifolia* Ait., *Myrica asplenifolia* L. in part.

Distribution and Use.—Occurring on dry sandy soils, chiefly in clearings, from Nova Scotia westward to Saskatchewan and southward to North Carolina, Indiana, and Minnesota, sweetfern is an attractive, sweet-scented, low, deciduous shrub valuable for ornamental purposes, and for game food. Because of its tendency to form extensive colonies, it is also useful for soil protection and wildlife cover. It was introduced into cultivation in 1714.

Seeding Habits.—The unisexual flowers are borne from April to May in catkins on different parts of the same plant. The staminate catkins are relatively long and cylindrical, the pistillate short and globose. The latter develop into a burlike fruit about one-half inch in diameter and contain about four nutlets or seeds. These are oblong, about one-fourth inch in length, olive brown and lustrous, and consist (fig. 85) of a thick, bony coat and a large embryo without endosperm. They ripen in August and fall from the branches soon thereafter.

Collection, Extraction, and Storage.—The "burs" may be collected as soon as they become brown in color and begin to open. A short period of drying is enough to release the seeds. One hundred pounds of fresh "burs" will produce about 4 to 12 pounds of cleaned seed of about 4 percent soundness. Soundness (4 samples) ranged from 3 to 38 percent and may inherently be low. Cleaned seed per pound (based on same 4 samples): low, 31,200; average, 43,800; high, 54,800.

Germination.—Germination figures are not available. Seed which had been stored dry in sealed bottles at 41° F. for almost 2 years and then stratified in moist sand for 60 days at 41° gave no germination in 60 days, although 4 percent of the seed (original cutting test, 5 percent sound) was found sound at the close of this period. Since tests show that the seed coat is very permeable to water, it is not likely that the dormancy is due to lack of moisture.

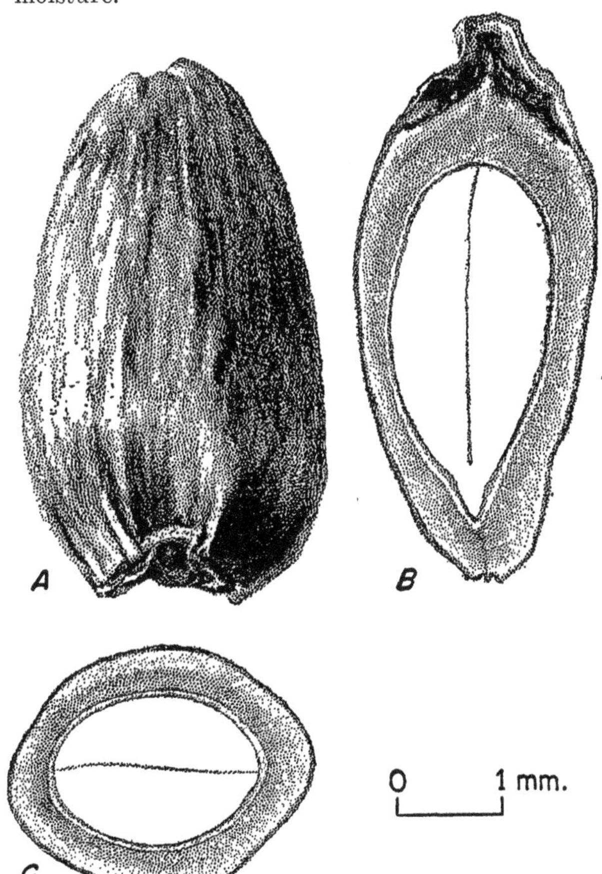

Figure 85.—Seed views of *Comptonia peregrina*: *A*, Exterior view; *B*, longitudinal section; *C*, cross section.

CORNUS L. Dogwood

(Dogwood family—Cornaceae)

DISTRIBUTION AND USE.—Including about 40 species of deciduous, rarely evergreen, trees and shrubs (2 are herbs), native to the temperate region of the Northern Hemisphere, Mexico, and Peru, the dogwoods are useful chiefly for their ornamental qualities—flowers, fruit, foliage, or color of twigs. The wood is hard, heavy and durable, and that of the tree species is used for turnery and charcoal; the bark of several species is medicinal; some produce edible fruits. Distribution data and chief uses of 8 American species of present or potential importance in conservation planting are given in table 70.

So far as known, none of the dogwoods is used for reforestation in the United States. *Cornus asperifolia,* however, has been used in shelter-belt planting in the southern Prairie-Plains region; *C. florida* is sometimes planted for game food and cover in the South and Central States; *C. nuttalli* has been planted to some extent in southern England where it is perfectly hardy. Most of the eight species listed in table 70 are used in horticulture to a more or less degree.

SEEDING HABITS.—The small, perfect flowers—white, greenish white, or yellow in color—are borne in terminal clusters in the spring. In some species,

TABLE 70.—*Cornus: Growth habit, distribution, and uses*

Accepted name	Synonyms	Growth habit	Natural range	Chief uses	Date of earliest cultivation
C. alternifolia L. f. (alternate-leaf dogwood).	blue dogwood, pagoda dogwood.	Small tree or tall shrub.	Nova Scotia to Minnesota, south to Florida, Alabama, and Missouri.	Ornamental planting; bird food.	1760
C. amomum Mill. (silky dogwood).	*C. sericea* L.	Small to large shrub.	Massachusetts to Georgia, west to southern Indiana and Tennessee.	Ornamental planting; game food and cover; medicinal bark.	1658
C. asperifolia Michx. (roughleaf dogwood).		Shrub, occasionally a small tree.	Southern Ontario to South Dakota, south to Florida, and eastern Texas.	Shelter-belt planting; game food.	1836
C. canadensis L. (bunchberry dogwood).	*Chamaepericlymenum canadense* (L.) Aschers. & Graebn., *Cornella canadensis* (L.) Rydb. (dwarf cornel).	Low herb	Labrador to Alaska, south to West Virginia, Minnesota, and California.[1]	Food and browse for birds and animals; ornamental planting.	
C. florida L. (flowering dogwood).	*Cynoxylum floridum* (L.) Raf. (boxwood, flowering cornel).	Small tree or shrub.	Massachusetts to Florida, west to southern Michigan, Kansas, and Texas; also in Mexico.	Shuttleblocks, engraver's blocks, wheel hubs, handles, pulleys, other turned articles; important ornamental species; game food; bark of this and next species contains quininelike substance.	1731
C. nuttalli Audubon (Pacific dogwood).	western dogwood	Tree to 80 feet.	British Columbia to southern California, west of Cascades and Sierra Mountains; also northwestern Idaho.	Wood used in turnery; ornamental planting; fair forage for sheep; game food and cover.	1835
C. racemosa Lam. (gray dogwood).	*C. paniculata* L'Hérit. (panicled dogwood).	Thicket-forming shrub.	Maine to Minnesota, south to Georgia and Nebraska.	Game food and cover; ornamental planting.	1758
C. stolonifera Michx.[2] (redosier dogwood).	redosier	do	Newfoundland to Yukon, south to Virginia, Nebraska, New Mexico, and California.	Landscape planting; fruit much eaten by birds; twigs browsed by game animals.	1656

[1] There is a species in eastern Asia which is very similar to or the same as *C. canadensis*—botanists are not in complete agreement.
[2] Includes several varieties.

notably *Cornus florida* and *C. nuttalli,* the clusters are surrounded by a conspicuous enlarged involucre[21] of four to six white to pinkish, petal-like, enlarged scales of terminal buds formed the previous season. The fruit, a globular to ovoid drupe 1/8 to 1/4 inch in diameter, with thin succulent or mealy flesh, and containing a single two-celled and usually two-seeded[22] bony stone (figs. 86 and 87) ripens in the late summer or fall. Commercial seed consists of dried fruit or clean stones. Such details as are available on the seeding habits of eight species discussed are given in table 71.

Data are not available on the age for commercial seed-bearing of any of the species listed in table 71. *Cornus florida* and *C. nuttalli* bear good seed crops about every other year; data are lacking for the others. Seed dispersal is largely by birds and other animals.

Nothing is known as to the development of geographic strains in *Cornus.* As is the case in other genera including species of wide distribution, it is

[21] An involucre is also present in *C. canadensis.*
[22] Very often only one will be fully developed.

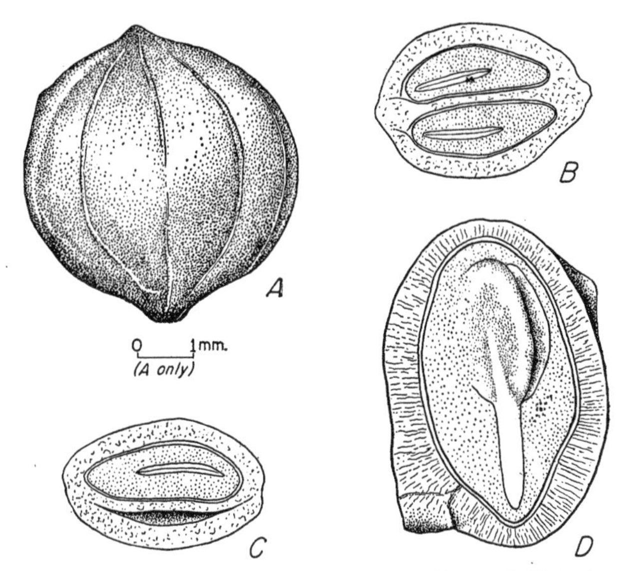

FIGURE 86.—Seed views of *Cornus stolonifera: A,* Exterior view; *B,* cross section of stone with two embryos; *C,* cross section of stone with single embryo; *D,* longitudinal section, showing embryo.

FIGURE 87.—Seed views of five species of *Cornus.*

CORNUS

TABLE 71.—*Cornus: Time of flowering, fruit ripening, and seed dispersal*

Species	Time of— Flowering	Fruit ripening	Seed dispersal	Color of ripe fruit
C. alternifolia	May–June	July–August	July–September	Dark blue.
C. amomum	June	September	September to mid-October.	Pale blue.
C. asperifolia	Mid-May to early September.	August–late October	August–late October	White.
C. canadensis	May–July	August	August–October	Scarlet.
C. florida	March–June	September–late October	October–late November	do.
C. nuttalli	May	September–October	September–October	Red.
C. racemosa	June–July	do	September–early winter	White.
C. stolonifera	May–July	Late July–October	Late July–October	White to lead color.

quite likely that such races have also developed in some of the dogwoods, particularly *C. stolonifera.*

COLLECTION, EXTRACTION, AND STORAGE.—As soon as ripe, the fruit should be collected by stripping or shaking from the branches to prevent its consumption by birds. In the case of the tree species, hand-picking may involve the use of ladders but ordinarily this can be done from the ground. Fruit of *Cornus florida* should not be collected from isolated trees, as these frequently show a very high percentage of empty stones. After collection the fruit can be sown or put into stratification. Seed that is to be stored should be cleaned to reduce bulk. If this cannot be done soon after collection, the ripe fruit should be spread out in shallow layers to prevent heating and fermentation. The drying process, however, should not be allowed to proceed too far, or removal of the pulp will be made rather difficult. The stones can be readily extracted[23] by macerating the fruit in water or running it through a hammer mill, allowing the pulp and empty stones to wash away. After drying, the clean stones are ready for sowing or for storage. Such data as are available on yield, size, purity, soundness, and cost of seed are given in table 72 for the eight species discussed here.

Practically no data are available on the longevity of *Cornus* seed. Seed of *C. racemosa,* however, has given good germination after having been stored for 28 months in sealed containers at 41° F. Until optimum methods are worked out, it is suggested that these conditions be used for storage of other *Cornus* species. One-year-old seed of *C. florida* had its germinability reduced from 15 to 0 percent when stored for 2 years at 41° and a moisture content of 12.6 percent.

GERMINATION.—Natural germination of *Cornus* seed in at least *C. florida* occurs in the spring after the seed have fallen and laid on the ground over winter. Some individual seeds of other species prob-

[23] Extraction of stones of *Cornus canadensis* appears to be considerably more difficult than that of other species. Soaking the whole fruit for 10 minutes in concentrated sulfuric acid prior to its maceration in water makes the latter operation much simpler. Fermenting fresh fruit for about 2 days is also said to be an effective aid to cleaning.

TABLE 72.—*Cornus: Yield of cleaned stones, and purity, soundness, and cost of commercial seed*[1]

Species	Cleaned stones: Yield per 100 pounds of fruit	Per pound: Low	Per pound: Average	Per pound: High	Basis, samples	Over-run[2]	Commercial seed: Purity	Soundness	Cost per pound
	Pounds	*Number*	*Number*	*Number*	*Number*	*Percent*	*Percent*	*Percent*	*Dollars*
C. alternifolia	----------	5,900	8,000	9,500	4	[3]3	[3]99	74	[4]1.50
C. amomum	17	10,900	--------	11,600	2	--------	--------	77 (60–96)	2.00
C. asperifolia	18–27	8,600	15,700	21,000	5	--------	[3]89	67 (33–85)	----------
C. canadensis	----------	59,000	--------	77,000	2	[3]6	[3]99	73	[4]3.00
C. florida	37 (22–46)	3,300	4,500	6,200	11	7 (2–14)	97	76 (46–94)	0.60–1.00
C. nuttalli	----------	4,000	--------	6,100	2	--------	[3]99	92	1.40
C. racemosa	18–25	10,200	12,100	15,300	9	15 (4–30)	99	83 (58–100)	1.75
C. stolonifera	15–20	13,800	18,700	26,700	8	20 (7–32)	99	85 (52–99)	1.50–2.50

[1]A single figure indicates the mean of observations; 2 figures separated by a dash indicate the range from low to high.
[2]Due to the fact that *Cornus* stones are typically double-celled, 100 stones invariably contain more than 100 seeds.
[3]1 sample.
[4]Dried fruit.

ably also germinate in this fashion, but others likely lie over until the second spring. There is evidence that seed collected somewhat green and sown immediately will germinate promptly the next spring. Germination is epigeous (fig. 88), the seed coat usually sloughing off and remaining in the soil. The best seedbed for some species is moist but well-drained, rich loams.

All of the eight species discussed here show delayed germination due to a dormant embryo and probably in most cases to the impermeability or the hardness of the pericarp as well. For those species showing both types of dormancy, warm stratification for about 60 days in moist sand or peat at temperatures alternating diurnally from 70° to 85° F., followed by a usually longer period at much lower temperatures, is required. Immersion in concentrated sulfuric acid and possibly mechanical scarification can be used in place of the warm stratification. In species showing embryo dormancy only, this can easily be broken by low-temperature stratification. Detailed information on pretreatment methods is given in table 73.

Adequate germination tests can be made in sand flats or peat mats, using 400[24] properly pretreated seeds per test. Temperatures alternating diurnally from 68° to 86° F. appear to be satisfactory for

[24] Probably about 1,000 should be used in the case of *Cornus canadensis*.

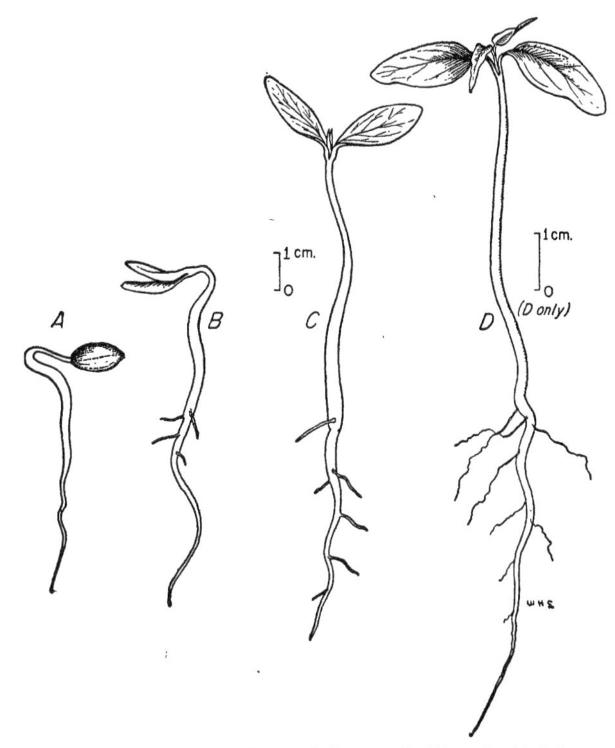

FIGURE 88.—Seedling views of *Cornus florida: A,* At 2 days; *B,* at 4 days; *C,* at 8 days; *D,* at 31 days.

TABLE 73.—*Cornus: Dormancy and methods of seed pretreatment*

Species	Kind of dormancy	Pretreatment: Medium	Pretreatment: Duration and temperature (Days at °F.)	Other methods of pretreatment
C. alternifolia	Hard pericarp and dormant embryo.	Sand, peat, sand and peat.	60 at [1]68–86 + 60 at 41	Mechanical scarification or acid in combination with cold stratification.
C. amomum	Dormant embryo and probably hard pericarp.	do.	90–120 at 41	Warm stratification prior to cold stratification.
C. asperifolia	Hard pericarp and dormant embryo.	Mechanical scarifier plus sand, peat, sand and peat.	[2]1 at 70–80 + 60+ at 41	do.
C. canadensis	do.	Sand or peat, sand and peat.	30–60 at 77 + 120–150 at 33	Acid plus cold stratification.
C. florida	Dormant embryo	do.	100–130 at 41	
C. nuttalli	Most fresh seeds not dormant.[3]	do.	120–200 at 33–41	Soak old seed 4 hours in H_2SO_4 and then stratify 90 days at 41° F.[4]
C. racemosa	Hard pericarp and dormant embryo.	do.	60 at [1]68–86 + 120 at 41	H_2SO_4 2 hours plus 120 days' stratification at 41° F.
C. stolonifera	Dormant embryo; some lots also show hard pericarp.	do.	90–120 at 41	Mechanical scarification plus 120 days' stratification at 41° F. for lots with hard pericarp.[5]

[1] Temperatures alternating diurnally.
[2] This period can likely be shortened markedly by using equipment of better design.
[3] Remaining seeds have dormant embryo and will germinate only after cold stratification.
[4] Stratification improves germination even of fresh seed.
[5] H_2SO_4 1 hour plus stratification for 90 to 120 days at 41° F. is better than stratification alone for lots with hard pericarp but seems to reduce potential germination.

all species. Results that may be expected under these conditions are given in table 74.

NURSERY AND FIELD PRACTICE.—Dogwood stones or fruits should be sown in the fall or stratified and sown in April or early May. Fresh seed of *Cornus nuttalli* is sown in the fall and begins to germinate about 70 days after sowing. Seeds having embryo dormancy only will germinate the first spring but those having also a hard pericarp will for the most part lie over until the second year. Seeds of such species can hence be sown in midsummer if desirable and the beds mulched until the next spring, at which time practically complete germination can be expected. In one case seeds of *C. alternifolia* and *C. racemosa* collected somewhat green in July and sown immediately gave full germination the next spring.

The seeds are usually sown in drills, sometimes broadcast, and are covered with 1/4 to 1/2 inch of nursery soil, depending on size. Forty viable seeds of the smaller-seeded species are sown per square foot. The beds are usually given a mulch of leaves or straw which is removed at the first signs of germination. Young seedlings of *Cornus nuttalli* are given half shade; this is usually omitted in the case of *C. florida*. Nursery germination of *C. florida* is 77 to 85 percent.

Dogwood seedlings appear to be generally resistant to damping-off and root rot. One-year-old seedlings of *Cornus racemosa* are sometimes badly defoliated in early summer by a leaf-spot disease caused by *Septoria cornicola. C. amomum* seedlings are apparently more resistant to this disease. Dogwood seedlings in nurseries in the Southeastern States are sometimes lightly defoliated by a leaf spot caused by *Cercospora cornicola.* Although an entirely satisfactory method of control for these leaf-spot diseases has not yet been worked out, applications of 4–6–50 bordeaux mixture at about 2-week intervals have been quite effective in checking their spread. Another fairly common foliar trouble, retarded and abnormal leaf development, is believed caused by the leafhopper, *Graphocephala versuta.*

One-year-old seedling stock is usually large enough for field planting. Most species should be planted on rather moist sites; however, *Cornus stolonifera* does well in wet places and *C. asperifolia* should be used only on dry soils. In addition to propagation from seed, most dogwoods can be grown from root cuttings, layering, and by division. Those with willowlike soft wood, such as *C. stolonifera* can be grown from cuttings of mature wood.

TABLE 74.—*Cornus: Recommended germination test duration, and summary of germination data*

Species	Test duration recommended[1]	Germination data from various sources							
		Germinative energy		Germinative capacity			Potential germination	Basis, tests
		Amount	Period	Low	Average	High		
	Days	*Percent*	*Days*	*Percent*	*Percent*	*Percent*	*Percent*	*Number*
C. alternifolia				0		13		2
C. amomum	35				10			1
C. asperifolia	50	14	30	1	25	60		3
C. canadensis		6	26	6		23	61	2
C. florida	20–25	14–45	15–20	0	35	70		7
C. nuttalli	40					100		1
C. racemosa	15–30	22–30	14	1	20	31	52–74	5
C. stolonifera	20–40	22–50	16	6	47	76	16–85	7

[1]Test duration for untreated seed of *Cornus racemosa* was 230+ days, and for *C. stolonifera* 180+ days; the percentage of germination for each of the species was only a small fraction of that obtained from pretreated seed.

CORYLUS L. Filbert, hazel

(Birch family—Betulaceae)

DISTRIBUTION AND USE.—The hazels include about 15 species of deciduous shrubs, or sometimes trees, which occur in the temperate portions of North America, Europe, and Asia. Some species are cultivated for their edible nuts or for ornament. Most of them furnish food for wild animals. *Corylus avellana* has been cultivated quite widely for commercial production of filberts, mostly in Europe but to some extent in the United States. The other species have not been grown extensively. Four species of present or potential usefulness are described in table 75.

SEEDING HABITS.—The male and female flowers are borne separately on 1-year-old lateral twigs or spurs of the same plant. They are formed late in the summer and open the next spring before the leaves appear. The male flowers, which appear first, are naked catkins bearing pollen only. The inconspicuous female flowers, blooming shortly afterwards, are clustered in scaly buds. By late summer or fall the fertilized flowers have developed into ripe fruits. The fruit is a round or egg-shaped, hard-shelled nut (fig. 89) surrounded or included in an involucre or husk consisting of two more or less united leafy and hairy bracts; these appear in clusters at the ends of the branchlets. Natural seed dispersal is chiefly by animals. Details of the seeding habits of the four species discussed here are given in table 76.

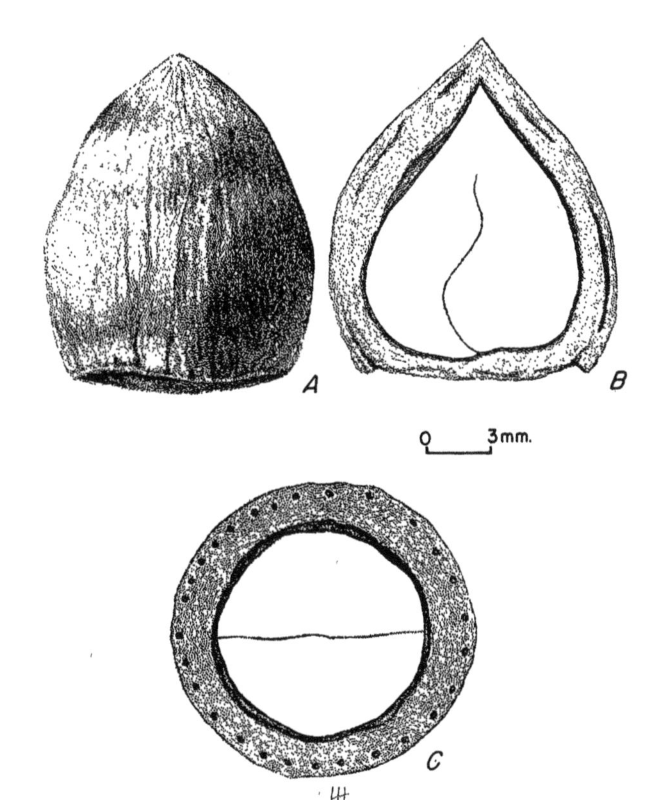

FIGURE 89.—Seed views of *Corylus californica*: *A*, Exterior view; *B*, longitudinal section; *C*, cross section.

TABLE 75.—*Corylus: Growth habit, distribution, and uses*

Accepted name	Synonyms	Growth habit	Natural range	Chief uses	Date of earliest cultivation
C. americana Marsh. (American filbert).	American hazel.	Medium-sized, thicket-forming shrub.	Maine to Saskatchewan and southward to Florida and Oklahoma.	Wildlife food and cover, ornamental, shelter-belt purposes.	1798
C. avellana L. (European filbert).	European hazel.	Medium to tall, thicket-forming shrub.	Europe	Human food, wildlife food and cover, shelter-belt purposes.	(1)
C. californica (A. DC.) Rose (California filbert).	*C. rostrata* var. *californica* A. DC. (California hazel).	do	British Columbia to central California.	Erosion control, wildlife food, human food, basket splints, medicinal.	1910
C. cornuta Marsh. (beaked filbert).	*C. rostrata* Ait. (beaked hazel).	Medium-sized, thicket-forming shrub.	Quebec to British Columbia and southward to Georgia and Missouri.	Wildlife food and cover	1745

[1] Long cultivated.

CORYLUS

TABLE 76.—*Corylus: Time of flowering, fruit ripening, and frequency of seed crops*

Species	Time of—			Seed year frequency	
	Flowering	Fruit ripening	Seed dispersal	Good crops	Light crops
				Years	*Years*
C. americana	March–April	July–October	July–winter	2–3	1+
C. avellana	Spring	Fall	Fall		1+
C. californica	February–March	do	do		
C. cornuta	April–May	July–September	July–winter	5±	1+

Scientific information as to racial variation within hazel species is lacking. Several varieties of *Corylus avellana* are known, but none of them seems to be associated with geographic origin. In view of the wide distribution of most of the species, racial variations may well have arisen.

COLLECTION, EXTRACTION, AND STORAGE.—Many nuts may be lost because of squirrels and other animals and through dropping out of the husks unless the fruits are gathered by hand from the bushes as soon as the edges of the husks begin to brown. The fruits should be spread out in a thin layer to dry for a short period and the husks afterwards removed by flailing. The nuts, which are the commercial seed, can then be sown, stratified, or stored. Yield, size, purity, soundness, and cost of commercial seed vary according to species as shown in table 77.

Storage of hazel nuts over winter can be accomplished satisfactorily by stratification in moist sand at temperatures just above freezing. Seed of *Corylus avellana* can be kept satisfactorily for 1 year in ordinary storage. Storage in sealed containers at 41° F. will retain a large part of the viability of *C. americana* seed and some viability of *C. cornuta* for at least 2 years.

GERMINATION.—Natural reproduction seems to be more common from sprouts than from seed. Germination is hypogeous (fig. 90). The seeds have a dormant embryo and, without pretreatment, germinate very slowly. Although the best means of pretreatment have not yet been worked out for any species, results so far obtained indicate that stratification in moist sand or peat for more than 90 days is needed. It is possible that alternations of warm and cold stratification may be necessary for good germination. Germination tests can be run in sand flats, using 100 to 200 seeds per test. Average results so far attained, with methods admittedly inadequate, are given in table 78.

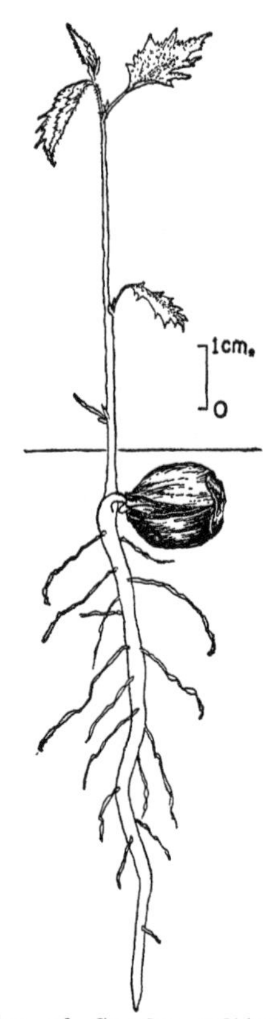

FIGURE 90.—Seedling of *Corylus californica* at 1 month.

NURSERY AND FIELD PRACTICE.—*Corylus* seed should be sown in drills in the fall or stratified seed sown in the spring. Ordinarily seed of *C. cornuta* germinate the second spring after sowing. The hazels will grow on a variety of soils but do best on moderately rich, well-drained soil. Horticultural varieties usually are propagated by suckers, layers, budding or grafting, and cuttings.

TABLE 77.—*Corylus: Yield of cleaned seed, and purity, soundness, and cost of commercial seed*

Species	Cleaned seed					Commercial seed		
	Yield per 100 pounds of fruit	Per pound			Basis, samples	Purity	Soundness	Cost per pound
		Low	Average	High				
	Pounds	*Number*	*Number*	*Number*	*Number*	*Percent*	*Percent*	*Dollars*
C. americana	25–30	197	476	736	9	96	87	0.70–1.20
C. avellana	60	176	480	530	5+	98	85	.65–1.20
C. californica		400		418	2	62		
C. cornuta		425	549	676	3	99	75	1.15

TABLE 78.—*Corylus: Method of seed pretreatment for dormancy, recommended germination test duration, and summary of germination data*

Species	Dormancy pretreatment	Test duration recommended		Germination data from various sources				
				Germinative capacity			Potential germination	Basis, tests
		Untreated seed	Pretreated seed	Low	Average	High		
		Days	*Days*	*Percent*	*Percent*	*Percent*	*Percent*	*Number*
C. americana	Stratify 60 days at 41° F., plus 67 days at 65° F., plus 30 days at 41° F.	346+	77		39		68	1
C. avellana	None	60					90	1
C. californica	Stratify in peat 90 days at 41° F.		96+			20		1+
C. cornuta	Stratify 60–90 days at 41° F.	346+	67	1		3	84	2

COTONEASTER B. Ehrh. Cotoneaster

(Rose family—Rosaceae)

Distribution and Use.—The cotoneasters comprise about 40 species of deciduous or evergreen shrubs, or rarely small trees, native to the temperate regions of Europe, northern Africa, and Asia. They are valued for ornamental purposes on account of their attractive fruits, for the white flowers in some species, and also for wildlife food. Three species, which are of value in the United States and for which reliable information is available, are described in table 79.

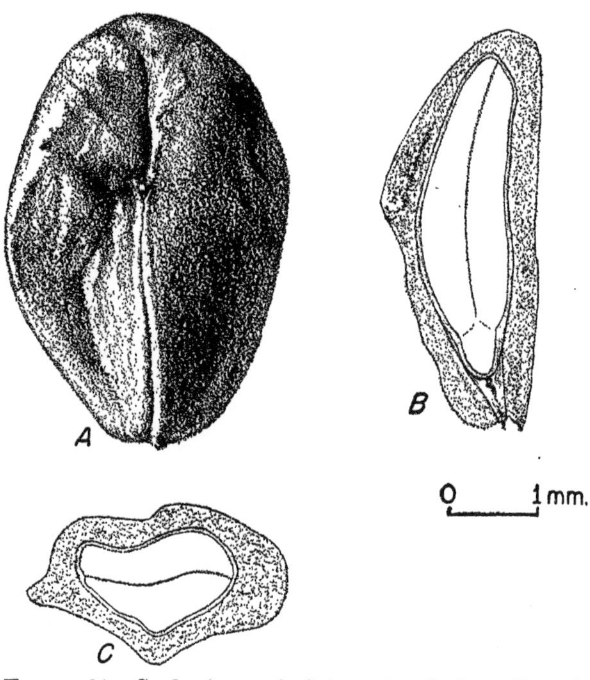

Figure 91.—Seed views of *Cotoneaster horizontalis: A*, Exterior view; *B*, longitudinal section; *C*, cross section.

Seeding Habits.—The perfect, white or pinkish flowers occur in few to many-flowered clusters at the ends of short, leafy lateral shoots and appear before the leaves are out. The fruits are small, black or red, pomaceous drupes which ripen in late summer or early fall and generally hang on the bushes for some time. Seed crops are borne annually. Each fruit contains from two to five stones or nutlets (fig. 91). The comparative seeding habits of the three species discussed here are given in table 80. No scientific information as to the development of climatic races in cotoneaster species is available.

Collection, Extraction, and Storage.—The ripe fruits should be collected by hand from the bushes from late summer to winter. There is a possibility that seed of fruit collected slightly green in midsummer will germinate more promptly. Seeds may be extracted by running the fruits through a macerator with water and skimming off or screening out the pulp; or the fruits may be dried and the seeds removed from the pulp by abrasion. The seed characteristics of the three species discussed here vary considerably as indicated in table 81. Cotoneaster seed should be stored dry. No information is available as to the longevity of the seed under storage conditions.

Germination.—Cotoneaster seed exhibits dormancy resulting apparently from a combination of impermeable seed coat and internal conditions of the embryo, and is difficult to germinate for testing purposes. The degree of dormancy seems to vary from crop to crop. For *C. horizontalis* dormancy can be overcome quite satisfactorily by first either stratifying the seed in moist peat for 90 to 120 days at 60° to 75° F. or soaking the seed for 1½ hours in concentrated sulfuric

Table 79.—*Cotoneaster: Growth habit, distribution, and uses*

Accepted name	Synonyms	Growth habit	Natural range	Chief uses	Date of earliest cultivation
C. acutifolia Turcz. (Peking cotoneaster).	*C. pekinensis* Zab.	Fairly tall shrub	Northern China	Ornamental, wildlife food.	1883
C. horizontalis Decne. (rock cotoneaster).	*C. davidiana* Hort.	Half evergreen, low shrub.	Western China	Ornamental (for rockeries), wildlife food.	1880
C. lucida Schlecht. (hedge cotoneaster).	*C. acutifolia* Lindl., not Turcz.	Upright, bushy shrub.	Altai Mountains and Lake Baikal region of central Asia.	Ornamental	1840

acid, followed by stratification in moist peat for 90 to 120 days at 41° to 50°. Although this procedure may also be effective with the other species, tests to prove it are not available. Untreated seed of *C. acutifolia* kept in flats over winter at 50° and then placed in a greenhouse at 70° germinated only 4 percent the first spring, but came up 79 percent the second spring. However, seed collected just prior to maturity (July 11) in Iowa germinated 80 percent the first season. Seed of *C. lucida* stored in the fruit for 15 days at 41° and then stored moist in a sealed bottle for 20 days at 36° gave no germination in 132 days of testing.

It is recommended that cotoneaster seed be tested in sand flats or peat mats for 100 days at 68° F. (night) to 86° (day), using 400 pretreated seeds per test. Seed of *Cotoneaster horizontalis* has a germinative capacity averaging about 30 percent, and varying from 0 to 76 percent (5+ tests). Comparable data for *C. acutifolia* and *C. lucida* are lacking.

Nursery and Field Practice.—Cleaned seed may be sown broadcast in the fall in board-covered coldframes, or properly pretreated seed may be sown in the spring. In the former case most germination will occur in the second spring, and even in the latter case there may be some germination the second spring. It is possible that seed collected slightly green and sown immediately will germinate promptly the first spring. Cotoneasters are reproduced by means of cuttings, layering, and grafting, as well as from seed. They grow well in most good soils, but do not do so well in particularly moist or shaded situations.

Table 80.—*Cotoneaster: Time of flowering, fruit ripening, and seed dispersal*

Species	Time of—			Color of—		Stones per fruit
	Flowering	Fruit ripening	Seed dispersal	Flowers	Fruits	
C. acutifolia	May–June	September–October	September–winter	Pinkish	Black	2
C. horizontalis	June	do	do	do	Bright red	3
C. lucida	May–June	September	do	do	Black	3–4

Table 81.—*Cotoneaster: Yield of cleaned seed, and purity, soundness, and cost of commercial seed*

Species	Cleaned seed per pound				Commercial seed		
	Low	Average	High	Basis, samples	Purity	Soundness	Cost per pound
	Number	*Number*	*Number*	*Number*	*Percent*	*Percent*	*Dollars*
C. acutifolia						99±	[1]3.50–6.50
C. horizontalis		64,000		1	95	60	4.00–6.00
C. lucida	[2]7,000		19,000	2		47	

[1]Dried berries cost $2 to $3 per pound.
[2]Moist seed.

891183°—50——11

COWANIA STANSBURIANA Torr. **Cliffrose**

(Rose family—Rosaceae)

Syns.: *C. mexicana* of U. S. authors, not D. Don; *C. davidsonii* Rydb.; quinine-bush.

Distribution and Use.—Native to dry, exposed, rocky situations such as mesas and canyon walls from southern Colorado and Utah west to southern California and south to northern Mexico, cliffrose is a spreading evergreen shrub to small tree (up to 25 feet) which furnishes, where abundant, an important and valuable browse to cattle, sheep, goats, and deer. Because of its ornamental qualities—conspicuous, fragrant flowers, feathery fruit, and aromatic foliage—it has been in occasional use for landscape planting since 1904.

Seeding Habits.—The white to sulfur-yellow, perfect flowers open from May to June; the fruit, consisting of an achene with a persistent feathery style about 2 inches long, is borne in clusters of 4 to 10 on a flat disk and ripens in October. Each achene (fig. 92) contains a single seed with membranous coat, thin endosperm, and oblong cotyledons. Seed of commerce consists of the dried achenes, because it is impractical to extract the true seeds. Good seed crops are borne at rather frequent intervals. Dispersal is largely by wind and to some extent by animals.

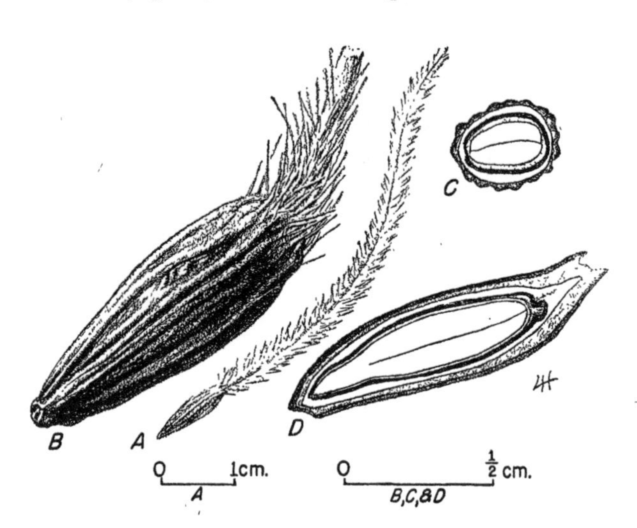

Figure 92.—Seed views of *Cowania stansburiana*: *A*, Exterior view; *B*, exterior view, style broken off; *C*, cross section; *D*, longitudinal section.

Collection, Extraction, and Storage.—*Cowania* seed can be collected in October by stripping from the branches. After a short period of drying they should be rubbed to break off the styles. This not only reduces bulk but also facilitates sowing.[25] No further treatment is necessary. Number of cleaned seed per pound: 92,000 (1 sample, some without styles); purity, 19 percent; soundness, 24 percent. No cost data are available. The seed can be stored at ordinary temperatures for at least 7 years.

Germination.—Little is known about the behavior of *Cowania* seed in nature. Since in the nursery the seeds germinate over a long period, it would appear that they are dormant. Whether or not stratification is beneficial is not yet certain. In one instance, untreated seed gave 10 percent germination in 21 days, while stratified seed did not germinate at all. Germinative capacity, 10 to 21 percent (2 samples).

Nursery and Field Practice.—The seed is usually sown in the nursery in the late fall or early spring (late December to mid-March in New Mexico). It should be sown thickly and covered with one-eighth inch of soil plus a one-fourth-inch mulch of sifted sand. Germination begins in about 12 to 40 days and is at its peak about 15 days later. The seedbeds should be covered with screens to prevent birds from digging up the seeds and eating the cotyledons.

[25] The styles are hygroscopic and twist and untwist with moisture changes. Hence, seed sown with styles attached are difficult to keep covered.

CRATAEGUS MOLLIS Scheele — Downy hawthorn, redhaw

(Rose family—Rosaceae)

Distribution and Use.—A thorny, deciduous tree, up to 30 or 40 feet, downy hawthorn occurs on low rich soils, usually bottom lands, from southern Ontario to Virginia and west through Kentucky and southern Minnesota, to southeastern South Dakota and Kansas. Although its chief value is for game food and shelter, it has been in occasional use for ornamental planting since 1683; the sweet mealy fruit is sometimes sold in markets.

Seeding Habits.—Downy hawthorn's showy, perfect flowers, white in color, open from April to June. The scarlet fruit, a pome about one-half inch in diameter and containing four to five bony one-seeded nutlets, ripens in late August and September and soon falls from the trees. Dispersal is chiefly by birds and animals. Commercial seed consists of the cleaned nutlets or dried pomes. Figure 93 illustrates a typical seed of the *Crataegus mollis* type with thick bony outer coat or pericarp, a thin membranous inner coat (the seed coat proper), a large embryo, and absence of endosperm. No data are available on frequency of seed crops, age of commercial seed production, or on the development of climatic races.

Collection, Extraction, and Storage.—The fruit may be hand-picked or shaken from the trees as soon as it has become scarlet in color and has yellow, mealy flesh. Since the number of empty seeds varies greatly from tree to tree, frequent cutting tests are advisable during the collecting period. Unless immediate extraction is planned, care should be taken to prevent heating of the fruit. Extraction is readily accomplished by maceration in water, allowing the pulp to float away. After superficial drying, the seed is ready for storage or for use. Cost of commercial seed: $1.25 to $2 per pound (clean); $0.50 (dried fruit). No data are available on number per pound, purity, or soundness of downy hawthorn seeds; soundness is probably not very high because of the presence, as in other species of *Crătaegus,* of many empty and abortive seeds. Optimum storage conditions are not known; it is likely, however, that the seed will keep for a year or two under ordinary storage or much longer if stored dry in sealed containers at about 40° F.

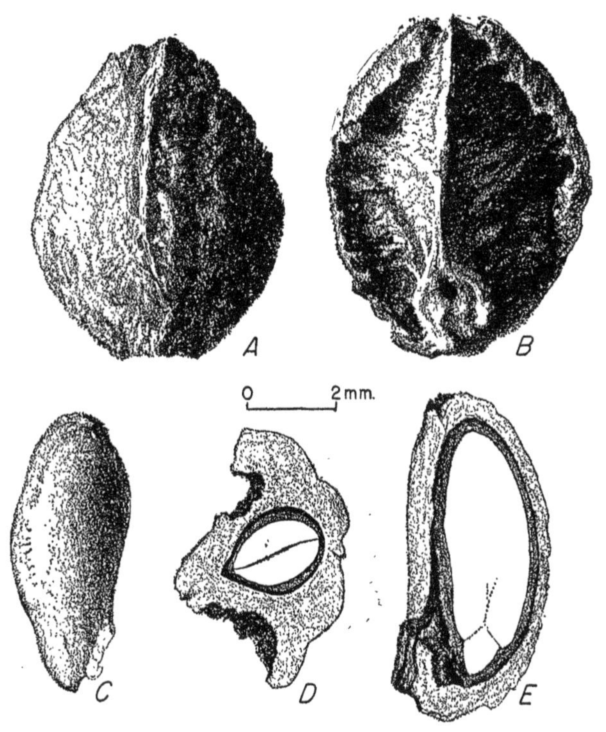

Figure 93.—Seed views of *Crataegus* spp.: *A* and *B,* Exterior views in two planes; *C,* cross section; *D,* longitudinal section; *E,* embryo.

Germination.—Natural germination, which is epigeous, takes place in the spring, occasionally 1 year after seed fall but more often after 2 to 3 years. Rich moist bottom-land soils form the best seedbeds. The seeds show pronounced dormancy due to a combination of an impermeable seed coat and conditions within the embryo, particularly the hypocotyl. Germination can be stimulated by stratification for several weeks in moist peat at 70° to 80° F., followed by stratification for an additional 75 to 90 days at 41°. Longer periods may be required if the seed are allowed to dry before stratification, or if old seed are used. Treatment with sulfuric acid or mechanical scarification can likely be substituted for the warm stratification. Germination tests can best be made in sand flats or peat mats and require about 400 properly pretreated seeds per test. Temperatures fluctuating diurnally from 60° to 80° appear to be satisfactory. Germinative capacity (2 samples): 50 to 64 percent.

NURSERY AND FIELD PRACTICE.—Downy hawthorn seed may be sown in well-prepared beds in the fall or spring. Fall sowing should be done as early as possible so as to have the seed exposed to a few weeks of warm moist weather before frost. A mulch of burlap, straw, or leaves held in place until germination time in the spring is necessary. Spring sowing may employ pretreated or untreated seed. If pretreated seed are used, surface soil should be kept moist until after germination period. If untreated seed are used, beds should be mulched and kept intact until germination time the following spring and then the mulch removed. Seed should be drilled in rows 8 to 12 inches apart and covered with about one-fourth inch firmed soil. Nursery germination: 40 to 65 percent. *Crataegus mollis* is one of a group of species in the genus with seed coats contributing to dormancy. Seeds of those species without this characteristic do not require the initial stratification at high temperatures, but germinate readily after fall sowing or low-temperature stratification. Most hawthorns soon develop a long taproot and hence should not be kept in the seedbeds more than 1 year.

CUPRESSUS L. Cypress

(Pine family—Pinaceae)

DISTRIBUTION AND USE.—The true cypresses, comprising about 12 species, are evergreen trees or shrubs widely distributed in the warmer temperate or subtropical regions of the Northern Hemisphere. Six species are native to North America, but none of them has yet been widely used in reforestation. Because of their durable wood several cypresses are valuable for lumber, posts, poles, and railroad ties; and the wood of some is used for making furniture and for fuel. Many of the cypresses are valuable for erosion control and wildlife food and shelter; some of them are used for windbreaks and hedges; and most of them are useful as ornamentals. Four species, which are of known or supposed value for reforestation in the United States and for which reliable information is available, are described in table 82.

Cupressus arizonica is planted in Europe where it grows quite thriftily. This species is also cultivated extensively for windbreaks in southern California. *C. macrocarpa* is planted extensively for windbreaks and hedges and is used as an ornamental on the Pacific coast, in the Southeastern States, and also in Europe, South America, and Australia.

SEEDING HABITS.—The male and female flowers are borne on different twigs of the same tree. The male flowers, bearing pollen only, are small yellow, or rarely red, cylindrical catkins borne on the ends of branchlets. The female flowers, which develop into cones, are small subglobose, erect greenish bodies, covered with scales. Cones of *Cupressus,* which mature at the end of the second season, are small globose or ellipsoid bodies, usually one-half to 1 inch in diameter, dark brown when mature, and are made up of 4 to 14 woody, shieldlike scales. Each scale has a 3-cornered protuberance at the center of its outer surface, and bears 15 to 20 seeds. The seeds have narrow, hard wings, and are small, compressed, and narrowly angled (figs. 94 and 95), and brown when ripe. Comparative seeding habits of 4 species are given in table 83. In *C. torulosa* cone bearing begins at about 15 years of age. Information for other species is lacking.

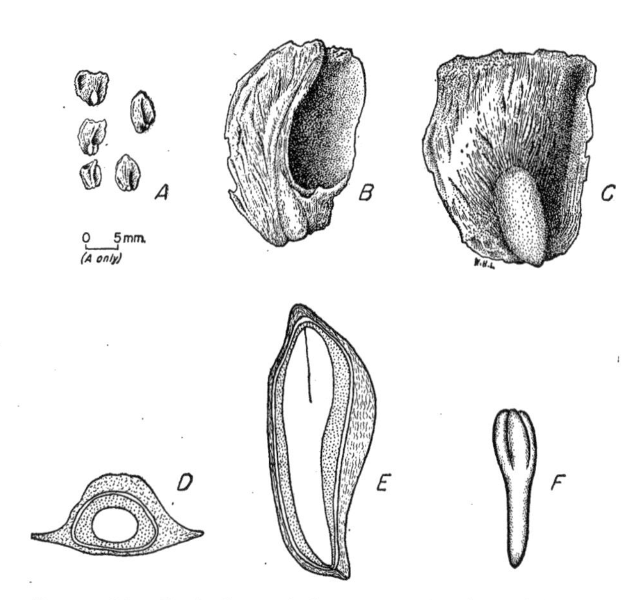

FIGURE 94.—Seed views of *Cupressus arizonica: A, B,* and *C,* Exterior views in several planes; *D,* cross section; *E,* longitudinal section; *F,* embryo.

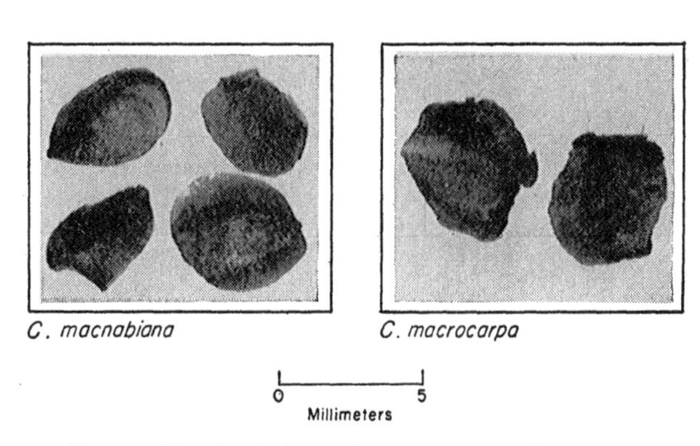

FIGURE 95.—Seed views of two species of *Cupressus.*

No information is available as to the development of climatic races in *Cupressus.* That there is variation within the individual species, however, is shown by the recognition of varieties of most of them.

CUPRESSUS

TABLE 82.—*Cupressus: Growth habit, distribution, and uses*

Accepted name	Synonyms	Growth habit	Natural range	Chief uses	Date of earliest cultivation
C. arizonica Greene (Arizona cypress).	*C. glabra* Sudw. (smooth cypress), *C. arizonica* var. *bonita* Lemm.	Small to medium-sized tree.	Mountains of central and southeastern Arizona, southwestern New Mexico, and northern Mexico.	Timber, erosion control, wildlife food, ornamental purposes, posts, fuel, windbreaks.	1882
C. macnabiana A. Murr. (McNab cypress).	California mountain cypress.	Shrub or small tree.	Dry slopes in northern California and southwestern Oregon.	Erosion control	1854
C. macrocarpa Hartw. (Monterey cypress).	*C. hartwegii* Carr.	Small to medium-sized tree.	Restricted area near Monterey, Calif.	Windbreaks, hedges, ornamental use.	1838
C. torulosa D. Don[1] (Bhutan cypress).	*C. nepalensis* Loud.	Tall tree	Outer ranges of western Himalayas and western Szechwan, China, at moderate elevations.	Timber, railroad ties, incense, ornamental purposes.	1824

[1] A tree sometimes cultivated in the United States as "*C. torulosa*," however, is the Chinese *C. duclouxiana* Hickel (syn.: *C. torulosa* Rehd. & Wils., not D. Don).

TABLE 83.—*Cupressus: Time of flowering and cone ripening, and frequency of seed crops*

Species	Time of—			Seed year frequency	
	Flowering	Cone ripening	Seed dispersal	Good crops	Light crops
C. arizonica	Spring	September	Intermittent for several years.	Almost annually.	Intervening years.
C. macnabiana	March–April	Summer	Late autumn	do.	do.
C. macrocarpa	February–March	August	Autumn		
C. torulosa	January–February	April	August–December	Frequent	

COLLECTION, EXTRACTION, AND STORAGE.—Cypress cones may be collected from 1 year to several years after ripening because many of them remain closed on the tree for long periods. They are ripe enough to collect when they have become dark brown, sometimes covered with a bluish bloom; older cones are grayish in color. The cones may be opened by drying in the sun or in a cone kiln—temperatures of more than 120° F. are probably unnecessary—and the seed separated from the cones by screening. Because of the small wings, no dewinging operation is necessary. Details as to time of collection and seed yield per cone are given in table 84. In *Cupressus torulosa* there are about 120 cones to the pound. The size, purity, soundness, and cost of commercial cleaned seed vary considerably by species as shown in table 85.

Few scientific data are on hand concerning the storage of cypress seeds, but it is known that seeds of this genus usually keep quite well. In the case of *Cupressus arizonica* sound and apparently germinable seed have been found in cones that had remained closed on the trees for 8 to 10+ years, and seed stored for 10 years in a loosely covered fruit jar kept at a temperature varying from 40° to 90° F. still had a germination of 13 percent. In the case of *C. torulosa* comparisons of the germinability of fresh seed with that shipped from India to Europe points to a rather rapid loss of viability. For most *Cupressus* species it appears that the seeds could be kept for 1 year to several years with little loss of viability if they were stored in sealed containers at temperatures of about 40°.

TABLE 84.—*Cupressus: Time of collection, and seed yield*

Species	Time of collection	Seeds per cone
C. arizonica	One year to several years following ripening.	48–112
C. macnabiana	September–December	60–96
C. macrocarpa	August–October	160–280
C. torulosa	April–September	45–80

GERMINATION.—Natural germination is best on moist, mineral soil; however, it is somewhat slow and irregular, and for *Cupressus macrocarpa* natural seedlings appear to be rare or absent in its native range but rather abundant in planted stands. Germination is epigeous. Although cypress seeds are usually germinated without pretreatment and the generally low germination is often considered to be due to the presence of poor or empty seed, limited tests on some species indicate the presence of embryo dormancy. In the latter case stratification in moist sand at low temperatures has improved germination. Two samples of *C. arizonica* seed when stratified in moist sand for 60 days at

TABLE 85.—*Cupressus: Yield of cleaned seed, and purity, soundness, and cost of commercial seed*

Species	Cleaned seed per pound				Commercial seed		
	Low	Average	High	Basis, samples	Purity	Soundness	Cost per pound
	Number	*Number*	*Number*	*Number*	*Percent*	*Percent*	*Dollars*
C. arizonica	27,000	40,000	59,000	8	82	[1]55	1.50–3.00
C. macnabiana	79,000	----------	130,000	2	75	92	3.75
C. macrocarpa	46,000	66,000	162,000	38	80	[2]33	1.75–2.50
C. torulosa	75,000	92,000	120,000	18	[3]50	----------	2.00–4.00

[1]2 samples. [2]11 samples. [3]3 samples.

TABLE 86.—*Cupressus: Recommended conditions for germination tests, and summary of germination data*

Species	Test conditions recommended				Germination data from various sources					
	Temperature		Duration		Germinative energy		Germinative capacity			Basis, tests
	Night	Day	Untreated seed	Stratified	Amount	Period	Low	Average	High	
	°F.	*°F.*	*Days*	*Days*	*Percent*	*Days*	*Percent*	*Percent*	*Percent*	*Number*
C. arizonica	68	86	75	30	28	8–20	17	26	40	9
C. macnabiana	70	90	30	----------	----------	----------	4	----------	27	2
C. macrocarpa	(1)	(1)	75	----------	11	30	2	14	37	37
C. torulosa	(1)	(1)	40	----------	10	20	1	58	70	19

[1]Temperatures of 68° (night) and 86° F. (day) suggested; experimental data not complete.

41° F. gave germination results of 26 and 30 percent in 30 days while unstratified seed gave 14 and 13 percent in 75 or more days. One test with *C. macrocarpa* stratified for 60 days at 32° germinated 22 percent. Germination of cypress seeds may be tested in sand flats or standard germinators, using 1,000 seeds per test. Methods recommended and average results for the 4 species discussed here are given in table 86.

NURSERY AND FIELD PRACTICE.—Seeds of *Cupressus* for which samples show dormancy should be fall sown, or nondormant seed may be sown broadcast in the spring or in drills covering them with one-eighth inch soil and one-fourth inch sand for a mulch. The seedbeds should be kept well moistened during germination and can be shaded, but that is probably not necessary. Germination usually takes place 2 to 3 weeks after spring sowing. Nursery information is lacking for *C. macnabiana*, but the methods described here are probably suitable for this species also. Seedlings of several species are attacked occasionally by cedar blight caused by *Phomopsis juniperovora* which can be controlled partially by frequent spraying with 4-4-50 bordeaux, colloidal sulfur, or normal Semesan. For field planting 2- or 3-year-old transplants of *C. torulosa* are recommended; similar stock of the other species would also probably be suitable. The cypresses grow well on many soils but do best on deep sandy loams. They may also be propagated by cuttings or veneer grafting.

DENDROMECON RIGIDA Benth. Stiff bushpoppy, treepoppy

(Poppy family—Papaveraceae)

Native to the chaparral forests of the coast range, Sierra Nevada, and mountains of southern California, stiff bushpoppy is a small to large, branching, evergreen shrub browsed by deer and useful as cover on burned-over areas.

The large, bright-yellow, poppylike, perfect flowers appear from April to June; the fruit, a linear grooved capsule 2 to 4 inches long, ripens from May to July, at which time the seeds may be collected. Stiff bushpoppy seeds are black in color, almost globular with a basal appendage which can be rubbed off, and a consist of a relatively thin coat, and a minute embryo embedded in an oily endosperm (fig. 96).

Cleaned seed per pound (2 samples): 42,000 to 50,000; purity, 77 percent; soundness, 97 percent. Cost of commercial seed, about $23 per pound.

Stiff bushpoppy is propagated from seeds which are slow to germinate. The best germination thus far obtained, 21 percent in 50 days, followed stratification for 2 months at temperatures alternating diurnally from 41° to 70° F. Germination begins in 50 days after sowing.

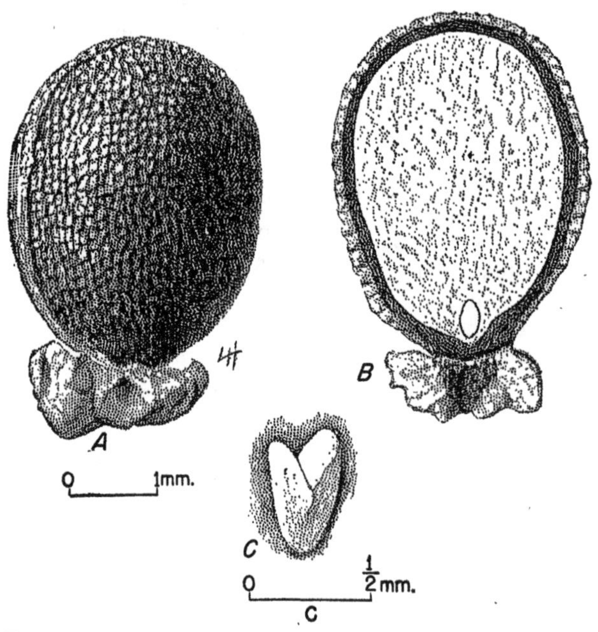

FIGURE 96.—Seed views of *Dendromecon rigida: A*, Exterior view; *B*, longitudinal section, showing minute embryo; *C*, embryo.

DIOSPYROS VIRGINIANA L. Common persimmon

(Ebony family—Ebenaceae)

Distribution and Use.—Occurring in open woods and as an invader of old fields from Connecticut west through southern Ohio to eastern Kansas and south to Florida and Texas, the common persimmon is a tree of medium to occasional large size; frequently it is a shrub forming dense thickets from stolons. Its wood is highly valued for weavers' shuttles, for golf-club heads, and for other uses which require hard, smooth-wearing wood. The nutritious fruit is relished by animals, birds, and, when thoroughly ripe, by man. It is a valuable honey plant. This species has been cultivated for its handsome foliage and fruit since 1629.

Seeding Habits.—The small, dioecious, axillary flowers are borne after the leaves from March to mid-June; the fruit, an orange or yellow—green in some forms, sometimes black—glaucous berry with conspicuous, persistent calyx and containing 3 to 8 seeds, ripens September to November. The seed, about one-half inch in length and much flattened, consists of a thick, light-brown, roughened seed coat, a large endosperm which is pearl gray in color, tough and flinty, and a rather small embryo (fig. 97). Dispersal occurs from ripening until late winter and is effected by birds and animals.

Commercial seed-bearing age: minimum, 10 years; optimum, 25 to 50 years; maximum, 60 to 80 years. Good seed crops are borne about every 2 years, with light crops in intervening years. Occasional trees bear fruits without seeds.

No information is available as to the existence of geographic or other strains. Since, however, there is considerable variation in the size, quality, and color of the fruit, it seems likely that such strains do exist. The development of horticultural varieties by selection and breeding of superior individuals seems promising.

Collection, Extraction, and Storage.—The fruit may be collected by picking or shaking from the trees as soon as it turns orange brown in color and becomes soft. It may also be picked from the ground after natural fall. The seeds are easily removed by running the berries with water through a macerator, allowing the pulp to float away, or by rubbing and washing pulp through one-fourth inch mesh hardware cloth. After cleaning, the seed should be spread out to dry for a day or two. Cleaned seed: yield per hundred pounds of fruit, 10 to 30 pounds; number per pound (8 samples): low, 665; average, 1,185; high, 1,764. Commercial seed averages 96 percent in purity, 90 percent in soundness, and costs $0.65 to $1 per pound. No data on optimum storage conditions are available; it is likely, however, that viability can be prolonged considerably by sealed storage of dry seed at temperatures of about 40° F.

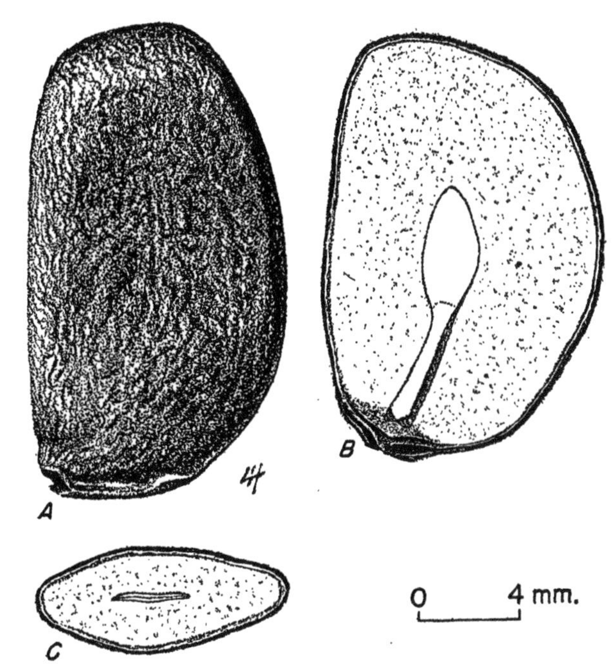

Figure 97.—Seed views of *Diospyros virginiana: A,* Exterior view; *B,* longitudinal section; *C,* cross section.

Germination.—Natural germination usually occurs in April or May[26] following seed ripening

[26] Some seeds apparently do not germinate for 2 or 3 years; these are probably from fruits which have hung on the trees until late in the season and have become dry.

and is best on moist but well-drained, loamy soils. Germination is epigeous (fig. 98). The seed is dormant; this condition is said to be due to a layer of cells which cap the radicle and prevent moisture absorption. Dormancy can be broken by stratification in sand or peat for 60 to 90 days at 50° F.[27] or in fresh seed by removing by hand the layer of impermeable cells. Adequate tests may be made in sand or peat flats at temperatures alternating diurnally from 68° to 86° and require 100 to 200 stratified seeds. Test duration: stratified seed, 40 to 60 days; untreated seed, 125 days or longer. Germinative energy, 53 to 94 percent in 20 to 34 days. Germinative capacity (10 tests): low, 0 percent; average, 61 percent; high, 100 percent.

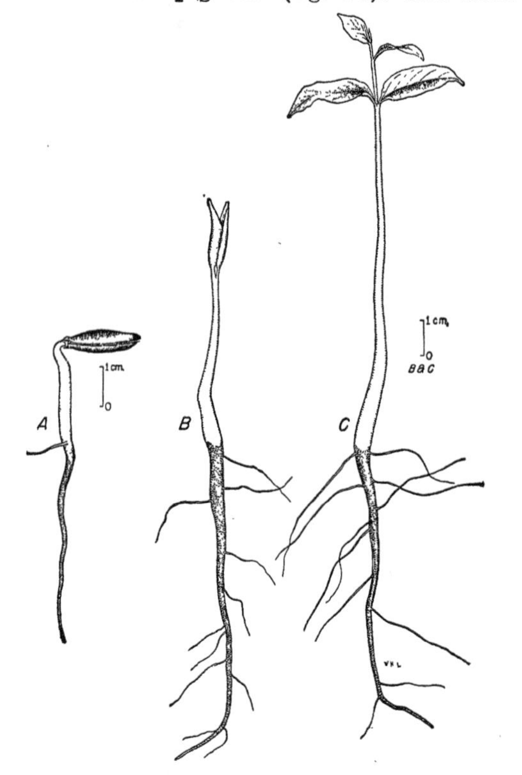

FIGURE 98.—Seedling views of *Diospyros virginiana*: *A*, At 4 days; *B*, at 6 days; *C*, at 8 days.

NURSERY AND FIELD PRACTICE.—Persimmon seed may be fall-sown or stratified and sown in the spring. Best results are obtained with seed sown in drills in deeply plowed beds and covered with one-half to three-fourths of an inch of firmed soil which is kept moist until germination begins. It is necessary to mulch fall-sown beds with burlap or straw, but the mulch should be removed at the first sign of germination in the spring. Twenty-five to thirty-three percent of the viable seed sown will produce usable seedlings. The seedlings have a strong taproot and should be field planted at the end of the first season. Such stock will average about 8 inches in height. Persimmon may also be propagated from root cuttings and by grafting.

[27] 41° F. can also be used but is somewhat less effective.

ELAEAGNUS L. Elaeagnus

(Elaeagnus family—Elaeagnaceae)

DISTRIBUTION AND USE.—*Elaeagnus* includes about 40 species of deciduous or evergreen shrubs or trees, some of which are spiny; they are native to southern Europe, Asia, Australia, and North America. Although they are grown chiefly as ornamentals, some species are useful for shelter-belt purposes, some produce edible fruits, and most of them are a source of wildlife food. Two species, which are valuable for planting and for which reliable information is available, are described in table 87. *E. commutata* has been planted in both Europe and North America, but less extensively than *E. angustifolia* which is grown widely and sometimes escapes from cultivation.

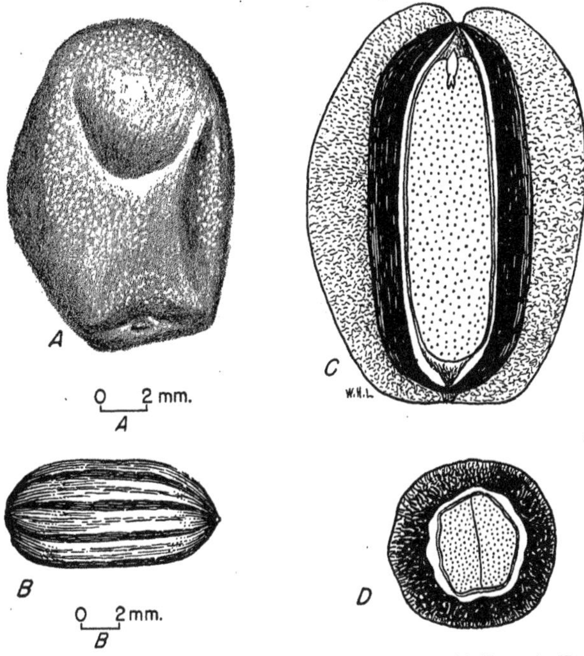

FIGURE 99.—Seed views of *Elaeagnus angustifolia*: *A*, Exterior view of dried fruit; *B*, exterior view of stone; *C*, longitudinal section of dried fruit; *D*, cross section of stone.

FIGURE 100.—Seed views of *Elaeagnus commutata*.

SEEDING HABITS.—The flowers of *Elaeagnus angustifolia* and *E. commutata* are small, perfect or polygamous and are borne singly or in clusters in the spring. The fruit, averaging 0.4 inch in length, ripens in the late summer or fall, is drupelike, fleshy or mealy, and encloses a single, ellipsoid, striate stone (figs. 99 and 100). Comparative seeding habits of these two species are as follows:

	E. angustifolia	*E. commutata*
Time of—		
Flowering	June	May–July.
Fruit ripening	August–October	July–September.
Seed dispersal	do	August–November.
Color of—		
Flowers	Silvery	Yellow.
Ripe fruit	Silvery yellow	Silvery.

TABLE 87.—*Elaeagnus: Growth habit, distribution, and uses*

Accepted name	Synonyms	Growth habit	Natural range	Chief uses	Date of earliest cultivation
E. angustifolia L. (Russian-olive).	*E. hortensis* Bieb. (oleaster).	Small tree or shrub, sometimes spiny.	Southern Europe and western Asia to the western Himalayas.	Ornamental, shelter-belt purposes, winter bird food, honey plant.	[1]
E. commutata Bernh. (silverberry).	*E. argentea* Pursh not Moench. (wolfberry).	Upright, thicket-forming, unarmed shrub.	Quebec to Yukon Territory, south to South Dakota, Utah, and British Columbia.	Shelter belts, bird food, erosion control.	1813

[1] Long cultivated.

ELAEAGNUS

Scientific information as to racial variations in *Elaeagnus* is lacking, but in view of the wide natural distribution of the species described the development of climatic races is likely. The small-seeded spiny form of *E. angustifolia* is considered hardiest for Prairie-Plains planting.

COLLECTION, EXTRACTION, AND STORAGE.—The ripe fruits may be picked from the plants by hand or beaten or stripped from the branches onto canvas. Collection usually is done from September to December. The fruits may then be spread out to dry—the usual practice with *Elaeagnus angustifolia*—or they may be run through a macerator with water and the pulp floated off or screened out—the usual practice with *E. commutata*. Accordingly, commercial seed may consist either of dried fruits or of cleaned stones. There is some evidence to indicate that moderate fermentation accelerates after-ripening and otherwise improves viability in *E. angustifolia*, so this might be desirable practice for that species. When dried fruits are used, leaves and twigs should be removed by fanning. The size, purity, soundness, and cost of commercial seed vary by species as shown below:

	E. angustifolia	*E. commutata*
Cleaned seed:		
Per 100 pounds of fruit..pounds..	15–60	--
Per pound:		
Low..number..	4,600	2,700
Average..do....	5,200	3,400
High..do....	6,100	3,800
Basis, samples..do....	4+	3
Dried fruits:		
Per 100 pounds of fruit[1]..pounds..	80–84	--
Per pound:		
Low..number..	1,800	--
Average..do....	2,900	2,000
High..do....	4,500	--
Commercial seed:		
Purity..percent..	97	96
Soundness..do....	92	100
Cost per pound..dollars..	0.20–0.80	0.75–1.25

[1] Basis, samples: 25.

Under ordinary storage conditions seed of *Elaeagnus commutata* retains its viability satisfactorily for 1 to 2 years and that of *E. angustifolia* up to 3 years. In one test of the latter, dried fruits which were stored in a sealed jar at 40° to 50° F. showed no loss in viability at the end of 5½ years.

GERMINATION.—*Elaeagnus* seed is slow to germinate, apparently because of embryo dormancy, unless sown in the fall or stratified. Germination is epigeous. Dormancy is overcome satisfactorily in *E. angustifolia* and *E. commutata* by 90 days' stratification in moist sand at 41° F. The stratification period for *E. angustifolia* could probably be reduced somewhat if cleaned seed were used instead of dried fruits. Germination tests should be made in sand flats at 68° (night) to 86° (day) using 200 stratified seeds per test. Average results for the 2 species discussed here are as follows:

	E. angustifolia	*E. commutata*
Basis, tests..number..	32	1
Test duration..days..	60	50
Germinative energy:		
Amount..percent..	7–76	52
Period..days..	10–32	13
Germinative capacity:		
Low..percent..	7	--
Average..do....	34	60
High..do....	79	--

NURSERY AND FIELD PRACTICE.—*Elaeagnus* seeds should be sown in drills in the fall, or stratified seed sown in the spring, at the rate of 40 viable seed per linear foot, and covered with three-fourths of an inch of soil. Some soil adheres to the pubescent leaves of newly emerged seedlings, often causing heavy losses. The beds should be mulched to cover the soil and prevent rain spattering. Since the leaves often drop late in the fall, some nurserymen prefer to dig the stock early in the spring. Field planting is usually done in the spring, using 1–0 stock. *Elaeagnus* species grow well in almost any well-drained soil, including limestone or alkaline soils. They may also be propagated by stem cuttings, root cuttings, layers, and grafting.

EPIGAEA REPENS L. Trailing-arbutus

(Heath family—Ericaceae)

DISTRIBUTION AND USE.—Native to woodlands on acid sandy soils from Newfoundland to Saskatchewan, south to Florida and Kentucky, trailing-arbutus is a prostrate, evergreen, creeping shrub of wide esthetic appeal. Although reputedly difficult to grow in cultivation, it has been in occasional use for ornamental planting since 1736. Its fruits are sometimes used for food by small game, and its flowers are often sold by street vendors—a practice which is to be deplored because of the damage caused to the plants by careless handling. The leaves are occasionally used in medicine.

SEEDING HABITS.—Trailing-arbutus' spicy, fragrant flowers, pink to white in color, bloom in April or May. They are usually unisexual, the two kinds of flowers occurring on different plants; sometimes, however, they are perfect. The fruit, a globular, dehiscent capsule about three-eighths of an inch in diameter and with white juicy flesh, ripens in late June to July. Each capsule contains 20 to 500, or an average of 250, tiny, shiny brown, hard seeds (fig. 101) embedded in the pulp. At the proper stage of ripeness the seeds are ejected from the opened capsule with some force; at other times they are separated with difficulty. Dehiscence and partial seed dispersal take place soon after ripening; seed dispersal, however, is apparently spread over a long period, as fruits containing some seed can still be found in late fall or winter.

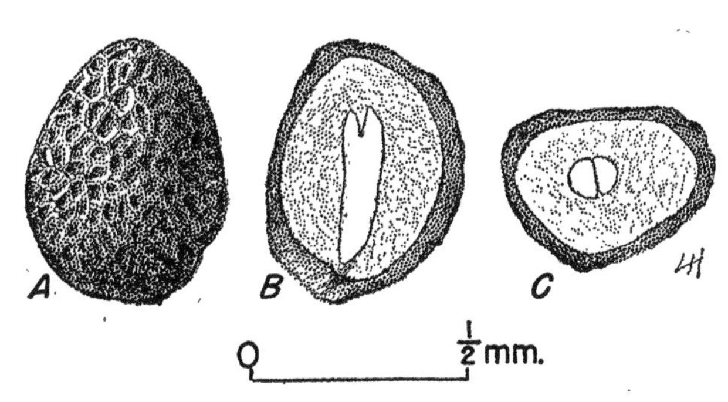

FIGURE 101.—Seed views of *Epigaea repens: A,* Exterior view; *B,* longitudinal section; *C,* cross section.

COLLECTION, EXTRACTION, AND STORAGE.—Collection and extraction of trailing-arbutus seed is done on a comparatively small scale by hand shortly after ripening. Cleaned seed per pound (1 sample): 10,300,000. Cost of commercial seed: 25 cents per packet. No formal storage tests have been reported but from the results after incidental storage, both at room temperature and in a refrigerator, it appears that viability declines rapidly after the seed are about 3 months old.

GERMINATION, PROPAGATION, AND FIELD PRACTICE.—Germination, which is epigeous, occurs readily without pretreatment. Germination tests and germination for plant production are made in the same manner—the essentials being an acid soil mixture of sandy peat and leaf mold, abundant moisture, and high humidity. Room temperatures are satisfactory. Covering the germination tray or pot with a pane of glass, or placing under a bell jar, will provide the desired humidity. The seeds are merely sprinkled on the soil surface without covering. For germination tests where the seedlings are not to be saved, sprinkling the seeds on filter paper laid over wet peat moss facilitates observation and counting of the minute plants. Germination of fresh seed takes place in 20 to 65 days, mostly within 30 days. Germinative capacity averages 45 percent.

As soon as the seedlings in the germination tray have developed three to five leaves above the cotyledons, perhaps 2 months after germination, they should be transplanted to flats or pots containing acid peaty soil to provide adequate growing space. These should be kept covered with glass until the seedlings are reestablished, after which the protection can be gradually removed. At the end of 1 year the plants should have grown to be 4- or 5-inch rosettes, and at 2 years they should provide a good display of bloom.

When placed in permanent locations, each plant should be set in a pocket of at least 1 cubic foot of prepared soil composed of 4 parts oak-leaf mold, 2 parts crumbled peat moss, and 2 parts sand, or some similar mixture. The acidity should be around pH 4.5 to 5.0. Plenty of moisture, good drainage, and partial shade are also essential.

Trailing-arbutus, in common with many other plants, thrives best in association with certain fungi which form mycorrhizae on the roots. It is, therefore, a wise precaution when preparing soil mixtures for germination and later growth to include soil collected near wild, healthy arbutus plants. This will almost certainly introduce the necessary fungus. Trailing-arbutus can also be propagated by cuttings if care is used to insure the introduction of the mycorrhizal fungus.

ERIOGONUM FASCICULATUM Benth. Flattop eriogonum

(Buckwheat family—Polygonaceae)

Also called California-buckwheat, flat-top, flat-top buckwheatbrush.

A low shrub, native to dry lower slopes and mesas of the Coast Range and other mountains of southern California, flattop eriogonum is a very valuable honey plant; its dry flower heads also provide food for deer in the fall and winter months.

The small, perfect flowers are borne in the spring and summer in small clusters or heads scattered along or near the ends of the shrub's branches. Its seed, a small three-angled achene, brown and shining and enclosed within the persistent calyx (fig. 102), ripens from June to August.

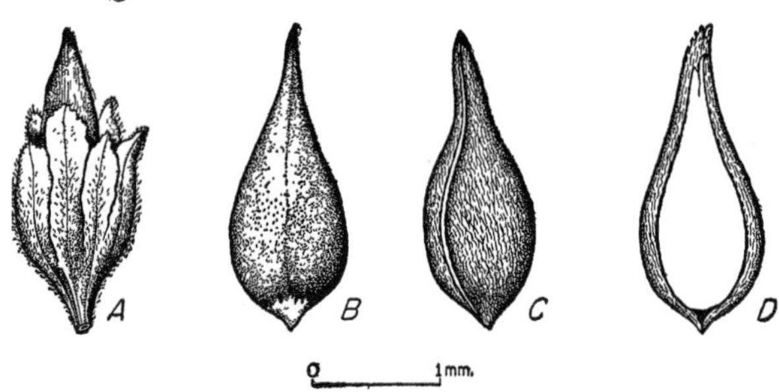

FIGURE 102.—Seed views of *Eriogonum fasciculatum: A,* Exterior view, fruit in calyx; *B* and *C,* exterior views in two planes; *D,* longitudinal section.

The seeds may be collected by stripping the flower clusters into sacks as soon as they have dried. They should then be run dry through a macerator, and through a fanning mill to separate the chaff from the seeds. Removal of the persistent calyx can probably be accomplished by rubbing the seeds between screens, but this is probably unnecessary since it is unlikely that the dry scales interfere with germination. Cleaned seed per pound (2 samples): 239,000 to 493,000. Soundness appears to be relatively low.

Germination is epigeous (fig. 103) and occurs about a week after sowing. No pretreatment is necessary. Germinative capacity (2 samples): 20 to 40 percent.

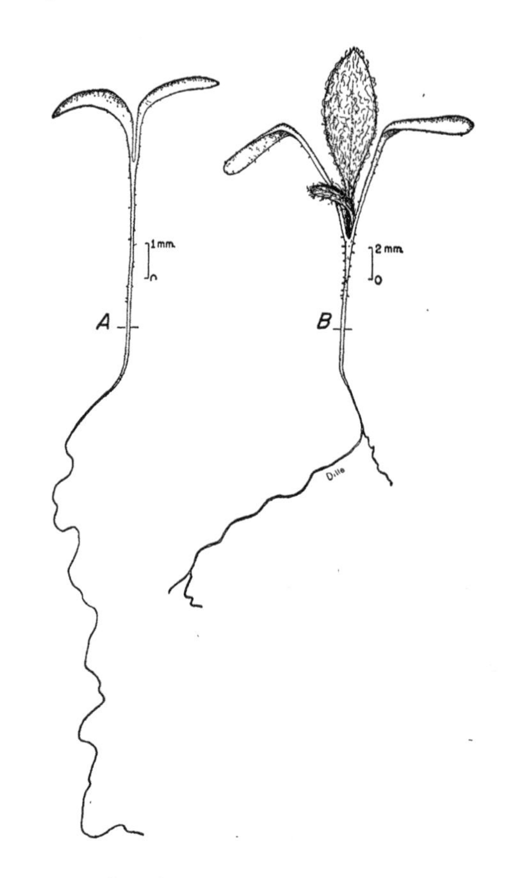

FIGURE 103.—Seedling views of *Eriogonum fasciculatum: A,* Very young seedling; *B,* older seedling.

EUCALYPTUS GLOBULUS Labill. Tasmanian blue eucalyptus, blue-gum

(Myrtle family—Myrtaceae)

Distribution and Use.—A tall forest tree, native to Tasmania, Tasmanian blue eucalyptus is the most widely planted *Eucalyptus* in America. It was extensively planted in California[28] for many years because it was thought to have considerable promise for timber production. Although it has been used there to some extent for piling, insulator pins, spokes and other specialty products, the quality of the wood is not very high and the chief uses of the tree are for ornamental purposes, shelter belts, and fuel. The leaves are distilled for oil and the flowers yield considerable honey, although of a rather strong flavor. *Eucalyptus* has also been planted to some extent in Florida, South America, Italy, Algeria, and India.

Seeding Habits.—The large, perfect axillary flowers, mostly solitary, appear from December to May; the fruit, a hemispherical woody capsule ¾ to 1 inch in diameter and opening at the apex by 3 to 6 valves, ripens from March to June and is borne in abundance. Capsules contain numerous small, brown, angular seeds (fig. 104), each consisting of a thin coat and twisted cotyledons without endosperm. Dispersal is effected largely by wind within a month or two after ripening. Good seed is produced by trees as young as 10 years of age. No data are available on the frequency of seed crops or the existence of geographic races.

[28] Tasmanian blue eucalyptus reproduces abundantly in California; seedlings may be found in nearly every grove, especially in shaded ravines.

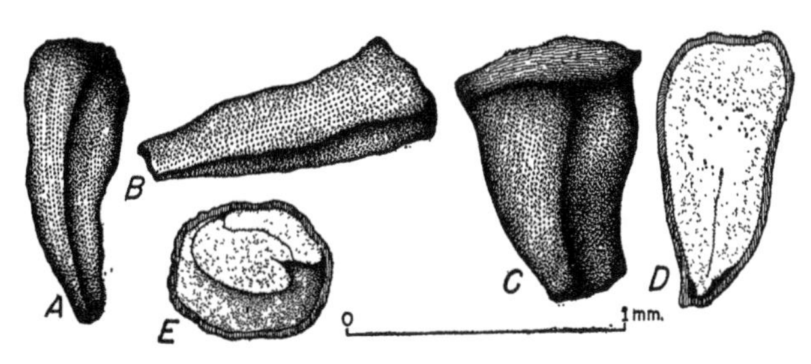

Figure 104.—Seed views of *Eucalyptus rudis: A, B,* and *C,* Exterior views in three planes; *D,* longitudinal section; *E,* cross section.

Collection, Extraction, and Storage.—The capsules can be picked from the trees in spring to early summer. However, since a single branch will often bear buds, flowers, empty capsules, and mature capsules about to open, care must be used to collect only large, well-developed fruit. After drying in the sun a few hours, capsules begin to open and release the seeds; the latter are cleaned by screening. Cleaned seed per pound (3 samples): low, 104,000; average, 138,000; high, 160,000. Soundness is rather low. Cost of commercial seed per pound: $6.50. Optimum storage conditions are not known. Since seed which had been stored in envelopes in a laboratory at room temperature showed 4.4 percent germination after 10 years' storage and 2.2 percent after 20 years, presumably sealed storage at low temperatures would give much better results.

Germination. — Germination is epigeous (fig. 105) and apparently occurs soon after seed fall. Tests may be run in sand flats, petri dishes, or standard germinators and require the use of 1,000 seeds per test. Since the seedlings are very susceptible to damping-off, sterilization of the seed or the germination medium is probably advisable. Temperatures of 60° to 70° F. appear to be satisfactory. Duration of test: 10 to 15 days. Germinative capacity (2 samples): 75 to 80 percent.

Nursery and Field Practice.—Blue-gum seed, like that of other eucalypts, is seldom sown direct in the nursery. Common practice is to sow the seed in June in flats about 18 inches square and 4 inches deep, filled with light, sandy loam. Enough seed to produce from 200 to 400 seedlings

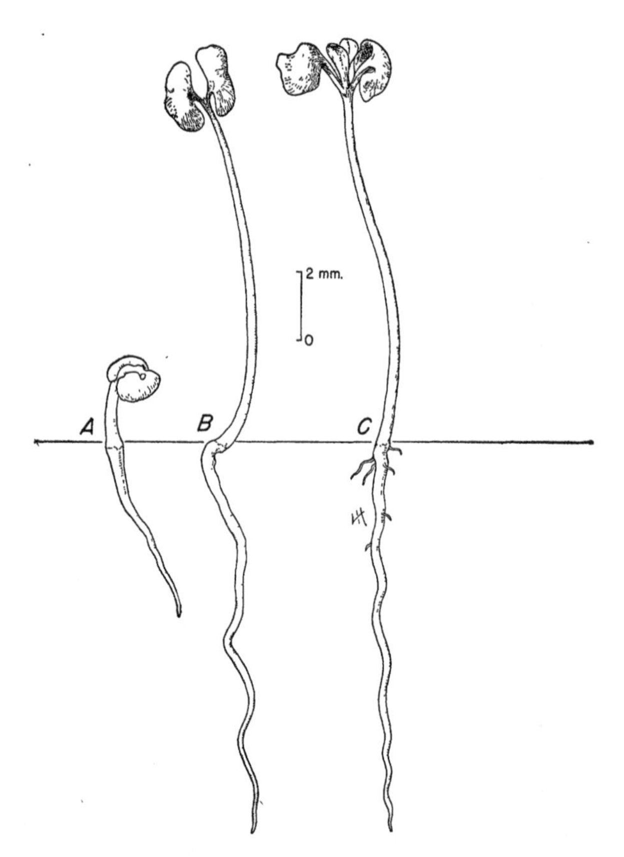

FIGURE 105.—Seedling views of *Eucalyptus: A, E. viminalis* at 1 day; *B*, at 8 days; *C, E. rudis* at 42 days.

is sown broadcast and covered with about one-eighth inch of fine sand or clean, fine sawdust. The flats are then placed in a shady spot, such as a lath house, until germination is complete. This usually takes about 2 weeks and varies from 5 to 30 percent. Since the young seedlings are very susceptible to damping-off, the soil must be watered sparingly during the germination period. Treatment of the soil with sulfuric acid is also suggested. If desired, the seed may also be sown in standard nursery seedbeds with board sides and covered with light cotton cloth. The beds are watered thoroughly after sowing and water applied only as needed until germination is at its height.

Because the roots of Tasmanian blue eucalyptus are very susceptible to drying, the usual practice is to use balled stock for field planting. Consequently, when the seedlings are 1 to 2 inches high (about 1 month old) they are transplanted from flats or seedbeds into paper pots in which they are set out in the field without disturbing the root system. Sandy loam containing some leaf mold forms a good medium for the growth of transplants. Field planting is ordinarily done after all danger from frost is past and when the young plants are 5 to 8 months old from seed. Plantations should be cultivated during the first 2 years to reduce competition from grass and weeds and also to prevent girdling of the young trees by mice. This species can also be grown from stump sprouts which should be thinned out in the early spring of the second year.

EUONYMUS L. Euonymus

(Staff-tree family—Celastraceae)

DISTRIBUTION AND USE.—Euonymus includes about 120 species of deciduous or evergreen shrubs or small trees, sometimes creeping or climbing, native to North and Central America, Europe, Asia, and Australia. Because of their attractive fruits, the euonymus species are planted chiefly for ornamental purposes, but they also have some value for wildlife food, shelter-belt purposes, and minor wood products; at least one species is a source of gutta. Three species, for which reliable information is available and which have potential value for conservation planting, are described in detail in table 88.

SEEDING HABITS.—The usually perfect flowers, borne in clusters, bloom in the spring. The fruit, which ripens in late summer or fall, is a four- to five-celled—occasionally two- to three-celled—capsule usually lobed and sometimes winged. Each fruit cell contains one to two seeds (fig. 106) enclosed in a fleshy, usually orange, aril. Good fruit crops are borne almost annually. The seeding habits of the three species discussed vary as shown in table 89. Scientific information as to the presence of climatic races among euonymus species is lacking.

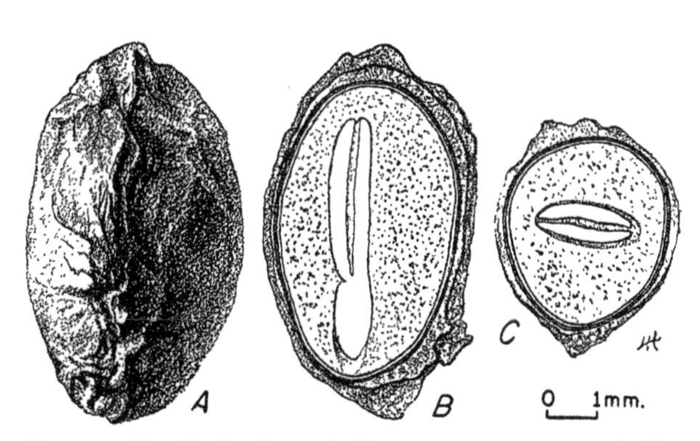

FIGURE 106.—Seed views of *Euonymus europaeus*: *A*, Exterior view of dried fruit; *B*, longitudinal section; *C*, cross section.

COLLECTION, EXTRACTION, AND STORAGE.—Seed of the euonymus species may be collected in the late summer or fall by picking the ripe fruits from the bushes by hand or shaking them onto outspread canvas. They should then be spread out to dry for several days in a warm room. The seeds can be extracted from the fruits by beating the latter in a canvas bag and then rubbing them through a coarse, round-hole grain screen

TABLE 88.—*Euonymus: Growth habit, distribution, and uses*

Accepted name	Synonyms	Growth habit	Natural range	Chief uses	Date of earliest cultivation
E. atropurpureus Jacq. (eastern wahoo).	eastern burning-bush, spindle tree.	Tall shrub or small tree.	Low wet sites from New York to Florida and west to Minnesota and Texas.	Ornamental, possible wildlife food; bark has medicinal properties.	1756
E. europaeus L. (European euonymus).	*Evonymus vulgaris* Mill. (European spindletree).	Upright tall shrub or small tree.	Europe to western Asia; sometimes escapes cultivation in eastern United States.	Ornamental, wood used for small articles, wildlife food, shelter-belt purposes.	(1)
E. verrucosus Scop. (wartybark euonymus).	warty spindletree.	Upright medium-tall shrub.	Southern Europe and western Asia.	do[2]	1763

[1] Long cultivated.

[2] Harvested for gutta in Russia.

891183°—50——12

EUONYMUS

TABLE 89.—*Euonymus: Time of flowering and fruit ripening, and fruit form and color*

Species	Time of—			Fruit form	Color of—			
	Flowering	Fruit ripening	Seed dispersal		Flower	Fruit	Seed	Aril
E. atropurpureus	May–July	Sept.–Oct.	Sept.–winter	Smooth, deeply 3- to 4-lobed, 4-celled.	Purple	Pink to crimson.	Brown	Scarlet.
E. europaeus	May–June	Aug.–Oct.	Aug.–Oct.	Smooth, 4-lobed, 3- to 5-celled.	Yellowish green.	Red to pink.[1]	White	Orange.
E. verrucosus	do.	August	August	Deeply 4-lobed	Brownish	Yellowish red.	Black	do.[2]

[1] Whitish in 1 variety.
[2] Seed not wholly covered by aril.

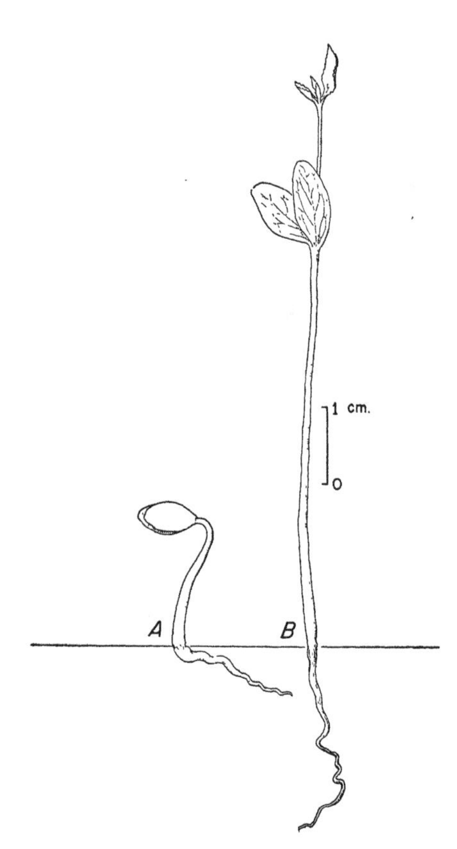

FIGURE 107.—Seedling views of *Euonymus europaeus: A*, At 1 day; *B*, at 12 days.

(10/64). Chaff can be removed by winnowing. Removal of the pulpy arils is accomplished by rubbing the seed through coarse, mesh-wire cloth after they have dried several weeks, but it is difficult to do without breaking the rather thin seed coat and injuring the seed. The yield, size, purity, soundness, and cost of commercial cleaned seed vary by species as shown in table 90.

Russian reports indicate that seed of *Euonymus europaeus* and *E. verrucosus* can be kept satisfactorily for 2 years in ordinary dry storage. It is likely that seed of these species and also *E. atropurpureus* could be kept for longer periods if stored in sealed containers at temperatures of about 41° F.

GERMINATION. — Germination is epigeous (fig. 107). The seeds have dormant embryos and although satisfactory germination in the nursery is obtained by spring sowing stratified seed, suitable methods of germinating the seed in the laboratory have not been developed. Warm plus cold stratification offers the best lead, but even with such treatment germinations of 2 to 12 percent only are obtained, with potential germinations ranging from 30 to 45 percent.

NURSERY AND FIELD PRACTICE.—Euonymus seed, preferably that which has been stratified 90 to 120 days at 32° to 50° F., should be sown in the spring. Nursery germinations of 30 to 70 percent are obtained by this method. The species may be planted satisfactorily on a variety of soils. Propagation by means of cuttings is also practiced, and varieties sometimes are propagated by grafting or budding.

TABLE 90.—*Euonymus: Yield of cleaned seed, and purity, soundness, and cost of commercial seed*

Species	Cleaned seed					Commercial seed		
	Yield per 100 pounds of fresh fruit	Per pound			Basis, samples	Purity	Soundness	Cost per pound
		Low	Average	High				
	Pounds	*Number*	*Number*	*Number*	*Number*	*Percent*	*Percent*	*Dollars*
E. atropurpureus	10–20	8,700	11,800	14,200	3	75	83	1.50–2.00
E. europaeus	15	8,000	10,900	15,500	8+	90	98	1.50–2.00
E. verrucosus	15	16,300	20,400	26,700	4+	90	98	4.00–5.00

EUROTIA LANATA (Pursh) Moq. Common winterfat

(Goosefoot family—Chenopodiaceae)

Also called lambs tail, sweet sage, white sage, feather sage.

DISTRIBUTION AND USE.—Native to dry sandy or shallow clay loams from Saskatchewan to Manitoba and south to Texas and California, common winterfat is a low shrub which is a very valuable browse plant for livestock and wildlife species, especially in winter when other plant food is scarce; it also has some value as an ornamental. The date of earliest cultivation is 1895.

SEEDING HABITS.—Common winterfat's small, greenish male and female flowers, borne separately on the same plant—occasional plants bear female flowers only—bloom from May to July. The fruit, enclosed by two bracts forming a long, hairy, two-horned coat, ripens from August to October. Each fruit contains a single vertical seed consisting of a tough outer testa which encloses a membranous endosperm and an almost ringlike embryo (fig. 108). The seed is dispersed chiefly by wind in the fall or winter soon after ripening. Plants begin to bear seed the first year, and they produce abundant crops almost annually. A 10-year-old stand has produced 70 to 80 pounds of seed per acre.

COLLECTION, EXTRACTION, AND STORAGE.—The fluffy white seed should be stripped from the bushes by hand in late summer to early winter as soon as it is ripe. If the seed are collected without too much debris, no cleaning is necessary; otherwise fruit fragments and debris should be removed. Unless they are to be sown immediately, the seed should be dried, as they heat and spoil badly when stored undried in bags. Cleaned seed per pound (4+ samples): low, 54,000; average, 90,000; high, 93,000. One sample of seed ran 52 percent in purity and 97 percent in soundness. Common winterfat seed does not ordinarily appear on the market. Under ordinary dry storage at room temperatures the seed loses about 10 to 50 percent of its initial viability in 1 year and most of it in 2 years, although a little remains up to 4 years. Sealed storage at low temperatures might maintain viability longer.

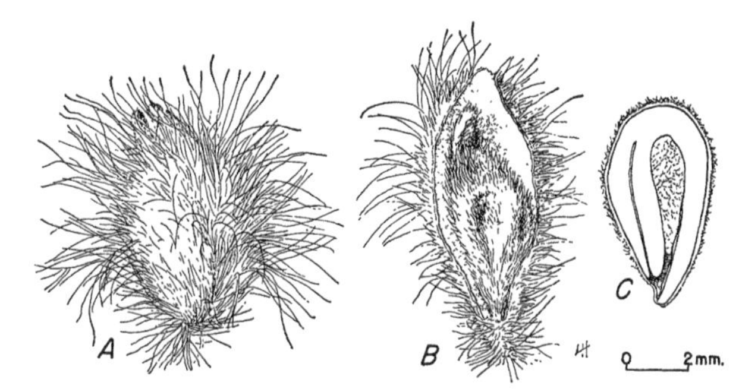

FIGURE 108.—Seed views of *Eurotia lanata*: *A*, Exterior view of fruit; *B*, longitudinal view of fruit showing seed; *C*, longitudinal view of seed showing embryo enclosing endosperm.

GERMINATION.—Natural germination occurs during cool or cold weather. The seeds take up moisture readily and germinate satisfactorily in the laboratory without pretreatment. Germination may be tested in sand flats for 10 to 30 days at about 50° (night) to 77° F. (day), using 400 to 1,000 seeds per test. Germinative capacity (29 tests): low, 5 percent; average, 58 percent; high, 99 percent.

NURSERY AND FIELD PRACTICE.—Common winterfat seed should be sown broadcast in the late fall or winter soon after collection, and worked into the soil with a hand spray if the weather is cloudy, or covered with one-eighth inch of sifted sand if the weather is sunny. Germination usually begins in 2 to 5 days after sowing. Because of its deep rooting habit, it is probably better to seed common winterfat directly in the field on plowed or otherwise well-prepared ground, using about 3 to 4 pounds of seed per acre. One-year-old stock is best for field planting. This species grows chiefly on subalkaline soils. It is quite drought-resistant. Rabbits are serious enemies of this plant.

FAGUS L. Beech

(Beech family—Fagaceae)

Distribution and Use.—The beeches include 10 species of medium-sized, deciduous trees native to the temperate regions of the Northern Hemisphere. They are valuable for their ornamental qualities and edible nuts, which are used as food by man, game, and domestic animals. An edible oil is extracted from the nuts. Some species are also important timber trees. Two species, which are of value for planting in the United States and for which reliable information is available, are described in table 91. *Fagus grandifolia* has been used to some extent for reforestation purposes in eastern United States, and this species and *F. sylvatica* L. have both been planted as ornamentals in that area.

Seeding Habits.—The male and female flowers of beech are borne separately on the same tree and bloom in the spring after the leaves unfold. The male flowers occur in long-stemmed heads. Occurring in clusters of two to four, the female flowers develop into two to three one-seeded nuts surrounded by a prickly bur which opens on the tree soon after maturity in the autumn, allowing the nuts to fall to the ground. Commercial seed consists of the ovoid, unequally three-angled, chestnut-brown, shining, thin-shelled nuts without endosperm (figs. 109 and 110). Natural dispersal is chiefly by gravity and animals. The comparative seeding habits of the two species of *Fagus* discussed here are as follows:

	F. grandifolia	*F. sylvatica*
Time of—		
Flowering	April–May	April–May.
Fruit ripening	Sept.–Nov.	Sept.–Oct.
Seed dispersal	After first heavy frost.	After first heavy frost

F. grandifolia bears good seed crops every 2 to 3 years, with light seed crops in most intervening years. Both *F. grandifolia* and *F. sylvatica* begin to bear seed in commercial quantities at 40 to 60 years. *F. sylvatica* bears good seed crops every 5 to 7 years—in some localities every 15 to 20 years—with light seed crops in most intervening years.

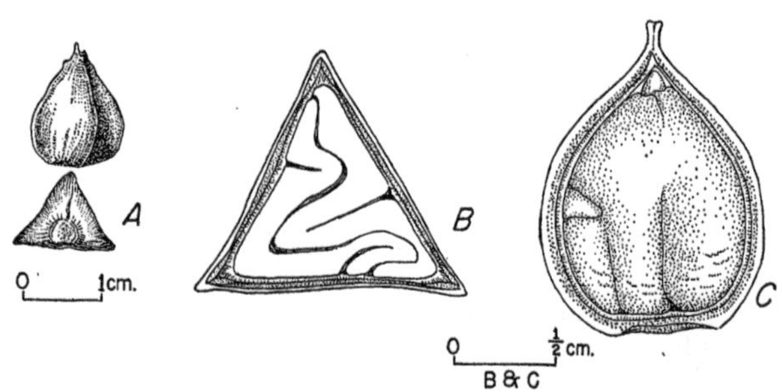

Figure 109.—Seed views of *Fagus grandifolia*: *A*, Exterior views in two planes; *B*, cross section; *C*, longitudinal section.

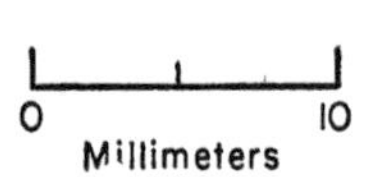

Figure 110.—Seed view of *Fagus sylvatica.*

Studies of *Fagus sylvatica* made in Switzerland, Germany, and Denmark have shown the presence of several climatic races which differ in phenological characteristics, frost hardiness, rate of growth, and form. Scientific information is not available as to the development of climatic races in *F. grandifolia*. However, in view of the wide range of the species and the recognition of a botanical variety, it seems likely that such races may have developed.

Collection, Extraction, and Storage.—The nuts may be shaken from the trees after frost has opened the burs, or they may be raked up from the ground after they have fallen, usually from mid-September to November. The closed burs may also be picked in the fall from trees recently felled in logging operations. In such cases the burs should be dried for a short period after which the nuts may be shaken or screened out. The cleaned nuts usually are sown or stratified as soon as possible after extraction. Yield, size, purity, soundness, and cost of commercial cleaned *Fagus* seed vary as shown below:

	F. grandifolia	*F. sylvatica*
Cleaned seed:		
Per bushel of burs ... pounds	9	--
Per 100 pounds of burs ... do	32	--
Per pound:		
Low ... number	1,300	1,800
Average ... do	1,600	2,100
High ... do	2,300	2,400
Basis, samples ... do	10	4
Commercial seed:		
Purity ... percent	97	[1]96
Soundness ... do	88	[1]45
Cost per pound ... dollars	0.85–1.40	0.75–2.25

[1] Based on 40 samples.

TABLE 91.—*Fagus: Distribution and uses*

Accepted name	Synonyms	Natural range	Chief uses	Date of earliest cultivation
F. grandifolia Ehrh. (American beech).	*F. americana* Sweet, *F. ferruginea* Ait., *F. atropunicea* (Marsh.) Sudw. (red beech).	New Brunswick to northern Michigan, south to Florida and Texas.	Lumber, handle and cooperage stock, fuel wood, ornamental; food for man, game animals, and livestock.	About 1800.
F. sylvatica L. (European beech).	------------------------	Central and southern Europe eastward to Crimea.	-----do------------------	(1)

[1]Long cultivated.

An acre of 150-year-old *Fagus sylvatica* high forest in a good seed year has been reported from France to yield 57 bushels of seed from which 409 pounds of oil could be extracted.

In Europe it is common practice to store *Fagus sylvatica* seed over winter in fairly dry sand at low temperatures, ordinarily in a good root cellar. Danish tests showed that the nuts could be stored at 34° to 36° F. in sealed containers for 1 year without loss of viability, but at 2 years no good seed were left. Although comparable information for *F. grandifolia* seed is not available, it is likely that it would behave similarly; it can be stored satisfactorily over winter at 41° in sealed containers.

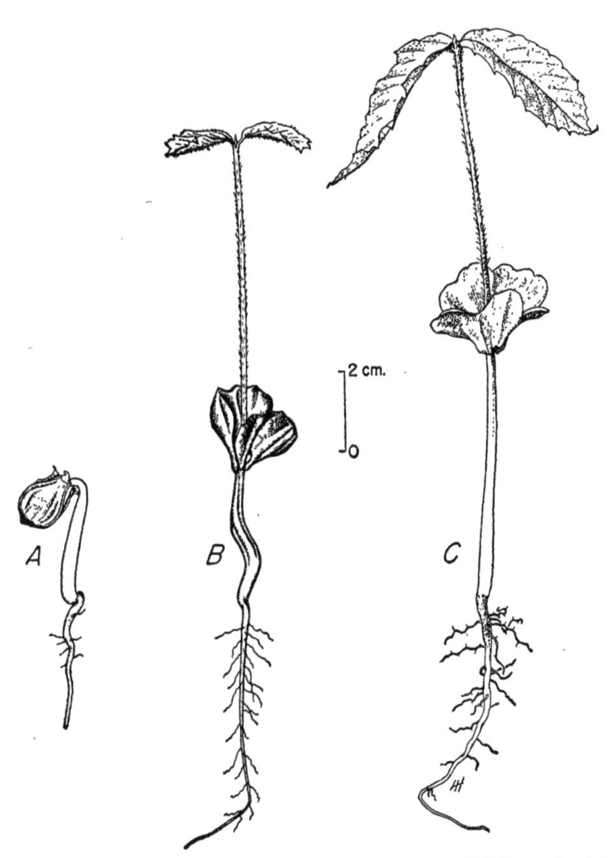

FIGURE 111.—Seedling views of *Fagus grandifolia*: *A*, At 2 days; *B*, at 5 days; *C*, at 7 days.

GERMINATION. — Natural germination usually takes place from early spring to early summer beneath the parent stand, the preferred seedbed for *Fagus sylvatica* being a humous loamy soil with a fair lime content. Germination is epigeous (fig. 111) and is slow because of a dormant embryo. Satisfactory germination results from fall sowing or stratification in moist sand at 41° F. A 90-day treatment is satisfactory for *F. grandifolia* and, although specific information is lacking, similar treatment (or possibly a 60-day treatment) should suffice for *F. sylvatica* seed also. The germination of beech seed should be tested in sand flats at 68° (night) to 86° (day) for 60 days using 100 to 200 stratified seeds per test. Average results are as follows:

	F. grandifolia	*F. sylvatica*
Stratified seed:[1]		
Basis, tests..............number..	6	--
Germinative energy:		
Amount..............percent..	84	--
Period..............days..	40–47	--
Germinative capacity:		
Low..............percent..	80	--
Average..............do....	85	--
High..............do....	96	--
Period..............days..	60	--
Potential germination....percent..	93	--
Untreated seed:		
Basis, tests..............number..	5	4
Germinative energy:		
Amount..............percent..	14	28
Period..............days..	131–145	20
Germinative capacity:		
Low..............percent..	14	7
Average..............do....	24	34
High..............do....	40	71
Period..............days..	150–160	60
Potential germination....percent..	58	89

[1] Stratified 90 days at 41° F.

NURSERY AND FIELD PRACTICE. — *Fagus* seed should be sown broadcast or in drills in the fall, or stratified seed sown in the spring, and covered with one-half inch of soil. Fall-sown seed germinates early in the spring, often by the end of April, and must be specially protected against rodents. The seedbeds should be given half-shade until past midsummer of the first year. One- or two-year-old seedlings are used in field planting. For best results, this species must be planted only on well-drained sandy loams or limestone soils. Horticultural varieties are propagated by grafting.

FALLUGIA PARADOXA (D. Don) Endl. Apacheplume

(Rose family—Rosaceae)

Syns.: *F. acuminata* (Woot.) Cockerell, *Sieversia paradoxa* D. Don; fallugie, ponil.

Distribution and Use.—An attractive, many-branched, often evergreen shrub occurring on a great variety of soils from western Texas and southwestern Colorado to southern Nevada, southeastern California and south into Mexico, Apacheplume is an important forage plant furnishing browse to both domestic and wild animals. Because of its conspicuous flowers and abundance of decorative plumelike seeds, it has been in occasional use for ornamental planting since 1877 and is perfectly hardy as far north as Massachusetts. It is also an excellent natural aid in erosion control in the dry lands of the Southwest and has been planted to a limited extent in that section for this purpose.

Seeding Habits.—The large, white, roselike, perfect flowers are borne on long stalks singly or in clusters in June to August; the fruit, consisting of a small hairy achene tipped with a persistent feathery style 1 to 2 inches in length, is borne in dense reddish clusters of 20 to 30 or more from August to October. Each achene (fig. 112) contains a single seed. Commercial "seed" consists of the ripe achenes. Sometimes another crop of seed is produced in June or July from flowers borne in April and May. Dispersal is largely by wind and to some extent by animals and occurs about a month after ripening.

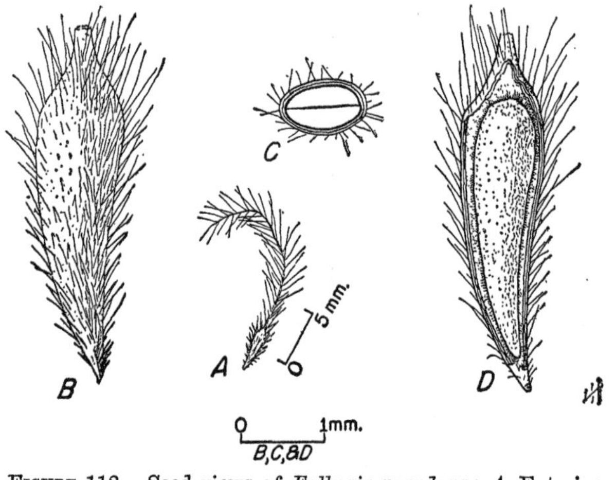

Figure 112.—Seed views of *Fallugia paradoxa: A,* Exterior view of fruit; *B,* exterior view, achene without style; *C,* cross section; *D,* longitudinal section, showing seed.

Collection, Extraction, and Storage.—The seeds may be collected when the pink hairy styles turn whitish and the plump seeds fall readily from the branches. The seeds may be stripped from the branches or collected from the

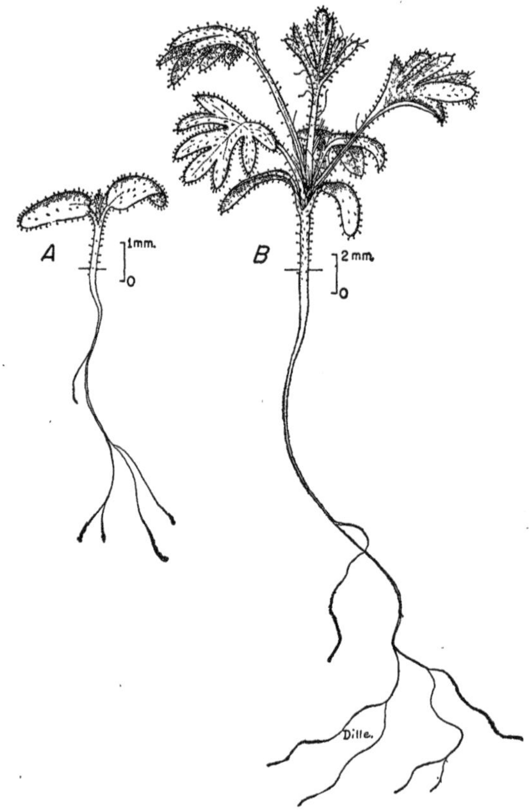

Figure 113.—Seedling views of *Fallugia paradoxa: A,* Very young seedling; *B,* older seedling.

ground among dense clumps of the plants by means of fine-toothed collecting rakes. No subsequent treatment is necessary other than rubbing to break off the styles, followed by fanning to remove chaff and other debris. Cleaned seed per pound (1 sample) : 420,000. No data are available on purity, soundness, cost, or optimum storage conditions.

GERMINATION. — Apacheplume seeds germinate readily without special treatment; germination is epigeous (fig. 113). Although definite information is lacking, tests can likely be made in sand flats or standard germinators at temperatures fluctuating diurnally from about 70° to 85° F. Number of seeds required: 1,000. Test duration: 20 days. Germinative energy: 42 percent in 14 days. Germinative capacity (4 samples): low, 19 percent; average, 45 percent; high, 62 percent.

NURSERY AND FIELD PRACTICE.—In the southwestern United States Apacheplume seed is sown broadcast either from July to October or from February to April with apparently good results. After sowing, the beds are rolled and the seed then covered with $\frac{1}{16}$ inch of soil and ⅛ to ¼ inch of sifted sand. Germination occurs in 4 to 10 days after sowing.

FRAXINUS L. Ash

(Olive family—Oleaceae)

DISTRIBUTION AND USE. — The ashes consist of about 65 species of deciduous trees—a few are shrubs—occurring in the Northern Hemisphere, in Asia south to Java, and in North America south to Mexico. Several are important timber trees with straight-grained and tough wood used for tool handles, wagons, furniture, interior finish, and fuel; others can be used in erosion control; still others are valuable ornamental trees or shrubs. Some Chinese species yield a white wax. The manna of commerce used in medicine is obtained from exudations from the trunk and leaves of *Fraxinus ornus* L. Data on distribution and chief uses of 8 important species and 1 variety are given in table 92.

Compared with the pines, none of the species of *Fraxinus* is planted to any great extent for reforestation in the United States. However, they are as widely used as any hardwoods. *F. americana* and *F. pennsylvanica* var. *lanceolata* are planted

TABLE 92.—*Fraxinus. Growth habit, distribution, and uses*

Accepted name	Synonyms	Growth habit	Natural range	Chief uses	Date of earliest culti-vation
F. americana L. (white ash).	American ash, cane ash.	Large forest tree.	Nova Scotia to southern Minnesota, south to Florida and Texas.	Handle stock, furniture, finishing, wagons, cooperage; landscape planting.[1]	1724
F. dipetala Hook. and Arn. (twopetal ash).	California shrub ash, flowering ash.	Large shrub	Canyons and lower slopes of Sierra Nevada and Coast Ranges of California.	Erosion control and game food.	1878
F. excelsior L.[2] (European ash).	--------	Large tree	Europe and Asia Minor	Automobile bodies, aircraft, sporting goods, handles, wheels, etc.; shelter-belt planting, landscape planting, game food and cover.	([3])
F. nigra Marsh. (black ash).	swamp ash, hoop ash, basket ash.	Medium-sized tree.	Newfoundland to Manitoba and south to West Virginia, Arkansas, and Iowa.	Interior finishing, a favorite basket wood, cooperage.	1800
F. oregona Nutt. (Oregon ash).	*F. Californica* Hort.	do	Lower slopes of Cascade, Coast and Sierra Nevada ranges from British Columbia to southern California.	Handle stock, furniture, finishing, wagons, cooperage; landscape planting.	1870
F. pennsylvanica Marsh. (red ash).	--------	do	Nova Scotia to North Dakota, south to Georgia and Mississippi.	do	1783
F. pennsylvanica var. *lanceolata* (Borkh.) Sarg. (green ash).	*F. lanceolata* Borkh., *F. viridis* Michx.	do	Maine to Saskatchewan and Montana, south to Florida and Texas.	Handle stock, furniture, finishing, wagons, cooperage; landscape planting; shelter-belt and ornamental planting.[4]	1824
F. quadrangulata Michx. (blue ash).	--------	Medium to large tree.	Southern Michigan, south and west to Arkansas and Texas.	Inner bark contains a blue dye.[5]	1823
F. velutina Torr.[6] (velvet ash).	*F. pistaciaefolia* Torr.	Small tree	Southern Nevada and Utah, southeastern California, southern and central Arizona and southern New Mexico; Mexico.	Occasionally for handles; shelter-belt and ornamental planting; game food.	([7])

[1] Wood very valuable.
[2] Includes over a dozen varieties in cultivation.
[3] This species has been cultivated for centuries.
[4] Wood similar to that of *F. americana*.
[5] Wood valuable but commercially less important than that of *F. americana* and other common ashes.
[6] Includes several varieties.
[7] Before 1890.

TABLE 93.—*Fraxinus: Time of flowering and fruit ripening, and frequency of seed crops*

Species	Time of—			Seed year frequency		Commercial seed-bearing age		
	Flowering	Fruit ripening	Seed dispersal	Good crops	Light crops	Minimum	Optimum	Maximum
						Years	*Years*	*Years*
F. americana	April–May	Mid-Aug. to Oct.	Sept.–early winter	Every 3 to 5 years	Intervening years	20	40–100	125–175
F. dipetala[1]	March–June	July–Oct.	Aug.–Nov.					
F. excelsior[1]	Mid-April to mid-May	Mid-Aug. to late Sept.	Winter–early spring	1 to 2 years	Intervening years	15	40–60	
F. nigra[1]	May	Aug.–Sept.	Oct.–early spring					
F. oregona	April–May	do	Sept.–Oct. (early winter)	Annually[2]				
F. pennsylvanica	May	Autumn	Autumn–spring					
F. pennsylvanica var. *lanceolata*	do	Sept.–Oct.	Oct.–May	Frequently				
F. quadrangulata[1]	April–early May	Sept.–early Oct.	Oct.–Nov.			25	40–125	200–300
F. velutina	March–May	Sept.						

[1] These species have perfect flowers; all others are dioecious and hence some trees never bear seed.
[2] In open stands.

for timber in the Northeast, the Ohio Valley States, and Michigan. Large quantities of the latter species are being used in shelter-belt planting in the Prairie-Plains States; like quantities of the former species are used for this purpose in Russia. *F. excelsior* is considerably used for timber and shelter-belt planting in Europe. *F. quadrangulata* is sometimes planted on dry calcareous uplands within its native range. Practically all of the ashes are used to some extent for landscape, park, and street planting, particularly *F. americana, F. pennsylvanica,* and *F. pennsylvanica* var. *lanceolata* in the Eastern and Central States, *F. oregona* in the Pacific Coast States, *F. velutina* in arid situations in the Southwest, and *F. excelsior* in Europe.

SEEDING HABITS.—The small flowers, rather inconspicuous in most species, showy in others, are dioecious or polygamous, sometimes perfect; they appear in early spring with or before the leaves in terminal or axillary clusters. The fruit, an elongated samara (figs. 114 and 115), borne in clusters, contains a single seed with small embryo and rather large endosperm. Ripening occurs in late summer or fall, sometimes in spring—as for *Fraxinus berlandieriana* A. DC.—and is indicated by the light yellowish-brown to brown color of the samaras, and also by their degree of crispness. Commercial seed consists of the ripe samaras. Detailed seeding habits of the 8 species and one variety discussed here are given in table 93. Seed dispersal is largely by wind; *F. nigra* and *F. pennsylvanica* var. *lanceolata* seed may also be disseminated by water.

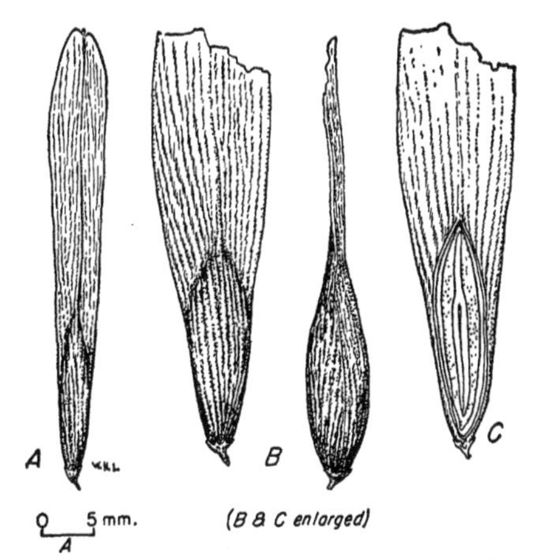

FIGURE 114.—Seed views of *Fraxinus pennsylvanica* var. *lanceolata: A,* Exterior view of fruit; *B,* exterior view of seed with part of wing, in two planes; *C,* longitudinal section.

Little information is available concerning the development of races in *Fraxinus.* Studies in Germany have shown the presence of at least two races of *F. excelsior* in that country: one, called lime ash, which is native to dry limestone areas; the other, called water ash, on various fresh to moist sites. The two races differ in leaf size, drought resistance, and height growth. That such races also exist in green ash is suggested by the fact that seed collected from trees in the northwest part of the Prairie-Plains region of the United States produces seedlings which are smaller, have foliage which is darker green,

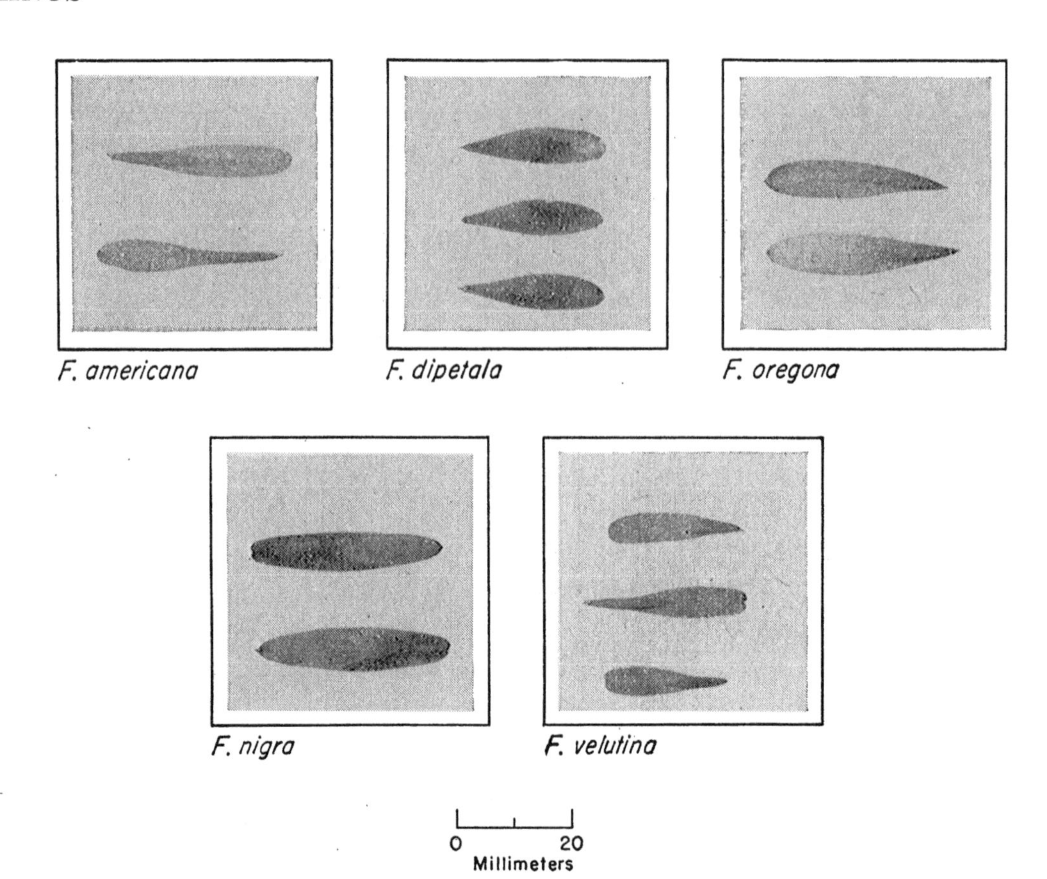

FIGURE 115.—Seed views of 5 species of *Fraxinus.*

and are more drought-resistant than those from seed collected from the southern and eastern parts of that region. Additional study would likely disclose the presence of other races in these and some of the other species of ash with a wide geographic range.

COLLECTION, EXTRACTION, AND STORAGE.—Ash seed may be collected in the fall or early winter by picking the seed clusters from standing trees or those recently felled in logging operations. This may be done by hand, or the clusters may be cut off with pruners or seed hooks. On days without wind, the seed may be shaken or whipped from the trees by means of light bamboo poles and collected on canvas spread on the ground. In such cases care should be used to avoid breaking the brittle twigs and smaller branches, especially of the red, white, and green ashes. Seed may sometimes be gathered from the ground in case of heavy fall and drifting.

When seed is collected early, it is usually somewhat moist and should be spread out to dry in shallow layers. Otherwise, no special treatment after collection is needed with the exception of breaking up the stems of the seed clusters so as to reduce bulk. This can be readily accomplished in the case of thoroughly dry seed by gentle flailing or by running the clusters dry through a macerator. The seed is then separated from twigs, stems, leaves, and other debris by fanning or screening.[29] No further cleaning is necessary. The wings can also be broken off to reduce bulk but this is not ordinarily done. The seed can either be sown after cleaning or stored. Detailed information on yield, size, purity, soundness, and cost of *Fraxinus* seed is presented in table 94.

Ash seeds are somewhat variable in keeping quality. Seeds with the greatest length of life are apparently those of the subsection *Bumelioides,* to which belong *Fraxinus excelsior, F. nigra,* and *F. quadrangulata.*[30] Seed of the first species will keep for 2 to 3 years under ordinary storage conditions and for 3 to 6 years when stored dry in sealed

[29] Screens with ⅜ x ⅜-inch openings are about right for seeds of *Fraxinus americana* and *F. pennsylvanica* and its var. *lanceolata.*

[30] This is believed due to the presence of a rather impermeable seed coat which is not typical of the other common species of *Fraxinus. F. nigra* seed will often lie on the ground for 2 years before germinating. *F. excelsior* seed have been found to remain dormant in soil up to 6 years.

TABLE 94.—*Fraxinus: Yield of cleaned seed, and purity, soundness, and cost of commercial seed*

Species	Cleaned seed					Commercial seed[1]		
	Yield per 100 pounds of fruit	Per pound			Basis, samples	Purity	Sound-ness	Cost per pound
		Low	Average	High				
	Pounds	*Number*	*Number*	*Number*	*Number*	*Percent*	*Percent*	*Dollars*
F. americana[2]		5,500	10,000	18,200	19	92 (70–100)	80 (38–98)	0.40–0.70
F. dipetala			8,000		1+			
F. excelsior[3]	75	4,000	5,900	7,000	10+	92	60–80	.85–1.20
F. nigra[4]		6,100	8,100	9,500	4	94	85 (64–100)	1.25
F. oregona		10,000		14,300	2		48	1.25
F. pennsylvanica[2]	75	9,300	11,800	19,000	6	93	65–85	.95
F. pennsylvanica var. *lanceolata*.	75	11,000	17,300	24,600	49	89 (60–99)	88 (55–100)	.25–0.75
F. quadrangulata[4]		5,900		7,000	2+	[5]74	[5]88	1.00–1.25
F. velutina		13,000	20,600	28,000	6	92	20–74	1.60–2.00

[1]First figure is average percent; figures in parentheses indicate range from lowest to highest.
[2]Weight of seed per bushel: *F. americana*, 12.4 lbs.; *F. pennsylvanica*, 6.5–8.2 lbs. Number of seed per bushel: *F. americana*, 139,000; *F. pennsylvanica*, 66,000–92,000.
[3]Seed collected after a wet summer are reported to be of low vitality.
[4]Seed of these species have a characteristic spicy odor which also is present in at least some of the other members of the subsection *Bumelioides*. This is not found in the seed of *F. americana* and its closely related species.
[5]1 sample.

containers at a low temperature. That of *F. nigra* will retain full viability for at least a year, and probably longer, if stored at a moisture content of about 7.5 percent and at temperatures as high as 75° F. On the other hand, seeds of the other species described keep rather poorly in ordinary storage unless kept at a low moisture content. Thus seed of *F. oregona* had its viability reduced to only one-fourth of the original after 3 years of ordinary storage, but that of *F. pennsylvanica* dried to a moisture content of 7.3 percent and then sealed showed no loss in germinability after 1 year's storage at temperatures as high as 85°. At a moisture content of about 10 percent, however, complete loss of viability occurred in the latter species.

Even when *Fraxinus* seeds are stored at low temperatures, they will probably keep best if their moisture content is low. For example, seed of *F. pennsylvanica* var. *lanceolata* stored in open containers at 40° to 50° F. and high humidity showed about 25 percent germinable seed at the end of 5 years, but that of *F. americana* showed little loss after 3 years at 34° to 38°, and that of *F. dipetala* no loss after 4 years at 35° to 40° when stored dry and sealed.

GERMINATION.—*Fraxinus* seeds vary in germination behavior. Those of most species germinate in the spring after lying on the ground over winter.[31] Seeds of *F. excelsior, F. nigra,* and *F. quadrangulata* and probably others of the subsection *Bumelioides* do not usually germinate until the second spring. Germination in all species, so far as known, is of the epigeous type (fig. 116). For most species moist, well-drained, fertile loams form the best natural seedbed; for *F. quadrangulata,* rich calcareous soils protected by partial cover and litter appear best.

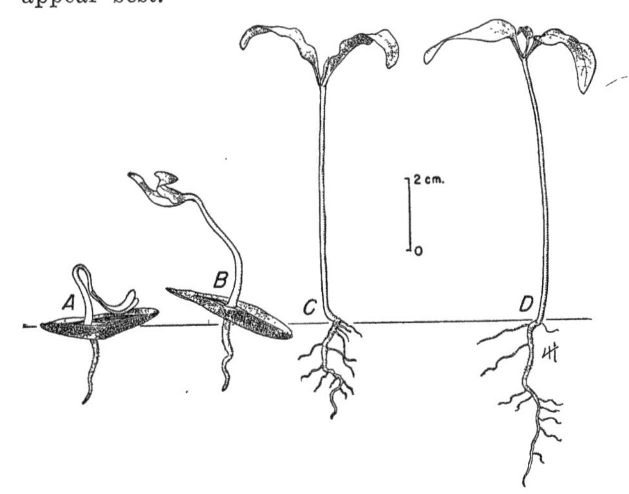

FIGURE 116.—Seedling views of *Fraxinus nigra: A,* At 1 day; *B,* at 2 days; *C,* at 8 days; *D,* at 14 days.

Seeds of the eight species and one variety discussed here are dormant to a greater or less degree. In some cases this seems to be associated with degree of freshness, the older lots usually containing a higher proportion of dormant seeds than fresher lots. In most species dormancy is apparently due to conditions within the embryo. In others, such as *Fraxinus nigra, F. excelsior,* and likely also *F. quadrangulata,* the embryo is apparently immature and its enlargement is somewhat retarded by the seed coat. Detailed data on

[31] Seeds which hang on the trees until late spring, as is common in *F. pennsylvanica* var. *lanceolata,* likely do not germinate until the second spring.

the type of dormancy shown by each of the eight species and one variety and how best to overcome it are given in table 95.

Germination tests are most easily made in flats using sand or peat for a medium and should use for a basis 400 to 800 properly pretreated seeds per test. Germinators can also be used for testing. For most species temperatures alternating diurnally from 68° to 86° F. appear to be satisfactory. Results that can be expected from such tests are given in detail in table 96.

NURSERY AND FIELD PRACTICE.—Seeds such as those of *Fraxinus americana* and *F. pennsylvanica* may be either fall-sown[32] or stratified and sown in the early spring. In the cases of *F. americana*, *F. pennsylvanica* var. *lanceolata*, and possibly other similar species, soaking the seed in aerated water at about 70° F. from 10 to 21 days prior to spring sowing is about as effective as stratification, particularly with fresh seed. The seed is sown preferably in drills or nursery rows, although it may also be sown broadcast, at the rate of 25 to 30 viable seeds per linear foot and is covered with 1/4 to 1/2 inch of nursery soil or by a layer of sand which is covered by a thin layer of soil. Fall-sown beds should be mulched with burlap or straw, and the mulch removed as soon as germination begins in the spring. Seedlings usually appear within 2 to 3 weeks in the spring. Nursery germination: *F. pennsylvanica*, 56 percent. Field planting of *F. americana, F.*

[32] Some lots of *Fraxinus americana* and *F. pennsylvanica* should be sown in September or early October; otherwise germination will not be complete the first spring.

TABLE 95.—*Fraxinus: Dormancy and methods of seed pretreatment*

Species	Dormancy		Stratification		Other methods
	Kind	Occurrence	Medium	Duration and temperature	
				Days at *°F.*	
F. americana	Embryo	Most seeds	Sand or peat	60–90 at 41	Water soaking 2 to 10 days.[1]
F. dipetala	do	All seeds	do	120–150 at 35–40	
F. excelsior	Seed coat and embryo.	Most seeds	do	60–90 at 68 + 60–90 at 41	
F. nigra	do	do	do	60–90 at 68 + 60–90 at 41	
F. oregona	Embryo	do	do	90 at [2]41	
F. pennsylvanica	do	do	do	60–90 at 41	
F. pennsylvanica var. *lanceolata.*	do	do	do	60–90 at 41	Water-soak at room temperature 10 to 21 days.[1]
F. quadrangulata	Seed coat and embryo.	do	Sand	60–90 at 68–86 + 120 at 41	
F. velutina	Embryo	Some, not all lots.	Sand or peat	90 at [2]41	

[1]Some seed lots of southern and eastern origin show better germination if stratified for 30 days at 68° to 86° F. prior to low-temperature stratification.
[2]Suggested; experimental data not complete.

TABLE 96.—*Fraxinus: Recommended germination test duration, and summary of germination data*

Species	Test duration recommended		Germination data from various sources						
	Untreated seed	Stratified seed	Germinative energy		Germinative capacity			Potential germination	Basis, tests
			Amount	Period	Low	Average	High		
	Days	*Days*	*Percent*	*Days*	*Percent*	*Percent*	*Percent*	*Percent*	*Number*
F. americana	60+	40–60	65–75	18	1	38	82		10
F. dipetala	240	30–40					86		1+
F. excelsior					50		70	98	
F. nigra[1]	[2]190+	30	7	18	1	20	58	80–100	6
F. oregona		15–20					54		1
F. pennsylvanica					49		56		2
F. pennsylvanica var. *lanceolata.*	60–90+	40–60	10–87	10–35	[3]2	[3]42	[3]93		48
F. quadrangulata[1]	[2]240+	30	11	16		12		77	1
F. velutina	60–90+	60			3	33	81		5

[1]Further tests needed for this species.
[2]No germination obtained in this period.
[3]Stratified seed, 5, 50, and 93 percent (20 samples); seed soaked in water 2 to 14 days, 2, 24, and 66 percent (28 samples).

pennsylvanica, and *F. pennsylvanica* var. *lanceolata* calls for 1–0 or sometimes 2–0 stock.

Practice with seeds such as those of *Fraxinus excelsior, F. nigra,* and *F. quadrangulata* is somewhat different. The seed should be sown in spring or summer months and the beds mulched until the following spring when germination will occur. Pretreated seed, i. e., seed given warm plus cold stratification, should be sown in spring and will germinate soon thereafter. The seed is sown usually in drills about 6 to 12 inches apart and covered with ½ to ¾ of an inch of nursery soil. Nursery germination: *F. excelsior,* 75 percent; *F. nigra,* 50 to 75 percent. The seedlings should be given half shade during the first summer. 1–1 or 2–0 stock of *F. excelsior* is ordinarily used for field planting, or in the case of unusually moist sites, 2–2 or 2–3 stock.

Ash seedlings are relatively free from insect and fungous pests in the nursery. However, seedlings of *Fraxinus pennsylvanica* var. *lanceolata* and other species are frequently subject to severe defoliation caused by the fungus *Marssonia fraxini,* particularly in northern nurseries. This can be controlled by spraying with bordeaux mixture (4–6–50) at 2-week intervals, or with a 2-percent solution of lime-sulfur. In some nurseries applications of nitrate of soda after the seedlings have emerged, or of organic fertilizers prior to seeding, seem to enable the stock to overcome the ill effects of this disease.

FREMONTODENDRON Cov. Fremontia

(Sterculia family—Sterculiaceae)

Distribution and Use.—The fremontias include about three species—sometimes grouped as one—of deciduous shrubs or small trees, often forming thickets, native to the foothill region of California, Arizona, and northern Mexico. They are sometimes cultivated for ornamental purposes and can be used for erosion-control purposes. Two species, for which reliable information is available, are described in table 97. *Fremontodendron mexicanum* is the more widely planted of the two species, chiefly for ornamental purposes.

Seeding Habits.—The perfect, yellow or orange, roselike flowers bloom in May to July *(Fremontodendron californicum)* or April to July *(F. mexicanum)*. The fruit is a densely woolly, egg-shaped, four- to five-celled capsule which ripens in August to September and splits open at its tip. It persists on the plant for several months longer. Upon opening, the capsule releases its numerous dark reddish-brown seeds (fig. 117) as the plants are shaken by the wind or browsing animals. Good seed crops are borne almost annually.

Collection, Extraction, and Storage.—The ripe pods of *Fremontodendron mexicanum* are gathered before they split open in July to August, and those of *F. californicum* are gathered in August to September. They should be spread out to dry and, after they split open, shaken to remove the seeds. Seed characteristics of the two species vary as shown below:

	F. californicum	*F. mexicanum*
Cleaned seed per pound:		
Low .... number	15,000	--
Average .... do	--	27,000
High .... do	18,000	--
Basis, samples .... number	2	1
Commercial seed:		
Purity .... percent	57	--
Soundness .... do	[1]53	--
Cost per pound .... dollars	20	28

[1] 5-year-old seed.

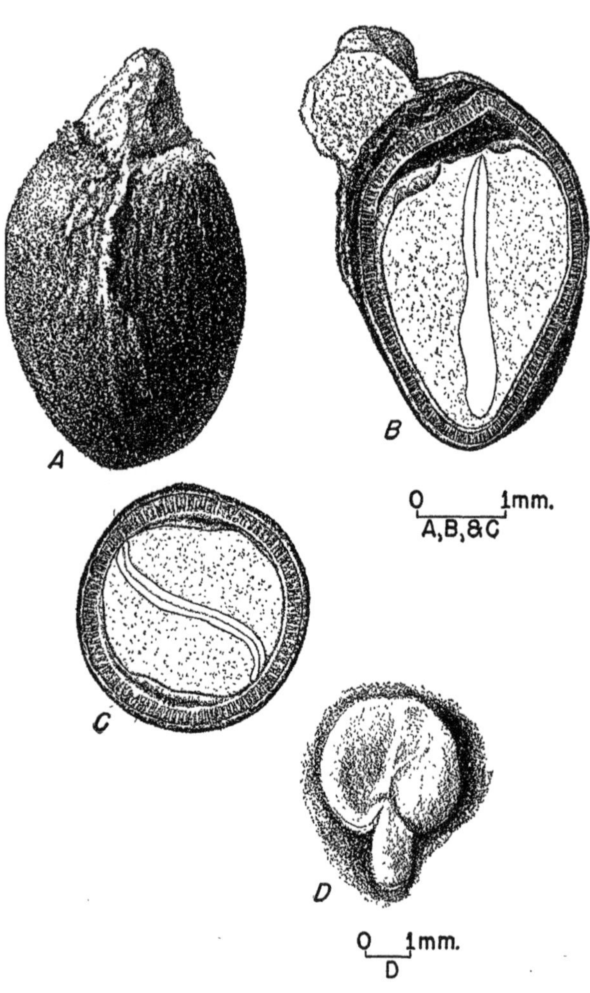

Figure 117.—Seed views of *Fremontodendron californicum*: *A*, Exterior view; *B*, longitudinal section; *C*, transverse section; *D*, embryo.

Table 97.—*Fremontodendron: Distribution and uses*

Accepted name	Synonyms	Natural range	Chief uses
F. californicum (Torr.) Cov. (California fremontia).	*Fremontia californica* Torr. (flannelbush).	Foothills and mountains of California, and Arizona.	Ornamental, possible erosion control, browse for cattle, inner bark for poultices.
F. mexicanum Davids. (Mexican fremontia).	*Fremontia mexicana* (Davids.) Macbr., *Fremontia californica* var. *mexicana* (Davids.) Jeps. (San Diego fremontia, southern fremontia).	Southern San Diego County, Calif., and adjacent Lower California.	do.

Seed of *F. californicum* stored 5 years in envelopes at room temperatures still showed apparent viability of 53 percent. The seed of both species could probably be kept satisfactorily for several years in sealed containers at temperatures of about 41° F.

GERMINATION.—Natural germination of fremontia is most common on exposed mineral soil where the seeds have been covered by washed soil. The seeds exhibit dormancy due probably to a combination of seed coat and embryo conditions. Soaking the seeds in hot water improves germination some, and stratification in moist sand at low temperatures gives still better results. Germination may be tested in sand or soil flats for 40 to 60 days at about 60° to 80° F., using 200 to 400 properly pretreated seeds per test. Germinative capacity, stratified seed (1 test): *F. californicum,* 50 percent; *F. mexicanum,* 55 percent. For the former species hot water treatment alone gave 26 percent germination in 37 days (1 test).

NURSERY AND FIELD PRACTICE.—Untreated fremontia seeds should be sown in the fall, or pretreated seeds in the spring. Propagation by softwood cuttings is also possible. Fremontias grow best on well-drained rather dry soils, where winter moisture is not too abundant, and in sunny, rather sheltered positions.

GARRYA FREMONTII Torr. Fremont silktassel, bearbrush

(Garrya family—Garryaceae)

Native to California, Oregon, and Washington, Fremont silktassel is an evergreen, medium-sized shrub, desirable as an ornamental and useful for erosion control. It was introduced into cultivation in 1842. The male and female flowers, borne separately on different trees, bloom from January to May. The fruits, rather dry, oval, dark purple, one- or two-seeded berries, ripen from August to December. They are often disfigured by larval holes, and care must be taken to collect only sound seeds. The oval seeds have abundant endosperm and minute embryos (fig. 118). One hundred pounds of dry berries yield about 50 pounds of cleaned seed. Seeds can be cleaned by rubbing the fruits through a fine-mesh screen and then floating off the pulp and empty seeds. There are about 26,000 to 33,000 seeds per pound (2 samples). One sample tested was 97 percent pure and 99 percent sound.

Seed does not germinate without pretreatment. Stratification for 3 months at 41° F. gave a maximum germination of 24 percent in 4 months after sowing (1 test). One test flat left in the greenhouse for 3 months and then transferred to a temperature of 41° for an additional 3 months gave a total germination of 88 percent. This occurred in the 3-week period immediately following removal from cold storage. Propagation by cuttings and layers is also feasible. For best results Fremont silktassel should be planted on a well-drained soil in a sunny sheltered position.

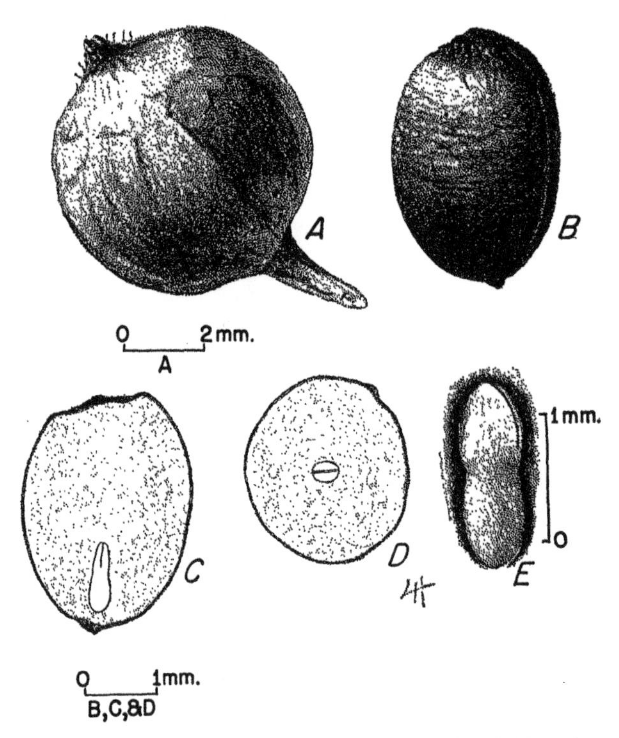

FIGURE 118.—Seed views of *Garrya fremontii*: *A*, Exterior view of fruit; *B*, exterior view of seed; *C*, longitudinal section, showing small embryo; *D*, cross section; *E*, embryo.

GAULTHERIA PROCUMBENS L. Checkerberry wintergreen

(Heath family—Ericaceae)

Also called checkerberry, teaberry, aromatic wintergreen.

DISTRIBUTION AND USE.—Native from Newfoundland to Manitoba and southward to Georgia and Michigan, checkerberry wintergreen is a low creeping, almost herbaceous, aromatic, evergreen shrub of value chiefly for game food and for ground cover. The leaves contain oil of wintergreen which is used in medicine, although most of this is now obtained from the sweet birch, *Betula lenta,* or is made synthetically. The edible fruit sometimes is marketed. Wintergreen was introduced into cultivation in 1762.

SEEDING HABITS.—Wintergreen's perfect flowers are borne from June to September. The bright red fruit, consisting mostly of the fleshy calyx surrounding a rather dry capsule, which contains many minute seeds (fig. 119), ripens in the fall. It persists on the plants until early the next summer, slightly increasing in size. Commercial "seed" consists of the dried fruits.

COLLECTION, EXTRACTION, AND STORAGE. — The fruits may be collected at any time in the fall after ripening, and the seed extracted by drying the fruit until it is brittle and powdery and then rubbing it through a fine screen (30-mesh). There are about 3,000 dried fruits per pound. Cleaned seed per pound (2 samples): 2,870,000 to 4,840,000; purity, 80 percent; soundness, 97 percent. Commercial "seed" costs $3.75 to $4.50 per pound. Wintergreen seed stored in sealed bottles showed 16 percent germination after 2 years at 41° F. The original germination was not known.

GERMINATION.—Wintergreen seed has a dormant embryo and will germinate much more rapidly and abundantly if stratified for 30 to 75 days at 41° F. Because of its minute size, the seed should be scattered on or pressed into peat and then covered with glass. Germinative capacity (1 sample): 16 percent.

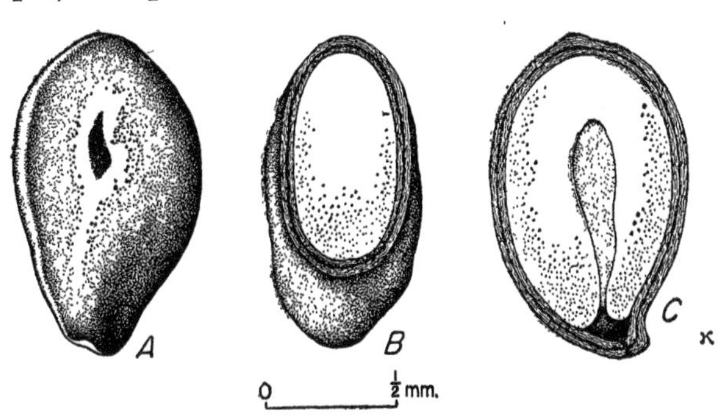

FIGURE 119.—Seed views of *Gaultheria procumbens: A,* Exterior view; *B,* transverse section; *C,* longitudinal section.

NURSERY AND FIELD PRACTICE.—Untreated seed should be sown in the fall, or stratified seed in the spring. Propagation by layers and suckers, division of older plants, and by cuttings is also possible. Results are best when wintergreen is planted on rather moist peaty or sandy soil in partial shade.

891183°—50——13

GAYLUSSACIA BACCATA (Wangenh.) K. Koch — Black huckleberry

(Heath family—Ericaceae)

Also called huckleberry. Botanical syn.: *G. resinosa* (Ait.) Torr. & Gray

DISTRIBUTION AND USE.—Native from Newfoundland to Manitoba, south to Georgia, Kentucky, and Iowa, black huckleberry is an upright, deciduous, low shrub, valuable for game food and cover. The berries are also eaten to considerable extent by man. It was introduced into cultivation in 1772.

SEEDING HABITS.—The perfect flowers are borne from May to June; lustrous fruit, a berrylike drupe, containing 10 one-seeded, bone-colored nutlets (fig. 120), ripens from July to September and persists on the bushes for a few weeks.

COLLECTION, EXTRACTION, AND STORAGE.—Black huckleberry fruit may be stripped from the branches with a blueberry rake any time after it is thoroughly ripe. The seeds are extracted by macerating the berries in water, allowing the pulp and empty seeds to float away. One hundred pounds of fruit yields about 3 pounds of cleaned seed. Cleaned seed per pound (4 samples): low, 282,000; average, 354,000; high, 412,000. Commercial seed (3 samples) averages 98 percent in purity, 45 percent in soundness and costs about $3 per pound. Seed has been stored for over 2 years in sealed bottles at 41° F. without loss in viability.

GERMINATION.—Black huckleberry seed is slow to germinate, apparently because of an impermeable seed coat. Germination is epigeous (fig. 121).

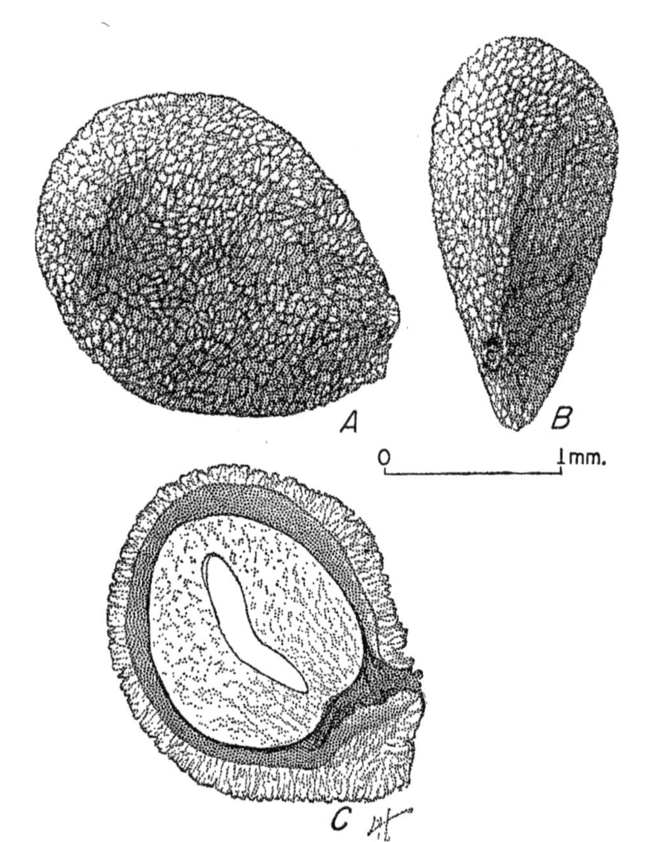

FIGURE 120.—Seed views of *Gaylussacia baccata*: *A* and *B*, Exterior views in two planes; *C*, longitudinal section.

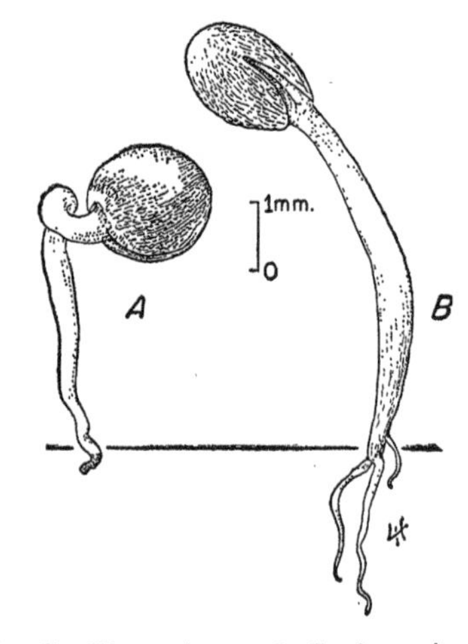

FIGURE 121.—Seedling views of *Gaylussacia baccata*: *A*, At 3 days; *B*, at 9 days.

Seed lots germinated thus far do not appear to have embryo dormancy. The best germination (24 out of a possible 25 percent) was obtained from 2-year-old seed which had been stratified in peat for 30 days at a temperature alternating diurnally from 68° to 86° F. During a subsequent stratification period of 50°, the seed germinated, germination beginning somewhere between the 8th and 27th days and being complete by the 47th day. Germinative energy: 20 percent in 27 days.

GLEDITSIA L. Honeylocust

(Legume family—Leguminosae)

DISTRIBUTION AND USE. — About 12 species of *Gleditsia* are found in North America, central and eastern Asia, tropical Africa, and South America. Only 3 of these species, attaining heights from 60 to 140 feet, are native to North America—the eastern temperate portion. The wood is hard, coarse-grained, red to red brown and is quite durable in contact with the soil; it is used for posts, poles, and fuel. Several of the Asiatic species have been introduced into the United States for ornamental planting. Further details of range and use of the 2 North American species and their natural hybrid are described in table 98.

SEEDING HABITS.—Polygamous flowers appear in single or clustered racemes in May and June in the axils of last year's leaves. The pod fruits ripen in the fall of the first year and contain in most species from one to many light-brown seed (fig. 122); pod fruits of *Gleditsia aquatica* contain from one to three seed. The pods open slowly and often remain on the trees until late fall or winter. *Gleditsia's* seed coat is thin and crustaceous but generally impermeable to water. The embryo is surrounded with a layer of horny, orange-colored endosperm, the source of a material essential to the processing of rubber. Seeds of the different species are similar in structure and in response to treatment, but differ in size and shape. Comparison of seed production of the two species and one hybrid studied is given in table 99.

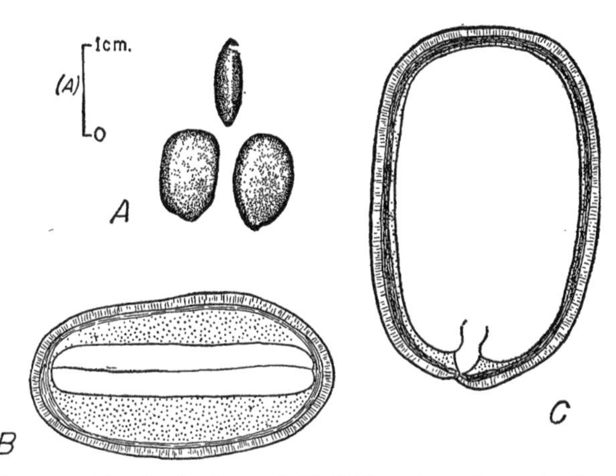

FIGURE 122.—Seed views of *Gleditsia triacanthos*: *A*, Exterior views in three planes; *B*, cross section; *C*, longitudinal section.

COLLECTION, EXTRACTION, AND STORAGE. — Pods may be gathered from the ground or from the trees from the time of ripening until they begin to disintegrate in late winter or spring. In localities where weevil infestations are common, it is advisable to make collections as soon as fruits mature. They should not be picked until brown and dry. Moist pods should be thoroughly dried after which seed extraction may be accomplished either by mechanical threshers or by hand flails. The special macerator and thresher developed by

TABLE 98.—*Gleditsia: Growth habit and uses*

Accepted name	Synonyms	Natural range	Chief uses	Date of earliest cultivation
G. aquatica Marsh. (waterlocust).	*G. monosperma* Walt., *G. inermis* Mill. (swamp honey locust).	In wet soils South Carolina to Florida, west to Texas, Mississippi Valley to Kentucky, Illinois, Indiana, and Missouri.	Game food, posts, and fuel	1723
X*G. texana* Sarg. (*G. aquatica* x *triacanthos*) (Texas honeylocust).	*G. brachycarpa* Nutt.	Southeastern Texas, Louisiana, Mississippi, and Indiana.	Game food, poles, posts, ties, fuel.	1900
G. triacanthos L.[1] (honeylocust).	sweet locust, thorny locust, thorn tree, three thorned acacia.	Moist, fertile soils from Ontario southwest to Nebraska, south to Florida, and west to Texas.	Game and livestock food, posts, poles, ties, honey plant, fuel, lumber, rubber processing, ornamental planting.	1700

[1]Includes *G. triacanthos* var. *inermis* Willd., a thornless variety often preferred for planting.

GLEDITSIA

TABLE 99.—*Gleditsia: Time of flowering and fruit ripening, and frequency of seed crops*

Species	Time of—			Commercial seed-bearing age			Seed year frequency	
	Flowering	Fruit ripening	Seed dispersal	Minimum	Optimum	Maximum	Good crops	Light crops
				Years	*Years*	*Years*	*Years*	*Years*
G. aquatica	May–June	Aug.–Oct.	Sept.–Dec.	10	25–75	100	1–2	Intervening.
X*G. texana*	April–May	Aug.–Sept.	do	10	25–75	100	------	------
G. triacanthos	May–June	Sept.–Oct.	Sept.–late winter	10	25–75	100	1–2	Intervening.

the Forest Service for extracting deciduous tree seeds is satisfactory for honeylocust; with its use, 400 to 600 pounds of clean seed can be extracted per day. Chaff may be separated from the seed by a fanning mill or by water flotation. Data for *Gleditsia triacanthos* are as follows: Yield per 100 pounds of fruit, 20–35 pounds. Cleaned seed per pound (36+ samples): low, 1,750; average, 2,800; high, 4,050. Purity, 95 percent; soundness, 98 percent; cost per pound, $0.35-$1.10. Because of the relative scarcity of the thornless variety, seed of *G. triacanthos* var. *inermis* cost approximately 50 percent more than those of the thorny species.

Although no storage data are available for seed of *Gleditsia aquatica* and X*G texana*, it is believed recommendations for *G. triacanthos* may be applied satisfactorily to them. Seed of honeylocust may be kept for several years in sealed containers at temperatures of 32° to 45° F. without appreciable loss of viability. In regions where weevil damage is common, it is advisable to store seed at slightly above 32° to retard activity.

GERMINATION.—The best natural seedbed is a moist fertile soil. It is believed that a high percentage of seed germinating under natural conditions has passed intact through the alimentary canals of animals or birds, since observations of seed have indicated high germination as against low germination for those undisturbed. Germination is epigeous. Processes of decay of the fruit pod may render the impermeable coats of some seed permeable to water, stimulating germination. In nursery practice, some kind of effective pretreatment is necessary. Details of recommended pretreatments are listed in table 100.

Adequate tests may be conducted on 200 pretreated seeds in flats of sand, peat, sand and peat, or a sandy loam soil at temperatures alternating from 70° F. (night) to 85° (day). Methods recommended and average results for *Gleditsia triacanthos* are as follows: Test duration: pretreated seed, 15 to 40 days; untreated seed, 120 to 180 days. Germinative energy: 45 to 99 percent in 9 to 20 days. Germinative capacity (22+ samples): low, 50 percent; average, 75 percent; high, 99 percent. No data for the other species are available.

NURSERY AND FIELD PRACTICE.—Spring sowing is much preferred to fall sowing. Pretreated seed should be drilled in rows 6 to 10 inches apart in well-prepared beds and covered to a depth of ½ to ¾ of an inch with firmed soil. Spacing is governed by methods of weeding and cultivation, but approximately 10 to 15 seed should be drilled per linear foot of row. Surface soil should be kept constantly moist until germination is complete. The seedlings reach suitable size for field planting in 1 year. Vegetative propagation by cuttings may be easily done. The species are relatively free from insect pests and diseases.

TABLE 100.—*Gleditsia triacanthos*:[1] *Methods of seed pretreatment for dormancy*

Pretreatment			Remarks
Medium	Temperature	Duration	
	°F.	*Hours*	
Scarification	------	------	A number of mechanical scarifiers have been developed to handle large quantities of seed.
Concentrated commercial sulfuric acid.	60–80	1–2	Acid must be thoroughly washed from seed which are immediately drilled or dried for limited storage. Avoid prolonged washing.
Water	185–195	------	Quantity of water used for soaking should be 3 or 4 times that of seed volume. Stir seed in water and leave to soak as water cools until majority of seed have swollen. Drill at once. The first 2 treatments are usually much more satisfactory.

[1]It is recommended that the methods of pretreatment for this species be employed for *G. aquatica* and X*G. texana*.

GYMNOCLADUS DIOICUS (L.) K. Koch Kentucky coffeetree

(Legume family—Leguminosae)

Also called coffee-nut, nicker-tree.

Distribution and Use.—Native to the Central States and adjacent Appalachian highlands of Tennessee, Kentucky and northward to Pennsylvania, New York, Ontario, and Minnesota, Kentucky coffeetree is a medium-tall, deciduous tree valuable for timber, fence posts, and game food. It is used to a limited extent as an ornamental in the eastern United States and in Europe. The seeds are said to have been used by the Kentucky pioneers as a substitute for coffee. It was introduced into cultivation before 1748.

Seeding Habits.—The fruit is an indehiscent or tardily dehiscent, stout, leguminous pod, 6 to 10 inches long and 1.5 to 2 inches wide, externally tough and woody, pulpy inside, and bears three to six seeds. The seeds are ovoid, about three-fourths of an inch long, hard, dull, dark brown, with thin endosperm (fig. 123). Kentucky coffeetree flowers are dioecious and appear in June. The seeds ripen in September or October of the same year and most of the pods remain unopened on the trees through the winter.

Collection, Extraction, and Storage. — The pods may be gathered from the trees after the seed ripen, or from the ground throughout the fall and winter. Pods on the tree often can be freed by vigorously shaking or flailing the branches. Small quantities of seed may be extracted by hand either dry or after maceration in water. For larger quantities, the special macerator and thresher developed by the Forest Service for extracting decidu-

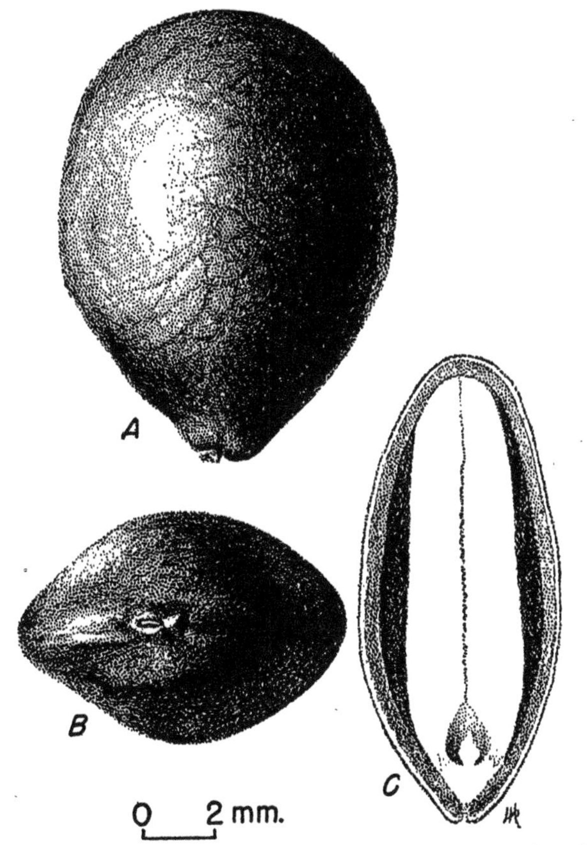

Figure 123.—Seed views of *Gymnocladus dioicus: A* and *B*, Exterior views of two planes; *C*, longitudinal section.

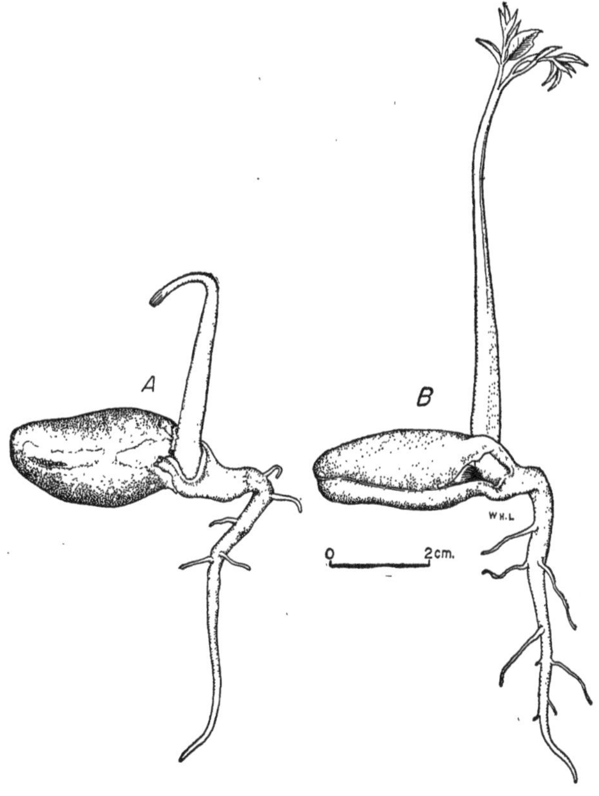

Figure 124.—Seedling views of *Gymnocladus dioicus: A*, At 2 days; *B*, at 5 days.

ous tree seeds may be used. After threshing, the seeds must be cleaned in a fanning mill or by water flotation. Cleaned seed per 100 pounds of pods: 30 to 50 pounds. Number per pound (13 samples): low, 210; average, 275; high, 365. The seed runs almost 100 percent in purity and is usually over 95 percent sound. Cost per pound: $0.75 to $1.40. Dried seed may be kept under ordinary dry-storage conditions at least over winter, and probably for several years. Cold dry storage is recommended for periods longer than 1 year.

GERMINATION. — Germination in this species is hypogeous (fig. 124). An impermeable seed coat either prevents or long delays germination. Two to four hours' immersion in concentrated sulfuric acid followed by thorough washing in water is the most effective treatment known for inducing prompt germination. Recent evidence indicates that the efficacy of the acid treatment is further enhanced, particularly for old seed, if the treatment is divided into 2 immersions of 1 to 2 hours each with an intervening wash and period of drying. An alternative, but generally less effective, treatment is immersion for 2 to 6 minutes in hot water at 185° to 195° F. For small lots of seed, scarification by hand with a file or emery wheel gives fair results. Test methods recommended: Sand flats at temperatures alternating diurnally between 68° and 86°. Seeds required: 100 to 200 properly pretreated. Duration: 10 to 40 days. Germinative energy: 24 to 90 percent in 10 to 25 days. Germinative capacity (14 tests): low, 29 percent; average, 75 percent; high, 92 percent. All data are for acid-treated seed.

NURSERY PRACTICE.—Acid-treated seed should be sown in the spring in drills and covered with 1 to 2 inches of firmed soil. Where ditch irrigation is used, the rows should be 21 to 27 inches apart, with sowing at the rate of 10 to 15 seeds per linear foot. With overhead irrigation, or where irrigation is not required, the rows may be closer together. Seedlings may be transplanted to the field after 1 year in the nursery, or carried over a second year. Propagation by cuttings is also practiced.

HALESIA CAROLINA L. Carolina silverbell

(Snowbell family—Styracaceae)

Also called bell tree, snowdrop tree. Botanical syns.: *H. tetraptera* Ellis; *Mohrodendron carolinum* (L.) Britt.

Distribution and Use.—Native to the Appalachian highlands from West Virginia southward, and west to Missouri, Oklahoma, and Texas, Carolina silverbell is a small deciduous tree or large shrub valued chiefly as a cultivated ornamental. It is of but minor importance as a source of game food, and the wood is marketed only in insignificant quantity for a few special uses. The species is cultivated in the Eastern States as far north as Massachusetts and in northern and central Europe. It was introduced into cultivation in 1756.

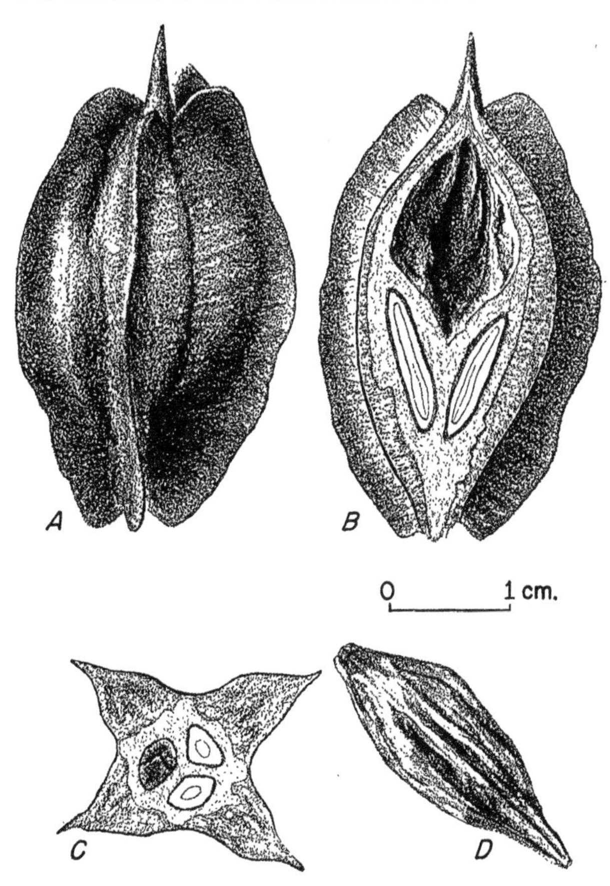

Figure 125.—Seed views of *Halesia carolina: A,* Exterior view of fruit; *B,* longitudinal section of fruit, showing filled and aborted seeds; *C,* cross section of fruit, showing filled and aborted seeds; *D,* cleaned seed.

Seeding Habits.—The fruit is an oblong or oblong-obovoid dry, 4-winged, corky drupe, 1 to 2 inches long. Though it develops from a 4-celled ovary and occasionally matures 2 to 4 seeds, it typically is 1-seeded by abortion (100 fruits yield 106 to 135 seeds, 4 samples). For practical purposes, the stone of the fruit is treated as the seed (fig. 125). Flowers appear in April or May; the fruit ripens in late autumn, persists well into the winter, and is dispersed chiefly by wind, gravity, and water.

Collection, Extraction, and Storage. — The fruits may be collected from the trees throughout late fall and early winter. Wings should be removed to reduce bulk and facilitate handling. Apparently the fruits may be stored dry temporarily; there are no data on longevity in storage. For seeds that are to be germinated the next spring after collection, no storage problem exists because the entire period is required for after-ripening treatments. There are 1,200 to 2,500 cleaned (dewinged) fruits

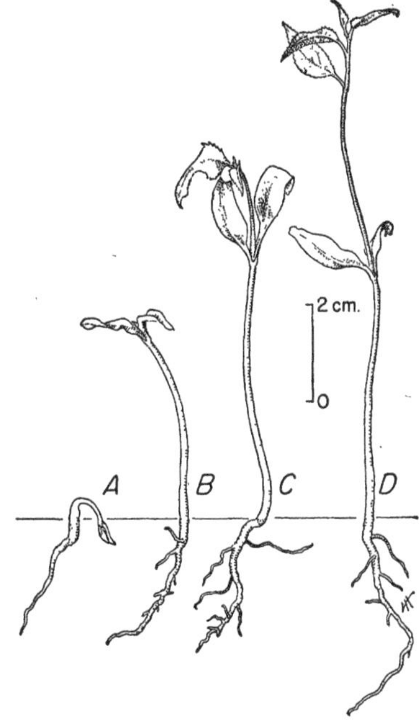

Figure 126.—Seedling views of *Halesia carolina: A,* At 1 day; *B,* at 4 days; *C,* at 16 days; *D,* at 49 days.

per pound (2 samples). The fruits can be purchased commercially at prices from $1 to $2.50 per pound.

GERMINATION.—Untreated seeds will not germinate satisfactorily until the second year after planting. Dormancy is broken best by stratifying the seed first at high temperatures (56° to 86° F.) for 60 to 120+ days and then at low temperatures (33° to 41°) for 60 to 90 days, although some lots germinate fairly well with cold stratification alone. Tests should be run in sand or sand and peat flats for 60 to 90 days at 68° F. (night) to 86° (day) using 100 to 200 properly pretreated seeds per test. Germinative capacity, pretreated seed (7 tests): low, 15 percent; average, 53 percent; high, 84 percent. Germination is epigeous (fig. 126).

NURSERY AND FIELD PRACTICE.—To obtain good germination the first year the fruits should be subjected to comparatively high temperatures (59° to 75° F.) for 2 to 3 months, followed by 2 to 3 months at 33° to 40°. The treatment may be applied to fruits either planted in soil in flats or stratified in moist sand or peat. In the more northern latitudes it suffices for the cold treatment to move the flats to an outdoor coldframe in January. They should then be covered with boards or mulch. Fair results can be obtained for some lots of fruits by stratifying cold for 6 or 7 months without previous high temperatures; this treatment, however, is more likely to result in uneven and prolonged germination.

Different collections of fruit seem to vary somewhat in their response to treatment. It appears also that after-ripening proceeds better with fluctuating than with constant temperatures. Well after-ripened seed show 40 to 80 percent germination. Thirty to forty-five days are required after planting for germination to start; 50 to 80 days are required for its completion. Seed after-ripened in outdoor flats germinate mostly in May and June. Plants are also propagated by layers, root cuttings or greenwood cuttings. Carolina silverbell grows well in a variety of soils but does best on rich, well-drained soils in sheltered positions.

HAMAMELIS VIRGINIANA L. Witch-hazel

(Witch-hazel family—Hamamelidaceae)

Distribution and Use.—Witch-hazel is a deciduous shrub or small tree native from Nova Scotia south to Florida and west to southern Ontario, Wisconsin, eastern Nebraska, and eastern Texas. Locally it occurs on rich, moist soils in partial shade of forests. It is used in ornamental planting in the Northern States and in western and northern Europe, but it has a very restricted value as a game food. The bark, leaves, and twigs are used medicinally, and a volatile oil is distilled from the plant. It was introduced into cultivation in 1736.

Seeding Habits.—Perfect and sometimes monoecious flowers open in September to mid-November, but the fruit does not ripen until the next autumn when the tree is in flower again. Fruits burst, each forcibly discharging two shiny black seeds (fig. 127). There is a limited dispersal beyond the vicinity of the plant by birds.

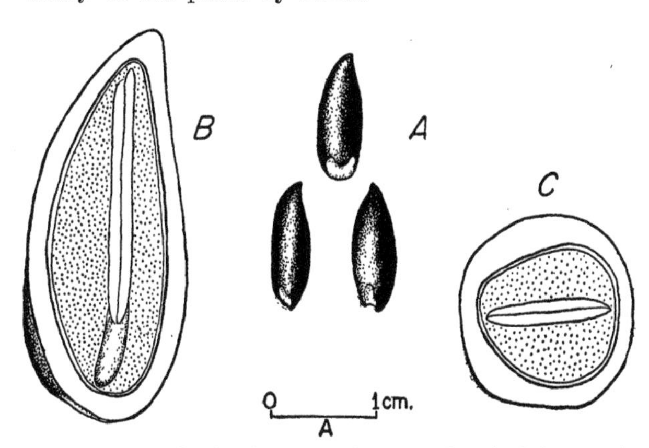

Figure 127.—Seed views of *Hamamelis virginiana: A,* Exterior view in three planes; *B,* longitudinal section; *C,* cross section.

Collection, Extraction, and Storage.—Witch-hazel fruits or capsules should be picked in the autumn shortly before they split and discharge the seeds. Ripe fruits are dull orange brown with blackened, adhering remnants of floral bracts at the base. They should be spread out to dry soon after collection; the seeds may then be separated from the capsules by screening. Seeds average 8,700 to 10,900 per pound (2 samples). Commercial seed averages 99 percent in purity, 95 percent in soundness, and costs $1.50 to $3.25 per pound. Seed stored dry in sealed containers at 41° F., soon after collection, can be kept for 1 year without loss of viability. For overwinter storage prior to spring sowing, the seed should be stratified in a mixture of sand and peat at 41°.

Germination.—Damp, rich soils in partial forest shade are the best natural seedbeds. Some seed germinate in the spring, but many lie over until the second spring. Germination is epigeous (fig. 128). Dormancy due to conditions in both seed coat and embryo is the rule. Satisfactory stratification methods for overcoming dormancy have not been worked out. Best results were obtained, however, in one series of tests by stratifying seed for 60 days at 68° F. (night) to 86° (day) plus 90 days at 41°. Soaking the seed in concentrated sulfuric acid for 15 minutes and then stratifying them for 90 days at 41°, the only other method to break dormancy, gave only 1 percent germination in 90 days. The stratification period is shortened if the seeds are first soaked in hot water. Germination tests should be run in sand or sand and peat flats for 60 days at 68° (night) to 86° (day) using 200 pretreated seeds per test. The germinative capacity of pretreated seed was 17 percent and potential germination 81 percent in one test.

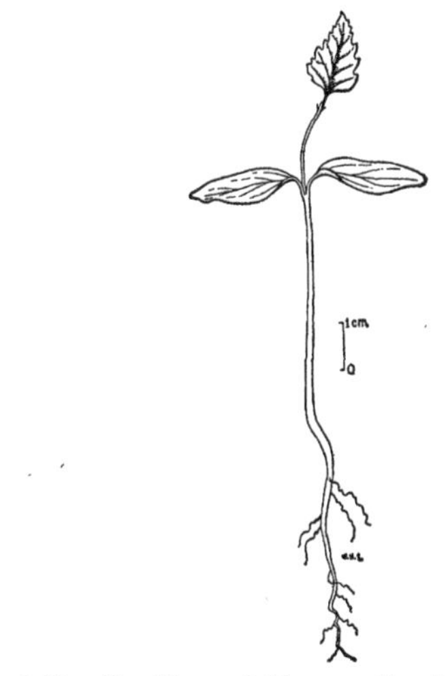

Figure 128.—Seedling of *Hamamelis virginiana* at 21 days.

HAMAMELIS

NURSERY PRACTICE.—Witch-hazel seed may be fall sown in the nursery or spring sown in well-prepared beds. Fall sowing is advised as soon as the seed are gathered. The beds should be mulched with burlap, straw, or leaves, and the mulch removed at germination time in the spring. In one case seed collected slightly green in August, and sown immediately, germinated 90 percent the following spring. In spring sowing, stratified seed are used in beds prepared as early as soil conditions permit. Surface soil should be kept moist until germination is complete. If dry seed are used, the beds should be mulched and kept intact until the following spring. Sowing may be broadcast or drill, the latter being advised, with 8- to 12-inch spacing to provide ease in weeding and cultivation. Propagation by layering is also possible. For best results witch-hazel should be planted on rather moist peaty or sandy soil.

HIPPOPHAE RHAMNOIDES L. Common seabuckthorn, swallow-thorn

(Elaeagnus family—Elaeagnaceae)

Distribution and Use. — Native from Europe through central Asia to the Altai Mountains, western China, and the northwestern Himalayas, common seabuckthorn is one of two species in this genus, the other of which is little cultivated. It is a very hardy, deciduous shrub to small tree which has long been cultivated for ornamental purposes, and is also useful for controlling shifting sands and for shelter-belt purposes. Although its berries are reported to be somewhat poisonous, they are eaten by birds and in Siberia are used for making jelly and jam.

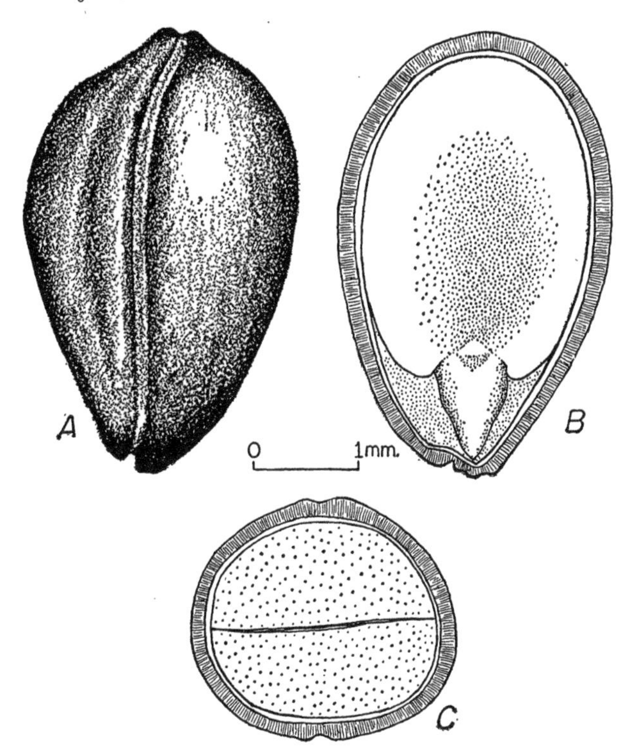

Figure 129.—Seed views of *Hippophae rhamnoides: A,* Exterior view; *B,* longitudinal section; *C,* cross section.

Seeding Habits.—The species is dioecious, and its very small, yellowish, pistillate flowers appear in March and April before the leaves. Acid, orange-yellow, drupelike fruits about the size of a pea ripen in September and frequently persist on the shrubs until the following March. Each fruit contains one bony, ovoid stone (fig. 129).

Collection, Extraction, and Storage.—Common seabuckthorn fruits may be collected by hand from the bushes, or flailed onto canvas or cloth, from late fall to early spring. The seed are probably most easily extracted by running the fruit wet through a macerator. One hundred pounds of fruit yields 10 to 30 pounds of cleaned seed. Cleaned seed per pound (10+ samples): low 25,000; average, 40,000; high, 59,000. Commercial seed averages 97 percent in purity, 85 percent in soundness, and costs about $2.50 per pound. The seed can be kept satisfactorily for 1 to 2 years in ordinary dry storage and perhaps longer if stored dry in sealed containers at low temperatures.

Germination. — Germination is epigeous (fig. 130). The seed exhibit dormancy, probably due to embryo conditions, and stratification in moist sand for 90 days at 41° F. is suggested to overcome it. Germination tests may be run in sand flats for 40 days at 68° (night) to 86° (day), using 400 stratified seeds per test. Three tests using untreated seed gave germinative capacities of only 8 to 60 percent at 60 days.

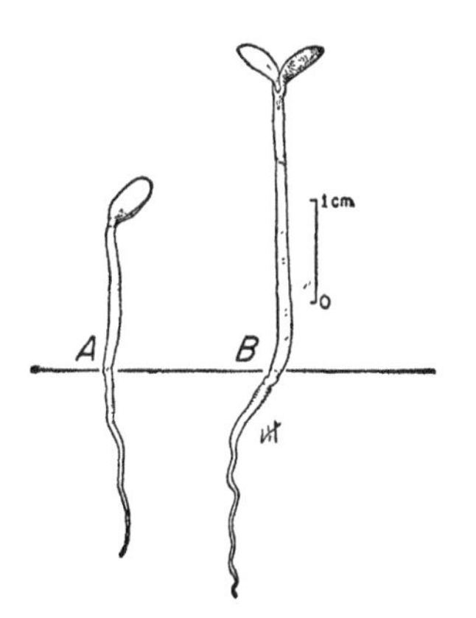

Figure 130.—Seedling views of *Hippophae rhamnoides: A,* At 1 day; *B,* at 7 days. (Seedlings typically covered with silvery scales.)

Nursery and Field Practice.—Untreated seed may be sown in the fall, or stratified seed in the spring, either broadcast or in drills, and covered with about one-fourth inch of soil. This species is propagated by layers, suckers, and root cuttings, as well as by seeds. It grows best on moist, sandy soils.

ILEX L. Holly

(Holly family—Aquifoliaceae)

DISTRIBUTION AND USE. — The hollies include about 300 species of deciduous or evergreen shrubs or trees occurring in the temperate and tropical regions of both hemispheres. Several species are cultivated for ornamental purposes and some evergreen species for the production of holiday decorations. The wood of some of the trees is light and tough and valued for cabinetmaking and interior finish. Several of the hollies are good honey plants, and the leaves of some are used for tea, while the leaves, bark, and fruit of others are used medicinally. About 20 species, both trees and shrubs, are native to eastern North America. Five species, which are of known or supposed value for planting in the United States and for which reliable information is available, are described in detail in table 101. *Ilex aquifolium* is the most widely cultivated holly; it is hardy in the United States only in the southeastern part and in the Pacific Northwest, where it is cultivated commercially for holiday decoration. *I. opaca* is the most commonly planted native holly.

SEEDING HABITS.—The white or greenish, axillary male and female flowers usually are borne separately on different plants, but bisexual flowers may occur. The fertile flowers usually are solitary, the sterile ones in clusters. By the end of the season the former develop into berrylike drupes (the fruits), each containing two to nine bony, one-seeded, flattened nutlets,[33] the seed of commerce (figs. 131 and 132). A membranaceous, pale brown seed coat covers a fleshy endosperm, at the apex of which is a very small embryo. Dispersal is chiefly by animals and gravity. The comparative seeding habits of the five species discussed here are given in table 102. In *Ilex aquifolium* seed bearing begins on trees 5 to 12 years old, but is best on trees over 20 years in age. Good crops are borne annually, although older trees in dense stands may produce only biennially. Comparable information for the other species is not available. Scientific information as to racial variations among the hollys is lacking.

[33] Smooth on the back in *Ilex verticillata,* ribbed on the back in *I. montana,* ribbed on the back and sides in *I. opaca.*

TABLE 101.—*Ilex: Growth habit, distribution, and uses*

Accepted name	Synonyms	Growth habit	Natural range	Chief uses	Date of earliest cultivation
I. aquifolium L. (English holly).	------------------	Short to medium-tall evergreen tree.	Western and southern Europe, northern Africa, and western Asia to China.	Holiday decorations, ornamental, wildlife food.	(1)
I. glabra (L.) A. Gray (inkberry).	*Prinos glaber* L. (gallberry, winterberry).	Medium-sized evergreen or semideciduous shrub.	Maine to Florida and west through Gulf States to eastern Texas and Arkansas.	Wildlife food, bee pasture, tea, and medicinal purposes.	1759
I. montana Torr. & Gray (mountain winterberry).	*I. monticola* A. Gray, *I. dubia monticola* (A. Gray) Loes. (mountain holly).	Deciduous shrub or slender small tree.	New York to South Carolina and west to Alabama.	Wildlife food.	1870
I. opaca Ait. (American holly).	holly, evergreen holly.	Short to medium-tall evergreen tree.	Massachusetts to Florida, westward through Gulf States to Texas and in Mississippi Valley to Indiana.	Holiday decorations, ornamental, turnery wood, wildlife food, medicinal purposes.	1744
I. verticillata (L.) A. Gray (common winterberry).	black-alder-------	Medium-sized deciduous shrub.	Swamps and low ground from Nova Scotia to Florida and west to western Ontario and Missouri.	Wildlife food, ornamental, winter decorations, tea, medicinal purposes.	1736

[1] Cultivated since ancient times.

TABLE 102.—*Ilex: Time of flowering and fruit ripening, and fruit characteristics*

Species	Time of—			Fruit		
	Flowering	Fruit ripening	Seed dispersal	Color	Diameter	Seeds
					Inches	*Number*
I. aquifolium	May–June	September	Winter–spring	Bright red	¼–⅜	1–4
I. glabra	March–June	Fall	do	Black	¼	6±
I. montana	June	September	do	Orange red	⅜	
I. opaca	April–June	Fall	do	Red or yellow	¼	4
I. verticillata	June–July	September–October	Fall–winter	Bright red or yellow	¼	4–6

However, varieties of most species are recognized, and some of these may be synonymous with climatic races.

COLLECTION, EXTRACTION, AND STORAGE. — Ripe holly fruits may be picked from the plants by hand or flailed onto canvas in the late fall to winter or early spring. They may then be spread out and dried with the pulp on or run wet through a macerator, and the pulp and empty seeds floated or skimmed off. Holly seed should be dried before storage or use, or they may be stratified immediately. The yield, size, purity, soundness, and cost of commercial cleaned seed vary considerably by species as shown in table 103. Storage for periods longer than over winter has not been reported for holly seed. For such storage cleaned seeds or dried berries may be placed dry in sealed containers at low temperatures, or they may be stratified in moist sand.

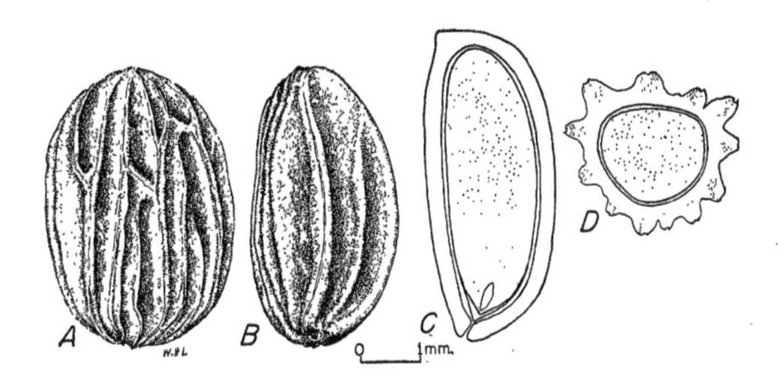

FIGURE 131.—Seed views of *Ilex montana*: *A* and *B*, Exterior views in two planes; *C*, longitudinal section, showing small embryo; *D*, cross section.

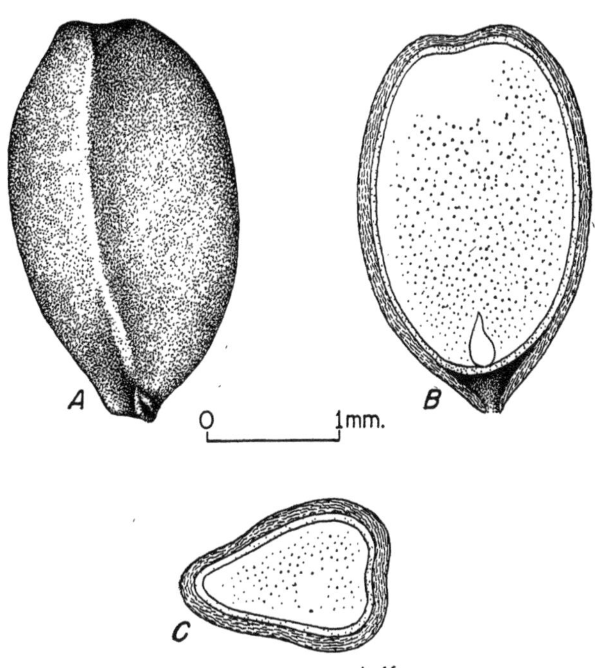

FIGURE 132.—Seed views of *Ilex verticillata*: *A*, Exterior view; *B*, longitudinal section, showing small embryo; *C*, cross section.

GERMINATION. — In nature, holly seeds usually germinate from 1 to 3 years after dispersal, and natural regeneration commonly is quite sparse. Germination probably is epigeous. The seeds have hard but not impermeable seed coats, and dormant or immature embryos; as a rule they are very difficult to germinate. For several species neither stratifica-

TABLE 103.—*Ilex: Yield of cleaned seed, and purity, soundness, and cost of commercial seed*

Species	Cleaned seed					Commercial seed			
	Yield per 100 pounds of fruit	Per pound			Basis, samples	Purity	Soundness	Cost per pound	
		Low	Average	High				Seed	Dry berries
	Pounds	*Number*	*Number*	*Number*	*Number*	*Percent*	*Percent*	*Dollars*	*Dollars*
I. aquifolium			57,000		1		87	1.25–2.00	
I. glabra	5				1			1.25–2.10	
I. montana			35,000		1	90	84	2.00	
I. opaca	20–27	22,000		31,000	2		70	2.00–3.70	0.50–1.70
I. verticillata	11–20	40,000	92,000	129,000	4	96	75	3.00–3.50	1.00

tion, boiling, nor various chemical treatments have been very successful in overcoming dormancy; it still takes 2 to 3 years to get fairly complete germination. In such cases the only reasonably rapid estimate of viability is by means of cutting tests. In one series of tests *Ilex opaca* seed with the pericarps removed germinated 60 percent in 150 days when tested in a 5-percent dextrose-treated substratum at 77° to 86° F. *I. verticillata* germinates quite satisfactorily (52 to 73 percent in one series of tests) in 60 days in sand flats if they are first stratified in moist sand for 60 days at 68° (night) to 86° (day) followed by 60 days at 41°.

NURSERY AND FIELD PRACTICE.—Holly seed may be sown broadcast or in drills in the fall, or stratified over winter and sown in the spring. Fall-sown beds should be mulched over winter. For *Ilex opaca* recommendations also include: stratification for 30 days at 75° to 80° F. prior to sowing; burying seeds mixed with sand or peat and sowing them after the second winter; sowing berries and covering them with a heavy mulch until the spring of the second year. The seed should be covered with about one-eighth inch of soil. For *I. aquifolium* it is recommended that half-shade be given the beds the first 2 summers, and that field planting be done with 2–2, 2–3, or 2–2–2 stock. Best results usually are attained if planting is done on rich, well-drained soil, although *I. verticillata* does best in moist places. The evergreen species do best on partly shaded situations. Germination occurs chiefly during the second and third season after sowing. Holly plants should be transplanted only when dormant, and the evergreen species only after being stripped of their leaves. Propagation by cuttings, layering, or grafting is also practiced to some extent.

JUGLANS L. Walnut

(Walnut family—Juglandaceae)

DISTRIBUTION AND USE.—The walnuts, consisting of about 15 species of deciduous trees, rarely shrubby, are native to the temperate regions of North America, northwestern South America, and from southeastern Europe to eastern Asia. Six species are native to the United States. The wood of several species is highly valued for cabinet work and interior trim, and the nuts of many species are valuable as human and animal food. Six species, which are of present or potential value for growing in the United States, are described in table 104. *Juglans regia* is the most extensively planted walnut, chiefly for nut culture, in the United States, and on an international basis also. Of the native species, *J. nigra* is most widely planted, and *J. cinerea, J. rupestris,* and *J. hindsii* are planted less extensively. *J. cinerea* and *J. rupestris* are also planted somewhat in Europe.

SEEDING HABITS.—The male and female flowers are borne separately on the same tree. The male flowers, bearing pollen only, develop from axillary buds on the outer nodes of the previous year. They are borne in slender, lateral catkins. The female flowers, which develop into the fruits, occur in few- to many-flowered short, terminal spikes and are borne on the current year's shoots. These flowers appear with, or shortly after, the leaves. The ovoid, globose, or pear-shaped fruit ripens in the first year and is a type of dry drupe with an indehiscent, thick husk. Each fruit contains one nut—the seed of commerce—which is incompletely two- to four-celled and indehiscent, or finally separates into two valves (figs. 133 and 134). The nut has a crustaceous or bony, rugose shell. The seed is two- to four-lobed; it remains within the shell in germination. Seeds of the various species are quite similar

TABLE 104.—*Juglans: Growth habit, distribution, and uses*

Accepted name	Synonyms	Growth habit	Natural range	Chief uses	Date of earliest cultivation
J. californica, S. Wats., (California walnut).	California black walnut, SPN.	Low tree[1]	Southern California coast region.	Veneer, erosion control, wildlife food.	------
J. cinerea L. (butternut).	--------------------	Tree of medium size.	New Brunswick to northern Georgia and west to North Dakota and northern Arkansas.	Cabinet work, human and animal food, hulls for dyes.	1633
J. hindsii Jeps. (Hinds walnut).	*J. californica* var. *hindsii* Jeps. (Hinds black walnut, SPN).	Large tree	Central California coast region.	Shade tree, stock for grafting Persian walnut.	1878
J. nigra L. (black walnut).	eastern black walnut	do.	Massachusetts to Florida, west to Minnesota and Texas.	Furniture, interior trim, gunstocks, human and wildlife food, shelter belts, ornamental, hulls for dye.	1686
J. regia L. (Persian walnut).	English walnut	do.	Southeastern Europe to the Himalayas and China.	Human food, furniture, cabinet wood, gunstocks, wildlife food.	(2)
J. rupestris Engelm. (little walnut).	Texas black walnut, SPN; desert walnut, western walnut.	Shrub or small tree.	Limestone banks of streams southwestern Oklahoma, central Texas, and northwest into southeastern New Mexico.	Veneer, wildlife food, ornamental and shelterbelt planting.	1868

[1] Sometimes shrubby.
[2] Long cultivated.

in behavior and differ mainly in size, thickness of shell, and palatability. Comparative seeding habits of the six species discussed here are given in table 105.

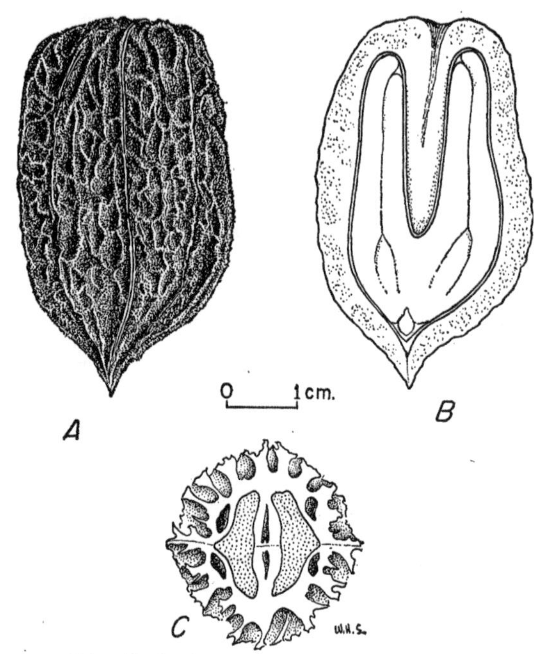

FIGURE 133.—Seed views of *Juglans cinerea: A,* Exterior view; *B,* longitudinal section; *C,* cross section.

There is little scientific information available as to the development of climatic races in *Juglans* other than for *J. regia,* but it is likely that most species of extensive range have developed such races. Three distinct climatic races of *J. regia* are recognized: Turkestanian, Himalayan, and Central Asian. They differ considerably in frost hardiness. Several horticultural varieties are also recognized.

COLLECTION, EXTRACTION, AND STORAGE.—Walnut fruits are collected in the fall or early winter from the ground, either after they have fallen naturally, usually after frosts, or after they have been knocked

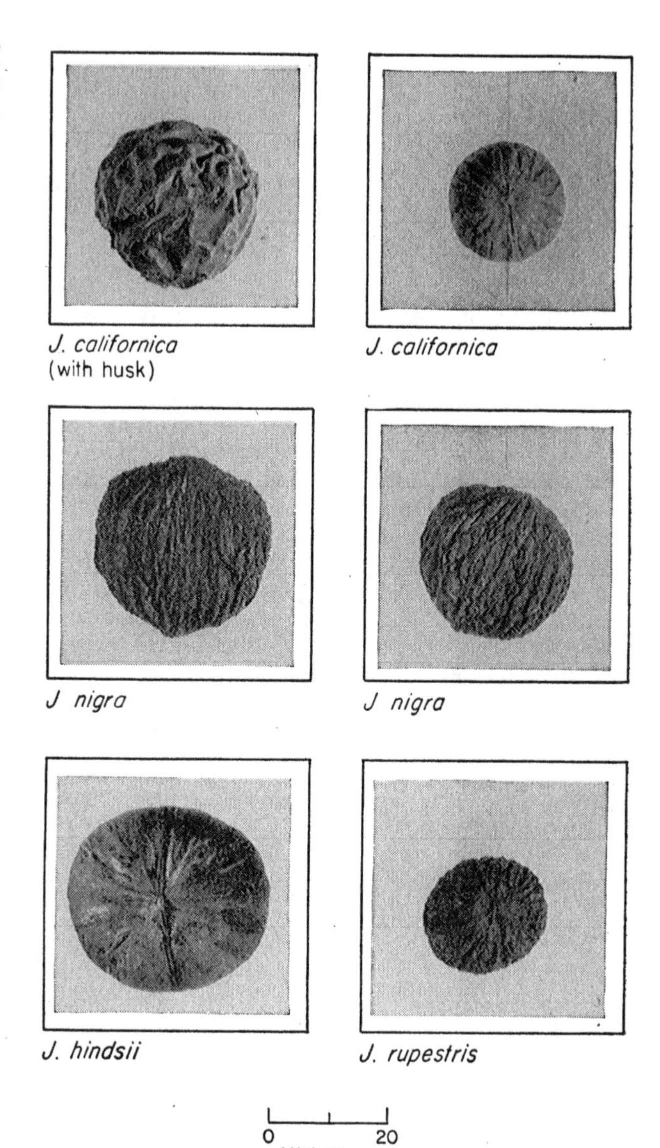

FIGURE 134.—Seed views of four species of *Juglans.*

TABLE 105.—*Juglans: Time of flowering and fruit ripening, and frequency of seed crops*

Species	Time of—			Commercial seed-bearing age			Seed year frequency	
	Flowering	Fruit ripening	Seed dispersal	Minimum	Optimum	Maximum	Good crops	Light crops
				Years	*Years*	*Years*	*Years*	*Years*
J. californica	April–May	Fall	Fall					
J. cinerea	do	Sept.–Oct.	Sept.–Oct.	20	30–60	80	2–3	Intervening.
J. hindsii	do	Fall	Oct.–Nov.					
J. nigra	May–June	Sept.–Oct.	Fall	12	30+		(1)	Intervening.
J. regia	March–May	Sept.–Nov.	do.	5			(1)	do.
J. rupestris	March–April	Aug.–Sept.	do.	20	50–125	150	2–3	do.

[1] *J. nigra* and *J. regia* bear good seed crops irregularly.

TABLE 106.—*Juglans: Yield of cleaned seed, and purity, soundness, and cost of commercial seed*

Species	Cleaned seed					Commercial seed		
	Yield per 100 pounds of fruit	Per pound			Basis, samples	Purity	Sound-ness	Cost per pound[1]
		Low	Average	High				
	Pounds	*Number*	*Number*	*Number*	*Number*	*Percent*	*Percent*	*Dollars*
J. californica			75		1			0.50–0.60
J. cinerea	20–30	15	30	40	13	100	96	.10– .30
J. hindsii		38		80	2			.25– .60
J. nigra	30–65	20	40	100	19+	99	87	.10– .30
J. regia		30	40	50	8+	([2])	([2])	.45– .65
J. rupestris	40–65	78		107	2	100	94	

[1]Seed of many species can be purchased locally much more cheaply by the bushel.
[2]Purity and soundness in *J. regia* are high.

down by flailing or shaking. Collection should begin promptly after the nuts fall, or many of them will be gathered by squirrels and other animals. The husks turn brownish when ripe, and the nuts are easier to extract if the husks have not dried. They may be removed either by hand or by running the fruits through a corn sheller. If the nuts are not removed from the husks immediately, the fruits should be spread out to dry before they are stored. Detailed experience in collection practice is available for three species only. In *Juglans cinerea* thrifty trees will yield from 1/4 to 1 bushel of clean nuts. Those of *J. nigra* may produce up to several bushels. Fruits of *J. nigra* and *J. rupestris* can be collected at a labor cost of $0.50 to $1 per bushel. Yield, size, purity, and soundness of commercial cleaned seed vary considerably by species as shown in table 106.

Storage information is not available for several *Juglans* species. In general they should either be sown in the fall soon after collection or stored over winter in moist sand, peat, or sandy loam at temperatures of 35° to 45° F. If outdoor storage is used, the fruits or nuts should be covered with at least 2 feet of soil or straw. The entire fruits may be so stored. Seed of *J. cinerea* will retain its viability for 4 or 5 years if stored either husked or unhusked in sealed containers at temperatures a little above freezing. It is recommended that clean nuts of *J. rupestris* be placed in sealed containers with a relative humidity of at least 85 percent and kept at 41° for long-time storage.

GERMINATION. — Natural germination usually takes place in the spring after seed fall. Many seed that germinate have been previously buried by rodents. Germination usually is best on a moist, rich, loamy soil with side but not overhead shade where the nuts have been covered by debris or litter. Squirrels, other rodents, some birds, and larger mammals destroy much of the seed in nature. Germination is hypogeous. Most, if not all, *Juglans* species have dormant seed. Dormancy is apparently caused by internal conditions of the embryo and probably also by a hard seed coat. This condition may be broken by stratification at low temperatures. Further information on pretreatment is given in table 107. The germination of walnut seed may be tested in sand flats or peat mats using 50 to 100 stratified seeds per test. Methods recommended and average results for 6 species are presented in table 108.

NURSERY AND FIELD PRACTICE.—The nuts, husked or unhusked, should be sown in the fall—or in the spring if stratified—in drills 8 to 12 inches apart,[34] using about 15 nuts per linear foot of row. A sandy soil is preferable because it promotes a more fibrous root system. The nuts should be covered with 1 to 2 inches of firmed soil. Seedbeds should be protected with screens against rodents, and those sown in the

[34] Spacings from 16 to 42 inches are sometimes used.

TABLE 107.—*Juglans: Method of seed pretreatment for dormancy*

Species	Stratification			Remarks
	Medium	Temperature	Duration	
		°F.	*Days*	
J. californica				Stratification probably necessary for good germination.
J. cinerea	Sand, sand and peat, sandy loam.	35–45	90–120	Germination may be hastened by preceding the low-temperature stratification with a 60-day period of room-temperature stratification.
J. hindsii				Stratification probably necessary for good germination.
J. nigra	Sand or peat	33–50	60–120	37° F. is most effective stratification temperature.
J. regia	do	[1]41	[1]30– 60	
J. rupestris	Sand, or sandy loam.	40–45	90	

[1]Suggested; experimental data not complete.

891183°—50——14

fall should be mulched with straw or leaves to prevent alternate freezing and thawing of the nuts. The mulch should be removed in the spring after late frosts, and the surface soil should be kept moist until after germination is complete. Root pruning at a depth of 8 to 10 inches in midsummer is helpful in producing stock with a more branchy, compact root system and in removing the characteristic deep taproot.

Juglans nigra seedlings in the nursery often are killed by a *Phytophthora* root rot. To avoid this disease, seeding should be done only on the better-drained parts of the nursery, taking care not to seed too deeply. A leaf disease believed to be caused by the fungus *Gnomonia leptostyla,* which does some damage, may be controlled by 5 to 7 applications of 3–4–50 bordeaux. Sometimes a powdery mildew, *Microsphaera alni,* causes serious damage. It may be controlled by dusting on ground sulfur mixed in the proportion of 1 to 10 with talc, hydrated lime, or other flux material when the disease first appears.

Field planting should be done with 1–0 stock; root-pruned 2–0 stock is sometimes satisfactory. Direct seeding in the fall or using stratified seed in the spring is often more satisfactory than using nursery stock. A deep, well-drained soil of good quality is best for field planting. Species such as *Juglans regia* are often propagated by means of grafting, budding, or layering.

TABLE 108.—*Juglans: Recommended conditions for germination tests, and summary of germination data*

Species	Test conditions recommended				Germination data from various sources					
	Temperature		Duration		Germinative energy		Germinative capacity			Basis, tests
	Night	Day	Stratified seed	Untreated seed	Amount	Period	Low	Average	High	
	°F.	°F.	Days	Days	Percent	Days	Percent	Percent	Percent	Number
J. californica	----------	----------	----------	30+	----------	----------	----------	----------	35	1+
J. cinerea	68	86	45–60	110+	36–82	30–60	30	65	90	7
J. hindsii	----------	----------	----------	30+	----------	----------	20	----------	30	2+
J. nigra	68	86	15–40	100–300	50–70	10–40	70	75	80	4+
J. regia	68	86	40	----------	----------	----------	70	----------	84	2+
J. rupestris	68	86	30–60	----------	68	14	17	36	72	3

JUNIPERUS L. Juniper

(Pine family—Pinaceae)

DISTRIBUTION AND USE.—The junipers include about 40 species of evergreen trees and shrubs which occur throughout the temperate and subtropical regions of the Northern Hemisphere. Practically all species are valuable ornamental plants and many varieties have been developed in horticulture. Tree species furnish close-grained, aromatic wood which is much used in interior finishing, chests, in the manufacture of small articles such as pencils, and, because of its extreme durability, for posts and poles. The fruits and also the young branchlets of some species contain an aromatic oil used in medicine and in the manufacture of gin, and the fruits of at least one species are eaten by man. Those of many others furnish valuable food for wildlife. Of about 15 species indigenous to the United States, 7 species and 1 variety are of present or potential value for conservation planting. Their distribution and chief uses are given in table 109.

Of the species listed in table 109, only *Juniperus virginiana* (Central and Southern States) and *J. scopulorum* (Colorado and Nebraska) are used to any extent for reforestation in the United States. These two species, together with *J. ashei*, are also used for shelter-belt planting in the Prairie-Plains States and to a lesser degree in planting for wildlife food and cover. Practically all of the eight species or

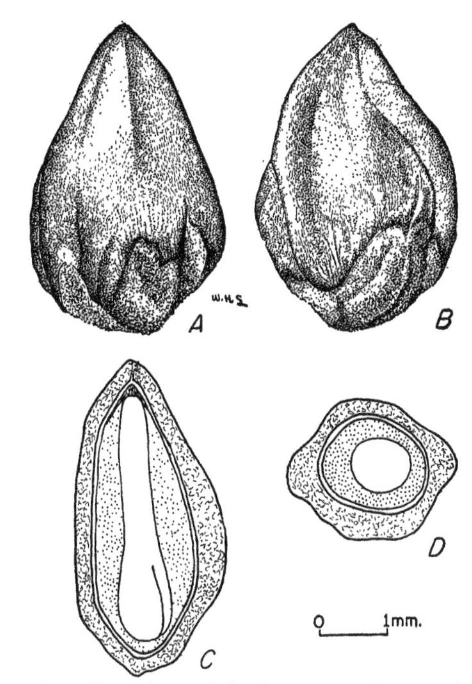

FIGURE 135.—Seed views of *Juniperus scopulorum*: *A* and *B*, Exterior views in two planes; *C*, longitudinal section; *D*, cross section.

TABLE 109.—*Juniperus: Growth habit, distribution, and uses*

Accepted name	Synonyms	Growth habit	Natural range	Chief uses	Date of earliest cultivation
J. ashei Buchholz (Ashe juniper).	*J. sabinoides* (H.B. K.) Nees not Griseb., *J. mexicana* Spreng. (mountain-cedar, rock-cedar, Mexican juniper).	Small to large tree, thicket-forming.	Limestone hills, Brazos River in Texas, south and west into Mexico; also in Ozark region of Missouri, Arkansas, and Oklahoma.	Posts and poles, ties, fuel; leaves and wood, distilled for oil, wax, and camphor; fruit eaten by birds.	1925
J. communis var. *depressa* Pursh[1] (old-field common juniper).	prostrate juniper____	Low procumbent shrub.	Newfoundland to Alaska, south to Connecticut, Indiana, Minnesota, and Colorado.	Ornamental planting; food for birds; possibly for erosion control.	------
J. monosperma (Engelm.) Sarg. (one-seed juniper).	*Sabina monosperma* (Engelm.) Rydb.	Tree or several-stemmed shrub.	Wyoming to western Texas, New Mexico, and Nevada.	Fence posts, fuel; fruit eaten by birds, browsed by goats; protection forest.	1900

[1] The extremely variable *J. communis* L., of wide distribution, also occurs in the United States as do others of its varieties.

TABLE 109.—*Juniperus: Growth habit, distribution, and uses—Continued*

Accepted name	Synonyms	Growth habit	Natural range	Chief uses	Date of earliest cultivation
J. occidentalis Hook. (Sierra juniper).	*Sabina occidentalis* (Hook.) Heller (western juniper).	Small to large tree.	Mountain slopes, western Idaho and eastern Washington, west to east slopes of Cascades, and south in the high Sierras to southern California.	Posts, fuel, ties, wildlife food.	1840
J. pachyphloea Torr.[2] (alligator juniper).	oakbark juniper	Tree often of massive size.	Dry mountain slopes, Arizona, New Mexico, southwestern Texas, and Mexico.	Posts, fuel; fruit eaten by birds and animals; browse for game animals.	1873
J. scopulorum Sarg.[2] (Rocky Mountain juniper).	*Sabina scopulorum* (Sarg.) Rydb., *J. virginiana* var. *scopulorum* Lemm. (Rocky Mountain red cedar).	Tree of medium size, sometimes shrubby.	Dry ridges in foothills of Rocky Mountains, Alberta to Texas and west to eastern Oregon, Nevada, and northern Arizona.	Posts, pencils, fuel, shelter-belt and erosion-control planting; wildlife food.	1836
J. utahensis (Engelm.) Lemm.[3] (Utah juniper).	*Sabina utahensis* (Engelm.) Rydb., *J. californica* var. *utahensis* Engelm. (big-berry juniper, desert juniper).	Bushy tree	Desert foothills and mountain slopes, Wyoming and southern Idaho to California, Arizona, and New Mexico.	Fuel, posts, interior finishing; game food; fruit eaten by Indians.	1900
J. virginiana L.[2 4] (eastern redcedar).	*Sabina virginiana* (L.) Ant. (red cedar, savin).	Small to large tree.	Dry slopes and ridges, often on limestone; New Brunswick to Georgia, west to eastern North Dakota, and eastern Texas.	Furniture and cabinetry, interior finishing, pencils, woodenware, posts, poles; shelter-belt planting; wildlife food.	Before 1664

[2] Several to many varieties are recognized in horticulture.
[3] Includes var. *megalocarpa* (Sudw.) Sarg.
[4] Includes *J. virginiana* var. *creba* the northern form; the type occurs from Virginia southward.

TABLE 110.—*Juniperus:*[1] *Time of flowering and fruit ripening, frequency of seed crops, and commercial seed-bearing age*

Species	Time of—			Color of ripe fruit	Seeds per fruit	Seed year frequency		Commercial seed-bearing age		
	Flowering	Fruit ripening	Seed dispersal			Good crops	Light crops	Minimum	Optimum	Maximum
					Number	*Years*	*Years*	*Years*	*Years*	*Years*
J. ashei	Jan.–April	Sept.–Nov.	Fall–winter	Deep blue	1–2			20	50–175	250
J. communis var. *depressa*	May	Aug.–Sept.[2]	Persists for 1–2 years	Blue black	3					
J. monosperma	March–April	do	do	Dark blue[3]	1					
J. occidentalis	Spring	Sept.[4]	do	Blue black	2–3					
J. pachyphloea	Feb.–March	Aug.–Oct.[4]	Persists for abt. a year	Dark red brown	4	([5])				
J. scopulorum	Spring	Nov.–Dec.[4]	Long persistent.	Bright blue	2	2–5	Intervening.	10	50–200	300
J. utahensis	do	Sept.[4]	do	Reddish brown	1	2	do			
J. virginiana	Mid-March to mid-May	Mid-Sept. to mid-Nov.[6]	Feb.–March	Dark blue	1–2	2–3	do	10–15	25–75	125–175

[1]All species listed here are dioecious except *J. utahensis.*
[2]Requires three seasons to mature its fruit.
[3]Sometimes copper colored.
[4]Requires 2 seasons to mature its fruit; all others, except *J. communis* var. *depressa,* require 1 season.
[5]*J. pachyphloea* bears good seed crops almost every year.
[6]There is evidence that the seed is mature before the berries appear ripe.

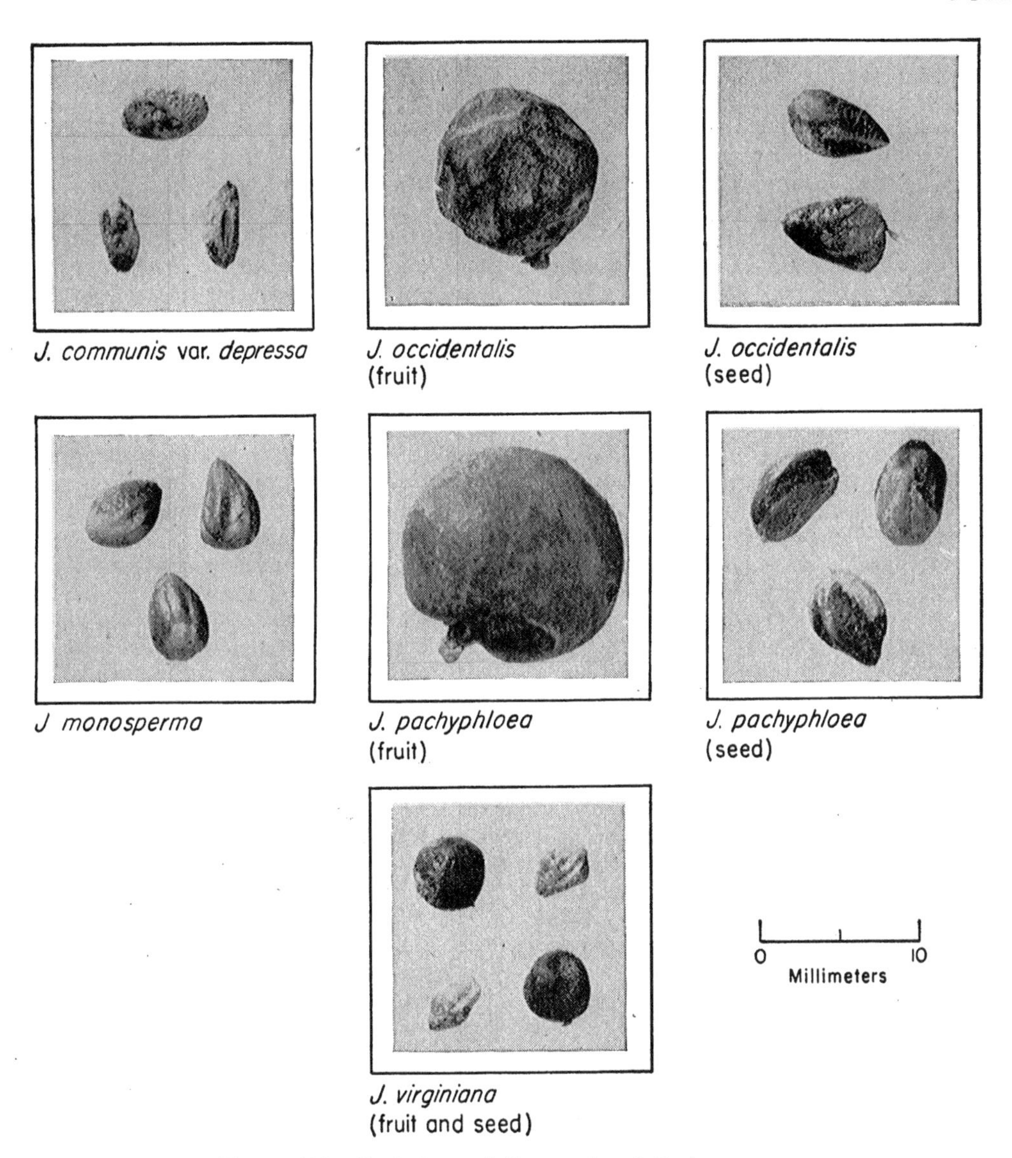

FIGURE 136.—Seed views of five species of *Juniperus.*

some of their horticultural varieties are used to more or less extent for ornamental planting.

SEEDING HABITS.—The small inconspicuous flowers of *Juniperus* are unisexual, the two kinds being borne in the spring on separate plants or occasionally in some species on the same plant. The male flowers are yellow and form a short catkin; the greenish female flowers are composed of 3 to 8 pointed scales, some or all of which bear 1 to 2 ovules. The scales gradually become fleshy and unite into a berrylike, indehiscent conelet or strobile, ripening the first, second, or third fall, depending on the species. Juniper berries—in some species resinous, in others sweet and nearly dry—are blue, blue black, or reddish in color with conspicuous bloom. They contain 1 to 6, rarely 12, brown seeds. These are rounded or variously angled and often show longitudinal pits caused by the pressure of resin cells within the flesh of the fruit (figs. 135 and 136). In section the seed shows a coat of two layers, the outer thick and bony, the inner thin and membranous, and embedded within a fleshy endosperm a straight embryo with 2 to 6 cotyledons. Commercial seed consists of the ripe dried berries or clean seed. Such details of the seeding habits as are known are given for 7 species and 1 variety in table 110. Seed dispersal is almost entirely by birds, but sometimes by mammals. Examples are known of dispersal of *J. utahensis* by sheep.

Little is known as to the existence of climatic races in *Juniperus*. In *J. communis,* several distinct geographic varieties are recognized; *J. virginiana* has a northern form (var. *creba*) which differs a little in growth habit and has slightly pitted seeds.

Further study would undoubtedly reveal the presence of such races in some of the other species which occur over a wide range.

COLLECTION, EXTRACTION, AND STORAGE.—The fruit is usually collected by stripping or picking by hand, or by shaking or flailing it from the trees or shrubs onto canvas. The large fruits of *Juniperus pachyphloea* may sometimes be picked up from the ground. In the case of those species which require more than one season for seed development, care should be taken to avoid the collection of 1-year fruit; its color is light to dark green in contrast to the blue to red-brown color of mature fruits. Although in some species collection can be prolonged over much of the winter, it is better to collect the fruit as soon as possible after ripening to prevent its consumption by birds. Cutting tests should be made at time of collection, since in most species the number of empty seeds may vary widely from tree to tree. Freshly collected fruit should be stored in shallow piles to prevent heating until extraction.

Juniper seed may be extracted by running the fruit with water through a macerator or a hammer mill,[35] allowing the pulp and empty seeds to float away. If the fruit is well dried, the seed can also be extracted by rubbing on screens. In the case of *J. virginiana* and *J. scopulorum* and others with resinous fruits, the pulp is more easily removed if the berries are first soaked in lye solution (one teaspoon per gallon of water) for a day or two. In addition to the first cleaning, another day's soaking in this solution will serve to remove any resin which may still adhere. After drying, the seed may be stored or sown. Data on the yield, size, purity, soundness, and cost of juniper seed are given in table 111.

Information on longevity of *Juniperus* seed is available for only three species. Seed of *J. ashei* stored in a bag at about 40° F. and high humidity showed about half of its original germinability at the end of 4 years. That of *J. pachyphloea* stored in the pulp in an unheated warehouse and open containers fell to about 30 percent germination in 5 years. Seed of *J. scopulorum* stored in a cool cellar (50° to 65°) both in the pulp and as cleaned seed showed about 30 percent germination after 3½ years. Since so much viability is retained under conditions which are more or less average, it seems likely that under conditions optimum for other species, i.e., dry storage in sealed containers at a low temperature,[36] germinability can be extended for even longer periods.

GERMINATION.—Natural germination of *Juniperus* seed occurs in the early spring. In some species, notably *J. monosperma,* this takes place during the first spring after dispersal; in others, such as *J. virginiana,* most of the seeds germinate the second year, some the third year, and only a few the first year.[37] For most species this information is not available. Germination is epigeous (fig. 137); the best natural seedbed is moist mineral soil on the loamy side, and, for at least *J. ashei* and *J. virginiana,* somewhat calcareous.

Most of the junipers[38] show delayed germination due to embryo dormancy and in some cases also to

[35] Seed of *Juniperus communis* var. *depressa* will apparently not stand as rough treatment during extraction as that of most other junipers.

[36] For *Juniperus virginiana,* a moisture content of 7 percent and a temperature of 20° F. have been recommended.

[37] Seed eaten by birds is said to germinate more promptly.

[38] *Juniperus pachyphloea* seed is apparently not always dormant.

TABLE 111.—*Juniperus: Yield of cleaned seed, and purity, soundness, and cost of commercial seed*

Species	Berries per pound	Cleaned seed: Yield per 100 pounds of fruit	Cleaned seed: Per pound: Low	Cleaned seed: Per pound: Average	Cleaned seed: Per pound: High	Cleaned seed: Basis, samples	Commercial seed: Purity[1]	Commercial seed: Soundness[1]	Commercial seed: Cost per pound (berries)
	Number	*Pounds*	*Number*	*Number*	*Number*	*Number*	*Percent*	*Percent*	*Dollars*
J. ashei	----------	13	----------	10,100	----------	1	98	81	0.75–1.75
J. communis var. *depressa.*	----------	16	37,600	46,800	54,500	3	----------	78 (50–96)	1.00–1.50
J. monosperma	----------	15	16,300	18,100	19,700	4	89	51 (38–70)	1.50
J. occidentalis	3,600	10	8,000	----------	12,900	2	----------	51	----------
J. pachyphloea	----------	36	----------	9,800	----------	1	97	82	[2].10
J. scopulorum	4,200– 8,600	22–25	17,900	28,600	42,100	19	94	59 (25–95)	[3].50–1.50
J. utahensis	800	[4]25	3,600	5,000	7,100	5	----------	(20–78)	.50
J. virginiana	7,000–10,000	[5]20–26	17,600	43,200	59,000	17	89 (55–99)	70 (16–100)	[3].65–1.50

[1]First figure is average percent; figures in parentheses indicate range from lowest to highest.
[2]Data contributed by Soil Conservation Service.
[3]Cost of cleaned seed per pound: *J. scopulorum,* $4.; *J. virginiana,* $2.25.
[4]Yield per tree may vary from 3 quarts to 3 bushels.
[5]Weight per bushel of fruit, about 35 pounds.

an impermeable seed coat. Germination of seeds showing embryo dormancy can be hastened by low temperature stratification in moist sand or peat. That of seeds or lots showing both types of dormancy can be improved by stratification in moist sand or peat for a period at warm (summer) temperatures followed by another period at low temperatures. Mechanical scarification is sometimes substituted for the warm stratification. The use of sulfuric acid has also been suggested for this purpose. Details of pretreatment that have proved successful or are believed to be beneficial are given in table 112.

Germination tests are best made in sand flats, although peat mats may also be used, using 400 to 800 seeds per test, these to be pretreated in the case of dormant lots. Other conditions in making tests and the results that may be expected are given in table 113.

Nursery and Field Practice.—The handling of *Juniperus* seed in the nursery varies according to species as follows: Those species which have embryo dormancy only should be fall sown or stratified. Germination occurs the next spring. These will include *J. ashei*, *J. monosperma*, and some lots of *J. pachyphloea*. *J. communis* var. *depressa*, *J. scopulorum*, *J. utahensis*, and *J. virginiana*, which in addition have an impermeable seed coat, may be handled in one of several ways: (1) store seed in fruit 1 year, clean, scarify, and fall sow; (2) store seed in fruit 1 year, clean, scarify and stratify in peat for 100 days at 41° F., and spring sow; or (3) stratify outdoors in the shade from May until sowing time in the fall. Germination in all 3 cases will occur the following spring. Occasionally the fruit of *J. virginiana* is stratified outdoors or sown

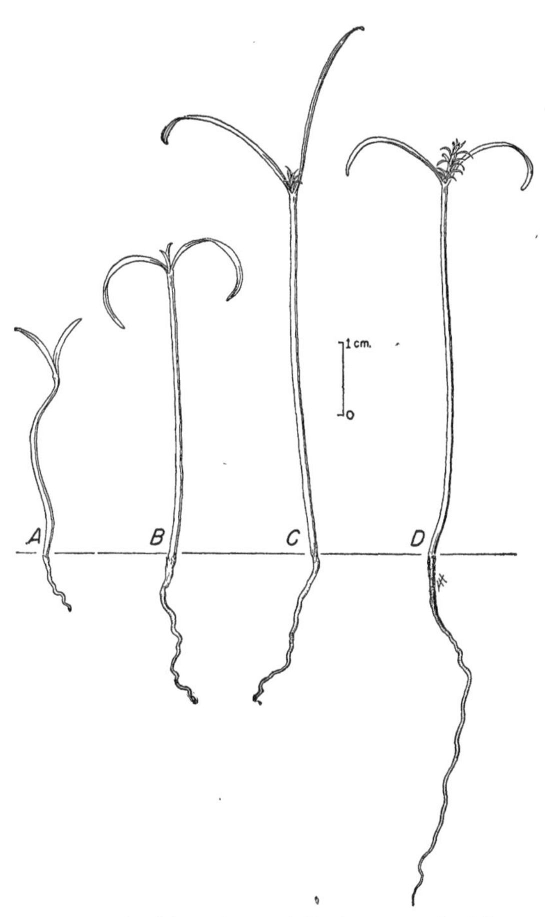

Figure 137.—Seedling views of *Juniperus pachyphloea: A*, At 2 days; *B*, at 17 days; *C*, at 43 days; *D*, at 96 days.

Table 112.—*Juniperus: Dormancy and method of seed pretreatment*

Species	Dormancy	Stratification		Remarks
		Medium	Duration and temperature	
			Days *°F.*	
J. ashei	Embryo	Sand or peat	120 at 41	Needs further study.
J. communis var. *depressa.*	Embryo and impermeable seed coat.[1]	do.	60–90 at [2]68–86 + 90+ at 41	do.
J. monosperma	do.[3]	do.	60 at 41	do.
J. occidentalis	Probably embryo and seed coat.	do.	60–90 at [4]68–86 + 90+ at [4]41	do.
J. pachyphloea	Embryo in some lots.[5]	do.	30–60 at 41	Water soaking sometimes of benefit.
J. scopulorum	Embryo and seed coat.	Sand and peat.	120± at [2]68–86 + 120 at 41	Mechanical scarification or acid treatment can likely be substituted for warm stratification.
J. utahensis	do.[1]	Sand or peat	120± at [4]68–86 + 120 at 41	do.
J. virginiana	Embryo and possibly seed coat.[6]	do.	100–120 at 41	Acid treatment or mechanical scarification are sometimes used prior to stratification; needs further study.

[1]Not definitely proved. [2]Alternated diurnally. [3]Some seeds not dormant.
[4]Treatment suggested; experimental data not complete. Temperatures alternated diurnally.
[5]Other lots apparently not dormant.
[6]In some lots either the seed coat is impermeable or the presence of a resinous film may prevent moisture absorption.

JUNIPERUS

TABLE 113.—*Juniperus: Recommended conditions for germination tests, and summary of germination data*

Species	Test conditions recommended[1]				Germination data from various sources						
	Temperature		Duration		Germinative energy		Germinative capacity			Potential germination	Basis, tests
	Night	Day	Stratified seed	Untreated seed[2]	Amount	Period	Low	Average	High		
	°F.	*°F.*	*Days*	*Days*	*Percent*	*Days*	*Percent*	*Percent*	*Percent*	*Percent*	*Number*
J. ashei	68	86	40	270+	30	10	2	----	38	42	2
J. communis var. *depressa*	68	86	20–30	([3])	----	----	2	----	7	88	2
J. monosperma	----	----	30	70+	----	----	4	----	33	----	2
J. pachyphloea	[4]50	[4]77	30–40	150	----	----	----	45	----	----	1
J. scopulorum	68	86	20–30	([3])	5–31	8–15	7	22	38	[5]45	7
J. virginiana	[6]50	[6]77	20–30	180+	6–74	9–24	1	32	76	----	16

[1]Considerable study of temperatures still needed on all species.
[2]Germination much lower than that obtained from stratified seed.
[3]More than 200 days.
[4]Alternations of 68° and 86° F. almost as effective.
[5]Lowest potential germination, 32 percent; highest, 58 percent.
[6]Alternations of 68° and 86° F. almost as effective. Germination will also occur at constant temperatures of 50° and 41° F but is considerably slower.

in the nursery in the fall and kept mulched until germination occurs during the second spring. Where stratified seed is spring sown, this should be done early enough so that germination will be practically complete before air temperatures go higher than 70°.

Juniper seeds are usually drilled in well-prepared soil in rows 6 to 8 inches apart and covered with about one-fourth inch of firmed soil or sand. Beds should be mulched with straw or burlap and protected with screens until just before germination begins. They should also be kept moist. If stratified seed is used, germination will commence in 6 to 10 days after spring sowing and be complete in 4 to 5 weeks. Nursery germination of *Juniperus monosperma,* 65 to 70 percent; *J. virginiana,* 30 percent. The young seedlings of a least *J. ashei, J. monosperma, J. scopulorum,* and *J. virginiana* should be given light to half shade during the first season. *J. pachyphloea* should be shaded only during the germination period.

Juniper seedlings are highly resistant to organisms causing damping-off. Those of most species are occasionally injured by the cedar blight disease caused by *Phomopsis juniperovora.* Spraying in the early spring and at 2- to 3-week intervals with 4-4-50 bordeaux mixture, colloidal sulfur, or semesan may have some control value. Each year's seedbeds should be placed on ground remote from any recent growth of junipers. All infested trees should be pruned or cut out and the diseased portions destroyed. Stock used for field planting: *Juniperus pachyphloea,* 1–1; *J. scopulorum,* 1–1 to 2–2; *J. virginiana,* 1–1, sometimes 2–0. In addition to propagation from seed, most junipers can also be grown from cuttings, and some from layers.

KALMIA LATIFOLIA L. Mountain-laurel, calico bush

(Heath family—Ericaceae)

Distribution and Use.—Native to the Atlantic Coastal Plain and Appalachian Mountains from Maine to Louisiana, mountain-laurel is one of our most attractive native shrubs—occasionally a small tree—in the wild, and is extensively cultivated as an ornamental in the Eastern States and in Europe. The foliage is an important winter food for deer. However, it may be toxic to deer if they are forced by scarcity of other browse to subsist on laurel exclusively. It is toxic also to sheep and cattle. The wood is utilized to a small extent in turnery and for small specialties; the root burls are used for making pipes. It was introduced into cultivation in 1734.

Seeding Habits.—Mountain-laurel's attractive, perfect flowers appear from March to July, depending on latitude and altitude, and its fruit ripens about September and is long persistent. The seeds are borne in clusters of dry, dehiscent, globose capsules about one-fourth inch in diameter. They are oblong, brown, numerous and minute (fig. 138), and are dispersed by splitting of the capsules.

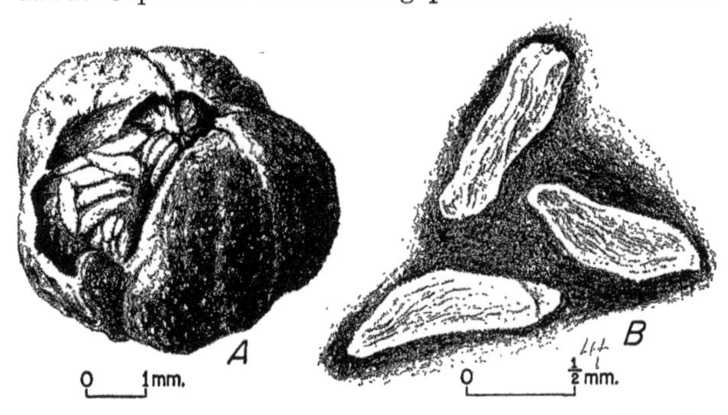

Figure 138.—Seed views of *Kalmia latifolia: A,* Exterior view of five-celled capsule open to show seeds; *B,* three seeds.

Collection, Extraction, and Storage.—To collect seed the capsules are picked from the plants at maturity, dried if necessary and rubbed or beaten to open them, after which the seeds can be shaken out. Seed can be purchased at prices ranging from $0.85 to $2 per ounce, or somewhat cheaper in larger quantities. The dry capsules also are offered at $0.25 per ounce to $1.60 per pound.

Germination.—Germination is increased fivefold or more, and hastened as compared to no treatment, by exposing the seeds after sowing to outdoor winter temperatures in a coldframe for 2 or 3 months. The seeds usually are germinated in flats or pans of sandy or peaty soil in a coldframe or greenhouse. Germination on live moss, which is a common occurrence in nature, also has been recommended.

Nursery and Field Practice.—The seedlings are lifted and replanted in boxes or pots as soon as large enough to handle. The following year they can be transplanted to outdoor beds for a year or more of additional growth before setting in permanent locations. The species grows over a considerable range of conditions as regards soil moisture and fertility but is rather strictly confined in nature to acid soils. Heavy clay soils and limestone soils are unfavorable planting sites unless texture and acidity are properly altered. Varieties often are propagated by side grafting or layers.

KOELREUTERIA PANICULATA Laxm. Panicled goldenraintree

(Soapberry family—Sapindaceae)

Also called Pride-of-India, China-tree, varnish-tree.

DISTRIBUTION AND USE.—Native to China, Korea, and Japan, the goldenraintree is a small deciduous tree which has been cultivated since 1763 chiefly for ornamental purposes.

SEEDING HABITS.—The irregular yellow flowers, occurring in broad, loose terminal panicles, bloom in July and August. The fruits, which are bladdery, triangular, three-celled capsules about 1½ to 2 inches long, ripen in September and October and change from a reddish color to brown. Within the papery walls of the ripe fruit are three roundish, black seeds (fig. 139). Good seed crops are borne almost annually.

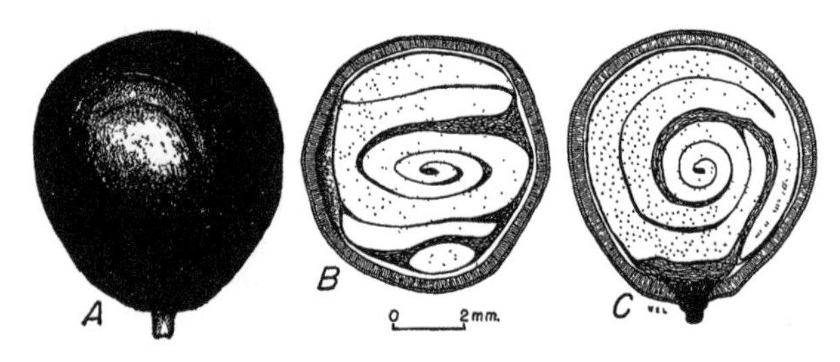

FIGURE 139.—Seed views of *Koelreuteria paniculata: A,* Exterior view of fruit; *B,* cross section; *C,* longitudinal section.

COLLECTION, EXTRACTION, AND STORAGE.—Collection of capsules from the trees should be made in September and October and the seed extracted and cleaned. Cleaned seed per pound (2 samples): 2,600 to 3,500. Commercial seed runs about 99 percent in purity, 92 percent in soundness, and costs about $1.25 per pound. One sample stored in fruit jars with loosely fastened lids and exposed to temperatures ranging from about 40° to 90° F. still germinated 15 percent at the end of 10 years.

GERMINATION.—Germination of goldenraintree is epigeous (fig. 140). The seeds exhibit dormancy which is apparently caused both by an internal condition of the embryo and by an impermeable seed coat. Sulfuric acid treatment for 1 hour, soaking in hot water, and stratification in moist sand at low temperatures (41° F. is suggested) have all been recommended. In one series of tests soaking in sulfuric acid for 1 hour plus stratification in moist sand for 90 days at 41° gave the best results. Germination should be tested in sand flats for 5 to 10 days at 68° (night) to 86° (day), using acid-treated plus stratified seed. One test using seed without pretreatment gave a germination of only 2 percent at 29 days, whereas seed of the same sample gave 52 percent germination in 3 days after having been soaked in concentrated sulfuric acid for 1 hour and then stratified 90 days at 41° before testing.

NURSERY AND FIELD PRACTICE.—Untreated seed may be sown in drills or broadcast in the fall, or stratified seed sown in the spring, and covered with one-half inch of soil. This species should be planted only in sunny locations; it is not particular as to soil. Propagation may also be by layers, cuttings, or root cuttings.

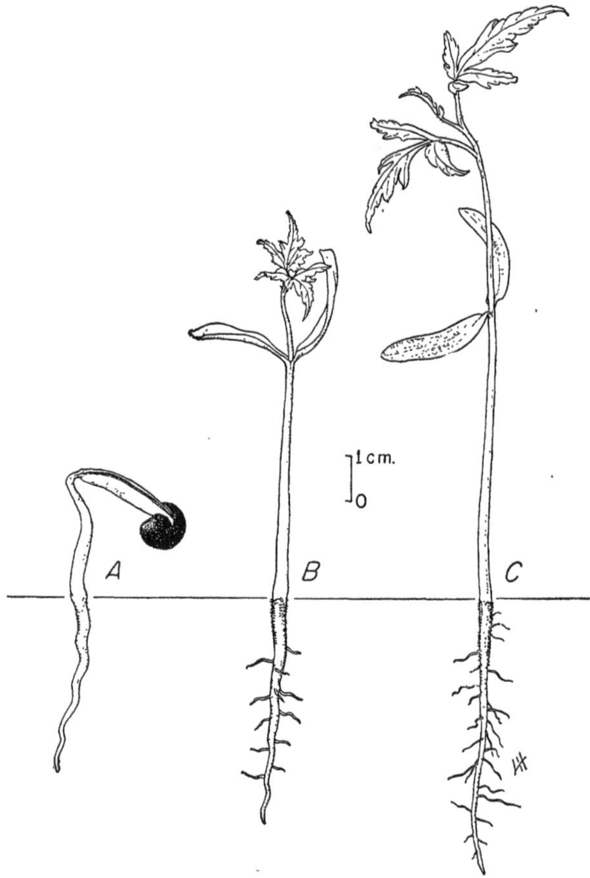

FIGURE 140.—Seedling views of *Koelreuteria paniculata: A,* At 1 day; *B,* at 3 days; *C,* at 5 days.

LARIX Mill. Larch

(Pine family—Pinaceae)

DISTRIBUTION AND USE.—The larches include 10 species of small to large, cone-bearing deciduous trees widely distributed over the cooler regions of the Northern Hemisphere. Three species are native to North America, but none of them has yet received any general use in reforestation. Because of its durability in contact with the ground larch wood is especially valuable for posts and poles, railroad ties, and mine props. Many larches are useful for watershed protection, and some of them for ornamental purposes. Venetian turpentine is produced from larches. Seven species are described in table 114.

Of the American species, *Larix laricina* has been planted a little and probably will come into further use. *L. europaea* has had its range extended into northern Europe by reforestation, and it has been planted rather extensively in the northeastern United States. *L. leptolepis* has been planted quite commonly in northern Asia and Europe and some in the eastern United States. *L. gmelini* and *L. sibirica* have been planted on a small scale in northern Europe and Siberia and may have possibilities for the colder, inland portions of the United States.

SEEDING HABITS.—The male and female flowers of the larches are borne separately on the same tree. Solitary male flowers appear on the side of twigs or branches. They are yellow, globose to oblong bodies which bear wingless pollen. The female flowers, which appear with the leaves early in the spring, are small red or greenish cones. These usually short-stalked erect cones ripen the first year and contain numerous brownish, woody scales, each of which bears two seeds at the base. The seeds are chiefly wind dispersed during the autumn or following spring, and the empty cones remain on the trees for an indefinite period. Mature larch seed

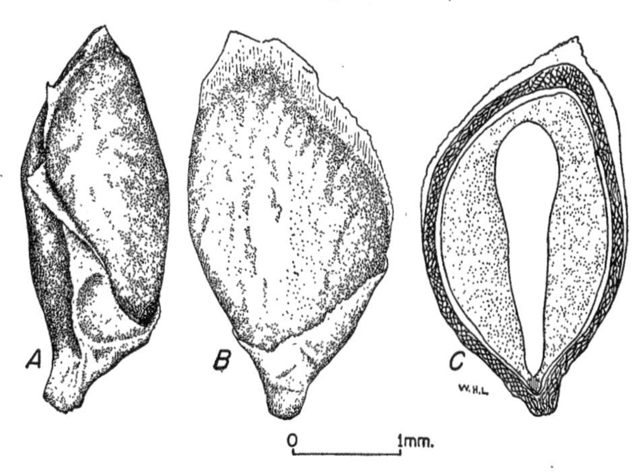

FIGURE 141.—Seed views of *Larix laricina*: *A* and *B*, Exterior views in two planes of seed showing wing fragments; *C*, longitudinal section.

TABLE 114.—*Larix: Distribution and uses*

Accepted name	Synonyms	Natural range	Chief uses	Date of earliest cultivation
L. decidua Mill. (European larch).	*L. europaea* DC., *L. larix* Karst.	Mountains of central Europe	Timber, tannin, watershed protection, Venetian turpentine, wildlife food and cover, ornamental.	1629
L. gmelini (Rupr.) Litvin. (Dahurian larch).	*L. dahurica* Turcz., *L. cajanderi* Mayr.	Eastern Siberia to northeastern China to Saghalien.	Posts, small poles, shipbuilding, furniture.	1827
L. laricina (Du Roi) K. Koch (tamarack; eastern larch, SPN).	*L. americana* Michx. (American larch, hackmatack).	Yukon River to Labrador, south to Maryland and Minnesota.	Lumber, ties, mine timbers, fence posts, poles, fuel, inner bark for medicinal purposes.	1737
L. leptolepis (Sieb. & Zucc.) Gord. (Japanese larch).	*L. kaempferi* Sarg., not Carr., *L. japonica* Carr.	Mountains of Japan	Timber, watershed protection, ornamental.	1861
L. lyallii Parl. (alpine larch).		High mountains of northwestern United States and adjacent Canada.	Watershed protection, wildlife food and shelter.	1904
L. occidentalis Nutt. (western larch).	None	British Columbia to Montana and Oregon.	Lumber, posts, poles, ties, wildlife food, ornamental, possible galactan and ethyl alcohol production.	1881
L. sibirica Ledeb. (Siberian larch).	*L. europaea* var. *sibirica* (Ledeb.) Loud. (Russian larch).	Northeastern Russia and western Siberia.	Poles, piling, ties	1806

is winged and nearly triangular in shape. It has a crustaceous, light-brown to reddish brown, outer coat; a membranaceous, pale chestnut brown, lustrous inner coat; a light-colored endosperm; and a well-developed embryo (figs. 141 and 142). During poor seed years much of the seed is often destroyed by weevils. Details of the seeding habits of six of the species discussed here are given in table 115.

Geographic strains probably have developed in most larches of wide range, but scientific evidence on this point is lacking for most species. Available information for four species is summarized here:

Larix decidua.—Provenience studies made in England, Switzerland, Austria, and Sweden have shown the existence of at least three distinct races which differ in size and viability of seed, survival, rate and time of growth, form, and resistance to diseases and insect enemies. The more local strains are generally superior. Scotch or Silesian races are preferred over the Tyrolean race for planting in the eastern United States.

Larix gmelini.—Tests made in Finland show marked differences in survival, rate of growth, frost resistance, and susceptibility to insect attack as between Korean and Saghalien seed origins.

Larix leptolepis.—In Great Britain three forms or races have been noted. They are distinguished by bark characteristics, and show differences in susceptibility to aphid *(Chermes)* atttacks.

Larix sibirica.—Stock grown from seed of the Altai region seems to be less frost-hardy than that from other parts of the range.

COLLECTION, EXTRACTION, AND STORAGE.—Collection of larch cones should be made in the fall as soon as they ripen. They may be picked by hand from standing trees, gathered from felled trees or slash, or squirrel caches may be used where available.[39] Ripe cones are brown, and collecting may begin as soon as tests show that the seed coats have become hard and the endosperm firm. Freshly collected cones should be spread out in thin layers to dry in the sun or in well-ventilated cone sheds. The cones may be opened by solar heat, by heating them in a cone kiln,[40] by placing them in a heated room, or by tearing them apart mechanically. After opening, the cones should be run through a shaker to remove the seed. The seed should then be dewinged. This may be done by using a dewinging machine, by treading them in a grain sack, or by

[39] In Tyrol, European larch seed are picked from the snow by hand. Or sometimes they are collected by placing canvas beneath the trees in late winter and shaking them to release the seed.

[40] *Larix laricina* cones open satisfactorily in 8 hours at 120° F. in a simple convection kiln. Kiln schedules for other species are not available.

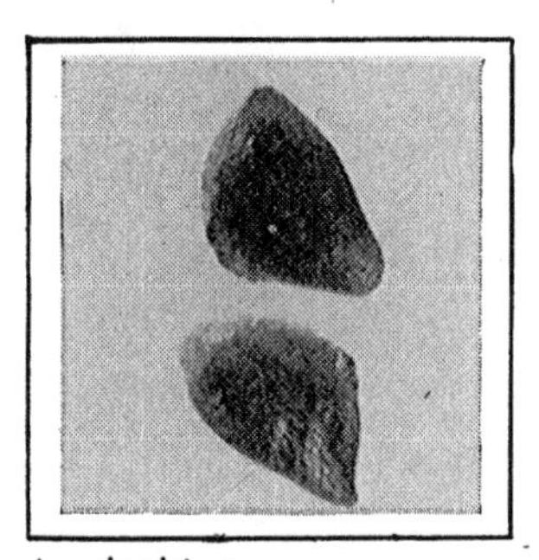

L. decidua

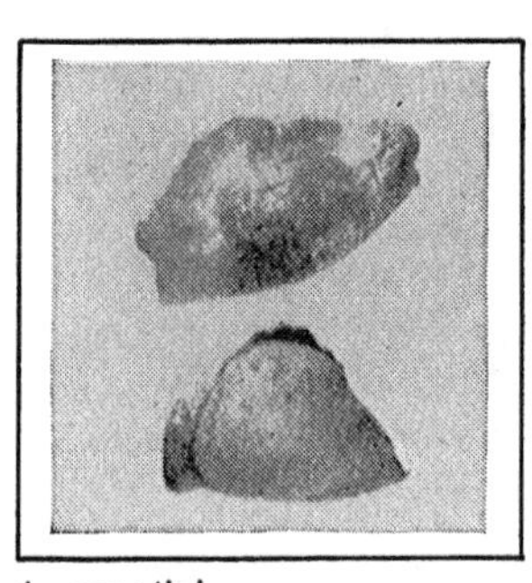

L. gmelini

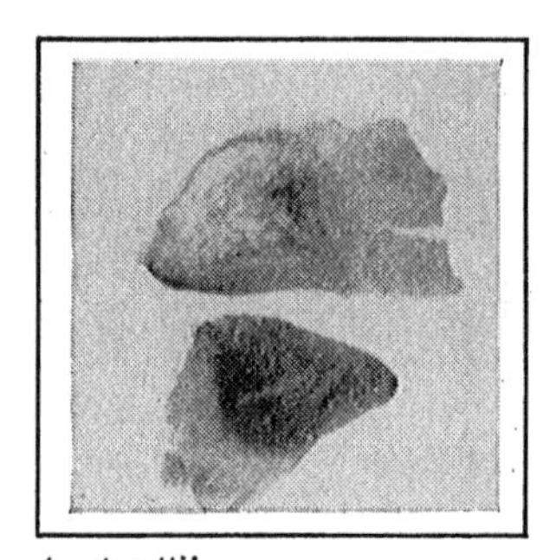

L. lyallii

L. occidentalis

L. sibirica

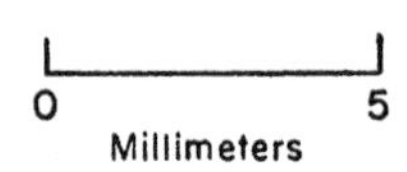

FIGURE 142.—Seed views of five species of *Larix*.

TABLE 115.—*Larix: Time of flowering and cone ripening, and frequency of seed crops*

Species	Time of—			Commercial seed-bearing age			Seed year frequency	
	Flowering	Cone ripening	Seed dispersal	Minimum	Optimum	Maximum	Good crops	Light crops
				Years	*Years*	*Years*	*Years*	*Years*
L. decidua	March–May	Sept. 15–Nov.	Sept. 15–spring	10–15	50	300	5–10	Intervening.
L. gmelini		Sept.–Nov.	Feb.–March					
L. laricina	May	Aug. 20–Sept. 10	Sept.	40	75+		5–6	Intervening.
L. lyallii		August	August				(1)	
L. occidentalis		Aug. 20–Sept. 10	Aug.–Sept.	40–50	60+		5–6	Most intervening.
L. sibirica	April–May	Sept. 15–Nov.	Sept. 15–March	12–15			3–5	Intervening.

[1] Good seed crops borne irregularly.

hand rubbing. And finally, the seed should be cleaned by a blower or fanning mill. Size, purity, and soundness of commercial cleaned seed vary considerably by species, as shown in table 116.

Although special storage studies have not been made for most species, it appears likely that larch seed can be kept for one to several years with little or no loss of viability if they are stored dry in sealed containers at temperatures between 32° and 50° F. Known results are as follows:

Larix decidua.—Seed stored dry in sealed containers kept in the dark at 32° to 50° F. retained its viability for 3 or 4 years, whereas that stored in open containers at room temperatures lost most of its viability in the same period. Seed keeps quite well if stored in the cones.

Larix laricina.—Seed stored dry in sealed containers in a Forest Service seed house retained most of its viability at the end of 6 years in one case. In another case viability was much impaired after 3 years.

Larix occidentalis.—Seed can be kept for 1 or 2 years in sealed containers at room temperature with an annual loss of about 6 percent of its germinative capacity. Results would probably be better with low temperature storage.

Larix sibirica.—Seed stored dry in sealed containers at 41° F. retains unimpaired viability for at least 1 year. Russian reports are that the seed retains viability satisfactorily for 2 to 3 years under ordinary storage.

GERMINATION. — Natural germination usually takes place the spring following seed dispersal, although some may occur the second spring. As a rule, it is not abundant. Germination is epigeous (fig. 143). The best seedbed is exposed, well-drained, mineral soil, although seedlings may appear on any moist site. North and east exposures are best for

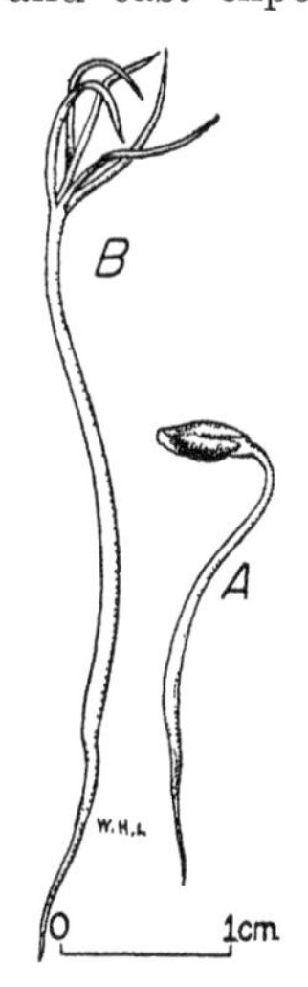

FIGURE 143.—Seedling views of *Larix laricina*: *A*, At 1 day; *B*, at 8 days.

TABLE 116.—*Larix: Yield of cleaned seed, and purity, soundness, and cost of commercial seed*

Species	Cleaned seed					Commercial seed		
	Yield per bushel of cones	Per pound Low	Per pound Average	Per pound High	Basis, samples	Purity	Soundness[1]	Cost per pound
	Ounces	*Number*	*Number*	*Number*	*Number*	*Percent*	*Percent*	*Dollars*
L. decidua	24–32	44,000	77,000	167,000	156	83	56	1.75–5.00
L. gmelini		80,000	120,000	211,000	21	91	79	2.50–6.50
L. laricina	9±	210,000	318,000	420,000	16	84	60	8.00
L. leptolepis		53,000	118,000	154,000	88	88	60	2.50–6.50
L. lyallii		140,000	154,000	163,000	3	86	20	
L. occidentalis	8	98,000	143,000	197,000	25+	88	12±	3.50–5.00
L. sibirica		31,000	43,000	74,000	60+	88	70	1.50–3.50

[1]The consistently low percentage of sound seed in larch may be attributable to a combination of the development of many seed which have not been fertilized, and the filling of a large number of such seed with resin. The latter hinders their removal in the cleaning operation.

LARIX

TABLE 117.—*Larix: Dormancy and method of seed pretreatment*

Species	Seed dormancy		Stratification			Remarks
	Kind	Occurrence	Medium	Temperature	Duration	
				°F.	*Days*	
L. decidua	Embryo	Probably variable.	Moist sand[1]	41	60	Stratification induces prompt germination.
L. gmelini						Mild dormancy probably occurs in some lots.
L. laricina	Embryo	General	Moist sand	41	30–60	Stratification induces prompt germination.
L. leptolepis	do	Probably variable.	do	41	30	Do.
L. lyallii[2]	Probably embryo.	do	do	41	60–90	No treatment tried has succeeded in inducing germination.
L. occidentalis[2]	do	do	do[3]	41	30	Stratification probably will induce better as well as prompt germination.
L. sibirica[2]	do	do	do	41	30	Do.
						Russian recommendations are to soak the seed in lye water.

[1]Other pretreatment method: soak in water 14 to 20 days.
[2]Pretreatment methods only suggested; experimental data not complete.
[3]Other pretreatment method: soak in tepid water 5 days.

Larix occidentalis. Most of the larches—other than *L. lyallii* which germinates poorly—germinate fairly well without pretreatment. However, there seems to be a mild dormancy which can be overcome by stratification in moist sand or peat at low temperatures. Details for seven species are given in table 117.

Germination of larch seed may be tested in sand flats or standard germinators, using 1,000 seeds per test. The seed may be stratified or untreated, depending on the time available for testing. Methods recommended and results for the 7 species are given in table 118.

NURSERY AND FIELD PRACTICE.—Larch seed should be sowed broadcast or in drills in the fall, or in the spring if less prompt germination is satisfactory, and covered with one-eighth to one-fourth inch of nursery soil. Fall sown beds should be covered with burlap or mulched with straw or litter over the first winter. The mulch should be removed before germination commences in the spring, and the seedbeds kept under half shade through the first summer. Sowing at a rate to produce 50 to 75 seedlings per square foot at the end of the second year is recommended. Tree percents average about 10 for European larch, 20 for Japanese larch, and 30 for Siberian larch. Depending upon the individual nursery, stock should be field planted as 2–0, 1–1, 2–1, or 1–2. Larches have no particular enemies in the nursery other than the common ones, such as white grubs. However, planted trees are threatened by the larch sawfly, a defoliating insect, which nearly exterminated tamarack between 1890 and 1920.

TABLE 118.—*Larix: Recommended conditions for germination tests, and summary of germination data*

Species	Test conditions recommended				Germination data from various sources					Basis, tests
	Temperature		Duration		Germinative energy		Germinative capacity			
	Night	Day	Untreated seed	Treated seed	Amount	Period	Low	Average	High	
	°F.	*°F.*	*Days*	*Days*	*Percent*	*Days*	*Percent*	*Percent*	*Percent*	*Number*
L. decidua	68	86	75	35	4–40	10–37	0	37	86	5,588
L. gmelini	68	86	60	[1]30	12–80	10–30	12	52	90	22
L. laricina	68	86	110	50	15–52	20–37	10	47	85	16
L. leptolepis	68	86	75	30	3–55	13–20	0	28	92	442
L. lyallii			[1]110	[1]50	0	0	0	0	0	2
L. occidentalis	68	86	60	[1]30	0–50	9–21	0	27	65	25+
L. sibirica	68	86	40	[1]20	10–63	7–36	1	40	71	65+

[1]Estimated figure.

LIBOCEDRUS DECURRENS Torr. California incense-cedar

(Pine family—Pinaceae)

Also called incense-cedar.

DISTRIBUTION AND USE.—California incense-cedar is native to the mountain slopes of southern Oregon, western Nevada, and California to Lower California. This tree is valuable for purposes for which wood resistant to decay is used. It is also used for making pencil slats, shingles, and other products. Because it is a very handsome tree, it is frequently cultivated as an ornamental in parks and gardens. Deer sometimes use it for food. It is planted in New England and the Middle Atlantic States.

SEEDING HABITS.—The seeds are small, have large, light wings, and contain resin glands (fig. 144). Male and female flowers occur separately on the same tree, or occasionally on different trees, and appear in January. The fruit is an oblong cone about 1 inch long. It becomes light reddish brown when ripe. The seed matures usually during the latter part of September or the first 3 weeks of October. California incense-cedar produces more or less seed annually, but good crops generally occur once every 3 years. The bulk of the seed comes from thrifty, mature trees growing in full sunlight.

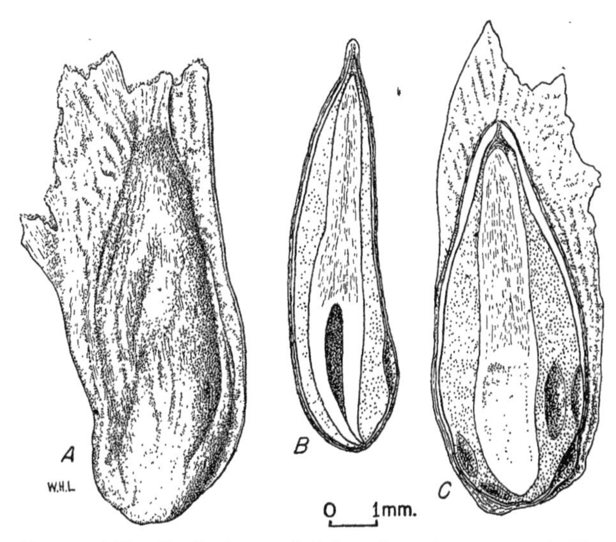

FIGURE 144.—Seed views of *Libocedrus decurrens: A,* Exterior view of seed with wing fragments; *B* and *C,* longitudinal sections in two planes; note resin cells along seed coat.

COLLECTION, EXTRACTION, AND STORAGE.—Cones may be collected in September or October either from standing or from recently felled trees, and the seed extracted later. When the cones are already open, the seed may be collected directly by spreading canvas beneath the trees and vigorously thrashing the branches. Cones from which seed are to be extracted should be spread on canvas in the sun until they open. This requires about 3 days, and the seed can then be thrashed out by hand. Wings may be taken off by rubbing the seed between the hands. Chaff and empty seed are removed by means of a fanning mill or by pouring the seed from one box to another in a current of air. Cleaned seed per bushel of cones: 2 or 3 pounds. Number of seed per pound (66 samples): low, 6,400; average, 15,000; high, 29,000. Commercial seed averages in purity about 80 percent; in soundness, about 65 percent; in cost per pound, $1.75 to $3. Under ordinary conditions of storage the seed loses its viability very rapidly, and 3-year-old seed usually does not germinate at all. Storage of dry seed in airtight containers at low temperatures apparently preserves viability, although exact information on this subject is lacking. It is advisable to collect fresh seed every year.

GERMINATION.—*Libocedrus* seed has the ability to germinate well in organic as well as in mineral soil. A fair percentage, 20 to 40 percent, of fresh seed germinates well without any treatment. This seems to indicate that some seed has no dormancy. On the other hand, stratification always improves the germination, doubling it on the average; this indicates that a certain portion of the seed has dormancy. Accordingly, in order to obtain maximum germination it is advisable to stratify the seed for 2 or 3 months at 35° to 40° F. Tests may be run in sand flats for 40 to 60 days in a greenhouse where temperature fluctuates diurnally between 60° and 80°, using 200 to 400 stratified seeds per test. Germinative capacity, stratified seed (5 tests): low, 18 percent; average, 50 percent; high, 76 percent; average of 15 tests with untreated seed was 21 percent. Both greenhouse and nursery germination average about 40 percent of cutting-test values.

NURSERY AND FIELD PRACTICE.—Fall sowing is preferable. The seed should be sown so as to produce 60 to 80 plants per foot of drill for 1–0 stock, or 35 to 40 per foot for 2–0 stock. Broadcast sowing can also be used. The seed should be sown ¼ to ½ inch deep. If spring sowing is used, burlap mulching is advocated. One-two or two-one stock should be used in field planting, and well-drained soils in the open make the best planting sites.

LIGUSTRUM VULGARE L. European privet

(Olive family—Oleaceae)

Also called common privet, prim. Includes many garden varieties.

DISTRIBUTION AND USE.—Native to Europe, northern Africa, and western Asia, the European privet is a deciduous or half-evergreen shrub which has been cultivated since ancient times chiefly for hedge purposes. It has some value as a producer of wildlife food and as a honey plant. This species has become naturalized in eastern North America.

SEEDING HABITS.—The small, perfect, white flowers, occurring in rather dense panicles, bloom in June and July. The fruit, a one- to four-seeded (fig. 145), lustrous black, subglobose to ovoid berrylike drupe, is about one-fourth inch long. It ripens in September and October and persists over winter.

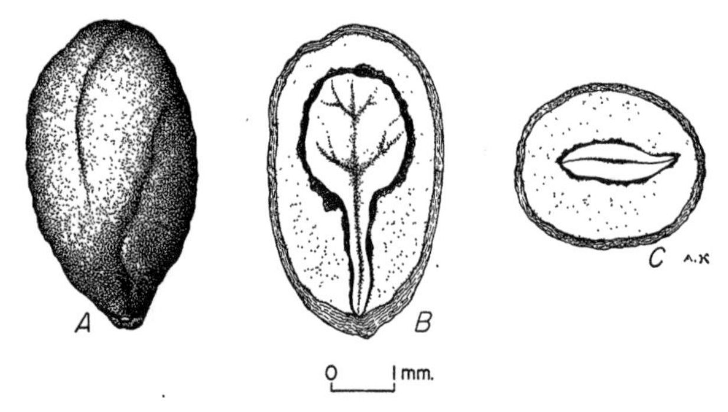

FIGURE 145.—Seed views of *Ligustrum sinense: A,* Exterior view; *B,* longitudinal section; *C,* cross section.

COLLECTION, EXTRACTION, AND STORAGE.—Privet berries should be collected from the bushes by hand in the late fall or winter and macerated in water to extract the seed. Cleaned seed per pound (8+ samples): low, 13,000; average, 20,000; high 37,000. Commercial seed averages 93 percent in purity, 85 percent in soundness, and costs about $1.25 per pound. After cleaning, the seed should be stored dry in sealed containers at low temperatures. Seed can be kept satisfactorily for 1 to 2 years under ordinary dry storage.

GERMINATION.—Germination is epigeous (fig. 146). The seeds exhibit dormancy which can be overcome by stratification in moist sand and peat for 60 to 90 days at 32° to 50° F. Germination may be tested in sand flats; 60 days at 70° (night) to 85° (day) with 800 stratified seeds per test is suggested. Reported germination is not high—5 to 27 percent in 2 tests.

NURSERY AND FIELD PRACTICE.—Privet seed may be sown broadcast or in drills in the fall, or stratified seed used in the spring. They should be covered with one-fourth inch of soil. One- or two-year-old seedling stock is preferred for field planting. The privet grows on almost any kind of soil and can withstand a moderate amount of shade. This species frequently is propagated by cuttings.

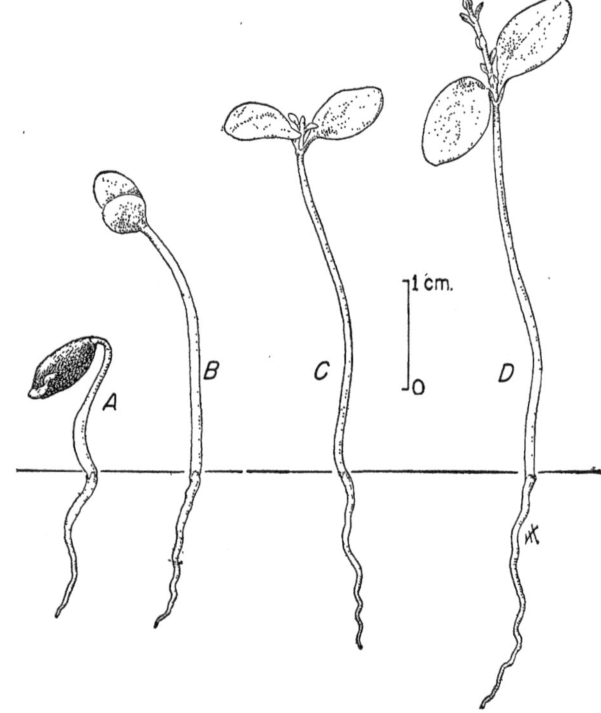

FIGURE 146.—Seedling views of *Ligustrum vulgare: A,* At 1 day; *B,* at 5 days; *C,* at 50 days; *D,* at 132 days.

LINDERA BENZOIN (L.) Blume Common spicebush

(Laurel family—Lauraceae)

Botanical syns.: *Benzoin aestivale* (L.) Nees, *B. benzoin* (L.) Coult.

Distribution and Use.—Common spicebush is native to the eastern United States from Maine to Ontario and Kansas, and south to Florida and Texas. It is a deciduous, medium to tall shrub valuable for game food and ornamental planting; the bark and fruit are used medicinally. This species was introduced into Washington and Oregon. It has been in cultivation since 1683.

Seeding Habits.—The greenish-yellow, perfect flowers open in March or April before the leaves unfold. Spicebush fruit is a fleshy, scarlet drupe about 0.4 inch long, which ripens in September or October of the first year. Each fruit contains a single oval seed, light violet brown in color, with flecks of darker brown (fig. 147).

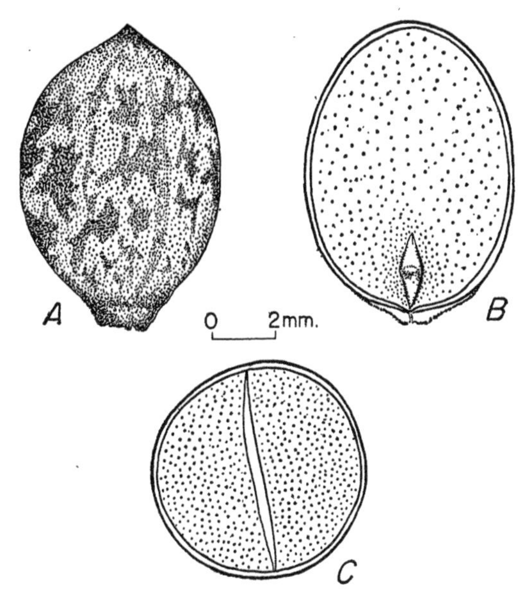

Figure 147.—Seed views of *Lindera benzoin: A,* Exterior view; *B,* longitudinal section; *C,* cross section.

Collection, Extraction, and Storage.—Spicebush fruit may be collected by picking it from the bushes in September or October. It should be pulped in water immediately, the pulp floated off, and the seeds thoroughly air dried. Small lots of seed may be dried with the pulp on. One hundred pounds of fresh fruit yields 15 to 25 pounds of cleaned seed. Cleaned seed per pound (2 samples): 4,500 to 4,600. Commercial seed runs about 95 percent sound and costs $1 to $1.25 per pound. Ordinarily spicebush seed loses its viability quickly, but it is possible that sealed storage in a cold room may keep it viable longer.

Germination.—Germination is hypogeous (fig. 148). Seed of this species has a dormant embryo which responds most promptly if stratified 15 to 30 days at 77° F. followed by 90 to 120 days at 34° to 41°. Nearly as good germination results from 120 days' stratification in moist peat or sand at 41° to 50°. Tests may be made in sand flats or petri dishes at temperatures alternating between 50° (night) and 86° (day). Satisfactory tests require use of 400 seeds and can be completed in 45 days for pretreated seed or 180 days with untreated seed. Germinative energy, 70 to 79 percent in 15 to 21 days; germinative capacity pretreated seed (6 tests): low, 78 percent; average, 85 percent; high, 90 percent.

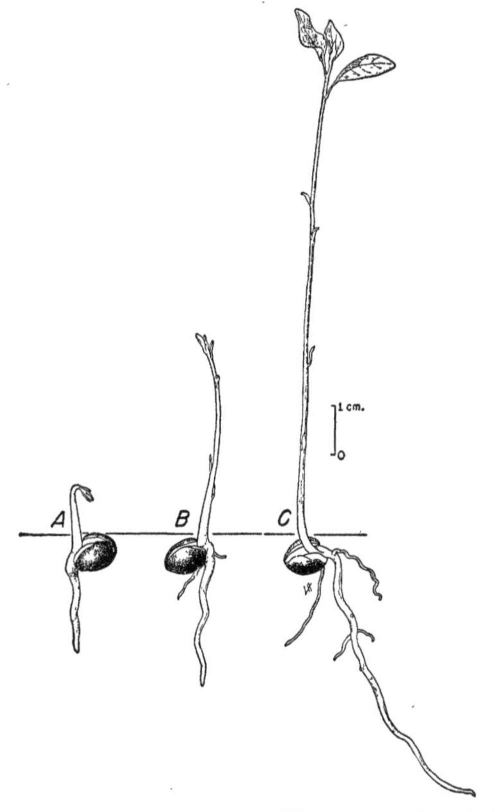

Figure 148.—Seedling views of *Lindera benzoin: A,* At 2 days; *B,* at 3 days; *C,* at 10 days.

Nursery Practice.—Spicebush seed should be sown in the fall and mulched over winter. The mulch should be removed in April or May before germination starts. In a climate where mulching does not prevent freezing, the seed may be stratified below the frost line and sown in early spring. Seventy to eighty percent of the viable seed sown may be expected to produce seedlings. Plants grow best on moist sandy or peaty soils. Propagation by layers or cuttings is also sometimes practiced.

891183°—50——15

LIQUIDAMBAR STYRACIFLUA L. Sweetgum

(Witch-hazel family—Hamamelidaceae)

Also called American sweetgum, SPN; bilsted, liquidambar.

Distribution and Use.—Sweetgum is a medium to large deciduous tree which grows in swamps and wet bottom lands from southern Illinois to eastern Texas, and eastward to Connecticut and Florida. It is very valuable for forest products and is often planted in the Eastern States as an ornamental. Sweetgum trees can be worked in much the same manner as the pines are for naval stores, and produce a compound known as storax (styrax). Storax is valued for use in perfuming such products as soap and glove powder, as well as for use in pharmaceutical preparations. The seed is an important game food. This species was introduced into cultivation in 1681.

Seeding Habits.—The seeds are borne in light brown, globose, aggregate heads 1 to 1½ inches in diameter, and are discharged through numerous openings which resemble small birds' beaks. They are angular, light brown, about one-fourth inch long, and have a wing slightly shorter than their total length (fig. 149). They are wind disseminated, and many are abortive. The flowers of this species open from March to May; its fruits ripen from September to November of the same year, and the seed are shed during this period. Empty fruits hang on the tree over winter. Seed bearing begins in 20 to 30 years and the tree seeds vigorously up to 150 years. Fair seed crops are borne every year with bumper crops about every 3 years.

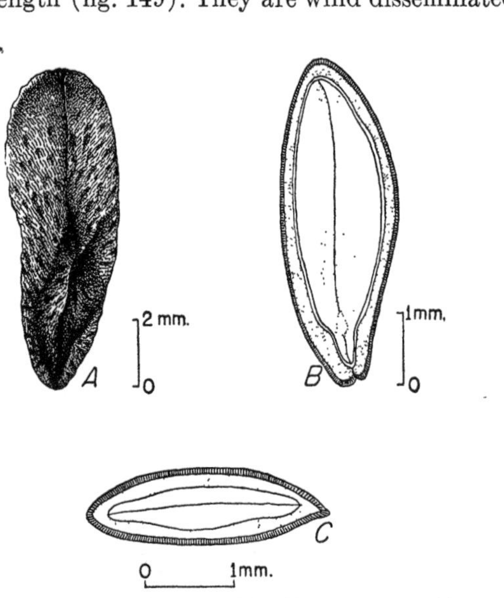

Figure 149.—Seed views of *Liquidambar styraciflua*: *A*, Exterior view; *B*, longitudinal view without wing; *C*, cross section.

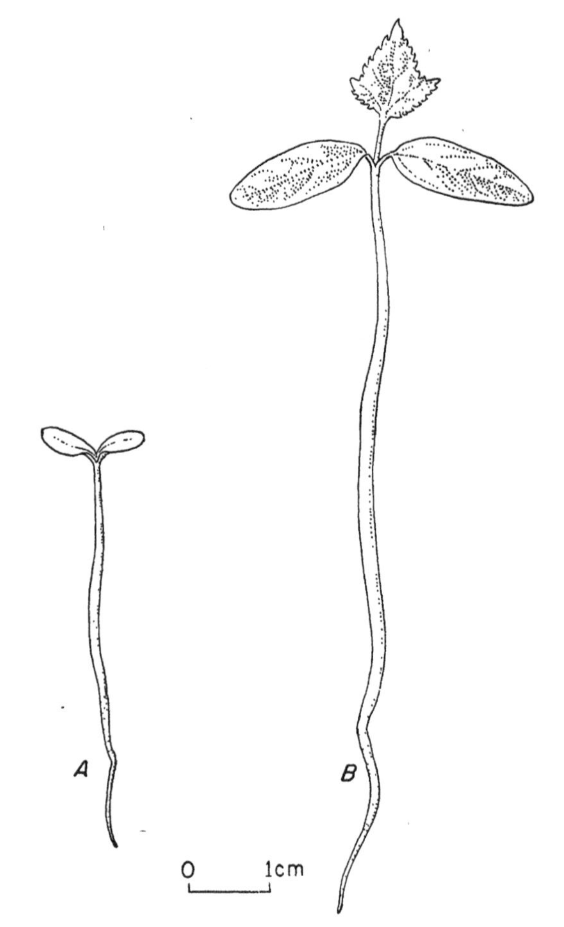

Figure 150.—Seedling views of *Liquidambar styraciflua*: *A*, At 2 days; *B*, at 30 days.

Collection, Extraction, and Storage.—Sweetgum fruit turns yellow from September to November and the seed matures during this period. Immediate collection is necessary since the fruits soon open and release the seed. In Louisiana October 5 to 25 is given as date of collection. Fruits may be picked from logged or standing trees, and they will open if exposed to air temperature for about 2 weeks. The seed can then be shaken out. Sawdust-like, abortive seeds may be removed in a fanning mill, or trash and light seed floated off in water. Dewinging is not necessary. Seeds per fruit, 7 or 8; seed per bushel of fruit, ¾ of a pound; seed per pound (10+ samples): low, 65,000; average, 82,000; high, 90,000. Commercial seed purity: 90 to 95 percent; soundness: 80 to 90 percent. Commercial seed has been quoted at $1.75 to $3.20 per pound. Seed of this species can be kept in dry storage at low temperature for at least 1 year with no loss in viability.

Germination.—Moist, bottom-land soil with good light is a good seedbed. Germination is epigeous (fig. 150). The seeds usually exhibit dormancy which can be overcome satisfactorily by stratification in moist sand for 30 to 90 days at 41° F.; in one case soaking the seed in water for 20 days gave good results. Germination may be tested in sand flats, peat mats, or standard germinators at 68° (night) to 86° (day) using 800 to 1,000 pretreated seeds per test. The germinative energy of pretreated seed is 40 to 75 percent in 10 to 30 days; the germinative capacity (12 tests): low, 50 percent; average, 70 percent; and high, 85 percent.

Nursery and Field Practice.—In the South, a germination of 50 percent in about 20 days has been secured by sowing sweetgum in January or February, covering the seed lightly, and mulching with leaves. Fall sowing has been tried in Illinois but difficulty was experienced in keeping the seed covering on the bed over winter. Sowing of stratified seed in the spring is recommended. Shading is not necessary.

LIRIODENDRON TULIPIFERA L. Yellow-poplar

(Magnolia family—Magnoliaceae)

Also called tuliptree, SPN; tulip poplar.

DISTRIBUTION AND USE.—Yellow-poplar[41] is native to the eastern United States from Massachusetts to Wisconsin and south to Florida and Mississippi. It is a large, deciduous tree highly prized for timber and also useful for shade and ornamental plantings. The wood is used for lumber, excelsior, and veneer, and the bark for medicines. It is a good honey plant. This species is not hardy outside its natural habitat. Attempts to grow it in forest plantations in Europe have met with little success. It has been cultivated since 1663.

SEEDING HABITS.—The seed of yellow-poplar are borne in a tan or light-brown cone composed of closely overlapped, dry, woody carpels. Commercial "seed" is the mature carpel with wing; 2 seeds, 1 usually aborted, are borne within. One hundred fruits usually contain from 100 to 120 seeds. The seeds have a large, oily endosperm and a small embryo (fig. 151). Yellow-poplar flowers open from April to June, its fruit ripens from September to November, and the seed falls from October to January. The minimum commercial seed-bearing age of this species is 15 to 20 years, and the maximum age is 200 years and older. Some seed are borne almost every year but good crops occur at irregular intervals. The existence of geographic strains has not been proved, but southern seed is reported to have a lower germinative capacity and to produce less frost-hardy plants than seed from New York State.

COLLECTION, EXTRACTION, AND STORAGE.—Cones may be collected in October and November by hand plucking from standing trees or slash, by cutting with pruning hooks from standing trees, or by gathering from squirrel caches. The latter are reported to yield the best quality seed. Cones are closed and less fragile in wet weather. By cutting the cones open in August prior to collection, trees with a large percentage of good seed can be selected. Seed may also be shaken onto canvas sheets in early winter. The best seed comes from the upper portion of the crowns. Mature trees yield 0.75 to 1.50 bushels of cones. Newly collected cones should be separated from leaves, twigs, and other debris and spread out to dry.

[41] Unpublished information on this species was supplied by Elmon Radway, formerly of the U. S. Forest Service.

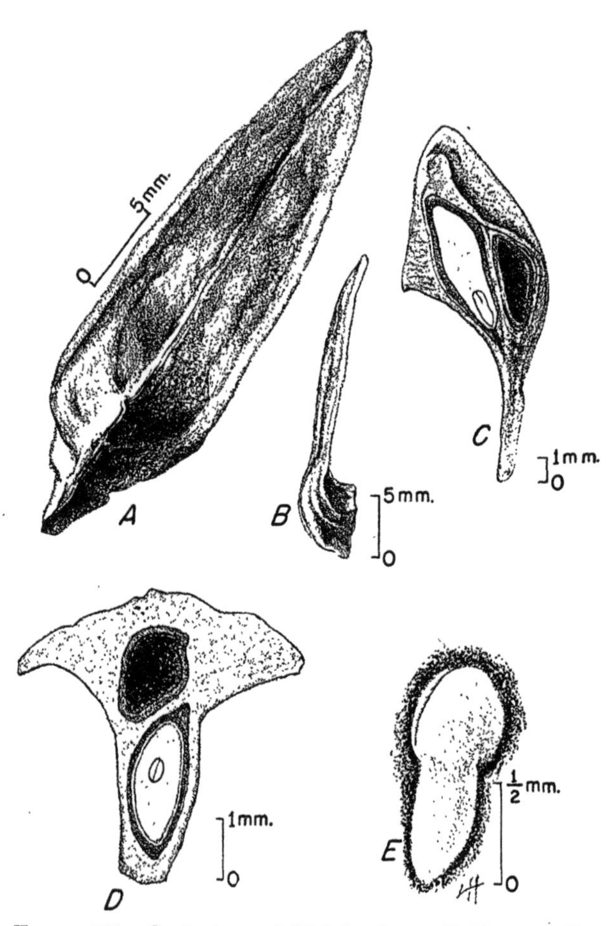

FIGURE 151.—Seed views of *Liriodendron tulipifera*: *A*, Exterior view of fruit; *B*, exterior view of fruit from different plane; *C*, longitudinal section showing filled and aborted seed; *D*, cross section; *E*, embryo.

Individual carpels can readily be separated from the axis of thoroughly dried cones by hand shucking, flailing, treading, or by running the fruits through a hammer mill. Further extraction is unnecessary and impractical. Cleaned seed: per bushel of cones, 7 to 13 pounds; per 100 pounds of cones, 30 to 80 pounds; per cone, 80 to 100; per pound (28 samples): low, 10,000; average, 14,000; high, 24,000. Commercial seed averages in purity 75 percent, in soundness 14 percent, and costs from $0.60

to $0.75 per pound. Seed to be sown the following spring should be stratified in peat or moist sand and placed below the frost line immediately following extraction. A decrease in germination from 22 to 4.5 percent in 3 years was noted in seed stored in the duff of the forest floor. Rapid deterioration is also reported to occur in dry storage unless seed is held at low temperatures, 32° to 45° F.

GERMINATION.—Natural seedlings occur most frequently in abandoned fields or other places where a mineral soil seedbed is available and competing vegetation sparse. Germination is epigeous (fig. 152). It is slow because of embryo dormancy and possibly because the seed coat is somewhat impermeable to water and oxygen. Good germination is reported from seed sown in the fall, from seed stratified over winter in the soil, and from seed stratified in moist peat for 70 days at temperatures varying daily or weekly between 32° and 50° F. Seed stored dry over winter and sown the following spring germinates the second season. Germination may be tested in sand flats, Jacobsen germinators, or petri dishes, for 45 to 60 days at 68° (night) to 86° (day) using 1,000 stratified seeds per test. Germination is low partly because dormancy may not be completely broken and because of the low soundness of the seed. Germinative energy of stratified seed: 1 to 11 percent in 30 to 45 days; germinative capacity (34 tests): low, 1 percent; average, 5 percent; high, 14 percent.

NURSERY PRACTICE.—Nursery germination of yellow-poplar averages about 50 percent of that found by laboratory test. A fertile nursery soil is required for satisfactory growth. The seed should be sown at a rate of 50 to 75 seeds per linear foot in drills 8 to 12 inches apart and covered to a depth of one-fourth inch. Mulching is desirable to prevent surface drying, but should be replaced by high shade when germination begins. The shade should be retained for 1 or 2 months. This species grows best on deep, rich, and rather moist soil. Varieties are frequently propagated by grafting or budding, and sometimes by layering.

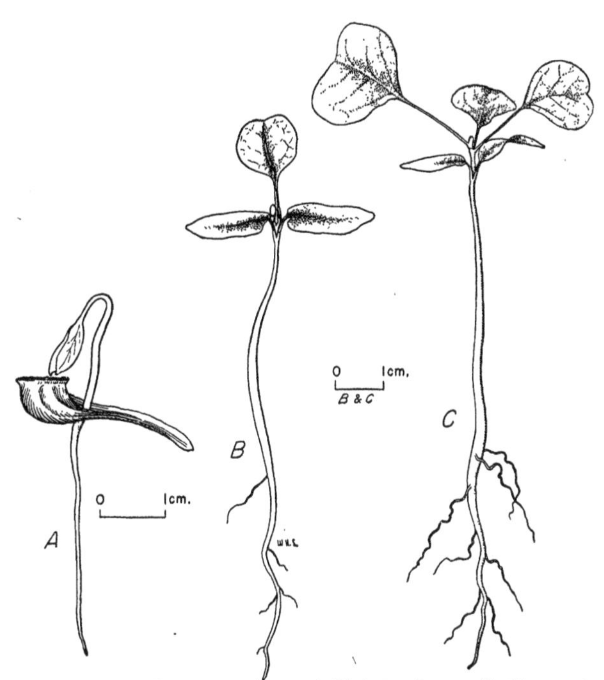

FIGURE 152.—Seedling views of *Liriodendron tulipifera: A*, At 1 day; *B*, at 18 days; *C*, at 48 days.

LITHOCARPUS DENSIFLORUS (Hook. & Arn.) Rehd. Tanoak

(Beech family—Fagaceae)

Also called tanbark oak. Botanical syn.: *Pasania densiflora* (Hook. & Arn.) Oerst.

Native to the Coast Range and the Sierra Nevada of California to southern Oregon, tanoak is a medium-sized evergreen tree. It is occasionally cultivated in parks and can be used for erosion control. The wood is largely used for fuel; the bark is exceedingly rich in tannin, which is used for tanning leather. This species was introduced into cultivation in 1874. The tree flowers in July and August, and its acorns (fig. 153) ripen the second autumn. They are very similar to oak acorns.

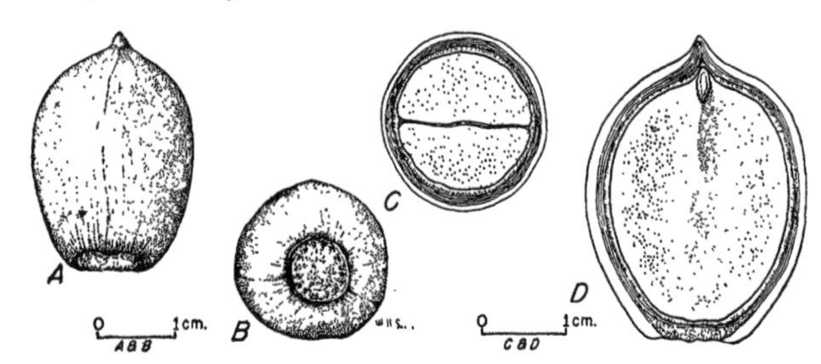

FIGURE 153.—Seed views of *Lithocarpus densiflorus: A* and *B,* Exterior views in two planes; *C,* cross section; *D,* longitudinal section.

Tanoak seed may be collected from standing trees and does not require any cleaning except removal of the cups. There are about 110 acorns to the pound (1 sample). The acorns should be planted immediately after collection in very light soil or peat, or stratified at a temperature just above freezing—not to improve their germination but rather to retard it until spring. Germination is hypogeous (fig. 154). Germinative capacities of 19 and 78 percent in 85 days have been recorded (2 tests). Seedlings (fig. 154) appear 3 weeks after planting. In field planting, acorns should be protected from rodents.

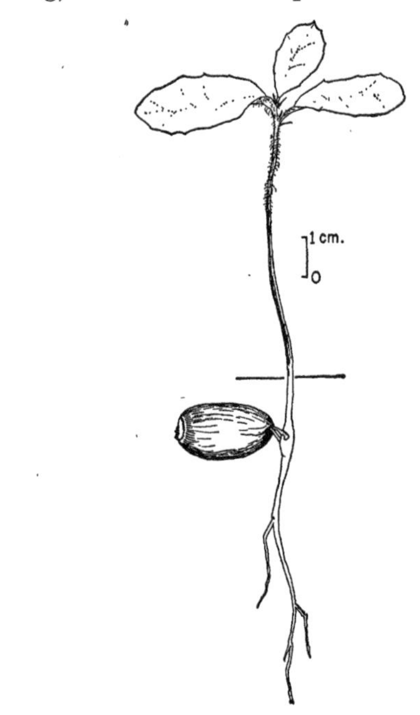

FIGURE 154.—Seedling of *Lithocarpus densiflorus* at 2 months.

LONICERA L. Honeysuckle

(Honeysuckle family—Caprifoliaceae)

DISTRIBUTION AND USE.—The honeysuckles are composed of about 180 species of deciduous, sometimes evergreen, upright shrubs or climbing vines which occur throughout the Northern Hemisphere, south to Mexico, north Africa, Java, and the Philippines. They are widely planted for their attractive, often fragrant flowers, and ornamental fruits. Some species furnish food and cover for birds and game animals; others are valuable for erosion control and shelter-belt planting. Of the 7 species now used or believed valuable in conservation planting, 2 are native to the Old World; the others are indigenous to this country. The distribution and use of these 7 species are given in table 119.

Only *Lonicera tatarica* has been used to any considerable extent in conservation planting, and this largely in connection with shelter belts in the Prairie-Plains region. This species has been widely used in landscaping over much of the northern United States and has escaped from cultivation in many areas. The remaining honeysuckles have all been used more or less for ornamental purposes. All are potentially valuable for wildlife planting and a few for erosion control.

SEEDING HABITS.—Honeysuckle's usually rather small, perfect flowers vary from white or yellow to pink, purple, or scarlet. They are borne in axillary pairs, or in stemless whorls, generally in spring but

TABLE 119.—*Lonicera: Growth habit, distribution, and uses*

Accepted name	Synonyms	Growth habit	Natural range	Chief uses	Date of earliest cultivation
L. canadensis Marsh. (American fly honeysuckle).	*L. ciliata* Muhl. (fly honeysuckle).	Shrub	Quebec to Saskatchewan, south to Pennsylvania, Indiana, and Minnesota.	Wildlife food and cover.	1641
L. dioica L. (limber honeysuckle).	*L. glauca* Hill, *L. media* Murr.	Slightly twining or bushy shrub.	Quebec to Saskatchewan, south to North Carolina, Ohio, and Iowa.	Wildlife food	[1]1636
L. glaucescens Rydb. (Donald honeysuckle).	*L. douglasii* Koehne not DC., *L. dioica* var. *g.* (Rydb.) Butters, *L. hirsuta* var. *g.* Rydb. (glaucous honeysuckle).	Twining vine	Quebec to Alberta, south to North Carolina, Ohio, and Nebraska.	do	[1]1890
L. hirsuta Eaton (hairy honeysuckle).	*L. pubescens* Sweet	High-climbing shrub.	Quebec to Saskatchewan, south to Pennsylvania, Michigan, and Nebraska.	Ornamental purposes, and wildlife food.	1825
L. maackii Maxim.[2] (Amur honeysuckle).		Shrub	Manchuria, China, and Korea.	Ornamental purposes, wildlife food, erosion-control and shelter-belt planting.	[1]1855
L. oblongifolia (Goldie) Hook.[2] (swamp fly honeysuckle).		do	New Brunswick to Manitoba and south to Pennsylvania, Michigan, and Minnesota.	Wildlife food and ornamental purposes.	1823
L. tatarica L.[2] (Tatarian honeysuckle).	twin honeysuckle	Large shrub	Southern Russia to Altai and Turkestan.	Ornamental purposes, shelter-belt and erosion-control planting, wildlife food.	1752

[1] Approximate date.
[2] Includes 1 or more varieties.

LONICERA

TABLE 120.—*Lonicera: Time of flowering and fruit ripening*

Species	Time of—		Color of ripe fruit
	Flowering	Fruit ripening	
L. canadensis	April–June	June–July	Red.[1]
L. dioica	May–June	July–October	Salmon red.[1]
L. glaucescens	do	do	Coral red.[1]
L. hirsuta	June–July	September–October	Red.
L. maackii	June	do	Dark red.
L. oblongifolia	May–June	July	Red to purplish.
L. tatarica	do	July–August	Red orange.[2]

[1]Berries contain 3 to 4 seeds.
[2]Berries contain many seeds.

sometimes in late summer. The attractive fruit—red, orange, blue, or black in color and often borne in coalescent pairs—ripens in the summer or early fall months; it is a berry containing few to many rather small seeds (figs. 155 and 156). Commercial seed may consist of dried berries or cleaned seed. Such detailed data as are available on seeding habits of seven species are given in table 120.

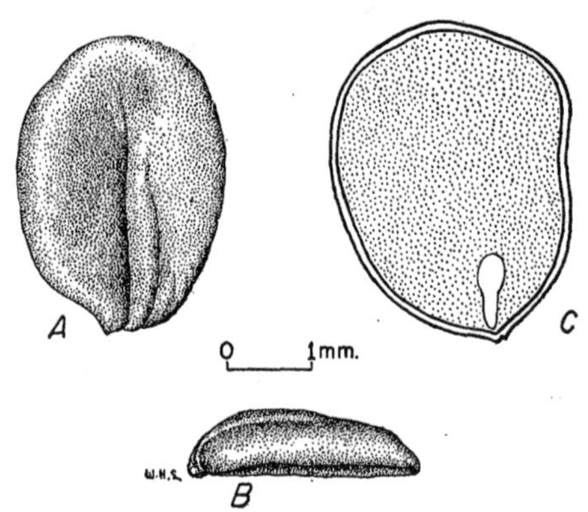

FIGURE 155.—Seed views of *Lonicera tatarica: A* and *B*, Exterior views in two planes; *C*, longitudinal section.

Good seed crops of *Lonicera maackii* and *L. tatarica* are borne almost every year. No data are available as to the age of commercial seed-bearing for any of the honeysuckles. Seed dispersal is primarily by birds and other animals. Dispersal in most cases probably occurs quite soon after ripening. Occasionally the dried fruit of *L. maackii* and of *L. tatarica* will remain on the bushes well into the winter. Information regarding the development of geographic strains is lacking. However, it is likely that such strains do exist, particularly in species which have a wide range and a great many forms that are recognized by horticulturists, such as *L. tatarica*.

COLLECTION, EXTRACTION, AND STORAGE.—Honeysuckle fruit should be hand-picked or stripped from the branches as soon after ripening as possible to prevent its consumption by birds. Since most species hybridize rather freely, it is better to collect seed

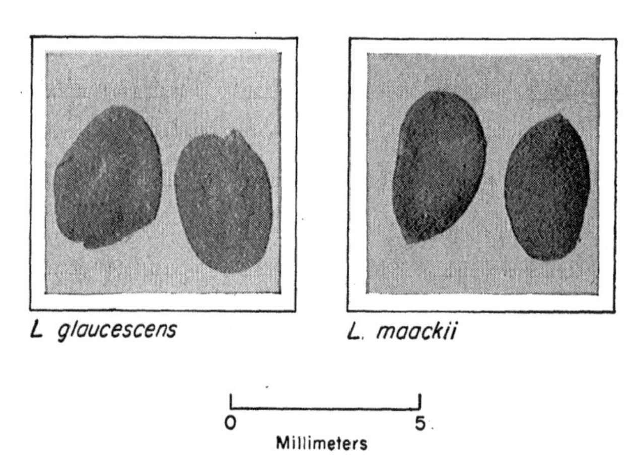

FIGURE 156.—Seed views of two species of *Lonicera.*

TABLE 121.—*Lonicera: Yield of cleaned seed, and purity, soundness, and cost of commercial seed*

Species	Cleaned seed					Commercial seed		
	Yield per 100 pounds of fruit	Per pound			Basis, samples	Purity[1]	Soundness[1]	Cost per pound
		Low	Average	High				
	Pounds	*Number*	*Number*	*Number*	*Number*	*Percent*	*Percent*	*Dollars*
L. glaucescens	----------	----------	86,000	----------	1	----------	100	----------
L. maackii	[2]14	116,000	148,000	194,000	3	94	80 (53–99)	3.50–5.00
L. oblongifolia	----------	----------	238,000	----------	1	----------	90	----------
L. tatarica	2–8	[3]116,000	142,000	198,000	21+	92 (84–100)	92	2.00–4.50

[1]First figure is average percent; figures in parentheses indicate range from lowest to highest.
[2]100 pounds of commercial seed (dried fruit) will yield about 30 pounds of cleaned seed.
[3]Seeds from orange-colored fruit appear to be larger; 2 samples contained 85,000 and 101,000 seeds per pound.

TABLE 122.—*Lonicera: Dormancy and method of seed pretreatment*

Species	Kind of dormancy	Stratification	
		Medium	Duration and temperature
			Days *°F.*
L. hirsuta[1]	Possibly embryo and seed coat	Sand or peat	60 at [2]68–86 + 60 at 41
L. maackii	Embryo	do	60–90 at 32–50
L. oblongifolia	Probably embryo and seed coat	do	60 at [2]68–86 + 90 at 41
L. tatarica[1]	Some lots embryo and possibly seed coat; others, none.[3]	do	[4]30–60 at 41

[1]Further study needed.
[2]Temperatures alternated diurnally.
[3]In such lots stratification appears to retard germination.
[4]Tentative recommendation.

only from isolated shrubs or groups. Unless the seed is to be extracted immediately, the berries should be spread out in rather thin layers to prevent heating. Extraction may be easily accomplished by macerating the fruit in water, allowing empty seeds and pulp to float away. After a short period of drying, the seeds are ready for storage or sowing. Data on yield, size, purity, soundness, and cost for four species are given in table 121.

Definite information on optimum storage conditions for *Lonicera* seeds is lacking. *L. oblongifolia* seed stored in a sealed container at 41° F. showed a decrease in germinability of only about 20 percent at the end of 1 year. Seed of *L. tatarica* stored in moist sand at 41° up to 11 months lost no viability but required 160 days to test germinative capacity. This species when stored under ordinary conditions without temperature or humidity controls is reported to have a longevity of 2 years.

GERMINATION.—Natural germination in most species of *Lonicera* is believed to occur in the spring following dispersal. However, in some lots of *L. tatarica* it occurs in part soon after ripening in midsummer, and in part the next spring. The proportions germinating during the two seasons fluctuate considerably between individual lots. Germination is of the epigeous type (fig. 157). All of the seven species discussed here show delayed germination more or less. In most species this is likely due to a dormant embryo; in others it appears that the seed coat may also retard germination. Stratification in sand or peat at a low temperature is recommended for seeds of the first type; for those of the second type low-temperature stratification preceded by a short period of warm stratification is suggested. Possibly light scarification can be substituted for the warm stratification. Such information as is available concerning type and duration of pretreatment is given in table 122 for four species.[42]

Germination tests are made most successfully in sand flats and require 400 to 800 seeds per test. Seed from dormant lots should be properly pretreated. Germination temperatures alternating diurnally from 68° to 86° F., or from 50° to 77°, appear to be equally good in tests of *Lonicera oblongifolia* and *L. tatarica.* A constant temperature of 50° retards germination of the latter species but does not seem to affect the amount. Available data on the results that may be expected under test conditions described here are given in table 123.

[42] *Lonicera canadensis* and *L. dioica* may have little or no seed dormancy. Outdoor stratification of these species for 83 winter days improved germination but little over no treatment, and required almost as long a period. Data for *L. glaucescens* are not available, but its behavior is probably similar to *L. dioica.*

TABLE 123.—*Lonicera: Suggested germination test duration, and summary of germination data*

Species	Duration of test	Germinative energy		Germinative capacity			Potential germination	Basis, tests
		Amount	Period	Low	Average	High		
	Days	*Percent*	*Days*	*Percent*	*Percent*	*Percent*	*Percent*	*Number*
L. canadensis	[1]90					100		1
L. dioica	[1]90–110					93		1
L. hirsuta	[1]100				43			1
L. oblongifolia	[1]60	32	24		37		78	1
L. tatarica	60–90	10–92	25–70	11	66	99		17+

[1]This period can likely be reduced once the proper pretreatment has been determined.

LONICERA

Nursery and Field Practice.—*Lonicera* seed of lots showing embryo dormancy should be sown broadcast or in drills in the fall, or stratified prior to early spring sowing. That of species or lots believed to have also an impermeable seed coat (these are mostly summer-ripening species) should be sown as soon as possible after collection to insure germination the next spring. Nondormant seeds can be sown in the spring without treatment. The seed should be covered with one-eighth to one-fourth inch of nursery soil. Germination of *L. tatarica* (60 to 85 percent) is usually complete in 40 to 60 days after spring sowing. This can be accelerated by soaking the seed for 24 to 48 hours prior to sowing or by using stratified seed. Ordinarily about 15 percent of *L. tatarica* seed sown produces usable 1-0 seedlings. One- or two-year-old seedlings of this species and also of *L. maackii* are suitable for field planting. In addition to propagation from seed, many honeysuckles may be easily grown from cuttings of ripened wood or from greenwood cuttings grown under glass in summer.

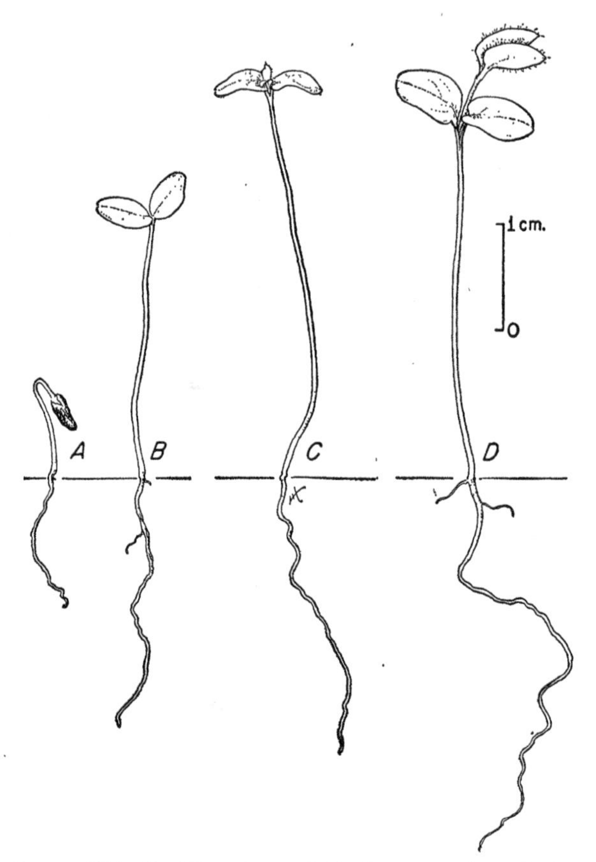

Figure 157.—Seedling views of *Lonicera tatarica: A,* At 1 day; *B,* at 3 days; *C,* at 13 days; *D,* at 31 days.

LUPINUS LONGIFOLIUS (S. Wats.) Abrams   Pauma lupine

(Legume family—Leguminosae)

Native to southern California, this erect shrubby lupine is valuable for ornamental plantings and erosion control. The closely related *Lupinus albifrons* Benth. (whiteface lupine), native to central and northern California, is similar in habit, uses, and germination. Pauma lupine flowers from January to July, and the seed ripen from May to August. Its pods pop open when dry. Therefore it is necessary to collect them somewhat green, or the seed (fig. 158) may be dispersed when the pods are touched. There are 18,000 to 24,000 clean seeds per pound (2 tests). In one sample, purity was 91 percent and soundness 76 percent. Freshly collected seeds of all the lupines tested germinated well without treatment. Stored seeds require hot-water treatment. Ninety-two percent is the highest germination recorded for Pauma lupine (1 test). After seed are sown, the first seedlings appear in 10 days.

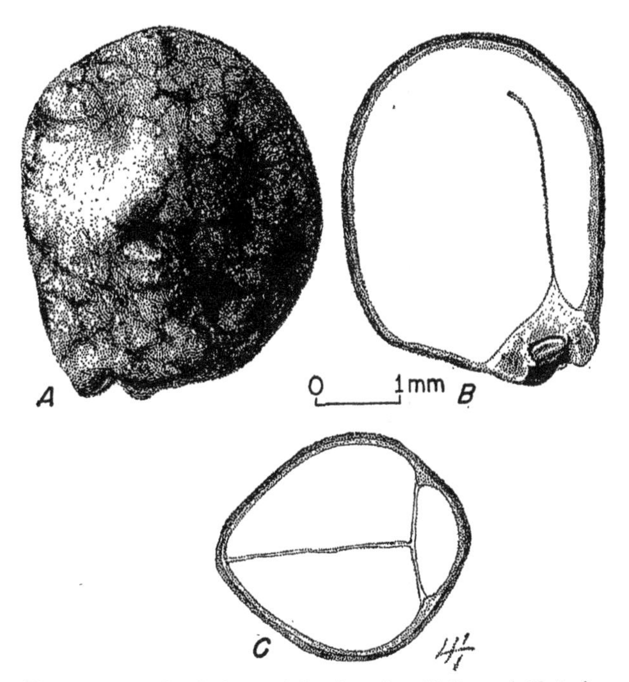

FIGURE 158.—Seed views of *Lupinus longifolius: A,* Exterior view; *B,* longitudinal section; *C,* cross section.

MACLURA POMIFERA (Raf.) Schneid. Osage-orange

(Mulberry family—Moraceae)

Also called bois d'arc, bow-wood. Botanical syns.: *Toxylon pomiferum* Raf., *Maclura aurantiaca* Nutt.

Distribution and Use.—Osage-orange is native from southern Missouri and southern Arkansas westward through southern Oklahoma and northern Texas. Locally it is a small, deciduous tree which occurs on rich soils, especially in bottom lands. Its best development is in the Red River Valley of Oklahoma. This species is chiefly valuable for fence posts, hedge fences, windbreaks, and a yellow dye produced from the root bark. Introduced into cultivation in 1818, it has been successfully planted throughout the eastern United States and southeastern Canada, and reproduces naturally in most of these extended ranges.

Seeding Habits.—The dioecious flowers of Osage-orange open from April to June. Its fruits ripen in September and October of the same year and fall to the ground soon after maturing. Seed (fig. 159) are often dispersed beyond the tree vicinity by cattle which eat the fruits sparingly. Good seed crops occur in most years. Commercial seed-bearing age: minimum, 10 years; optimum, 25 to 65 years; maximum, 75 to 100 years.

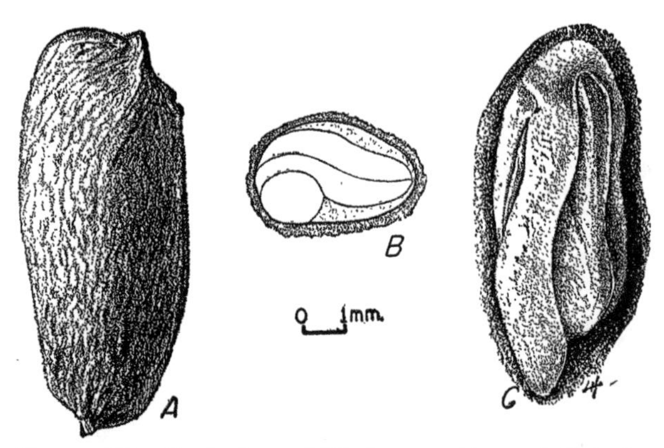

Figure 159.—Seed views of *Maclura pomifera: A,* Exterior view; *B,* cross section showing large embryo partly surrounded by endosperm; *C,* embryo.

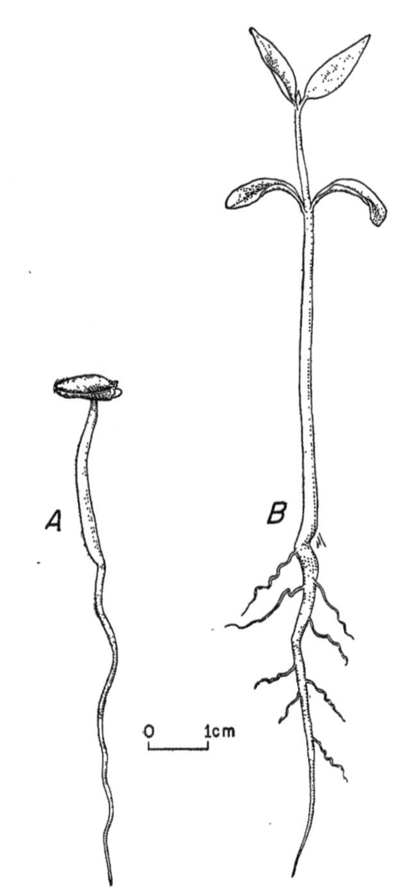

Figure 160.—Seedling views of *Maclura pomifera: A,* At 1 day; *B,* at 8 days.

Collection, Extraction, and Storage. — The light greenish-yellow fruits of this species may be collected from the ground in the autumn or winter. They may be crushed, or placed in moist storage and allowed to ferment, after which the seed may be rubbed from the pulp, further macerated, and washed clean. Eighty average-size fruits make a bushel and yield approximately 24,500 seeds which weigh about 2 pounds. Seed per pound (22 samples): low, 7,000; average, 14,000; high, 16,000. Commercial seed average about 96 percent pure, 95 percent sound, and cost from $1.25 to $2.60 per pound. Viability may be retained for 3 or more years by sealing air-dry seed in containers soon after cleaning, and storing them at 41° F. For spring sowing, seed may be stratified in sand or peat at 41° soon after cleaning.

Germination.—Although Osage-orange seed may germinate under a wide range of natural seedbed conditions, well-drained, moist, loamy soils are the best seedbeds. Germination takes place in the spring following seed fall. The seed do not often remain viable a second year. Germination is epigeous (fig. 160). Very limited dormancy exists in the seed coat and embryo. This may be overcome by soaking the seed in water for 48 hours, by stratifying it in sand or peat for 30 days, or by holding it over winter in sealed containers at 41° F. Adequate tests may be made on 400 pretreated seeds in sand or soil flats in 40 days at 68° (night) to 86° (day). Average germinative energy, pretreated seed: 20 to 79 percent in 14 to 34 days. Germinative capacity (13 tests): low, 23 percent; average, 58 percent; high, 85 percent.

Nursery Practice.—Seed of this species may be drilled in beds in the fall, but it is more economical to drill stratified seed, or dry-stored seed soaked for 48 hours, in early prepared beds in the spring. Fall-sown beds should be mulched; spring-sown beds need not. It is preferable to drill the seed in rows 8 to 12 inches apart to provide for ease of cultivation and weeding. The seed should be covered with one-fourth to one-half inch of firmed soil, and the surface soil kept moist during the germination period.

MAGNOLIA L. Magnolia

(Magnolia family—Magnoliaceae)

Distribution and Use.—The magnolias consist of about 35 species of deciduous or evergreen trees or shrubs which occur in North and Central America, eastern Asia, and the Himalayas. Some of the species produce valuable lumber, and the bark and fruit of some are used occasionally in medicine. Many magnolias are used in ornamental planting, because of their showy flowers and fruit, and attractive foliage. Their seeds are eaten by some of the larger birds. Of the 9 species native to the United States, *Magnolia acuminata* and *M. grandiflora* are used or are potentially valuable for conservation planting. Their distribution and chief uses are given in table 124. Practically all plantings made of both species thus far have been for ornamental purposes. *M. acuminata* has been often planted as an ornamental tree in the Eastern States[43] and in northern and central Europe. *M. grandiflora* is used to a great extent in landscape planting from Maryland southward. It is also used in the temperate parts of Europe where many horticultural varieties, which differ largely in leaf form and duration of the flowering period have been recognized.

Seeding Habits.—The large solitary perfect flowers (greenish yellow in *Magnolia acuminata* and white in *M. grandiflora*) are borne singly at the ends of the branches in the spring or early summer. The red or rusty brown, conelike fruit, which consists of several to many coalescent one- to two-seeded fleshy follicles, ripens in late summer to early fall. At maturity the seeds are red or scarlet and drupelike, the outer portion of the outer seed coat being fleshy, oily, and soft, the inner portion stony. The inner seed coat is thin and membranous and encloses a large fleshy endosperm in which is embedded a minute embryo (fig. 161). The seed usually is suspended from the open follicle for some time by a slender, elastic thread. Seed dispersal is largely by wind and birds, and occurs soon after ripening. Squirrels sometimes cut off the ripe fruits which eventually open up on the ground. Commercial seed consists of ripe, dried seeds or clean seeds from which the fleshy part of the outer coat has been removed. Details of seeding habits follow:

	M. acuminata	*M. grandiflora*
Time of—		
Flowering	April–June	April–August.
Fruit ripening	Late August–October.	Late July–October.
Seed dispersal	September–October.	August–October.
Seeds per fruit	10–60	40–60.

[43] This species is perfectly hardy and produces viable seed at St. Paul, Minn.

Table 124.—*Magnolia: Growth habit, distribution, and uses*

Accepted name	Synonyms	Growth habit	Natural range	Chief uses	Date of earliest cultivation
M. acuminata L. (cucumbertree).	cucumbertree magnolia, SPN, mountain magnolia.	Forest tree of pyramidal habit; deciduous.	Scattered individuals in rich woods from western New York to Georgia, west to southern Ohio, Illinois, and Missouri, to eastern Oklahoma and Louisiana.	Durable, light wood valuable for lumber and cabinet materials; ornamental planting; used as stock for scions of exotic magnolias; game food.	1736
M. grandiflora L.[1] (southern magnolia).	*M. foetida* Sarg. (evergreen magnolia, bull bay).	Large forest tree of pyramidal habit; evergreen with very large, fragrant flowers.	Rich moist soils along coast from North Carolina to Florida, west and north through Gulf Coast region to Texas and southern Arkansas.	Ornamental planting; wood used for baskets and crates; game food; fuel.	1734

[1] Read, A. D. Seed Handling Practice With Little-used Species. 1941. (Unpublished manuscript. Copy on file U. S. Forest Serv., South. Forest Expt. Sta., New Orleans (Pollock), La.)

Good seed crops are borne by *M. acuminata* every 4 to 5 years, but they are less frequent at the margins of its geographic range; light seed crops are borne in intervening years. Commercial seed bearing age of *M. acuminata* extends from 30 to 250 years and is optimum from 50 to 200 years. No data are available as to the existence of climatic races in *M. acuminata* and *M. grandiflora*. It is likely that such races have developed, particularly in *M. acuminata* which occurs over a wide range.

COLLECTION, EXTRACTION, AND STORAGE. — The fruits may be picked from standing trees by hand or by pruners, or from trees recently felled on logging operations. Collection can be started as soon as the cones turn bright red in *Magnolia acuminata* or rusty brown in *M. grandiflora*, but may be delayed until mid-September or October when the cones have begun to open. Early collections should be spread out to dry in shallow layers. The fruits will open in a few days and the seeds can be shaken out. Seed to be used in the near future should have the fleshy outer part of the outer seed coat removed by maceration in water or by rubbing on hardware cloth. After a short period of superficial drying it is ready for use. Seed to be kept for any length of time should be stored in the pulp which has been dried enough to lose its fleshy characteristics. Sometimes the pulp is removed from seed that is to be stored. Data on yield, number of cleaned seed per pound, purity, soundness, and cost for *M. acuminata* and *M. grandiflora* are as follows:

	M. acuminata	*M. grandiflora*
Fresh cones per pound	38	------
Cleaned seed (fleshy portion of outer seed coat removed):[1]		
Basis, samples ------ number	10	1
Per bushel of fruit --- pounds	3.7	------
Per pound:		
Low ------ number	2,900	------
Average ------ do	4,600	5,800
High ------ do	[2]6,600	------
Commercial seed:		
Purity ------ percent	97–100	94–100
Soundness ------ do	[3]91(11–100)	95
Cost per pound ------ dollars	1.75–3.00	1.10–2.50

[1]Cleaned seed of *M. acuminata* are dark brown to black in color, and those of *M. grandiflora* are straw-colored.
[2]Fresh seeds with pulpy coat, 1,675 to 2,200 per pound; air dried with pulp, 3,680 to 4,990 per pound.
[3]First figure is average percent; figures in parentheses indicate range from lowest to highest.

Little information is available as to the proper storage conditions for magnolia seed or its longevity. Seed of some species can apparently be kept either cleaned or in the dried pulp for several years with little loss if stored in sealed containers at 32° to 41° F. Seed stored at appreciably higher temperatures should not be cleaned. *Magnolia grandiflora* seed loses its viability if stored over winter at air temperature.

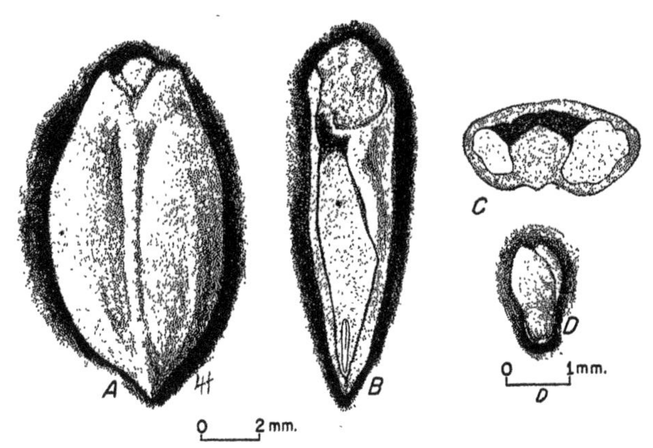

FIGURE 161.—Seed views of *Magnolia grandiflora: A,* Exterior view, outer part of outer coat missing; *B,* longitudinal section showing small embryo; *C,* cross section; *D,* embryo.

GERMINATION.—The best natural seedbed for magnolia is a rich, moist, woodland soil protected by litter. Germination, which occurs in the first or second spring following seed fall, is epigeous (fig. 162). Reproduction in nature of at least *M. acuminata* is scarce. This is probably due to the destruction of seeds by birds and rodents, the species' high susceptibility to injury by freezing, and the exacting conditions required for germination. Dormancy, which is due to conditions within the embryo, can be overcome by several months of low-temperature stratification. Details of pretreatment that have proved beneficial for the two species discussed here are as follows:

	M. acuminata	*M. grandiflora*
Stratification:		
Medium	Peat,[1] sand and peat.	Peat
Temperature ------ °F.	41	41
Duration ------ days	[2]150–180	[3]90–120

[1]Acid medium appears to give best results.
[2]A somewhat longer period may be required if the pulpy outer coat has not been removed from the seed. Seed stratified too long will likely show secondary dormancy.
[3]Stratification for 120 to 150 days at 50° F. also recommended.

Germination tests for magnolia are best made in sand flats, or in peat mats, using 400 or more stratified seeds per test. Other requirements and the results that may be expected are as follows:

	M. acuminata	*M. grandiflora*
Basis, tests ------ number	3	2
Test conditions:		
Night temperature ------ °F.	59	[1]59
Day temperature ------ do	79	[1]79
Duration:		
Stratified seed --- days	----	60
Untreated seed ---- do	----	210+
Germinative capacity:		
Low ------ percent	8	8
Average ------ do	55	----
High ------ do	86	90

[1]Suggested temperature; experimental data not complete.

Sterilization of seed prior to germination tests has proved decidedly harmful. A fairly close estimate of viability in *M. acuminata* can be made by removing the seed coats, lightly scratching the surface of the endosperm, and then placing the seeds at germinating temperatures on moist cotton or blotting paper in covered dishes. Viable seeds produce green pigment in the vicinity of the scratches in 2 to 3 days.

NURSERY AND FIELD PRACTICE.—Magnolia seed may either be fall sown or stratified over winter and sown early the following spring.[44] In fall sowing, clean seed or seed in the pulp should be drilled in rows 8 to 12 inches apart and covered with about one-fourth inch of firmed soil. The beds must be mulched with a layer of straw or leaves heavy enough to prevent freezing of the seed. This is not removed until all danger from late spring frosts is past. Seedbeds must be kept moist until germination is complete. The seedlings require half shade during much of the first summer and do best in rich soil. Nursery germination of *M. grandiflora* is 60 to 70 percent and begins in 20 days. Spring sowing following stratification appears to be the best method in areas where the rodent problem is serious. Magnolias may also be propagated from cuttings and by grafting.

[44] Sowing should be done in February or March in the deep South.

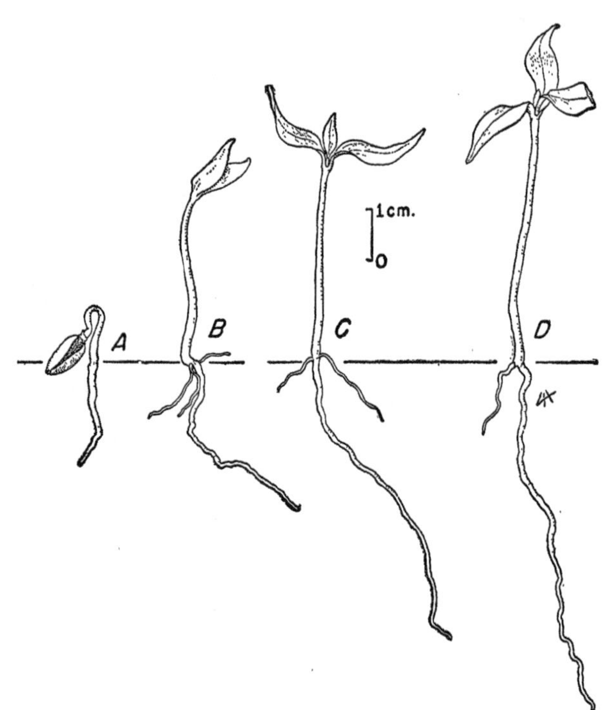

FIGURE 162.—Seedling views of *Magnolia grandiflora: A,* At 1 day; *B,* at 5 days; *C,* at 13 days; *D,* at 31 days.

MAHONIA HAEMATOCARPA (Woot.) Fedde Red mahonia

(Barberry family—Berberidaceae)

Also called red hollygrape, algerita. Botanical syns.: *Odostemon haematocarpus* (Woot.) Heller, *Berberis haematocarpa* Woot.

Red mahonia is an evergreen shrub which occurs in New Mexico, Arizona, and Colorado. This species was introduced into cultivation in 1916. Its fruit, much used for food by game and canned or preserved by man, is a red berry about 8 mm. in diameter which contains several small seeds (fig. 163). The bushes begin to bear fruit when they are about 4 years old. Ripening of the fruit occurs in June and the seed are dispersed by birds and animals. After collection the berries should be macerated and washed to remove the pulp, and the seed dried and stored in a cool place. There are about 103,000 seeds per pound (1 test). The seed, when sown in the spring, germinate within a few days to a month after sowing. Dormancy is apparently not present. It should be noted, however, that seed of some other western species of this genus do require stratification. When sowing, seed of this species should be covered with one-eighth inch of soil, then one-fourth inch of sand.

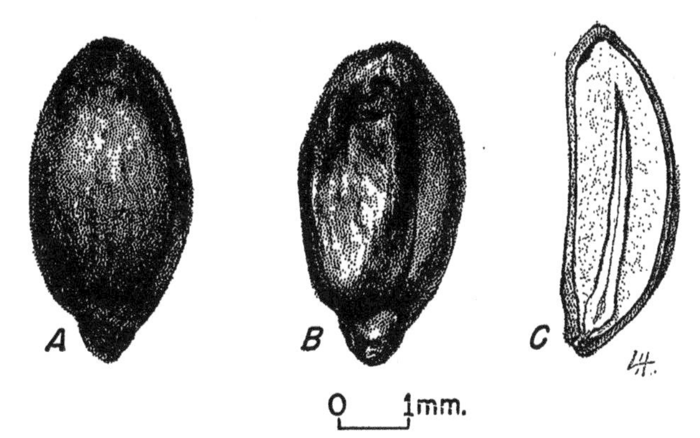

FIGURE 163.—Seed views of *Mahonia aquifolium*: *A* and *B*, Exterior views in two planes; *C*, longitudinal section.

MALUS Mill. Apple

(Rose family—Rosaceae)

Distribution and Use.—The apples include about 25 species of deciduous, or rarely half-evergreen, trees or shrubs native to the temperate regions of North America, Europe, and Asia. Some species are among the most important fruit trees and others are some of our most valuable ornamentals. Most of the apples are of value for wildlife food. Some of them are useful for shelter-belt planting. Four species, which are of present or potential value for conservation planting and for which accurate information is available, are described in table 125. *Malus pumila* is the parent of most of our cultivated apples and *M. baccata* is the primary source from which all the cultivated crabs have been developed. *M. coronaria* and *M. ioensis* are cultivated chiefly for ornamental purposes.

Seeding Habits.—The showy pink or white, perfect flowers appear in the spring with or before the leaves. The fruit, which is a pome, ripens in the summer and becomes yellowish to red. It is with or without some grit cells and contains from 4 to 10 small- to medium-sized dark seeds. The seeds (figs. 164 and 165) have a thick outer coat with an open hilum and a thin, translucent, inner coat of very dense structure without openings. Usually adhering to the inner coat is a fairly thick, white, apparently endospermous tissue which surrounds the embryo. The comparative seeding habits of the 4 species discussed here are given in table 126.

Scientific evidence as to the development of climatic races among the apple species is lacking. The commonly cultivated species, particularly the commercial-fruit producers, have been developed into a great number of forms and varieties. They differ considerably in hardiness, form, phenological characteristics, germination behavior, and kind and amount of fruit borne. In view of that fact, and also their extensive range, it seems likely that climatic races occurred at least among the wild plants of *Malus baccata* and *M. pumila*.

Collection, Extraction, and Storage. — The ripe apples may be collected from standing trees or gathered from the ground. Seed may be extracted by running the fruits through a macerator[45] with water and screening out or floating off the pulp. The seed should then be dried. The yield, size, purity, soundness, and cost of commercial seed vary considerably by species as

[45] The fruits may also be run through a cider mill, but this often results in injury to the seed.

Table 125.—*Malus: Growth habit, distribution, and uses*

Accepted name	Synonyms	Growth habit	Natural range	Chief uses	Date of earliest cultivation
M. baccata (L.) Borkh. (Siberian crab apple)	*M. baccata* var. *sibirica* (Maxim.) Schneid., *Pyrus baccata* L. (Siberian crab).	Small tree	Siberia, Manchuria, northern China, and the Himalayas.	Shelterbelts, wildlife food, human food, ornamental.	1784
M. coronaria (L.) Mill. (sweet crab apple)	*M. fragrans* Rehd., *Pyrus coronaria* L. (wild sweet crabapple, SPN).	do	New York to Alabama, and west to Missouri.	Wildlife food and cover, ornamental.	1724
M. ioensis (Wood) Britton (prairie crab apple).	*M. coronaria* var. *ioensis* (Wood) Schneid., *Pyrus ioensis* (Wood) Bailey.	do	Minnesota and Wisconsin to Nebraska, Kansas, and Missouri.	Wildlife food, ornamental.	1885
M. pumila Mill. (apple).	*M. communis* DC., *M. malus* Britton, *Pyrus malus* L. in part (common apple, SPN; wild apple).	Small to medium tree.	Europe and western Asia; often naturalized in eastern North America.	Human food, wildlife food.	(1)

[1]Cultivated since ancient times.

shown in table 127. Apple seed may be stored satisfactorily over winter by placing it in stratification in moist sand held at low temperatures. For longer storage the seed should be stored dry in sealed containers at temperatures just above freezing. Seed of *Malus pumila* will keep satisfactorily up to 2½ years, and possibly longer, by the latter method.

GERMINATION. — Seed of these species are disseminated in nature chiefly by gravity and animals. Germination is epigeous (fig. 166). The seeds display dormancy which can be overcome by stratification in moist sand, peat, or a mixture of the two for 30 to 90 days at 32° to 50° F. The best practice varies according to the species and variety. Further information on pretreatment is given in table 128. The germination of apple seeds may be tested in sand flats, using 200 to 400 properly pretreated seeds per test. Methods recommended and the average results for the 4 species discussed are given in table 129.

NURSERY AND FIELD PRACTICE.—Ordinarily stratified apple seeds are sown in the spring in drills, although fall sowing may also be practiced and the seeds covered with one-fourth inch of soil. Seed of *Malus ioensis* collected slightly green and sown immediately germinated 100 percent the following spring. About 15 percent of the *M. baccata* seed sown produces usable 1-0 seedlings. Comparable information for other species is lacking. Field planting is usually done with 1-0 or 2-0 stock. Most of the commercial apple varieties are propagated by budding or grafting onto hardy rootstocks. Apples grow over a wide variety of soils and conditions, but do best when grown in a moderate, temperate climate on a clay-loam soil.

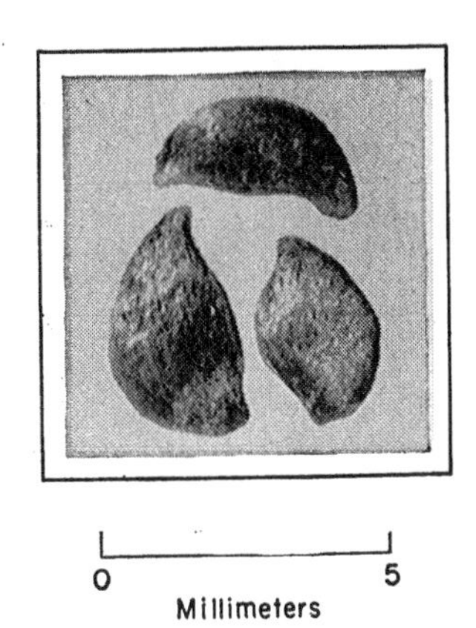

FIGURE 165.—Seed views of *Malus baccata*.

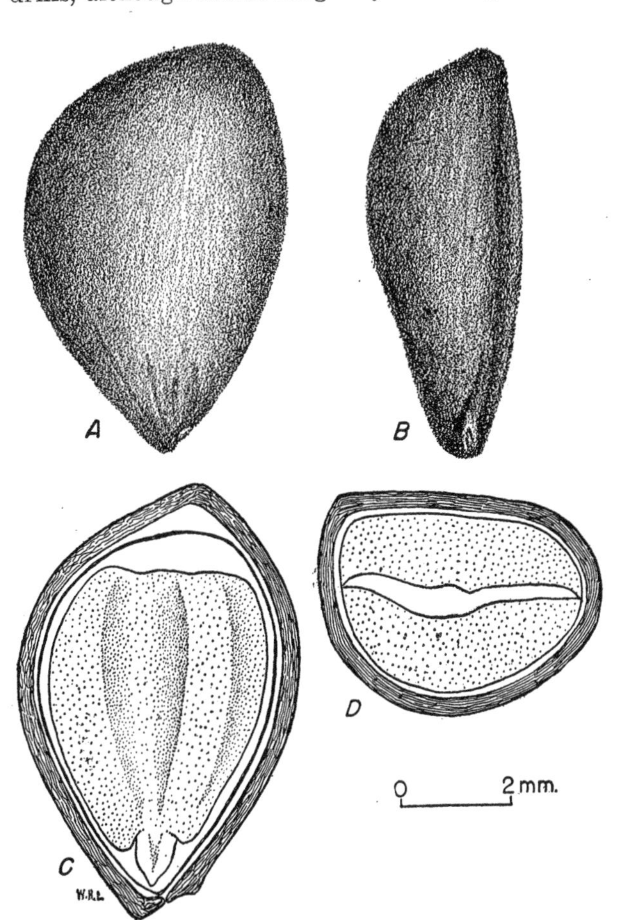

FIGURE 164.—Seed views of *Malus coronaria: A* and *B*, Exterior views in two planes; *C*, longitudinal section; *D*, cross section.

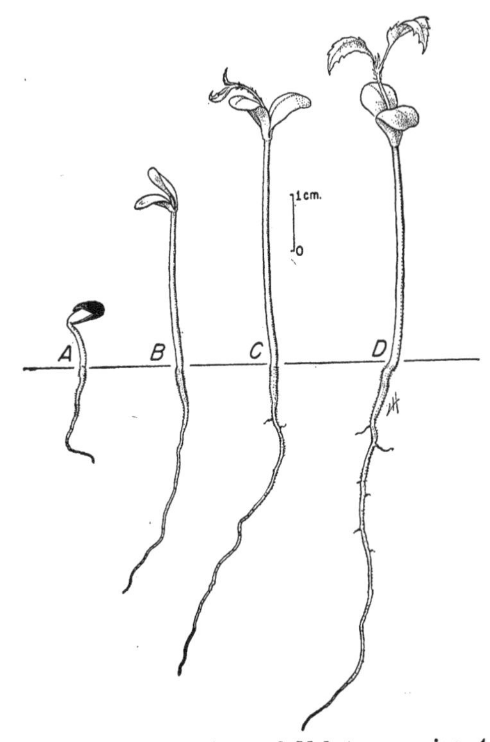

FIGURE 166.—Seedling views of *Malus coronaria: A*, At 1 day; *B*, at 3 days; *C*, at 9 days; *D*, at 16 days.

MALUS

TABLE 126.—*Malus: Time of flowering and fruit ripening, and frequency of seed crops*

Species	Time of—			Color of—		Diameter of ripe fruit	Seed year frequency	
	Flowering	Fruit ripening	Seed dispersal	Flowers	Ripe fruits		Good crops	Light crops
						Inches		
M. baccata	May	Late summer-early fall.	Late summer-early fall.	White	Red or yellow.	¼–½	Every 2 or more years.	Intervening years.
M. coronaria	Spring	do	do	White, flushed pink.	Yellow-green.	1–1¼	Every 2-4 years.	do.
M. ioensis	May–June	September–October.	do	do	Greenish, waxy.	1–1¼	About every 2 years.	do.
M. pumila	May	August–October.	do	White, pinkish, or rose.	Yellowish to red.	¾+	Every 2 or more years.	do.

TABLE 127.—*Malus: Yield of cleaned seed, and purity, soundness, and cost of commercial seed*

Species	Cleaned seed					Commercial seed		
	Yield per 100 pounds of fruit	Per pound			Basis, samples	Purity	Soundness	Cost per pound
		Low	Average	High				
	Pounds	*Number*	*Number*	*Number*	*Number*	*Percent*	*Percent*	*Dollars*
M. baccata	2–3	22,000	66,000	85,000	5	90	90	[1]3.50–5.25
M. coronaria	1		14,000		1	93	99	4.00
M. ioensis			30,000		1	86	94	10.00
M. pumila	0.6–0.9	7,000	20,000	27,000	5+	90	85	.75–1.50

[1]Dried fruits cost $2.50 per pound.

TABLE 128.—*Malus: Method of seed pretreatment for dormancy*

Species	Stratification			Remarks
	Medium	Temperature	Duration	
		°F.	*Days*	
M. baccata	Moist sand or peat	41	30	Longer stratification and combination of warm plus cold stratification gave no better germination.
M. coronaria	do	41	120	Satisfactory germination occurs with 60 days' stratification, but only after 104 days as compared to 24 days for the 120-day treatment.
M. ioensis	do	41	60	Longer stratification might improve germination slightly. Seed collected slightly green germinated 100 percent in the nursery (1 sample).
M. pumila	Moist acid or neutral peat	41	75	

TABLE 129.—*Malus: Recommended conditions for germination tests, and summary of germination data*

Species	Test conditions recommended			Germination data from various sources						
	Temperature		Duration	Germinative energy		Germinative capacity			Potential germination	Basis, tests
	Night	Day		Amount	Period	Low	Average	High		
	°F.	*°F.*	*Days*	*Percent*	*Days*	*Percent*	*Percent*	*Percent*	*Percent*	*Number*
M. baccata	68	86	30	48	8	52		55	78–86	2
M. coronaria	50	50	30	93	19		96		98	1
M. ioensis	68	86	10	48	4		58		64	1
M. pumila	68	86	[1]60				65			1+

[1]A more rapid test can be obtained by placing excised embryos in a standard germinator.

MELIA AZEDARACH L. Chinaberry

(Mahogany family—Meliaceae)

Also called China-tree, bead tree, Indian lilac. Includes var. *umbraculiformis* Berckmans.

Distribution and Use.—Native to the Himalayas, the Chinaberry is a short-lived, moderate-sized, half-evergreen tree. It is one of about 10 species of *Melia* which are confined to southern Asia and Australia. This tree has become naturalized in most tropical and subtropical countries, including the southern United States. Although it has been cultivated since the sixteenth century chiefly for ornamental purposes, it is also useful as a producer of wildlife and livestock food, and as a honey plant. Its bark is used in medicines. In India, furniture, agricultural implements, and cigar boxes are made of Chinaberry wood, and the fruit stones are used for beads.

Seeding Habits.—Borne in axillary panicles, the showy, lilac, perfect flowers bloom in March to May. The fruit is a subglobose, yellow, berrylike drupe with a fleshy outer coat about one-half inch across. It ripens in September and October, and persists on the trees well into the winter. Each fruit contains one fluted, light-brown stone which includes four to five pointed, smooth, black seeds (fig. 167). The seed are dispersed largely by birds and mammals. Abundant crops are borne almost annually.

Collection, Extraction, and Storage.—Chinaberry fruits should be collected by hand from the trees after the leaves have fallen in late fall or early winter. They may either be run wet through a macerator and the pulp floated off or screened out, or the entire fruits may be planted immediately. The number of fruits per pound averages about 640 (7 samples). Commercial seed —the stones—costs from $0.80 to $1.25 per pound. Seed of this species retains unimpaired viability for at least 1 year under ordinary dry storage.

Germination. — Chinaberry seed germinates freely in nature the spring following dispersal. Despite reports to the contrary, germination is epigeous. One fruit produces 1 to 4 seedlings. Seedlings frequently come in on cut-over oak lands in the South where lime is abundant. The seed of this species apparently are not dormant. Germination may be tested in sand flats. It is suggested that temperatures of 70° F. (night) to 85° (day) be used for 60 days, with 200 seeds per test. Average germinative capacity is about 65 percent (5 tests).

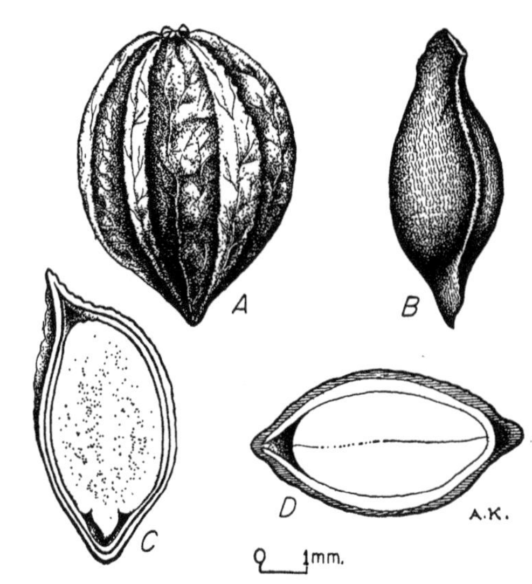

Figure 167.—Seed views of *Melia azedarach:* *A*, Exterior view of stone; *B*, exterior view of seed; *C*, longitudinal section; *D*, cross section.

Nursery and Field Practice.—The fruits may be sown immediately after collection in the fall, or they may be sown in the spring. They should be sown 2 to 3 inches apart in drills and covered with about 1 inch of soil. Germination takes place in the spring about 3 weeks after planting. One-year-old seedling stock is preferred for field planting. Older stock should be top- and root-pruned. Direct seeding in the field may be used. This species may also be propagated from cuttings and root suckers.

MENODORA SCABRA A. Gray **Rough menodora, twinberry**

(Olive family—Oleaceae)

Rough menodora is native to the Southwest and southern California. This low shrub is valuable as browse for cattle, horses, sheep, and goats. Its seed (fig. 168) ripen in September and October and are dispersed during October and November. Collection should be made from September to November. Good seed crops usually occur every year. Cleaned seed per pound (2 samples): 102,000 to 112,000. In one sample purity was 41 percent and soundness 98 percent. Storage should be in a dry place. Seed of this species germinate freely and apparently do not need stratification. Germinative capacity (2 tests): 70 to 99 percent.

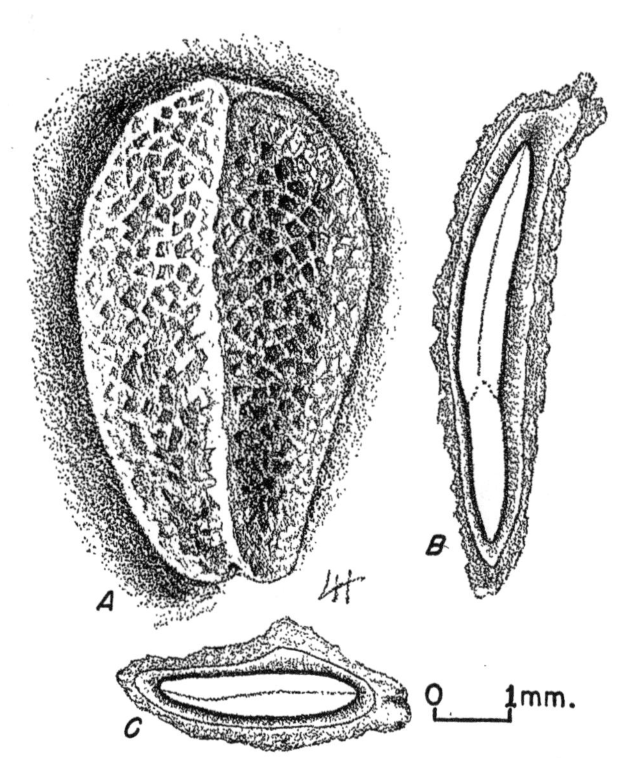

FIGURE 168.—Seed views of *Menodora scabra: A,* Exterior view; *B,* longitudinal section; *C,* cross section, showing large embryo.

MORUS L. Mulberry

(Mulberry family—Moraceae)

Distribution and Use.—The mulberries consist of about 12 species of deciduous trees or shrubs native to temperate and subtropical regions of North America, South America, and Asia. Some are valuable for timber products and shelter belts in semiarid regions, and the fruits are edible by both man and wildlife. Certain species, particularly *Morus alba,* provide the best known food for the silkworm. Two species and one variety, which are grown in the United States, are described in table 130. *M. alba* var. *tatarica* probably is the most widely planted mulberry in the United States, particularly in the prairie region where it was introduced by the Russian Mennonites in 1875–77.

Seeding Habits.—The male and female flowers—the former bearing pollen and the latter developing into fruit—are borne on the same or on different trees. They are very small, occur in elongated clusters, and bloom early in the spring. The fruit is an aggregate composed of many small, closely appressed drupes, the whole ½ to 1½ inches long, juicy, and white, red, or black when ripe. They are dispersed soon after ripening, chiefly by birds and animals and sometimes by water. Each fruit contains several minute seed which have thin, membranous coats and endosperms (figs. 169 and 170). Comparative seeding

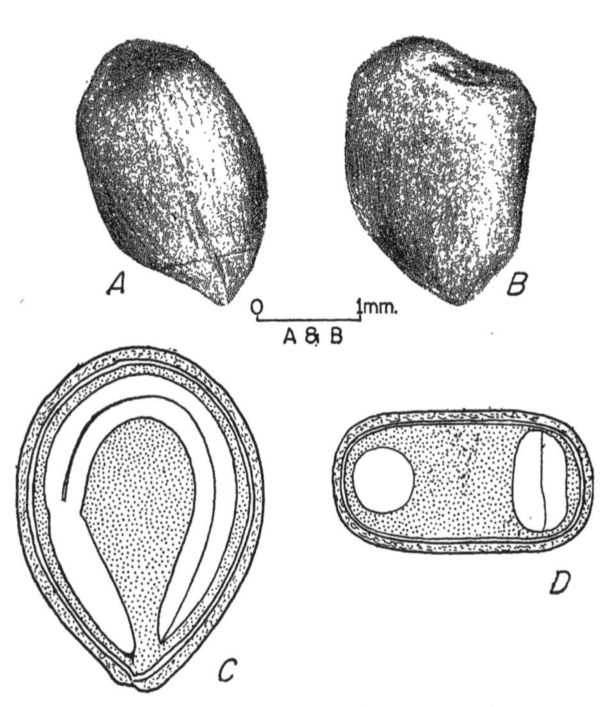

Figure 169.—Seed views of *Morus alba* var. *tatarica: A* and *B,* Exterior views in two planes; *C,* longitudinal section showing curled embryo; *D,* cross section.

Table 130.—*Morus: Growth habit, distribution, and uses*

Accepted name	Synonyms	Growth habit	Natural range	Chief uses	Date of earliest cultivation
M. alba L. (white mulberry).	----------------	Small, short-lived tree.	China; introduced in North America, Europe, and other parts of Asia.	Silkworm food, wildlife food and cover, human food; furniture, turnery, boats.	(1)
M. alba var. *tatarica* (L.) Ser. (Russian mulberry).	*M. tatarica* L.	-----do--------	-----do----------------	Hedges, shelter belts, fence posts, fuel, wildlife and livestock food and cover, silkworm food.	(1)
M. rubra L. (red mulberry).	----------------	-----do--------	Massachusetts south to Florida, west to southern Ontario and Texas.	Fence posts, boat building; human and wildlife food.	1629

[1] Long cultivated.

MORUS

TABLE 131.—*Morus: Time of flowering and fruit ripening, and commercial seed-bearing age*

Species	Time of—		Color of fruit	Fruit length	Commercial seed-bearing age[1]		
	Flowering	Fruit ripening			Minimum	Optimum	Maximum
				Inches	*Years*	*Years*	*Years*
M. alba	May	July-August	White[2]	½-1	5		
M. alba var. *tatarica*	do.	June-August	Dark red, sometimes white	⅓-½			
M. rubra	do.	do.	Dark red	¾-1¼	10	30-85	125

[1]*Morus alba* var. *tatarica* bears good seed crops almost annually; *M. rubra* bears good crops every 2 to 3 years with light crops in the intervening years.
[2]Sometimes pinkish or purplish violet.

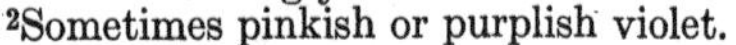

FIGURE 170.—Seed views of *Morus rubra.*

habits of the two species and one variety discussed are given in table 131.

Little scientific information is available as to the development of climatic races among the mulberries. In the case of *Morus alba* many varieties are recognized, and the fact that there are differences in drought resistance and chromosome number among the varieties indicates the likelihood that climatic races have developed. *M. alba* var. *tatarica,* if not a separate species, may be a climatic race of *M. alba;* it is distinctly more drought resistant than *M. alba.* In the case of var. *tatarica,* one series of tests showed that large seed produced taller and better rooted seedlings than did medium or smaller-sized seed. This brings up the possibility of some racial differentiation on the basis of seed size.

COLLECTION, EXTRACTION, AND STORAGE.—Ripe mulberry fruits may be collected by stripping, shaking, or flailing them from the trees into a tarpaulin. If collection is not made as soon as they are ripe, they may be destroyed by birds or other animals. To extract the seed, mash and soak the berries in water for 24 hours and then run them wet through a macerator, skimming or floating off the pulp. Fermentation of the fruits for 1 day before extraction improves germination of *Morus alba* var. *tatarica* seed. A 1-percent lye solution is reported to aid the extraction and cleaning process. Small samples may be pulped by rubbing them gently through a No. 6 screen and then floating off the pulp and empty seed. The cleaned seed should be air-dried in the shade for a short period before storage or use. The size, yield, purity, soundness, and cost of commercial cleaned seed vary by species as shown in table 132.

Detailed information as to the longevity of *Morus* seed under storage is generally lacking. *M. alba* var. *tatarica* seed is reported to maintain its viability satisfactorily for 3 years under ordinary storage conditions. In general it is recommended that the seed be stored dry in sealed containers at about 41° F., or unsealed under seed warehouse conditions. *Morus* seed can be kept satisfactorily over winter in moist, cold stratification.

GERMINATION.—The best natural seedbed is a moist, rich, loamy soil, at least partially protected by litter or vegetation. Germination occurs in the spring following seed fall; it is epigeous. In the laboratory, germination of untreated seed usually is poor. Individual seed lots may vary consid-

TABLE 132.—*Morus: Yield of cleaned seed, and purity, soundness, and cost of commercial seed*

Species	Cleaned seed					Commercial seed		
	Yield per 100 pounds of fruit	Per pound			Basis, samples	Purity	Soundness	Cost per pound
		Low	Average	High				
	Pounds	*Number*	*Number*	*Number*	*Number*	*Percent*	*Percent*	*Dollars*
M. alba	1-6	130,000	235,000	350,000	18+	80-95	90+	1.50-2.50
M. alba var. *tatarica*	2	245,000	300,000	370,000	12	80	85	1.40-2.50
M. rubra	2-3	200,000	360,000	500,000	4	85	90	2.00-3.50

erably, but ordinarily part of each lot consists of seed with dormancy due to embryo conditions and probably also to impermeability of the seed coat. Stratification in moist sand helps overcome the dormancy. Further information on pretreatment is given in table 133. The germination of mulberry seed may be tested in sand, or sand and peat, flats at 68° F. (night) to 86° to 95° (day) for 30 to 45 days, using 1,000 properly pretreated seeds per test. Average results for the 2 species and 1 variety discussed here are given in table 134.

TABLE 133.—*Morus: Method of seed pretreatment for dormancy*

Species	Stratification	
	Medium	Duration and temperature
		Days °*F.*
M. alba	Moist sand	60 at [1]68–86 + 60 at 41
M. alba var. *tatarica*	do	30– 60 at [2]41
M. rubra	do	90–120 at 41

[1]Fairly good results may also be obtained by 90 days' stratification at 41° F.
[2]Seed soaked in water 1 week germinates equally well.

NURSERY AND FIELD PRACTICE.—Mulberry seed mixed with sand or sawdust may be sown broadcast or in drills in the fall; or seed which has been stratified or water-soaked for 1 week may be sown in the spring and covered with one-fourth inch of soil. Drills 8 to 12 inches apart, using about 50 seed to the linear foot of row, are satisfactory. Fall-sown beds should be mulched with straw or leaves and protected by bird or shade screens until germination begins in the spring. Spring-sown beds should be mulched with burlap and kept moist until germination begins. Seedbeds should be given half-shade for a few weeks after germination. Germination usually takes place 1 to 2 weeks after spring sowing. In the case of *Morus alba* var. *tatarica,* about 12 percent of the viable seed sown produce usable seedlings.

M. alba var. *tatarica* has been attacked by a bacterial canker in nurseries in Oklahoma and elsewhere. This disease can be controlled to some extent by disinfecting the soil with formaldehyde, or by weekly spraying with 0.06 percent corrosive sublimate plus 1.5 percent slaked lime, or 0.2 percent potassium permanganate with 2 percent starch. To prevent the spread of this disease, mulberry seed should be sown in areas remote from old mulberry plantings and the badly diseased seedlings destroyed. Some leaf-spot diseases caused by the fungi *Cercospora* spp. and *Phleospora maculans* (?) cause damage sometimes. Early applications of 4–4–50 bordeaux probably will reduce the spread of other diseases. For protection against insects, the seeds sometimes are soaked in water mixed with camphor and the beds sprinkled with a mixture of lime, ashes, and white arsenic. The seedlings should be hardened off in late summer to prevent frost injury. One-year-old seedling stock is used for field planting. Propagation by budding is sometimes practiced.

TABLE 134.—*Morus: Summary of germination data from various sources*

Species	Stratified seed						Untreated seed					
	Germinative energy		Germinative capacity			Basis, tests	Germinative energy		Germinative capacity			Basis, tests
	Amount	Period	Low	Average	High		Amount	Period	Low	Average	High	
	Percent	*Days*	*Percent*	*Percent*	*Percent*	*Number*	*Percent*	*Days*	*Percent*	*Percent*	*Percent*	*Number*
M. alba	73–84	8–12	75		87	2	33	13	34		46	3
M. alba var. *tatarica*	4–79	8–25	4	48	82	8	2–55	11–30	3	33	76	10
M. rubra			20	50	80	3+						

MYRICA

MYRICA CERIFERA L. Southern waxmyrtle

(Waxmyrtle family—Myricaceae)

Also called bayberry. Botanical syn.: *M. caroliniensis* Mill.

Distribution and Use.—Southern waxmyrtle is an evergreen shrub or small tree native from New Jersey to Florida and Texas. The wax obtained from its fruits is important in the manufacture of candles which give off a pleasant fragrance when burning. Its bark, berries, and leaves have medicinal properties. This species was introduced into cultivation in 1699.

Seeding Habits.—The male and female flowers, which are borne separately on different trees, open in the spring from March to April. Waxmyrtle fruit is a one-seeded, grayish-white, wax-covered drupe (fig. 171). It ripens in September or October and persists until late winter or spring. The seed are dispersed by birds, rodents, and water.

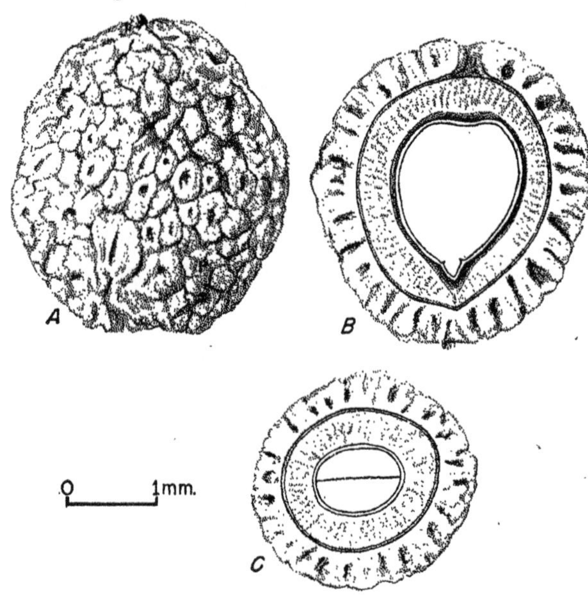

Figure 171.—Seed views of *Myrica cerifera: A,* Exterior view of waxy-coated fruit; *B,* longitudinal section; *C,* cross section.

Collection, Extraction, and Storage.—Fruits of this species may be collected from October until spring by stripping them from the branches by hand or by whipping them onto canvas. The waxy covering of the seed is usually removed by agitation of the dry fruits in a stirring machine or between surfaces. One pound of fruit yields about 9 or 10 ounces of clean seed, which number 52,000 to 58,000 per pound (3 samples). Average purity and soundness of commercial seed are, respectively, 98 percent and 90 percent. The cost of commercial seed ranges from $1.25 to $2.50 per pound. For prolonged storage, one to several years, seed still retaining wax should be sealed in containers at 41° F. For storage periods up to 1 year, the seed should be cleaned and either stored in sealed containers or stratified in sand, or a mixture of sand and acid peat, or peat alone, at 41°.

Germination.—Optimum natural seedbed conditions for southern waxmyrtle are found in moist, silty, acid sand. Epigeous germination (fig. 172) occurs in spring in those seed which have fallen in late fall or early winter and become covered. Those remaining on the shrubs until spring may fall and germinate later in the growing season or the next spring. Dormancy, caused chiefly by conditions in the embryo but prolonged somewhat by the waxy seed covering, may be most quickly broken by removing the wax and stratifying the seed in acid peat for 3 months at 34° to 50° F. Adequate germination tests may be made on 100 or 200 pretreated seeds in acid sand and peat flats in 30 days at 70° to 80°. Fresh, clean seed give better germination results than waxy seed that have been stratified for 3 months, but clean seed deteriorate more rapidly. Germinative capacity (20 tests): low, 36 percent; average, 70 percent; high, 92 percent.

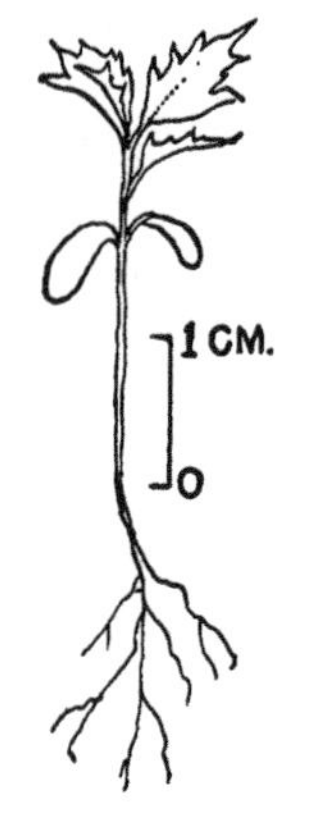

Figure 172.—Seedling of *Myrica californica* at 1 month, actual size.

Nursery Practice.—Seed of this species may be sown in nursery beds in the fall or spring. Drilling in rows 8 to 12 inches apart is preferable to broadcasting. The seed should be covered with about one-fourth inch of firmed soil. It is desirable to mulch fall-sown beds with straw or leaves and protect them with bird or shade screens until after late spring frosts. Seeding must be done late in the fall to avoid germination during that season and seedling mortality during the winter. Spring seeding should employ stratified seed. Propagation by layering is possible.

NEMOPANTHUS MUCRONATUS (L.) Trel. Mountain-holly

(Holly family—Aquifoliaceae)

Mountain-holly is a deciduous, branchy shrub which occurs in swamps from Newfoundland to Minnesota and south to Virginia and Indiana. It was introduced into cultivation in 1802. This species provides food and cover for game. Its flowers open from May to June. The fruit, which ripens from August to September of the same year, is a dull red drupe 6 to 8 mm. in diameter containing four to five bony nutlets. The latter are somewhat crescent shaped and are bone-colored with one rib on the back (fig. 173). Because the fruit is somewhat persistent, it may be collected as late as mid-October.

Approximately 10 pounds of cleaned seed are produced from 100 pounds of fresh fruit. There are about 1,600 berries in a pound (1 sample). Cleaned seed per pound (3 samples): low, 31,000; average, 45,000; high 66,000. Commercial seed averages 96 percent in purity (1 sample), 80 percent in soundness (4 samples), and costs $3.50 per pound. Small lots of seed can be extracted by rubbing the fruits through a No. 10 screen and then floating off the pulp and empty seed. Very little information is available on germination. It is very slow, probably because of the combination of an impermeable seed coat and dormant embryo. Stratification alone does not appear to be of much value—the seeds begin to germinate only after several months and continue to germinate for almost 2 years to reach a germinative capacity of 14 to 66 percent (3 tests). Possibly some treatment to render the seed coat permeable—acid soaking or scarification—in conjunction with stratification or warm plus cold stratification would be of benefit. Propagation by green-wood cuttings is sometimes practiced. Mountain-holly should be planted on moist soil; it will do well under partial shade.

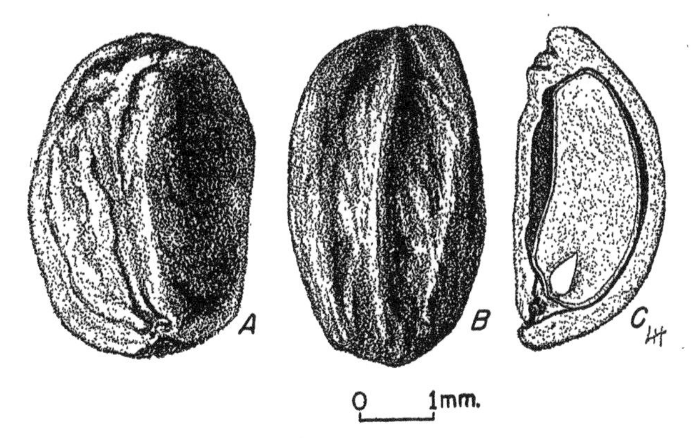

FIGURE 173.—Seed views of *Nemopanthus mucronatus: A* and *B,* Exterior views in two planes; *C,* longitudinal section.

NYSSA L. Tupelo

(Tupelo family—Nyssaceae)

DISTRIBUTION AND USE.—The tupelos include six species of short to medium-sized, deciduous trees found in North America and Asia, usually on wet sites. They are useful as producers of timber, wildlife food, bee pasture, and to some extent as ornamentals. Four species are native to eastern North America. Two of these, for which reliable information is available, are described in table 135. So far none of the tupelos have been widely planted for conservation purposes, although they offer possibilities for use on wet sites.

SEEDING HABITS.—The minute, greenish-white flowers appearing in early or midspring may be perfect or the male and female flowers may be borne separately on different trees. The fruit is a red, blue, or purple, oblong, thin-fleshed drupe one-third to 1 inch long. It ripens in the fall of the first year and is dropped soon afterward. Each fruit contains a bony, ribbed or winged, usually 1-seeded stone, the seed of commerce (fig. 174). The stones of *Nyssa aquatica* and *N. sylvatica* are light brown to almost white in color. Natural seed dispersal is largely by animals. The comparative seeding habits of these two species are as follows:

	N. aquatica	*N. sylvatica*
Time of—		
Flowering	Mar.–April	April–June
Fruit ripening	Sept.–Oct.	Sept.–Oct.
Seed dispersal	Fall	Sept.–Oct.

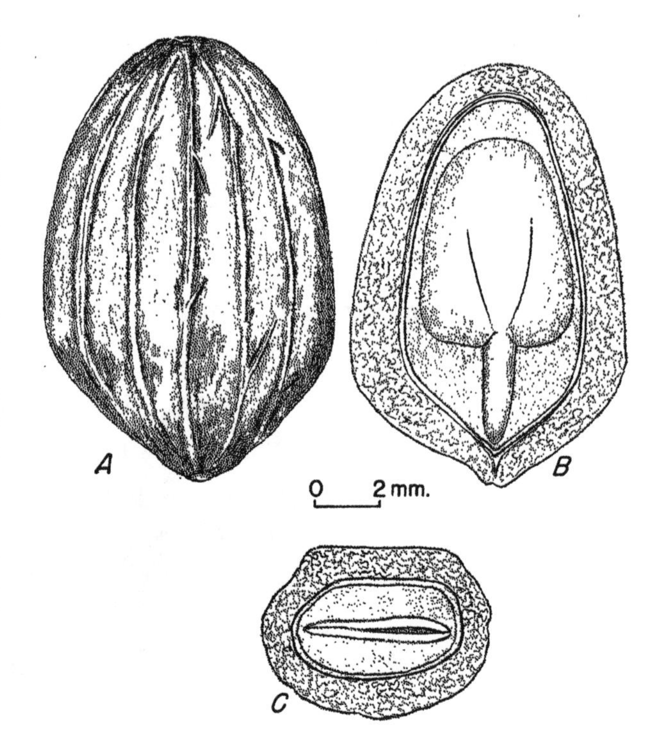

FIGURE 174.—Seed views of *Nyssa sylvatica: A,* Exterior view; *B,* longitudinal section; *C,* cross section.

TABLE 135.—*Nyssa: Growth habit, distribution, and uses*

Accepted name	Synonyms	Growth habit	Natural range	Chief uses	Date of earliest cultivation
N. aquatica L. (water tupelo).	tupelo gum, cotton gum, tupelo.	Medium to tall tree.	Deep swamps from Virginia to southern Illinois, south to Florida and Texas.	Timber products, wildlife food, bee pasture.	Before 1735.
N. sylvatica Marsh. (black tupelo).	blackgum, sour gum, pepperidge.	---do---	Southern Main to Ontario and Michigan, south to Florida and Texas.	Timber and ties, wildlife food, ornamental, bee pasture.	Before 1750.

Fruits of *N. aquatica* are dark purple, ¾ to 1 inch long, and the seeds have 10 prominent longitudinal ribs. Fruits of *N. sylvatica* are dark blue, ¼ to ½ inch long, and the seeds have 10 to 12 slightly accented ribs. Scientific information as to racial variations among the tupelos is lacking. The variety *biflora* of *N. sylvatica* is confined to a definite geographic zone and may be a climatic race.

COLLECTION, EXTRACTION, AND STORAGE.—Ripe tupelo fruits may be picked by hand from standing or freshly felled trees. To extract the seed, the fruits should be run through a macerator with water and the pulp floated off or screened out. Small samples may be depulped by rubbing the fruits over a submerged screen. The yield, size, soundness, and cost of commercial seed vary by species as follows:

	N. aquatica	*N. sylvatica*
Cleaned seed:		
Per 100 pounds of fruit...pounds..	----	25
Per pound:		
Low..................number..	----	1,850
Average..................do....	[1]456	3,300
High..................do....	----	4,000
Basis, samples..................do....	1	5
Commercial seed:		
Soundness..................percent..	----	87
Cost per pound..................dollars..	1.70	0.85–1.60

[1]Average number of *N. aquatica* fruits per pound, 215 to 250.

Nyssa seed can be kept satisfactorily over winter by stratifying them in moist sand at low temperatures. No other methods of storage have been reported, except a variation of stratification in which *N. aquatica* seed were kept over winter in sand submerged in water.

GERMINATION.—Germination is epigeous (fig. 175). Tupelo seeds have dormant embryos and do not germinate readily unless they are first stratified in moist sand at 30° to 50° F. for several months. Stratification for 60 to 90 days is recommended for *Nyssa sylvatica* and would probably be suitable for *N. aquatica.* The germination of tupelo seed may be tested in sand flats at temperatures of 68° (night) and 86° (day) for 30 to 60 days using 100 to 200 properly stratified seeds per test. Average results for *N. aquatica* and *N. sylvatica* are as follows:

	N. aquatica	*N. sylvatica*
Germinative capacity:		
Low..................percent..	45	3
Average..................do....	----	30
High..................do....	50	55
Basis, tests..................number..	2	4

NURSERY AND FIELD PRACTICE.—Untreated seed may be sown in the fall, or seed that have been stratified 90 days or longer may be sown in the spring. Sowing should be at the rate of 15 seeds per linear foot for *Nyssa aquatica,* and probably about 30 per linear foot for *N. sylvatica.* The seed should be covered with ½ to 1 inch of soil. After sowing, the seed must not be allowed to dry out; if drying occurs, they will not germinate until the following year. Spring sowings of stratified *N. aquatica* seed have given a germination of 38 to to 50 percent which began 17 to 26 days after sowing. Since the root systems of 1-year seedlings are usually rather sparse, root-pruning or transplanting is probably necessary to produce good stock for field planting. The tupelos sometimes are increased by layering, although not very satisfactorily.

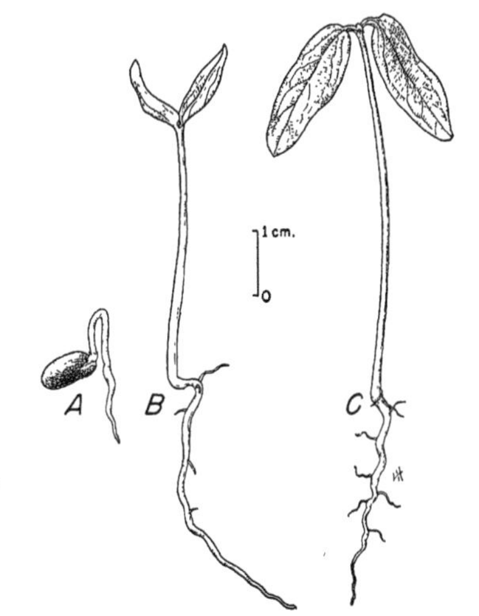

FIGURE 175.—Seedling views of *Nyssa sylvatica: A,* At 1 day; *B,* at 4 days; *C,* at 39 days.

OLNEYA TESOTA A. Gray Tesota

(Legume family—Leguminosae)

Also called ironwood, Sonora ironwood, palo de hierro.

This evergreen tree, which is native to southern Arizona, southern California, and northwestern Mexico, is valuable chiefly for fuel and for manufacture into small objects. It also furnishes fair browse for cattle and can be planted in arid, treeless localities within its climatic range. Tesota flowers appear in June, and its fruit—a 1- to 5-seeded, light-brown, rounded, hairy pod—ripens before the end of the following August. The seeds are chestnut brown, shiny, ovoid, and about 8 mm. long (fig. 176). There are about 2,000 to 2,200 seeds in a pound (2 samples). In 1 sample tested, purity was 84 percent. Sometimes it is necessary to fumigate the seed with hydrocyanic acid in order to kill beetles which may infest it. Fresh seed germinates well without any treatment. Sowing is usually done broadcast in the spring, and it is advisable to soak the seeds in water for 24 hours before they are sown. After sowing, the beds should be rolled and then covered with one-fourth inch of sifted sand. Care must be taken not to water the beds too heavily, or the seeds will rot in the ground. Germination is prompt, sometimes taking place in 18 hours. Seedlings appear in 6 days after sowing.

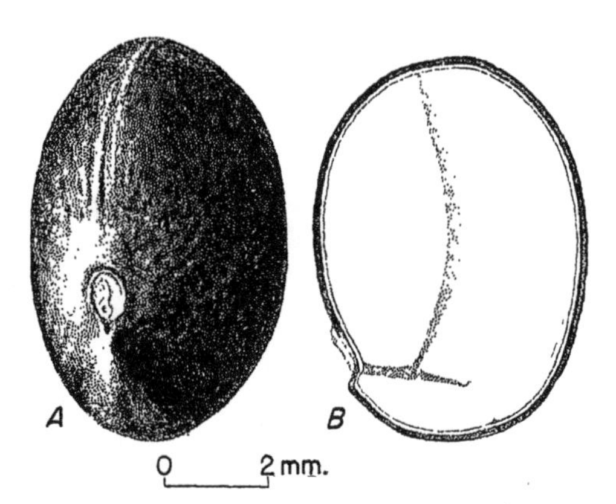

FIGURE 176.—Seed views of *Olneya tesota*: *A*, Exterior view; *B*, longitudinal section showing large embryo.

OSTRYA VIRGINIANA (Mill.) K. Koch Eastern hophornbeam

(Birch family—Betulaceae)

Also called American hophornbeam, SPN; leverwood, ironwood.

Distribution and Use.—Eastern hophornbeam is a small, deciduous tree native from Nova Scotia westward to southern Manitoba, northern Minnesota, the Black Hills of Dakota, eastern Nebraska and Kansas, and south to Florida and west to eastern Texas and Oklahoma. Its best development is in southern Arkansas and eastern Texas. Locally it occurs in rich woods as part of the understory. Because the wood is hard, strong, and durable, it is commercially valuable for fence posts, handles, mallets, golf club heads, and other products having durability requirements. It provides food and cover for many birds and some mammals. An attractive ornamental, this species was introduced into cultivation in 1690.

Seeding Habits.—The monoecious flowers—the staminate in long, clustered catkins formed the preceding year, and the pistillate in small open clusters appearing at the end of small leafy

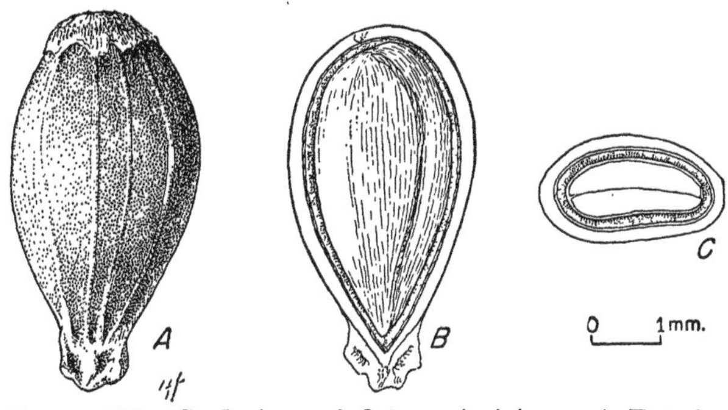

Figure 177.—Seed views of *Ostrya virginiana: A,* Exterior view of nutlet; *B,* longitudinal section; *C,* cross section.

branches—open from April to June. The fruit is a hard-shelled nutlet (fig. 177) containing a single seed without endosperm. A bladderlike sac, i. e., an involucre, clothed with stinging hairs covers the fruit. The involucres occur in strobiles which dry and gradually fall apart from soon after ripening, August through September, until early winter. These seed-bearing involucres are dispersed largely by the wind, but occasionally by birds. *Ostrya* is slow-growing and does not produce seed in quantities until about 25 years old. One variety, *glandulosa,* is confined to the northern part of the range and may be a climatic race.

Collection, Extraction, and Storage. — The strobiles should be hand-picked from the trees when they are a pale greenish brown in color, and before they are dry enough to shatter. When ripe, they are light gray to greenish brown. After the fruit has been dried in bags or trays, the seed may be beaten or rubbed from the involucres and separated from the chaff by fanning. One bushel of fruit will yield approximately 2 pounds of seed, or 100 pounds will yield about 20 pounds of cleaned seed. Cleaned seed per pound (5 samples): low, 25,000; average, 30,000; high, 35,-000. Commercial seed averages about 97 percent in purity, 80 percent in soundness, and costs from $2 to $3.80 per pound. It is recommended that the seed be stratified in sand or peat at 41° F. for over-winter storage. Although no reliable information

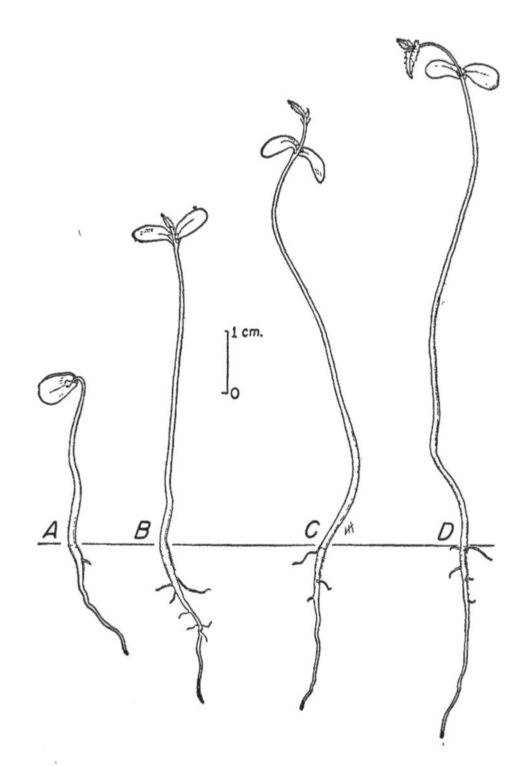

Figure 178.—Seedling views of *Ostrya virginiana*: *A,* At 2 days; *B,* at 4 days; *C,* at 23 days; *D,* at 27 days.

on dry storage is available, it is believed that such storage is feasible, particularly at low temperatures.

GERMINATION.—The best natural seedbed for *Ostrya* is moist, well-drained, loamy, mineral soil partially protected by vegetational cover. Germination, epigeous (fig. 178), probably takes place in the spring following autumn seed fall. Dormancy apparently lies in the embryo, since the seed coats are known to be somewhat water-permeable. However, the only successful method of overcoming dormancy so far found is stratification in moist sand or peat for 60 days at 68° to 86° F. plus 140 days at 41°, or for 6½ months at 50° to 77° plus 90 days at 41°. Reliable tests on 200 to 400 pretreated seeds in sand flats at 50° to 77° may be made in 30 to 40 days. Using this test method, a germinative capacity of 27 to 65 percent (2 tests), and a potential germination of 85 to 90 percent (3 tests) were obtained.

NURSERY PRACTICE.—Either fall or spring sowing in well-prepared nursery beds is practicable. Preferably, fall sowing should be done in drills soon after seed collection. In one case seed collected slightly green in August and sown immediately germinated 100 percent the following spring. Fall-sown beds should be mulched with burlap, straw, or other available material, to be held in place until germination begins in spring and then removed. It is recommended that spring sowing of stratified seed be done as soon as the soil can be worked; the beds should be mulched or their surfaces kept moist until germination starts. A sowing depth of one-fourth inch with firmed soil is recommended.

OXYDENDRUM ARBOREUM (L.) DC. Sourwood

(Heath family—Ericaceae)

Also called sorrel-tree.

Distribution and Use.—Sourwood, the only species in its genus, is native throughout the Eastern States from Indiana to Pennsylvania south to Louisiana and Florida. This small tree is of minor commercial importance. However, its wood, which is fairly hard and strong, is occasionally used for small articles such as tool handles. Locally the flowers are an important source of honey. This species is used to some extent as an ornamental for both flowers and colorful autumn foliage as far north as Massachusetts. It was introduced into cultivation in 1747.

Seeding Habits.—The fruit is an ovoid-pyramidal, dry, dehiscent capsule one-fourth to one-third inch long, borne in profuse, panicled clusters. Its seeds are brown, reticulate, elongate, minute, and numerous (fig. 179). Sourwood flowers appear in copious masses from June to August, depending on location, and the fruit ripens in September or October. It is long persistent, and the seeds are dispersed gradually through the winter by dehiscence of the capsules. Seed of this species are produced in abundance nearly every year.

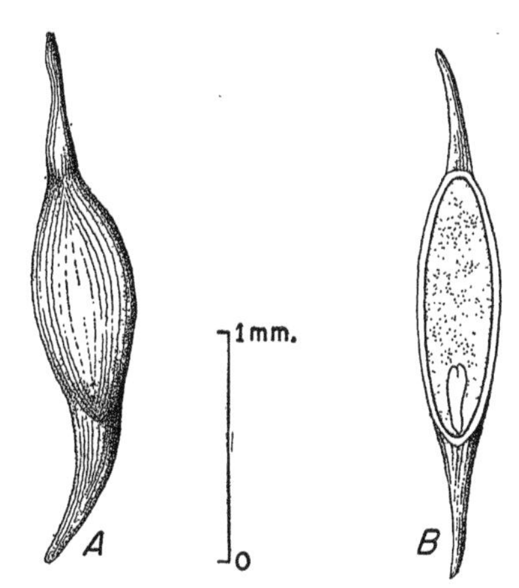

Figure 179.—Seed views of *Oxydendrum arboreum: A,* Exterior view; *B,* longitudinal section, showing small embryo.

Collection, Extraction, and Storage.—Collection of the fruits may be made from the trees during the fall and early winter. Some capsules containing seed can be found as late as February or March. The capsules should be dried if necessary and beaten or rubbed in a bag to break or completely open them. Their seed can then be shaken out and cleaned by screening and careful fanning. One lot treated in this manner showed 33 percent purity, with 96 percent of the seed sound, and ran about 5,500,000 cleaned seeds per pound. Dealers list the seed at $0.75 to $1.25 per ounce, and the dried capsules at $0.15 to $0.20 per ounce.

Germination.—Sourwood seeds germinate reasonably well without special after-ripening treatments. One test on seed sown 1 month after collection in September gave 13.8 percent germination in 33 days, and 16.2 percent in 63 days. In another instance, seed collected and immediately tested in February showed good germination (percent not determined); the first seedlings appeared in 11 days. Tests may be run in petri dishes or similar glass-covered receptacles by sprinkling the seed on moist pulverized peat or peat and sand and covering them very lightly with the same material. Germination is epigeous (fig. 180).

Nursery and Field Practice. — To propagate sourwood from seed it is recommended that the same methods be followed as described for rhododendron and mountain-laurel, viz, germination in acid sandy peat or sand and pulverized, decayed oak litter in covered flats in a greenhouse or coldframe; subsequent transplanting to seedling flats in the coldframe, followed by transplanting to nursery beds in the spring of the second year for additional growth before moving to permanent locations.

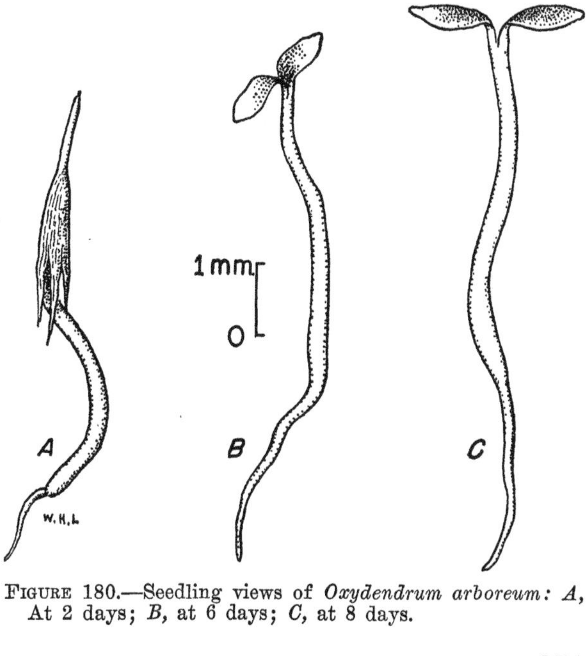

Figure 180.—Seedling views of *Oxydendrum arboreum: A,* At 2 days; *B,* at 6 days; *C,* at 8 days.

891183°—50——17

PARTHENOCISSUS Planch. Creeper, five-leaf ivy

(Grape family—Vitaceae)

DISTRIBUTION AND USE.—The creepers consist of about 10 species of deciduous, rarely evergreen, woody, climbing and trailing vines with tendrils, and are distributed in North America, eastern Asia, and the Himalaya region. They are of value chiefly for ornamental purposes, because of their handsome foliage, which turns scarlet to crimson in the fall, or attractive fruits. They are also useful as game food and cover plants and to some extent for erosion control. The two species described in table 136 are occasionally planted in the United States.

SEEDING HABITS.—The flowers in *Parthenocissus* are small, greenish in color, usually perfect, sometimes polygamo-dioecious, and are borne in the early summer in rather inconspicuous long-stemmed clusters. Its fruit, a dark blue to bluish-black berry containing one to four seeds, ripens in the autumn. The seed (fig. 181) has a small embryo and a large endosperm. Details of seeding habits are as follows:

Time of—	*P. inserta*	*P. quinquefolia*
Flowering	June–July	July–Aug.
Fruit ripening	July–early Sept.	Sept.–Oct.
Seed dispersal	Late Aug.–late autumn	Sept.–midwinter.

No data are available on the age for commercial seed bearing or on the regularity of seed crops although it is likely that good crops are borne at frequent intervals. Dispersal is effected largely by birds and mammals, and to some extent by gravity. Information as to the existence of geographic strains in the genus is not available, but owing to the wide geographic range of some species, particularly those described here, and also to the recognition of botanical varieties, it is quite likely that such strains do occur.

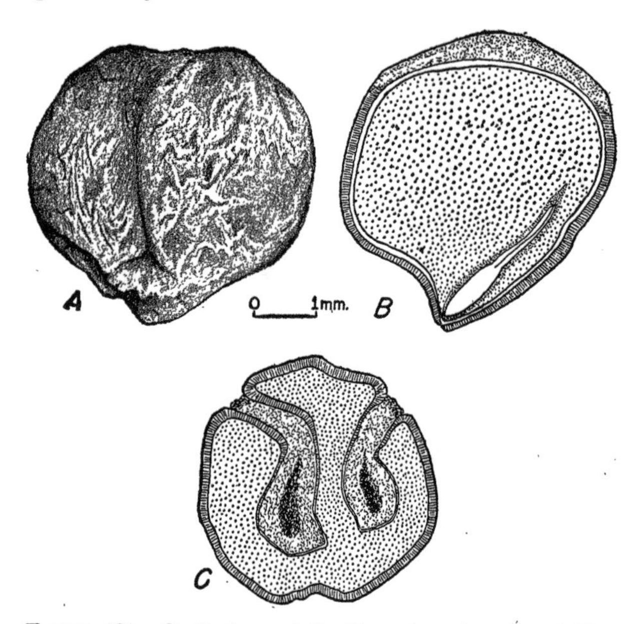

FIGURE 181.—Seed views of *Parthenocissus inserta*: *A*, Exterior view; *B*, longitudinal section; *C*, cross section.

COLLECTION, EXTRACTION, AND STORAGE.—*Parthenocissus* berries can be collected by stripping the clusters from the vines as soon as they have turned blue black in color. After collection, leaves and

TABLE 136.—*Parthenocissus: Growth habit, distribution, and uses*

Accepted name	Synonyms	Growth habit	Natural range	Chief uses	Date of earliest cultivation
P. inserta (Kern.) K. Fritsch[1] (thicket creeper).	*P. vitacea* (Knerr) Hitchc., *Psedera vitacea* (Knerr) Greene.	Low, rambling vine.	Maine to Alberta, south to Pennsylvania, Texas, and Arizona.	Game food and cover, erosion control, ornamental purposes.[2]	(3)
P. quinquefolia (L.) Planch.[1] (Virginia creeper).	*Ampelopsis quinquefolia* (L.) Michx. *Psedera quinquefolia* (L.) Greene.	Climbing vine.	New England west to Minnesota, south to Florida, Texas, and Mexico.	Ornamental purposes, game food and cover.	1622

[1]Includes several varieties.
[2]Does not cling to walls, hence suitable only for covering arbors, or as bushes.
[3]Before 1800.

other debris should be removed by screening or fanning. The seed may be extracted by running the fruit with water through a macerator or a hammer mill, allowing the pulp and empty seeds to float away. Extraction should be carried on rather carefully because the seed coats are often soft and easily injured. After a short period of drying, the seed is ready for sowing or for storage. Commercial seed may consist of dried berries or cleaned seed. Data on yield, seed size, soundness, and cost for the two species discussed here follow:

	P. inserta	*P. quinquefolia*
Cleaned seed:		
Per 100 pounds of fruit...pounds..	----	25
Per pound:		
Low.....number..	14,100	12,000
Average.....do....	18,800	------
High.....do....	23,300	[1]26,200
Basis, samples.....do....	3	2
Commercial seed:		
Average soundness.....percent..	92	92
Cost per pound.....dollars..	([2])	[3]1.50

[1]Moisture content 8.15 percent.
[2]Apparently not sold in commerce.
[3]Dried berries, 85 cents per pound.

No data on longevity or optimum storage conditions are available for this genus. However, *Vitis riparia* seeds have been stored in a sealed container at 41° F. for over 2 years without loss in germinability. It seems likely therefore that *Parthenocissus* seeds, which are very close to *Vitis* seeds morphologically, may also be kept under these conditions.

GERMINATION.—Natural germination is believed to take place during the spring following dispersal and is of the epigeous type (fig. 182). Most seed lots show delayed germination due to a dormant embryo; this can be readily overcome by stratification in moist sand or peat for about 60 days at 41° F. Germination tests are most feasibly made in flats and should be based on 200 to 400 properly pretreated seeds. Such information as is available on recommended procedures and the results to be expected is as follows:

	P. inserta	*P. quinquefolia*
Test conditions:		
Night temperature.....°F...	68	68
Day temperature.....do..	86	86
Duration:		
Stratified seed.....days..	30	30–40
Untreated seed.....do..	90+	150+
Germinative energy:		
Amount.....percent..	72	90
Period.....days..	16	15
Germinative capacity:		
Low.....percent..	----	45
Average.....do....	78	69
High.....do....	----	94
Basis, tests.....number..	1	3

NURSERY AND FIELD PRACTICE. — *Parthenocissus* seeds may be fall sown or stratified prior to spring sowing. The seed should be sown in drills and covered with about three-eighths of an inch of nursery soil. Propagation can also be effected by hardwood cuttings or layering, and in some species by green-wood cuttings.

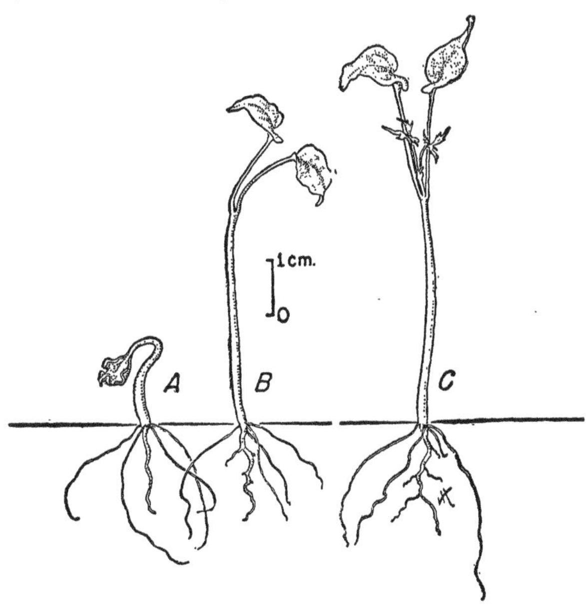

FIGURE 182.—Seedling views of *Parthenocissus quinquefolia*: *A*, At 1 day; *B*, at 3 days; *C*, at 22 days.

PHOTINIA ARBUTIFOLIA Lindl. Christmasberry

(Rose family—Rosaceae)

Also called toyon, tollon. Botanical syns.: *Photinia salicifolia* Presl, *Heteromeles arbutifolia* (Lindl.) Roem.

Distribution and Use.—Native to the Sierra Nevada and the Coast Range of California, this large shrub is extensively cultivated for its evergreen foliage and red berries. One variety, *cerina,* has yellow berries. Christmasberry can be used for erosion control on dry slopes, and it is a useful honey plant. The berries of this shrub provide food for birds and sometimes for humans, and its foliage is browsed by deer.

Seeding Habits.—Seed of this species are small, rigid, brown, and dotted. One or two are contained in each of the two cells of the fruit (fig. 183). The perfect, white flowers of Christmasberry occur in clusters and bloom from June to July. Its berries ripen from October to December, and abundant crops are produced nearly every year.

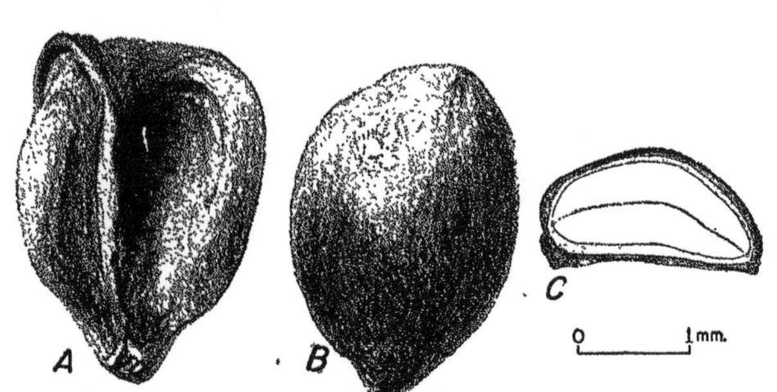

Figure 183.—Seed views of *Photinia arbutifolia: A* and *B,* Exterior views from two planes; *C,* cross section.

Collection, Extraction, and Storage.—Clusters of berries may be harvested with pruning shears from October to December. They should be soaked in water and the seeds carefully separated by hand; large lots probably could be run through a macerator. Prolonged soaking, which causes fermentation of the berries, is harmful to the seed. Cleaned seed per pound (1 sample): 24,000. Cost of commercial seed: $1.25 to $1.50 a pound. Storage at low temperatures in airtight containers is recommended.

Germination.—Sand or soil flats are recommended for greenhouse germination. Fresh seed germinates well without any pretreatment, and germination is epigeous (fig. 184). Stored seed develops dormancy and should be stratified for 3 months. Germination begins about 10 days after sowing; the highest germination obtained was 73 percent (1 sample).

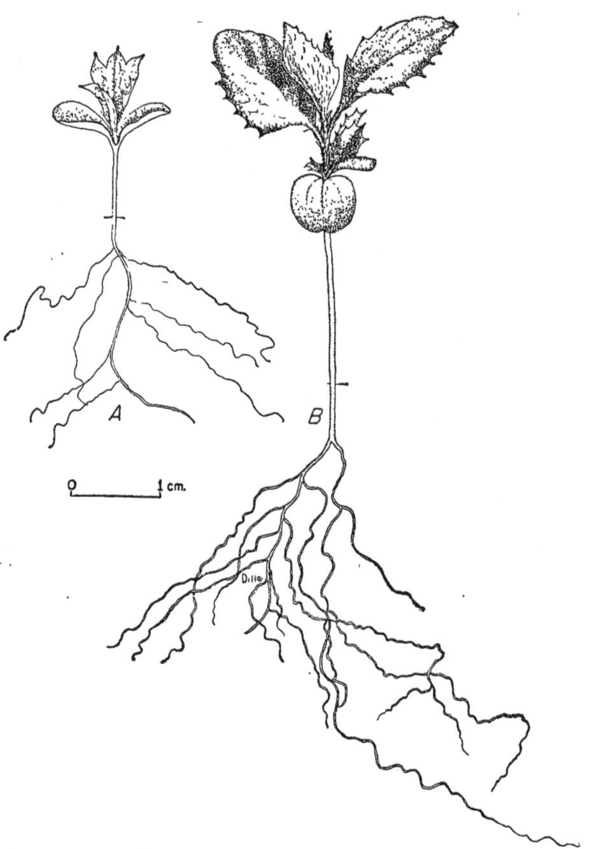

Figure 184.—Seedling views of *Photinia arbutifolia: A,* Young seedling; *B,* older seedling.

Nursery and Field Practice.—Untreated seed may be sown in the fall or stratified seed used in the spring. Christmasberry will grow on a variety of soils, but does best on a rather light, sandy loam. Propagation by cuttings and grafting is also practiced.

PICEA A. Dietr. Spruce

(Pine family—Pinaceae)

Distribution and Use.—The spruces, consisting of some 40 species, are medium to tall evergreen forest trees distributed over the cooler parts of the Northern Hemisphere. Eight species are native to North America, and 8 species are used there to some extent in reforestation. All spruces are valuable, or potentially so, for the production of lumber and pulpwood, and nearly all are useful for ornamental purposes. Nine species and one variety, which are of known or supposed value for reforestation in the United States and for which reliable information is available, are described in table 137.

In the United States the spruces which are used most in reforestation are, in descending order of importance, as follows: *Picea glauca, P. rubens, P. engelmanni, P. abies, P. sitchensis,* and *P. glauca* var. *albertiana. P. abies* is the most widely planted spruce in Europe. While most of these species have been planted within their native range, some have also been planted extensively elsewhere: *P. abies* has been widely planted in the northeastern quarter of the United States and adjacent Canada; *P. glauca* and *P. sitchensis* are planted quite commonly in parts of northern Europe; *P. glauca* var. *albertiana* is planted on the Plains and farther

Table 137.—*Picea: Distribution and uses*

Accepted name	Synonyms	Natural range	Chief uses	Date of earliest cultivation
P. abies (L.) Karst. (Norway spruce).	*P. excelsa* Link (European spruce).	Northern and central Europe	Lumber, pulpwood, ornamental, windbreaks, Christmas trees.	1548
P. breweriana S. Wats. (Brewer spruce).	weeping spruce	Mountains of northern California and southern Oregon.	Watershed protection, ornamental.	1893
P. engelmanni Parry (Engelmann spruce).		Mountains of British Columbia, the Rocky Mountains, and the Cascades.	Lumber, poles, ties, interior trim, watershed protection, wildlife food and cover.	1862
P. glauca (Moench) Voss (white spruce).	*P. canadensis* (L.) B. S. P. not Link; *P. alba* Link.	Northern Lake States to New York, northward to Labrador and west to Alaska.	Lumber, pulpwood, ornamental, windbreaks.	1700
P. glauca var. *albertiana* (S. Brown) Sarg. (western white spruce).	*P. albertiana* S. Brown (Black Hills spruce, Alberta white spruce, SPN).	Alaska and British Columbia, south to Montana and the Black Hills).	Shelter belts, ornamental, lumber.	1904
P. mariana (Mill.) B. S. P. (black spruce).	*P. nigra* (Ait.) Link	Alaska to Labrador, south to Northeastern and Lake States and into mountains of northern Virginia.	Pulpwood, Christmas trees, twigs used for beverage, spruce gum.	1700
P. pungens Engelm. (blue spruce).	*P. parryana* Sarg. (Colorado spruce).	Rocky Mountains	Ornamental, windbreaks, posts, poles, fuel, watershed protection, wildlife food and cover.	1862
P. rubens Sarg. (red spruce).	*P. rubra* (Du Roi) Link, not A. Dietr.	Prince Edward Island and St. Lawrence Valley south through Northeastern States into southern Appalachians at higher elevations.	Pulpwood, lumber, ornamental, Christmas trees, twigs used for beverage.	(1)
P. sitchensis (Bong.) Carr. (Sitka spruce).		Pacific coast from Alaska to northern California.	Lumber, pulpwood, airplane timber, ornamental.	1831
P. smithiana Boiss. (Himalayan spruce).	*P. morinda* Link, *P. khutrow* Mast.	Cool, temperate zone of western Himalayas.	Matchwood, pulpwood, lumber, shingles, ornamental.	1818

[1]Before 1750.

eastward; *P. pungens* is extensively cultivated both in the United States and in Europe chiefly as an ornamental; *P. rubens* and *P. smithiana* are planted somewhat in Europe, chiefly as ornamentals.

SEEDING HABITS.—The male and female flowers are borne separately on the same tree, on twigs of the previous years growth. Bearing pollen only, the male flowers are drooping, yellow, bright purple, or rose red with short to long cylindrical bodies. The female flowers, which develop into cones, are erect, yellowish green or bright red, somewhat cylindrical bodies from one-fourth to three-fourths inch in diameter. Spruce cones mature in the fall of the first season, are cylindrical or egg-shaped, and hang drooping or bent downward. The persistent cone scales are thin and without prickles, each bearing two seeds on the under side. When ripe, the cones open and the seeds are dispersed by wind. The seeds (figs. 185 and 186) are oblong, acute at the base, much shorter than their wings, and are brown to black when ripe. Comparative seeding habits of nine species of *Picea* are given in table 138.

For most of the spruces there is little scientific information available as to the development of geographic strains; it is likely, however, that most species of extensive range have developed such strains. A summary of available information for four species is as follows:

Picea abies.—Norway spruce seed vary in size and also in weight from about 4 to 13 grams per 1,000 seed according to the locality of origin. Seed from northern Europe is lighter than that from central Europe; seed from high elevations is considerably smaller than that from the lowlands. It is reported but not proved conclusively that in trees growing at high altitudes (or those whose buds open early), the young cones are reddish in color, while in trees from low altitudes (or those whose buds open late), the young cones are yellowish green. Trees of different origins vary considerably in rate of growth, phenology, and form of crowns. In Sweden 5 races are recognized on the basis of branching habit; the "comb-type" is preferred in forest management.

Picea glauca.—Differences in rate of growth and frost resistance associated with origin have occurred in Lake States nurseries—northern sources slower growing, more frost hardy.

Picea rubens.—Seed from southern part of range heavier than from northern collections.

Picea sitchensis.—Canadian strains more frost resistant than those from California and Oregon; Alaskan sources proved hardier in Norway than Washington sources.

COLLECTION, EXTRACTION, AND STORAGE.—Spruce cones should be collected in the fall beginning just before the cone scales start to open; they may be picked by hand from standing trees, gathered from felled trees or slash, or collected from squirrel caches. Cones are ripe enough to collect when the seed coats darken and the endosperm becomes firm. Usually they turn brownish at this time, but in *Picea mariana* they become purple. Freshly collected cones should be spread out in thin layers to dry in the sun or in well-ventilated cone sheds to prevent heating and molding. They may be opened

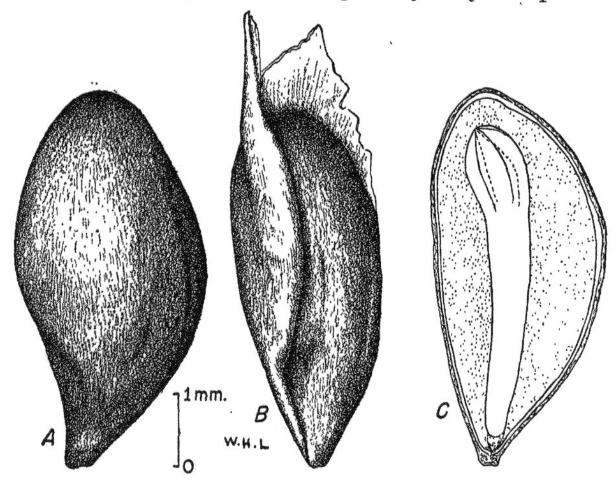

FIGURE 185.—Seed of *Picea breweriana*: *A* and *B*, Enlarged views of two seed surfaces, one with wing attachments; *C*, longitudinal section of seed.

TABLE 138.—*Picea: Time of flowering and cone ripening, and frequency of seed crops*

Species	Time of—			Commercial seed-bearing age			Seed year frequency	
	Flowering	Cone ripening	Seed dispersal	Minimum	Optimum	Maximum	Good crops	Light crops
				Years	*Years*	*Years*	*Years*	*Years*
P. abies	May–June	Sept.–Nov.	Sept.–April	30–50	100		4–8	Intervening.
P. breweriana			Sept.–Oct.					
P. engelmanni	June–July	Aug.–Sept.	do.	16–25	150–200		2–3	Intervening.
P. glauca	May	do.	Aug.–Nov.	30	60+		2–6	do.
P. mariana	May–June	Sept.	Oct.[1]	30–40	50–150	250	4–5	do.
P. pungens	April–May	Fall	Fall and winter.				1	
P. rubens	do.	Sept.	Sept.	30–40	50+		3–8	Intervening.
P. sitchensis	do.	Early fall	Early fall	35			3–4	do.
P. smithiana	do.	Oct.–Nov.	Oct.–Nov.	20			5±	do.

[1] *P. mariana* appears to be unique among spruces in retaining its cones for 2 to 3 years in a state of active seed dispersal.

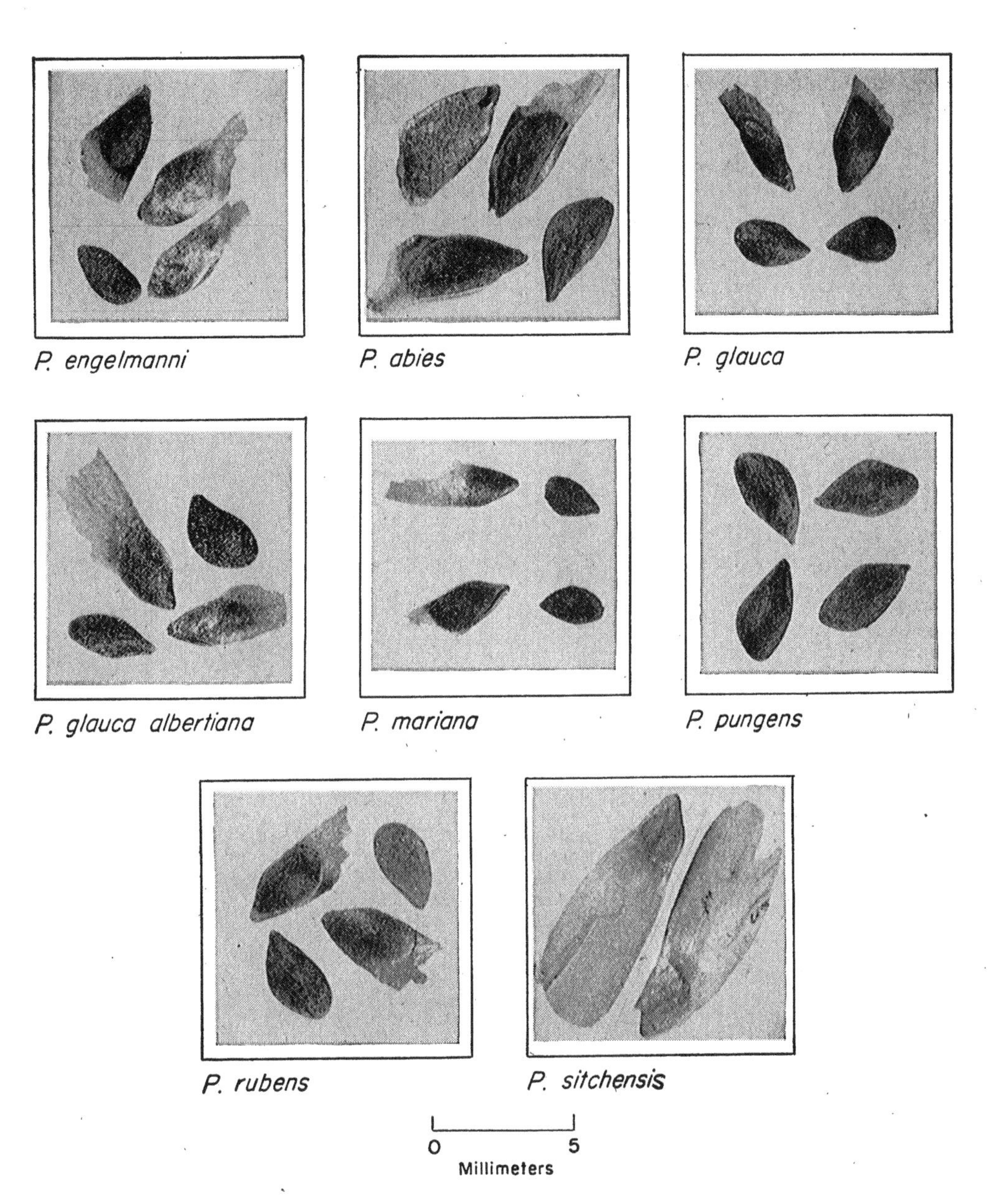

FIGURE 186.—Seed of several spruces.

by further exposure to the sun, or in cone kilns. Ordinarily extraction is not difficult, but in *P. mariana* a kiln is necessary and the cones may have to be soaked in water before they will open fully. After opening, a shaking will remove the seed; the seed should then be dewinged either by hand or machine and cleaned by a blower or fan. Detailed data on experience in collection and extraction practice are available only for three spruces (table 139), but similar procedures would probably be suitable for other species. The size, purity, and soundness of commercial cleaned seed vary considerably by species as shown in table 140.

Seed of *Picea glauca* and *P. rubens* have kept their viability well for 10 years when stored dry in sealed containers at temperatures just above freezing—36° to 40° F. in one case and 40° to 50° in the other. *P. abies* and *P. mariana* seed have retained their viability satisfactorily for 5 years in sealed containers kept in cool cellars. *P. abies* seed also has retained satisfactory viability for 5 years when stored in the cones in a dry loft. *P. engelmanni* seed in sealed containers stored in cool cellars retained its viability unimpaired for 3 years and lost about 20 percent of it at the end of 5 years. It is probable that spruce seed kept under proper conditions will retain its viability unimpaired for 10 or more years.

GERMINATION.—Natural germination of spruce seed usually takes place the spring or summer following dispersal. Sometimes seed may lie over and germinate the second year. Germination is best on mineral soil with side shade and overhead light but may occur on most any moist surface. Germination is epigeous (fig. 187). The natural seed supply is reduced considerably by insects, birds, and rodents. For instance, the spruce seed moth (*Laspeyresia youngana* Kerf.) larvae attack the cones of *Picea glauca, P. sitchensis,* and *P. pungens;* the spruce seed midge (*Dasyneura canadensis*

TABLE 139.—*Picea: Time of collection, cone yield, and extraction method*

Species	Time for collection	Yield per tree	Cones per bushel	Cost per bushel of cones	Kind of kiln	Kiln schedule		
						Duration	Temperature	Relative humidity
		Bushels	*Number*	*Dollars*		*Hours*	*°F.*	*Percent*
P. abies	October–March	----------	150– 400	0.50	Simple convection	6–24	105–120	--------
P. glauca	August–September	1.0 –2.5	6,500– 8,000	2.00	do	4– 6	130	38
P. mariana	September–Dec. 1	.05– .40	7,000–16,000	--------	do	6– 8	130–145	--------

TABLE 140.—*Picea: Yield of cleaned seed, and purity, soundness, and cost of commercial seed*

Species	Cleaned seed					Commercial seed		
	Yield per bushel of cones	Per pound			Basis, samples	Purity	Soundness	Cost per pound
		Low	Average	High				
	Ounces	*Number*	*Number*	*Number*	*Number*	*Percent*	*Percent*	*Dollars*
P. abies	8–12	47,000	64,000	99,000	200+	95	96	1.00– 2.25
P. breweriana	----------	51,000	61,000	74,000	8	91	90	18.00–22.00
P. engelmanni	6–16	69,000	135,000	200,000	22+	90	67	2.50– 3.00
P. glauca	6–20	142,000	240,000	398,000	75	83	80	2.75– 3.50
P. glauca var. *albertiana*	[1]24–32	173,000	188,000	196,000	6	88	87	3.75– 6.00
P. mariana	2– 5	335,000	404,000	510,000	49	81	90	4.50– 6.00
P. pungens	12–20	80,000	106,000	163,000	48+	88	93	2.50– 3.25
P. rubens	17–24	100,000	140,000	289,000	22	85	85	1.50– 4.00
P. sitchensis	8–20	155,000	210,000	400,000	80+	92	50±	4.00– 4.50
P. smithiana	----------	24,000	34,000	38,000	160	73	----------	7.50–20.00

[1]Yield per 100 pounds of cones.

TABLE 141.—*Picea: Dormancy and method of seed pretreatment*

Species	Seed dormancy		Stratification			Remarks
	Kind	Occurrence	Medium	Temperature	Duration	
				°F.	*Days*	
P. abies	Embryo	Occasionally	----------	----------	----------	Pretreatment not required.
P. breweriana	do	Probably variable.	Moist sand	41	30–90	Fair germination without pretreatment.
P. engelmanni	do	Occasionally	----------	----------	----------	Pretreatment not required.
P. glauca	do	General	Moist sand	41	60–90	Pretreatment needed.
P. glauca var. *albertiana.*	None	----------	----------	----------	----------	Pretreatment not required.
P. mariana	Embryo	General	Moist sand	41	30–60	Fair germination without pretreatment.
P. pungens	do	Variable	do	32–41	30–90	About 50 percent of seeds are dormant.
P. rubens	do	General	do	41	30–45	Fair germination without pretreatment.
P. sitchensis	do	Variable	do	41–50	60–90	Pretreatment needed only on some lots.
P. smithiana	do	Probable	do	41	30–60	Treatment suggested.

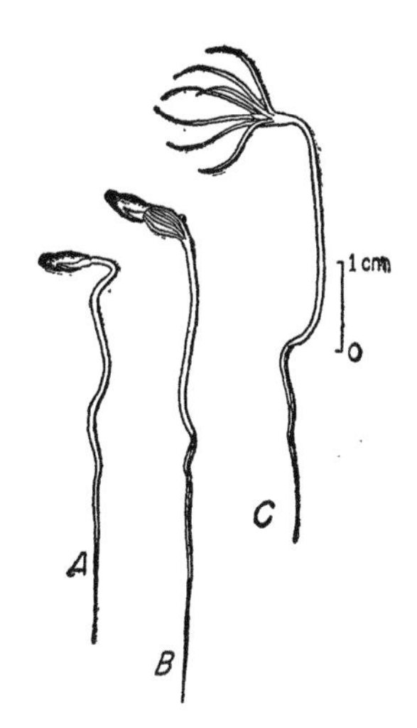

FIGURE 187.—Seedling views of *Picea pungens: A,* At 2 days; *B,* at 5 days; *C,* at 7 days.

Felt) destroy the seed of *P. glauca* without deforming the cones; birds, chiefly nutcrackers, and flying squirrels devour the cones of *P. smithiana.* As a general rule most spruces have dormant seed, but in several cases it is known that different lots of seed of the same species may or may not be dormant. Only 1 of the 10 spruces discussed here appears to have consistently nondormant seed, and this finding may be changed after an adequate number of tests have been made. Dormancy may be broken by stratification at low temperatures. In all cases tested, stratification has improved the rate of germination. Further information on seed dormancy and pretreatment is given in table 141.

The germination of spruce seed should be tested in sand flats or standard germinators, using 1,000 seeds per test, either stratified or untreated as indicated in table 141. Test conditions recommended and the average results that may be expected are given in table 142.

NURSERY AND FIELD PRACTICE.—Spruce seed of dormant lots should be sown in the fall, or if the seed are nondormant or stratified they may be sown in the spring. Broadcast sowing or sowing in drills so as to produce about 75 usable seedlings per square foot is the recommended practice. The seed should be covered with about one-fourth inch (one-eighth inch for *Picea mariana*) of nursery soil. A well-drained sandy loam makes the best seedbed. Fall-sown beds should be mulched over winter. Both fall- and spring-sown beds should be protected against birds and rodents by covering with 2-mesh wire screen. Seedbed soil should be treated with sulfuric acid or aluminum sulfate to prevent damping-off, and until August of the first year should be given half shade. Germination begins 15 to 30 days after spring sowing. The tree percent is apt to be about 50. Transplant stock—usually 2–1 or 2–2—is preferred for field planting.

Spruce seedlings often are attacked by white grubs or by weevils (*Hylobius pales* Herbst.). *Picea mariana* and *P. pungens,* and to a less extent *P. glauca,* are sometimes attacked in the nursery by needle rust fungi of the genus *Chrysomyxa.* Control is by eradication of alternate hosts—members of the heath family—or by spraying with 4–4–50 bordeaux just as the needles attain full length. Tips of *P. rubens* sometimes are killed by fungi of the genus *Rhizoctonia;* control is by decreasing moisture or spraying with Semesan solution or 4–4–50 bordeaux.

TABLE 142.—*Picea: Recommended conditions for germination tests, and summary of germination data*

Species	Test conditions recommended				Germination data from various sources					
	Temperature		Duration		Germinative energy		Germinative capacity			Basis, tests
	Night	Day	Untreated seed	Stratified seed	Amount	Period	Low	Average	High	
	°F.	*°F.*	*Days*	*Days*	*Percent*	*Days*	*Percent*	*Percent*	*Percent*	*Number*
P. abies	68	86	[1] 20–30	---	18–95	10–22	0	70	99	3,888
P. breweriana	---	---	100	30	---	---	11	54	73	7+
P. engelmanni	70	80	[1]50	---	58–76	10–20	42	69	97	45+
P. glauca	68	86	55–60	[1]45	37	20–30	7	49	91	140
P. glauca var. *albertiana*	68	86	---	---	43–65	16–25	44	54	70	5
P. mariana	68	86	[1]50	---	65	33	14	61	89	49
P. pungens	68	86	60	[1]20	13–75	10–29	14	73	97	18+
P. rubens	68	86	[1]50	30	57	---	32	60	94	22
P. sitchensis	75	75	[1]60	60	---	---	1	60	96	305+
P. smithiana	---	---	[1]30	30	40	20	20	41	66	154+

[1]Indicates the condition reported on for germinative energy and germinative capacity.

PINUS L. Pine

(Pine family—Pinaceae)

DISTRIBUTION AND USE.—The pines include about 80 species of evergreen trees, or rarely shrubs, distributed throughout the Northern Hemisphere from the Arctic Circle to the West Indies, Central America, northern Africa, and Malaysia. They include some of the most valuable timber trees in the world. Some species are useful in the protection of steep slopes from erosion; a few provide the great bulk of naval stores; some are used for shelter belts; many of them provide cover and food for birds and other animals, and sometimes for man; and some are used for ornamental planting. More than 30 species are native to North America (north of Mexico) and these include the species most widely used there in reforestation. In addition, many exotic pines have been planted in North America. Forty-four species and four additional varieties, which are of known or supposed value for reforestation in the United States and for which reliable information is available, are described in table 143.

TABLE 143.—*Pinus: Growth habit, distribution, and uses*

Accepted name	Synonyms	Growth habit	Natural range	Chief uses	Date of earliest cultivation
P. albicaulis Engelm. (whitebark pine).	----------	Small to medium tree.	High mountains from British Columbia, south to California and Wyoming.	Possible erosion control, wildlife food.	1852
P. aristata Engelm. (bristlecone pine).	*P. balfouriana* var. *aristata* (Engelm.) Engelm. (hickory pine).	Shrub to low, bushy tree.	High elevations of Rocky Mountains from Colorado to Arizona and to desert regions of California and Nevada. Introduced into Maine and England.	Possible erosion control, ornamental.	1861
P. attenuata Lemm. (knobcone pine).	*P. tuberculata* Gord., not D. Don.	Small to medium tree.	Barren, rocky mountain slopes of southern Oregon and California.	Windbreaks, erosion control.	1847
P. banksiana Lamb.[1] (jack pine).	*P. divaricata* (Ait.) Dum.-Cours. (scrub pine, gray pine, Banksian pine).	-----do-----	Mackenzie River and Rocky Mountains to Nova Scotia and Maine, southward to northern New York, southern tip of Lake Michigan and central Minnesota.	Pulpwood, mine timber, lumber, windbreaks.	Before 1783.
P. canariensis C. Smith (Canary pine).	Canary Island pine--	Medium-sized tree.	Dry exposed slopes on Canary Islands.	Locally for lumber, foliage for packing bananas.	-------
P. caribaea Morelet[2] (slash pine).	*P. bahamensis* Griseb. (Cuban pine).	Medium to large tree.	South Carolina, Georgia, Florida to eastern Louisiana, Cuba, and Central America.	Lumber, naval stores, pulpwood, erosion control, wildlife food.	-------
P. cembra L. (Swiss stone pine).	cembran pine, arolla pine, Siberian cedar.	Small to tall tree.	Alps and Carpathian Mountains of central Europe, northeastern Russia, and Siberia.	Edible seed, cabinet wood, possible erosion control.	(3)
P. clausa (Engelm.) Vasey (sand pine).	Florida spruce pine--	Low, spreading tree.	Florida and sandy coasts of Alabama.	Locally for pulpwood, lumber; some wildlife food.	1832
P. contorta Dougl. (shore pine).	beach pine---------	Low, stunted tree.	Pacific coast from Alaska to California.	Locally for minor timber products.	1855

Footnotes on page 263.

TABLE 143.—*Pinus: Growth habit, distribution, and uses*—Continued

Accepted name	Synonyms	Growth habit	Natural range	Chief uses	Date of earliest cultivation
P. contorta var. *latifolia* Engelm. (lodgepole pine).	*P. murrayana* Grev. & Balf.	Small to large tree.	Rocky Mountains and Sierras from Colorado and Lower California north to Yukon.	Pulpwood, ties, lumber, mine timbers, erosion control.	--------
P. coulteri D. Don (Coulter pine).	nut pine, bigcone pine.	Small to medium tree.	California and Lower California.	Erosion control, ornamental.	1832
P. densiflora Sieb. & Zucc. (Japanese red pine).	--------------------	Medium-sized tree.	Medium elevations on mountains in Japan and Korea.	Lumber----------------	1854
P. echinata Mill. (shortleaf pine).	*P. mitis* Michx.------	-----do-------	New York to Florida, west to Illinois and Texas.	Lumber, pulpwood, erosion control, wildlife food.	1726
P. edulis Engelm. (pinyon).	*P. cembroides* var. *edulis* (Engelm.) Voss (Colorado pinyon pine, SPN).	Small tree----	Colorado south to Mexico and west to Nevada and Lower California.	Edible nuts, locally for secondary forest products.	1848
P. flexilis James (limber pine).	--------------------	Short to medium tree.	Rocky Mountains from Canada to Arizona and New Mexico, and west to California.	Locally for rough lumber, possible erosion control.	1851
P. flexilis var. *reflexa* Engelm. (border limber pine).	*P. reflexa* (Engelm.) Engelm., *P. strobiformis* auth., not Engelm.	Medium-sized tree.	New Mexico, Arizona, and adjacent Mexico.	Locally for lumber-------	1840
P. glabra Walt., (spruce pine).	cedar pine---------	-----do-------	Coast region from South Carolina to central and northwestern Florida and west to southeastern Louisiana.	Locally for fuel and rarely for lumber, wildlife food.	--------
P. griffithii McClelland (Himalayan pine).	*P. excelsa* Wall., not Lam., *P. nepalensis* DeChambr., not Forbes (Bhotan pine, blue pine).	Large tree----	Moist temperate portions of the Himalayas and Afghanistan.	Lumber, naval stores----	1827
P. halepensis Mill. (Aleppo pine).	*P. alepensis* Poir. (Jerusalem pine), *P. maritima* Lamb. (1803), not R. Br. in Ait. (1813).	Small to medium tree.	Mediterranean Region of Europe and Asia Minor.	Locally for lumber and naval stores, erosion control and windbreaks in hot, dry regions.	1683
P. heldreichii var. *leucodermis* (Ant.) Markgraf (palebark Heldreich pine).	*P. leucodermis* Ant., *P. nigra* var. *leucodermis* (Ant.) Rand. (Balkan pine, Bosnian pine, graybark pine).	-----do-------	Dry limestone formations at high elevations in the Balkans.	Possible erosion control on high, dry limestone areas; ornamental.	1864
P. jeffreyi Grev. & Balf. (Jeffrey pine).	*P. ponderosa* var. *jeffreyi* (Grev. & Balf.) Engelm.	Large tree----	Mountains of southern Oregon, California, and Lower California.	Lumber, production of heptane.	1853
P. lambertiana Dougl. (sugar pine).	big pine------------	Largest pine--	Western Oregon, California, and Lower California.	Lumber----------------	1827
P. latifolia Sarg. (Apache pine).	*P. apacheca* Lemm., *P. mayriana* Sudw. (Arizona longleaf pine).	Medium-sized tree.	Southern Airzona, New Mexico, and adjacent Mexico.	Locally for lumber, possibly erosion control.	--------
P. leiophylla var. *chihuahuana* (Engelm.) Shaw (Chihuahua pine).	*P. chihuahuana* Engelm. (smooth-leaved pine).	-----do-------	Mountains of southern Arizona, New Mexico, and northern provinces of Mexico.	Fuel------------------	--------
P. monophylla Torr. & Frém. (singleleaf pinyon).	*P. cembroides* var. *monophylla* (Torr. & Frém.) Voss (singleleaf pinyon pine, SPN).	Small tree----	Great Basin region to southern California and Lower California.	Edible nuts, minor timber products.	1848
P. monticola Dougl. (western white pine).	Idaho white pine, mountain white pine.	Medium to large tree.	Southern British Columbia to northern Montana and south to California.	Lumber, interior trim, etc.	1851

Footnotes on page 263.

TABLE 143.—*Pinus: Growth habit, distribution, and uses*—Continued

Accepted name	Synonyms	Growth habit	Natural range	Chief uses	Date of earliest cultivation
P. mugo Turra (Swiss mountain pine).	*P. montana* Mill. (Mugho pine).	Low shrub to medium tree.	Mountains of central and southern Europe.	Ornamental, erosion control, sand dune stabilization.	1779
P. muricata D. Don (bishop pine).	prickle-cone pine	Small to medium tree.	Coastal region of California and Lower California.	Locally for lumber, windbreaks.	1846
P. nigra Arnold (Austrian pine).	(4)	Medium to large tree.	Central and southern Europe.	Lumber, naval stores, ornamental, windbreaks, sand dune control, erosion control.	1759
P. palustris Mill. (longleaf pine).	*P. australis* Michx. f. (southern pine).	Medium-sized tree.	Southeastern United States.	Lumber, naval stores, poles and piling, wildlife food, decorations.	1727
P. peuce Griseb. (Balkan pine).	*P. excelsa* var. *peuce* (Griseb.) Beissn. (Rumelian pine, Macedonian pine).	do	Mountains of western Bulgaria, eastern Albania, and southern Yugoslavia.	Locally for lumber, ornamental.	1863
P. pinaster Ait. (cluster pine).	*P. maritima* Lam. (1778), not Mill. (1768) nor R. Br. (1813) (maritime pine).	do	Coastal portions of Mediterranean Region extending to Atlantic Ocean in France and Portugal.	Naval stores, rough lumber, poles, posts, mine timbers, ties, sand dune reclamation.	Before 1660.
P. ponderosa Laws. (ponderosa pine).	western yellow pine.	Small to large tree.	British Columbia south to Mexico and east to Nebraska, Colorado, and western Texas.	Lumber, interior trim, boxboards, ties, mine timbers, fuel, shelter belts.	1827
P. ponderosa var. *arizonica* (Engelm.) Shaw (Arizona pine).	*P. arizonica* Engelm. (Arizona ponderosa pine, SPN).	Medium-sized tree.	Mountains of southern Arizona, New Mexico, and adjacent Mexico.	Locally for lumber, possibly erosion control.	--------
P. pungens Lamb. (Table-Mountain pine).	mountain pine, hickory pine, poverty pine.	Small tree	Appalachian region from New Jersey to Georgia and Tennessee.	Fuel, possibly erosion control.	1804
P. quadrifolia Parl. (Parry pinyon).	*P. parryana* Engelm., *P. cembroides* var. *parryana* (Engelm.) Voss (Parry pinyon pine, SPN).	do	California and Lower California.	Edible nuts	1885
P. radiata D. Don (Monterey pine).	*P. insignis* Dougl. (insular pine).	Medium to large tree.	California coast	Rough lumber	1833
P. resinosa Ait.[5] (red pine).	Norway pine	do	Nova Scotia to Manitoba, south to Pennsylvania, Michigan, and Minnesota.	Lumber, poles and piling, mine props, ties, and pulpwood.	1756
P. rigida Mill. (pitch pine).	hard pine	Small to medium tree.	New Brunswick to Georgia, west to Ontario, and Tennessee.	Rough lumber, ties, fuel, wildlife food.	About 1743.
P. rigida var. *serotina* (Michx.) Loud. (pond pine).	*P. serotina* Michx. (marsh pine).	do	Low, wet flats and swamps from New Jersey to northern Florida and central Alabama.	Occasionally for lumber	1713
P. roxburghii Sarg. (chir pine).	*P. longifolia* Roxb., not Salisb. (longleaf Indian pine).	Large tree	Outer ranges and principal valleys of the Himalayas.	Naval stores, lumber, boxboards, ties, ornamental.	--------
P. sabiniana Dougl. (Digger pine).	Sabines pine	Small tree	Dry foothills of California	Shelter belts, erosion control, heptane, food for Indians.	1832
P. strobus L. (eastern white pine).	*Strobus weymouthiana* Opiz (white pine, Weymouth pine).	Medium to large tree.	Newfoundland to Manitoba, south to Georgia, Illinois, and Iowa.	Lumber, interior trim, sash and door, pattern making, boxboards, matchwood, boat spars.	1705
P. sylvestris L. (Scotch pine).	Scots pine, yellow deal.	do	Europe to western and northern Asia.	Lumber, interior trim, flooring, posts and poles, paving block, ties, naval stores.	(3)

Footnotes on page 263.

TABLE 143.—*Pinus: Growth habit, distribution, and uses*—Continued

Accepted name	Synonyms	Growth habit	Natural range	Chief uses	Date of earliest cultivation
P. taeda L. (loblolly pine).	old field pine	Medium to large tree.	New Jersey south to Florida, west to Arkansas and Texas.	Lumber, pulpwood, erosion control, leaf mulch, wildlife food.	1713
P. thunbergii Parl. (Japanese black pine).	*P. massoniana* Sieb. & Zucc., not Lamb.	do	Japan	Lumber, sand dune reclamation, ornamental, naval stores.	1855
P. torreyana Parry (Torrey pine).	*P. lophosperma* Lindl. (Soledad pine).	Small tree	San Diego County and Santa Rosa Island, California.	Ornamental	
P. virginiana Mill.[6] (Virginia pine).	*P. inops* Ait. (scrub pine, stickbark pine, old field pine).	Small to medium tree.	Southern New York south to Georgia, west to Ohio and Alabama.	Erosion control, secondary forest products, wildlife food.	Before 1739.

[1]Valuable information was obtained from a typewritten thesis prepared by H. S. Shen at the University of Minnesota.

[2]Valuable information on this species was contributed by M. B. Wilder of Lake City, Fla. Correspondence on file at the Southeastern Forest Experiment Station, Asheville, N. C. 1941.

[3]Long cultivated.

[4]Austrian pine (*Pinus nigra* Arnold) has its typical form in the so-called var. *austriaca* (Hoess) Aschers. & Graebn. (of which *P. austriaca* Hoess, *P. laricio* var. *austriaca* (Hoess) Loud , and *P. nigricans* Host are synonyms). *P. maritima* R. Br. (1813), not Mill. (1768) nor Lam. (1778) is a synonym of *P. nigra*. Corsican pine, on the other hand, is *P. nigra* ("black pine") var. *poiretiana* (Ant.) Aschers. & Graebn. (whereof *P. laricio* Poir., *P. laricio* var. *corsicana* Loud., and *P. laricio* var. *calabrica* Loud. are synonyms).

[5]CONZET, G. M. A QUALITATIVE AND QUANTITATIVE STUDY OF THE SEED PRODUCTION AND REPRODUCTION OF PINUS RESINOSA. 1913. [Thesis. Minn. Univ.]

[6]STONE, L. H. A CONE SEED AND SEEDLING STUDY OF VIRGINIA PINE. 1933. [Thesis. Pa. State Col., Dept. of Forestry.]

The pines most widely used in reforestation in the United States are, in descending order of importance, about as follows: *Pinus resinosa, P. banksiana, P. strobus, P. caribaea, P. palustris, P. sylvestris, P. ponderosa, P. echinata, P. taeda,* and *P. monticola*. The most widely used species in Europe is *P. sylvestris*. Many species have been planted outside their native range, the most notable examples probably being *P. sylvestris* and *P. nigra* in the United States; *P. strobus* in Europe; *P. radiata* in South Africa and New Zealand; *P. banksiana* in Nebraska; *P. caribaea* in Oregon, southern Africa, Australia, and New Zealand; *P. echinata* in South Africa and Australia; and *P. palustris* and *P. taeda* in southern Africa, Australia, and New Zealand. Within the United States most western species have been tried in the East and most eastern species in the West, generally without conspicuous success.

SEEDING HABITS.—Male and female flowers are borne separately on the same tree, and appear during late spring and early summer. The yellow or reddish male flowers occur in groups of cylindrical catkins, around the base of the young shoot, from which the wind blows pollen to the tiny female flowers. The latter are small, green to purplish conelets usually borne at or near the tips of new shoots. After pollination the female flowers develop slowly; at the end of the first summer their dimensions are only about one-seventh those of ripe cones. Most pines mature their cones in the fall of the second year; in three, *Pinus leiophylla* Schlecht. & Cham., *P. pinea* L., and *P. torreyana* Parry, an extra season is required.

The ripe cones, which vary greatly in size, form, and other characteristics among the many species, consist of brown to yellowish overlapping woody scales, each of which if fully developed bears 2 seeds at the base. The seeds are usually winged, but in a few species the wings are absent or rudimentary[46] or often remain attached to the cone scale upon dispersal.[47] The mature seed, usually brown or black, has a comparatively hard seed coat which is relatively simple in structure and usually permeable to water. Inside the brown, papery, inner coat there is a whitish endosperm containing stored food in the form of carbohydrates, proteins and fats, and a well-developed embryo (figs. 188, 189, and 190). In most species the cones open on

[46] *Pinus koraiensis* Sieb. & Zucc., *P. cembra* L., *P. albicaulis* Engelm., *P. flexilis* James, *P. armandi* Franch.

[47] *Pinus monophylla* Torr. & Frém., *P. edulis* Engelm., *P. quadrifolia* Parl., *P. cembroides* Zucc., *P. pinceana* Gord., *P. nelsonii* Shaw, *P. bungeana* Zucc., *P. gerardiana* Wall.

the trees shortly after ripening and the seed are dispersed chiefly by wind and gravity; in some species all or part of the cones may remain unopened on the tree over a period of several years; in a few species, notably *Pinus albicaulis* and *P. cembra*, the entire cone is shed and the seeds released only by disintegration of the cone or by the action of birds and rodents. The comparative seeding habits of 40 species and 4 additional varieties of *Pinus* are given in table 144.

Information on seeding habits is lacking for *Pinus canariensis*, *P. glabra*, *P. heldreichii* var. *leucodermis*, and *P. quadrifolia*. Natural hybrids of *P. palustris* and *P. taeda*, which flower at the same time, are found quite commonly in nature and are known as *P. sondereggeri* H. H. Chapm.

Scientific information as to the development of climatic races among the pines is lacking for many of them, but it is likely that most species of extensive range have developed such races. Available information for the pines discussed here is summarized as follows:

Pinus banksiana.—Within the Lake States, seeds from southern sources average larger than those from northern sources. Seedlings from widely separated seed origins grown in northeastern Minnesota showed decided differences in fall coloration, length of needles, and branching habits.

Pinus caribaea.—One-year-old trees from Cuban and British Honduran seed sown in Texas in 1929 were much less frost-resistant than stock from Florida seed; some botanists claim that more than one species is included, however, in *P. caribaea* as considered here.

Pinus contorta var. *latifolia*.—Seed varies in size in different parts of its range, the largest seed being found to the south (Sierra Nevada). In Sweden, seed from Alberta, Canada, produced better trees than did Colorado or Montana seed. In Finland, seed of nine Canadian sources varied in form, growth rate, and disease resistance. After storage at low temperatures, seed from warmer climates germinates more slowly than that from colder climates.

Pinus echinata.—Among several lots of shortleaf pine stock of diverse origins grown in a Louisiana nursery, those from Pennsylvania and Texas showed marked differences in period and amount of growth. The Pennsylvania stock ceased growth much earlier and grew less.

Pinus jeffreyi.—Seed of different origins varies in color and size.

Pinus monticola.—Seed from California is heavier than that from Washington, which in turn runs heavier than that from northern Idaho.

Pinus mugo.—Swiss mountain pine has three recognized varieties varying from a sprawling shrub to a medium-sized tree, which may be considered as races. Within the varieties, differences in development due to seed origin have been recognized.

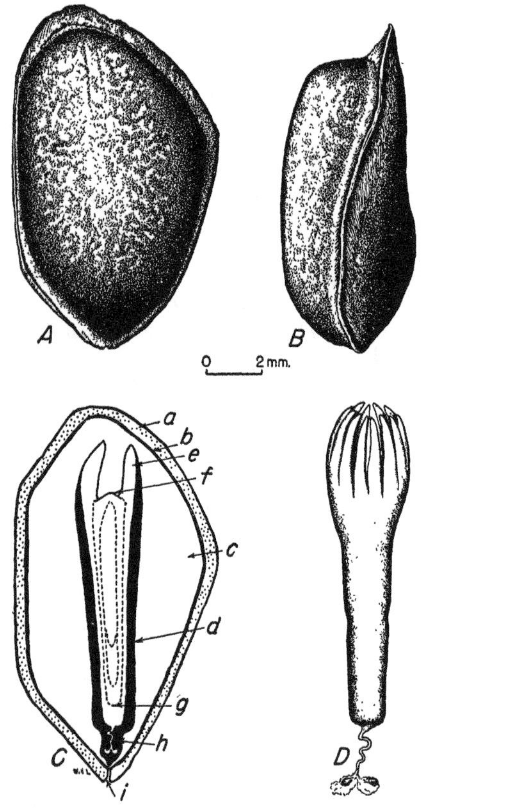

FIGURE 188.—Seed views of *Pinus lambertiana*: *A* and *B*, Exterior view in two planes. *C*, Longitudinal section: *a*, Seed coat; *b*, nucellus; *c*, endosperm; *d*, embryo cavity; *e*, cotyledons; *f*, plumule; *g*, radicle; *h*, suspensor; *i*, micropyle. *D*, Embryo.

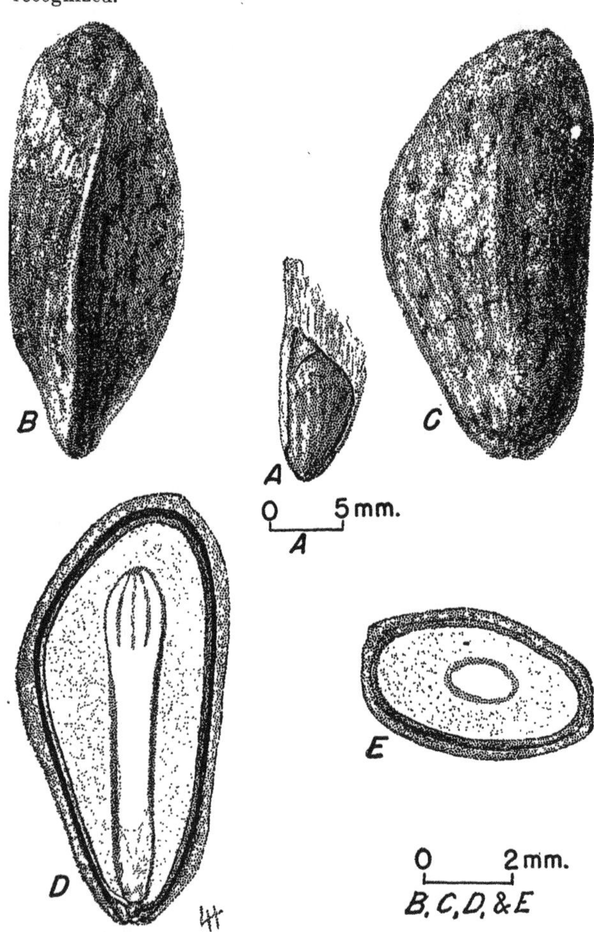

FIGURE 189.—Seed views of *Pinus ponderosa*: *A*, Exterior view of seed with wing fragment; *B* and *C*, exterior views of seed in two planes; *D*, longitudinal section; *E*, cross section.

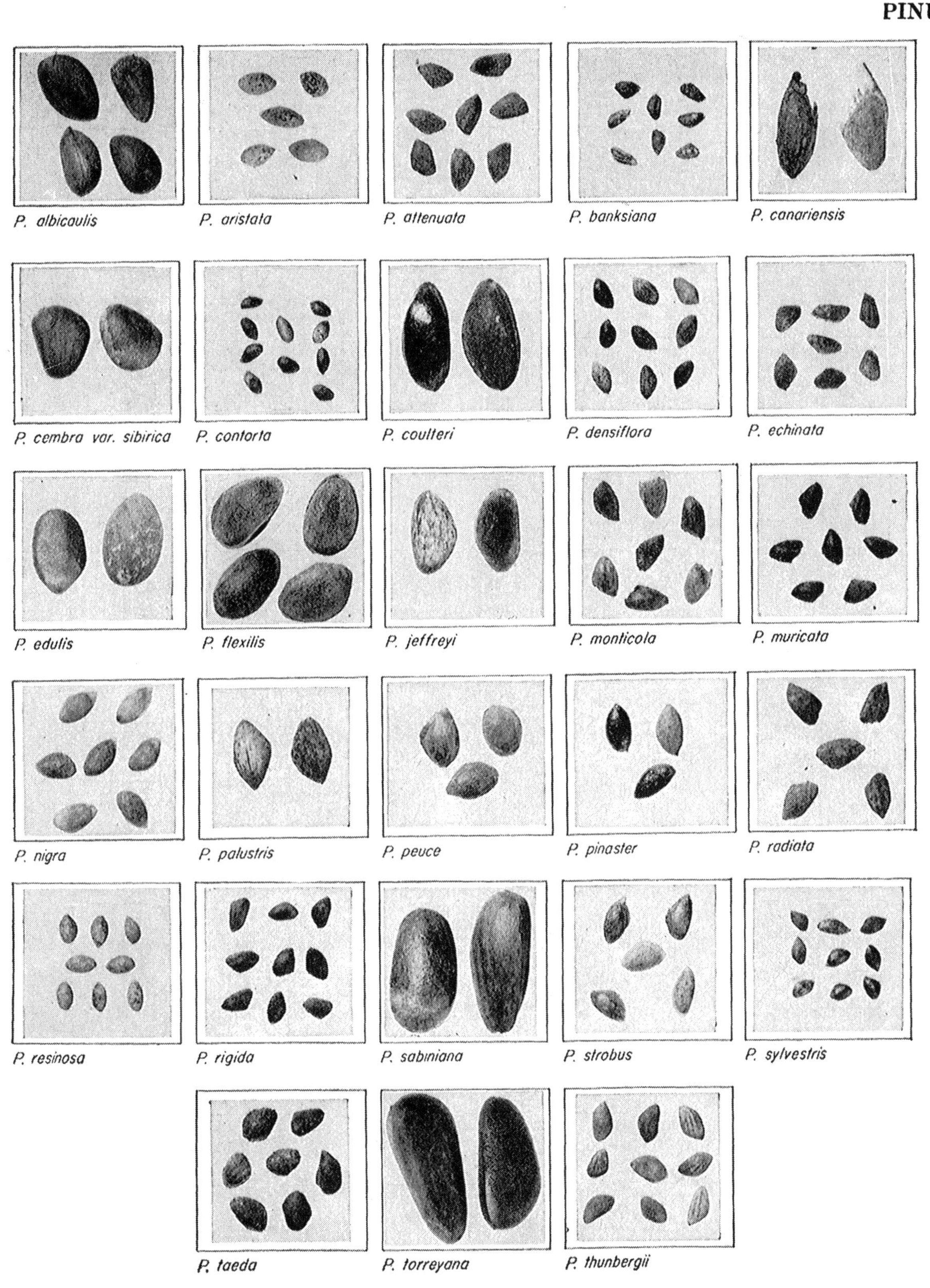

FIGURE 190.—Seed views of 27 species and 1 variety of *Pinus.*

Pinus nigra.—Several distinct varieties are recognized, of which at least four (var. *austriaca* (Hoess) Aschers. & Graebn., var. *caramanica* (Loud.) Rehd., *poiretiana* (Ant.) Aschers. & Graebn., and var. *cebennensis* (Gren. & Godr.) Rehd.) are geographical races.

Pinus palustris.—Seedlings in the nursery have shown differences in root habit, growth, and needle color and character associated with origin.

Pinus peuce.—Seed from the Rila and Pirin Mountains in Bulgaria are considered to produce the best trees.

Pinus pinaster.—Studies of plantations in Belgium, France, and South Africa indicate at least three distinct races.

Pinus ponderosa.—Two main forms, the Rocky Mountain (var. *scopulorum* Engelm.) and the west coast, have long been recognized by botanists. The former has smaller and quicker germinating seed than the latter. Stock grown from different seed origins shows considerable variation in seed size, cone size, number and length of needles, seed germination, rate of growth, resistance to mistletoe, and deer damage. At least four distinct races are present.

Pinus resinosa.—Red pine seed from the southern part of its range is larger than that of northern sources and produces larger 1–0 seedlings. In northeastern Minnesota, stock grown from local or near local seed has shown greater frost hardiness than that grown from Lower Michigan or New England seed.

Pinus rigida.—Pitch pine varies considerably in form and development from locality to locality. Although there is no definite proof of climatic races, it is considered unwise to move the seed more than 150 miles from its source.

Pinus strobus.—Seed from the western part of the range averages lighter than that from the eastern part.

Pinus sylvestris.—Scotch pine has been studied longer and more intensely than any other forest tree as to effects of seed origin. Seed size increases from south (65,000 seed per pound in Poland) to north (127,000 per pound in Lapland). Studies indicate at least five distinct races, four of them named: Riga, Scottish, Lapland, and Tyrolean. There are also a great number of forms or subraces differing in frost hardiness, disease resistance, rate of growth, phenological characteristics, needle size and color, bud characteristics, properties of the wood, bark characteristics, and chemical composition of needles and young shoots. Generally local seed has done best, but in some localities the East Prussian or Baltic strain has produced better trees.

Pinus taeda.—Differences in rate of growth and infection by *Cronartium* cankers have been found associated with origin among trees grown in Louisiana from seed collected in Louisiana, Texas, Georgia, and Arkansas. The local source was generally superior with other sources progressively inferior as their distance of origin from Louisiana increased. Indications of climatic races have also shown up in plantations in the Union of South Africa.

Pinus thunbergii.—The best formed trees are found inland; this indicates some racial variation.

TABLE 144.—*Pinus: Time of flowering and cone ripening, and frequency of seed crops*

Species	Time of—			Commercial seed-bearing age			Seed year frequency	
	Flowering	Cone ripening	Seed dispersal	Minimum	Optimum	Maximum	Good crops	Light crops
				Years	*Years*	*Years*	*Years*	*Years*
P. albicaulis	July	Aug.–Sept.	Sept.–Oct.[1]				(2)	
P. aristata	do	Sept.–Oct.	Fall	20			1	
P. attenuata		Sept.	After fires[3]	(4)			1	
P. banksiana	May	do	Fall—several years.	5–15	40–50	80+	3– 4	Intervening.
P. caribaea	Jan.–March	do	Sept.–Oct.	12–18	20+		1–10	Most intervening.
P. cembra	May	Aug.–Oct.	Next spring[1]	25			6– 7	
P. clausa		Autumn	Sept.—several years.					
P. contorta		Sept.–Oct.	Fall—several years.					
P. contorta var. *latifolia.*	June	Aug.–Sept.	do	5–20	50–200	200+	1– 3	Intervening.
P. coulteri		do	Oct.—several years.	8–20			3– 6	
P. densiflora				20–30				
P. echinata	March–April	Oct.–Nov.	Nov.–Dec.	16	40–50	280+	5–10	Most intervening.
P. edulis	June	Sept.	Sept.–Oct.	25	75+	150–375	{2–5 or 10+}	Intervening.
P. flexilis	June	Sept.	Sept.–Oct.				3+	do.
P. flexilis var. *reflexa.*		do	Oct.			180	3	
P. griffithii	April–June	Aug.–Oct.	Sept.–Nov.	15–20			1– 2	Intervening.
P. halepensis		Sept.	Fall—several years.	15–20				
P. jeffreyi	June	Aug.–Sept.		8	150		2– 4	1–3.
P. lambertiana	May–June	do	Aug.–Oct.	40–50	125–175	(5)	3– 5	Most intervening.
P. latifolia		Nov.	Winter–spring				2– 3	
P. leiophylla var. *chihuahuana.*		do	Dec.–Jan.					
P. monophylla		Aug.	Sept.			100+		
P. monticola	Spring	Aug.–Sept.	Fall–spring	10	45	300+	4– 6	3.
P. mugo			Winter	10			1	

Footnotes on page 267.

TABLE 144.—*Pinus: Time of flowering and cone ripening, and frequency of seed crops*—Continued

Species	Time of—			Commercial seed-bearing age			Seed year frequency	
	Flowering	Cone ripening	Seed dispersal	Minimum	Optimum	Maximum	Good crops	Light crops
				Years	*Years*	*Years*	*Years*	*Years*
P. muricata			Fall—several years.	5– 6				
P. nigra	May	Sept.–Oct.	March–April	30	60–90		3– 4	Most intervening.
P. palustris	Feb.–April	do	Sept.–Dec.	20	40–60	350+	3– 7	do.
P. peuce			Fall	12	40–50	300	3– 4	do.
P. pinaster		Late summer	Spring–summer	15	25–50		1+	
P. ponderosa	April–June	Aug.–Sept.	Fall–spring	20	150		2– 5	2–3.
P. ponderosa var. *arizonica.*		Oct.–Nov.	Nov.–Dec.	20			2– 3	
P. pungens	April	Fall	Fall—several years.	5				
P. radiata	Feb.–March	Fall	Spring	10			1	
P. resinosa[6]	April–June	Sept.–Oct.	Fall to next summer.	25	50–150	200+	3– 7	Most intervening.
P. rigida	April–May	Fall	Nov.—several years.	8–12		200	3	
P. rigida var. *serotina.*		Early fall	Fall—several years.					
P. roxburghii	Feb.–April	March–July	March–July	30–40			2– 3	Most intervening.
P. sabiniana	March	Fall	Oct.–Dec.					
P. strobus	April–June	August–Sept.	Sept.–Oct.	15–20	50–150	250	3– 5	Most intervening.
P. sylvestris	May–June	Sept.–Oct.	Winter–April	9–20	40–80	150	2– 5	do.
P. taeda	March–April	Sept.–Nov.	Fall–spring	12–15	35–60+		3–10	do.
P. thunbergii		Nov.	Nov.–Dec.	6	15–40			
P. torreyana		June–July		12–15		(7)	1	
P. virginiana[8]		Oct.	Nov.	5	50		1	

[1]Seed dispersal within cone.
[2]Interval between good seed crops ranges from a few to many years.
[3]Cones remain unopened for 15 or more years, usually until opened by fires or logging.
[4]*P. attenuata* begins to bear seed in commercial quantities at an early age.
[5]*P. lambertiana* bears seed in commercial quantities until it is very old.
[6]CONZET, G. M. A QUALITATIVE AND QUANTITATIVE STUDY OF THE SEED PRODUCTION AND REPRODUCTION OF PINUS RESINOSA. 1913. [Unpublished thesis. Copy on file Univ. Minn. Minneapolis.]
[7]*P. torreyana* continues to produce seed in commercial quantities until death.
[8]STONE, L. H. A CONE SEED AND SEEDLING STUDY OF VIRGINIA PINE. 1933. [Unpublished thesis. Copy on file Dept. of Forestry, Pa. State Col. State College.]

TABLE 145.—*Pinus: Specific gravity and color of ripe cones*

Species	Ripe cones			Remarks
	Specific gravity	Will float in—	Color	
P. banksiana			At least half of cone definitely brownish.	Specific gravity not a reliable index.
P caribaea[1]	0.90	SAE 20 motor oil		Test within 10 minutes of picking sound cones. When a majority of 5 freshly picked cones from a tree float, collection from that tree is recommended.
P. nigra			Cones yellowish green	
P. palustris	.90	SAE 20 motor oil		See *P. caribaea.*
P. ponderosa	.80– .86	Kerosene		When majority of 5 freshly picked cones from a tree float, collection from that tree is recommended.
P. resinosa	.80– .94	do	Cones deep purple with reddish-brown scale tips.	do.
P. strobus	.92– .97	Linseed oil	Cones yellowish green with brown-tipped scales.	do.
P. taeda	.90	SAE 20 motor oil		See *P. caribaea.*
P. thunbergii			Cones deep purple	

[1]It is believed the same test is suitable for cones of *P. echinata* also.

891183°—50——18

PINUS

Collection, Extraction, and Storage.—Ripe pine cones may be collected from standing trees, from newly felled trees or slash, and sometimes from animal caches. Trees growing in the open or in understocked stands generally yield more abundant and regular seed crops than those grown in dense stands. Before making collections it is advisable to cut open a few cones from several trees to see whether they are reasonably well filled with sound seed. The cones of most species are ripe enough to pick when they begin to turn brownish. More definite evidences of ripening, based on specific gravity or color, have been determined for a few species, table 145.

As a rule, young trees produce larger seeds than older trees. Depending on the species, abundance of crop, and skill of the worker, one man can collect from about 1 to 20 bushels of cones per day by climbing trees. For most species, freshly collected cones should be spread out in thin layers to dry in the sun or in a well-ventilated shed from 2 to 12 weeks. Cones of some species open satisfactorily with such treatment while others must be run through artificially heated kilns or extraction drums; in a few species they must be torn apart. The opened cones should then be shaken to remove the seed. To save storage space and facilitate sowing, the seed wings usually are re-

Table 146.—*Pinus: Time of collection, collection costs, and extraction practices*

Species	Time for collection	Closed cones per bushel	Collection cost per bushel of cones	Collection cost per pound clean seed	Extraction practice	Extraction schedule—cured cones[1]		
						Duration	Temperature	Relative humidity
		Number	*Dollars*	*Dollars*		*Hours*	*°F.*	*Percent*
P. albicaulis	Sept.	----	----	1.23	None needed	----	----	----
P. banksiana	do[2]	2,500–3,200	1.00	3.00–4.00	{Convection	2–4	145–150	----
					{Forced draft	6	170	30
P. caribaea	do	200–220	0.60–1.69	2.00	{Sun dry	400–1,200	----	----
					{Convection	6–48	120	[3]
P. clausa	do[2]	700–1,200	----	----	do	2+	140–150	25+
P. contorta var. *latifolia.*	Sept.[2]–Oct.	----	.75–1.00	2.75	do	6–8	140	----
P. echinata	Oct.–Nov.	1,400–2,500	1.89–25.00	2.00–4.50	{Sun dry	190–1,900	----	----
					{Forced draft	6–8	130	----
P. glabra	Oct.	----	----	----	None needed	264	Room	----
P. halepensis	Sept.[2]	----	----	----	Convection	----	----	----
P. jeffreyi	Aug.–Sept.	----	----	.50–1.00	Sun dry	92–144	Sun	----
P. lambertiana	do	20	----	----	Sun dry or convection.	----	----	----
P. latifolia	Nov.–Dec.	162	1.50	----	Convection	60+	110	----
P. leiophylla var. *chihuahuana.*	Nov.	1,300	1.00	----	----	----	----	----
P. monticola	Aug.–Sept.	----	----	----	Sun dry or convection.	10+	110	----
P. muricata	Winter[2]	----	----	----	Convection	----	170—	----
P. nigra	Oct.–spring	400–600	2.00	2.50	Extraction drum	----	120	----
P. palustris	Oct.	60–120	----	1.10	{Forced draft	8–16	115	20–30
					{Convection	12–72	120	20–30
P. ponderosa	Sept.–Nov.	200–300	----	1.00–3.00	{Sun dry	96–144	----	----
					{Convection	3	120—	----
P. ponderosa var. *arizonica.*	Oct.–Nov.	235	1.50	----	do	60	110	----
P. pungens	Fall	----	----	----	do	----	90+	----
P. radiata	Dec.[2]–Mar.	----	----	----	do	----	110	----
P. resinosa	Sept.–Oct.	1,300–1,800	1.50–2.00	6.50	{do	24–72	130–140	----
					{Forced draft	5	170	21
P. rigida	do	800	----	----	Convection	----	----	----
P. rigida var. *serotina.*	Sept.[2]	350	----	----	Convection, or soak in water ½ hour, then place on radiator.	----	130	----
P. strobus	Sept.–Oct.	500–700	.50–1.50	3.50–8.00	{Sun dry, convection.	----	110–120	----
					{Forced draft	8	140	40
P. sylvestris	Oct.–Mar.	1,100–3,000	----	----	{Convection	5–24	130	----
					{Forced draft	4–8	130	----
P. taeda	Oct.–Nov.	400–1,100	----	2.80	{Air dry	400–1,200	----	----
					{Convection	6–48	120–140	20–30

[1]Green cones usually require a lower temperature and longer treatment.
[2]Older cones at any time of year.
[3]Relative humidity, low.

moved by dewinging machines or by moistening followed by gentle kneading and thorough redrying.[48]

[48] Exceptions occur for *Pinus palustris,* the persistent wings of which are partly broken but not completely removed, and those species which have seed without wings or with wings adhering to the cone scales. Seed of some species also appear to be injured by standard dewinging practice. For instance, hand-dewinged lots of *P. banksiana* seed almost invariably germinate significantly better than those of the same collection dewinged by commercial methods. This is also apparently true for seed of *P. caribaea* and *P. echinata.*

The broken wings and cone fragments are removed by screening and fanning. Detailed experience in collection and extraction practice is available only for the 24 species and 2 additional varieties listed in table 146, but similar procedures probably would be suitable for other pines. The cones of *Pinus attenuata* should either be scalded, or scorched by open fire, and then allowed to open in a kiln or a warm, dry place. Seed of *P. edulis, P. monophylla,* and *P. quadrifolia* may be removed from the cones by shaking the tree, or

TABLE 147.—*Pinus: Yield of cleaned seed, and purity, soundness, and cost of commercial seed*

Species	Cleaned seed						Commercial seed		
	Per bushel of cones	Per 100 pounds of fresh cones	Per pound			Basis, samples	Purity	Soundness	Cost per pound
			Low	Average	High				
	Ounces	*Pounds*	*Number*	*Number*	*Number*	*Number*	*Percent*	*Percent*	*Dollars*
P. albicaulis	--	--	2,800	3,600	4,200	5	92	79	6.00
P. aristata	--	2.0	16,000	23,000	42,000	7	91	65+	2.75– 3.00
P. attenuata	--	--	22,000	29,000	34,000	3	86	82	5.35–11.50
P. banksiana	4–11	1.0	71,000	131,000	250,000	423	91	94	2.50– 4.00
P. canariensis	--	--	3,500	3,900	4,300	9	98	--	1.40– 3.50
P. caribaea	8–24	1–2	13,000	14,500	16,000	175±	97	95	2.00– 3.00
P. cembra	--	--	1,600	1,800	2,000	15+	99	75	.40– 1.75
P. clausa	10–15	3.5	65,000	75,000	85,000	10+	--	--	--
P. contorta	--	--	111,000	135,000	165,000	28	97	70	5.00– 8.50
P. contorta var. *latifolia*	5–8	3.0	38,000	102,000	160,000	89	93	97	3.00– 4.75
P. coulteri	--	--	1,300	--	1,400	2	90	96	2.50– 4.50
P. densiflora	9–12	2.0	31,000	45,000	60,000	39	96	93	.75– 1.50
P. echinata	5–24	2–3	36,500	48,000	62,500	30+	96	95	4.75– 7.00
P. edulis	--	--	1,500	1,900	2,500	9	100	94	.25– 1.25
P. flexilis	--	--	3,300	4,400	5,800	9	99	80	2.75– 5.30
P. flexilis var. *reflexa*	--	--	--	2,700	--	1+	--	--	--
P. glabra	--	--	65,000	--	80,000	2+	--	--	--
P. griffithii	--	--	7,200	9,100	12,400	176+	93	78	1.00– 3.00
P. halepensis	--	--	23,000	25,000	30,000	11+	96	99	.40– 2.40
P. heldreichii var. *leucodermis.*	--	--	20,000	--	25,000	2	96	--	8.00
P. jeffreyi	33	--	3,100	4,000	5,400	53+	96	77	1.75– 4.30
P. lambertiana	25–32	3.0	1,500	2,100	3,200	53	99	83	1.25– 2.00
P. latifolia	6	1.2	--	10,000	--	1	--	82	12.00
P. leiophylla var. *chihuahuana.*	12–14	--	--	40,000	--	1+	--	--	--
P. monophylla	--	--	--	1,200	--	1	--	--	3.50
P. monticola	12	--	14,000	27,000	32,000	99+	95	75	4.00– 8.00
P. mugo	--	--	48,000	62,000	92,000	141+	97	50+	2.00– 4.00
P. muricata	--	--	18,000	50,000	60,000	14	94	--	6.00– 7.80
P. nigra	7–12	2	14,000	26,000	39,000	159+	96	88	1.25– 5.00
P. palustris[1]	8–19	1–3	3,800	4,200	6,000	500±	92	90	1.00– 3.50
P. peuce	--	--	10,000	11,000	14,000	6	98	84	3.50– 5.00
P. pinaster	--	3.5–5.5	6,800	9,500	13,000	74+	97	75	1.15– 2.00
P. ponderosa	9–32	2–7	6,900	12,000	23,000	185	96	81	1.50– 2.50
P. ponderosa var. *arizonica.*	16	2.3	--	11,000	--	1	--	85	8.00–10.00
P. pungens	--	3	29,000	36,000	41,000	9	85	80	3.25
P. quadrifolia	--	--	--	1,200	--	1	--	--	--
P. radiata	--	--	13,000	16,000	23,000	43	98	95	2.50– 3.80
P. resinosa	9–12	1–2	30,000	52,000	71,000	497	97	96	5.00– 6.00
P. rigida	12	2–3	36,000	62,000	83,000	48	97	84	3.50– 5.70
P. rigida var. *serotina*	7	--	47,000	54,000	63,000	4+	--	--	--
P. roxburghii	--	--	3,100	4,600	11,300	136	98	--	3.60– 4.50
P. sabiniana	--	--	700	--	800	2	--	--	1.25– 4.30
P. strobus	5–28	2	20,000	27,000	53,000	162	92	92	1.50– 4.00
P. sylvestris	7–9	2	52,000	78,000	111,000	312	94	96	1.25– 7.50
P. taeda	8–24	2–3	16,000	18,400	25,000	50±	97	90	2.75– 4.00
P. thunbergii	[2]4–12	--	26,000	34,000	50,000	50	96	96	.85– 2.10
P. torreyana	--	--	400	500	800	7	100	74	5.00–11.00
P. virginiana	--	3	40,000	53,000	75,000	9	90	88	4.70

[1]Seed with wings; partially dewinged seed runs 6.5–16 ounces per bushel of cones and 4,200–6,700 per pound.
[2]Uncleaned.

seed already fallen may be gathered from the ground. Cones of *P. albicaulis* and *P. cembra* must be torn apart to release the seed. The size, purity, soundness, and cost of commercial cleaned seed vary considerably by species as shown in table 147.

The seed of some pines will remain viable for several years even when stored at ordinary temperatures. For others the seed must be dried and held at low temperatures to maintain its germinability as long as 1 year. The viability of most species can be prolonged greatly by storing the seed dry in sealed containers at temperatures

Table 148.—*Pinus: Viability maintenance under various storage methods*

Species	Viability maintained under—				Period of viability	Remarks
	Ordinary storage[1]	Sealed storage at: Ordinary temperature	Sealed storage at: 32°–41° F.	Other storage methods		
	Years	*Years*	*Years*		*Years*	
P. attenuata		10±				
P. banksiana			5+	In cones on trees[2]	15	
P. caribaea			9			A maintained 7 to 9 percent moisture content recommended.
P. cembra				In moist acid peat or sand	1+	
P. contorta var. *latifolia*.		5	[3]7+	In cones on trees	30+	
P. coulteri	2–3			do	2–4+	
P. densiflora			1+			
P. echinata			7–9	Some indications that storage under partial vacuum at 5° to 23° F. better for periods over 5 years.		A maintained moisture content of 7 to 9 percent recommended.
P. flexilis	6					
P. halepensis	2+					
P. jeffreyi	3		10+			
P. lambertiana			10+			
P. monticola	2					
P. mugo	1–2					
P. muricata				In cones on trees	20+	
P. nigra				In unopened cones	2+	
P. palustris			[4]2	Some indications that storage under partial vacuum at 5° to 23° F. gave good results.		A maintained moisture content of 7 to 9 percent recommended.
P. pinaster	3–4					
P. ponderosa			[5]3–4			Coast form.
P. ponderosa var. *scopulorum*.			[5]23+			
P. pungens				In cones on trees	9±	
P. resinosa		3–5	[3]10+			A maintained moisture content of 4.7 percent recommended.
P. rigida			2+			
P. roxburghii	1	2				
P. strobus			8+			
P. sylvestris		4	13			Seed from northern sources keep better than those from southern sources.
P. taeda			7–9	Some indications that storage under partial vacuum at 5° to 23° F. better for periods over 2 years.		A maintained moisture content of 7 to 9 percent recommended.
P. thunbergii			1+			
P. virginiana			5+			

[1]Unsealed, no temperature control.

[2]Shen, H. S. viability of the seed of jack pine from cones retained on the tree for various periods of time. 1936 [Unpublished thesis. Copy on file Minn. Col. of Agr., Forestry, and Home Econ., St. Paul.]

[3]Temperatures of 32° to 50° F.

[4]Maintained fairly well for 5 to 9 years.

[5]In cool cellar.

slightly above freezing.[49] Seed of the species listed in table 148 can be stored with little or no loss of viability under the conditions indicated. The best storage practices and consequently the maximum longevity for many of these species have not yet been determined. In most cases some viability is retained for much longer periods than shown. For species which are not listed, the safest practice is to store dried seed in sealed containers at temperatures of 32° to 41° F. Seed of *Pinus palustris* spoils rather easily after collection and should be extracted promptly and either sown or stored properly without delay. After cold storage, also, it should be sown promptly to avoid deterioration.

GERMINATION.—Natural germination of pine seed usually takes place the first spring following dispersal. However, under favorable conditions seed may germinate during the summer or early fall, and seed of some species may germinate the second or even third year after dispersal. Some species have cones which open slowly over a period of years, or suddenly after fires or logging. Chief among these are *Pinus attenuata, P. banksiana, P. contorta* var. *latifolia, P. muricata,* and *P. pungens.* Ordinarily germination and early survival is best on mineral soil with side shade and overhead light, but may occur on almost any moist surface. For some species, such as *P. cembra* and *P. edulis,* germination is best in the dark, such as is provided by a cover of litter or fine soil. Germination is epigeous (figs. 191 and 192). The natural seed supply often is reduced considerably both before and after maturity by insects, birds, and rodents, and for some species by larger animals, including man, and fungi.

Pine seed exhibits variable germination behavior. Seeds of many species germinate readily when sown, while those of other species show embryo dormancy and hence germinate much better if fall sown or stratified prior to spring sowing. Seed

[49] Since this was written, reanalysis of an earlier study has indicated that lower storage temperatures are advantageous. The following note by Philip C. Wakeley of the Southern Forest Experiment Station appeared in the January 1947 mimeographed Southern Forestry Notes:

"Seed of the principal southern pines can be stored for years at 5° F., without injury from cold and usually with better results than at 35°-41° F., the temperatures hitherto recommended. Relatively few cold storage warehouses operate at 35°-41° F., but many between freezing and 5° F. This finding therefore greatly increases the storage space available to southern nurserymen and seed collectors.

"If the air of the storage chamber is moist, sealing the containers is the obvious way of keeping the seed dry. If the relative humidity of the chamber is low, however, it makes practically no difference whether southern pine seed is stored in open, sealed, or vacuum containers. Aside from their relation to seed moisture content, seals and vacuums appear to have no particular merit.

"The laboratory work for these advances was done by Dr. Lela V. Barton of Boyce Thompson Institute, the originator of "stratification" of southern pine seed, in one of the most comprehensive forest-tree-seed storage studies ever undertaken. By applying new analytical methods to her data, published in 1935 (Contrib. Boyce Thompson Inst. 7: 379-404), the Southern Station has been able to show that her results on temperatures and containers, as summarized in this note, were much farther reaching and conclusive than was originally realized. This reanalysis has secured important information for the postwar planting program about 7 years sooner than could have been done by repeating Dr. Barton's excellent study."

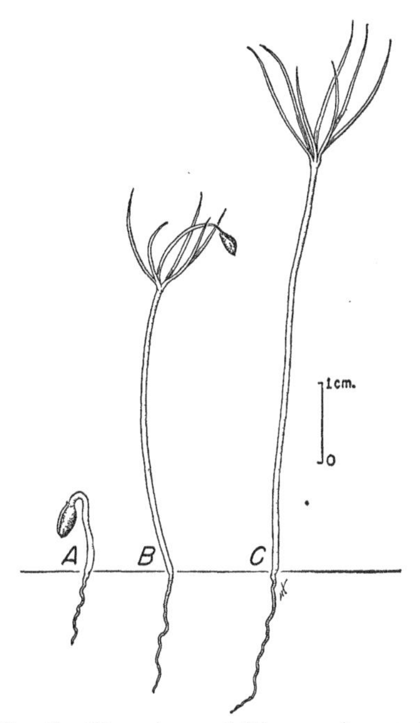

FIGURE 191.—Seedling views of *Pinus resinosa:* A, After 1 day; *B,* after 7 days; *C,* after 30 days.

FIGURE 192.—Seedling views of *Pinus palustris: A,* after 1 day; *B,* after 2 days; *C,* after 5 days; *D,* after 32 days.

of a few species have thick seed coats and germination is hastened and improved if the seed are first scarified or acid-treated. For many species of pine, dormancy varies considerably from lot to lot and even within lots. Seed of the following species ordinarily germinate satisfactorily without pretreatment: *Pinus aristata, P. banksiana, P. canariensis, P. caribaea, P. clausa, P. contorta, P. edulis, P. griffithii, P. halepensis, P. heldreichii* var. *leucodermis* (probably), *P. jeffreyi, P. latifolia, P. mugo, P. nigra, P. palustris, P. pinaster, P. ponderosa, P. ponderosa* var. *arizonica, P. pungens* (probably), *P. radiata, P. resinosa, P. roxburghii, P. sylvestris, P. thunbergii, P. virginiana.* However, some of these species give somewhat better germination results if pretreated seed is used, and for most of them stratification is decidedly beneficial when it is necessary to use seed which has been in storage for some time. Information as to the germination characteristics of *P. leiophylla* var. *chihuahuana* and *P. quadrifolia* is lacking. Details of seed dormancy and its pretreatment are given in table 149 for 28 species and 2 additional varieties of pine.

TABLE 149.—*Pinus: Dormancy and method of seed pretreatment*

Species	Seed dormancy		Stratification			Remarks
	Kind	Occurrence	Medium	Temperature	Duration	
				°F.	Days	
P. albicaulis	Embryo, possibly seed coat.	Variable	Moist sand	41	60+	Some lots not benefited by stratification. In one test, cracking the seed coat improved results.
P. aristata	Probably embryo.	Occasional	do	41	60+	Some lots benefited by stratification; ordinarily no pretreatment needed.
P. attenuata	do	Probably general.	do	32–41+	15–40	Full germination does not result even from this treatment; longer stratification probably necessary.
P. banksiana	do	Occasional	do	41	30	Stratification usually not needed, but does hasten and improve germination some.
P. caribaea	do	do	Moist sand or peat.	35–40	15–45	Dormancy frequent in seed stored over winter or longer; not general in fresh seed. Stratification over 60 days appears injurious.
P. cembra	Embryo, possibly seed coat.	General	Moist sand or acid peat.	36	270	Fair results obtained by soaking in concentrated sulfuric acid for 3–5 hours followed by 90 days' stratification.
P. contorta var. *latifolia.*	Probably embyro.	Occasional	Moist sand	41	30–90	See remarks for *P. banksiana.*
P. coulteri	do	General	do	41	30–90	Freshly collected seed germinates well without pretreatment.
P. densiflora	do	Probably variable.	do	41–50	30–60	Stratification hastens and somewhat improves germination, but fairly good results are obtained without.
P. echinata	do	Quite general	Moist acid peat or sand.	35–38	30–45	Dormancy common in fresh seed and very frequent in stored seed.
P. flexilis	Embryo, possibly seed coat.	Probably variable.	Moist sand or peat.	41	30–90	Fair results usually attained without stratification. Best results (90%) occurred after scarification plus 90 days' stratification.
P. flexilis var. *reflexa.*[1]	Probably embryo.	do	Moist acid peat or sand.	35–50	90	Germination only fair without pretreatment.
P. glabra	do	do	Moist sand or peat.	30–35	30±	In one series of tests nursery sowing in the spring gave much better results when stratified seed were used.
P. jeffreyi	do	Occasional	do	41	30±	Stratification usually unnecessary; beneficial in some lots.
P. lambertiana	do	Variable	do	35–50	90	Most lots require stratification to give satisfactory germination in a reasonable time.
P. monophylla	Embryo, possibly seed coat.	General	do	32–40	60–90	Germination poor without pretreatment.
P. monticola	Seed coat, probably embryo.	Variable	do	32–40	90	Dormancy overcome satisfactorily in many cases by soaking 45 minutes in concentrated sulfuric acid. Some lots need no treatment.

[1]Treatment only suggested for this species; experimental data not complete.

TABLE 149.—*Pinus: Dormancy and method of seed pretreatment*—Continued

Species	Seed dormancy		Stratification			Remarks
	Kind	Occurrence	Medium	Temperature	Duration	
				°F.	Days	
P. *muricata*[1]	Probably embryo.	Probably variable.	Moist sand or peat.	41	25	Stratification probably will hasten and improve germination.
P. *nigra*	do	do	do	41	30–60	Stratification hastens germination; results at 30 days nearly as good with untreated seed.
P. *palustris*	Probably embryo.	Rare	Moist acid peat.	35–40	15–30	Occasionally slight to moderate dormancy in dried or stored seed. Soaking seed in water at 80° F. or lower for 6–12 hours often improves germination.
P. *peuce*[1]	do	Probably variable.	Moist acid peat or sand.	32–41	90	Germination fairly good without pretreatment; probably can be hastened and improved by stratification.
P. *ponderosa*	do	Occasional	do	41	30–60	Stratification usually not needed but hastens and improves germination of some lots, particularly of coast form.
P. *rigida*	do	Probably variable.	do	41	30	Germination fairly good without pretreatment, stratification hastens and somewhat improves it in some lots at least.
P. *rigida* var. *serotina*.	do	do	do	41–46	30–60	do.
P. *sabiniana*	Embryo, possibly seed coat.	Quite general	do	35–40	60–90	Occasional lots germinate well without stratification.
P. *strobus*	Embryo	do	do	50	30	Stratification 60 days at 40° F. also effective.
P. *sylvestris*	Probably embryo.	Occasional	do	41	60–90	Seed from extreme northern parts of range require stratification; from other localities it is not needed.
P. *taeda*	do	Quite general	do	35–38	30–90	To avoid heating, large seed lots (50–100 lbs.) should not be stratified longer than 40–50 days as a rule.
P. *thunbergii*	do	Probably variable.	do	41	30–60	Stratification hastens but does not improve germination.
P. *torreyana*	do	Quite general	do	32–40	30–90	Stratification hastens and improves germination as a rule.

The germination of pine seed may be tested in sand flats, peat mats, or standard germinators, using 800 to 1,000[50] seeds per test, either pretreated or untreated, as indicated in table 149. Methods recommended and average results for 42 species and 4 additional varieties are given in table 150. Information for *Pinus leiophylla* var. *chihuahuana* and *P. quadrifolia* is lacking.

NURSERY AND FIELD PRACTICE.—Pine seed may be sown in the nursery or in the field in the fall or in the spring if nondormant seed is used. It should be sown either broadcast or in drills, usually 4 to 6 inches apart. For most species a well-drained sandy loam makes the best seedbed. Ordinarily the seeds are covered with one-fourth inch of soil, although one-half inch is recommended for such large-seeded species as *Pinus albicaulis.* This depth is probably also advisable in general for species having less than 4,000 seeds per pound. A depth of three-eighths of an inch is commonly recommended for *P. lambertiana, P. monticola,* and *P. strobus,* and one-eighth inch is sometimes used for such species as *P. glabra* and *P. virginiana.* Ordinarily seed of *P. caribaea, P. echinata, P. palustris,* and *P. taeda* are covered only to the extent caused by rolling the seedbed, not over one-eighth to one-fourth inch.

Seedlings of some species are field planted directly from the seedbed. In such cases the seed should be sown at a rate to produce 40 to 70 usable seedlings per square foot of seedbed. For other species a year or two in transplant beds is required, and the seed may be sown at a rate to produce 80 to 100 usable seedlings per square foot of seedbed. The rate of sowing varies with the species and locality but in general 6 to 10 ounces of seed per 100 square feet of seedbed is sufficient, although in some cases 20 or more ounces are required. Germination is complete for most species in

[50] 400 seeds sufficient for the following species: *Pinus cembra, P. flexilis* var. *reflexa, P. jeffreyi, P. lambertiana, P. monophylla, P. quadrifolia.* Two hundred seeds sufficient for *P. sabiniana* and *P. torreyana.*

PINUS

TABLE 150.—*Pinus: Recommended conditions for germination tests, and summary of germination data*

Species	Test conditions recommended				Germination data from various sources					
	Temperature		Duration		Germinative energy		Germinative capacity			Basis, tests
	Night	Day	Untreated seed	Pretreated seed	Amount	Period	Low	Average	High	
	°F.	*°F.*	*Days*	*Days*	*Percent*	*Days*	*Percent*	*Percent*	*Percent*	*Number*
P. albicaulis			[1]120	60+			2	[2]20	75	4+
P. aristata	68	95	30				66	86	99	7+
P. attenuata				30			22	57	80	14
P. banksiana	68	86	15– 60		2–90	7–39	3	68	94	328
P. canariensis			30		86	20	61	86	88	9
P. caribaea	65	85	[1]45	20–40	10–87	19	12	61	89	41
P. cembra	68	80	350+	[1]200	30–50	90	35	50	72	6
P. clausa			35					72		1
P. contorta	68	86	50				51	80	98	29
P. contorta var. *latifolia*	60	80	[1]60	30	23–62	5–40	36	64	95	465
P. coulteri	60	80	60–150	[1]50	70–96	14–19	50	89	100	20
P. densiflora	75	75	[1]60	[1]30	21–70	12–40	4	84	99	47
P. echinata	60	80	[1]60–120	35–45	10–85	46	11	68	88	35
P. edulis	60	70	30		53	15	72	83	96	8
P. flexilis	68	86	90+	[1]40+	10–84	8–14	35	70	100	7
P. flexilis var. *reflexa*								47		1+
P. glabra			26+	[1]10+				64		[3]1
P. griffithii	68	86	[1]60	30	44	20	30	67	95	100+
P. halepensis	68	71	30		50–66	20	61	73	92	43
P. heldreichii var. *leucodermis.*			40					69		1
P. jeffreyi	60	80	60– 90		12–89	30–80	25	68	93	48+
P. lambertiana	60	70	120	[1]40			30	56	92	8
P. latifolia			30				18	39	88	3+
P. monophylla				90			20	47	90	22
P. monticola	60	80	[1]200	[1]60–90	8–75	15–168	0	48	95	122+
P. mugo	68	71	45				34	77	98	425+
P. muricata			90				14	45	78	23
P. nigra	68	86	[1]30	30	12–70	5–18	0	69	98	2,499+
P. palustris	55	75	[1]30– 40	35	10–94	22	8	54	94	201
P. peuce			30				35	69	89	4
P. pinaster	60	80	30				6	63	98	168
P. ponderosa	65	85	[1]60	30	14–87	7–29	5	59	97	186
P. ponderosa var. *arizonica.*			20					60–70		2+
P. pungens	75	75	40				26	65	85	9
P. radiata	60	80	[1]60	30	46–56	20–26	18	60	89	42
P. resinosa	68	86	30		25–75	7–25	9	75	100	551
P. rigida	65	80	50	30	16–87	5–38	19	77	99	127
P. rigida var. *serotina*			[1]60	20			42		97	3+
P. roxburghii			30		79	10	16	83	100	237
P. sabiniana			60				10	44	86	19
P. strobus	68	86	[1]60–100	[1]30–40	7–90	12–39	0	64	96	1,840
P. sylvestris	65	85	30– 40		15–93	7–26	0	72	100	19,122
P. taeda	65	85	[1]70	35–45	17–75	52	11	60	76	22
P. thunbergii	68	86	60		16–93	10–45	45	77	99	45
P. torreyana	65	80	120	[1]60			30	81	100	21
P. virginiana	68	86	30		25–54	10–32	31	65	90	5

[1]Indicates the condition reported under germinative energy and germinative capacity.
[2]In 1 test stratified seed germinated 30 percent as compared to 2 percent for untreated seed.
[3]In nursery.

from 10 to 40 days after spring sowing with nondormant seed, although it may straggle over 2 or 3 years for some dormant-seeded species if proper pretreatment is not used. For the few species for which such information is available, nursery germination usually is from 50 to 75 percent of laboratory germination. About 50 percent of the viable seed sown produce usable 1–0 seedlings as a rule, but the value ranges from 10 to 60 percent for various species.

The seedbeds should be mulched with coarse cloth, burlap, or litter until after germination is well under way, and they should be protected against birds and rodents by a 2-mesh wire screen. Ordinarily pine seedlings do not require artificial shading after germination is completed. However, it is recommended throughout the first season for such species as *Pinus lambertiana* and *P. thunbergii*, and in dry localities where it will reduce the amount of necessary watering.

The most suitable stock for field planting varies with the species, and within the species according to such factors as seed origin, local climate, nursery soil and watering conditions, and conditions under which planting is intended. For instance, *Pinus banksiana* stock suitable for field planting is produced in some nurseries as 1–0, in most nurseries as 2–0, but in a few as 2–1. As a general rule, however, satisfactory planting stock, which should have a sturdy stem at least 4 inches to the tip of the bud, healthy foliage, and a well-developed root system, is as follows:

Seedling Stock

P. banksiana, P. canariensis, P. caribaea, P. clausa, P. contorta, P. contorta var. *latifolia, P. echinata, P. glabra, P. palustris, P. pinaster, P. pungens, P. radiata, P. rigida, P. rigida* var. *serotina, P. taeda, P. virginiana.*

Transplant Stock

P. albicaulis, P. cembra, P. densiflora, P. flexilis, P. griffithii, P. heldreichii var. *leucodermis, P. jeffreyi, P. lambertiana, P. monticola, P. mugo, P. nigra, P. peuce, P. ponderosa, P. resinosa, P. roxburghii, P. strobus, P. sylvestris, P. thunbergii.*

Information is lacking for *P. attenuata, P. aristata, P. coulteri, P. edulis, P. flexilis* var. *reflexa, P. halepensis, P. latifolia, P. leiophylla* var. *chihuahuana, P. monophylla, P. muricata, P. ponderosa* var. *arizonica, P. quadrifolia, P. sabiniana, P. torreyana.*

The seedlings of many species of pine are susceptible to damping-off and are protected against this disease by acidifying the beds with sulfuric acid or aluminum sulfate prior to sowing. Other important sources of injury are as follows:

1. White pine blister rust caused by *Cronartium ribicola* Fischer and affecting chiefly *Pinus lambertiana, P. monticola,* and *P. strobus;* it may be controlled by removing currant and gooseberry plants (*Ribes*) and infected pines within 1,500 feet of the nursery.

2. Western blister rust caused by *Cronartium filamentosum* and affecting *Pinus ponderosa* and other hard pines; it may be controlled by cutting out infected host plants, either pine or painted cups (*Castilleja* spp.) and owlclover (*Orthocarpus*).

3. Sweetfern blister rust caused by *Cronartium comptoniae* and affecting *Pinus ponderosa, P. sylvestris, P. rigida, P. contorta* var. *latifolia, P. nigra,* and other hard pines; control is by removing infected pines or plants of sweetfern (*Comptonia*) and sweet gale (*Myrica*).

4. Brown-spot needle disease caused by *Scirrhia acicola* (Dearn.) Siggers and the closely related red-spot needle blight, the former affecting chiefly *Pinus palustris* and the latter *P. nigra* and *P. pinaster;* control by spraying with 4–4–50 bordeaux 2 to 6 times beginning between late May or early June and December 30.

5. Rust-gall diseases caused by *Cronartium cerebrum* and *C. fusiforme* Hedgc. & Hunt and affecting *Pinus nigra, P. caribaea, P. taeda,* and other hard pines; control is by culling out infected stock or possibly by lime sulfur or bordeaux spray.

6. European needle disease caused by *Lophodermium pinastri* and affecting *Pinus sylvestris, P. banksiana, P. resinosa,* and *P. rigida;* it may be controlled by spraying with 8–8–50 bordeaux when the needles are half grown and repeating every 3 to 4 weeks until growth ceases and eradicating infected plants within 100 to 300 yards of the nursery.

7. Dwarf mistletoe caused by *Arceuthobium* spp. and affecting *Pinus ponderosa* and other western pines; it may be controlled by pruning or cutting out infected pines.

8. Phytophthora root rot caused by *Phytophthora cinnamomi* and affecting chiefly *Pinus resinosa* and *P. sylvestris;* control by not sowing when fungus is known to be present.

9. White grubs chiefly of the genus *Phyllophaga,* which attack many pines; control is by a combination of roto-tilling the soil and using an arsenate spray on surrounding June berries (*Amelanchier*) and other plants on which the adults feed.

10. Pales weevil (*Hylobius pales* Herbst.) which attacks chiefly *Pinus strobus* and *P. sylvestris;* best prevention is to avoid fresh cuttings near nursery.

11. Ants sometimes destroy seed of *Pinus roxburghii* and other species; control is by coating seed with red lead before sowing.

12. Pine tip moths and snout beetles sometimes attack *Pinus sylvestris* and other species; control must depend chiefly on hand methods.

Although pines generally are propagated by seed, varieties and the rarer kinds sometimes are grafted on their types or allied species; some species, such as *Pinus radiata,* are occasionally grown from cuttings.

Despite the fact that individual species vary considerably in their requirements, the pines as a rule prefer well-drained or even dry soils, most commonly sandy loams. A few grow in swamps.

PLATANUS L. Sycamore, planetree

(Sycamore family—Platanaceae)

DISTRIBUTION AND USE.—Sycamore, comprised of about 11 species, is confined to North America, southeastern Europe, and southwestern Asia. About eight species are distributed in Canada, the United States, Mexico, and Central America. All develop into sizable trees. The wood is heavy and hard and valuable for interior construction and trim and for furniture. A further description of two species is given in table 151. *Platanus occidentalis* is one of the largest deciduous trees of eastern North America. This species is easily established in plantations on bottom lands and grows rapidly. It is becoming more important as a primary cover on stripped coal lands in many areas.

SEEDING HABITS.—Sycamore's minute staminate and pistillate flowers occur in separate, dense, globular heads on the same tree, and appear with the leaves in the spring. The staminate dark-red flowers are usually borne along the branchlets, while the pistillate flowers are light-green tinged with red and occur at the tips. In *Platanus occidentalis* one to two pistillate heads arise from a single stem as compared to two to seven in *P. racemosa.* The head (ball) is a syncarp or multiple fruit which ripens in the first year and usually persists on the tree over winter. This species has an elongated-oblong, hairy seed—an achene—with a light chestnut-brown, thin but firm seed coat (figs. 193 and 194). The embryo is erect and surrounded by a thin endosperm. Descriptions of seed development for two species are as follows:

Time of—	*P. occidentalis*	*P. racemosa*
Flowering	May	
Fruit ripening	Sept.–Oct.	June–Aug.
Seed dispersal	Sept.–May	June–Dec.

The minimum commercial seed-bearing age for *P. occidentalis* is 25 years; the optimum, 50 to 200 years; the maximum, 250 years. It bears good seed crops from 1 to 2 years, with light crops in the intervening years.

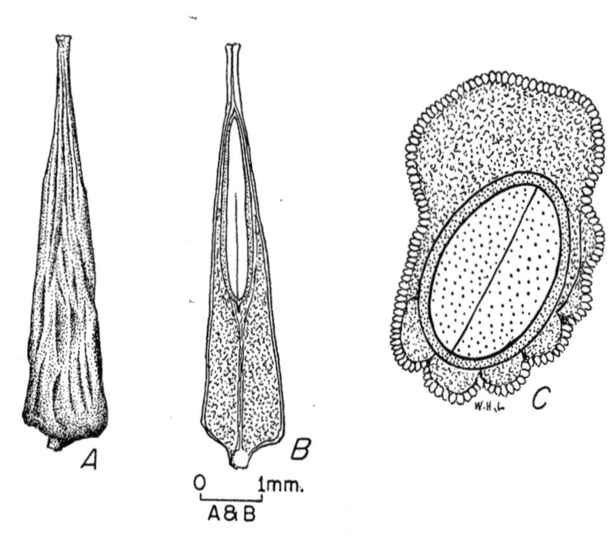

FIGURE 193.—Seed views of *Platanus occidentalis: A*, Exterior view; *B*, longitudinal section; *C*, cross section, enlarged.

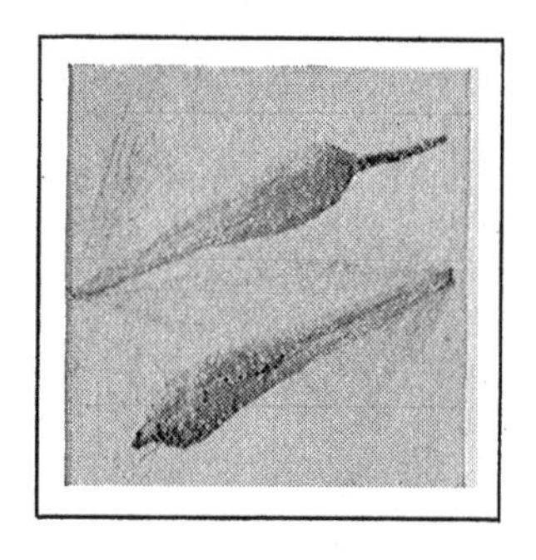

FIGURE 194.—Seed views of *Platanus racemosa.*

TABLE 151.—*Platanus: Distribution and uses*

Accepted name	Synonyms	Natural range	Chief uses	Date of earliest cultivation
P. occidentalis L. (American sycamore).	American planetree, planetree, buttonball-tree, buttonwood.	Rich bottom-land soils Maine west to Kansas, south to Florida, Texas, and northeastern Mexico.	Furniture, interior finishing, lumber, fuel, tobacco boxes, butcher's blocks, bird food.	1640
P. racemosa Nutt. (California sycamore).	California planetree, western sycamore.	Rich valley soils northern California to Lower California.	Stream-bank protection, ornamental planting, fuel.	1870

COLLECTION, EXTRACTION, AND STORAGE.—Sycamore fruits (seed balls) may be collected most easily in late fall or during the winter after the leaves have fallen. Occasionally they may be collected from the ground. They should be spread to dry in well-aerated trays after which the seed may be easily rubbed from the heads. Removal of the tawny hairs, accomplished by rubbing over screen, will facilitate sowing. For *Platanus occidentalis* 100 pounds of fruit yields about 7 pounds of cleaned seed (including hairs) which run in number per pound (15+ samples) from low, 151,000, through average, 204,000, to high, 228,000; commercial seed averages 85 percent in purity, 66 percent in soundness, and costs $0.60 to $1.30 per pound. Comparable information for *P. racemosa* is lacking. If prolonged storage is planned, the seed should be sealed in containers and kept at 35° to 41° F. Probably the best procedure for overwintering seed is to allow the fruits to remain on trees. Pretreatment is then unnecessary. Seed may also be stored by stratification for short periods of 60 to 90 days.

GERMINATION.—Germination takes place in the spring after seed fall and is epigeous. Moist to wet alluvial soils, or sand or gravel, form the best natural seedbeds. When the fruit matures, the seed are dormant, likely due to conditions in the embryo. A brief period of stratification at somewhat above freezing generally will break this dormancy. Further details of pretreatment to hasten and increase germination in two species of *Platanus* are as follows:

	P. occidentalis[1]	*P. racemosa*
Pretreatment:		
Medium	Sand, peat, sand and peat, or sandy loam.	Sand, peat, sand and peat, or sandy loam.
Temperature (°F.)	35–41	40.
Duration (days)	45–60	60–90.

[1]If fruits are collected in the spring, no pretreatment is required. For dry stored seed, soaking in water is somewhat less desirable than stratification.

Adequate germination tests may be made in sand flats on 400 to 800 pretreated or spring-collected seeds. Test conditions recommended for *P. occidentalis,* and the average results to be expected, are as follows:

Test conditions:	
Night temperature °F.	70
Day temperature do.	85
Duration:	
Stratified seed days	15–20
Untreated seed do.	30–60
Germinative energy:	
Amount percent	34
Period days	14
Germinative capacity:	
Low percent	5
Average do.	35
High do.	69
Basis, tests number	15+

P. racemosa has shown a germinative capacity of 12 percent in 1 test.

NURSERY AND FIELD PRACTICE.—Although satisfactory results may be obtained from well-prepared and protected fall-sown beds, spring seeding with late winter or spring-collected seed is preferred. *Platanus* seed are preferably sown in rows 6 to 8 inches apart and should be covered with about one-fourth inch of firmed soil. The spacing depends somewhat upon methods of seeding and cultivation. Fall beds require a mulch of leaves, or a covering of burlap or straw held in place by bird screens or bed shades. Spring beds are benefited by a burlap cover until germination begins and thereafter by partial shade until seedling establishment. This serves to prevent drying of surface soil and damage by birds and rodents. Seedlings are field planted as 1–0 stock. *Platanus* usually has a sprawly root system; root pruning about midseason will effect a more compact root system and simplify field planting. Stock of these species is among the easiest of the hardwoods to produce, and field survival in moist bottom-land soils is generally good.

POPULUS L. Poplar

(Willow family—Salicaceae)

Distribution and Use. — The poplars include about 30 species of medium to large, rapidly growing, usually short-lived, deciduous trees which occur in North America south to northern Mexico, in Europe and North Africa, and in Asia south to the Himalayas. Some form extensive forests, especially on recently burned lands; others occur commonly along stream bottoms and other low-lying areas. The wood of several species is used in large quantities for paper making, and others are a source of lumber and such wood products as excelsior and match sticks. Many species are cultivated widely for shelter-belt and ornamental purposes. Some species furnish wildlife food and cover. Seven species of poplars, which are of known or supposed value for reforestation in the United States and for which reliable information is available, are described in detail in table 152. So far *Populus deltoides* is the only poplar which has been used extensively in actual reforestation practice in the United States. It is also planted in Europe. *P. sargentii* is used in shelter belts. Besides the true species, a great many forms and hybrid poplars have been developed and some of them have been planted more or less widely in the United States.

Seeding Habits.—The male and female flowers are borne on separate trees early in the spring, usually before the leaves, and occur as rather large, conspicuous catkins. The male flowers bear pollen only, while the female flowers develop into fruits. These fruits, which usually ripen from 1 month to 6 weeks after flowering, are one-celled capsules borne in long pendulous clusters (catkins), and each capsule contains many small, brown seeds (fig. 195).[51]

[51] In *Populus tremula* there are 1 to 12 seeds per capsule and 150 capsules per catkin.

Table 152.—*Populus: Growth habit, distribution, and uses*

Accepted name	Synonyms	Growth habit	Natural range	Chief uses	Date of earliest cultivation
P. deltoides Bartr. (eastern cottonwood).	*P. balsamifera* var. *virginiana* Sarg., *P. monilifera* Ait. (necklace poplar, cotton tree, eastern poplar, SPN).	Large tree	Banks of streams and lake shores from Quebec to North Dakota, south to Texas and Florida.	Box lumber, pulpwood, excelsior, veneer, cooperage, fuel, shelter belts, ornamental.	Prior to 1750
P. grandidentata Michx. (bigtooth aspen).	largetooth aspen, largetooth poplar.	Small tree	Nova Scotia to northeastern North Dakota, south to Iowa and Pennsylvania, and along the mountains to North Carolina.	Locally used for pulpwood, boxboards, matchsticks along with *P. tremuloides*.	1772
P. nigra L.[1] (black poplar).		Medium-sized tree.	Europe, western Asia, northern Africa.	Ornamental and windbreak planting.	([2])
P. sargentii Dode (plains cottonwood).	*P. deltoides* var. *occidentalis* Rydb. (Great Plains cottonwood, plains poplar, SPN).	Medium to large tree.	Saskatchewan and Alberta to Nebraska, New Mexico, and western Texas.	Locally used for fuel wood and posts, shelter belts, veneer, and baskets.	1908
P. tacamahaca Mill.[3] (balsam poplar).	*P. balsamifera* auth. (tacamahac poplar, balm-of-Gilead).	Medium-sized tree.	Alaska to Labrador, south to New York and Oregon.	Ornamental and windbreak planting; buds medicinal.	1689
P. tremula L. (European aspen).		Small to medium tree.	Europe, northern Africa, western Asia and Siberia.	Matchwood, pulpwood	([2])
P. tremuloides Michx. (quaking aspen).	aspen, trembling aspen	do	Labrador to Alaska, south to Pennsylvania, Missouri, northern Mexico, and Lower California.	Locally used for pulpwood, boxboards, matchwood, excelsior; bee pasture; bark medicinal.	About 1812

[1]Var. *italica* Muenchh. is the well-known Lombardy poplar.
[2]Long cultivated.
[3]Includes var. *michauxii* (Dode) Farwell.

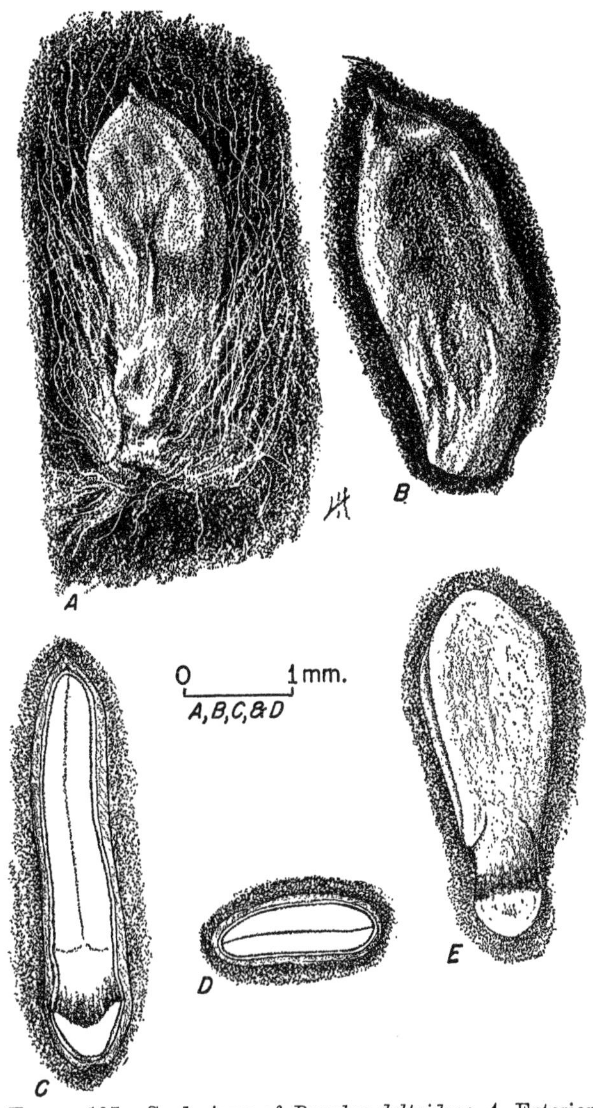

FIGURE 195.—Seed views of *Populus deltoides: A,* Exterior view seed, with "cotton"; *B,* exterior view seed without "cotton"; *C,* longitudinal section; *D,* cross section; *E,* embryo.

Each seed is surrounded by tufts of long, white, silky hairs attached to the short stalks of the seed, which enable it to be dispersed over great distances by the wind. Water dispersal is also effective. Ordinarily the seeds are dispersed within a few days after ripening. Comparative seeding habits of the seven species discussed here are given in table 153.

There is little scientific information available as to the development of climatic races within the poplars. Since many of them are of wide distribution, however, the development of such races is likely. In *Populus tremuloides,* var. *aurea* (Tidestr.) Daniels is confined to the West, and var. *vancouveriana* (Trel.) Sarg. is confined to the vicinity of Vancouver Island and Puget Sound. Both of these varieties are probably climatic races. Three variations of *P. deltoides*—var. *virginiana* (Castiglioni) Sudw., *missouriensis* (Henry) Henry, and f. *pilosa* (Sarg.) Sudw.—which may be climatic races, are recognized by some authors.

COLLECTION, EXTRACTION, AND STORAGE.—Because one or a few trees will provide enough seed for extensive planting, seed trees should be selected very carefully. Poplar seed should be collected from standing or felled trees as soon as the capsules begin to break open, showing the hairs within. At this time the capsules are quite green and should be well aerated to prevent heating. The correct time for collecting seed can be determined by placing a few fruiting twigs in a glass of water in a room. These will ripen 1 to 2 days ahead of those in the field. If spread out to dry for a few hours, the capsules will open completely and the "cottony" seed may be stripped off easily. The hairs can be removed safely in some species by running the opening capsules through a thresher, hammer mill, or macerator, or forcing them through screens. In the smaller-seeded species, this is probably not feasible. The cotton may also be removed by igniting it, but this usually results in injury to the seed. In commercial seed the "cotton" usually is not removed. The size, purity, and soundness of commercial cleaned seed vary considerably by species as shown in table 154 for 5 species. Individual trees of *Populus tremula* may yield from 8,000 to 54,000,000 seed per year. In *P. deltoides* var. *virginiana* 1 bushel of fresh fruit yields about 2 pounds of seed.

Poplar seeds are usually sown at once because they lose their viability rapidly under ordinary con-

TABLE 153.—*Populus: Time of flowering and fruit ripening, and frequency of seed crops*

Species	Time of—		Commercial seed-bearing age			Seed year frequency	
	Flowering	Fruit ripening	Minimum	Optimum	Maximum	Good crops	Light crops
			Years	*Years*	*Years*	*Years*	*Years*
P. deltoides	February–May	April–June	10	40	(1)	(1)	Intervening.
P. grandidentata	April–May	May–June	20	50–70		4–5	Most intervening.
P. nigra	Early spring	Late spring					
P. sargentii		June–August					
P. tacamahaca	April–May	May–June					
P. tremula	April	do	8	40–50	60+	4–5	Most intervening
P. tremuloides	April–May	do	20	50–70		4–5	do.

[1]Seed borne on commercial scale almost annually until death.

POPULUS

TABLE 154.—*Populus: Yield of cleaned seed, and purity, soundness, and cost of commercial seed*

Species	Average seeds per capsule	Cleaned seed per pound[1]				Commercial seed		
		Low	Average	High	Basis, samples	Purity	Soundness	Cost per pound
	Number	*Number*	*Number*	*Number*	*Number*	*Percent*	*Percent*	*Dollars*
P. deltoides	10–30	200,000	350,000	590,000	6+	40	95	----------
P. grandidentata	----------	----------	3,032,000	----------	1	20	----------	----------
P. sargentii	27	429,000	----------	479,000	2	94	92	----------
P. tremula	1–12	2,660,000	3,670,000	7,550,000	30	----------	----------	1.00
P. tremuloides	----------	----------	3,600,000	----------	1	50	----------	----------

[1]Cotton removed.

ditions. However, seed of some species can be kept from several months to 3 years and over without appreciable loss in germinability if it is stored in sealed containers at a temperature not far from freezing. Details of storage for a few species are as follows:

Seed of *Populus tacamahaca* stored in stoppered bottles showed 90 percent germination after 10 weeks at 23° F., but only 15 percent at the end of 30 weeks.

Populus deltoides seed, which was allowed to dry for 3 days at room temperature and then stored in a stoppered bottle at 36° to 41° F. dropped from 98 to 96 percent germination in 2 months, but lost all viability by the end of a year. Seed of the same lot dried only 1 day before similar storage lost all vitality during the first 2 months. Seed stored at 40° in sealed flasks from which air had been exhausted by an ordinary faucet vacuum showed 60 percent germination at the end of 8 months. Uncleaned seed packed tightly in vials at the same temperature retained partial viability for 6 months.

Seed of *Populus grandidentata* dried for 5 days and stored in sealed containers at 41° F. showed 44 percent germination after 45 months, but seed stored without preliminary drying deteriorated completely in 1 month.

Seed of *Populus nigra* stored in sealed containers under partial vacuum (1 mm. pressure) and low temperature (34° to 39° F.) retained its viability from several months to 1 year. Low temperatures alone are more effective than partial vacuum alone.

Dried seed of *Populus tremula* lost its viability in 2 to 6 months when stored in sealed containers at room temperature; when stored in a refrigerator it had lost no viability after 4½ months; and when stored at low atmospheric pressure (1 and 40 mm.) in a cool cellar it lost only 10 percent of its viability in 22 months.

Populus tremuloides seed maintained reasonable viability for 8 weeks when stored in an open dish out of direct sunlight at 71° F. and a 40- to 50-percent relative humidity; it showed signs of life and a 10- to 20-percent germination after 105 days, but no true germination after 326 days when stored at 68° in desiccators with relative humidities of 10 to 50 percent. In another study, seed dried for 3 days and then stored in sealed containers at 41° germinated 97 percent at the end of 1 year.

GERMINATION.—Natural germination takes place within a day or two after seed dispersal, provided suitably moist seedbeds are reached. Germination is epigeous (fig. 196). Fungi often cause high mortality among the succulent newly germinated seedlings, which are also highly susceptible to heat and drought damage. Poplar seeds are not dormant and need no pretreatment. Using 400 seeds per test, germination may be tested in petri dishes or standard germinators, or in sand flats for some of the larger-seeded species. Tests should be run from 4 to 7 days for fresh seed, or double that time for stored

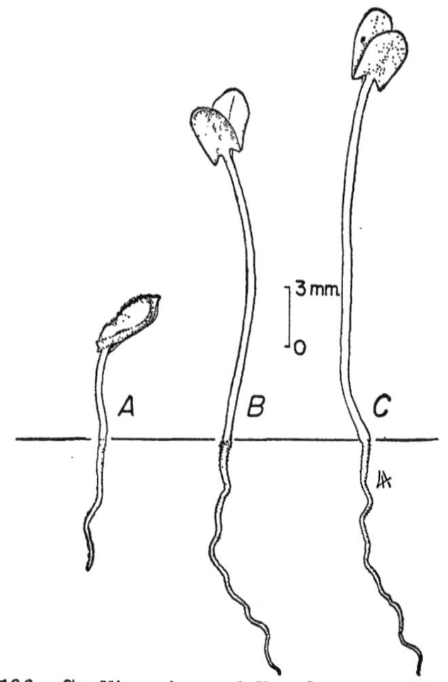

FIGURE 196.—Seedling views of *Populus sargentii*: *A*, At 1 day; *B*, at 4 days; *C*, at 8 days.

seed, at temperatures ranging from 68° (night) to 86° F. (day). Average results for 6 species are given in table 155.

NURSERY AND FIELD PRACTICE.—(1) Poplar seeds may be sown broadcast or in drills in the spring as soon after collection as possible; or (2) branchlets gathered from the trees just before the capsules open may be inserted at an angle in the soil of the seedbed—they will disperse the seeds naturally on the seedbed; or (3) freshly collected catkins may be spread over the seedbeds and allowed to shed their seeds naturally. The seedbeds should be kept thoroughly moist from 5 to 15 days after germination, or until fairly good roots are established. Ordinarily the seed are not covered; at the most, they should be covered with only one-eighth inch of soil. Germination begins within a few days after sowing. The beds should be shaded during germination and at least during the hottest part of the day through midsummer. Shade frames should be kept 3 to 5 feet above ground surface to provide good aeration. Seedlings of this genus are very susceptible to damage from heat, drought, and fungi. Heat damage can be controlled by proper shading and watering.

In the nursery poplars are sometimes affected by leaf rusts due to fungi (*Melampsora* spp.). These may be controlled by sprays or by the removal of certain alternate hosts such as firs, larches, hemlocks, and Douglas-fir. Sometimes also poplar stock has areas of bark-kill caused by *Cytospora* and other fungi. Stock so affected should be destroyed and not taken into the field. Much damage can be prevented by spraying the nursery soil with a 2-percent solution of Uspulun and raking it into the ground a few days before sowing. Ants destroy considerable seed, but they can be controlled by placing carbon bisulfide in their runways. Earthworms occasionally pull the catkins into the ground. This can be avoided by using cleaned seed. Poplar leaf beetle and caterpillars may cause partial or complete defoliation of small trees in the nursery or in plantations, especially in late summer. Aphids often retard the growth of weak trees. Control usually is not needed but can be obtained by spraying with 40-percent nicotine sulfate or a kerosene emulsion. One-year-old seedling stock of most species of poplar is suitable for field plantings; for others, however, 2–0 stock is best. Wildings may be used to propagate some of the poplars, particularly the cottonwoods. Such seedlings should not be over 1 year old. Most species can also be propagated by stem (except the aspens) or root cuttings.

TABLE 155.—*Populus: Recommended germination test duration, and summary of germination data*

Species	Test duration recommended	Germination data from various sources			
		Germinative capacity			Basis, tests
		Low	Average	High	
	Days	*Percent*	*Percent*	*Percent*	*Number*
P. deltoides	2–6	80–	88	100	9
P. grandidentata	4–7	34	99	100	4+
P. nigra	2–3	67	89	99	5
P. sargentii[1]	5	----	98	----	1
P. tremula	5	0	90	100	76
P. tremuloides[2]	7	17	----	100	2

[1]This species showed a germinative energy of 73 percent in 3 days.
[2]*P. tacamahaca* seed is reported to behave similarly.

PROSOPIS JULIFLORA (Sw.) DC. *Mesquite*

(Legume family—Leguminosae)

Including var. *glandulosa* (Torr.) Cockerell, and var. *velutina* (Woot.) Sarg. Syn.: *P. chilensis* auth., not (Mol.) Stuntz

Distribution and Use.—Mesquite is a deciduous thorny shrub or small tree which occurs from southwestern United States south to northern South America. In the United States it diverges into two extreme forms: (1) var. *glandulosa,* honey mesquite, occurs in eastern Texas to southern Kansas and reappears with rather shorter and more crowded leaflets in Arizona and southern California, and (2) var. *velutina,* velvet mesquite, occurs in the hot valleys of southern Arizona. Mesquite produces hard, durable wood, particularly valuable as fuel, and its pods are used as fodder and for human food. Flowers of var. *glandulosa* are the source of excellent honey.

Seeding Habits.—Mesquite's perfect flowers open from May to the first half of July. Its fruit is a pod containing several seeds (fig. 197), which ripen from August to September.

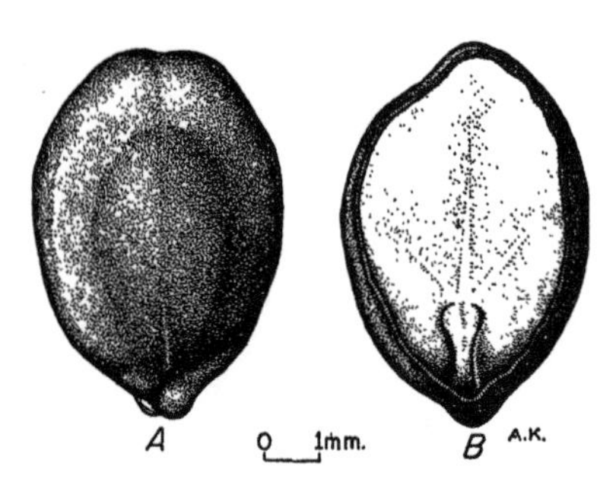

Figure 197.—Seed views of *Prosopis juliflora: A,* Exterior view; *B,* longitudinal section.

Collection, Extraction, and Storage. — The ripe pods may be collected from the ground under trees in stock-excluded areas, and can be broken into longitudinal segments and stored in a dry place. Apparently cold storage is not necessary for keeping mesquite seed. Cleaned seed per pound (8 samples): low, 10,000; average, 14,000; high, 16,000. In one sample purity and soundness were both 91 percent. Mesquite seeds are often infested with larvae of a weevil. Although it has been reported that weevil-infested seed germinates better than the sound ones, it is advocated that the pods be fumigated before storage. Seed of this species do not ordinarily appear on the market.

Germination. — Germination is epigeous (fig. 198). The seeds exhibit some dormancy due to seedcoat conditions; this can be overcome best by soaking the seeds in concentrated sulfuric acid for 15 to 30 minutes, or by placing them in boiling water and allowing them to cool as they soak for 24 hours. Soaking the seeds in ether is helpful but not as efficient as the acid or hot-water treatments. Germination tests should be run in sand flats for 15 days at 68° F. (night) to 86° (day), using 200 to 400 properly pretreated seeds per test. Under these test conditions, germinative energy was 78 to 88 percent in 5 to 6 days, and germinative capacity was as follows (11 tests): low, 86 percent; average, 89 percent; and high, 90 percent.

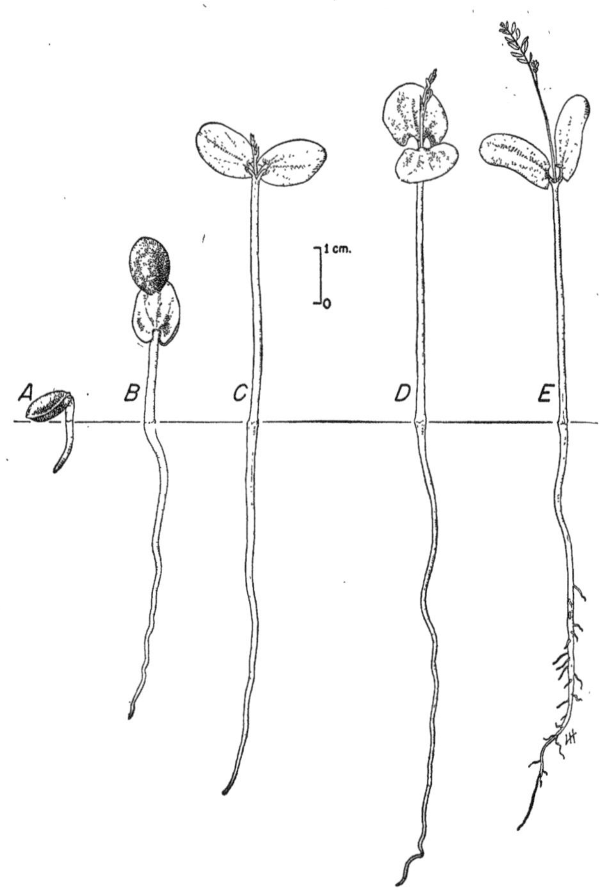

Figure 198.—Seedling views of *Prosopis juliflora: A,* At 1 day; *B,* at 2 days; *C,* at 5 days; *D,* at 10 days; *E,* at 25 days.

PRUNUS L. Plum and cherry

(Rose family—Rosaceae)

DISTRIBUTION AND USE.—The plums and cherries, comprising nearly 200 species, are deciduous, or sometimes evergreen, shrubs or small- to medium-sized trees distributed chiefly in the temperate regions of the Northern Hemisphere, although some species do range south into the Andes Mountains of South America. About 22 of the 25 to 30 species native to the United States are trees. Some species are important producers of commercial fruit; most of them are valuable as sources of wildlife food and as honey plants; many of them are used for ornamental planting; the wood of some species is a valuable cabinet material; several species are useful for shelter-belt and erosion-control planting. Fifteen species and two varieties, which are useful for planting in the United States, are described in detail in table 156. *Prunus serotina* is the only species used at all widely for regular reforestation

TABLE 156.—*Prunus: Growth habit, distribution, and uses*

Accepted name	Synonyms	Growth habit	Natural range	Chief uses	Date of earliest cultivation
P. alleghaniensis Porter (Allegheny plum).	sloe	Shrub or straggling tree.[1]	Connecticut to North Carolina; Michigan.	Wildlife food and cover, human food.	1889
P. americana Marsh. (American plum).	wild plum	Small tree[1]	Massachusetts to Manitoba, south to Georgia, New Mexico, and Utah.	Human food, wildlife food and cover, shelter belt.	1768
P. amygdalus Batsch (almond).	*Amygdalus communis* L., *P. communis* Arcang., not Huds.	Short tree	Western Asia, possibly North Africa.	Edible nuts, ornamental, flavoring extracts, prussic acid, wildlife food.	(2)
P. angustifolia Marsh. (Chickasaw plum).	*P. chicasa* Michx. (mountain cherry).	Small tree or low shrub.[1]	Texas and Oklahoma, naturalized throughout southeast.	Wildlife food and cover, human food, shelter belt, erosion control.	(3)
P. angustifolia var. *watsoni* (Sarg.) Waugh (sand Chickasaw plum).	*P. watsoni* Sarg. (sandhill plum).	Small shrub	Sand dunes from Kansas and Texas to New Mexico.	Wildlife food and cover, shelter belt, human food.	1879
P. armeniaca L. (apricot).	*Armeniaca vulgaris* Lam.	Small tree	Western Asia	Human food, shelter belt, wildlife food.	(2)
P. avium (L.) L. (mazzard).	*Cerasus avium* Moench. (sweet cherry, mazzard cherry, SPN).	Small to medium tree.	Europe and western Asia, naturalized in North America.	Human food, ornamental, wildlife food.	(2)
P. besseyi Bailey (Bessey cherry).	*P. pumila* var. *besseyi* (Bailey) Waugh, *P. prunella* Daniels (sand cherry, western sand cherry).	Small shrub	Manitoba to Wyoming, south to Kansas and Colorado.	Wildlife food and cover, erosion control,[4] ornamental, human food.	1892

[1]Form thickets.
[2]Long cultivated.
[3]Prior to 1874.
[4]Valuable, but not widely used for this purpose.

891183°—50——19

TABLE 156.—*Prunus: Growth habit, distribution, and uses*—Continued

Accepted name	Synonyms	Growth habit	Natural range	Chief uses	Date of earliest cultivation
P. caroliniana (Mill.) Ait. (Carolina laurelcherry).	laurelcherry, mockorange, wildorange.	Shrub or small tree.	Coast region and islands from North Carolina to Florida, west to Texas.	Ornamental, wildlife food.	--------
P. cerasus L. (sour cherry).	*Cerasus caproniana* DC. (pie cherry, Morello cherry).	Small tree	Asia Minor and southeastern Europe.	Human food, ornamental, wildlife food.	(2)
P. padus L. (European birdcherry).	*P. racemosa* Lam., *Padus racemosa* Schneid., *Cerasus padus* (L.) DC.	do	Europe and northern Asia to Korea and Japan.	Ornamental, wildlife food.	(3)
P. pensylvanica L. f. (pin cherry).	*Cerasus pensylvanica* (L. f.) Loisel. (fire cherry, bird cherry, pigeon cherry, wild red cherry).	Small tree or shrub.[1]	Newfoundland to British Columbia, south to North Carolina and Colorado.	Reclothing cut-over or burned-over areas, nurse crop, wildlife food, human food, ornamental.	(5)
P. persica (L.) Batsch (peach).	*Amygdalus persica* L.	Short tree	China	Edible fruit, wildlife food.	(2)
P. pumila L. (sand cherry).	----------------------	Low shrub	Sandy areas from western New York to Manitoba, south to Indiana and Illinois.	Wildlife food, erosion control,[4] ornamental.	1756
P. pumila var. *susquehanae* (Willd.) Jaeg. (Appalachian sand cherry).	*P. cuneata* Raf., *P. pumila* var. *cuneata* Bailey (dwarf cherry).	Shrub	Rocky and sandy areas from Maine to Manitoba, south to Pennsylvania and Wisconsin.	Wildlife food and cover, erosion control.[4]	1805
P. serotina Ehrh. (black cherry).	wild black cherry, rum cherry.	Medium to tall tree.	Nova Scotia to Lake Superior, south to Florida and Texas.	Backing for electrotyping blocks, furniture, finishing material, wildlife food.	1629
P. virginiana L. (common chokecherry).	*P. nana* Du Roi, *Padus virginiana* Roem.	Large shrub to small tree.[1]	Newfoundland and Hudson Bay to British Columbia, south to North Carolina and southern California.	Wildlife food and cover, shelter belt, erosion control, livestock and game browse, ornamental, human food.	1724

[1]Form thickets.
[2]Long cultivated.
[3]Prior to 1874.
[4]Valuable, but not widely used for this purpose.
[5]About 1773.

purposes in this country, but *P. americana, P. angustifolia, P. angustifolia* var. *watsoni, P. armeniaca*, and *P. virginiana* are used in shelter-belt planting. *Prunus avium* is the source of most of the commercial sweet cherries, and *P. cerasus* the source of most of the commercial sour cherries; *P. amygdalus* is the almond of commerce, and *P. persica* the peach.

SEEDING HABITS.—The perfect flowers of *Prunus* are solitary, or in fascicles or racemes, and appear in the spring before, with, or after the leaves. The pistil has two ovules, and the fruit is a one-seeded drupe with thick fleshy pulp (dry in *P. amygdalus*), yellowish to black when ripe. The stone (figs. 199, 200, and 201), which is the "seed" of commerce, actually is a bony, smooth, rugose, or pitted and indehiscent, matured endocarp enclosing the true seed. The latter is covered with a brown membranous coat. Comparative seeding habits of 15 species and 2 varieties are given in table 157. Commercial seed-bearing in *P. serotina* begins at 10 years of age, is best from 25 to 75 years, and is usually over at 100 to 125 years. Detailed information for the other species is lacking. Seed dispersal is almost entirely by animals, especially birds.

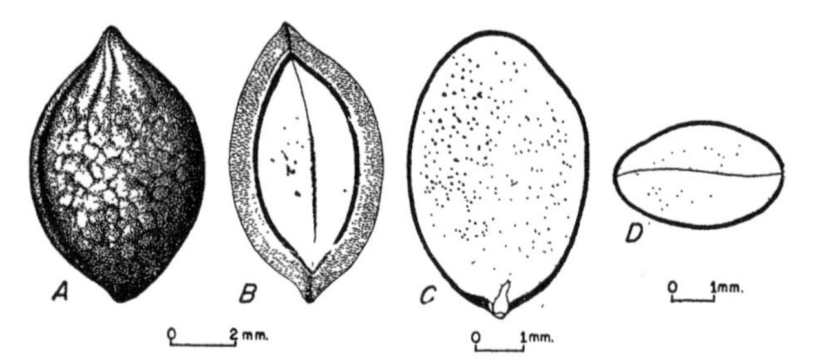

FIGURE 199.—Seed views of *Prunus alleghaniensis*: *A*, Exterior view of stone; *B*, longitudinal section of stone; *C*, longitudinal section of seed; *D*, cross section of seed.

The meager scientific information available as to the development of geographic strains in *Prunus* species is summarized here:

In *P. americana* it has been found that seed from northern Minnesota germinates much better at a temperature of 50° F. than at higher temperatures, whereas seed from Nebraska germinated as well and more rapidly at 70° (night) to 80° (day).

In *P. armeniaca* it is known that the variety called Russian apricot is hardier than the typical form. Horticulturists distinguish five races of *P. persica*.

In view of their wide range and in some cases the recognition of a number of varieties, it is likely that the following species also have developed geographic strains: *P. alleghaniensis, P. avium, P. besseyi, P. padus, P. pensylvanica, P. pumila, P. serotina,* and *P. virginiana.*

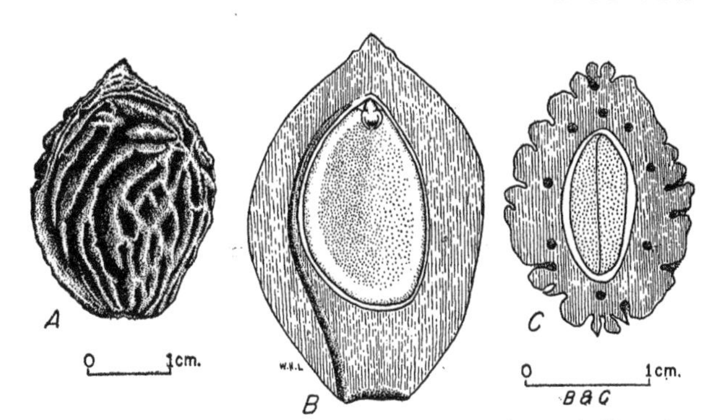

FIGURE 200.—Seed views of *Prunus persica: A,* Exterior view of stone; *B,* longitudinal section; *C,* cross section.

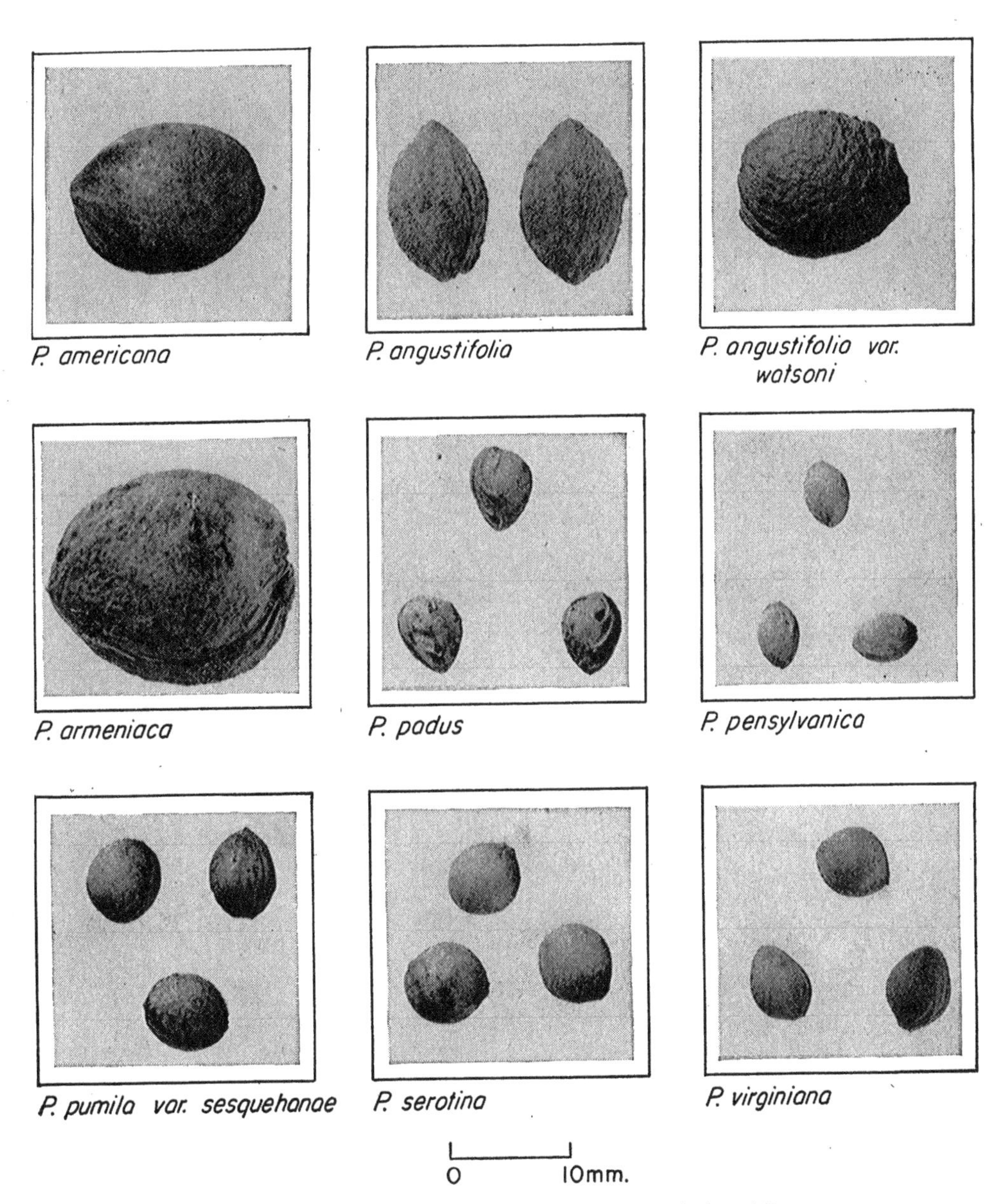

FIGURE 201.—Seed views of seven species and two varieties of *Prunus.*

TABLE 157.—*Prunus: Time of flowering and fruit ripening, and frequency of seed crops*

Species	Time of—			Color of ripe fruit	Seed year frequency	
	Flowering	Fruit ripening	Seed dispersal		Good crops	Light crops
P. alleghaniensis	May	Aug. to Sept. 15	Aug. to Sept. 15	Dark purple	Usually annually	
P. americana	March to May	June to Oct.	June to Oct.	Red or yellowish	do	
P. amygdalus	Jan. to April	Aug. to Oct.	Aug. to Oct.	Brownish	do	
P. angustifolia	Mar. to April 15	May to July	May to July	Bright yellow or red.		
P. angustifolia var. *watsoni.*		Aug. to Sept.	Aug. to Sept.			
P. armeniaca	Early spring	Early summer	Early summer	Yellow to reddish	Periodically	Intervening years.
P. avium	April	June to Aug.	June to Aug.	Yellow to red	Frequently	do.
P. besseyi	April to May	July to Sept.	July to Sept.	Purple to black		
P. caroliniana	Feb. to April	Autumn	Autumn to spring.	Black		
P. cerasus	May	July	July	Red	Almost annually	
P. padus	April 15 to June	July to Oct.	July to Oct.	Black	do	
P. pensylvanica	April to June	July to Aug.	July to winter	Red		
P. persica	Early spring	Early to late summer.	Summer	Yellow to reddish	Almost annually	Intervening years.
P. pumila	May to June	July to Aug.	July to Aug.	Purple black		
P. pumila var. *susquehanae.*			Aug.	Purple to black		
P. serotina	Mar. to June	June to Oct.	June to Oct.	Black	Almost annually	
P. virginiana	April to July	July to Sept. 15	July to Sept. 15	Red to black		

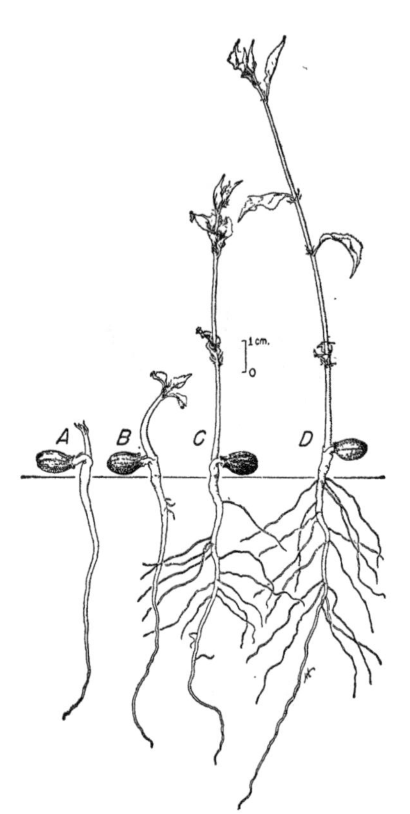

FIGURE 202.—Seedling views of *Prunus americana: A,* At 1 day; *B,* at 3 days; *C,* at 5 days; *D,* at 9 days.

COLLECTION, EXTRACTION, AND STORAGE.—Plum and cherry fruits should be collected in the summer or fall after, or in some cases just before, the fruits have assumed the color which denotes ripeness. They may be picked by hand from standing trees or gathered from the ground after falling, but usually the most economical method is to shake, strip, or flail them onto canvas or cloth. Somewhat green fruits can probably be collected safely if tests show that the seed coats within the stones have become tan to brown. After collection, the fruits should be spread out in shallow layers to prevent heating and fermentation. In *Prunus virginiana,* however, there is some evidence that a few days, fermentation may be beneficial. Plum and cherry seeds may be extracted by running the fruits through a macerator or hammer mill with water and floating or skimming off the pulp. In *P. caroliniana* and *P. pensylvanica* and possibly some of the other small-fruited cherries, the entire fruit may be dried. Seed of *P. amygdalus* may need to be separated from adhering hulls by shaking. After extraction, the seed should be dried for a short time before storage or sowing. Too much drying appears to increase dormancy in some species. The yield, size, purity, and soundness of commercial cleaned seed vary considerably by species as indicated in table 158.

Storage of *Prunus* seed for long periods is not commonly attempted, but with optimum moisture contents and low temperatures, seed of most species could probably be kept with high viability for several years. *P. americana* seed has been stored dry in open vessels at room temperature for 2½ years without loss in viability and for nearly 4 years with little loss. Seed of *P. pensylvanica* will retain its viability at least 1 year if stored in sealed containers at 41° F. Seed of *P. serotina* has kept well for 2 years in sealed containers held at 41°. Seed of *P. virginiana* stored dry for 1 year at 26° showed relatively little loss of viability, but at higher temperatures the losses were progressively greater; seed which was fermented during extraction still germinated more quickly and better than seed from fresh berries after 1 year's storage. Seeds of *P. armeniaca* and *P. avium* also keep satisfactorily for 1 to 2 years under ordinary storage conditions. Seed of *P. amygdalus* should be stored dry.

GERMINATION.—Natural germination takes place the spring following seed dispersal. Moist, exposed mineral soil makes a favorable seedbed. Germination is epigeous in some species and hypogeous in others (figs. 202 and 203). All species have delayed germination due to dormant embryos, and in several cases probably also due to hard seed coats. Dormancy may be overcome in most species by stratification in moist sand, peat, or a combination of the two, at low temperature. Further information on seed dormancy and pretreatment is given in table 159. As is evident, however, the best practices have not yet been worked out for several of the species listed, and for *P. padus* not even a fairly satisfactory procedure has been developed. Germination tests should be made in sand flats using up to 200 properly pretreated seeds per test. Recommended procedures and results vary with the species as shown in table 160. Results can be obtained in a much shorter time if the seeds are first removed from the stones. This practice involves considerable labor, however.

NURSERY AND FIELD PRACTICE.—Plum and cherry seeds should be sown in the fall in drills spaced 8 to 12 inches apart or in the spring if pretreated seeds are used. *Prunus pensylvanica* seed should be

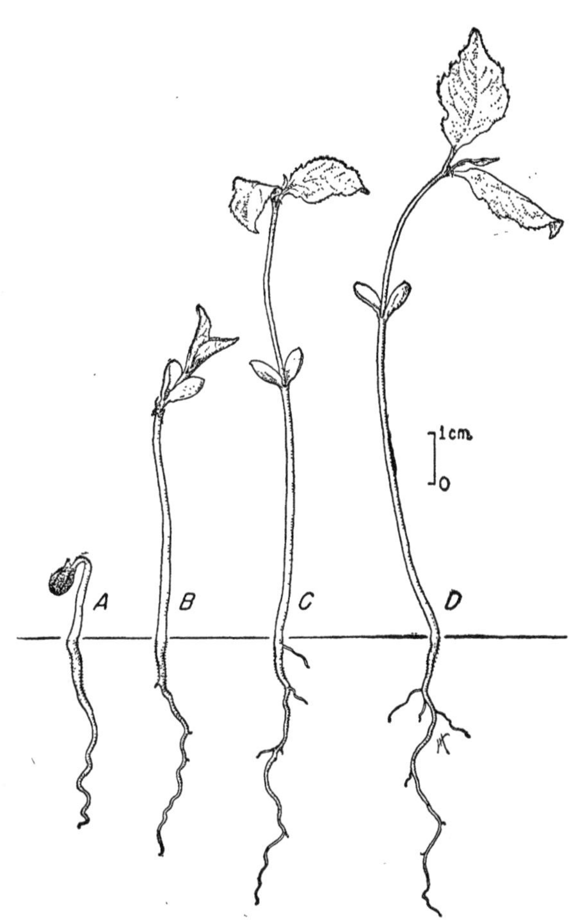

FIGURE 203.—Seedling views of *Prunus virginiana: A*, at 1 day; *B*, at 3 days; *C*, at 7 days; *D*, at 11 days.

TABLE 158.—*Prunus: Yield of cleaned seed, and purity, soundness, and cost of commercial seed*

Species	Cleaned seed					Commercial seed		
	Yield per 100 pounds of fruit	Per pound			Basis, samples	Purity	Soundness	Cost per pound
		Low	Average	High				
	Pounds	*Number*	*Number*	*Number*	*Number*	*Percent*	*Percent*	*Dollars*
P. alleghaniensis	----	----	2,951	----	1	----	96	----
P. americana	15–30	550	840	1,380	16	96	94	0.50–1.25
P. amygdalus	----	----	126	----	1	100	100	.25–1.00
P. angustifolia	15–30	770	1,060	1,530	6	----	96	1.50
P. angustifolia var. *watsoni*	15–30	----	1,000	----	1	----	----	----
P. armeniaca	30–40	250	300	360	5	----	99	.75– .85
P. avium	8	2,200	2,600	3,000	3	98	88	.60– .75
P. besseyi	15–28	1,500	2,060	2,410	3	----	94	1.00–2.25
P. caroliniana	----	----	----	----	----	----	----	.75–1.70
P. padus	----	6,600	6,800	7,045	2	----	----	.50– .90
P. pensylvanica	±16	11,900	15,700	21,800	4	----	84	1.25–1.75
P. persica	20	----	165	----	1	99	88	.15– .80
P. pumila	----	2,500	----	2,740	2	----	----	----
P. pumila var. *susquehanae*	----	3,780	4,750	5,970	3	98	99	----
P. serotina	20–40	3,100	4,800	8,100	18	99	96	.40–1.30
P. virginiana	15–20	3,000	5,800	8,400	21	97	94	1.25–1.75

sown in July or August. The following sowing depths are recommended: *P. persica,* 2 inches; *P. armeniaca,* 1½ inches; *P. angustifolia* and *P. angustifolia* var. *watsoni,* 1 inch; other species listed in table 156, one-half inch. The plum species should be sown at the rate of about 18 seeds per linear foot. With *P. virginiana* seed, and probably those of the other cherries, about 25 seeds per linear foot should be sown. The seedbeds should be mulched until germination begins, and protected from birds and rodents. For *P. angustifolia* and *P. armeniaca* one-third to one-half of the viable seed sown produce usable seedlings; similar information for the other species is lacking.

Some *Prunus* species are attacked by a leaf blight in the nursery. A "shot-hole" or leaf spot disease caused by the fungus *Coccomyces lutescens* sometimes causes considerable damage to *P. serotina, P. virginiana,* and *P. melanocarpa* in nurseries. The disease may be controlled by spraying with bordeaux mixture (4–6–50 or 3–4–50) or a 2-percent solution of lime sulfur. Such treatment may, however, cause considerable leaf burning. Another leaf spot disease affecting wild cherries in the nursery is caused by bacteria (*Bacterium pruni*); it probably can be controlled by the measures recommended above. A powdery mildew caused by the fungus *Podosphaera oxyacanthae* occasionally causes some losses in wild cherry stock. It may be controlled by dusting with sulfur or spraying with bordeaux. *P. americana* seedlings may be attacked by a brown rot caused by *Monilinia fructicola.* The spread of the disease can be held down by raking and burning the fallen leaves and tips, by rotation of the beds and keeping them away from infected hosts, and by applications of 4–4–50 or 3–4–50 bordeaux.

Field planting of *Prunus* species is usually done with 1–0 stock. Ordinarily planting should be done on deep, well-drained soils in sunny locations free of frost pockets. However, *P. cerasus* is one of the few fruit trees which can thrive in shade, and *P. serotina* in the Central States makes best growth under partial shade of black locust and sassafras. Horticultural varieties of most *Prunus* species are propagated chiefly by budding or grafting.

TABLE 159.—*Prunus: Method of seed pretreatment for dormancy*

Species	Stratification			Other methods	Remarks
	Medium	Duration and temperature			
		Days	*°F.*		
P. alleghaniensis	Moist sand	90+	41		Stratification longer than 90 days probably better.
P. americana	Moist sand or peat.	150	41		
P. angustifolia	Moist sand	120+	41		150 days or more stratification probably better.
P. angustifolia var. *watsoni.*[1]	do	120+	41		
P. armeniaca	do	60	41		
P. avium	Moist sand or peat.	90–120	32–41		
P. besseyi	Moist sand	90+	41		
P. caroliniana	Soil	60	Winter		Stratify outdoors in January and February in Iowa.
P. cerasus	Moist peat	90–120	32–50		
P. padus	Moist sand	90 at + 60 at	68–86 + 41	None satisfactory	Meager germination given by recommended method; further work necessary.
P. pensylvanica	do	60 at + 90 at	68–86 + 41	Clip off one end of stone and stratify 120 days at 41° F.	Better methods need to be worked out; about one-third of nongerminated seed is still sound.
P. persica	Moist peat	45– 90	35–45		Mature seed from green fruits reported nondormant.
P. pumila[1]	Moist sand	90+	41		Stratification in coldframe during winter gave 18 percent germination.
P. pumila var. *susquehanae.*	do	60 at + 120 at	68–86 + 41	Soak in H_2SO_4 for 1 hour, then stratify 60–90 days.	Stratification for 120 days at 50° F. has been relatively ineffective.
P. serotina	Moist sand or peat.	90–120	41	Soak dry seed in H_2SO_4 30 minutes.	Some seed germinate in stratification when 150-day period is used. Best results occur when seed is soaked in H_2SO_4 for 30 minutes and then stratified in moist sand at 41° F. for 120 days.
P. virginiana	do	90–160	41		Seed germinate in stratification if held too long.

[1]Treatment suggested; experimental data not complete.

TABLE 160.—*Prunus: Recommended conditions for germination tests, and summary of germination data*

Species	Test conditions recommended			Germination data from various sources							Type of germination
	Temperature		Duration	Germinative energy		Germinative capacity			Potential germination	Basis, tests	
	Night	Day		Amount	Period	Low	Average	High			
	°F.	*°F.*	*Days*	*Percent*	*Days*	*Percent*	*Percent*	*Percent*	*Percent*	*Number*	
P. alleghaniensis	50	50	60	12–27	18–25	2	25	37	73–100	7	Hypogeous.
P. americana (northern seed)	50	50	60	9–92	4–28	12	60	97	45–100	21	do.
P. americana (southern seed)	70	80									
P. amygdalus	[1]68	[1]86	60	—	—	—	—	—	—	—	—
P. angustifolia	68	86	60	13+	20+	21	39	70	—	3	—
P. angustifolia var. *watsoni.*	68	86	60	—	—	—	2	—	100	1	—
P. armeniaca	68	86	40	38–84	8–21	44	—	89	—	3	—
P. avium	[1]68	[1]86	[1]60	—	—	56	76	94	—	4	—
P. besseyi	68	86	60	7+	13+	8	—	72	—	2	—
P. caroliniana	—	—	30	—	—	—	—	—	—	1	Hypogeous.
P. cerasus	[1]68	[1]86	[1]60	—	—	—	—	—	—	—	—
P. padus	[1]68	[1]86	[1]90	—	—	—	—	—	—	1	Epigeous.
P. pensylvanica	50	77	60	45	10	50	—	73	72–97	2	do.
P. persica	[1]68	[1]86	60	—	—	15	50	95	—	6+	—
P. pumila	[1]68	[1]86	[1]60	—	—	—	18+	—	—	1	Probably hypogeous.
P. pumila var. *susquehanae.*	50	77	40	44–70	5–19	59	67	78	86–99	3	Hypogeous.
P. serotina	50 68	77 86	30	51–83	5–12	21	63	87	68–100	7	do.
P. virginiana	50	77	40	14–66	11–20	10	43	66	58–100	7	Epigeous.

[1]Suggested treatment.

PSEUDOTSUGA Carr. Douglas-fir

(Pine family—Pinaceae)

Distribution and Use.—The Douglas-firs include five species of medium to very large evergreen trees native to western North America and eastern Asia. Of the two North American species, one (*Pseudotsuga taxifolia*) is widely distributed and the most important timber tree in the United States; the other (*P. macrocarpa*) is of restricted distribution and minor importance. These two species are described in table 161. *P. taxifolia* is extensively used in reforestation in the western United States and Canada and has been widely planted in Europe; for ornamental purposes it has been planted successfully in parts of Europe and northern and eastern United States.

Seeding Habits.—Flowers of both sexes are borne singly on branchlets formed the previous year on different parts of the same tree. The male flowers, arising from the axils of the leaves, are small-stalked, cylindric bodies. The female flowers, occurring at the tips of branchlets, are small, bristly, scaly, conical bodies which develop into cones at the end of the first season. Two larchlike seeds, partly enveloped in large rounded wings (figs. 204 and 205), are borne under each cone scale. The seeds are dispersed chiefly by wind. Comparative seeding habits of two species of *Pseudotsuga* are as follows:

	P. macrocarpa	*P. taxifolia*[1]
Time of—		
Flowering	----------	Spring–summer.
Cone ripening	Aug.	Aug.–Sept.
Seed dispersal	Aug.–Sept.	do.
Cone length inches	3½–7	2–4.

[1]Commercial seed-bearing age: Minimum, 9 years; optimum, 100 to 200 years; maximum, 600 years.

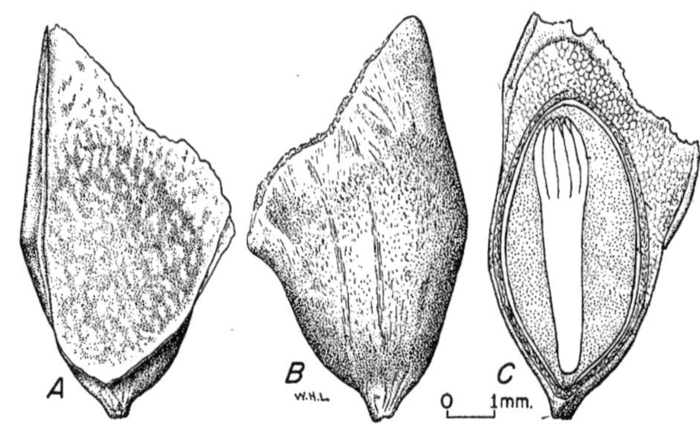

Figure 204.—Seed views of *Pseudotsuga taxifolia: A* and *B*, Exterior views in two planes, showing attached wing base; *C*, longitudinal section.

0 5mm.

Figure 205.—Seed views of *Pseudotsuga macrocarpa.*

Table 161.—*Pseudotsuga: Distribution and uses*

Accepted name	Synonyms	Natural range	Chief uses	Date of earliest cultivation
P. macrocarpa (Vasey) Mayr (bigcone-spruce).	bigcone Douglas-fir, SPN	Mountain slopes of southern California.	Locally for lumber; can be used for erosion control.	1910
P. taxifolia (Poir.) Britton (Douglas-fir).	*P. douglasii* Carr., *P. mucronata* (Raf.) Sudw. (Douglas-spruce, red pine, Oregon pine).	Mountainous regions from British Columbia to California, and Montana to Colorado, western Texas, and northern Mexico.	Lumber, structural timbers, posts and poles, pulpwood, ties, plywood, veneer, sash work; erosion control; windbreaks; ornamental; wildlife food.	1827

Good seed crops are borne infrequently by *P. macrocarpa* with light crops in the intervening years. In *P. taxifolia* good crops are borne every 3 to 7 years with light crops in the intervening years. The cones of *P. taxifolia* usually drop off the trees within a few months after seed dispersal while those of *P. macrocarpa* may remain on the branches for a year or more.

Information is not available as to the development of climatic races in *Pseudotsuga macrocarpa*. However, in view of its restricted range, it is probable that no racial differentiation has taken place. In the case of *P. taxifolia,* two forms—the Rocky Mountain or blue and coast or green—sometimes considered as distinct species, have long been recognized. The Rocky Mountain form is hardier and its seed usually germinate quicker and more easily in the laboratory than do those of the coast form which is faster growing. Seed from the southwest appear to have little or no dormancy, as contrasted to that from most other localities. Average seed size appears to be independent of race but seems to decrease from south to north. Studies in the West further indicate the presence of races within the coastal form. In Europe, foresters recognize three forms of *P. taxifolia:* green (coastal), gray (inland), and blue (Rocky Mountain), and some intermediate forms. Studies show that these differ in color and length of foliage, rate of growth, form, frost hardiness, and resistance to needle-cast disease. The coastal form is recommended for use in milder conditions and the inland form for use on drier and colder sites.

Collection, Extraction, and Storage.—Ripe, unopened *Pseudotsuga* cones may be collected from standing or felled trees or squirrel caches in August or September. Ripe cones are brownish in color. Ripe seed of *P. macrocarpa* are dark chocolate brown and shiny on the upper side and dull, very slightly reddish brown on the under side; those of *P. taxifolia* are a dull russet brown with areas of white. In *P. taxifolia* an average tree produces about 2½ bushels of cones; an average bushel contains about 1,000 cones.

The following extraction and cleaning methods are recommended for seed of *Pseudotsuga taxifolia,* and similar practices probably can be employed for *P. macrocarpa:* (1) Spread the freshly collected cones out to dry in the sun or place them in a simple convection type cone kiln for 10 to 15 hours at 104° F. (for fresh cones) or 110° (for precured cones) and gradually increase the temperature to 130°; (2) run the cones through a shaker or flail them to remove the seed; (3) dewing the seed by running them through a dewinger, by kneading them in a sack, or by rubbing them over a one-sixth-inch mesh screen; (4) clean the seed by fanning or blowing off the chaff. The yield, size, purity, soundness, and cost of commercial cleaned seed vary according to the species, as follows:

	P. macrocarpa	*P. taxifolia*
Cleaned seed:		
Per bushel fresh cones____pounds__	____	⅓–1⅓
Per pound:		
Low________________number__	4,000	20,000
Average________________do____	5,000	[1]42,000
High___________________do____	6,000	68,000
Basis, samples______________do____	3	148
Commercial seed:		
Purity________________percent__	88	93
Soundness________________do____	____	72
Cost per pound__________dollars__	6.00	2.50–4.50

[1]Seed from different localities average about as follows: California, 30,000–35,000; central and northern Rocky Mountains, 40,000–45,000; Washington (coast), 44,000; British Columbia (coast), 49,000.

Under open storage at ordinary temperatures, seed of *Pseudotsuga macrocarpa* loses its viability completely in 4 years as compared to 10 years for *P. taxifolia.* When stored in sealed containers at ordinary temperatures, the germination of *P. taxifolia* seed dropped from 52 to 30 percent in 4 years. Seeds of both species stored in sealed containers at 40° F. still retained their original viability at the end of 4 years.

Germination.—Although natural germination of *Pseudotsuga macrocarpa* ordinarily is quite scanty, it is best in leaf litter under shade near seed trees or under live oaks. That of *P. taxifolia* is best on warm, moist, pure mineral soil or a mixture of it with humus. Much seed is destroyed by animals. Germination is epigeous (fig. 206). In both *P. macrocarpa* and *P. taxifolia* dormancy occurs in the embryo and possibly in the seed coat. However, seed dormancy in *Pseudotsuga* is a variable thing. In *P. macrocarpa* part of the seed in some lots is dormant; cold stratification in moist sand does not increase germination, but speeds it up considerably. In *P. taxifolia* some seed lots are nondormant,

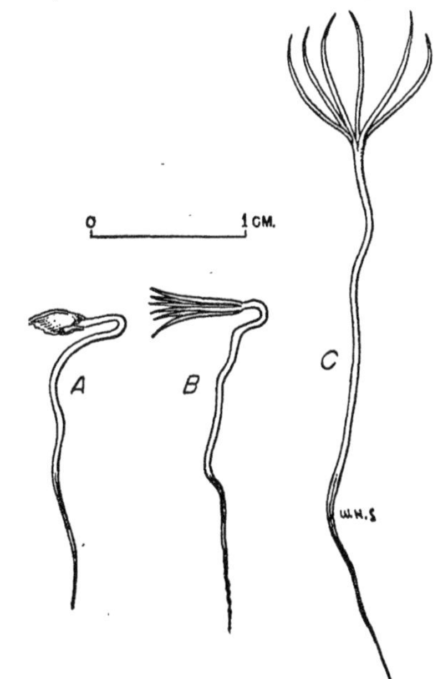

Figure 206.—Seedling views of *Pseudotsuga taxifolia: A,* At 2 days; *B,* at 5 days; *C,* at 8 days.

whereas others are dormant and will not even begin to germinate after more than 60 days unless pretreated. Most lots of this species contain some dormant seed which can be stimulated to germinate by cold; moist stratification. In one test high germination temperatures fluctuating diurnally from 57° to 95° F. stimulated untreated seed into good germination, but germination was not as rapid as with stratified seed. To overcome dormancy in *P. macrocarpa* and *P. taxifolia*, stratification in moist sand at 41° is recommended; the former species should be stratified for 30+ days and the latter 30 to 60 days. One series of tests for *P. taxifolia*, in which seed were soaked in water at 59° for 2 days, gave fairly good results, and higher germination temperatures improved the germination of unstratified seed.

The germination of *Pseudotsuga* seeds may be tested in sand or soil flats or standard germinators, using 400 to 1,000 stratified or nondormant seeds per test. Methods recommended and average results for 2 species are as follows:

	P. macrocarpa	*P. taxifolia*
Test conditions:		
Night temperature ----------- °F.	[1]70	60
Day temperature ----------- do.	[1]80	95
Duration:		
Stratified seed ----------- days.	20–30	15–30
Untreated seed ----------- do.	60–90	60–90
Germinative energy:		
Amount ----------- percent.	----	80
Period ----------- days.	----	10–22
Germinative capacity:		
Low ----------- percent.	15	24
Average ----------- do.	30+	85
High ----------- do.	57	90
Basis, tests ----------- do.	5+	[2]50

[1]Suggested.
[2]Stratified seed only; 664 tests of untreated seed averaged 54 percent germinative capacity, ranging from 0 to 95 percent.

Nursery and Field Practice.—Seed of *Pseudotsuga taxifolia*[52] should be sown in drills in the fall, or nondormant or stratified seed in the spring, at a rate of about 5 to 10 ounces per 50 square feet of seedbed to produce about 80 to 120 usable 1–0 seedlings per square foot of bed. The lower density of sowing is preferred if the stock is to be field planted as 2–0, and the higher density if the stock is to be transplanted. The seed should be covered with about one-fourth inch of soil and the seedbeds mulched until germination begins. Nursery germination starts about 20 days after spring sowing and is complete about 40 days later. Half-shade should be provided for the seedbeds the first season, and sometimes for the second. For field planting, 1–1 stock or root-pruned 2–0 stock grown at a density of 75 per square foot is preferred. In the nursery seedlings may be injured by damping-off, which can be minimized by acid or sulfate (aluminum or ferrous) treatment of the beds. The seedlings are sometimes attacked by a root rot fungus, *Phytophthora cinnamomi*, for which no adequate control measures are known. Sometimes, too, the seedlings are attacked by the needle-cast disease caused by the fungus *Rhabdocline pseudotsugae*. This may be controlled by spraying with bordeaux mixture and by cutting out infected trees near the nursery.

[52] The nursery practices discussed here refer to *Pseudotsuga taxifolia*, but they may be used for *P. macrocarpa* since detailed information for this species is lacking.

PTELEA TRIFOLIATA L. Common hoptree

(Rue family—Rutaceae)

Also called wafer ash, three-leaved hop-tree, shrubby trefoil; includes var. *mollis* Torr. & Gray.

DISTRIBUTION AND USE.—Hoptree occurs on rocky slopes and alluvial banks from Ontario and New York to Florida and west to Iowa, Kansas, and Arizona. It is a deciduous shrub or small tree with a disagreeable odor, and is of some value for game food and for shelter-belt and ornamental planting. The bark is occasionally used as a tonic and the fruit as a substitute for hops. This species was introduced into cultivation in 1724.

SEEDING HABITS.—The small, greenish, polygamous flowers, borne in terminal clusters, open from March to July after the leaves. The fruit is about 1 inch in diameter and is a two-celled, often two-seeded, rounded samara with a thin wing (fig. 207). It is straw-colored to dark brown, ripens from August to September, and may persist on the branches through much of the winter. Hoptree is an abundant seeder and dispersal is by wind.

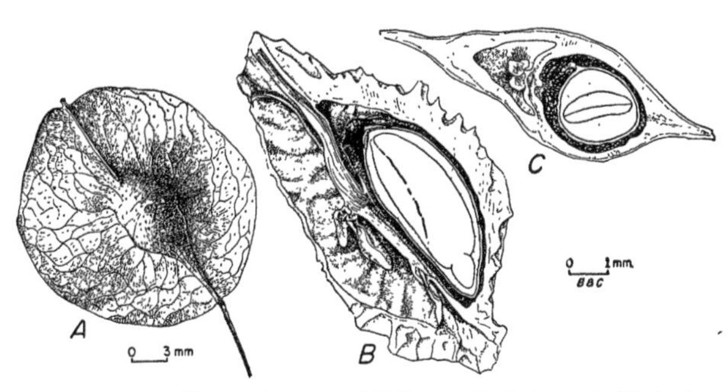

FIGURE 207.—Seed views of *Ptelea trifoliata: A,* Exterior view of fruit; *B,* longitudinal section showing seed; *C,* cross section showing seed.

COLLECTION, EXTRACTION, AND STORAGE.—Commercial "seed" consists of the ripened samaras which may be picked from the branches. Fresh fruit may require a few days drying before it is stored. No extraction is necessary. Cleaned samaras per pound (5 samples): low, 9,000; average, 12,000; high, 18,000. One hundred samaras will contain about 97 sound seeds. Commercial seed averages 98 percent in purity, 90 percent in soundness, and costs about $1.25 per pound. Seed of this species will retain most of its viability for at least 16 months if stored at 41° F. in sealed containers.

GERMINATION.—Hoptree seed germinates slowly, probably because of embryo dormancy. Germination, which is epigeous (fig. 208), can be greatly hastened by stratification in sand or peat for 3 to 4 months at 41° F. Germination tests can be made in sand flats at temperatures alternating diurnally from 50° to 77°. Germinative-energy period, 15 to 20 days; germinative capacity (6 tests): low, 10 percent; average, 28 percent; high, 91 percent.

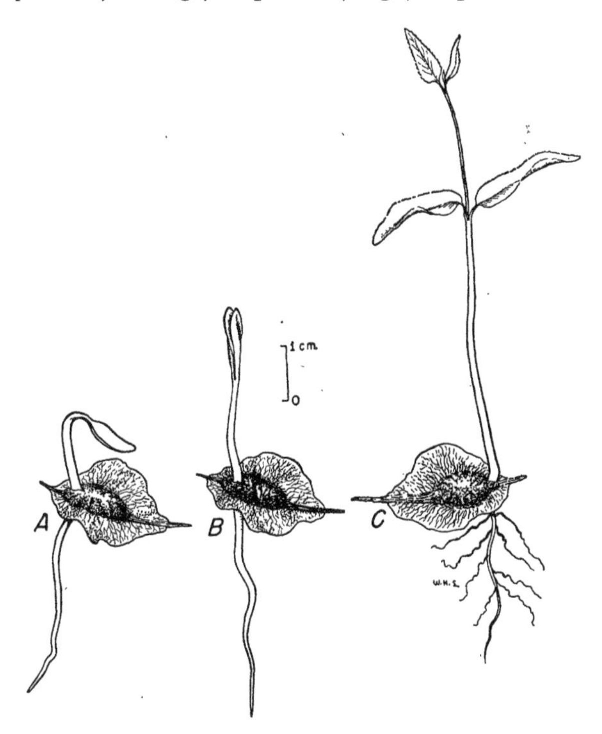

FIGURE 208.—Seedling views of *Ptelea trifoliata: A,* At 1 day; *B,* at 2 days; *C,* at 10 days.

NURSERY PRACTICE.—Seed of this species should be fall sown or stratified over most of the winter and sown in the spring. If fall sown, beds must be mulched to prevent extreme freezing and thawing. Propagation by layers, grafting, or budding is also feasible. For best results, common hoptree should be planted on porous, moderately moist soils in partial shade.

PURSHIA TRIDENTATA (Pursh) DC. Antelope bitterbrush

(Rose family—Rosaceae)

Botanical syn.: *Kunzia tridentata* (Pursh) Spreng.

Native to the arid flats of Oregon, California, and eastward throughout the Rocky Mountains, bitterbrush, the only species in its genus, is a deciduous spreading shrub. It is one of the most important plants of the West for game food and cattle and sheep browse. This species can be used for erosion control. It was introduced into cultivation in 1826.

Bitterbrush seed is actually a fruit, an oblong pubescent achene about one-fourth inch to one-half inch long (fig. 209). The perfect, yellowish flowers bloom from April to July, and the fruit ripens from July to September. At this time it can be collected, rubbed in a sack, winnowed, and stored in airtight containers at low temperatures (35° to 40° F.). By this method the original viability of the seed can be preserved for at least 3 years. One pound contains from 17,000 to 23,000 seeds (3 samples). Purity of 1 sample was 77 percent, and soundness of 2 samples averaged 82 percent. This species is not listed in commercial seed catalogs.

Because about 85 percent of the bitterbrush seed in a lot is dormant, this species requires from 2 to 3 months' stratification. A germination of 69 percent in 5 days was obtained after such treatment (1 test). Bitterbrush has been neither grown in nurseries on a large scale nor sown in the field. It should be planted on well-drained soils in sunny locations. Propagation by layering is also possible.

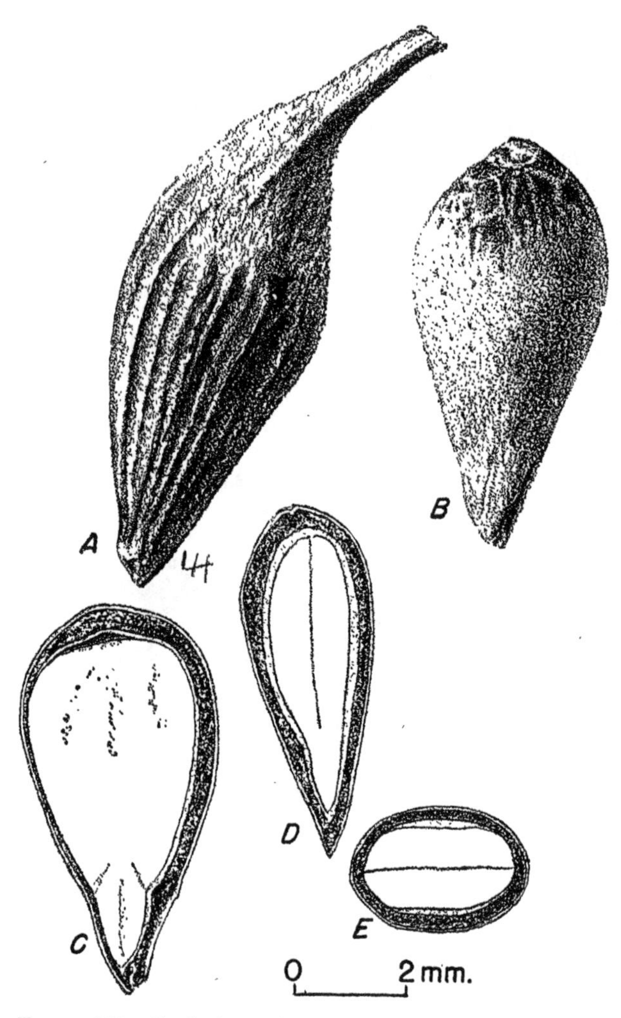

FIGURE 209.—Seed views of *Purshia tridentata*: *A*, Exterior view of achene; *B*, exterior view of seed; *C* and *D*, longitudinal section in two planes; *E*, cross section.

PYRUS COMMUNIS L. Common pear

(Rose family—Rosaceae)

Includes var. *pyraster* L., var. *cordata* (Desv.) Schneid., var. *sativa* DC., and some others.

Distribution and Use.—Native to Europe and western Asia, the common pear is the best known representative of a genus containing about 20 species. It is a small to medium-sized, deciduous tree which has long been cultivated chiefly for fruit and ornamental purposes, and also to some extent for shelter-belt and wildlife food purposes. In many places it has escaped from cultivation and become naturalized.

Seeding Habits.—Large, perfect, white flowers appear in May with or before the leaves. The fruits, turbinate or subglobose, yellow to reddish pomes up to 2 inches long, and larger in many horticultural varieties, ripen from late August to early October; some varieties ripen from early July, on. Each fruit usually contains from 4 to 10 rather small, smooth black, or nearly black, seeds (fig. 210). The greater hardiness of the Russian forms indicates the existence of geographic strains.

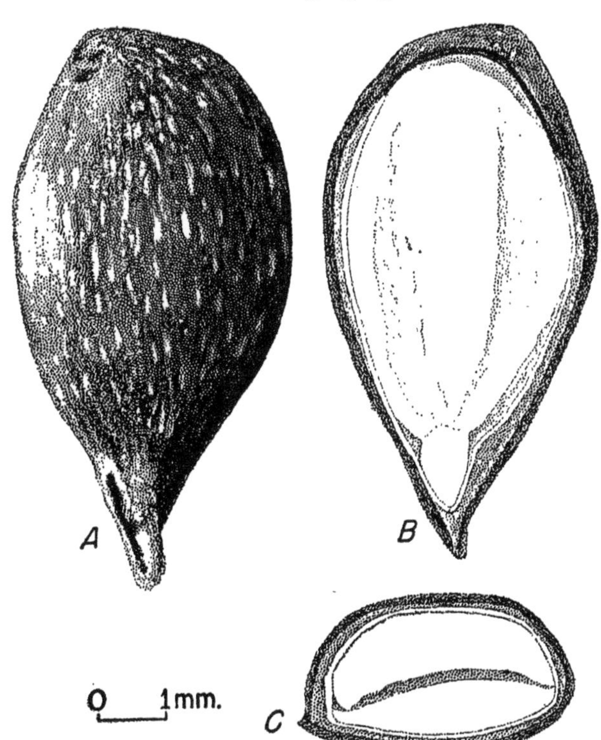

Figure 210.—Seed views of *Pyrus communis*: *A*, Exterior view; *B*, longitudinal section; *C*, cross section.

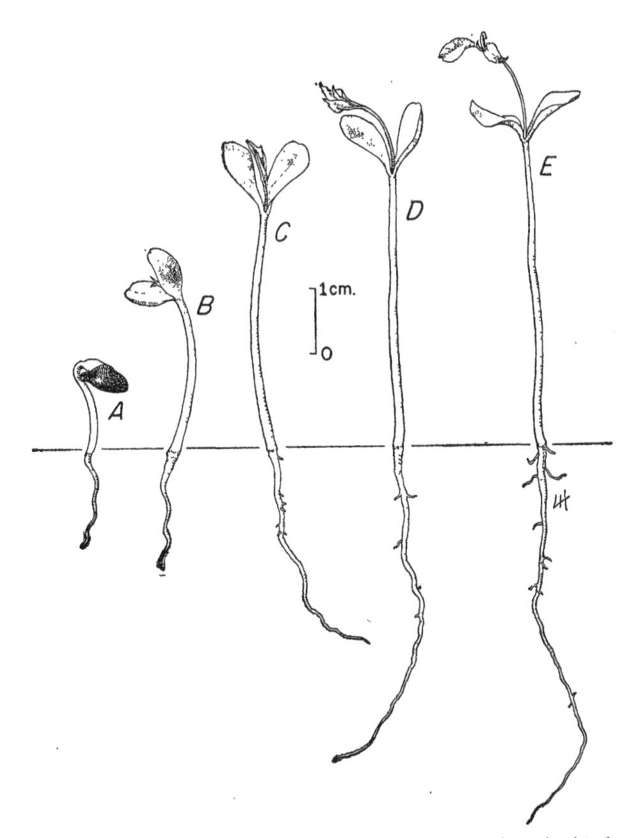

Figure 211.—Seedling views of *Pyrus communis*: *A*, At 1 day; *B*, at 2 days; *C*, at 3 days; *D*, at 6 days; *E*, at 12 days.

Collection, Extraction, and Storage.—The ripe fruits may be picked from standing trees, or shaken down. It is suggested that the fruits be run wet through a macerator with water, and the pulp floated off or screened out, or that they be run through a cider mill, and the seeds dried. One hundred pounds of fruit yields about 1 pound of cleaned seed. Cleaned seed per pound (11+ samples): low, 12,000; average, 15,000; high, 24,000. Commercial seed averages about 98 percent in purity (3 samples), 96 percent in soundness (2 samples), and costs about $0.75 per pound for domestic seed and $2 to $3.25 for that of "French pears." Cleaned seed should be stored dry at room temperature before they are sown or tested, or in moist sand at room temperature or lower. Seed viability is maintained satisfactorily for 2 to 3 years in ordinary dry storage. No information as to long-time storage practices is available.

GERMINATION.—Germination is epigeous (fig. 211). The seeds exhibit dormancy due to internal conditions of the embryo. This can be overcome by stratification in moist sand and peat, or peat alone, for 60 to 90 days at 32° to 45° F. Various chemical treatments have failed to break dormancy satisfactorily. Germination can be tested in sand flats for 45 days, using 400 stratified seeds per test; a temperature of 50° is suggested. Germinative capacity of untreated seed averages about 1 percent; of stratified seed (15 tests): low, 41 percent; average, 65 percent; high, 98 percent.

NURSERY PRACTICE.—Common pear seeds should be sown in drills in the fall, or stratified seeds used in the spring, and covered with about one-half inch of soil. Seedlings of 1–0 stock can be field planted, or they can be root pruned at a depth of 6 to 8 inches and transplanted for 1 year. They are frequently subject to attack by mildew. Horticultural varieties of this species usually are propagated by budding or grafting. Pears do best on clay loam soils, not too moist.

QUERCUS L. Oak

(Beech family—Fagaceae)

DISTRIBUTION AND USE.—Approximately 275 species of oak have been identified. They occur in the North Temperate Zone and tropics as far south as Colombia in South America, the Malay States in southeast Asia, and the Indian islands. There are both deciduous and evergreen trees and shrubs among the many species. About 54 of the North American species are large or small trees. The oaks provide some of the most valuable hardwood timber. Some are chief sources of essential bark products and bird and animal food. Many small tree and shrub species have their greatest value in watershed protection and soil stabilization. Of the two broad groups of oaks, white and black, the white oaks have more durable wood and a greater economic value. The natural range and chief uses of 31 of the more valuable species and 2 additional varieties are described in table 162.

TABLE 162.—*Quercus: Growth habit, distribution, and uses*

Accepted name	Synonyms	Growth habit	Natural range	Chief uses	Date of earliest cultivation
Q. agrifolia Née (California live oak).	coast live oak	Evergreen tree attaining 95-foot height, often shrubby.	Valleys, rocky hills, canyon walls of California.	Watershed cover, soil stabilization, ornamental planting.	1849
Q. alba L. (white oak).	stave oak	Large, deciduous tree 80 to 125 feet, occasionally 150 feet.	Well-drained bottom lands, rich uplands, sandy plains, gravelly ridges Maine to Florida, west to Minnesota and Texas.	Lumber, flooring, furniture, shakes, tight cooperage, fuel, ties, posts, animal food, shade.	1724
Q. arizonica Sarg. (Arizona white oak).		Deciduous tree attaining 50- to 60-foot height.	Southern New Mexico, Arizona, northern Mexico.	Fuel, game food and cover.	
Q. bicolor Willd. (swamp white oak).	*Q. platanoides* (Lam.) Sudw.	Deciduous tree attaining 100-foot height.	Rich soil, bottom lands and swamps Maine to Ontario, Minnesota to Missouri, south to Virginia, west to Arkansas.	Tight cooperage, flooring, furniture, lumber, fuel, animal food.	1800
Q. borealis Michx. f. (northern red oak).	red oak	Deciduous tree 60 to 80 feet high.	Rich, moist soils of protected sites Nova Scotia to Ontario, south to North Carolina, Ohio, Michigan, Wisconsin, Minnesota, and Iowa.	Lumber, interior finishing, furniture, cross ties, posts, game food, forest planting, shade.	1800
Q. borealis var. *maxima* (Marsh.) Ashe (eastern red oak).	*Q. rubra* Du Roi, not L. (red oak).	Deciduous tree 70 to 80 feet high, occasionally 150 feet.	Rich moist soils Nova Scotia to Florida, west to Minnesota and Texas.	Flooring, lumber, cross ties, fuel, posts, forest planting, game food.	1724
Q. chrysolepis Liebm. (canyon live oak).	maul oak, California live oak.	Medium-sized tree 40 to 50 feet high. Evergreen, leaves deciduous after 3 and 4 years.	Canyon bottoms, moist ridges and flats southern Oregon, California, Arizona, and New Mexico.	Watershed protection, soil stabilization, ornamental planting, fuel, agricultural implement parts, pack saddles, tool handles.	1877

TABLE 162.—*Quercus: Growth habit, distribution, and uses*—Continued

Accepted name	Synonyms	Growth habit	Natural range	Chief uses	Date of earliest cultivation
Q. coccinea Muenchh. (scarlet oak).	----------	Deciduous tree 70 to 100 feet high.	Maine to Ontario, south to Florida and Oklahoma.	Rough lumber, cross ties, fuel, ornamental planting.	1691
Q. douglasii Hook. & Arn. (blue oak).	California blue oak, mountain white oak.	Medium-sized, deciduous tree 50 to 60 feet high, rarely 90 feet.	Sierra Nevada foothills and inner coastal ranges of California.	Fuel, watershed protection, soil stabilization.	--------
Q. dumosa Nutt. (California scrub oak).	scrub oak --------	Small tree, rarely 20 feet high.	Southern California, the Coast Ranges, and Sierra Nevada.	Watershed protection, soil stabilization.	--------
Q. ellipsoidalis E. J. Hill (northern pin oak).	jack oak, Hill's oak.	Medium-sized, deciduous tree 60 to 70 feet high.	Dry, acid soils of uplands Michigan and Indiana, northwest to Manitoba.	Rough lumber, fuel, game food.	1902
Q. falcata Michx.[1] (southern red oak).	*Q. rubra* L., not Du Roi, *Q. digitata* (Marsh.) Sudw. (red oak, Spanish oak).	Deciduous tree 70 to 80 feet high, occasionally 100 feet.	Dry uplands New Jersey to southern Illinois, south to central Florida and Texas.	Rough lumber, cross ties, fuel, tannin, game food.	1763
Q. falcata var. *pagodaefolia* Ell. (swamp red oak).	*Q. rubra* var. *pagodaefolia* (Ell.) Ashe, *Q. pagoda* Raf. (Spanish oak, swamp Spanish oak, red oak).	Deciduous tree 80 to 90 feet high, occasionally 120 feet.	Rich bottom lands Maryland to northern Florida, west to Texas and Mississippi Valley to Illinois.	Rough lumber, cross ties, posts, fuel.	1904
Q. garryana Dougl. (Oregon white oak).	white oak, Pacific post oak, Garry oak.	Medium-sized, deciduous tree 60 to 70 feet high.[2]	Pacific coast from Vancouver Island to northern California.	Flooring, furniture, lumber, fuel.	1873
Q. imbricaria Michx. (shingle oak).	laurel oak --------	Medium-sized, deciduous tree, 50 to 60 feet high, rarely 100 feet.	Poor to moderately drained soils Pennsylvania to Florida, west to Missouri and Kansas.	Structural timber, fuel.	1724
Q. kelloggii Newb. (California black oak).	black oak --------	Medium-sized, deciduous tree, occasionally 100 feet high.	Lower and middle mountain slopes southern Oregon and California.	Fuel ----------------	1878
Q. laevis Walt. (turkey oak).	*Q. catesbaei* Michx.	Small tree 20 to 35 feet high.	Dry, sandy soils of Coastal Plain southeastern Virginia to Florida and west to eastern Louisiana.	Fuel, nurse tree -------	(3)
Q. lobata Née (California white oak).	valley oak --------	Large, deciduous tree often 100 feet high and 10 feet in diameter.	Foothills and valleys of California.	Fuel ----------------	1874
Q. lyrata Walt. (overcup oak).	swamp white oak.	Medium-sized, deciduous tree 70 to 80 feet high, rarely 100 feet.	Bottom lands New Jersey to Florida, west through the Gulf region to Texas and up Mississippi Valley to southern Indiana.	Lumber, posts --------	1786
Q. macrocarpa Michx.[4] (bur oak).	mossy cup oak, overcup oak.	Large, deciduous tree usually 80 to 100 feet high, occasionally 170 feet.	Rich bottom lands Nova Scotia to West Virginia and Tennessee west to Minnesota and Texas.	Construction lumber, tight cooperage, furniture, cross ties, fuel, game food, shelter belts, ornamental planting.	1811
Q. montana Willd. (chestnut oak).	*Q. prinus* Engelm., not L. (rock oak).	Medium-sized tree 60 to 70 feet high, rarely 100 feet.	Well-drained soils and rocky slopes Maine to Ohio, along Appalachians to Georgia.	Rough lumber, fuel, posts, tannin, game food.	1688
Q. nigra L. (water oak).	spotted oak, pin oak, red oak.	Medium-sized tree occasionally 80 feet high.	Rich, sandy bottom lands Delaware to Florida, Gulf States and Mississippi Valley to Missouri.	Rough lumber, cross ties, fuel, shade and ornamental planting.	1723

Footnotes on page 299.

TABLE 162.—*Quercus: Growth habit, distribution, and uses*—Continued

Accepted name	Synonyms	Growth habit	Natural range	Chief uses	Date of earliest cultivation
Q. nuttallii Palmer (Nuttall oak).	smooth-barked red oak, tight-barked red oak, yellow-butt oak, striped oak, Red River oak, swamp red oak, Mississippi Valley red oak.	Medium-sized tree occasionally 80 feet high.	Tight soils in bottom lands Mississippi Valley, Missouri to Louisiana, Alabama, and Texas.	Rough lumber, cross ties, fuel.	1923
Q. palustris Muenchh. (pin oak).	swamp Spanish oak.	Large, deciduous tree usually 70 to 90 feet high, occasionally 120 feet.	Low, wet ground Massachusetts to Virginia, west to Kansas and Oklahoma.	Rough lumber, cross ties, fuel, shade planting, game food.	(5)
Q. petraea (Mattuschka) Lieblein (durmast oak).	*Q. sessiliflora* Salisb., *Q. robur* Mill., not L. (sessile oak).	Deciduous tree 60 to 120 feet high.	Europe and western Asia. (Hardy in central and northeastern United States.)	Woodworking, cabinet work, tannin, fuel, game and livestock food.	(6)
Q. phellos L. (willow oak).	swamp oak, water oak, pin oak.	Large, deciduous tree usually 70 to 90 feet high, occasionally 120 feet.	Borders of streams and lakes, rich sandy soils New York to Florida along maritime plain, west to Texas and southern Illinois.	Rough lumber, cross ties, fuel, shade and ornamental planting, game food.	1723
Q. prinus L. (swamp chestnut oak).	*Q. michauxii* Nutt. (cow oak, basket oak).	Large, deciduous tree often 100 feet high, occasionally 125 feet.	Bottom lands New Jersey to Florida, west to Texas, and north to Arkansas and southern Indiana.	Lumber, flooring, cross ties, tight cooperage, posts, fuel, game food.	1737
Q. robur L. (English oak).	*Q. pedunculata* Ehrh. (pedunculate oak).	Deciduous tree 75 to 150 feet high.	Europe, northern Africa, western Asia.	Structural timber, fuel, charcoal, tannin.	(6)
Q. stellata Wangenh. (post oak).	*Q. minor* (Marsh.) Sarg.	Deciduous tree 50 to 60 feet high, rarely 100 feet.	Well-drained uplands Massachusetts to Florida, west to Iowa and Texas.	Rough lumber, flooring, tight cooperage, cross ties, fuel, animal food.	1819
Q. suber L.[7] (cork oak).	------------------	Evergreen tree 60 to 70 feet high.	Southwestern Europe and Algeria.	Cork, ornamental planting.	[8]1699
Q. velutina Lam. (black oak).	yellow oak, yellow bark oak.	Large, deciduous tree 70 to 80 feet high, occasionally 150 feet.	Well-drained uplands Maine to Florida, west to Ontario and Texas.	Rough lumber, cross ties, fuel, tannin, yellow dye.	1800
Q. virginiana Mill. (live oak).	------------------	Evergreen tree usually 40 to 50 feet high, occasionally 60 to 70 feet high and 6 to 7 feet in diameter.	Coastal areas Virginia to Florida and Louisiana, inland to Texas.	Rough lumber, fuel, ornamental and shade planting.	1739
Q. wislizeni A. DC. (interior live oak).	highland live oak.	Evergreen tree 70 to 80 feet high, leaves deciduous the second year.	Lower mountain slopes, foothills, and valleys in California and Lower California.	Watershed cover, fuel, ornamental planting.	1874

[1]ERAMBERT, F. G. UNPUBLISHED DATA. 1941. [Copy on file U. S. Forest Service, Glenn Bldg., Atlanta, Ga.]
[2]Var. *semota* Jeps. and var. *breweri* Jeps. are shrubs.
[3]About 1834.
[4]Includes var. *olivaeformis* (Michx. f.) A. Gray.
[5]Before 1770.
[6]Long cultivated.
[7]METCALF, W., and WALTZ, R. S. EXPERIMENTS WITH CORK OAK IN CALIFORNIA. Calif. Agr. Ext. Serv., Berkeley. 1941. [Typewritten.]
[8]Cultivated in California since 1858.

Relatively little use is made of the oaks in conservation planting at the present time. *Quercus alba, Q. borealis,* and *Q. borealis* var. *maxima* are occasionally planted for timber; *Q. macrocarpa* is used to some extent in shelter-belt planting in the Prairie-Plains region; small amounts of *Q. suber* are being planted in the Southwest for the production of cork. Many species, because of their handsome appearance, are used to a more or less degree in ornamental planting, including *Q. agrifolia, Q.*

891183°—50——20

borealis and its var. *maxima, Q. chrysolepis, Q. coccinea, Q. macrocarpa, Q. nigra, Q. palustris, Q. phellos, Q. suber, Q. virginiana,* and *Q. wislizeni.*

SEEDING HABITS. — The staminate flowers are borne in catkins and the pistillate flowers solitary or in 2- to many-flowered spikes on the same tree in the spring before or with the leaves. Staminate flowers develop from leaf axils of the previous year whereas pistillate flowers develop from axils of leaves of the current year. The fruit, an acorn or nut, ripens in 1 year in the white oak group[53] and in 2 years in the black oak group, with one known American exception, *Quercus agrifolia.* Oak seeds, ¼ to 1½ inches long, are subglobose to oblong, short-pointed at the apex, and marked with a circular scar at their base which is covered by a scaly cup or ripened involucre (figs. 212 and 213). The seed coats are a thick, hard outer shell and an inner brown membrane. The embryo has two large fleshy cotyledons. There is no endosperm. Acorns of the several species vary widely in size and palatability. Available details of seed bearing in 31 species and 2 varieties are given in table 163.

Many variations occur among the oaks. There is little authentic, specific information available on climatic races. Because hybridization occurs frequently, identification is difficult. To this confusion is added great fluctuations in growth habit due to differences in site factors. However, *Quercus robur* appears to have several races which differ in rate and time of growth and in resistance to frost, drought, and disease. *Q. macrocarpa* var. *olivaeformis,* which has much smaller acorns than the typical species occurring farther south, is probably a climatic race.

[53] Certain species native to the Eastern Hemisphere, such as *Quercus acutissima, Q. castaneaefolia, Q. cerris, Q. libani, Q. macrolepis, Q. trojana, and Q. variabilis,* ripen their fruit in 2 years.

TABLE 163.—*Quercus: Time of flowering and fruit ripening, and frequency of seed crops*

Species	Time of—			Commercial seed-bearing age			Seed year frequency	
	Flowering	Fruit ripening	Seed dispersal	Minimum	Optimum	Maximum	Good crops	Light crops
				Years	*Years*	*Years*	*Years*	*Years*
Q. agrifolia	April–May	Sept.–Oct.[1]	Sept.–Oct.					
Q. alba	do	do[1]	do	20	75–200	300	4–10	Intervening.
Q. arizonica	do	do[1]	Oct.–Nov.	30	50–100	150	3	do.
Q. bicolor	do	do[1]	Sept.–Oct.	35	75–200	300	3–5	do.
Q. borealis	do	do	do	25	50–125	200	2–3	do.
Q. borealis var. *maxima.*	do	do	Sept.–Nov.	30	50–200	250	3–5	do.
Q. chrysolepis	May–June	do[1]	Sept.–Oct.					
Q. coccinea	April–May	do	do	20	50–125	150	(2)	Irregular.
Q. douglasii	Feb.–May	Aug.–Oct.[1]	Aug.–Oct.					
Q. dumosa	April–May	Sept.–Nov.[1]	Sept.–Nov.					
Q. ellipsoidalis	do	Sept.–Oct.	Sept.–Oct.					
Q. falcata	do	do	do	25	50–75	125	1–2	Intervening.
Q. falcata var. *pagodaefolia.*	March–April	do	Sept.–Nov.					
Q. garryana	April–May	do[1]	Sept.–Oct.					
Q. imbricaria	do	Sept.–Nov.	Sept.–Nov.	25	30–75	125	2–4	Intervening.
Q. kelloggii	do	Sept.–Oct.	Sept.–Oct.					
Q. laevis	March–April	do	do					
Q. lobata	Feb.–April	do[1]	do					
Q. lyrata	March–April	do[1]	do					
Q. macrocarpa	April–May	Aug.–Sept.	Aug.–Sept.	35	75–250	400	2–3	Intervening.
Q. montana	do	Sept.–Oct.[1]	Sept.–Nov.	20	50–100	150	1–2	do.
Q. nigra	Feb.–April	Sept.–Oct.	Sept.–Oct.	25	50–125	175	1–2	do.
Q. nuttallii	March–April	do	Sept.–Feb.					
Q. palustris	April–May	do	Sept.–Nov.	20	40–75	125	1–2	Intervening.
Q. petraea	May	Oct.[1]	Nov.		50–200		8–15	1–3.
Q. phellos	Feb.–May	Aug.–Oct.	Sept.–Nov.					
Q. prinus	April–May	Sept.–Oct.[1]	Sept.–Oct.					
Q. robur	do	do[1]	Sept.–Nov.	20	60–100		2–4	Intervening.
Q. stellata	March–May	Sept.–Nov.[1]	do	25	50–150	250	2–3	do.
Q. suber[3]	Aug.	Nov.–Dec.[1]	Dec.		30–40		(4)	
Q. velutina	April–May	Sept.–Oct.	Sept.–Nov.	20	40–75	100	2–3	Intervening.
Q. virginiana	March–April	do[1]	Sept.–Oct.					
Q. wislizeni	April–May	do	do					

[1]Seed ripen first year.
[2]Good seed crops borne irregularly.
[3]Var. *occidentalis* (Gay) Arcang. seed ripen second year.
[4]Good seed crops borne periodically.

Collection, Extraction, and Storage.—Ripe acorns can be collected from early fall to early winter from the ground or flailed or shaken from branches onto canvas after ripening. To retard early germination, it is particularly necessary to collect acorns of the white oak group—generally those maturing their fruits in one season—immediately after they have fallen. No extraction is required, but any cups clinging to the seeds after a brisk stirring should be removed. Trees may yield from 1/4 to 1 1/2 bushels of clean acorns each. Labor costs per bushel range from $0.75 to $5 and occasionally more depending upon labor supply and size of crop. Light crops are usually heavily infested by weevils. Available details of yield, size, purity, soundness, and cost are given in table 164 for 30 species and two varieties of oak.

Although it is possible to store acorns of many oaks for 3 to 4 years without too great a loss in viability, it is usually impracticable to store them for more than 6 months or from the time of seed fall to sowing time in the spring. For short periods between collection and late fall seeding, they can be stored satisfactorily in cool, humid cellars.[54] For longer periods, dry storage in sealed containers at 32° to 36° F. is probably best.[55] Viability decreases rapidly, however, at higher temperatures when seed is so stored. Acorns can also be stored over winter

[54] Acorns of *Quercus macracarpa* var. *olivaeformis* stored in sacks under such conditions (40°–65° F.) showed 65 percent germination at the end of the first winter but only 22 percent during the second spring.

[55] *Quercus robur* acorns show about half of their original viability after 3 years under such storage conditions.

Table 164.—*Quercus: Yield of cleaned seed, and purity, soundness, and cost of commercial seed*

Species	Cleaned seed					Commercial seed		
	Yield per 100 pounds of fruit	Per pound[1]			Basis, samples	Purity	Soundness	Cost per pound
		Low	Average	High				
	Pounds	*Number*	*Number*	*Number*	*Number*	*Percent*	*Percent*	*Dollars*
Q. agrifolia			200		1	100		1.00
Q. alba	[2]60–90	70	150	210	13+	96–100	92	[3]0.05– .25
Q. bicolor	60–75	90		175	2	100	98	.05– .40
Q. borealis	[2]70–80	80	[4]140	255	17+	100	80	[3].05– .40
Q. borealis var. *maxima*	42–48	75	105	150	10	100	85–90	.05– .40
Q. chrysolepis			150		1			.45
Q. coccinea	[2]40–50	155	280	405	3+	100	85	.05– .50
Q. douglasii			180		1			
Q. dumosa			100		1			
Q. ellipsoidalis		205	245	290	11+	100	80	
Q. falcata	[2]75–86	390	595	785	5+	98	60	.05– .50
Q. falcata var. *pagodaefolia*			745		1			
Q. garryana			100+		1+			
Q. imbricaria	40–55	315	415	795	11+	99	95	.05– .60
Q. kelloggii			115					.45
Q. laevis			395		1			
Q. lobata			75					
Q. lyrata			130		1			.35– .50
Q. macrocarpa	65–75	40	75	[5]135	8	86	85	[3].05– .50
Q. montana	[2]60–75	55	75	100	4	100	98	.05– .50
Q. nigra		280		635	3+	100	95	
Q. nuttallii			104		1	100		
Q. palustris	50–70	320	410	540	3	100	90	.05– .15
Q. petraea		60	115	225	3+	98	60	.05– :35
Q. phellos		595		695	2			.05– .50
Q. prinus	40–50	55	100	195	5			
Q. robur		90	130	225	10+	72		.05– .35
Q. stellata	100	240	400	635	6	98	68	.05– .50
Q. suber		55	70	90	3+			.40–1.40
Q. velutina	[2]40	125	250	400	6	100	85	.05– .50
Q. virginiana		335	390	510	3+			
Q. wislizeni			100		1+			

[1]Number of seed per pound varies considerably with moisture content which for freshly collected acorns is very high, particularly in those of white oak group. One sample of *Q. macrocarpa* var. *olivaeformis* had a moisture content of 82 percent when collected right after falling.

[2]Weight per bushel of cleaned seed: *Q. alba*, 64–70 lbs.; *Q. borealis*, 42–56 1/2 lbs.; *Q. coccinea*, 39–49 lbs.; *Q. montana*, 50–53 lbs.; *Q. falcata*, 38 lbs.; *Q. velutina*, 41–49 lbs.

[3]Cost per bushel: *Q. alba*, $2.50–$3.50; *Q. borealis* and *Q. macrocarpa*, $2.

[4]Number per bushel, 4,480–8,040.

[5]Number per pound var. *olivaeformis* (7 samples): low, 160; average, 270; high, 440; weight per bushel, 46 pounds.

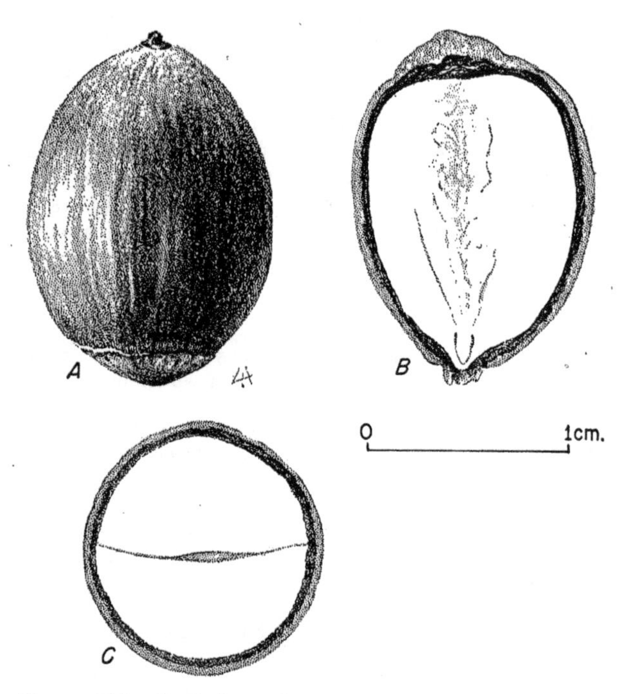

FIGURE 212.—Seed views of *Quercus macrocarpa* var. *olivaeformis: A,* Exterior view; *B,* longitudinal section; *C,* cross section.

in well-drained outdoor pits below the frost line, or they can be stratified in sand at low temperatures. Storage at slightly above freezing retards weevil activity in seed lots which at higher temperatures might be destroyed. Acorns which germinate in nature soon after falling should be stored immediately after collection. In general, seed of the white oaks loses its viability more rapidly in dry storage over winter than does that of the black oak.

GERMINATION.—A wide variation in germination exists between oak seed, particularly between the white oak and black oak groups. With a few exceptions, seed of the white oak group have little or no dormancy and will germinate soon after falling,[56] while seed of the black oak group are dormant and germinate the following spring under natural conditions. Germination is hypogeous (fig. 214). The best natural seedbed is a moist, well-aerated soil with an inch or more of leaf litter. Dormancy is largely due to conditions in the embryo and may be broken by stratification at low temperatures. The following species do not require pretreatment to break dormancy: *Quercus agrifolia* (a black oak), *Q. alba, Q. arizonica, Q. bicolor, Q. chrysolepis, Q. douglasii, Q. dumosa, Q. garryana, Q. lobata, Q. macrocarpa,*[57] *Q. montana, Q. petraea, Q. prinus, Q.*

[56] This is not true of at least *Quercus macrocarpa* var. *olivaeformis* which germinates the following spring.

[57] Acorns of at least var. *olivaeformis,* however, appear to be dormant and are benefited by stratification.

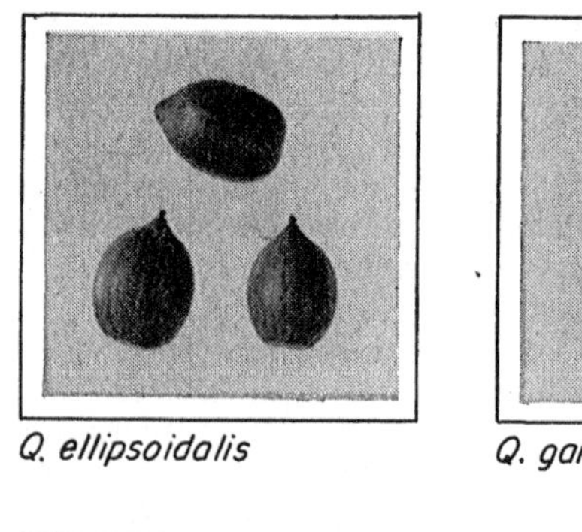

FIGURE 213.—Seed views of 4 species of *Quercus.*

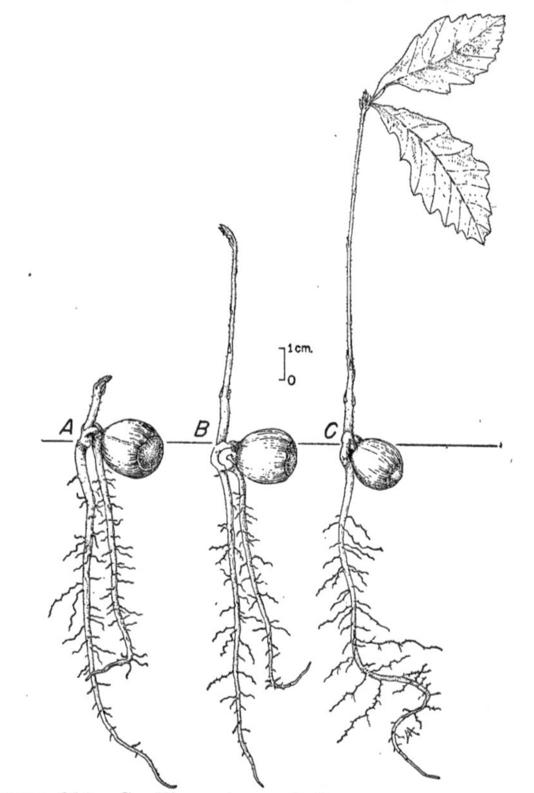

FIGURE 214.—Seedling views of *Quercus macrocarpa* var. *olivaeformis: A,* At 1 day; *B,* at 5 days; *C,* at 12 days.

Table 165.—*Quercus: Method of seed pretreatment for dormancy*

Species	Stratification[1]		Remarks
	Temperature	Duration	
	°F.	*Days*	
Q. borealis	32–38	30–45	In spring beds, better results obtained from stratified acorns than from dry-stored acorns. Stratification medium should be kept moist and well drained. Retain good viability in stratification up to 9 months.
Q. borealis var. *maxima*	32–38	30–45	do.
Q. coccinea	32–41	30–60	Stratify from collection time to seeding if spring nursery sowing is practiced. Nuts must not thoroughly dry.
Q. ellipsoidalis	41	60–90	do.
Q. falcata	32–38	30–45	do.
Q. falcata var. *pagodaefolia*	30–35	30–60	do.
Q. imbricaria	32–41	30–60	do.
Q. kelloggii	32–41	30–45	do.
Q. macrocarpa var. *olivaeformis*	41	30–60	
Q. nigra	30–35	30–60	Stratify from collection time to seeding if spring nursery sowing is practiced. Nuts must not thoroughly dry.
Q. nuttallii	30–35	30–60	do.
Q. palustris	32–41	30–45	do.
Q. phellos	30–35	30–60	do.
Q. velutina	33–40	30–60	do.
Q. wislizeni	32–41	30–60	do.

[1]Medium: Sand, sand and peat, or sandy loam.

Table 166.—*Quercus: Recommended conditions for germination, and summary of germination data*

Species	Test conditions recommended				Germination data from various sources					
	Temperature		Duration		Germinative energy		Germinative capacity			Basis, tests
	Night	Day	Untreated seed	Stratified seed	Amount	Period	Low	Average	High	
	°F.	*°F.*	*Days*	*Days*	*Percent*	*Days*	*Percent*	*Percent*	*Percent*	*Number*
Q. agrifolia			15–40	(1)					73	1
Q. alba	68	86	30–50	(1)	62–93	18–40	31	78	99	9
Q. bicolor	55	75	60–80					99		1
Q. borealis	68	86	300+	40–60	39–84	13–42	11	58	89	11+
Q. borealis var. *maxima.*	68	86	60+	30–40	60–85	19–26	87		100	3
Q. chrysolepis	[2]68	86	60	(1)					56	1+
Q. coccinea	68	86	240+	60			24	62	94	4
Q. douglasii	[2]68	86	30	(1)					72	1+
Q. dumosa	[2]68	86	90						80	1+
Q. ellipsoidalis	70	85	340+	30–60	80–93	18–26	32	73	95	5
Q. falcata	60	75	240+	30–40					91	2
Q. falcata var. *pagodaefolia.*								38		1
Q. garryana	70	85	90				46	77	95	3+
Q. imbricaria	60	75	240+	30			28		66	2
Q. kelloggii	70	85		30–40					95	1+
Q. lobata									24	1+
Q. lyrata			110					84		1
Q. macrocarpa	68	86	[3]130–225+	[3]40	28–85	25–45	7	45	99	11
Q. montana	65	80	60	(1)	72–78	40	66	82	94	3+
Q. nigra								61		1+
Q. nuttallii	[2]68	86						82		1+
Q. palustris	[2]68	86						68		1+
Q. petraea	[2]68	86		(1)				74		1
Q. phellos	[2]68	86	240+			22		46		2
Q. prinus	60	75	110			30		87		2
Q. robur	60	77	30–60	(1)			46	81	100	4+
Q. stellata	68	86	60	(1)			54		62	2
Q. suber	80	80	20–30				90		100	2+
Q. velutina	65	80	240+	30–50			7	47	88	5
Q. virginiana	[2]68	86		(1)			41		65	5+
Q. wislizeni	[2]68	86	160+						40	1

[1]Pretreatment not needed.
[2]Suggested temperatures; no data available.
[3]Dormant lots.

robur, Q. stellata, Q. suber, and *Q. virginiana.* Germination of some of these species can be hastened and increased somewhat by stratification at low temperatures (32° to 41° F.) for 30 to 90 days. Such pretreatment is impractical, however, since either fall seeding or proper over-winter storage accomplishes the same effect. Germination of *Q. virginiana* is peculiar in that the radicle, soon after it appears, becomes enlarged just below the surface of the ground because of the transfer of food from the cotyledons. Information for 13 species and 2 additional varieties of oaks which require pretreatment for dormancy is given in table 165.

Adequate germination tests can be made by using 50 to 100 acorns in flats of sand, sand and peat, or sandy loam. The test conditions recommended and available average results for 29 species and 2 varieties are given in table 166.

Nursery and Field Practice.—When practicable, it is better to practice fall seeding of oaks than spring seeding. This is particularly advisable for the white oaks since their seed are more susceptible to injury in winter storage than those of the black oaks. For spring seeding, acorns should be stratified over winter. Drilling in rows 8 to 12 inches apart in well-prepared soil is preferred, and the seed should be covered with one-half to 1 inch of firmed soil. Fall beds should be mulched with leaves or straw that is held in place by hardware-cloth covers or other effective materials; the covering also serves as a protection against rodents. In the spring, after the frost-danger period, the mulch should be removed and the surface soil kept moist until after germination is complete. Germination: *Quercus alba,* 66 percent; *Q. borealis,* 72 percent; *Q. montana,* 82 percent; *Q. velutina,* 90 percent. Tree percent: 80 to 90.

A collar rot has killed seedlings in patches in some nurseries. The disease is spread through the soil by a gray mycelium. It may be controlled by treating the patches with 1½ fluid ounces of formaldehyde in 2 pints of water to each square foot of bed. Powdery mildew, although not often destructive, may be controlled by dusting with sulfur.

It is seldom necessary to produce older stock than 1–0 seedlings for average field planting. Direct seeding is in general more desirable than planting of nursery stock, but it is a hazardous practice if a heavy population of rodents is present.

RHAMNUS L. Buckthorn

(Buckthorn family—Rhamnaceae)

Distribution and Use.—The buckthorns comprise about 100 species of deciduous, or sometimes evergreen, shrubs or small trees native chiefly to the temperate regions of the Northern Hemisphere; a few species are found in Brazil and South Africa. Some species are grown for hedges and shelter belts, and most of them produce wildlife food. Several buckthorns yield yellowish-green dyes, and the fruits and bark of some species are used medicinally. Six species, which are of present or potential use for planting in the United States, are described in table 167. *Rhamnus purshiana* is sometimes cultivated in the eastern United States and in western Europe.

Seeding Habits.—The perfect or polygamous, small, greenish flowers, which bloom in the spring or early summer, are borne in heads or clusters along the sides or at the tips of the branches. Buckthorn fruits, which ripen in late summer or early fall, are oblong or globose berrylike drupes, each containing two to four nutlike seeds (figs. 215 and 216). Typical seeds are small, roundish, with a flat side marked by a slight central rib, and a rounded side with a small terminal knob (probably the hilum). The comparative seeding habits of six species are given in table 168. *Rhamnus purshiana* begins to produce fruit when it is 5 to 7 years old; comparable information for other species is lacking. Good seed crops of *R. davurica* are common; in *R. purshiana* good seed crops are borne almost annually with light crops in intervening years. Comparable information for the other species is lacking. No scientific information is available as to the development of climatic races within the buckthorns.

Collection, Extraction, and Storage.—The ripe fruits should be picked from the bushes by hand in the late summer or fall. Ordinarily collection should begin soon after ripening, or much of the fruit will be eaten by birds. To extract the seeds, the fruits should be run through a macerator with water, immediately after collection,

Table 167.—*Rhamnus: Growth habit, distribution, and uses*

Accepted name	Synonyms	Growth habit	Natural range	Chief uses	Date of earliest cultivation
R. alnifolia L'Hérit. (alder buckthorn).	Dwarf alder	Low, spreading, compact shrub.	Moist sites from New Brunswick to British Columbia, south to West Virginia and northern California.	Wildlife food, sheep browse, limited horticultural use.	1778
R. californica Eschsch. (California buckthorn).	California coffeeberry.	Variable evergreen shrub.	California and adjacent parts of Oregon and Arizona.	Wildlife food, including browse.	1871
R. cathartica L. (European buckthorn).	Common buckthorn, SPN; Hart's thorn, waythorn rhineberry.	Shrub or small tree.	Europe and northern and western Asia. Naturalized in eastern United States.	Hedges and shelter belts, wildlife food and cover.	(1)
R. davurica Pall. (Dahurian buckthorn).		Large, spreading shrub or small tree.	Dahuria to northern China, Manchuria, and Korea.	do	1817
R. frangula L. (glossy buckthorn).	alder buckthorn	Shrub or small tree.	Europe, western Asia, northern Africa. Naturalized in eastern United States and Illinois.	Ornamental, wildlife food and cover, charcoal for gunpowder.	(1)
R. purshiana DC. (cascara buckthorn).	cascara sagrada	Tall shrub or small tree.	British Columbia to central California and eastward to Idaho.	Source of drug, cascara; wildlife food and cover, ornamental.	1870

[1]Long cultivated.

RHAMNUS

Table 168.—*Rhamnus: Time of flowering and fruit ripening, and ripe-fruit characteristics*

Species	Time of—		Ripe-fruit characteristics		
	Flowering	Fruit ripening	Color	Diameter	Seeds per fruit
				Inch	*Number*
R. alnifolia	May–June	July–October	Black	1/5	3
R. californica	April–July	June–October	Purplish black	1/4	2
R. cathartica	April–June	September–October	Black	1/5	3–4
R. davurica	May–June	do	do	1/5	
R. frangula	May–July	July–October	Red to dark purple	1/4	2
R. purshiana	April–June	July–September	Purplish black	1/4	2–3

and the pulp skimmed or floated off. The seed should then be dried before sowing or storage. Sometimes the fruits are dried in the sun and used in that form. The size, purity, and soundness of commercial seed vary considerably by species as shown in table 169. Seed storage practices have not been worked out for most buckthorns, but it appears likely that satisfactory results will be obtained if the seed are kept in sealed containers at low temperatures, probably about 41° F. Such storage information as is available is as follows: *Rhamnus alnifolia* seed will retain most of its viability for at least 2 years if stored in sealed containers at 41°; *R. davurica* seed can be stored safely over winter in sealed containers at low temperatures; *R. cathartica* will retain its viability for 2 years in ordinary storage according to Russian experience.

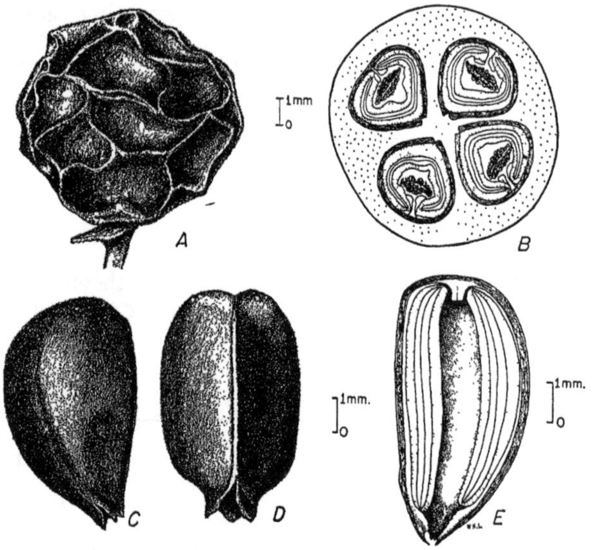

Figure 215.—Seed views of *Rhamnus cathartica*: *A*, Exterior view of dried fruit; *B*, cross section of fruit showing four seeds; *C* and *D*, exterior views of seeds in two planes; *E*, longitudinal section of seed showing clasping endosperm.

Germination.—Natural germination usually occurs in the spring following seed dispersal. *Rhamnus frangula* seedlings come in quite abundantly on peat soils. In some species germination is epigeous and in some hypogeous (fig. 217). Although there appears to be considerable variability between species and perhaps within species, most of the buckthorns seem to have some seed dormancy due to either or both seed-coat and embryo conditions. This can be overcome by moist, cold stratification or scarification. Further information on seed dormancy and pretreatment is given in table 170. Test results are not available for *R. californica* but it is likely that stratification is needed for good germination, since untreated seed germinate rather poorly. As is evident, the best pretreatments are still to be worked out for most of the species listed. The germination of buckthorn seeds can be tested in sand flats at 68° (night) to 86° F. (day), using 1,000 properly pretreated[58] seeds per test. Average results for 6 species are given in table 171.

[58] No pretreatment needed for *Rhamnus cathartica*.

Table 169.—*Rhamnus: Yield of cleaned seed, and purity, soundness, and cost of commercial seed*

Species	Cleaned seed					Commercial seed		
	Yield per 100 pounds of fruit	Per pound			Basis, samples	Purity	Soundness	Cost per pound
		Low	Average	High				
	Pounds	*Number*	*Number*	*Number*	*Number*	*Percent*	*Percent*	*Dollars*
R. alnifolia		62,000		69,000	2	96	92	
R. californica			[1]12,000		1			
R. cathartica		13,000	19,000	28,000	6	90	93	1.50–2.50
R. davurica	20–30	18,000	27,000	39,000	3	97	83	4.50
R. frangula	9	18,000		20,000	2	98	98	1.50
R. purshiana	20	5,000	12,000	19,000	5+	99	90	1.50–3.25

[1]Dried fruits with 2 seeds each.

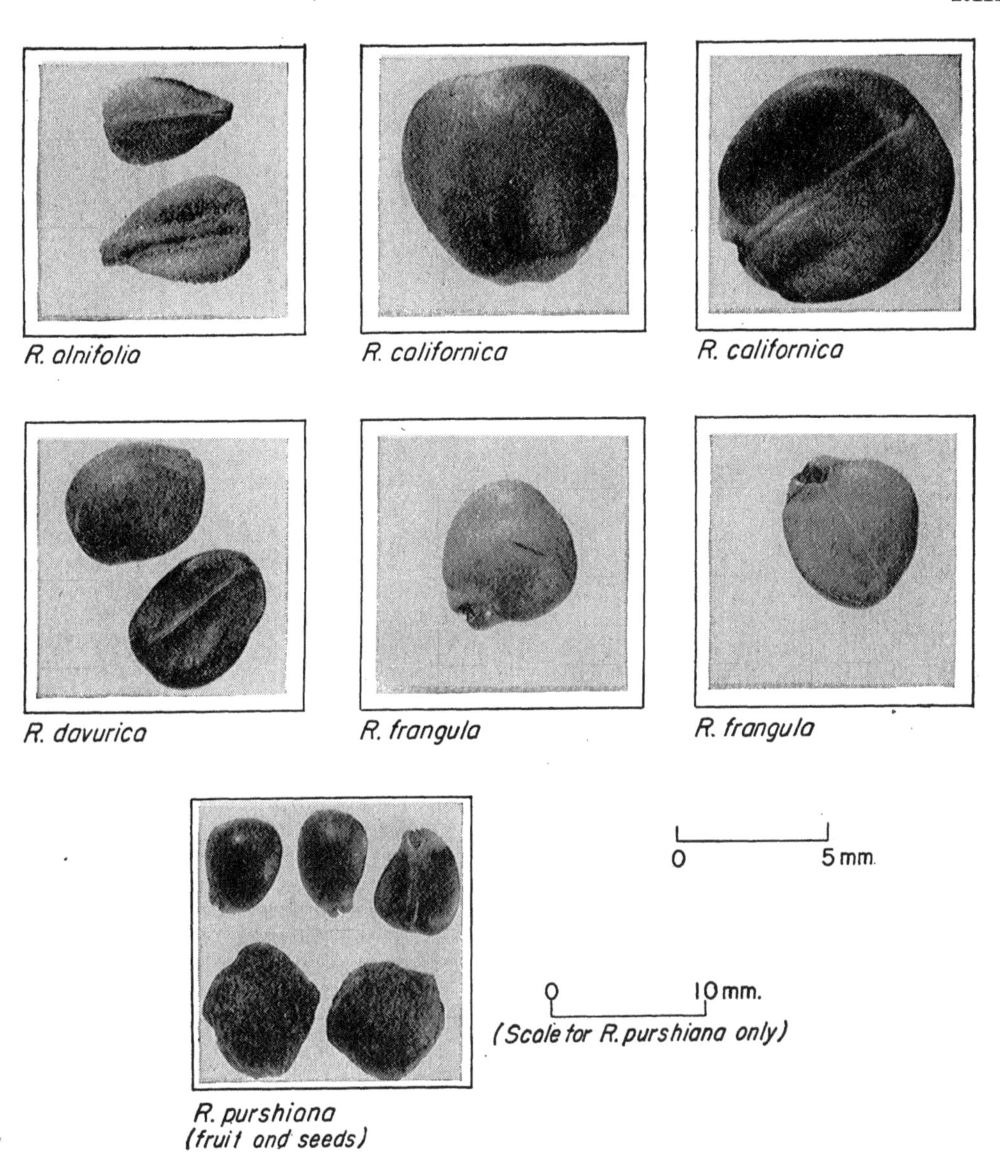

FIGURE 216.—Seed views of five species of *Rhamnus*.

TABLE 170.—*Rhamnus: Dormancy and method of seed pretreatment*

Species[1]	Seed dormancy		Stratification			Remarks
	Kind	Occurrence	Medium	Temperature	Duration	
				°F.	Days	
R. alnifolia	Embryo	General	Moist sand	41	60+	Sixty days' stratification is of considerable benefit, but does not seem to overcome dormancy completely.
R. davurica	Seed coat	do	do	41	90	Scarification or water soaking might be helpful.
R. frangula	Probably embryo.	do	do	41	60	
R. purshiana	do	do	do	41	60+	Stratification longer than 60 days probably would improve results.

[1]Seed of *Rhamnus cathartica* are apparently nondormant. Neither water soaking, 30-minute scarification, 30- to 60-day stratification, or 2-minute soaking in concentrated sulfuric acid improves germination; the latter is definitely harmful.

RHAMNUS

TABLE 171.—*Rhamnus: Summary of germination data from various sources*

Species	Germinative energy		Germinative capacity			Basis, tests	Type of germination
	Amount	Period	Low	Average	High		
	Percent	*Days*	*Percent*	*Percent*	*Percent*	*Number*	
R. alnifolia	50	15	51	63	80	3	Epigeous.
R. californica	----------	----------	----------	[1]30	----------	1+	----------
R. cathartica	24–84	18–41	35	54	92	5	Epigeous.
R. davurica	----------	----------	----------	40±	----------	1+	do.
R. frangula	55	14	54	----------	[2]66	2	Hypogeous.
R. purshiana	----------	----------	30	36	[3]40	3+	Epigeous.

[1]Untreated seed.
[2]Untreated seed germinated only 5 percent in 60 days.
[3]In 15 days.

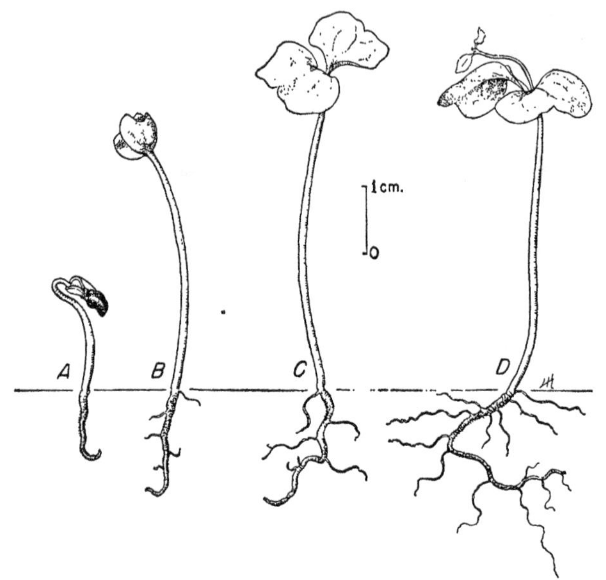

FIGURE 217.—Seedling views of *Rhamnus cathartica: A,* At 1 day; *B,* at 4 days; *C,* at 19 days; *D,* at 28 days.

NURSERY AND FIELD PRACTICE.—Detailed nursery practice for most species has not been worked out, but the following practice should insure reasonable success. The seed should be sown in the fall soon after they are extracted—or seed stratified 90+ days in the spring can be used—in drills about 8 inches apart and covered with one-half inch of soil. A sowing depth of 1 inch and shading of the seedbed are recommended for *Rhamnus purshiana.* About 25 percent of the viable *R. davurica* seed sown produce usable seedlings. In general, buckthorns like a rather moist soil and do well under shade. *R. cathartica* and *R. davurica,* however, grow well on dry soils. Some species of buckthorn, especially those originating in Europe, are alternate hosts for the oat rust, *Puccinia coronata.* Because of this, buckthorns and oats should not be planted in the same vicinity. Although buckthorns are usually propagated by seed, they can also be propagated by layers and some of the rarer species by cuttings (sometimes used for *R. purshiana*) or by grafting (such as some forms of *R. frangula*).

RHODODENDRON L. Rhododendron

(Heath family—Ericaceae)

DISTRIBUTION AND USE.—The rhododendrons include over 600 species of evergreen or deciduous shrubs, or rarely trees, native to the colder and temperate regions of the Northern Hemisphere, the high mountains of southern Asia and Malaysia and extending to New Guinea and Australia. Many of them are cultivated for their beautiful flowers, and the evergreen species for their foliage. Some species are valuable for wildlife browse; the wood of some tree species is used occasionally. Many hybrids have originated in cultivation and this often makes classification as to species difficult. About 26 species are known in North America. However, only the 3 for which reliable information is available, and which have value for conservation purposes, are described in table 172. *Rhododendron maximum*, both the natural type and various horticultural selections and hybrids, is used extensively in landscaping throughout the eastern United States and in Europe. *R. catawbiense*, though the natural species has not been much cultivated, enters into the parentage of many artificially created hybrids and varieties of great horticultural merit. It is considerd one of the most beautiful of our native shrubs. *R. macrophyllum* has not been cultivated extensively.

SEEDING HABITS.—The large, brightly colored, perfect flowers of rhododendron usually occur in clusters at the tips of the branches and bloom in the spring or early summer. The fruit, which ripens in the fall, is an egg-shaped or oblong capsule which becomes greenish brown on ripening and later turns brown. It splits along the sides soon after ripening and releases numerous seeds which are dispersed by the wind, the opened capsule remaining on the plant for several months. The seeds are minute, brown, compressed, oblong bodies with a fringed tuft at either end (fig. 218). Seed

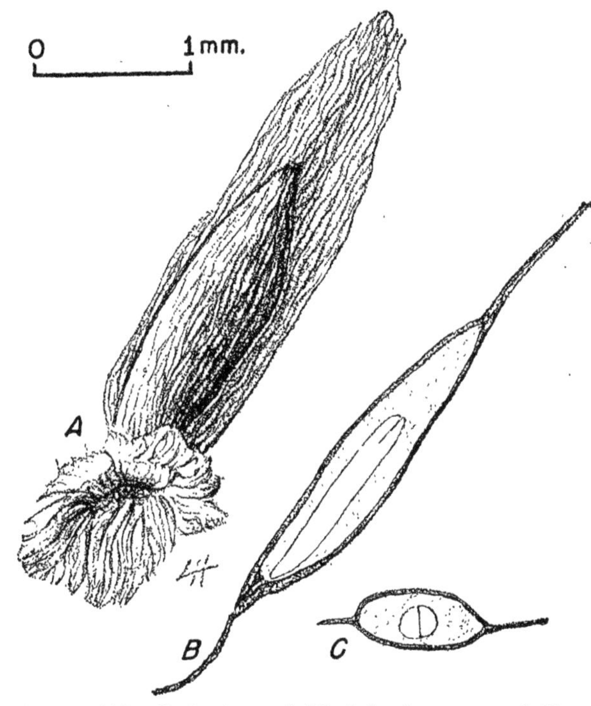

FIGURE 218.—Seed views of *Rhododendron macrophyllum*: *A*, Exterior view of winged seed; *B*, longitudinal section; *C*, cross section.

TABLE 172.—*Rhododendron: Growth habit, distribution, and uses*

Accepted name	Synonyms	Growth habit	Natural range	Chief uses	Date of earliest cultivation
R. catawbiense Michx. (catawba rhododendron).	mountain rosebay	Low to medium, spreading evergreen shrub.	Appalachian Mountains from Virginia to Georgia and Alabama.	Ornamental	1809
R. macrophyllum D. Don (Pacific rhododendron).	*R. californicum* Hook. (coast rhododendron).	Medium to tall, erect evergreen shrub.	West slope of the Cascades and in coastal regions from British Columbia to northern California.	do	1850
R. maximum L. (rosebay rhododendron).	Rosebay, great laurel, great rhododendron.	Large evergreen shrub or small tree.	Nova Scotia to Ontario and south to Ohio; through the Appalachian Mountains to Alabama and Georgia.	Ornamental, deer browse, small wood specialties.	1736

RHODODENDRON

TABLE 173.—*Rhododendron: Time of flowering, fruit ripening, and seed dispersal*

Species	Time of—			Flower color
	Flowering	Fruit ripening	Seed dispersal	
R. catawbiense	May–June	Fall	Fall	Lilac, purple.
R. macrophyllum	do	August–September	Late summer–fall	Pale rose, rose, purple.
R. maximum	May–July	Fall	Fall	Rose, purple, pink.

crops are borne annually in fair abundance. The comparative seeding habits of the three species discussed here are given in table 173. Scientific information as to racial variation among the rhododendrons is lacking. The three species show great variation within themselves as to leaf, flower, and growth characteristics, and it may be that some of these are racial variations.

COLLECTION, EXTRACTION, AND STORAGE. — The greenish-brown to brown capsules of rhododendron should be picked from the plants by hand before they have begun to open, and spread out in thin layers to dry. After drying, they can be rubbed or beaten if they have not already split open, and the seed shaken out. The size, purity, and soundness of rhododendron seed vary with the species as shown in table 174. Seed from wild plants ordinarily do not appear on the market. Seed of cultivated hybrids are usually sold in small packets and range in price from $2.50 per ounce upward. Detailed information on seed storage is not available, although it is reported that seed of rhododendrons, if properly stored, will maintain its viability for a great length of time.

GERMINATION.—Natural germination is reported to be best on moist mineral soil under partial shade. Germination is epigeous (fig. 219). In tests, germination ordinarily is low. Since there are no reports mentioning seed dormancy, this may be a reflection of the high proportion of empty seed which cannot be removed by ordinary commercial means because of their lightness. Germination can be tested in quartz sand in petri dishes using 1,000 seeds per test. In one series of tests with *Rhododendron macrophyllum* seed temperatures of 68° (night) to 86° F. (day) were a little better than temperatures of 50° (night) to 77° (day), and quartz sand gave much better results than peat—27 to 28 percent as compared to 3 to 8 percent—as a germination medium. Test methods recommended and the average results for 3 species are given in table 175.

NURSERY AND FIELD PRACTICE. — In propagating from seed, the seed are germinated in flats or pans of acid, sandy peat, or sand mixed with pulverized, decayed, oak litter. Sowings can be made in a cool greenhouse (45° to 50° F.) during the winter, or about April in a coldframe. The seed are sown on the surface of the soil and covered lightly with pulverized sphagnum. Some growers recommend covering the flats with panes of glass during the period of germination and watering from below.

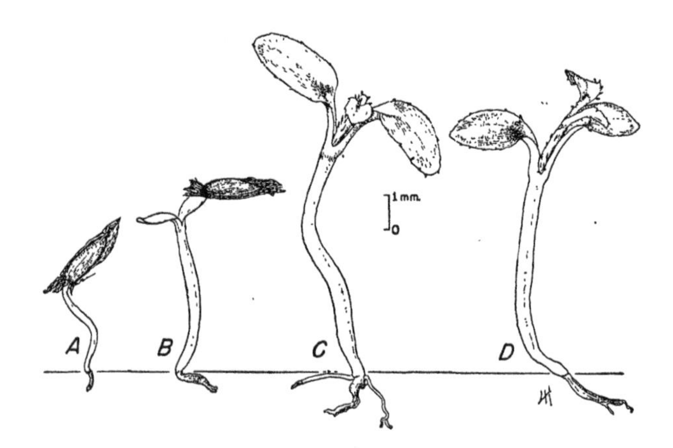

FIGURE 219.—Seedling views of *Rhododendron macrophyllum*: *A*, At 1 day; *B*, at 9 days; *C*, at 40 days; *D*, at 60 days.

With a properly prepared soil mixture the flats can be kept in shallow trays of water without the soil becoming waterlogged. Germination requires 2 to 4 weeks for completion. The tiny plants should be transplanted to seedling flats, as soon as they are large enough to handle, and carried through the first year in outdoor pits or coldframes under partial shade. The following spring they can be moved to nursery beds to grow one or more years before transplanting to permanent locations.

TABLE 174.—*Rhododendron: Yield of cleaned seed, and its purity and soundness*

Species	Cleaned seed per pound				Purity	Soundness
	Low	Average	High	Basis, samples		
	Number	*Number*	*Number*	*Number*	*Percent*	*Percent*
R. catawbiense	5,000,000	5,600,000	5,700,000	7	57	
R. macrophyllum	2,000,000		2,100,000	2	80	77
R. maximum	5,000,000	5,000,000	5,700,000	4	51	

Partial shade in the nursery is desirable. The keynotes to success may be summarized as (1) careful handling of the tender and delicate young seedlings; (2) an acid soil with a high content of organic matter, porous and well drained in the root zone; and (3) ample moisture at all times. The requirements with regard to soil properties and moisture must also be fulfilled by the permanent sites, or the plantings will be almost certainly foredoomed to failure.

TABLE 175.—*Rhododendron: Recommended conditions for germination tests, and summary of germination data*

Species	Test conditions recommended			Germination data from various sources					
	Temperature			Germinative energy		Germinative capacity			Basis, tests
	Night	Day	Duration	Amount	Period	Low	Average	High	
	°F.	*°F.*	*Days*	*Percent*	*Days*	*Percent*	*Percent*	*Percent*	*Number*
R. catawbiense			30+	0–54	20	8	48	70	7
R. macrophyllum	68	86	90	12–26	18–57	27		90	3
R. maximum			30+	4–14	20	5	13	31	4

RHODOTYPOS SCANDENS (Thunb.) Mak. Black jetbead

(Rose family—Rosaceae)

Botanical syn.: *R. kerrioides* Sieb & Zucc.

Distribution and Use.—Native to Japan and central China, black jetbead is an upright, spreading, deciduous shrub which was introduced into cultivation chiefly for ornamental purposes in 1866. It may also be of value for wildlife food and cover. This shrub represents the only species in the genus.

Seeding Habits.—The showy, white, perfect flowers (1 to 2 inches across) bloom from May to June. Black jetbead fruits are shiny, black, dry drupes, obliquely short-ellipsoid in shape. They ripen in October and persist on the plant well into the winter; each contains one stone. The seed is a small, stubby, ellipsoidal stone, about one-fourth inch long, characteristically sculptured much in the manner of leaf venation with the "midrib" extending around the longest periphery; it is a dull, tan color (fig. 220).

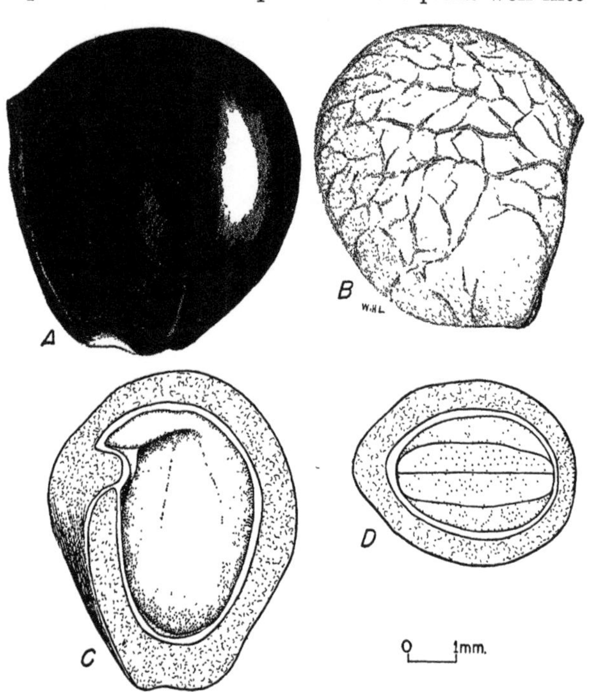

Figure 220.—Seed views of *Rhodotypos scandens*: *A*, Exterior view of fruit; *B*, exterior view of stone; *C*, longitudinal section of stone; *D*, cross section of stone.

Collection, Extraction, and Storage. — The fruits can be collected from the bushes by hand from October to midwinter. It does not appear necessary to extract the stones from the fruits. Cleaned seed per pound in 1 sample averaged 5,210; purity was 89 percent, and soundness 86 percent. Commercial seed costs $1.50 to $2.75 per pound. Seed of this species can be stored air-dry in open containers at 34° to 50° F. up to 9 months without loss (actually an increase) in viability. Storage in sealed containers and in vacuum at various humidities does not improve results.

Germination.—The seed exhibit dormancy due both to embryo and seed coat conditions; this can be overcome by stratification in moist peat for 30 days at 86° F. followed by 90 days' stratification at 41°. Partially after-ripened seed subjected to high temperature go into secondary dormancy. Germination tests can be made in sand flats using 200 to 400 pretreated seeds per test. A temperature range of 68° (night) to 86° (day) for 90 days is suggested. In one series of tests, stratified seed had a germinative capacity of 86 percent as compared with 16 percent for untreated seed.

Nursery and Field Practice.—Seed of this species should be sown in the fall in mulched or board-covered coldframes. A sowing depth of one-half inch is suggested. Some germination will take place the second year. Presumably properly stratified seed could be sown in the spring. In one case seed collected slightly green in August and sown immediately germinated 100 percent the following spring. Jetbead thrives on any good soil. It may also be propagated by cuttings.

RHUS L. Sumac

(Cashew family—Anacardiaceae)

Distribution and Use.—The sumacs include about 150 species of deciduous or evergreen shrubs or trees which are distributed throughout the temperate and subtropical regions of the world. A milky or resinous juice is a characteristic of these species. Because of their handsome foliage, which often assumes brilliant coloring in the fall, and showy red fruits, many species have decided ornamental value. The foliage and bark are often rich in tannin which is used in tanning leather, particularly sheepskin. Some Asiatic species yield valuable waxes and lacquers; certain others produce timber. Because of their suckering habit and edible fruit, many species are useful in plantings for erosion control and wildlife food and cover. In some areas several species, which are poisonous to the touch, are troublesome pests. Only 6 species, all of native origin, are used or are potentially useful for conservation planting in this country. Their distribution and uses are given in table 176. Relatively little use has thus far been made of the sumacs for planting in this country. *Rhus glabra* and *R. trilobata* are used to a minor extent in shelter-belt planting in the Prairie-Plains region; the former species and *R. typhina* are planted occasionally for erosion control and wildlife food and cover. All species are used to a limited extent in ornamental planting.

Seeding Habits.—The flowers in *Rhus* are dioecious or polygamous; they are small and rather inconspicuous and are borne in terminal or axillary clusters in the spring. The fruit is a small, smooth or hairy drupe with thin, dry flesh, and it contains a single, bony seed (figs. 221 and 222) without endosperm. It is frequently borne in dense clusters which ripen in the fall and often remain on the plant over winter. Commercial seed usually consists of the dried fruit. Such data as are available on the seeding habits of six species are given in table 177. Seed dispersal is almost entirely by birds or other animals. No information is available as to the frequency of seed crops,[59] the age for commercial seed production, or the existence of climatic races.

[59] Some seed is probably borne every year.

Table 176.—*Rhus: Growth habit, distribution, and uses*

Accepted name	Synonyms	Growth habit	Natural range	Chief uses	Date of earliest cultivation
R. glabra L. (smooth sumac).	*R. glabra* var. *cismontana* (Greene) Daniels, *R. glabra* var. *occidentalis* Torr.	Thicket-forming shrub or small tree.	Dry soils Maine to British Columbia, south to Florida and Arizona.	Shelter belt, erosion control and wildlife planting; ornamental purposes.	1620
R. integrifolia (Nutt.) Benth. & Hook. f. (mahogany sumac).	*R. integrifolia* var. *serrata* (Nutt.) Engler (lemonade berry, lemonade sumac).	Thicket-forming evergreen shrub or small tree.	Dry soils, coastal and southern California, and Lower California.	Erosion control, wildlife food, fuel.	--------
R. laurina Nutt. (laurel sumac).	--------------------	Evergreen shrub	Valleys, southern California and Lower California, Arizona.	Erosion control, wildlife food.	--------
R. ovata S. Wats. (sugar sumac).	sugarbrush	Tall shrub	Dry hills, southern California.	Ornamental purposes, wildlife food.	--------
R. trilobata Nutt. (skunkbush sumac).	*R. canadensis* var. *trilobata* (Nutt.) A. Gray, *Schmaltzia trilobata* (Nutt.) Small (skunkbush, quailbush, lemonade sumac).	Shrub	Dry, rocky hills, Alberta to Illinois, west and south to Oregon, California, and Mexico.	Shelter-belt and erosion-control planting, browse for livestock, wildlife food and cover; shoots used in basketry, fruits sometimes eaten by man.	1877
R. typhina Torner (staghorn sumac).	*R. hirta* (L.) Sudw. (hairy sumac).	Thicket-forming shrub or tree.	Dry soils New Brunswick to North Dakota, south to Georgia, Indiana, and Iowa.	Erosion control, wildlife planting, ornamental purposes.	1629

RHUS

TABLE 177.—*Rhus: Time of flowering, fruit ripening, and seed dispersal*

Species	Time of— Flowering	Time of— Fruit ripening	Time of— Seed dispersal	Color of ripe fruit
R. glabra	June–August	September–October	Persistent until early summer	Dark red.
R. integrifolia	February–April	July–October	September	do.
R. laurina	May–July (December)	August–September	Persistent	Whitish.
R. ovata	March–May	do.	do.	Dark red.
R. trilobata	March–April	August	Persists several months	Red.
R. typhina	June–July	August–September	Persists until early summer	Dark red.

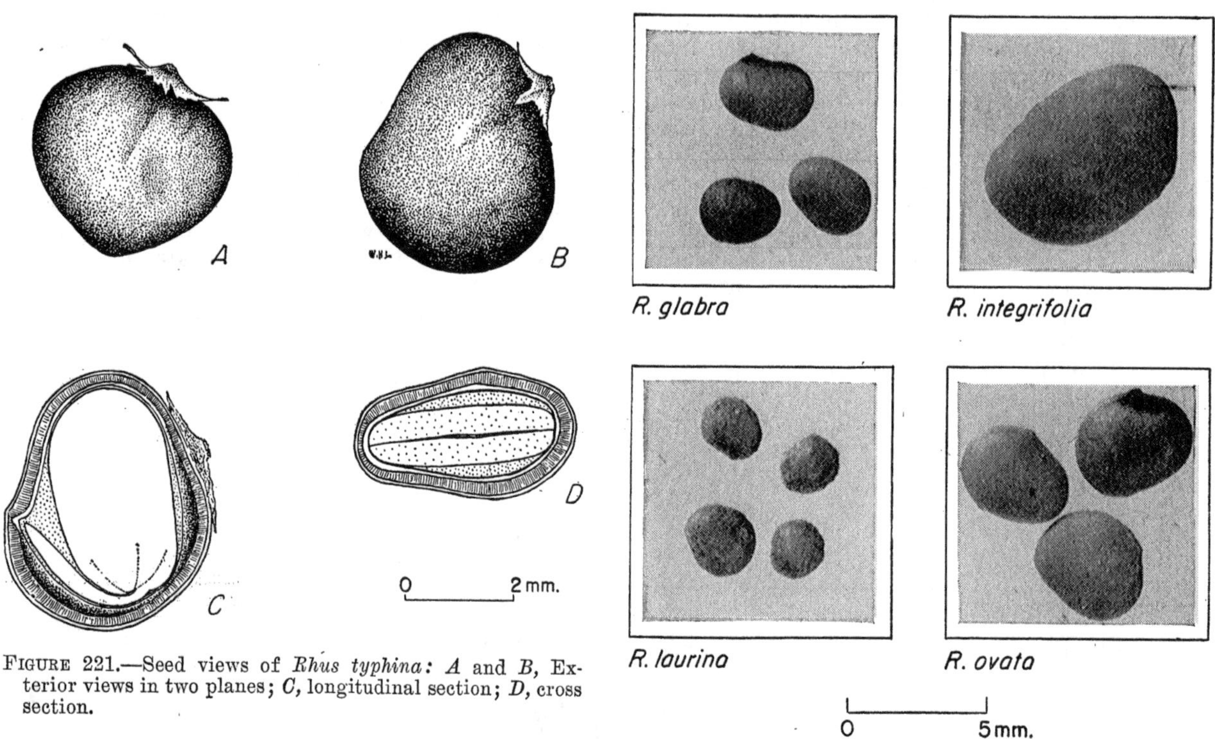

FIGURE 221.—Seed views of *Rhus typhina: A* and *B*, Exterior views in two planes; *C*, longitudinal section; *D*, cross section.

FIGURE 222.—Seed views of four species of *Rhus*.

COLLECTION, EXTRACTION, AND STORAGE. — The fruit clusters can be picked by hand from the shrubs as soon as ripe and until as late in the year as sound fruits can be found in quantity. Early collections, particularly of the fruits of *Rhus glabra* and *R. typhina* which occur in very dense clusters, are likely to require drying and should, therefore, be spread out in shallow layers. Fruits collected in the late fall and winter will have dried sufficiently on the shrubs. When thoroughly dry the clusters can be broken up into individual fruits by rubbing or beating in canvas sacks; they should then be screened or fanned to remove the debris. Although much of the thin pulp will be removed by this treatment, to obtain a thorough job of cleaning it is necessary to run the partially cleaned seed through a macerator with water, allowing the remaining pulp and the empty seeds to float away. Such complete cleaning, however, is not odinarily practiced, except in the case of *R. trilobata*. The seeds are usually sown with such parts of the pulp as may still cling to them. Data on yield of cleaned seed, its size, purity, soundness, and cost are presented in table 178.

Rhus seed can be stored dry over winter without special treatment. Little is known regarding storage methods for longer periods. However, clean seed of both *R. glabra* and *R. typhina* have been stored in sealed containers at 41° F. for 2½ years with no loss in viability. That of *R. ovata* is said to retain its germinability for a long time when kept under ordinary storage conditions.

GERMINATION.—The time of natural germination of *Rhus* seed is not known. Some seeds which

TABLE 178.—*Rhus: Yield of cleaned seed, and purity, soundness, and cost of commercial seed*

Species	Fruits per pound	Clean seed					Commercial seed		
		Yield per 100 pounds of clusters	Per pound			Basis, samples	Purity[1]	Soundness[1]	Cost per pound
			Low	Average	High				
	Number	*Pounds*	*Number*	*Number*	*Number*	*Number*	*Percent*	*Percent*	*Dollars*
R. glabra	23,000–48,000	50–60	62,800	68,600	74,200	3	80–94	73 (23–97)	0.50–0.90
R. integrifolia	3,000	----------	6,800	--------	8,000	2	--------	--------	3.50
R. laurina	90,000	----------	--------	129,600	--------	1	--------	--------	3.50
R. ovata	17,000	----------	18,700	--------	26,000	2	--------	--------	4.00
R. trilobata	7,000– 9,000	50	10,600	20,300	30,000	9	--------	62 (26–96)	.50–1.50
R. typhina	30,000	----------	48,700	53,300	67,600	5	89 (77–96)	71 (49–91)	.50–1.00

[1]First figure is average percent; figures in parentheses indicate range from lowest to highest.

have lain on the ground all winter probably germinate the first spring; others, likely not until the second spring. *Rhus* seed is often infested with larvae of some of the Chalcid flies. Germination so far as known is epigeous in all species (figs. 223 and 224). Seeds of the six species described here show delayed germination; practically none will germinate unless pretreated. Dormancy in most cases appears to be due mainly to the hard, impervious seed coats, although some species and some lots of other species appear to have embryo dormancy as well. Seed coat impermeability can be readily overcome by soaking the seeds in concentrated sulfuric acid. Seeds also having dormant embryos should be stratified at low temperatures in moist sand or peat following acid treatment. Details of treatment for the six species are given in table 179. Germination tests are most easily made in flats, using sand or peat as a medium, and 400 to 800 properly pretreated seeds. For most species, temperatures alternating diurnally from 68° to 86° F. are believed to be satisfactory. The results that may be expected under such test conditions are given in table 180.

NURSERY AND FIELD PRACTICE.—The usual practice with at least *Rhus trilobata* is to sow the seed in the fall, or to stratify them and sow in the spring. Better results would be obtained if the seed were treated with concentrated sulfuric acid prior to fall sowing or stratification. This same technique should be used with seed lots of other species thought to have embryo dormancy. Those known to have seed-coat impermeability only, should be treated with acid and sown in the spring. The seed should be sown in nursery rows at a depth of one-half inch and at the rate of about 25 viable seeds to the linear foot. Germination of properly pretreated seed will begin in the nursery in 10 to 20 days and is complete in about 30 days. Seedlings of *R. trilobata*, and possibly other species, are very susceptible to damping-off. *Rhus* can be propagated from root cuttings.

TABLE 179.—*Rhus: Dormancy and methods of seed pretreatment*

Species	Dormancy		Pretreatment recommended	Other methods of pretreatment	Remarks
	Kind	Occurrence			
R. glabra	Impermeable seed coat.	All lots[1]	Soak in H_2SO_4 60–80 minutes.	Mechanical scarification might prove helpful.	Needs further study.
R. integrifolia	do	do	Soak in H_2SO_4 4+ hours.[2]	Mechanical scarification might prove helpful. Hot water sometimes used.	----------
R. laurina	do	do	Expose seed 5 minutes to temperatures of 200°–240° F.	Soak in hot water	Acid treatment deserves trial.
R. ovata	do	do	Soak in H_2SO_4 1–6 hours.[2]	Hot-water soaking; exposure of seed to temperature of 180°–200° F.	Embryo dormancy apparently not present.
R. trilobata	Impermeable seed coat and dormant embryo.	do	Soak in H_2SO_4 1 hour; stratify 60 days at 34°–40° F.	Soak in H_2SO_4 3 hours; stratify 120 days at 41° F.	----------
R. typhina	Impermeable seed coat.	do[1]	Soak in H_2SO_4 60–80 minutes.	Mechanical scarification might prove helpful.	Needs further study.

[1]Some lots appear to have embryo dormancy also; these should be stratified 30–60 days at 41° F. following acid treatment.
[2]Duration depends on age of seed; older seed usually require longer soaking.

891183°—50——21

RHUS

TABLE 180.—*Rhus: Recommended germination test duration, and summary of germination data*

Species	Test duration recommended		Germination data from various sources					
			Germinative energy		Germinative capacity			Basis, tests
	Untreated seed	Treated seed	Amount	Period	Low	Average	High	
	Days	*Days*	*Percent*	*Days*	*Percent*	*Percent*	*Percent*	*Number*
R. glabra[1]	190+	50–60	18–43	10–34	1	11	47	9
R. integrifolia	----------	30	----------	----------	----------	----------	84	1+
R. laurina	----------	----------	----------	----------	32	----------	52	2
R. ovata	----------	----------	----------	----------	32	----------	39	2
R. trilobata	----------	15–20	18–78	8–13	1	30	86	9
R. typhina	120+	30	9–74	10–16	4	40	89	4

[1]This species was found to have a potential germination of 22 to 98 percent.

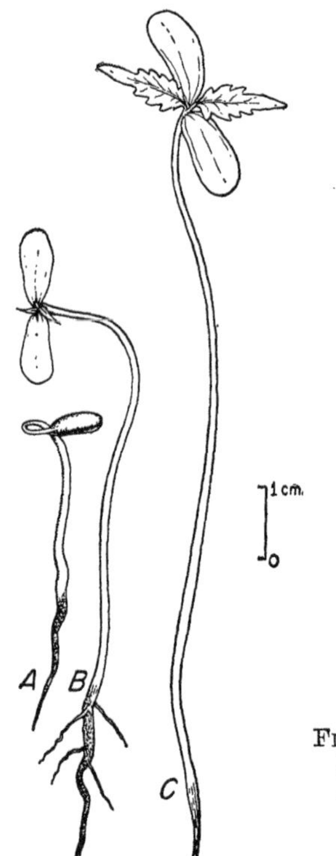

FIGURE 223.—Seedling views of *Rhus typhina: A,* At 2 days; *B,* at 4 days; *C,* at 17 days.

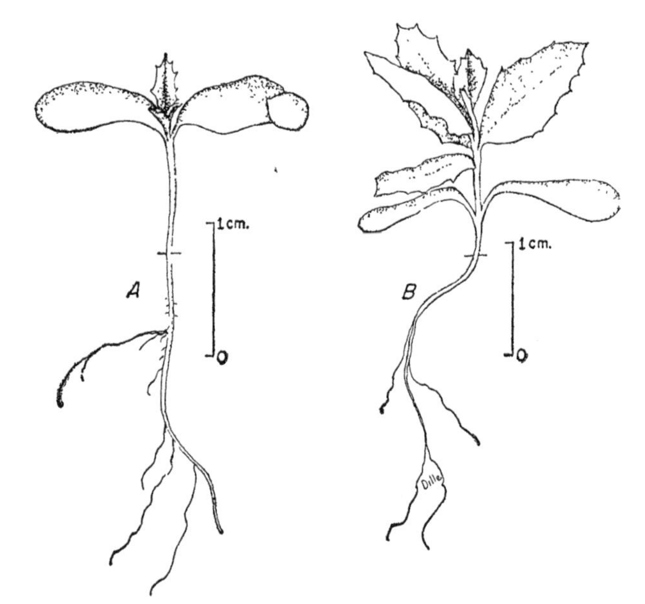

FIGURE 224.—Seedling views of *Rhus ovata: A,* Very young seedling; *B,* older seedling.

RIBES L. Currant, gooseberry

(Saxifrage family—Saxifragaceae)

Distribution and Use.—*Ribes* includes about 150 species of deciduous, rarely evergreen, shrubs, occurring in the colder and temperate portions of North America, Europe and Asia, and in the Andes of South America, south to Patagonia. The unarmed species are known as currants and the prickly species as gooseberries. About half of the species have been introduced into cultivation and are valued for their edible, often colorful, fruit, their attractive flowers or their foliage. Other species provide food and cover for game animals and browse for livestock; still others are useful in shelter-belt and erosion-control planting. Many species are alternate hosts for white pine blister rust and should not be planted where five-needled pines are commercially important. Six species, all native to North America, are used at present or have potential value for conservation planting. Their distribution and chief uses are given in table 181. Relatively little use is made of any of these species for conservation planting at the present time. *R. aureum* and *R. odoratum* are used to some extent in shelter-belt planting in the Prairie-Plains region, and the former species is also planted for gully control. Small amounts of *R. cynosbati* and *R. missouriense* have been planted to furnish food and cover for wildlife, particularly in Iowa. *R. aureum* and *R. odoratum,* because of their attractive foliage and rather conspicuous clove-scented golden flowers, have been widely used for ornamental purposes.

Seeding Habits.—The flowers in *Ribes* are perfect (dioecious in some species) usually small and greenish, but in some species larger and red to yellow in color. They are borne in the spring with the leaves in many- to few-flowered clusters or are solitary. The fruit, a many-seeded berry with calyx remnants attached, ripens in early to late summer

Table 181.—*Ribes: Growth habit, distribution, and uses*

Accepted name	Synonyms	Growth habit	Natural range	Chief uses	Date of earliest cultivation
R. americanum Mill.[1] (American black currant).	*R. floridum* L'Hérit.	Unarmed shrub.	Moist woods and swamps, Nova Scotia to Alberta, south to Virginia, Nebraska, and New Mexico.	Good bird food; has ornamental value.	1727
R. aureum Pursh (golden currant).	*Chrysobotrya aurea* (Pursh) Rydb. (slender golden currant, flowering currant).	Attractive shrub.	Along watercourses, Assiniboine to Washington, Montana, New Mexico, and California.	Useful for shelter belts, erosion control, and wildlife planting. Sometimes planted for its flowers and fruit.	1806
R. cynosbati L.[1] (pasture gooseberry).	*Grossularia cynosbati* (L.) Mill.; *R. gracile* Michx. (prickly gooseberry).	Prickly shrub.	Woods and thickets, New Brunswick to Manitoba, south to North Carolina, Alabama, and Missouri.	Wildlife food and cover	1759
R. missouriense Nutt. (Missouri gooseberry).	*R. gracile* Pursh, not Michx., *G. missouriensis* (Nutt.) Cov. & Britt.	do	Woods and thickets, Wisconsin to North Dakota, south to Illinois, Tennessee, and Kansas.	do	1907
R. odoratum Wendl.[1] (clove currant).	*Chrysobotrya odorata* (Wendl.) Rydb. (buffalo currant, Missouri currant, golden currant, flowering currant).	Handsome shrub.	Prairies and plains, South Dakota to Arkansas and Texas.	Widely planted for ornamental purposes; shelter-belt planting; wildlife food and cover.	1812
R. rotundifolium Michx. (roundleaf gooseberry).	*G. rotundifolia* (Michx.) Cov. & Britt.	Low shrub	Massachusetts to New York and North Carolina.	Game food	1809

[1]Includes 1 or more varieties.

and is varicolored—scarlet, purple, black, or yellowish. The seed of commerce is usually clean and contains a large endosperm in which is embedded a minute rounded embryo (fig. 225). Detailed seeding habits of six species are given in table 182. No data are available as to the frequency of seed crops for any of the six species; it is likely, however, that good crops are borne every year or two. Seed dispersal is almost entirely by birds and mammals. It is not known whether geographic strains occur in these species.

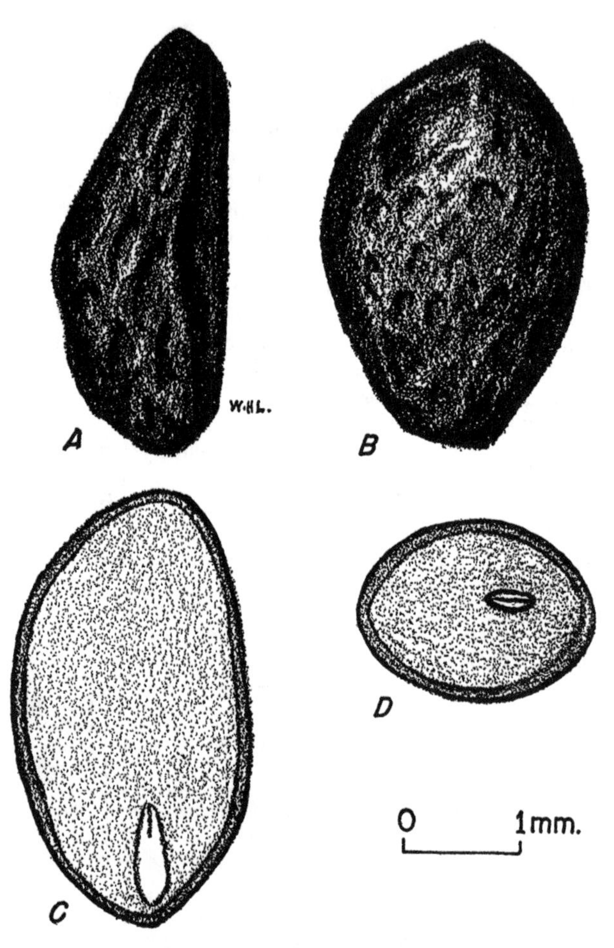

FIGURE 225.—Seed views of *Ribes missouriense:* *A* and *B*, Exterior views in two planes; *C*, longitudinal section; *D*, cross section.

COLLECTION, EXTRACTION, AND STORAGE. — The fruit should be picked or stripped from the branches as soon after ripening as possible to minimize consumption by birds. Unless the seed is to be extracted immediately, it should be spread out to dry in shallow layers to prevent heating. Extraction is easily accomplished by macerating the berries in water and allowing the pulp and empty seeds to float away. After drying, the filled seed should be run through a fanning mill to remove any remaining debris. Such data as are available on seed size, purity, soundness, and cost are given for five species in table 183. *Ribes* seed will probably keep best if stored with a low initial moisture content in sealed containers and at a low temperature. Seed of *R. americanum* with a moisture content of 16.4 percent showed a potential germination of only 38 percent after 3½ years' storage at 41° F. Seed of *R. cynosbati* gave 8 percent germination after 7 years' storage at room temperature.

GERMINATION.—Natural germination of most *Ribes* seed probably occurs in the spring following dispersal. The best seedbed appears to be moist mineral soil well supplied with humus. Germination is epigeous (fig. 226). The seeds of most species, with the exception of *R. rotundifolium*, are slow to germinate probably because they have dormant embryos. In some lots of other species, some seeds also seem to have impermeable seed coats. Germination can be hastened and increased by stratification at low temperatures in sand or peat, or a mixture of these media. In seed lots having seed-coat dormancy, the low-temperature stratification should be preceded by stratification at warm temperatures. Possibly mechanical scarification or a short period of acid treatment could be substituted for the latter period. Details of suggested treatment for dormancy are given in table 184. Germination tests are easily made in

TABLE 182.—*Ribes: Time of flowering, fruit ripening, and seed dispersal*

Species	Time of—			Color of ripe fruit
	Flowering	Fruit ripening	Seed dispersal	
R. americanum	April-early June	July-September	August-September	Black.
R. aureum	April-May	June-July	June-July	Yellow, sometimes red or black.
R. cynosbati	April-early June	Late July-September	August-September	Reddish purple (prickly).
R. missouriense	April-May	July-September	July-September	Black (smooth).
R. odoratum	do	July	July	do.
R. rotundifolium	do	July-September	July-September	Purplish (smooth).

TABLE 183.—*Ribes: Yield of cleaned seed, and purity, soundness, and cost of commercial seed*

Species	Cleaned seed				Commercial seed		
	Per pound			Basis, samples	Purity	Soundness	Cost per pound
	Low	Average	High				
	Number	*Number*	*Number*	*Number*	*Percent*	*Percent*	*Dollars*
R. americanum	247,000		333,000	2	97	95	
R. aureum[1]	200,000	217,000	231,000	3	97	98	[2]10.00–12.00; [3]1.00
R. cynosbati	189,000		221,000	2	99	86	
R. missouriense	156,000		168,000	2	98	97	
R. odoratum	106,000		155,000	2	98	97	

[1] Fruits yield 4 pounds of seed per 100 pounds of fresh fruit. [2] Cleaned seed. [3] Dried berries.

TABLE 184.—*Ribes: Dormancy and method of seed pretreatment*

Species	Dormancy		Stratification		Remarks
	Kind	Occurrence	Medium	Duration and temperature	
				Days °*F.*	
R. americanum	Embryo[1]	All lots	Sand or peat.	200 at 41–45	Suggest stratification 60 days 68°–86° F. plus 90–120 days at 41° for lots also having seed-coat dormancy.
R. aureum	do	do	Sand	90 at 41	
R. cynosbati	do	do	do	90+ at 41	
R. missouriense	do	do	do	90 at 41	
R. odoratum	do[1]		do	60 at [2]68–86 + 60–90 at 41	Needs further study; possibly acid treatment could be substituted for warm-temperature stratification.
R. rotundifolium	None	All lots			

[1] Some lots also appear to have impermeable seed coats.
[2] Temperatures alternated diurnally.

TABLE 185.—*Ribes: Recommended conditions for germination tests, and summary of germination data*

Species	Test conditions recommended				Germination data from various sources					
	Temperature		Duration		Germinative energy		Germinative capacity			Basis, tests
	Night	Day	Untreated seed	Stratified seed	Amount	Period	Low	Ave.	High	
	°*F.*	°*F.*	*Days*	*Days*	*Percent*	*Days*	*Percent*	*Percent*	*Percent*	*Number*
R. americanum	68	86	[1]580+	30			4	34	74	3
R. aureum	68	86	[2]60+	30			8		60	2
R. cynosbati	68	86	[2]130+	40	60	20	37	61	84	5
R. missouriense	68	86	[1]220+	30	85	15	42		90	2
R. odoratum	68	86	[2]60+	40	11	12		[3]20		1
R. rotundifolium	50	77	60				70		100	2

[1] Germination much lower than that of stratified seed.
[2] No germination in period indicated.
[3] Potential germination, 96 percent.

flats or germinators, using 400 or more seeds per test. The seeds should be previously pretreated in the case of lots or species showing delayed germination. Testing procedures recommended and results that may be expected are given for 6 species in table 185.

NURSERY AND FIELD PRACTICE.—*Ribes* seed is usually sown in the fall although it can be stratified and sown in the spring. Lots believed to have seed-coat dormancy should be sown as soon after collection as possible, thus insuring germination the following spring. About 40 viable seeds should be sown per linear foot of nursery row and covered to a depth of one-eighth to one-quarter inch. Most species can be propagated readily from hardwood cuttings taken in the autumn.

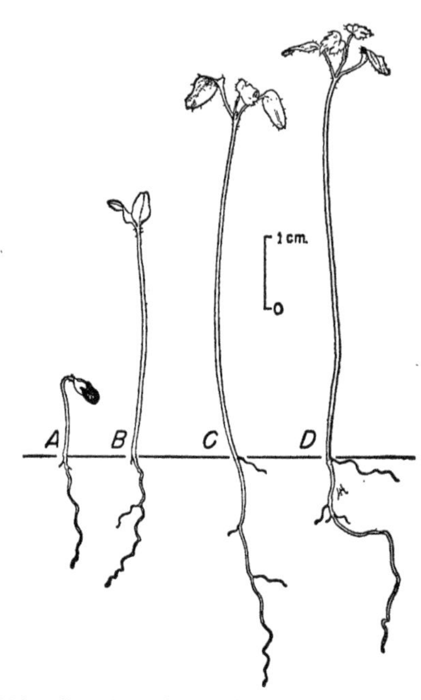

FIGURE 226.—Seedling views of *Ribes missouriense*: *A*, At 2 days; *B*, at 7 days; *C*, at 23 days; *D*, at 44 days.

ROBINIA L. Locust

(Legume family—Leguminosae)

Distribution and Use.—*Robinia* includes about 20 species which are native to the United States and Mexico. Three or four are deciduous trees and the rest are shrubs. Two species are described in table 186. *R. pseudoacacia* is native to the southeastern United States, but it grows best in the Appalachian region. This species has been widely introduced into cultivation both in the Western Hemisphere and in Europe. Its rapid growth on good sites and its durability make it the most valuable species.

Seeding Habits.—The perfect flowers of *Robinia,* which occur in racemes originating in the axils of leaves of the current year, appear in spring. The many-seeded fruit or pod ripens in the fall of the first year and opens while on the tree from winter to early spring in black locust and soon after ripening in New Mexican locust. The seed has a thin, hard outer coat that is generally impermeable to water, and it has a large embryo but no endosperm (fig. 227). Comparative seed developments of black and New Mexican locust are as follows:

	R. neo-mexicana	*R. pseudoacacia*[1]
Time of—		
Flowering	April–August	May–June.
Fruit ripening	Aug.–Oct.	Sept.–Oct.
Seed dispersal	Aug.–Dec.	Sept.–April.

[1]Commercial seed-bearing age: minimum, 6 years; optimum, 15–40 years; maximum, 60 years. Good seed crops borne every 1–2 years with light crops in intervening years.

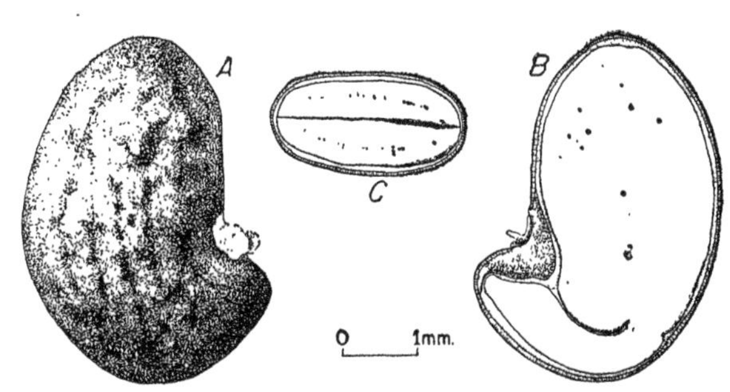

Figure 227.—Seed views of *Robinia pseudoacacia: A,* Exterior view; *B,* longitudinal section, showing large embryo; *C,* cross section.

A number of geographic strains or races of black locust have been reported but without experimental evidence. It is believed that many of the so-called races represent only differences in response to various site factors. The recognized *R. pseudoacacia* var. *rectissima* Raber (shipmast locust) develops excellent vegetative growth in parts of the eastern seaboard but is reported not to produce seed. In the Central States region numerous experimental tests have shown it to be decidedly inferior to local black locust, including a higher susceptibility to locust borer damage; in this region it is a prolific seed bearer.

Collection, Extraction, and Storage.—Pods of *Robinia pseudoacacia* can be picked from the

Table 186.—*Robinia: Growth habit, distribution, and uses*

Accepted name	Synonyms	Growth habit	Natural range	Chief uses	Date of earliest cultivation
R. neo-mexicana A. Gray (New Mexican locust).	----------	A deciduous shrub, usually only a few feet high.	Stream banks Colorado, New Mexico to Arizona and southern Utah.	Erosion control, game cover, browse, ornamental planting.	1881
R. pseudoacacia L. (black locust).	yellow locust, false acacia.	Medium-sized deciduous tree 70-80 feet high, occasionally 100 feet.	Rich soils, slopes of Appalachian mountains Pennsylvania to Georgia, west to Ozarks of Missouri, Arkansas, and Oklahoma.	Posts, poles, erosion control, soil improvement, shelter belts, ornamental planting.	1635

trees by hand or flailed or stripped onto canvas from late August throughout the winter. They should be spread out to air dry and later threshed by flailing them in a bag or by running them through a grain separator or a macerator. Chaff and light seed can be removed by fanning or flotation in water. *R. neo-mexicana* pods should be collected soon after ripening before seed are released. Yield, size, purity, soundness, and cost of commercial seed of the two species discussed here are as follows:

	R. neo-mexicana	*R. pseudoacacia*
Cleaned seed:		
Per 100 pounds of fruit...pounds..	20	[1]15–33
Per pound:		
Low...number..	21,000	16,000
Average...do....	----	24,000
High...do....	22,000	35,000
Basis, samples...do....	2+	50+
Commercial seed:		
Purity...percent..	98	97
Soundness...do....	96	90–99
Cost per pound...dollars..	----	0.20–1.50

[1]One bushel of clean seed weighs 64 pounds.

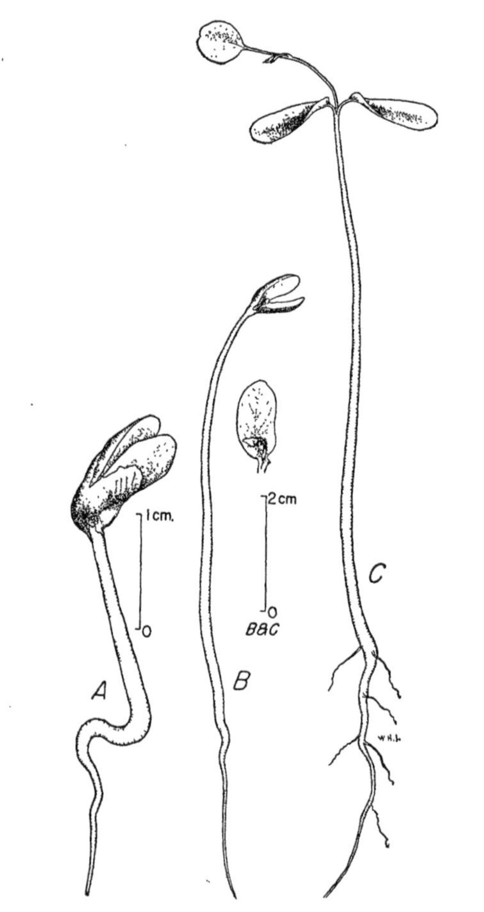

FIGURE 228.—Seedling views of *Robinia pseudoacacia*: *A*, At 1 day; *B*, at 3 days; *C*, at 8 days.

In prolonged storage, dry seeds retain their viability for 10 years or more if placed in closed containers at 32° to 40° F. Black locust seed buried in soil in 1902 was 27 percent viable in 1921. For periods of 3 to 4 years, open storage in a cool, dry place can be practiced.

GERMINATION.—The best natural seedbed for locust seeds is fresh, moist mineral soil, although seedling reproduction is rather rare; natural regeneration is largely by root suckers. Germination is epigeous (fig. 228). Dormancy is probably entirely due to the water impermeable seed coat. The differences in germination percents between lots of seed are often due to the degree of ripeness at time of collection and to scarification effected during extraction. Prompt germination can be induced by proper scarification. A modified Ames type, the barrel type, and the revolving disk type of scarifiers have all been used successfully to treat locust seed. The seed coats can also be rendered permeable by immersing the seed in concentrated sulfuric acid. Soaking in hot water is a third method, but it is less effective and less commonly used. Further details on pretreatment to overcome dormancy are given in table 187. Germination tests can best be made on 200 to 400 pretreated seeds in sand, peat, or sand and peat flats, or in standard germinators. Before the tests are made, the seed should be dusted with cuprous oxide or Semesan to prevent damping-off. Recommended methods and average results of such tests are as follows:

	R. neo-mexicana[1]	*R. pseudoacacia*
Basis, tests...number..	3	250+
Temperature:		
Night...°F...	60	60
Day...do..	80	80
Test duration...days..	10–25	10–25
Germinative energy:		
Amount...percent..	----	65
Period...days..	----	8–12
Germinative capacity:		
Low...percent..	10	51
Average...do....	29	68
High...do....	50	93

[1]Test conditions for this species recommended in absence of specific data.

NURSERY AND FIELD PRACTICE—The nursery soil should be a fertile, well-drained loam or sandy loam and well prepared. It is a preferred practice to drill seed of black locust from May to June, and New Mexican locust from March to April, in rows 9 to 12 inches apart; the seed should be covered with about one-fourth inch of firmed soil. Nursery beds should be kept moist but not wet since excessive watering favors damping-off fungi. Damping-off may be modified by dusting the seed with cuprous oxide or other fungicide prior to drilling. No mulch or shade is generally required. The tree percent is about 25. Field planting should be confined to fertile, moist, well-aerated soils, and it is best to use 1-year-old seedlings. *Robinia pseudoacacia* grows best on calcareous soils.

TABLE 187.—*Robinia: Methods of seed pretreatment for dormancy*

Species	Pretreatment			Remarks
	Medium	Temperature	Duration	
		°F.	*Mins.*	
R. neo-mexicana	Sulfuric acid	60–80	15–30	Handle same as *R. pseudoacacia* except time of treatment.
	Hot water	212—	0.2– 5	Handle same as *R. pseudoacacia*. Probably will require less severe scarification.
R. pseudoacacia[1]	Sulfuric acid	60–80	20–120	Avoid this treatment if seed are not to be sown immediately. Thickness of seed coat, ripeness of seed when collected, scarification effected in extraction are factors determining length of treatments. (Run preliminary tests.) Immediately after treatment wash seed thoroughly in water and spread to dry.
	Hot water	212—	0.2– 5	Subsequent soaking in water at about room temperature for 8 to 10 hours is necessary.

[1] For late spring sowing only: effective scarification is preferred to other treatments if seed are to be sown immediately or stored for a short time. This method provides ease in handling.

ROSA BLANDA Ait. Meadow rose

(Rose family—Rosaceae)

Also called smooth wild rose. Botanical syn.: *R. subblanda* Rydb.

Distribution and Use.—Meadow rose occurs in thickets at the edge of woods or on prairies from Newfoundland to Manitoba and south to Pennsylvania, Missouri, and North Dakota. It is a deciduous shrub, unarmed or with a few scattered prickles when young, and is valuable chiefly for game food and cover. Because of its habit of forming colonies, it may also have possibilities for erosion-control planting. The fruits may have some emergency value as a source of vitamin C. This species was introduced into cultivation in 1773.

Seeding Habits.—The bright pink, fragrant, perfect flowers are borne from May to June. The fruit (hip) which becomes fleshy and berrylike at maturity in the autumn, is scarlet in color, about one-half inch in diameter, and encloses several small, bony achenes or seeds (fig. 229); it often persists on the bushes until spring, sometimes until the second spring. Some fruit is borne every year. Dispersal is mostly by birds and mammals. Commercial seed consists of the dried hips or clean seed.

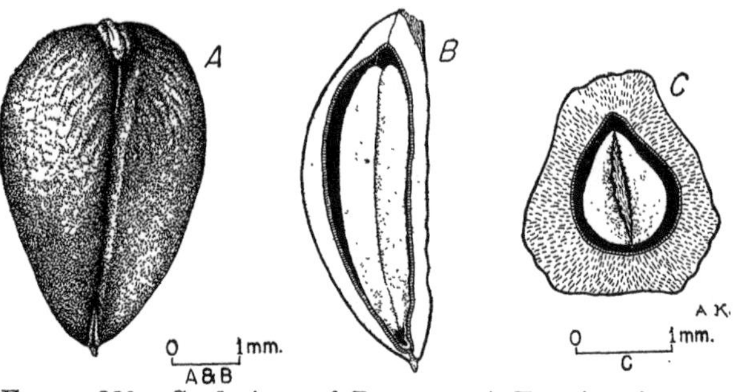

Figure 229.—Seed views of *Rosa* sp.: *A*, Exterior view; *B*, longitudinal section; *C*, cross section.

Collection, Extraction, and Storage. — The fruit can be collected at any time after ripening. Seed extraction is accomplished by macerating the hips in water and allowing the pulp and empty seeds to float away. After drying, it is ready for storage or for use. Clean seed per pound (3 samples): 37,000 to 53,000; soundness, 96 percent. Cost of commercial seed ranges from $1 to $1.25 per pound for dried fruit to $4 per pound clean. Closely related species yield about 20 pounds of clean seed per 100 pounds of fruit. Seed of this species has shown considerable germination after dry storage for 3 years.

Germination.—Germination is epigeous (fig. 230). Meadow rose seed, like that of most other roses, is dormant and will germinate only after prolonged stratification at 41° F. In addition to embryo dormancy, it is believed that the seeds have an impermeable seed coat. This is borne out by tests in which seed that had been soaked in sulfuric acid and then stratified for 60 days at 41°, and seed stratified for 60 days at 68° to 86° (temperatures alternated diurnally) plus 60 days at 41°, gave considerably higher germination than seed which had only been stratified at 41° for 90 days. Further evidence of seed-coat dormancy consists of the fact that seed which had been stratified after passage through the digestive tracts of ring-necked pheasants and quail also gave better germination than seed which had only been stratified. Suggested treatment: 1 to 2 hours' soaking in concentrated sulfuric acid plus stratification for 60 to 120 days at 41°. The optimum germination temperature for meadow rose seed is believed to be approximately 41°. Although the emergence of seedlings was found to be greatest at temperatures fluctuating diurnally from 68° to 86°, it is believed that actual germination may have begun before the seed was removed from the stratification chamber. Germinative energy, 32 percent in 10 days; germinative capacity, 36 percent; potential germination, 74 percent (1 sample).

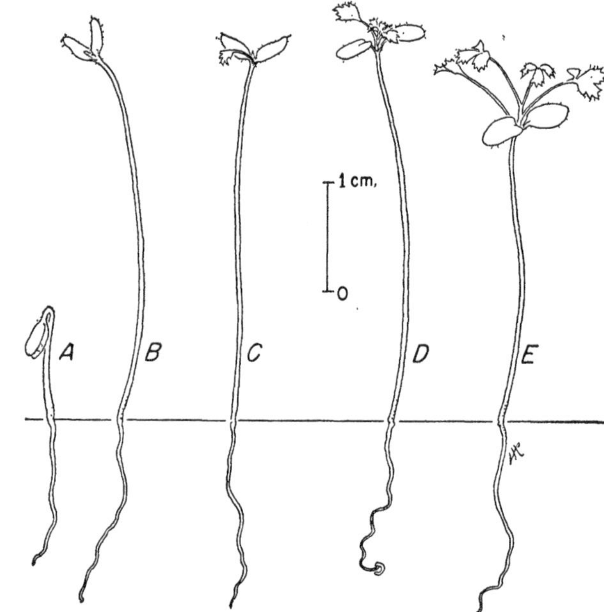

Figure 230.—Seedling views of *Rosa blanda*: *A*, At 1 day; *B*, at 3 days; *C*, at 6 days; *D*, at 26 days; *E*, at 41 days.

Nursery Practice.—Acid treatment of the seed followed by fall sowing or by stratification over winter prior to spring sowing are suggested practices. A sowing depth of about one-half inch is recommended. Propagation by cuttings is also possible.

RUBUS L. Blackberry, raspberry

(Rose family—Rosaceae)

Distribution and Use.—*Rubus* includes about 400 species of deciduous or evergreen, often prickly, erect or trailing shrubs or vines. They are native chiefly to the colder and temperate regions of the Northern Hemisphere, with a few in the tropics and the Southern Hemisphere. Many species are cultivated for their excellent fruit, their flowers, or foliage. Others provide food and shelter for wild animals and birds. Many are useful in establishing cover on denuded forest lands and have considerable possibilities for erosion-control planting. The bark of the roots, and to some extent the fruits, have medicinal properties. They are useful honey plants. Many species hybridize freely, thus frequently making identification difficult. Two species which are of importance for conservation planting, and for which reliable information is available, are described here in table 188. *R. idaeus* or its varieties has been the source of most of our cultivated red raspberries. *R. occidentalis* is also the source of several cultivated varieties.

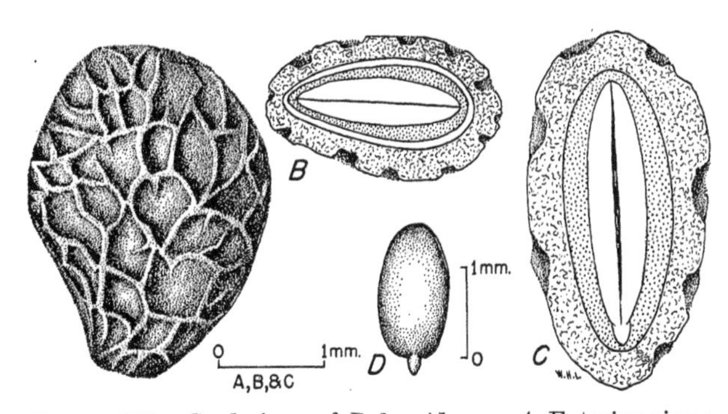

Figure 231.—Seed views of *Rubus idaeus: A,* Exterior view; *B,* cross section; *C,* longitudinal section; *D,* embryo.

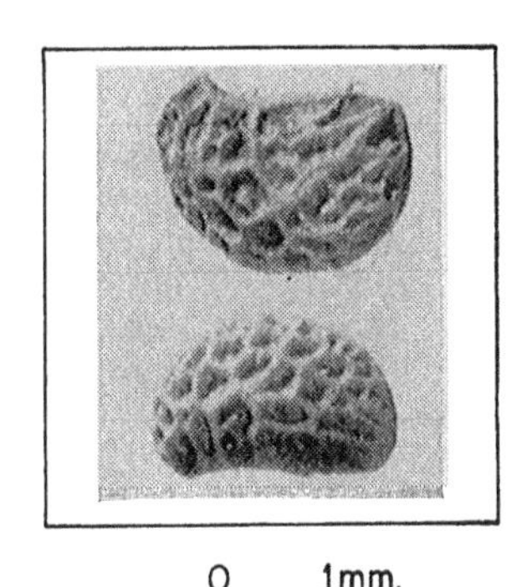

Figure 232.—Seed views of *Rubus occidentalis.*

Seeding Habits.—The perfect flowers bloom in the spring or early summer. The fruit, which ripens unevenly in the summer or early fall, is an aggregate of small, usually succulent drupelets, each containing a single hard, pitted seed (figs. 231 and 232). Natural seed dispersal is mostly by birds and mammals. *Rubus idaeus* and *R. occidentalis* flower from May to June; their fruits, which are about three-eighths of an inch in diameter, ripen from July to August. Good seed crops are borne almost annually. Fruits of *R. idaeus* are light red in color and white to yellow in some varieties; *R. occidentalis* fruits are purple black, but one variety is amber yellow. Although it has

Table 188.—*Rubus: Growth habit, distribution, and uses*

Accepted name	Synonyms	Growth habit	Natural range	Chief uses	Date of earliest cultivation
R. idaeus L. (red raspberry).	*R. melanolasius* Focke; *R. strigosus* Michx.; *R. idaeus* var. *strigosus* (Michx.) Maxim. (American red raspberry); *R. idaeus* var. *canadensis* Richards (Canada red raspberry); *R. carolinianus* Rydb.; *R. subarcticus* Rydb.	Thicket-forming, thorny, medium tall shrub.	Circumpolar: In North America from Newfoundland and Labrador to Alaska and south to North Carolina, Colorado, and Oregon.	Wildlife food and cover, soil protection; human food, honey plant.	Before 1894.
R. occidentalis L. (blackcap raspberry).	common black cap, wild black raspberry, thimble berry.	-----do---------	New Brunswick to Minnesota, south to Georgia and Colorado.	-----do---------	1696.

not been proved experimentally, it is likely that climatic races have developed within these two raspberry species because of their wide natural ranges and because of the great variety of cultivated forms. It is known, for instance, that the American varieties of *R. idaeus* are hardier in cultivation than the European forms.

Collection, Extraction, and Storage.—*Rubus* fruits should be picked from the bushes by hand soon after they are ripe to prevent loss caused by birds and other animals. Although they may then be spread out in thin layers to dry, the preferred practice is to run them through a macerator with water, floating off or screening out the pulp and empty seeds. The seed should then be dried, after which it can be sown or stored. *Rubus* seed is handled by very few seed dealers; many species do not appear on the market at all. The yield, size, purity, soundness, and cost of commercial cleaned seed vary by species, as follows:

	R. idaeus	*R. occidentalis*
Cleaned seed:		
Per 100 pounds fruit...pounds..	3	3–8
Per pound:		
Low...number..	305,000	286,000
Average...do....	328,000	334,000
High...do....	364,000	384,000
Basis, samples...do....	6	4
Commercial seed:		
Purity...percent..	99	98
Soundness...do....	86	83
Cost per pound...dollars..	8–9	----

Seed of *R. idaeus* stored for 1 year in a sealed container at 41° F. showed 78 percent germination out of a possible 80 percent. It probably could be stored satisfactorily for several years by such means. Although seed of *R. occidentalis* has been stored for 1 year in moist sand at 41° F. without apparent loss in viability, dry storage at a similar or lower temperature is believed to be a safer procedure.

Germination.—Germination is epigeous (fig. 233). Seed of many *Rubus* species are slow to germinate because they have a hard, impermeable coat (endocarp) combined with a dormant embryo. Such seeds can be stimulated into germination by warm plus cold stratification or by soaking them in concentrated sulfuric acid followed by cold stratification. Seed of *R. idaeus* and *R. occidentalis* require pretreatment. Stratification in moist sand at 68° to 86° F. for 90 days plus 90 days at 41° is recommended. Probably 2 hours' soaking in concentrated sulfuric acid, or mechanical scarification, could be substituted satisfactorily for the warm stratification. The germination of *Rubus* seed can be tested in sand flats at temperatures of 68° (night) to 86° (day)[60] for 30 days using 400 to 1,000 properly pretreated seeds per test. Average results for *R. idaeus* and *R. occidentalis* are as follows:

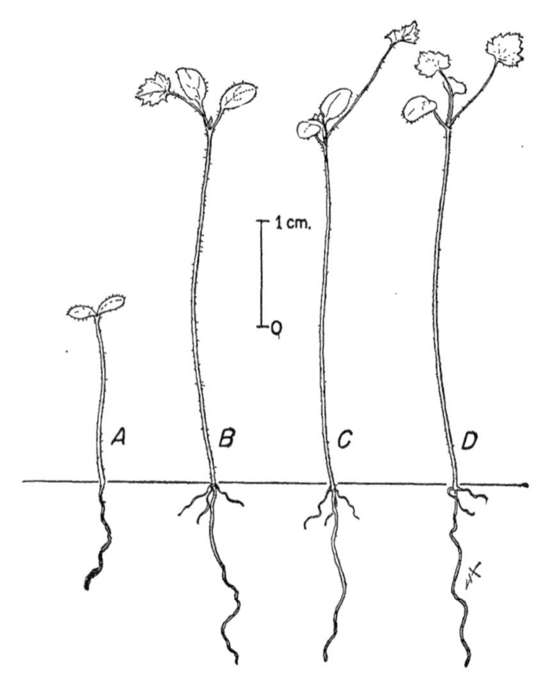

Figure 233.—Seedling views of *Rubus occidentalis*: *A*, At 1 day; *B*, at 13 days; *C*, at 22 days; *D*, at 36 days.

	R. idaeus	*R. occidentalis*
Basis, tests...number..	6	3
Test duration:		
Untreated seed...days..	230+	230+
Pretreated seed...do..	30	30
Germinative energy:		
Amount...percent..	63–78	44–78
Period...days..	7–13	7–8
Germinative capacity:		
Low...percent..	63	50
Average...do....	69	64
High...do....	78	80
Potential germination...do....	78–96	50–94

Nursery and Field Practice.—Results of laboratory tests indicate that sowing of *Rubus* seed immediately after collection in midsummer is the best practice. This insures prompt and complete germination the following spring. The seed should be sown in drills and covered with 1/8 to 3/16 inch of soil, and the seedbeds mulched to prevent drying. In horticultural practice *Rubus* plants usually are propagated by dividing clumps, by the use of natural stolons, or sometimes by root cuttings.

[60] A range of 50° (night) to 77° F. (day) is equally suitable for *Rubus idaeus*, but lower temperatures are not satisfactory.

SALIX L. Willow

(Willow family—Salicaceae)

Distribution and Use.—The willows consist of about 300 species of deciduous trees or shrubs widely distributed in both hemispheres from the Arctic region to South Africa and southern Chile. Hybrids are numerous. Of the approximately 70 species recognized in North America, some 30 attain tree form and size. Many of the tree species are valuable for their wood products, but only 3 of the more common ones are described in detail in table 189. Coppicing is often practiced to produce basket-weaving materials.

Seeding Habits.—Staminate and pistillate flowers are borne in catkins on separate trees, usually appearing before or with the leaves. The fruit, a capsule, occurring in elongated clusters, contains many minute, hairy seeds (fig. 234) which usually ripen in early summer, but in some species not until fall. The chief disseminating agents are wind and water. Comparative seeding habits of three species are given in table 190. Many horticultural varieties and hybrids have been developed.

Collection, Extraction, and Storage.—Willow seed must be collected as soon as the fruits ripen; ripeness is indicated when the capsules have turned from green to a yellowish color. Frequent observations are necessary to determine maturity, at which time capsules can be collected in bags; otherwise, if in the vicinity of water, seed may be gathered from drifts at the water's edge. It is unnecessary to separate the seed from the opened capsules. There are approximately 2 million to 3 million seed per pound. Because willow seed are viable for only a few days, commercial seed are not available. The maximum period of storage is from 4 to 6 weeks. For storage periods up to 10 days, the seed should be placed in closed containers at about room temperature. Relative humidity of surrounding air should not drop below 50 percent.

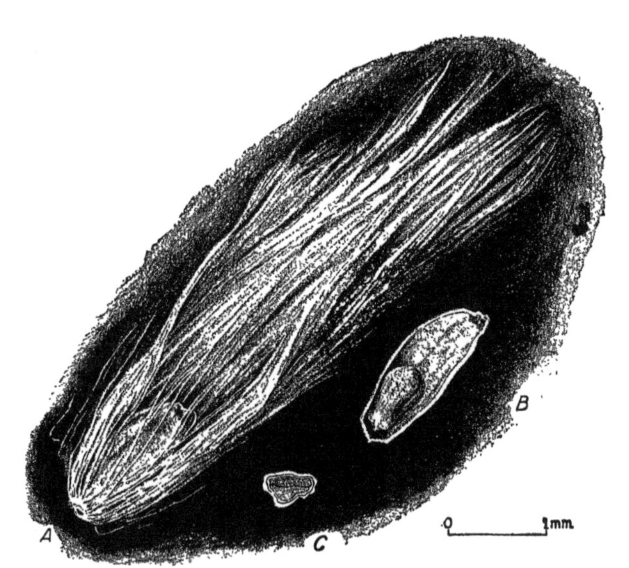

Figure 234.—Seed view of *Salix fragilis: A*, Exterior view of seed with cotton, showing corona; *B*, longitudinal section of seed showing embryo; *C*, cross section of seed.

Table 189.—*Salix: Growth habit, distribution, and uses*

Accepted name	Synonyms	Growth habit	Natural range	Chief uses	Date of earliest cultivation
S. bebbiana Sarg. (Bebb willow).	*S. rostrata* Richards., not Thuill. (beaked willow).	Bushy tree forming extensive thickets in the North.	Moist sandy or gravelly, rich soils Newfoundland to Alaska, south to Pennsylvania, Nebraska, New Mexico, Arizona and Oregon.	Erosion planting, game cover, browse, ornamental planting.	1889
S. discolor Muhl. (pussy willow).	glaucous willow	Tall shrub or small tree.	Moist meadows and along streams and lakes Nova Scotia to Manitoba, south to Delaware, west to Missouri and Black Hills.	Ornamental planting, erosion planting, game cover.	1809
S. nigra Marsh. (black willow).	swamp willow	Tree of medium size.	Moist rich soils of flats, stream and lake borders New Brunswick and Great Lakes to Florida, west to the Dakotas, Arizona, into Mexico.	Packing boxes, baskets, excelsior, woodenware, furniture drawers, stream-bank protection.	1809

SALIX

Table 190.—*Salix: Time of flowering and fruit ripening, and commercial seed-bearing age*

Species	Time of—			Commercial seed-bearing age[1]		
	Flowering	Fruit ripening	Seed dispersal	Minimum	Optimum	Maximum
				Years	*Years*	*Years*
S. bebbiana	May–June	May–June	May–June	4	10–30	45
S. discolor	March–April	April–May	April–May	2	8–25	35
S. nigra	April–May	May–June	May–June	10	25–75	125

[1] Good seed crops occur nearly every year with light crops in intervening years.

Germination.—Under natural conditions *Salix* seed usually germinate in 12 to 24 hours after falling on moist or wet sand, or alluvium of stream bottoms, lakes or swamps. Germination is epigeous. No dormancy is known to occur in any species. Adequate germination tests can be made on 200 to 400 seeds firmed on the surface of constantly moist sand or sandy loam in flats. It is recommended that the flats be kept at temperatures of 70° (night) and 85° F. (day) for 12 to 36 hours when testing *S. bebbiana, S. discolor,* and *S. nigra.* Under these test conditions *S. bebbiana* showed a germinative capacity of 28 percent (low) and 100 percent (high) in 2 tests; 1 test of *S. discolor* showed its germinative capacity to be 95 percent (high).

Nursery and Field Practice.—Seed must be sown immediately after collection. Broadcasting of seed on well-prepared beds should be followed by firming of the soil. The seedbeds should be kept constantly moist until after the seedlings are well established. To conserve soil moisture and maintain a high relative air humidity immediately above the bed surface, it is recommended that 4- to 6-inch side boards be placed along beds and shades and burlap rested on them. Usually a higher percentage of plantable stock is obtained if seedlings are transplanted at 3 to 4 weeks. The transplants, 1 year from seed, are generally large enough for field planting the next spring.

Nursery stock of the willows should be sprayed to control leaf rusts caused by *Melampsora* spp. if they are known to be present. *Larix, Abies, Ribes,* and species of the Saxifrage family are alternate hosts of these rusts. Two diseases, a fungus scab and black canker, recently introduced in the Northeastern States are destructive to leaves and shoots of some willow species. The scab kills trees of susceptible species. Infected nursery stock should not be field planted. Also seedlings or cuttings with dead bark, caused by *Cytospora* or other fungi usually more prevalent in stock growing under adverse conditions, should not be field planted. Root and stem cuttings are often used in field planting.

SAMBUCUS L. Elder

(Honeysuckle family—Caprifoliaceae)

DISTRIBUTION AND USE.—Including about 20 species of deciduous shrubs or small trees, rarely herbs, native to the temperate and subtropical regions of both hemispheres, the elders are useful chiefly for their fruit which serves as food for birds, mammals, and man. Some species have medicinal properties; others are desirable for ornamental use because of their handsome foliage, attractive flowers, and colorful fruit. The pith is used for various purposes. Three American species, which are useful for planting in the United States and for which reliable information is available, are described in table 191. *Sambucus canadensis* is probably the most commonly cultivated of the 3, with *S. pubens* less commonly used; *S. glauca* so far is seldom cultivated.

SEEDING HABITS.—The small, white or yellowish white, perfect flowers, blooming in the spring or summer, are borne in rather large clusters. The fruit is a berrylike drupe, with three to five one-seeded nutlets or stones (figs. 235 and 236), which ripens in summer. Dispersal is usually by birds or mammals. Comparative seeding habits of the three species discussed here are given in table 192. Scientific information as to racial variation among the elders is lacking. Both *Sambucus canadensis* and *S. pubens* have developed varieties none of which, however, appear to be climatic races. *S. glauca* has developed two varieties which have definite geographic limits and may be climatic races.

COLLECTION, EXTRACTION, AND STORAGE.—Elder fruits are collected by stripping the clusters from the branches. Long-handled pruning shears are convenient for this purpose. Collection should be made

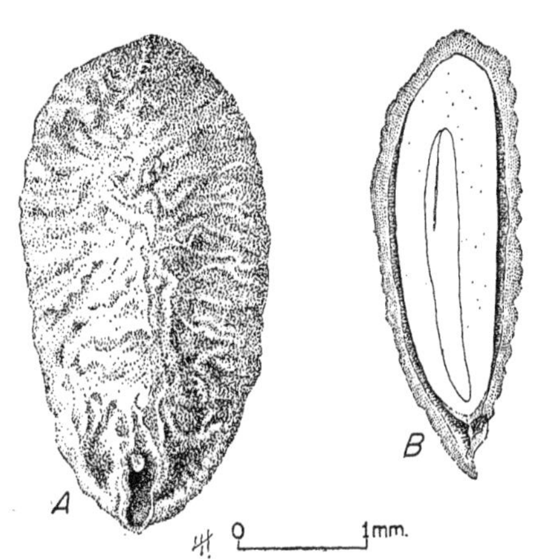

FIGURE 235.—Seed views of *Sambucus pubens: A,* Exterior view; *B,* longitudinal section.

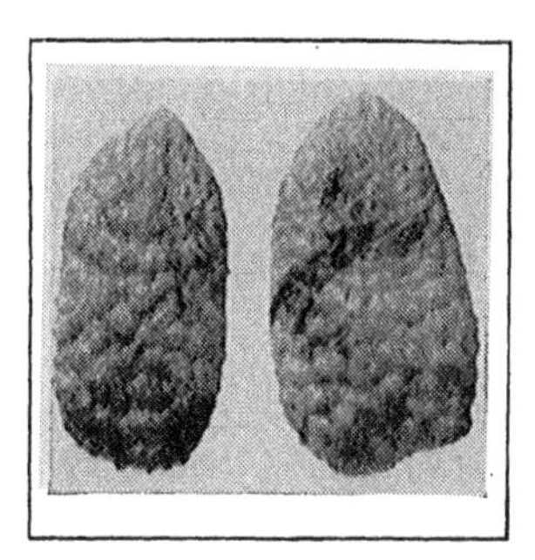

0 1mm.

FIGURE 236.—Seed views of *Sambucus canadensis.*

TABLE 191.—*Sambucus: Growth habit, distribution, and uses*

Accepted name	Synonyms	Growth habit	Natural range	Chief uses	Date of earliest cultivation
S. canadensis L. (American elder).	-------------------	Medium-tall, stoloniferous shrub.	Moist, rich soils from Nova Scotia to Manitoba, south to Florida and Texas.	Wildlife food, human food, shelter-belt planting, ornamental.	1761
S. glauca Nutt. (blueberry elder).	*S. cerulea* Raf.------	Large shrub or small tree.	Open woods and stream banks from British Columbia to California, east to Montana and Utah.	Wildlife food----	1850
S. pubens Michx. (scarlet elder).	*S. racemosa* of A. Gray, not L. (red-berried elder, red elder).	Medium to tall shrub.	Rocky banks, moist woods and waste places from New Brunswick to Manitoba, south to South Dakota, Ohio, and the mountains of Georgia.	Wildlife food, ornamental.	1812

SAMBUCUS

TABLE 192.—*Sambucus: Time of flowering and fruit ripening, and frequency of seed crops*

Species	Time of—			Ripe fruit			Seed year frequency	
	Flowering	Fruit ripening	Seed dispersal	Color	Average diameter	Average seed	Good crops	Light crops
					Inches	*Number*	*Years*	*Years*
S. canadensis	June–Aug.	July–Sept.	July–Sept.	Purple black[1]	⅛–3/16	4	1	----------
S. glauca	May–Aug.	Aug.–Sept.	Aug.–Sept.	Blue black	----------	----------	----------	----------
S. pubens	April–May	June–Aug.	June–Aug.	Scarlet[2]	3/16	3	1+	Intervening.

[1] Fruits of varieties vary in color from greenish to cherry red.
[2] Fruits of varieties vary in color from white to amber yellow.

as soon as the fruits ripen, since birds consume them quite rapidly. If the seed is not to be extracted immediately, the fruits should be spread out in thin layers to avoid heating. Otherwise, the fruits may be (1) dried, (2) run through a macerator with water and the pulp and empty seeds floated off, or (3) crushed, dried, and used without further cleaning. After a short period of drying, freshly extracted seed may be fanned or screened to remove debris. Excessive drying may be harmful to some lots and should be avoided. Commercial seed consists either of dried fruit or cleaned seed. The size, purity, soundness, and cost of commercial seed vary according to species as shown in table 193. Ordinarily dry storage of elder is recommended. Seed of *Sambucus canadensis* and *S. pubens* stored for almost 2 years in sealed jars at 41° F. showed little or no loss in viability. Seed of *S. pubens* has also been kept satisfactorily for 1 year in moist sand at 41°. Longevity of *S. glauca* seed under storage is not known.

GERMINATION.—Elder seed usually are rather difficult to germinate because of dormant embryos and hard seed coats. Best results so far have come from warm stratification followed by cold stratification. Germination is epigeous (fig. 237). Details as to dormancy and pretreatment are given in table 194. Germination tests can be made in sand flats at 68° (night) to 86° F. (day) for 60 to 100 days, using 1,000 properly pretreated seeds per test. Methods recommended and average results for the 3 species discussed here are given in table 195.

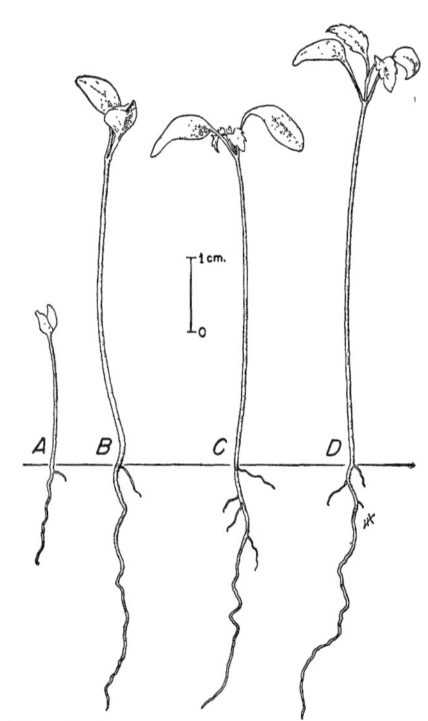

FIGURE 237.—Seedling views of *Sambucus canadensis: A,* At 2 days; *B,* at 20 days; *C,* at 33 days; *D,* at 45 days.

NURSERY AND FIELD PRACTICE.—Ordinarily elder seed is sown in the fall or stratified seed is used in the spring. In either case germination often is not complete until the second spring. On the basis of

TABLE 193.—*Sambucus: Yield of cleaned seed, and purity, soundness, and cost of commercial seed*

Species	Cleaned seed					Commercial seed		
	Yield per 100 pounds fresh fruit	Per pound			Basis, samples	Purity	Soundness	Cost per pound dried fruit
		Low	Average	High				
	Pounds	*Number*	*Number*	*Number*	*Number*	*Percent*	*Percent*	*Dollars*
S. canadensis	7–18	175,000	232,000	324,000	14	92	80	0.80–1.25
S. glauca	6	117,000	(1)	126,000	2	----------	----------	2.50
S. pubens	4	192,000	286,000	377,000	6	98	97	----------

[1] About 27,500 dried berries per pound.

laboratory results, it is believed that sowing in the summer shortly after collection should give good germination the following spring. For *Sambucus canadensis* it is recommended that the seed be sown in drills, using 35 viable seed per linear foot, and covered with one-fourth inch of soil. Fall-sown beds should be mulched. Beds of *S. pubens* seedlings should be given half shade. Nursery germinations of 85 percent for *S. canadensis* and 80 percent for *S. pubens* have been attained. One-year-old seedlings usually are large enough for field planting. Elders can also be propagated from cuttings.

TABLE 194.—*Sambucus: Dormancy and method of seed pretreatment*

Species	Dormancy		Stratification		Remarks
	Kind	Occurrence	Medium	Duration and Temperature	
				Days °*F.*	
S. canadensis	Embryo, sometimes seed coat.	Variable	Moist sand	60 at 68–86 + 90–150 at 41	Stratification for 90 days at 41° F. is satisfactory for some lots; some lots need no pretreatment. Sulfuric acid plus cold stratification might be a suitable substitute for warm plus cold stratification.
S. glauca	Embryo, possibly seed coat.	Probably regular.	do	90 at 41	The treatment recommended does not give very good germination. Warm plus cold stratification or acid treatment plus cold stratification are probably necessary for satisfactory germination.
S. pubens	Embryo and seed coat.	do	do	30–60 at 68–86 + 90–150 at 41	

TABLE 195.—*Sambucus: Recommended conditions for germination tests, and summary of germination data*

Species	Test conditions recommended				Germination data from various sources					
	Duration		Temperature		Germinative energy		Germinative capacity			Basis, tests
	Untreated seed	Pretreated seed	Night	Day	Amount	Period	Low	Average	High	
	Days	*Days*	°*F.*	°*F.*	*Percent*	*Days*	*Percent*	*Percent*	*Percent*	*Number*
S. canadensis	270+	[1] 60	68	86	18–47	12–20	13	63	95	7
S. glauca		[1] 105						17		1+
S. pubens	270+	[1] 60	68 or 50	86 or 77	47–54	21–34	32	47	63	6

[1] Indicates condition for which germination is reported.

891183°—50——22

SAPINDUS DRUMMONDI Hook. & Arn. Western soapberry

(Soapberry family—Sapindaceae)

Also called chinaberry, wild China-tree, soapberry, Indian soap-plant.

Distribution and Use. — Western soapberry occurs on moist clay soil or on dry limestone uplands from southwestern Missouri to Louisiana and westward to southern Colorado, southern Arizona, and northern Mexico. It is a deciduous tree useful for shelter-belt planting, ornamental purposes, and to some extent for game food. The ornamental fruit contains saponin and was formerly used in making soap. The heavy, strong, close-grained wood splits easily into thin strips and is used in basketry, for making frames of pack saddles, and for boxes. This species was introduced into cultivation in 1900.

Seeding Habits.—The small, white, polygamo-dioecious flowers, borne in rather large clusters, open from May to July; the fruit, a yellow, translucent, globular drupe, one-half inch in diameter and usually containing a single dark brown seed (fig. 238), ripens from September to October and persists on the branches until late winter or spring. Its pulp is yellow and turns black in drying.

Collection, Extraction, and Storage.—Western soapberry fruit may be collected at any time during the late fall or winter months by hand picking or by flailing it from the trees onto canvas. Although the fruit is fairly dry by this time, care should be taken to keep it from heating; spreading it out in shallow layers is usually sufficient. Prior to extraction, the fruit should be sprinkled with water twice daily until the pulp softens. The pulp can then be removed and floated away by running the fruit through a macerator with water. After drying, the seed is ready for storage or for use. One hundred pounds of fruit, as collected, will yield about 30 pounds of clean seed. Fruits per pound: 430 to 650. Clean seeds per pound (7 samples): low, 1,430; average, 1,620; high, 1,980. Soundness: 77 percent (12 samples). No cost data are available. Little evidence as to seed longevity is available. Seed stored dry for 18 months, then pulped and stored for 18 additional months in moist sand at about 40° F., showed 16 percent germination. It therefore seems likely that seed can be stored dry at a low temperature for several years without much loss in viability.

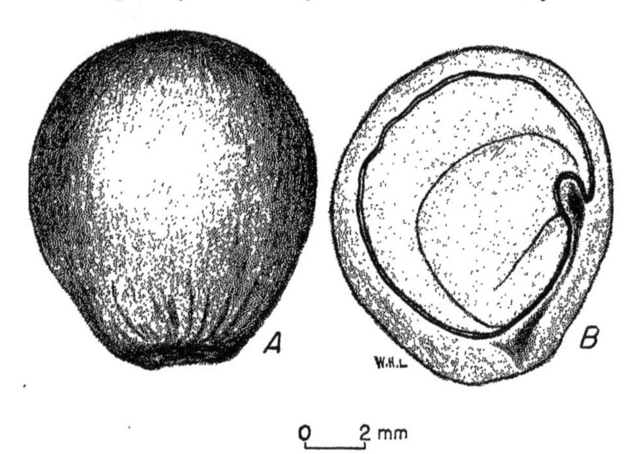

Figure 238.—Seed views of *Sapindus drummondi: A,* Exterior view; *B,* longitudinal section.

Germination.—Western soapberry seed germinate rather slowly, probably because of an impermeable seed coat and possibly a dormant embryo. Germination can be much improved by treating the seed with concentrated sulfuric acid for 2 to 2½ hours. Ninety days' stratification also gives fairly good results. Perhaps even better results would be obtained from a combination of the 2 treatments. Germination tests can be run in sand flats at temperatures alternating diurnally from 68° to 86° F. Germinative capacity (11 samples): low, 7 percent; average, 31 percent; high, 68 percent.

Nursery Practice.—Seed may be either fall sown or given acid treatment or stratification prior to spring sowing. Twenty viable seeds are usually sown per square foot at a depth of three-quarters of an inch. This species has a strong taproot; top growth is slow.

SASSAFRAS ALBIDUM (Nutt.) Nees Sassafras

(Laurel family—Lauraceae)

Common sassafras, SPN. Botanical syns.: *Sassafras variifolium* (Salisb.) Ktze., *S. officinale* Nees & Eberm., *S. sassafras* (L.) Karst.

Distribution and Use.—Sassafras is a short to medium-tall, deciduous tree native from Maine south to Florida and west to the Brazos River in Texas, eastern Oklahoma, Kansas, Iowa, and southern Ontario. Locally, its best development is in open woods in moist, rich, sandy loam soils, but it is found in pure, dense stands in old abandoned fields as a pioneer tree cover. The light-brown wood is soft, light in weight, and very durable, and it is used for fence posts, rails, cooperage, interior finish, and fuel. Bark of the roots is commercially important for tea, sassafras oil, and perfume for soaps and other articles. Gumbo filet, a powder made from the leaves, gives consistency to gumbo soup. With some difficulty, sassafras may be artificially regenerated from seed and root cuttings. It has been cultivated since 1630.

Seeding Habits. — Dioecious, greenish-yellow flowers appear in early spring with the leaves in loose, drooping, few-flowered racemes from between bracts of terminal buds. The pistillate flower buds are much larger than the staminate ones. Sassafras fruits are ovoid, dark-blue drupes one-half inch long; they ripen from August to September. The flesh or pericarp is pulpy, and it covers a hard, thin endocarp which encloses the seed (fig. 239). The seed coats are difficult to separate, and are membranous. Cotyledons are large and fleshy. The stones are the seed of commerce. Dispersal is chiefly by birds. Ages for commercial seed bearing are: minimum, 10 years; optimum, 25 to 50 years; maximum, 75 years. Good crops are produced at intervals of 1 to 2 years, with light intervening crops.

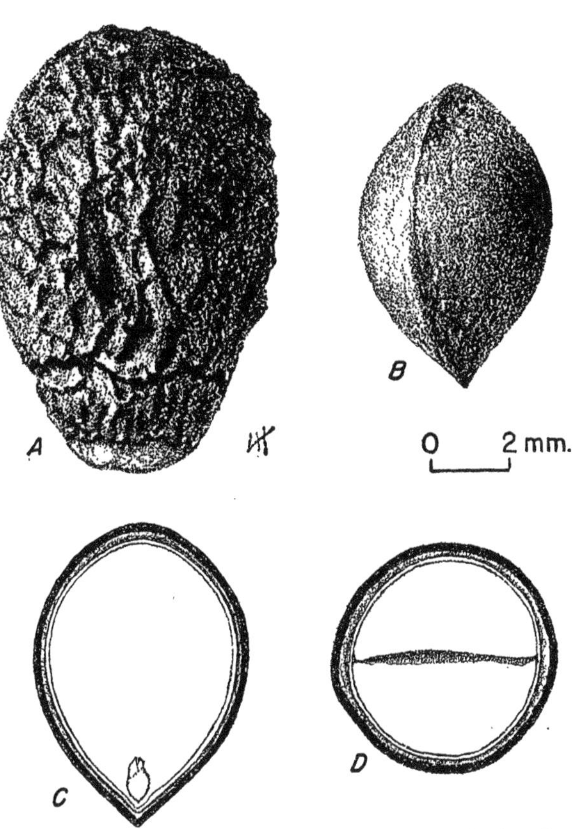

Figure 239.—Seed views of *Sassafras albidum: A*, Exterior view of fruit; *B*, exterior view of seed; *C*, longitudinal section, showing embryo with large cotyledons; *D*, cross section.

Collection, Extraction, and Storage.—Sassafras fruits may be collected by picking them from the trees soon after they ripen, or by shaking or whipping them from the branches onto canvas. Ripe fruits are dark blue and soft. After collection, they may be kept for a few days before cleaning, or cleaned at once. The pulp can be removed by rubbing the fruits over hardware cloth of mesh fine enough to hold the seeds, and the debris washed away with water. The number of seeds per pound varies from about 4,000 to 6,000 (2 samples) or about 3,000 dried fruits (1 sample). Commercial seeds have an average purity and soundness, respectively, of 98 and 85 percent. Costs range from $1 to $2.25 per pound of clean seed. Seed should not be allowed to air-dry for long; therefore it is necessary to complete storage soon after collection and cleaning. For prolonging viability, storage should be in sealed containers at 35° to 41° F. Overwinter storage can be accomplished by stratifying the seed in sand or sand and peat at 35° to 41°.

Germination.—The best natural seedbed for sassafras is a moist, rich, loamy soil partially protected by vegetative cover or litter. Germination takes place chiefly in April and May after seed fall. With favorable conditions, however, partial germination is common in early-maturing seed in the same fall. Sassafras seed are dormant for a short period due to conditions in the embryo; the dormancy may be broken by stratification in sand, or in a mixture of sand and peat, for 30 days at 41° F. Seed from the northern part of the species range, however, do not respond to this treatment. Adequate germination tests may be made on 100 pretreated seeds in sand or sand and peat flats in 30 to 45 days at temperatures of 55° to 75°. Forty-five to seventy-five days are sufficient for untreated seed.

Nursery Practice.—Sowing may be done in the fall or spring, and cleaned or uncleaned seed used. However, cleaned seed usually give better results. If seed with pulp are to be sown, it is recommended that they be stratified. The beds should be well prepared and drilled in rows 8 to 12 inches apart, the seed covered with one-fourth to one-half inch of firmed soil. Burlap, straw, or leaf mulch is necessary, and should be held in place by bird or shade screens until after late frosts in the spring. Fall seeding should be done as late in the fall as possible since early-sown seed often germinate before cold weather. It may be necessary to cold-store seed for a short period between collection and fall seeding. Spring beds can be prepared and drilled with stratified or dry cold-stored seed as early as soil conditions permit. It is necessary to keep the surface soil moist until germination is complete. From 80 to 85 percent germination may be expected in a nursery from properly handled sound seed. No shading is required.

SEQUOIA Endl. Sequoia

(Pine family—Pinaceae)

DISTRIBUTION AND USE.—The sequoias consist of two species of very tall evergreen trees largely confined to California. One species, redwood, is commercially important; the other, giant sequoia, is of little present importance in forestry, but bears promise of increased value in the future. The redwood is known as the tallest of trees, and the giant sequoia is the most massive of trees. Both species are described in detail in table 196. *Sequoia sempervirens* has been cultivated quite widely outside its natural range, including parts of Europe, and it thrives in cool, moist places.

SEEDING HABITS.—The minute and inconspicuous male and female flowers are borne separately on different branches of the same tree. The female flowers ripen into small egg-shaped cones composed of numerous closely packed, woody, persistent, thick scales. The cones[61] mature at the end of the first season in *Sequoia sempervirens* and at the end of the second season in *S. gigantea.* They are persistent for many years after seed fall in the latter species. Under each scale are several closely packed seeds. The ripe seeds are light brown, flattened, and each has two wings (narrower than the seed in *S. sempervirens* and broader than the seed in *S. gigantea*) (fig. 240). Good seed crops occur every year or so in both species, with light crops in intervening years. The comparative seeding habits of the sequoias are as follows:

	S. sempervirens	*S. gigantea*
Time of—		
Flowering	Nov.–March	Feb.–March
Seed dispersal	Fall	Fall
Seed-bearing age:		
Minimum ... years	20	125
Optimum ... do	60–100	200+
Scales per cone ... number	15–20	25–40
Seeds per scale ... do	2–5	3–9
Cone length ... inches	5/8–1 1/8	1 3/4–2 3/4

[61] These contain a considerable amount of dry, granulated pigment which consists chiefly of tannic acid.

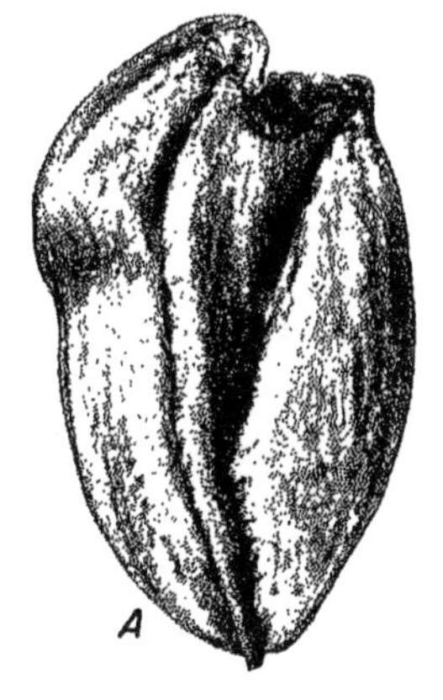

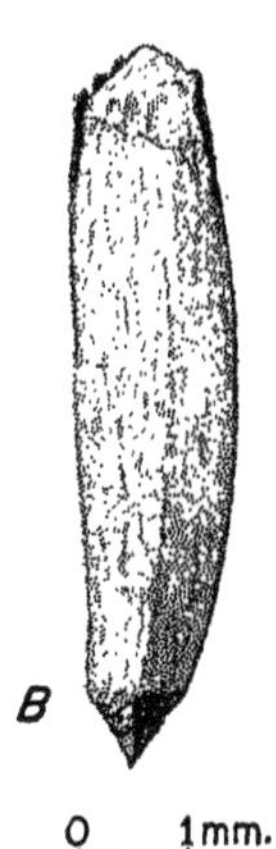

FIGURE 240.—Seed views of *Sequoia gigantea: A,* Exterior view; *B,* seed with outer coat removed; *C,* embryo.

TABLE 196.—*Sequoia: Distribution and uses*

Accepted name	Synonyms	Natural range	Chief uses	Date of earliest cultivation
S. sempervirens (D. Don) Endl. (redwood).		Fog belt in the California coast ranges, north to southwestern Oregon.	Lumber, ornamental	1843
S. gigantea (Lindl.) Decne. (giant sequoia).	*S. washingtoniana* (Winslow) Sudw.; *S. wellingtonia* Seem.; *Sequoiadendron giganteum* (Lindl.) Buchholz; *Wellingtonia gigantea* Lindl. (bigtree).	California on the western slope of the Sierra Nevadas at 4,500 to 6,500 feet above sea level.	Possibilities for lumber, erosion control, and ornamental.	1853

SEQUOIA

No information is available as to the development of climatic races in *Sequoia*. In view of the restricted range of both species, it is quite likely that no racial differentiation has taken place. There is some indication, however, that average seed size increases from south to north in *S. sempervirens*. Several varieties of *S. gigantea* are known.

COLLECTION, EXTRACTION, AND STORAGE.—Cones may be collected from August to December, usually in October (they may be collected as soon as they turn yellowish in color and the scales begin to separate), from (1) squirrel caches or cuttings, (2) from fresh logging slash, and (3) by climbing young trees. *Sequoia sempervirens* cones can be dried in shallow layers in the sun or at room temperature, using a fan to circulate the air. The cones should be stirred occasionally to promote uniform drying. Extraction requires 10 to 30 days by this process. By heating the cones in a forced air kiln at 120° to 130° F., extraction can be completed in half a day. The seeds usually are separated from debris by screening, although good results could doubtless be obtained by running them through a blower. Seed of *S. gigantea* very likely could be extracted by the same method. The yield, size, purity, soundness, and cost of commercial cleaned seed vary according to the species as follows:

	S. sempervirens	*S. gigantea*
Yield of cleaned seed:		
Per 100 pounds cones...pounds..	11	------
Per pound:		
Low..........number..	59,000	54,000
Average..........do....	122,000	91,000
High..........do....	300,000	132,000
Basis, samples..........do....	182	46
Commercial seed:		
Purity..........percent..	80	81
Soundness..........do....	23	41
Cost per pound..........dollars..	3.25–4.50	5.00–10.00

Seed of *S. sempervirens* kept in a sealed bottle at 26° to 30° F. lost about half its viability at the end of 1 year and deteriorated rapidly upon removal from cold storage. Another test showed fair viability at the end of 10 years' storage. Seed of *S. gigantea* is reported to retain good germinability for 8 to 24 years in storage.

GERMINATION.—Natural germination of the sequoias is best where mineral soil is exposed. Although sequoia seed do not as a rule require pretreatment to induce germination, the presence of slight but variable dormancy is indicated by the improved germination of some lots of seed after stratification for 60 days at 41° F. The germination of sequoia seed may be tested in sand flats or standard germinators at 60° (night) to 70° (day), or at a constant temperature of 68° for 40 to 60 days, using 1,000 seeds per test. Average results for the two species are as follows:

	S. sempervirens	*S. gigantea*
Germinative energy:		
Amount..........percent..	8–76	5–34
Period..........days..	20–35	20–35
Germinative capacity:		
Low..........percent..	1	2
Average..........do....	10	25
High..........do....	36	83
Basis, samples..........number..	46	30

The poor average germination of the sequoias is due largely to the high percentage of empty or nonviable seed rather than to dormancy.

NURSERY AND FIELD PRACTICE.—Sequoia seed should be sown in the fall in well-worked seedbeds in drills, and covered with one-eighth inch of soil. A mulch of burlap, to be removed as soon as germination starts, is required. The seedbeds should be screened and given half shade for the first 2 months. A density of 75 to 100 seedlings per square foot of seedbed is recommended. About 20 percent of the viable seed sown produce 1–0 seedlings in *Sequoia sempervirens* and about 25 percent of the 1–0 seedlings will be lost before removal from the nursery as transplants. One-one stock is desirable in field planting. *S. sempervirens* is also reproduced by sprouts.

SHEPHERDIA Nutt. Buffaloberry

(Elaeagnus family—Elaeagnaceae)

DISTRIBUTION AND USE.—The buffaloberries consist of three species of deciduous or evergreen shrubs confined to the western and northern parts of North America. The two cultivated species are deciduous. They are useful chiefly for wildlife food, ornamental purposes, and shelter belts, or erosion control; their roots bear nodules of bacteria and are hence able to fix nitrogen. These two species are described in table 197. *Shepherdia argentea* has been planted quite extensively in shelter belts. *S. canadensis* has been used less commonly.

SEEDING HABITS.—The small, yellowish male and female flowers, blooming in the spring, are borne separately on different plants, either solitary or in clusters on the branchlets. By summer the latter develop into drupelike, ovoid fruits about one-eighth to one-fourth inch long, each containing a single achene or nutlet, the seed (fig. 241). Natural seed dispersal is chiefly by animals. The comparative seeding habits of the two species discussed here are as follows:

	S. argentea	*S. canadensis*
Time of—		
Flowering	April–June	April–June.
Fruit ripening	June–Aug.	June–August.
Seed dispersal	June–Dec.	June–September.
Ripe fruit—		
Color	Scarlet[1]	Yellowish.
Flavor	Sour; edible	Bitter; unpalatable.

[1]Yellow in one variety.

Both species produce good seed crops almost every year. No scientific information is available concerning racial variation among the buffaloberries.

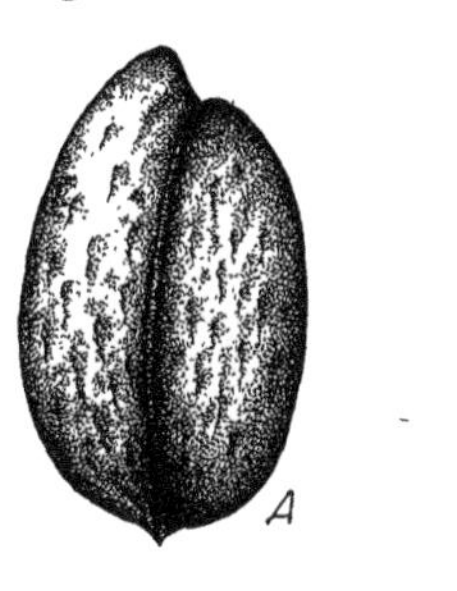
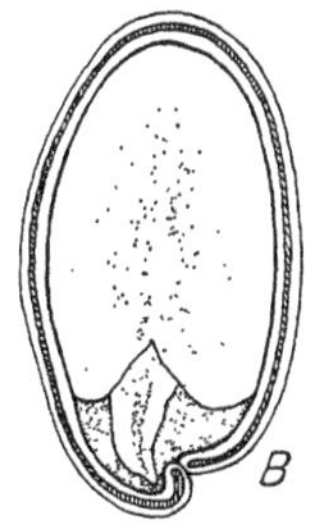
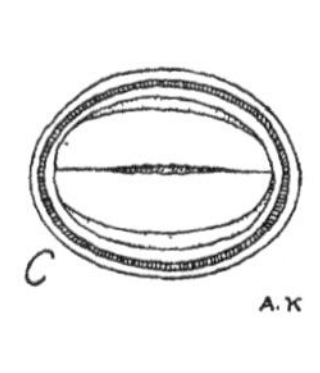

FIGURE 241.—Seed views of *Shepherdia argentea*: *A*, Exterior view; *B*, longitudinal section; *C*, cross section.

COLLECTION, EXTRACTION, AND STORAGE.—The fruits may be gathered any time after ripening by stripping or flailing them from the bushes onto canvas, or they may be picked by hand using heavy gloves to avoid injury from thorns in the case of *Shepherdia argentea*. Care should be taken to avoid heating of the collected fruits. Twigs, leaves, and other debris should be sifted out, and then the fruits should be run through a macerator with water and the pulp floated off or screened out; or the fruits may be spread out in a thin layer and dried. Commercial seed consists of dried fruits or cleaned seed. The size, purity, soundness, and cost

TABLE 197.—*Shepherdia: Growth habit, distribution, and uses*

Accepted name	Synonyms	Growth habit	Natural range	Chief uses	Date of earliest cultivation
S. argentea (Pursh) Nutt. (silver buffaloberry).	*Lepargyrea argentea* (Pursh) Greene (buffaloberry, red berry, bullberry).	Thorny shrub or small tree.	Moist situations from Minnesota and Manitoba to Saskatchewan, Kansas, and California.	Shelter belt, ornamental, livestock browse, wildlife food, gully control, preserves.	1818
S. canadensis (L.) Nutt. (russet buffaloberry).	*L. canadensis* (L.) Greene (Canadian buffaloberry, thornless buffaloberry, nannyberry, soapolallie, wild oleaster).	Thornless, spreading, medium-sized shrub.	Moist, open sites from Newfoundland to Alaska, South to Nova Scotia, Ohio, Minnesota, Black Hills, New Mexico and Oregon.	Wildlife food, sheep browse, erosion control.	1759

of commercial seed of *S. argentea* are as follows—Yield of cleaned seed per 100 pounds fruit, 5 to 10 pounds. Number per pound (13 samples): low, 18,000; average, 41,000; high, 67,000. Commercial seed: purity, 87 percent; soundness, 90 percent. Cost per pound: clean seed, $2 to $4; dried fruits, $0.75 to $1. Commercial clean seed of *S. canadensis* costs $3.50 per pound and dried fruits $2 per pound. Seed of *S. argentea* stored with a moisture content of 13.1 percent showed 97 percent germination after 3½ years at 41° F.

GERMINATION.—Germination is epigeous (fig. 242). Buffaloberry seed has variable embryo dormancy. Since some seeds in most lots may have dormancy, the seeds should be stratified in moist sand at 41° F. for 60 to 90 days before testing or spring sowing. Germination can be tested in sand flats for 30 to 60 days at 68° (night) and 86° (day), using 400 properly pretreated seeds per test. Average results for *Shepherdia argentea,* using pretreated seed, are as follows (7 tests)—Germinative energy: amount, 26 to 93 percent; period, 9 to 18 days. Germinative capacity: low, 27 percent; average, 57 percent; high, 96 percent. Pretreated seed of *S. canadensis* had an average germinative capacity of 14 percent, 1 test.

NURSERY AND FIELD PRACTICE.—Seed of *Shepherdia argentea* should be sown in the fall or stratified 60 to 90 days and sown in the spring, using 30 viable seed per linear foot of row. They should be covered with one-fourth inch of soil. About 12 percent of the viable seed sown produce usable 1–0 seedlings. Field planting is usually done with 2–0 stock. Similar practices probably would prove suitable for *S. canadensis.* The buffaloberries grow on a variety of soils and are very cold- and drought-hardy. *S. canadensis* is especially well suited for planting on dry, rocky banks where few other shrubs can thrive. *S. argentea* can also be propagated by means of wild root sprouts dug up and transplanted, although results are better if they are first transplanted in a nursery.

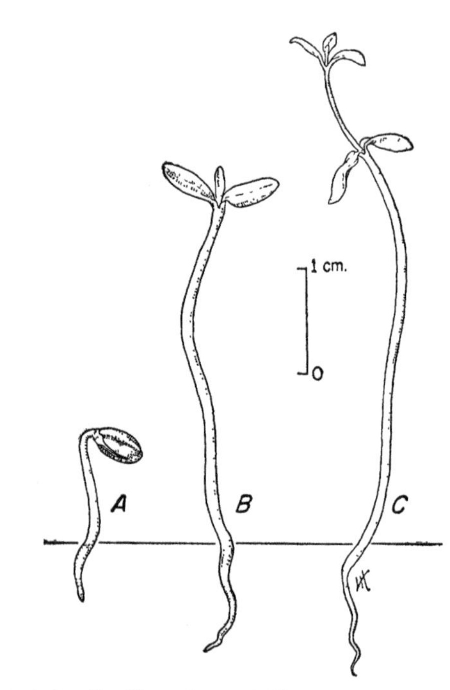

FIGURE 242.—Seedling views of *Shepherdia argentea: A,* At 1 day; *B,* at 9 days; *C,* at 38 days.

SIMMONDSIA CHINENSIS (Link) Schneid.[62] Jojoba

(Box family—Buxaceae)

Also called Californiajojoba, SPN; goat-nut, coffeeberry. Botanical syn.: *Simmondsia californica* Nutt.

DISTRIBUTION AND USE.—This evergreen desert shrub is native to arid hills of Arizona, southern California, and adjacent parts of Mexico. Its outstanding value, not yet fully appreciated, is that its seed oil is not a fat, as in *all* other plants so far analyzed, but a liquid wax. Probably in the future many uses will be found for the unique seed oil of this plant. The leaves and young twigs are much browsed by sheep, cattle, and goats. The seeds are edible for man and are used extensively by American Indians for food. The seed oil alone apparently cannot be digested and merely acts as a lubricant.

SEEDING HABITS.—The flowering period of jojoba is from March to May, in some places as early as the end of January. Jojoba is dioecious. The small, yellowish, male flowers are borne in dense, rounded, axillary clusters, while the larger, greenish, female flowers are solitary and on different plants. Seeds, which are the size of peanuts (fig. 243), ripen from June to the latter part of the fall. They are dispersed when the capsules split open at maturity.

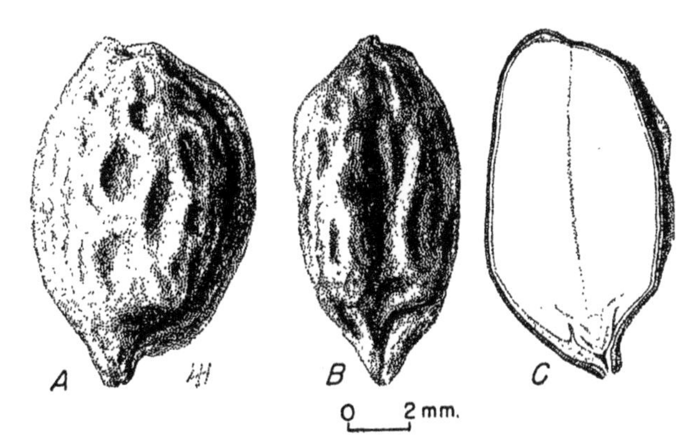

FIGURE 243.—Seed views of *Simmondsia chinensis: A* and *B*, Exterior views, 2 planes; *C*, longitudinal section.

COLLECTION, EXTRACTION, AND STORAGE.—The best time for seed collection is before the winter rains. The capsules can be stripped from bushes and the seeds removed by hand. Commercial seed are collected by the Arizona Indians. One pound contains from 1,100 to 1,700 cleaned seeds (2 samples). Purity and soundness ordinarily are very high. Storage of the seed for 1 or 2 years in an ordinary storage room is satisfactory. For a longer period airtight containers and low temperatures are advisable.

GERMINATION.—The best natural seedbed is deep soil, shaded by shrubs and lightly covered with litter. Jojoba seeds do not need pretreatment, and germinative capacity is (6+ samples): low, 60 percent; average, 90 percent; high, 100 percent. In a greenhouse, germination begins in 5 to 7 days after the seed is sown. Germination is hypogeous (fig. 244). Experiments have indicated that jojoba might be a good species for direct seeding in the field, were it not that small rodents dig up the seeds after planting. It is advisable to soak the seeds in water overnight, press them into the soil, and cover with sifted sand for a depth of one-fourth inch. In the nursery, seedlings appear in 15 to 20 days after sowing.

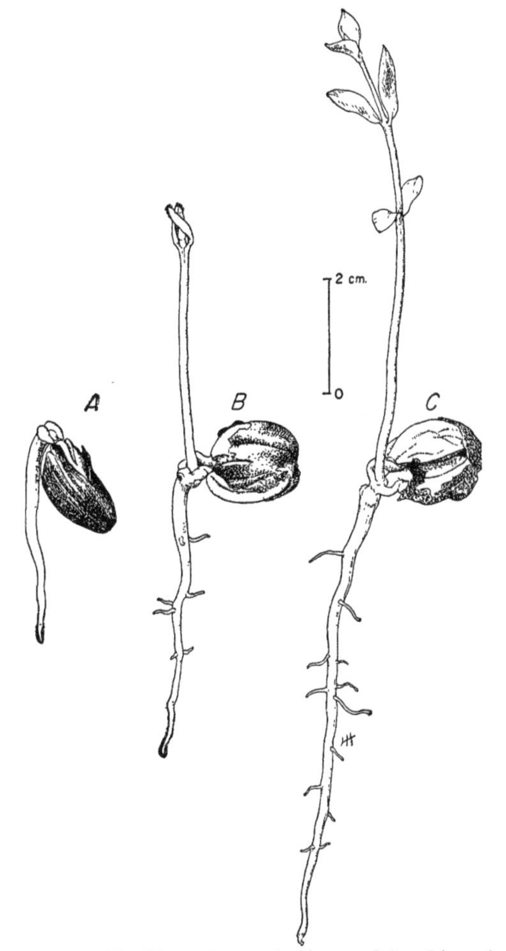

FIGURE 244.—Seedling views of *Simmondsia chinensis: A*, At 3 days; *B*, at 7 days; *C*, at 14 days.

[62] This inaccurate but unfortunately legal name is discussed in Standardized Plant Names, p. 588, 1942 ed.

SOLANUM DULCAMARA L. Bitter nightshade

(Nightshade family—Solanaceae)

Also called bittersweet, climbing nightshade. Includes at least 4 varieties.

Distribution and Use.—Native from Europe and northern Africa to eastern Asia, and often naturalized in North America, the bitter nightshade is a woody climber which has been cultivated since 1561 chiefly for ornamental purposes, but it is also of value for wildlife food and cover. It is one of some 1,200 species in the genus, most of which occur in the tropical and subtropical regions of both hemispheres.

Seeding Habits.—The violet flowers, occurring in long peduncled cymes, bloom from July to August; the poisonous fruits,[63] ovoid scarlet berries about one-half inch long, ripen from August to October. The seeds are small (about one-sixteenth of an inch across), flesh colored, irregular disks, dully glistening as if coated with fine sugar (fig. 245). Good seed crops are borne almost annually.

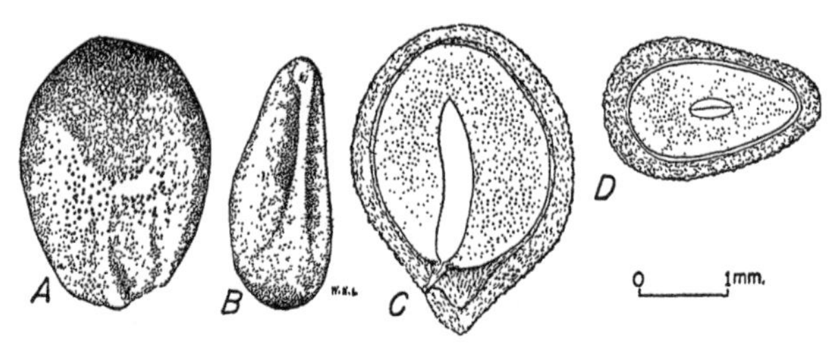

Figure 245.—Seed views of *Solanum dulcamara: A* and *B*, Exterior views in two planes; *C*, longitudinal section; *D*, cross section.

Collection, Extraction, and Storage.—Collection can be made from July to September by picking the ripe berries from the vines by hand. The fruits may then be rubbed through a No. 10 mesh screen, the fine pulp and empty seeds floated off, and the seed dried. On a larger scale, the use of a macerator would probably be advisable. In one collection the number of cleaned seeds per pound averaged about 350,000; purity averaged 99 percent, and soundness 92 percent (2 samples). No experimental evidence concerning methods of storage is available.

Germination. — Natural germination of bitter nightshade is best on moist soils. Germination is epigeous (fig. 246). The seeds of some lots exhibit dormancy, apparently due to conditions in the embryo, which can be overcome by stratification in moist sand outdoors over winter or for 30 days at 41° F. Germination tests can be run in sand flats at 68° (night) to 86° (day) for 60 days using 400 to 1,000 stratified seeds per test. In one series of tests germinative energy averaged 46 percent in 20 days for untreated seed and 71 percent in 25 days for stratified seed. Germinative capacity was 46 to 87 percent in 104 to 134 days for untreated seed (3 tests) and 78 percent in 53 days for stratified seed (1 test). In another case fresh seed germinated 52 to 70 percent (2 tests) in 9 days without pretreatment.

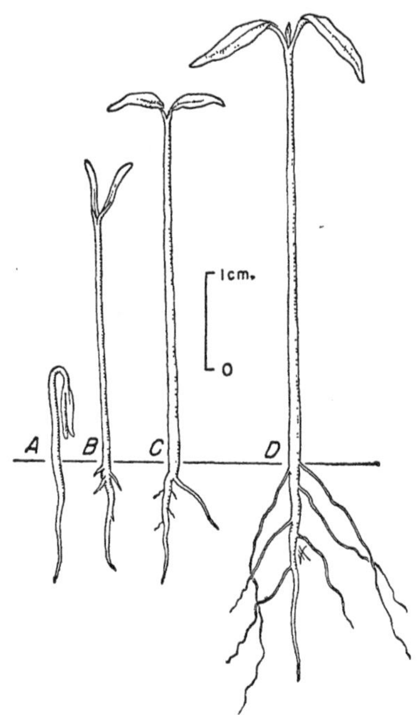

Figure 246.—Seedling views of *Solanum dulcamara: A*, At 1 day; *B*, at 2 days; *C*, at 6 days; *D*, at 12 days.

Nursery and Field Practice.—It is suggested that the seed be sown in the fall, or stratified seed sown in the spring, and covered with about one-eighth inch of soil. This species can also be propagated by root or stem cuttings.

[63] The fresh berries are poisonous to most people, and are fatal to rabbits, but some birds and other wildlife eat them with impunity.

SORBUS L. Mountain-ash

(Rose family—Rosaceae)

DISTRIBUTION AND USE.—Consisting of more than 80 species of deciduous trees or shrubs distributed through the Northern Hemisphere, the mountain-ashes, because of their graceful foliage, showy flowers, and handsome fruit, are used chiefly for ornamental purposes. The fruit is an important food for birds and that of some species is made into preserves; the twigs are used as browse by some of the larger game animals; the strong and close-grained wood is sometimes made into tool handles. Three species, which are grown in the United States, are described in detail in table 198. *Sorbus aucuparia* is probably the most widely planted of all species.

SEEDING HABITS.—The perfect, white flowers of mountain-ash are borne in the spring in large, rather flattened clusters. Its attractive fruit, which ripens in late summer or early fall, is a two- to five-celled, berrylike pome, each cell containing one or two small, brown seeds (figs. 247 and 248). Natural seed dispersal is chiefly by birds. Comparative seeding habits of the three species discussed here are given in table 199. In the case of *Sorbus aucuparia* seed bearing begins at about 15 years of age. It bears good seed crops almost annually, with light crops in intervening years. Similar information for the other species is lacking.

There is no scientific information available as to the development of climatic races in *Sorbus americana* and *S. decora*. In the case of *S. aucuparia*, although experimental evidence is largely lacking, it is likely that climatic races have developed in view of the wide range of the species and the recognition of several varieties. Furthermore, studies in Germany showed that seed of one Bavarian origin gave a much better and more uniform germination than that from another.

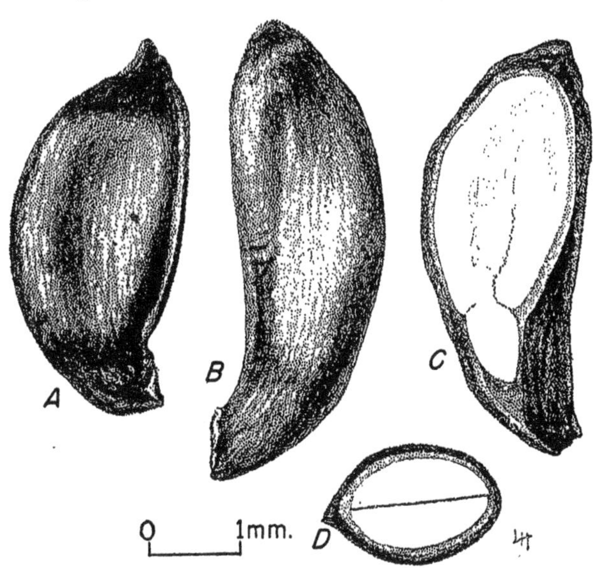

FIGURE 247.—Seed views of *Sorbus aucuparia*: *A* and *B*, Exterior views in two planes; *C*, longitudinal section; *D*, cross section.

TABLE 198.—*Sorbus: Growth habit, distribution, and uses*

Accepted name	Synonyms	Growth habit	Natural range	Chief uses	Date of earliest cultivation
S. americana Marsh. (American mountain-ash).	*Pyrus americana* (Marsh.) DC. (small-fruited mountain-ash, mountain-sumac).	Tall shrub or small tree.	Rich moist soils from Minnesota to Newfoundland and south in the mountains to North Carolina and Tennessee.	Wildlife food, ornamental; bark and berries have some medicinal value.	1811
S. aucuparia L. (European mountain-ash).	*Pyrus aucuparia* (L.) Gaertn. (rowantree).	Small tree	Europe to western Asia and Siberia. Naturalized in North America.	Ornamental purposes, wildlife food, nurse crop for spruce; preserves, tool handles.	[1]
S. decora (Sarg.) Schneid. (showy mountain-ash).	*S. americana* var. *decora* (Sarg.) Sarg. (large-fruited mountain-ash).	Tall shrub or small tree.	Labrador to western Ontario, south to Maine and Minnesota.	Wildlife food, ornamental	1636

[1] Long cultivated.

SORBUS

TABLE 199.—*Sorbus: Time of flowering, fruit ripening, and seed dispersal*

Species	Time of—			Fruit characteristics	
	Flowering	Fruit ripening	Seed dispersal	Color	Diameter
					Inches
S. americana	May-July	August-October	August-late winter	Bright red	¼
S. aucuparia	April-May	August-September	August-winter	Bright red to orange yellow.	¼
S. decora	May-July	do	do	Orange red	½

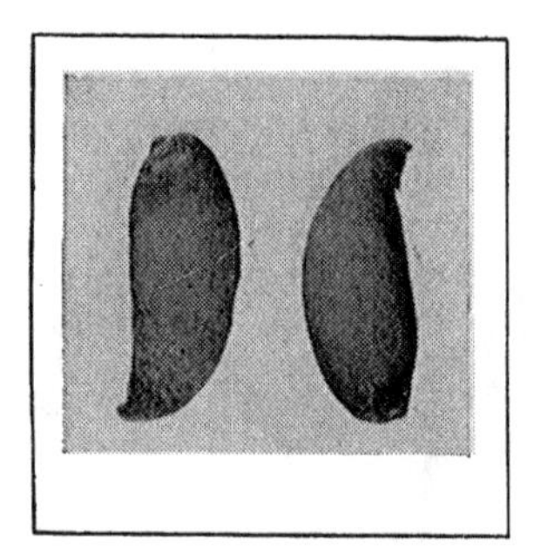

FIGURE 248.—Seed views of *Sorbus americana.*

COLLECTION, EXTRACTION, AND STORAGE.—The fruit should be picked by hand or shaken from the trees as soon as it is ripe to prevent destruction by the birds. Seeds may be extracted by running the fruits wet through a macerator or putting them through a fruit press. In the former method the pulp can be floated or skimmed off; in the latter, the dried pulp can be broken up and separated from the seeds in a blower, or the dried pulp and seed broken by hand as it is sown. After drying, the seed should be fanned to remove debris and flat, empty, or partly filled seeds. In some cases the fruits are spread out to dry, and used in that form. Commercial seed consists of the dried druit or cleaned seed. The yield, size, purity, soundness, and cost of commercial cleaned seed vary considerably by species as shown in table 200.

Seed of *Sorbus americana* stored in a tightly closed can at 30° to 50° F. lost no viability in 8 years. Cleaned seed or intact berries of *S. aucuparia* have been stored dry in unsealed containers at various temperatures from 17° to 70° for 2 years without significant loss in viability. Higher temperatures were injurious. Relative humidities much below or above 25 percent are unfavorable to long storage at the higher temperatures; ordinary room conditions are usually satisfactory. Storage in sealed containers or under vacuum showed no advantages. Seed may be stored over winter in outdoor stratification pits. Storage information for *S. decora* is lacking.

GERMINATION.—Natural seedlings are most commonly found near the seed trees where mineral soil is exposed and some protective shade is available. Germination is epigeous (fig. 249). Most mountain-ash seed is slow to germinate because of a dormant embryo; in some species the seed coat also appears to delay germination. Dormancy can be overcome at least in part by moist stratification; warm plus cold stratification appears advisable. Partially after-ripened seed exposed to higher temperatures goes into secondary dormancy. Further information on seed dormancy and pretreatment is given in table 201. As is evident, suitable methods of pretreatment still remain to be worked out for the American species. The germination of mountain-ash seed can be tested in sand flats using 400 to 1,000 suitably pretreated seeds per test. Methods recommended and average results for the 3 species discussed here are given in table 202. When proper methods of pretreatment are worked out, better germination should be obtained from the American species.

TABLE 200.—*Sorbus: Yield of cleaned seed, and purity, soundness, and cost of commercial seed*

Species	Cleaned seed					Commercial seed			
	Yield per bushel of fruit	Per pound			Basis, samples	Purity	Soundness	Cost per pound	
		Low	Average	High				Clean	Dried fruits
	Pounds	*Number*	*Number*	*Number*	*Number*	*Percent*	*Percent*	*Dollars*	*Dollars*
S. americana	1-2	84,000	160,000	235,000	8	80	86		0.60-1.50
S. aucuparia	3-7	[1]104,000	130,000	170,000	10+	96	[2]90	4.00-6.00	.85-2.00
S. decora			127,000		1	100	93		

[1]2,400 fruits per pound.
[2]Average number filled seeds per fruit, 2 to 3.

TABLE 201.—*Sorbus: Dormancy and method of seed pretreatment*

Species	Dormancy		Stratification		Remarks
	Kind	Occurrence	Medium	Duration and temperature	
				Days °F.	
S. americana	Embryo and seed coat	General	Moist sand	90 at 68–86 + 90 at 41[1]	Technique giving high germination in the laboratory has not yet been worked out. Sulfuric acid treatment for 10 minutes might possibly be substituted for the warm stratification.
S. aucuparia	Embryo	Probably variable	Moist acid (pH4) peat	90 at 33	Seed stored dry for 6 months requires only 60 to 80 days to break dormancy.
S. decora	Embryo and seed coat	General	Moist sand	90 at 68–86 + 90 at 41[1]	See remarks for *S. americana*.

[1]This treatment has not been tried but is based on experience with other species.

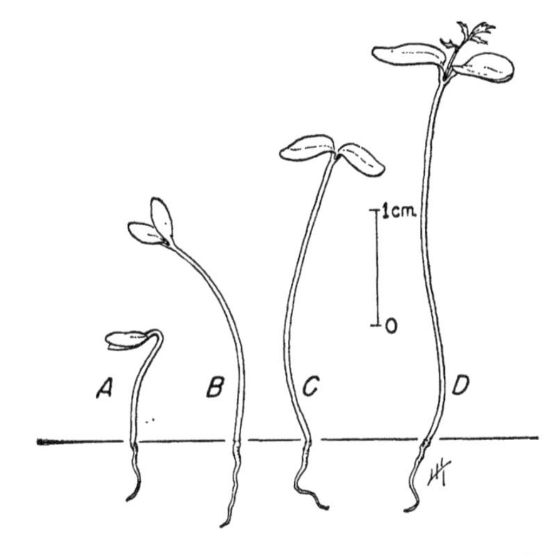

FIGURE 249.—Seedling views of *Sorbus americana: A,* At 1 day; *B,* at 3 days; *C,* at 7 days; *D,* at 24 days.

NURSERY AND FIELD PRACTICE.—Mountain-ash seed may be sown broadcast or in drills in the fall or later on the snow, or stratified seed may be sown in the spring. The seed should be covered with about one-sixteenth of an inch of soil. Sowing the seed in the pulp is said to help overcome seed dormancy in *Sorbus americana.* The seed may also be sown in board-covered coldframes; poorer results occur if the entire fruits are sown. In the case of *S. aucuparia* germination is largely complete in the second spring after sowing; most of the germination in *S. americana* and *S. decora* occurs during the second and third springs. The seeds of *S. aucuparia* are subject to attack by several species of chalcid flies. Mountain-ash seedlings are quite hardy and little susceptible to insects or diseases. In the field, seedlings are frequently nipped by deer. For field planting 1–1 stock is best, but 2–0 is often suitable. Best results are obtained when planting is done on cool, moist sites.

TABLE 202.—*Sorbus: Recommended conditions for germination tests, and summary of germination data*

Species	Test conditions recommended		Germination data from various sources				
			Germinative capacity				Potential germination
	Temperature	Duration	Low	Average	High	Basis, tests	
	°F.	Days	Percent	Percent	Percent	No.	Percent
S. americana	41–50	60		[1]15		1	3–71
S. aucuparia	32–50	90–100	27	68	96	12	80
S. decora				9		1	46

[1] Seed that had been stratified 150 days at 41° F., reached 33 percent after 330 days.

SYMPHORICARPOS Duham. Snowberry

(Honeysuckle family—Caprifoliaceae)

Distribution and Use.—The snowberries occur in North America, south to Mexico, with one species in western China, and include about 15 closely related species of deciduous upright or sometimes prostrate shrubs. They are grown chiefly for their attractive fruits, but they also provide food and cover for game birds and animals and considerable forage for livestock. Some species also have possibilities for erosion-control planting. Only 3 species, all native to North America, are of much importance from the standpoint of present or potential use in conservation planting. Their distribution and chief uses are given in table 203. All 3 of these species are used in wildlife planting and to some extent, particularly *Symphoricarpos occidentalis* and *S. orbiculatus,* in planting to control erosion. *S. albus,* especially its var. *laevigatus* which is taller and has larger leaves and fruit clusters than the type, and *S. orbiculatus* are much used in ornamental planting. The latter species is used because of its hardiness, long-persistent foliage, and handsome fruit.

Seeding Habits.—Snowberry's perfect flowers, pinkish to yellowish white in color, are borne in dense axillary or terminal clusters in the spring or summer months, and in some species in the late summer. The fruit, an attractive berry also borne in clusters, is usually white in color—in some species dark red, pink, or bluish black — and ripens in late summer or early fall of the same year. Each fruit contains two nutlets or seeds. These are flattened on one side and are composed of a tough, bony coat, a fleshy endosperm, and a small embryo (figs. 250 and 251). Commercial seed consists of dried berries or clean seed. Comparative seeding habits of the three species discussed here are given in table 204.

No data are available as to the frequency of seed production; it is likely that some seed is produced every year. Seed dispersal is largely by birds and mammals. Nothing is known regarding the existence of climatic races but it seems probable that these may have developed, particularly in such species as *Symphoricarpos albus* which is distributed over

Table 203.—*Symphoricarpos: Growth habit, distribution, and uses*

Accepted name	Synonyms	Growth habit	Natural range	Chief uses	Date of earliest cultivation
S. albus (L.) Blake[1] (common snowberry).	*S. racemosus* Michx. (waxberry).	Dwarf to medium-sized thicket-forming shrub.	Dry and rocky soil, Quebec and Hudson Bay to Alaska, south to Massachusetts, North Carolina, Kentucky, Minnesota, Colorado, and California.	Important forage plant in western United States, ornamental planting, game food and cover, erosion-control planting.	1806
S. occidentalis Hook.[2] (western snowberry).	wolfberry, buck brush.	Low, thicket-forming shrub.[3]	Dry soils, particularly on prairies, Michigan to British Columbia, south to Illinois, Kansas, and Colorado.	Valuable forage for livestock, wildlife planting, erosion-control planting, ornamental planting.	1880
S. orbiculatus Moench (Indian-currant coralberry).	Indian currant, coralberry.	Thicket-forming shrub.	New Jersey to South Dakota, south to Georgia and Texas, often along streams.	Wildlife and erosion-control planting, ornamental planting.	1727

[1] Includes var. *laevigatus* (Fern.) Blake (garden snowberry) and var. *pauciflorus* (A. Gray) Blake.
[2] Includes var. *heyeri* Dieck.
[3] Forms dense colonies from root suckers.

TABLE 204.—*Symphoricarpos: Time of flowering, fruit ripening, and seed dispersal*

Species	Time of—			Color of ripe fruit
	Flowering	Fruit ripening	Seed dispersal	
S. albus	May–September	August–October	Persists until following spring.	Waxy white.
S. occidentalis	June–July	September	do.	Dull white.
S. orbiculatus	July–August	September–frost	Persists until late winter	Purplish red.

a wide geographic range and in which at least one geographic variety is recognized.

COLLECTION, EXTRACTION, AND STORAGE.—The berries can be collected by stripping the clusters from the branches at any time after ripening. Those collected in the early fall contain considerable moisture and therefore require careful handling to prevent heating. The seed can be readily extracted by macerating the berries in water and allowing pulp and empty seeds to float away. Dried fruit should be soaked for several hours prior to maceration to make extraction easier. After drying and fanning, the seed is ready for storage or for use. Data on yield, size, purity, soundness, and cost of seed are given in table 205.

From such data as are available, it appears that *Symphoricarpos* seed will retain considerable germinability for long periods if kept at low temperatures. Dried seed of *S. albus* stored in a sealed container at 41° F. gave 45 percent germination after 2 years, with an additional 35 percent still sound at the conclusion of the test. Another lot of this seed which had been stratified in moist peat at 34° for 4 years showed very little germination but no loss in germinability. Seed of this species will also remain viable for at least 2½ years even at room temperatures. Nothing is known regarding optimum storage conditions for the other species, but it is probable that their behavior would be quite similar.

GERMINATION.—Germination behavior of *Symphoricarpos* seedlings in nature is not known. However, judging from the extreme resistance they show to the commoner types of pretreatment used to

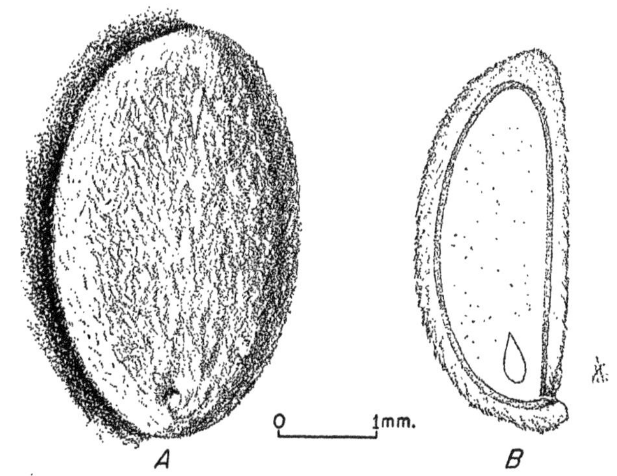

FIGURE 250.—Seed views of *Symphoricarpos albus: A,* Exterior view; *B,* longitudinal section, showing small embryo.

TABLE 205.—*Symphoricarpos: Yield of cleaned seed, and purity, soundness, and cost of commercial seed*

Species	Cleaned seed					Commercial seed		
	Yield per 100 pounds of fruit	Per pound			Basis, samples	Purity[1]	Soundness[1]	Cost per pound
		Low	Average	High				
	Pounds	*Number*	*Number*	*Number*	*Number*	*Percent*	*Percent*	*Dollars*
S. albus	[2]3	[3]39,000	75,000	113,000	10	92 (71–99)	78 (54–100)	[4]0.75–1.25 [5]2.30–3.50
S. occidentalis	5–10	52,000	73,000	94,000	4		85 (77–99)	
S. orbiculatus	[6]18–22		144,000		1	97	91	[4]1.10–1.25

[1]First figure is average percent; figures in parentheses indicate range from lowest to highest.
[2]Number of dried fruit per pound, 18,000.
[3]This is for var. *laevigatus* which appears to have larger seed; another sample showed 54,000 per pound.
[4]Dried fruit.
[5]Cleaned seed.
[6]This is apparently based on dried fruit.

overcome dormancy in the laboratory, it seems likely that many seeds do not germinate until the second spring after ripening; some probably require even longer. Germination is epigeous (fig. 252). Dormancy is due to a combination of a hard tough seed coat with a rudimentary or partially developed embryo. The seeds can be germinated by exposing them to warm plus cold stratification, or by acid treatment followed by low-temperature stratification. Such details of pretreatment as have been worked out are given in table 206. Germination may be tested in sand flats or peat mats, using 400 or more properly pretreated seeds per test. Temperatures fluctuating diurnally from 68° to 86° F. appear to be satisfactory for *S. albus* and *S. orbiculatus.* Results that may be expected are given in table 207. Germinability of *S. albus* may be tested by soaking the seeds in concentrated sulfuric acid until the coats have softened—75 minutes or more will be required—and then placing them in moist peat at 41° until the embryos have developed sufficiently that they may be removed from the seeds. The embryos are then placed in petri dishes at room temperature to germinate.

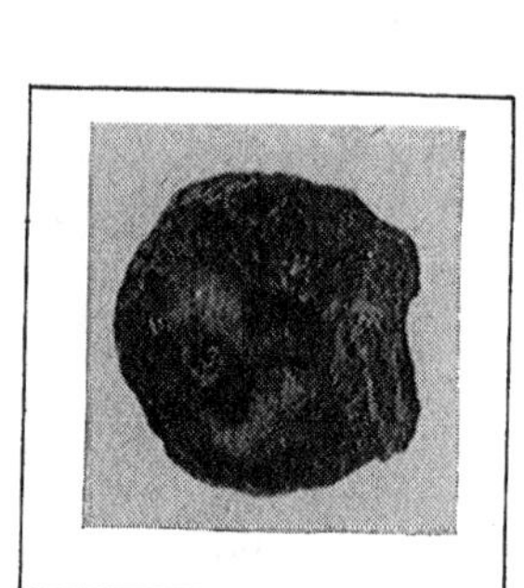

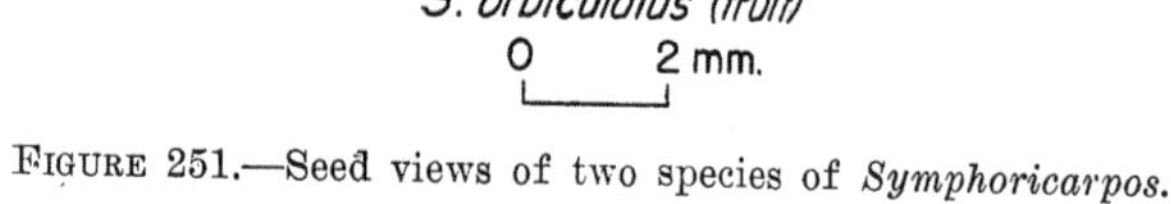

FIGURE 251.—Seed views of two species of *Symphoricarpos.*

FIGURE 252.—Seedling views of *Symphoricarpos albus: A,* At 5 days; *B,* at 7 days; *C,* at 13 days; *D,* at 20 days.

TABLE 206.—*Symphoricarpos: Dormancy and methods of seed pretreatment*

Species	Kind of dormancy	Stratification		Other methods of pretreatment
		Medium	Duration and temperature	
			Days / *°F.*	
S. albus[1]	Hard seed coat and immature embryo.	Sand or peat	90–120 at [2]68–86 + 180 at 41	Soak in sulfuric acid 75 minutes[3] and then stratify for 180 days at 41° F. Soak in sulfuric acid 60 minutes, stratify several weeks in peat at 77° then stratify 180 days at 41°.
S. orbiculatus	do[4]	do.	120 at [2]68–86 + 120 at 41	Stratify for 3 to 4 months at 77° F. and then for 5 to 6 months at 50°. Soak in sulfuric acid for 30 to 40 minutes and then stratify for 2 to 4 weeks at 77° plus 5 to 6 months more at 50°.

[1]No successful results available for *S. occidentalis;* probably should be handled like *S. albus.*
[2]Temperatures alternated diurnally; a constant temperature of 77° F. may also be used.
[3]Freshly collected seed may not require as long acid treatment as older seed.
[4]Some lots of fresh seed are reported to show only embryo dormancy which can be broken by low-temperature stratification.

Table 207.—*Symphoricarpos: Recommended germination test duration, and summary of germination data*

Species	Test duration for pretreated seed	Germination data from various sources						
		Germinative energy		Germinative capacity			Potential germination	Basis, tests
		Amount	Period	Low	Average	High		
	Days	*Percent*	*Days*	*Percent*	*Percent*	*Percent*	*Percent*	*Number*
S. albus	30	10–43	15	0	35	80	----	6
S. occidentalis	----	----	----	0	----	1	13–90	3
S. orbiculatus	10–25	83	8	74	81	96	----	3

Nursery and Field Practice.—Seed should be sown in seedbeds or nursery rows in the spring and these kept mulched until the following spring, at which time germination will occur. Sulfuric acid treatment, followed by either fall sowing or stratification over winter followed by spring sowing, should give good results the first spring. Fresh seed of *Symphoricarpos orbiculatus* can sometimes be fall-sown or stratified over winter prior to spring sowing with complete germination taking place the first season. About 30 viable seeds should be sown per linear foot of nursery row and covered to a depth of about one-fourth inch. Nursery germination: *S. albus,* 50 to 70 percent; *S. orbiculatus,* 90 percent. *Symphoricarpos* can also be propagated from stem and root cuttings.

891183°—50——23

SYRINGA VULGARIS L. Common lilac

(Olive family—Oleaceae)

Distribution and Use.—Native to southeastern Europe, the common lilac (includes numerous horticultural forms) has been cultivated since 1563 chiefly for ornamental purposes. It frequently escapes from cultivation. Hardy under a great variety of conditions, it is also useful for shelterbelt planting. It is the most commonly planted of some 28 species in this genus.

Seeding Habits.—The showy flowers, ranging from white through various shades of violet and purple to a deep reddish purple, bloom from mid-April to late June; the fruit, oblong, smooth, leathery, brown, two-celled capsules, about one-half inch long, ripens from early September to mid-October. Each capsule contains four thin, flat, lozenge-shaped seeds about ½ inch long and 3/16 inch wide, and a bright brown in color (fig. 253). Fair to good seed crops are produced annually.

Collection, Extraction, and Storage.—Common lilac fruits should be picked from the shrubs by hand in the early fall and spread out to dry. They can then be run dry through a macerator and fanned to remove impurities; the fanning must be carefully done or good seed will be lost. Cleaned seed per 100 pounds of fruit, approximately 1 to 4 pounds. Cleaned seed per pound (13 samples): low, 55,000; average, 90,000; high, 130,000. Commercial seed averages 60 percent in purity, 85 percent in soundness, and costs $2 to $3 per pound. Seed stored in moist peat in sealed containers at 41° F. after 20 months showed little or no loss in viability but was somewhat slower to germinate. The seed may be kept in ordinary dry storage for 1 to 2 years with little loss of viability.

Germination.—Most lots of common lilac seed appear to have dormant embryos and possibly impermeable seed coats; they can be stimulated into germinating by stratification in moist sand at 41° F. for 30 to 90 days. Tests may be run in sand flats for 60 days, or in Jacobsen germinators for 30 days at 68° (night) to 86° (day) using 400 to 1,000 stratified seeds per test. Germinative energy of stratified seed ranges from 20 to 50 percent in 18 to 30 days. Germinative capacity (11 tests): low, 30 percent; average, 58 percent; high, 84 percent.

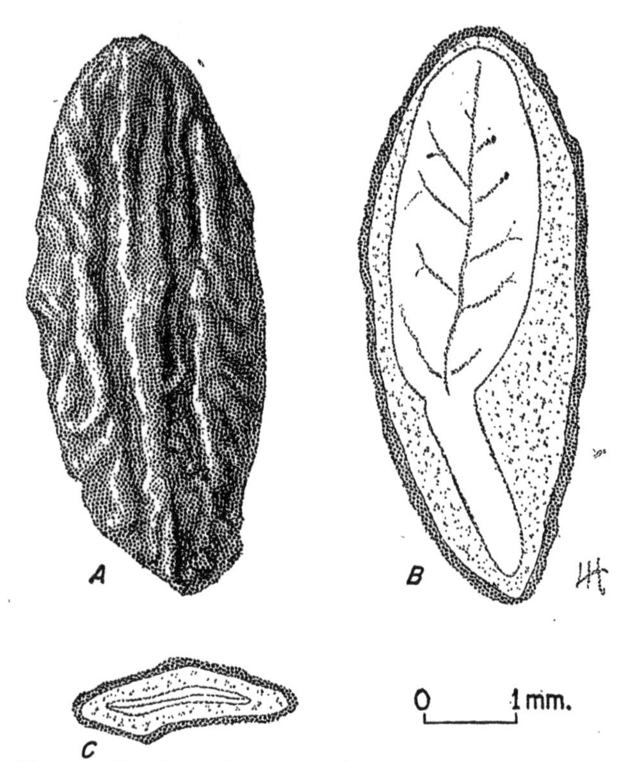

Figure 253.—Seed views of *Syringa vulgaris; A,* Exterior view; *B,* longitudinal section; C, cross section.

Nursery and Field Practice.—The seed may be sown in the fall, or stratified seed used in the spring; they should be covered with one-fourth inch of soil. An average of about 12 percent of the viable seed sown produces usable 1–0 seedlings; 1–1 stock is best for field planting. Suitable stock can be grown in one year from root cuttings. Common lilac grows in most soils but does best on moderately rich and moist ones.

TAXODIUM DISTICHUM (L.) Rich.[64] Baldcypress

(Pine family—Pinaceae)

Also called southern cypress, common baldcypress.

Distribution and Use.—Baldcypress grows in swamps and moist bottom lands from southern New Jersey to Florida, westward through the Gulf coast region to eastern Texas, and up the Mississippi Valley to southern Illinois and Indiana. Planted specimens are hardy as far north as Massachusetts, New York, and Michigan. Baldcypress was introduced into cultivation in Europe in 1640 and has since been planted in many parts of the world. It is an important timber tree. Because of its freedom from shrinkage and its durability in wet places, the wood is used for such purposes as building construction, deck and bridge timbers, sash and door, boxes and crates, and tanks and silos. Baldcypress is used as an ornamental and furnishes some food for wild ducks. However, it has not been successfully planted for timber production to any great extent.

Seeding Habits.—The male and female flowers, appearing with the leaves in March and April, are borne separately on the same tree. The male flowers, bearing pollen only, occur in slender, purplish, tassellike clusters at the end of the preceding year's shoots. The female flowers are scattered, small, rounded conelets which appear near the ends of the preceding year's branchlets, and occur singly or in groups. By late September to November of the first season the conelets develop into round, purplish cones ½ to 1¼ inches in diameter; they consist of a relatively few four-sided scales which break away irregularly as soon as they are ripe from mid-October to the end of December. Each scale bears two irregularly triangular seeds—both not always developed—which have thick, horny, warty coats and irregular projecting flanges along the edges (fig. 254). The flanges are often called "wings." Natural seed dispersal is chiefly by water. Good seed crops occur every 3 to 5 years but some seed are borne every year; very little seed mature in the northern part of the species' range. No scientific information is available as to racial variation in baldcypress.

[64] Some of the material reviewed almost certainly refers to *Taxodium ascendens* Brongn., pondcypress. As far as can be ascertained, the seeds of the two species can be handled alike.

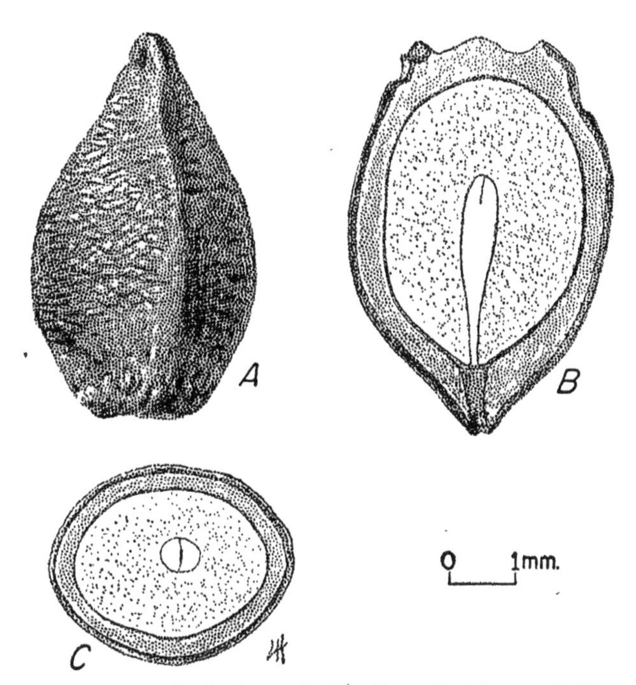

Figure 254.—Seed views of *Taxodium distichum*: *A*, Exterior view; *B*, longitudinal section; *C*, cross section, showing separation line in seed coat.

Collection, Extraction, and Storage.—When the cones of baldcypress begin to dry in October and November, they should be picked from logged or standing trees and immediately spread out to finish drying. Cones can be opened by air temperature and small lots of seed can be picked out by hand as the scales fall apart. When larger amounts are desired, the dried cones must be trampled or flailed. No cleaning is necessary. About 50 pounds of seeds are obtained from 100 pounds of fresh cones. Seeds per pound (17 samples): low, 1,300; average, 4,800; high, 9,100. Seeds per cone: 18 to 30. Cones per bushel: 2,600 to 3,500. Green weight per bushel: 40 to 50 pounds. Commercial seed averages 45 percent in purity, 40 percent in soundness, and costs $0.75 to $2.50 per pound. Authorities differ as to whether cypress seed should be sown immediately after collection, be stratified over winter, be stored dry over winter and stratified 30 days at 41° F. immediately prior to sowing, or be stored dry over winter and soaked in water for 4 to 8 weeks before sowing. Dry cold storage has been suggested for longer periods.

GERMINATION.—Wet muck forms a good seedbed. Seeds covered with water for as long as 30 months may germinate when the water drains off. Natural reproduction usually is sparse. Germination is epigeous (fig. 255). Baldcypress seed have dormant

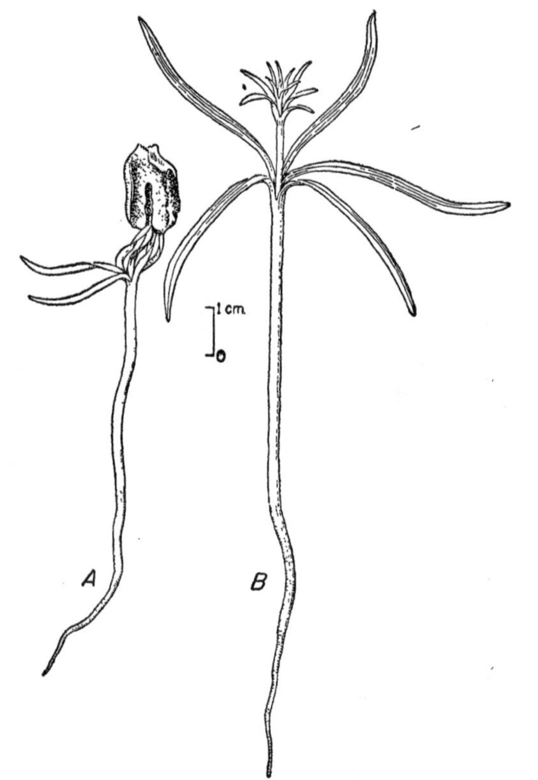

FIGURE 255.—Seedling views of *Taxodium distichum: A,* At 3 days; *B,* at 8 days.

embryos and there is a possibility that the rather thick resinous seed coats may also hinder germination. In one test, untreated seed soaked in distilled water for 96 hours absorbed 28 percent of its weight in water while seed of the same sample first washed in ethyl ether to remove surface pitch absorbed 59 percent of its weight in water in the same period. At any rate, untreated seed germinates slowly and poorly without pretreatment; stratification for 30 to 60 days in moist sand or peat at 41° F. speeds up germination and improves it somewhat. However, even such treatment stimulates only about one-third of the sound seed to germinate, and better methods should be sought. Germination may be tested in soil flats or peat mats —sand flats or standard germinators are a little less desirable—for 30 to 50 days (60 to 110 days for untreated seed) using 400 to 1,000 stratified seeds per test. Temperatures of 68° (night) and 86° (day) are suitable. Germinative capacity of stratified seed (21 tests): low, 1 percent; average 12 percent; high, 39 percent. Potential germination runs from 9 to 63 percent.

NURSERY AND FIELD PRACTICE.—In the South, sowing in December and early spring have both given good results. Sowing in autumn and mulching the beds has proved successful farther north where it has been found difficult to secure fully stocked beds with spring sowing. Seeds and scales are sown together either broadcast or in drills, a large quantity of seed being used to insure full stands. The seed should be covered with one-fourth to one-half inch of sand, loam, or peat moss, and mulched with leaves. In the South, shade may be needed from June to September. Seedbeds must be well watered. The germination period is ordinarily from 40 to 90 days but may be as short as 15. One reporter notes that tree percent is good, estimated at 40 to 50 percent. The resinous seeds are not eaten to any extent by rodents or birds.

TAXUS L. Yew

(Yew family—Taxaceae)

DISTRIBUTION AND USE.—The yews consist of about eight species—sometimes considered as geographical varieties of a single species—of nonresinous evergreen trees or shrubs occurring in the North Temperate Zone. They are useful chiefly for ornamental purposes, especially in hedges. The wood of the tree species is much prized for making bows and is also used in cabinetmaking and turnery. The "fruit" is eaten by birds; the foliage ordinarily is poisonous to horses and cattle. Three species, which are of known or supposed value for planting in the United States, and for which reliable information is available, are described in detail in table 208. *Taxus baccata* is the most widely cultivated yew.

SEEDING HABITS.—The male flowers, bearing pollen only, and the female flowers, which develop into "fruits," are rather inconspicuous and are borne on the same or on separate trees. The fruit, which ripens in late summer or autumn, consists of a scarlet, fleshy, cuplike aril bearing a single, hard seed (figs. 256 and 257). The seed has a large, very oily, white endosperm and a minute embryo. Comparative seeding habits of the species described here are given in table 209. Little information is available as to the occurrence of seed crops among the yews. *Taxus canadensis* bears some seed almost annually. *Taxus baccata* begins to bear seed at 30 to 35 years. Information as to the seed-bearing age for *T. brevifolia* and *T. canadensis* is not available. It is not known whether climatic races develop in *Taxus,* but it seems likely in view of the extensive range of each of the three species described here that such races have developed. In the case of *T. baccata* the variability within the species is further indicated by the recognition of some 40 varieties, some with common names, such as Irish yew and Dovaston yew.

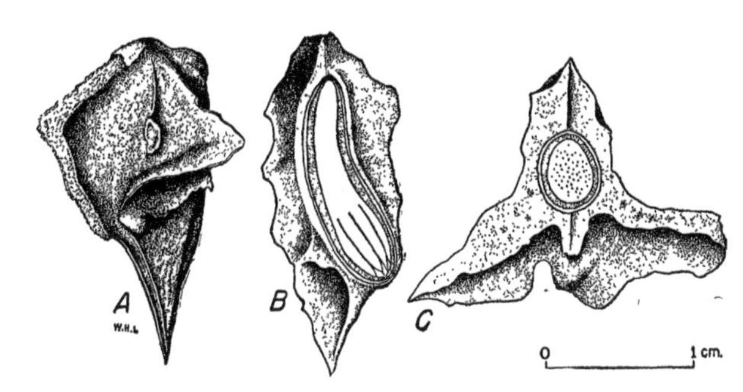

FIGURE 256.—Seed views of *Taxus brevifolia: A,* Exterior view; *B,* longitudinal section, showing small embryo; *C,* cross section.

TABLE 208.—*Taxus: Growth habit, distribution, and uses*

Accepted name	Synonyms	Growth habit	Natural range	Chief uses	Date of earliest cultivation
T. baccata L. (English yew).	----------	Shrub or tree	Throughout Europe and in Algeria, northern Persia and the Himalayas.	Ornamental, bow wood, furniture, posts, turnery; in Asia as cattle fodder.	(1)
T. brevifolia Nutt. (Pacific yew).	*T. baccata* var. *brevifolia* (Nutt.) Koehne	------do------	Deep shady canyons of Pacific coast and eastward to Idaho and Montana.	Erosion control, ornamental, bird food.	1854
T. canadensis Marsh. (Canada yew).	*T. baccata* var. *minor* Michx., *T. baccata* var. *canadensis* (Marsh.) A. Gray (American yew, ground hemlock).	Low, prostrate shrub, often in clumps.	Shady ravines and moist coniferous woods from Newfoundland to Manitoba and south to Virginia and Iowa.	Wildlife food and cover, deer and moose browse, ornamental.	1800

[1] Cultivated since ancient times.

TAXUS

TABLE 209.—*Taxus: Time of flowering, "fruit" ripening, and seed dispersal*

Species	Time of—		
	Flowering	Fruit ripening	Seed dispersal
T. baccata	March–May	September–November	Fall
T. brevifolia	June	August–October	do.
T. canadensis	April–May	July–September	do.

FIGURE 257.—Seed views of two species of *Taxus*.

COLLECTION, EXTRACTION, AND STORAGE.—*Taxus* "fruit" should be collected from the branches by hand as soon as it is ripe. The seed can be extracted by macerating the arils in water and floating off the pulp and empty seed. In some species the membranous, outer seed coat is partially destroyed during extraction; in others it remains tightly fixed to the bony inner coat. After extraction, the seeds may be dried and stored, or used. The size, purity, soundness, and cost of commercial seed vary considerably by species as shown in table 210. Little information is available as to proper storage practices for *Taxus* seed. Viability can be maintained for 4 years by storing *T. baccata* seed in moist sand or acid peat at low temperatures. Similar practices might be suitable for the other species.

GERMINATION.—Yew seeds are slow to germinate, natural germination usually not taking place until the second year. Germination is epigeous (fig. 258). In *Taxus baccata* natural seedlings develop satisfactorily beneath other trees but make more rapid and denser growth in the open; most natural germination is from seed which have passed through the digestive tracts of birds (nutcrackers). There appears to be dormancy due both to an impermeable seed coat and internal conditions of the embryo.

Satisfactory methods for laboratory germination have not yet been worked out. Moist, cold stratification as long as 90 days does not overcome dormancy. It is probable that the seed must first be treated with sulfuric acid, or possibly hot water, to overcome seed-coat resistance, and then stratified in moist sand or peat at temperatures of 32° to 50° F. for 3 months or more; or it may be that the seed should first be stratified at temperatures of 68° (night) to 85° (day) for 2 months or more, followed by stratification at 32° to 50° for a similar period. Such treatments remain to be tested. When proper methods of pretreatment are worked out, germination tests probably will best be made in sand flats using 1,000 seeds per test.

NURSERY AND FIELD PRACTICE. — *Taxus* seed should be sown broadcast or in drills in the fall, or stratified seed can be used in the spring. Germination will take place during the following 2 years, some during the third year. The yews can also be propagated from cuttings which are taken early in the fall and wintered in a cool greenhouse or coldframe, or by grafting. *T. baccata* will grow well on lime soils.

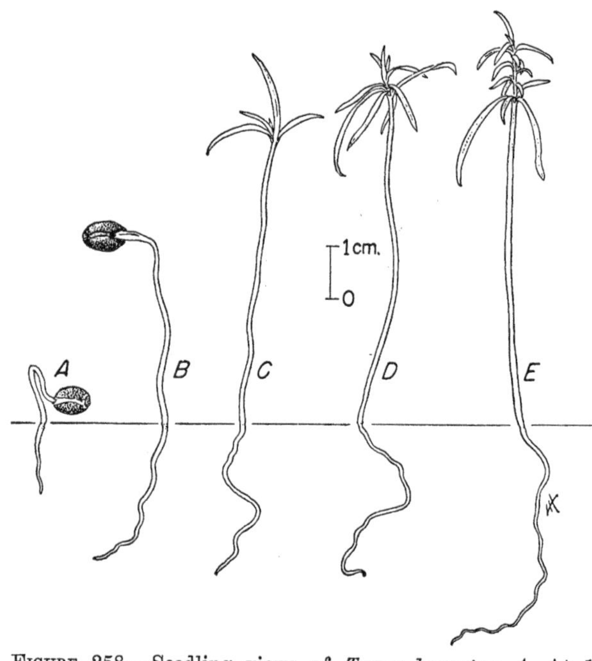

FIGURE 258.—Seedling views of *Taxus baccata: A,* At 1 day; *B,* at 8 days; *C,* at 12 days; *D,* at 22 days; *E,* at 39 days.

TABLE 210.—*Taxus: Yield of cleaned seed, and purity, soundness, and cost of commercial seed*

Species	Cleaned seed per pound				Commercial seed		
	Low	Average	High	Basis, samples	Purity	Soundness	Cost per pound
	Number	*Number*	*Number*	*Number*	*Percent*	*Percent*	*Dollars*
T. baccata		8,000		1	96	78	1.50–3.20
T. brevifolia	14,000		16,000	2+	99	98	
T. canadensis	15,000	21,000	32,000	3	100	99	7.20

THUJA L. Thuja, arborvitae

(Pine family—Pinaceae)

Distribution and Use.—The arborvitaes consist of six or seven species of medium to large evergreen trees native to China, Japan, Formosa, and North America. Some species are valuable timber trees because of their durable wood, and most of them are valuable as browse plants and ornamentals. The two species native to North America, both of which have been used to some extent in reforestation work, are described in detail in table 211. *Thuja occidentalis* has seldom been planted for forestry purposes, but it is coming into considerable use in game food plantings. It has been planted extensively both in North America and Europe for hedges and ornamental purposes. *T. plicata* has been planted for forestry purposes and as an ornamental to some extent both in western North America and northern and western Europe.

Seeding Habits.—The male and female flowers are borne separately on different branches of the same tree. The reddish, cylindric or globular male flowers, which bear pollen only, are borne on branchlets near the base of the shoot. The very small, conelike, green or purple-tinged female flowers, arising from short terminal branches, are composed of opposite pairs of leaflike scales. By the end of the season the female flowers develop into solitary cones one-third to two-thirds inch long, made up of 3 to 10 pairs of usually thin and flexible scales, of which only the middle 2 or 3 pairs ordinarily are fertile. Each fertile scale bears from two to five small, light chestnut brown, winged[65] seeds with a membranous coat (figs. 259 and 260). The comparative seeding habits of the two American species of *Thuja* are as follows:

	T. occidentalis	*T. plicata*
Time of—		
Flowering	April–May	April
Fruit ripening	August–October	August
Seed dispersal	do	August–October
Commercial seed-bearing age:		
Minimum ..... years	30	16
Optimum ..... do	75+	--
Average fertile cone scales ..... number	4	6
Average seeds per scale ..... do	2	2–3

T. occidentalis bears good seed crops every 5 years, while *T. plicata* bears good crops every 2 to 3+ years; light crops are borne by both species in the intervening years. No scientific information is available concerning the development of climatic races in *T. occidentalis*. Because of its wide distribution and the recognition of more than 30 horticultural varieties, it seems likely that such races have developed. In the case of *T. plicata*, tests in Norway brought out wide differences in frost hardiness of several sources from western United States and Canada, with the inland sources hardier than those from coastal areas.

Collection, Extraction, and Storage.—Arborvitae cones should be picked by hand from standing or recently felled trees as soon as the cones begin to turn brown and the seed has become firm. In the case of *Thuja occidentalis* the period between cone ripening and cone opening is short, only 7 to 10 days; for *T. plicata* the period is somewhat

[65] One foreign species is without wings.

Table 211.—*Thuja: Growth habit, distribution, and uses*

Accepted name	Synonyms	Growth habit	Natural range	Chief uses	Date of earliest cultivation
T. occidentalis L. (northern white-cedar).	*T. obtusa* Moench., *T. odorata* Marsh. (eastern arborvitae, SPN; American arborvitae).	Medium-sized tree.	Nova Scotia to Manitoba and south to Minnesota, Indiana, New Jersey, and in the Appalachian Mountains to North Carolina and Tennessee.	Shingles, posts and poles, canoe building, wreaths and decorations, ornamental and hedges, medicinal purposes, wildlife browse.	About 1536
T. plicata Donn (western red-cedar).	*T. gigantea* Nutt., *T. menziesii* Dougl., *T. lobbii* Hort. (western arborvitae, canoe cedar, giant arborvitae).	Large tree.	Alaska to northern California and eastward from British Columbia to Idaho and Montana.	Shingles, posts, poles, piling, ties, boxes, house and boat building, Indian baskets, wildlife browse.	1853

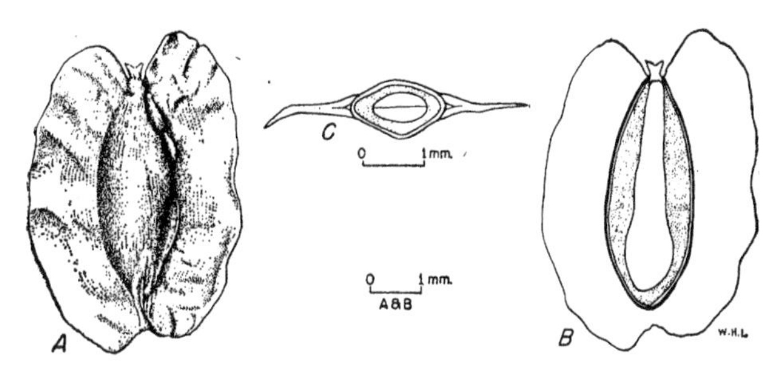

FIGURE 259.—Seed views of *Thuja occidentalis: A,* Exterior view, seed with wings; *B,* longitudinal section; *C,* cross section.

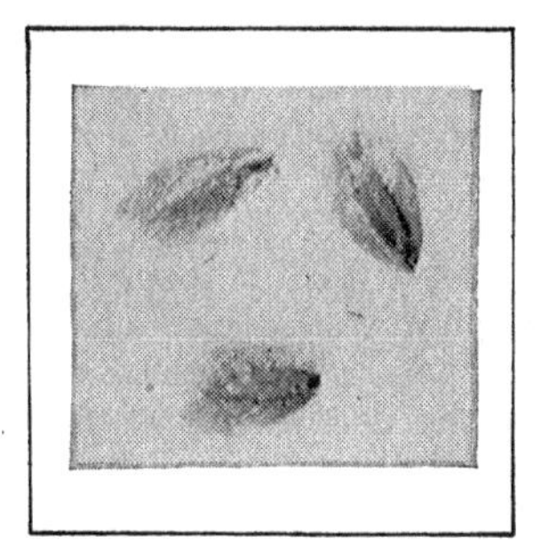

FIGURE 260.—Seed views of *Thuja plicata.*

longer. The freshly collected cones should be spread out on drying racks in the sun or in a dry, warm, well-ventilated room for several weeks. Ordinarily the cones will open with this treatment. If they fail to do so, they should be placed in a simple convection kiln or extracting drum at 100° F. It is necessary to put the cones through a shaker to sort out the seeds and then clean the latter of debris by blowing or fanning. The seeds should not be dewinged. The yield, size, purity, soundness, and cost of commercial cleaned seed vary as follows:

	T. occidentalis	*T. plicata*
Yield of cleaned seed:		
Per bushel of cones------ounces--	11–48	12
Per pound:		
Low----------------number--	184,000	203,000
Average---------------do----	346,000	414,000
High-----------------do----	568,000	504,000
Basis, samples-----------do----	66	60+
Commercial seed:		
Purity----------------percent--	79	75
Soundness----------------do----	73	81
Cost per pound---------dollars--	2.25–5.00	4.00–4.25

Seed of *T. occidentalis* stored dry in airtight containers at 35° to 40° retained high viability for 5 years. The seed deteriorates rapidly, however, at higher temperatures or when exposed to moisture. Seed of *T. plicata* stored in sealed containers at room temperature maintained fair viability for 2 years, but deteriorated rapidly thereafter. At low temperatures, viability probably would have remained higher and for a longer period.

GERMINATION.—Natural germination is best in shady places on moist materials such as rotten wood, decayed litter, peat, moss, or mineral soil. Germination is epigeous (fig. 261) and occurs in May or June of the year following dissemination in *Thuja occidentalis* and often in the same autumn as dispersal in *T. plicata. Thuja* seed, as a rule, show only slight dormancy, although occasional lots may be markedly dormant. Stratification in moist sand or peat at 32° to 50° F. for 30 to 60 days is sufficient to stimulate germination of dormant seed. The germination of arborvitae seed may be tested in flats of sand, soil, or mixtures of the two with peat, or in standard germinators at temperatures of 68° (night) to 86° (day), or constant temperatures of 75° for 20 to 50 days, using 1,000 seeds—they should be stratified in lots having marked dormancy—per test. Methods recommended and average results for the 2 species discussed here are as follows:

	T. occidentalis	*T. plicata*
Duration of test:		
Untreated seed-------------days--	50	40
Stratified seed---------------do--	30	20
Germinative energy:		
Amount----------------percent--	26–36	3–54
Period--------------------days--	15–30	20–30
Germinative capacity:		
Low--------------------percent--	0	3
Average--------------------do--	[1]46	[2]51
High-----------------------do--	92	91
Basis, tests--------------number--	59	66

[1]3 tests of stratified seed averaged 50 percent germination.
[2]4 tests of stratified seed averaged 59 percent germination.

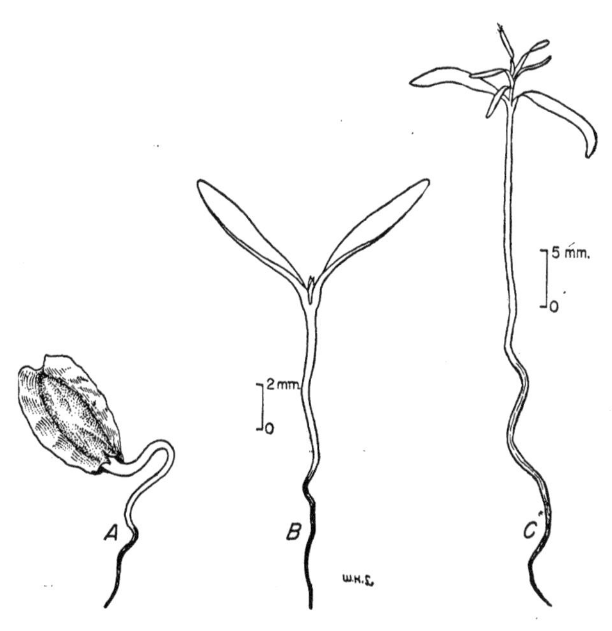

FIGURE 261.—Seedling views of *Thuja occidentalis: A,* At 1 day; *B,* at 5 days; *C,* at 25 days.

THUJA

Nursery and Field Practice.—The seed should be sown broadcast in the fall for *Thuja occidentalis,* or in the spring for *T. plicata,* at a rate to produce about 50 seedlings per square foot of seedbed. The seeds must be covered with one-eighth to one-fourth inch of soil, and fall-sown beds should be mulched until germination commences. Half shade for the beds the first season and mulching of the seedlings over the first winter are recommended. Field planting is best done with 2-0, 3-0, or 2-1 stock. The arborvitaes are fairly free of diseases and insect pests in the nursery. Under dry conditions *T. occidentalis* sometimes is attacked by scale insects which can be controlled by spraying weekly during May and early June with a paraffin wash. *T. plicata* seedlings sometimes are killed by a fungus, probably *Botrytis cinerea. T. occidentalis* should be grown on neutral or only slightly acid soil for best results, but will grow on a slightly basic soil. The arborvitaes can also be propagated vegetatively.

TILIA L. Basswood; linden, SPN[66]

(Linden family—Tiliaceae)

DISTRIBUTION AND USE.—The lindens include about 30 species of small to medium-large deciduous trees native to the temperate regions of the Northern Hemisphere in eastern North America south to Mexico; in Europe, and in Asia south to central China and southern Japan. They are valuable as timber trees, chiefly for lumber, excelsior, veneer, cooperage, drawing boards, and woodenware; for shade and ornamental planting; as bee pasture; and for wildlife food and cover. Many species hybridize freely in nature, making identification difficult. Two species, which are valuable for conservation planting and for which reliable information is available, are described in table 212. *Tilia cordata* has been grown successfully in northeastern United States in ornamental plantings. *T. americana* has been planted successfully on the Pacific slope.

SEEDING HABITS.—The perfect, yellowish flowers, which open in early summer, are borne in drooping clusters attached to large bracts. The fruits, which ripen in late summer or early fall of the first year, are round or egg-shaped capsular or nutlike structures; each consists of a crustaceous or woody pericarp which encloses usually a single

[66] *Tilia* is the linden of horticulture and literature, and the basswood of the U. S. Forest Service and lumber trade.

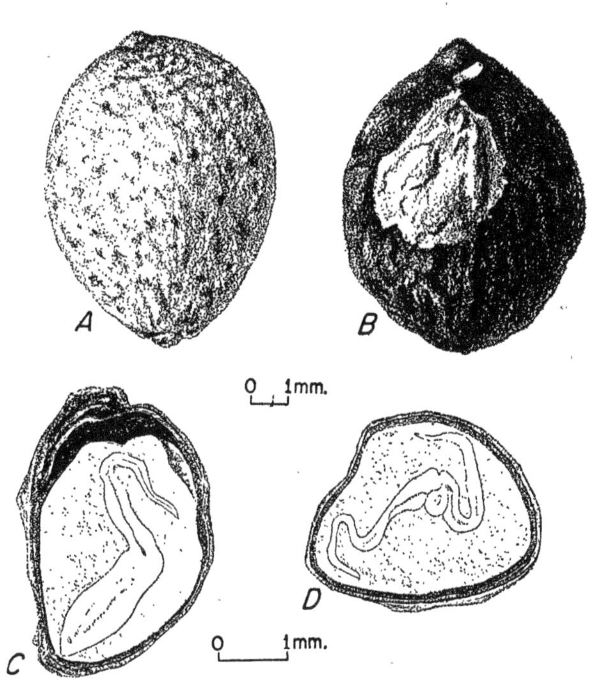

FIGURE 262.—Seed views of *Tilia americana*: *A*, Exterior view of fruit; *B*, exterior view of seed; *C*, longitudinal section; *D*, cross section.

TABLE 212.—*Tilia: Growth habit, distribution, and uses*

Accepted name	Synonyms	Growth habit	Natural range	Chief uses	Date of earliest cultivation
T. americana L.[1] (American basswood).	*T. glabra* Vent. (American linden, SPN; whitewood).	Medium to tall tree.	Northeastern United States and adjacent Canada.	Lumber, ornamental, shade tree, wildlife cover, bee pasture, flowers medicinal.	1752
T. cordata Mill. (littleleaf linden).	*T. europaea* L. in part, *T. ulmifolia* Scop., *T. parvifolia* Ehrh. (small-leaved linden, winter linden).	Medium-sized tree.	Europe and Siberia	Lumber, shade tree, ornamental, inner bark for tying material.	(2)

[1] TRIST, A. R. STUDIES OF AFTER-RIPENING AND GERMINATION AND PROPAGATION BY MEANS OF SHOOT AND ROOT CUTTINGS IN BASSWOOD. 1929. [Unpublished thesis. Copy on file Yale Forest School, New Haven, Conn.]

[2] Cultivated since ancient times.

seed, but sometimes two to four (figs. 262 and 263). The seeds have a crustaceous seed coat, a fleshy yellowish endosperm, and a well-developed embryo. Natural seed dispersal is chiefly by wind and animals. The comparative seeding habits of the two species of *Tilia* discussed here are as follows:

	T. americana	*T. cordata*
Time of—		
Flowering	June–July	June–July.
Fruit ripening	Sept.–Oct.	Sept.–Oct.
Seed dispersal	Fall–spring	Fall–winter.
Commercial seed-bearing age:		
Minimum years	15	20.
Maximum do	100+	
Fruit characteristics	Large, thick shelled	Small, thin shelled.

These two species bear good seed crops every 1+ years, and light seed crops in the intervening years. Scientific information concerning racial variation among the lindens is lacking, but in view of their wide distribution such variation is likely.

Collection, Extraction, and Storage.—After frosts, linden fruits[67] should be shaken from the branches onto canvas and spread out to dry. When thoroughly dried the persistent bracts may be removed by flailing, or by rubbing or running the fruits through a dewinging machine; sieving or fanning will remove the debris. The pericarps can also be removed by flailing in the case of *T. cordata,* but fruits of *T. americana* must be run through a coffee grinder or similar device, or acid treated to accomplish this. The yield, size, purity, soundness, and cost of commercial cleaned fruits vary by species as follows:

	T. americana	*T. cordata*
Yield of cleaned fruit:		
Per 100 pounds of fruit pounds	75	80
Seeds per pound:		
Low number	3,000	11,000
Average do	5,000	[1]14,000
High do	8,000	22,000
Basis, samples do	15+	25
Commercial seed:		
Purity percent	99	96
Soundness do	81	65
Cost per pound dollars	1.00–1.40	[2]2.00

[1]20,000 to 22,000 with pericarps removed.
[2]100-pound lots in Europe cost approximately 25 cents per pound.

Linden seed should be stored dry at 40° F. if it is to be kept any length of time. *T. cordata* with 10 to 12 percent moisture content will keep satisfactorily for 2 to 3 years under ordinary dry storage. *T. americana* will keep satisfactorily for 2 years if stored dry at room temperatures; for longer storage low temperatures are necessary.

Germination.—Natural germination of linden seed is best on mineral soil. Seed of *Tilia cordata* under natural conditions is reported to germinate

[67] Russian tests indicate that the best seed of *Tilia cordata* come from the upper part of the crown.

Figure 263.—Seed and fruit of *Tilia cordata.*

from 8 months to 7 years after dispersal, and seed of *T. americana* sometimes remains dormant for 4 years in nature. Germination is epigeous (fig. 264). Linden seed shows more or less delayed germination due to an impermeable seed coat, a dormant embryo, and, in some species, a tough pericarp. Methods which consistently will give prompt and fairly complete germination have not yet been worked out. Fair germination results sometimes from such practices as: (1) for *T. americana,* removing the pericarp by soaking in concentrated nitric acid for ½ to 2 hours, or removing it by mechanical means, followed by etching the seed coat with concentrated sulfuric acid for 15 to 20 minutes, and then stratifying in moist sand or peat for 3 to 5 months at 34° to 40° F.; (2) for *T. cordata,* stratification in moist peat or sand for 4 to 5 months at 59° to 77°, followed by 4 to 5 months stratification at 34° to 41°.

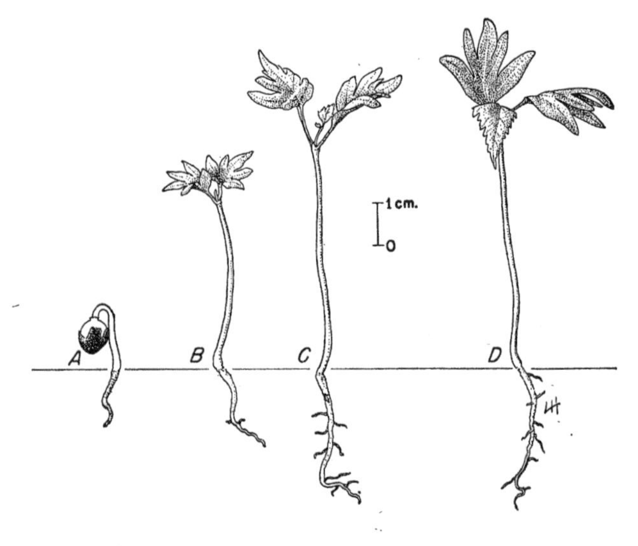

Figure 264.—Seedling views of *Tilia americana: A,* At 1 day; *B,* at 3 days; *C,* at 16 days; *D,* at 19 days.

Germination may be tested in sand flats or peat for 30 to 60 days at temperatures of 65° F. (night) to 85° (day), using 500 properly pretreated seeds per test. Average results for the two species discussed here, using seed pretreated as recommended, are as follows:

	T. americana	*T. cordata*
Germinative capacity:		
Low..................percent..	0	2
Average..................do....	29	25
High.....................do....	63	46
Potential germination........do....	26–100	50–98
Basis, tests..............number..	23	21

A quick test of the viability of stored or treated seed can be made within a few days by placing excised embryos or kernels with freed hypocotyls on moist cotton or filter paper. The best quick test of fresh seed is the cutting test.

NURSERY AND FIELD PRACTICE.—The usual practice is to sow the seed in drills in the fall, or to use stratified seed in the spring, and cover them with one-fourth to one-half inch of soil. This practice usually results in seedling production straggling over 2 or 3 years, much of it occurring in the second year. One series of tests with *Tilia cordata* indicated that the best seedling production (22 to 25 percent) occurred mostly the first spring after sowing dry-stored seed in June or July. It has been recommended that the following method be used with *T. americana* for good first-year seedling production: (1) Hold fruits in dry storage until 5 months before sowing time; (2) soak dry fruits in concentrated nitric acid for 30 to 120 minutes to remove the pericarps; (3) wash away digested pericarps, rinse, and dry seed thoroughly; (4) soak seed for 15 to 20 minutes in concentrated sulfuric acid and rinse thoroughly; (5) stratify seed in moist peat at 34° to 40° F. until time for sowing. This method, however, is not uniformly successful. Linden usually is field-planted as 1–0 or 2–0 stock. The lindens grow well on a variety of soils, so long as the soils are not dry. They are also propagated by layers and some of the varieties or rarer species by budding or grafting.

TORREYA CALIFORNICA Torr. California torreya

(Yew family—Taxaceae)

Also called California nutmeg. Botanical syn.: *Tumion californicum* (Torr.) Greene

Native to cool canyons in the central Coast Ranges and Sierra Nevada Mountains of California, this medium-sized evergreen tree is occasionally cultivated for a park tree. The "fruit" is drupelike and contains a single seed with a woody outer coat enclosed in a fleshy outer layer (fig. 265).

Flowering occurs in March to May; the fruits ripen from August to September and are ready for picking from September to November. There are approximately 110 to 140 seeds per pound (2 samples). It is not certain whether or not stratification is necessary. *Torreya* seeds germinate without stratification, but very slowly. The maximum germination recorded was 92 percent in 276 days (1 test).

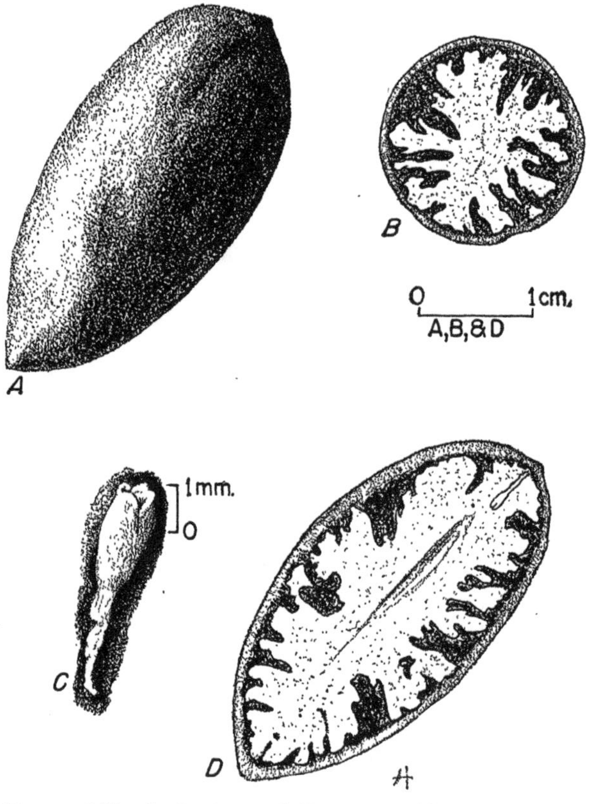

FIGURE 265.—Seed views of *Torreya californica: A,* Exterior view; *B,* cross section; *C,* embryo; *D,* longitudinal section.

TSUGA (Endl.) Carr. Hemlock

(Pine family—Pinaceae)

Distribution and Use.—The hemlocks comprise about 10 species of medium to large evergreen trees native to the temperate portions of North America, Japan, China, and the Himalayas. They are valuable as producers of timber, pulpwood, and tanbark, for wildlife food, and as ornamentals. Of the 4 American species, 2 are commercially important timber trees. The following 3 species, which are of known or supposed value for reforestation and for which reliable information is available, are described in detail in table 213. *Tsuga canadensis* and *T. heterophylla* are planted to some extent in reforestation work; *T. mertensiana* very rarely.

Seeding Habits. — Hemlock male and female flowers are borne on different branches of the same tree. The male flowers, bearing pollen only, occur in catkins borne in the axils of the previous year's shoots; small, greenish female flowers, which develop into cones, are borne at the ends of previous year's lateral shoots. The small, solitary, drooping cones are composed of thin, overlapping scales; they ripen during the first year, but remain on the tree after the release of the seeds until the summer or autumn of the second year. The cones change from yellowish green to purplish or brown when they ripen. Each scale bears two seeds, but only the seeds in about the central half of the cones are fertile. Hemlock seeds are brown, ovate, oblong, compressed, and nearly surrounded by their much longer wings (figs. 266 and 267). They are dispersed by the wind. The comparative seeding habits of the three species discussed here are given in table 214.

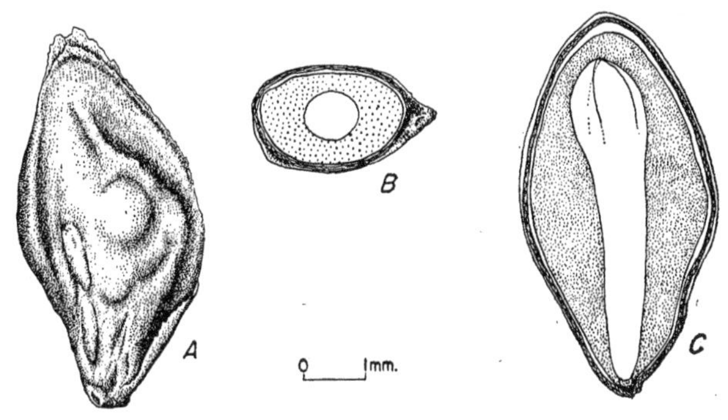

Figure 266.—Seed views of *Tsuga mertensiana: A,* Exterior view showing resin vesicles and adhering wing parts; *B,* longitudinal section; *C,* cross section.

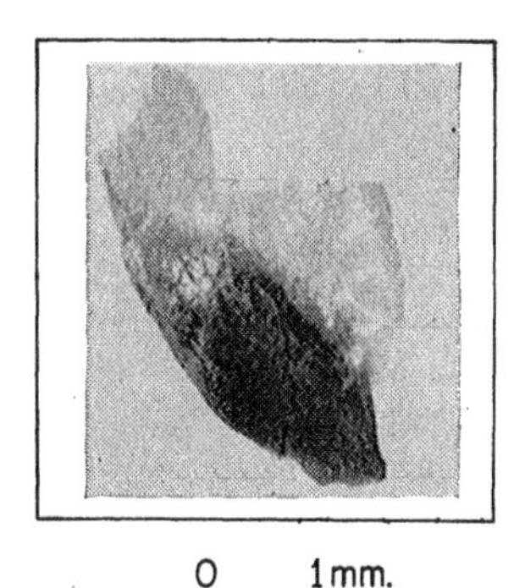

Figure 267.—Seed of *Tsuga heterophylla.*

Table 213.—*Tsuga: Growth habit, distribution, and uses*

Accepted name	Synonyms	Growth habit	Natural range	Chief uses	Date of earliest cultivation
T. canadensis (L.) Carr. (eastern hemlock).	*T. americana* (Mill.) Farwell. (Canada hemlock, SPN).	Medium to tall tree.	Nova Scotia to eastern Minnesota and Illinois, and south along mountains to northern Georgia and Alabama.	Lumber, poles, piling, ties, boxboards, pulpwood, tanbark; wildlife food and cover; ornamental.	1736
T. heterophylla (Raf.) Sarg. (western hemlock).	*T. mertensiana* auth., not (Bong.) Carr. (Pacific hemlock, SPN).	Tall tree.	Pacific coast from Alaska to northern California and in mountains of northern Idaho and northwestern Montana.	Lumber, ties, shingles, pulpwood, tanbark; wildlife food, ornamental.	1851
T. mertensiana (Bong.) Carr. (mountain hemlock).	*T. hookeriana* Carr.; *T. pattoniana* Senecl.	Medium to tall tree.	Southern Alaska to northern Montana, Idaho, and California.	Erosion control, wildlife food, ornamental.	1854

TSUGA

TABLE 214.—*Tsuga: Time of flowering, cone ripening, and seed dispersal, and frequency of seed crops*

Species	Time of—			Cone length	Commercial seed-bearing age			Seed year frequency		Color of ripe cone
	Flowering	Cone ripening	Seed dispersal		Minimum	Optimum	Maximum	Good crops	Light crops	
				Inches	*Years*	*Years*	*Years*	*Years*	*Years*	
T. canadensis	May-June	September-October	September-winter	1/2–1	30	-------	400+	2–3	Intervening	Purple to brown.
T. heterophylla	Spring	August	September	3/4–1¼	25–30	-------	-------	2–5	do.	do.
T. mertensiana	do.	September	September-October	1½–3	20	-------	-------	-------	-------	Yellowish green or bluish purple.

In view of their wide distribution and the recognition of several varieties, it is likely that climatic races have developed among the three hemlocks described here. However, only the following meager evidence to that effect is available: *Tsuga canadensis* seed from eastern origins is larger on the average than that from western origins. In *T. heterophylla,* stock grown from Alaskan seed is more frost-hardy than stock grown from Washington seed.

COLLECTION, EXTRACTION, AND STORAGE.—Hemlock cones should be picked by hand from standing trees or fresh slash as soon as they ripen, i. e., when they have changed from green to purplish or brown. After the cones have been spread out to dry for 2 days to several weeks in a cone shed, they should be run through a simple convection kiln for 2 days at 100° to 110° F. The seed can then be cleaned by running them through a dewinger and then fanning off the debris. The yield, size, purity, soundness, and cost of commercial cleaned seed vary considerably by species as shown in table 215. Scientific data concerning the storage of hemlock seed is mostly lacking. One lot of *Tsuga canadensis* seed stored dry in sealed containers at 41° germinated well (46 percent) after 4 years, but retained little viability (0.8 percent) at the end of 6 years. Seed of *T. heterophylla* can be stored sealed at ordinary room temperatures for 2 or 3 years without appreciable loss of germinative capacity.

GERMINATION.—Natural seedlings come in chiefly under shade on moist decaying logs, stumps, moss, or mineral soil. Germination is epigeous (fig. 268). Dormancy is variable in the three hemlocks discussed here; the seed of some lots may have dormant

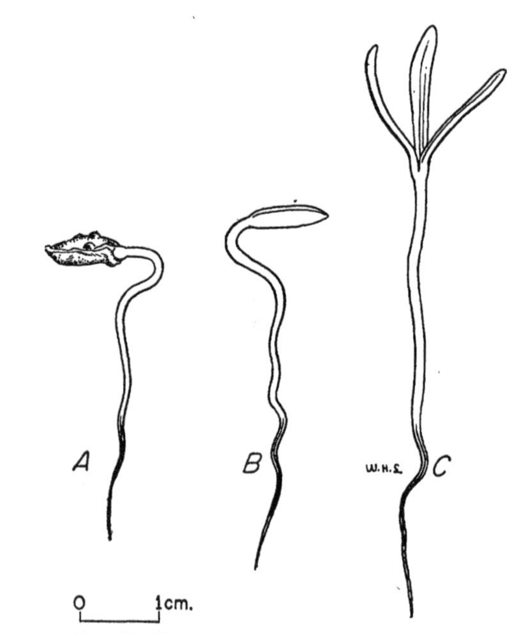

FIGURE 268.—Seedling views of *Tsuga canadensis: A,* At 2 days; *B,* at 4 days; *C,* at 7 days.

TABLE 215.—*Tsuga: Yield of cleaned seed, and purity, soundness, and cost of commercial seed*

Species	Cleaned seed					Commercial seed		
	Yield per 100 pounds fresh cones	Per pound			Basis, samples	Purity	Soundness	Cost per pound
		Low	Average	High				
	Pounds	*Number*	*Number*	*Number*	*Number*	*Percent*	*Percent*	*Dollars*
T. canadensis	3–6	132,000	187,000	360,000	59	85	80	5.00–7.00
T. heterophylla	3	220,000	297,000	508,000	22	85	56–74	[1]4.50–6.50
T. mertensiana	----------	60,000	114,000	208,000	6	83	----------	[1]6.00–8.50

[1] Forest Service collection.

embryos while other lots may not. Stratification in moist sand at 41° F. for 60 to 120 days will improve germination. Pretreatment for 60 to 120 days is recommended for *Tsuga canadensis.* Although seed of *T. heterophylla* usually germinate well without stratification, some lots need pretreatment. A stratification period of 90 days is recommended for this species and for *T. mertensiana.* Germination may be tested in sand flats, peat mats, or standard germinators at 68° (night) to 86° (day) for *T. canadensis* and *T. mertensiana,* 52° to 60° for *T. heterophylla,* using 1,000 stratified or nondormant seeds per test. Methods recommended and average results for the three species are given in table 216.

Nursery and Field Practice.—Hemlock seeds should be sown broadcast or in drills in the fall, or stratified or nondormant seeds used in the spring, and covered with about one-eighth inch of soil or a mixture of equal parts of sandy loam and screened humus. In *Tsuga heterophylla* nursery germination is about 50 percent of that obtained in the laboratory, and it occurs from 4 to 6 weeks after spring sowing. The seedbeds should be given half shade at the first sign of germination and throughout the first season; shade can be retained the second season. Hemlock seedlings are comparatively free from injurious pests but may be attacked by damping-off if the beds are not given previous acid or alum treatment. 2-2 stock is used in field planting. The hemlocks may also be propagated by layering.

Table 216.—*Tsuga: Recommended germination test duration, and summary of germination data*

Species	Test duration recommended		Germination data from various sources					
			Germinative energy		Germinative capacity			Basis, tests
	Untreated seed	Treated seed	Amount	Period	Low	Average	High	
	Days	*Days*	*Percent*	*Days*	*Percent*	*Percent*	*Percent*	*Number*
T. canadensis	200	60	10–55	15–30	12	38	58	15
T. heterophylla	90+	[1]30	38	20–30	12	56	96	25
T. mertensiana		[1]25–30	62–75	16–20	1	47	78	4

[1] Stratified or nondormant seed.

891183°—50——24

ULMUS L. Elm

(Elm family—Ulmaceae)

Distribution and Use.—About 20 species of elm are native to the Northern Hemisphere, exclusive of western North America. In North America, they extend south to northern Mexico. Most of them are trees valuable for their hard, heavy, tough wood. Because of their tall, spreading form, the elms are also highly valued for shade and ornament. The natural range and uses for 6 of the most important species are given in table 217. *Ulmus americana* develops to the largest size and has been extensively planted for its shade. In recent years, the Dutch elm disease, caused by a fungus (*Graphium ulmi*), and phloem necrosis, caused by a virus, have killed so many trees that park and street planting of this species has practically ceased in most sections of its range. *U. parvifolia* and *U. pumila* have been used considerably for shelter-belt planting in the Prairie-Plains region.

Seeding Habits. — The perfect, rather inconspicuous, flowers are borne in dense clusters[68] in the early spring generally before the leaves, or, in a few species, in late summer or early fall. The fruit, a samara, consisting of a compressed nutlet surrounded by a membranous wing (figs. 269 and

[68] An exception is *Ulmus thomasi;* its flowers are borne in racemes.

Table 217.—*Ulmus: Distribution and uses*

Accepted name	Synonyms	Natural range	Chief uses	Date of earliest cultivation
U. americana L. (American elm).	*U. floridana* Chapm. (white elm).	Moist soils of bottom lands and upland flats Great Plains east to Atlantic coast.	Lumber, slack cooperage, wheel hubs, crates, ornamental planting, fuel.	1752
U. fulva Michx.[1] (slippery elm).	*U. pubescens* Walt.; *U. rubra* Michx. f. (red elm).	Moist bottom lands and upland flats Quebec to North Dakota, south to Florida and Texas.	Lumber, furniture, slack cooperage, veneer, vehicle parts, sporting goods, fuel, posts. Inner bark medicinal.	1830
U. laevis Pall. (Russian elm).	*U. pedunculata* Foug. (spreading elm, European white elm).	Central Europe to west Asia.	Fuel, crate wood, shade and ornamental planting, animal food.	([2])
U. parvifolia Jacq. (Chinese elm).	*U. chinensis* Pers. (leatherleaf elm).	Northern and central China, Korea, and Japan.	Shade and ornamental planting, shelter belts, wildlife cover and food.	1794
U. pumila L. (Siberian elm).	Chinese elm, dwarf Asiatic elm.	Turkestan, eastern Siberia, northern China.	Shelter-belt, shade, ornamental planting; game cover.	1860
U. thomasi Sarg. (rock elm).	*U. racemosa* Thomas, not Borkh. (cork elm).	Dry, gravelly uplands and limestone soils Quebec to Tennessee, west to Minnesota and and Kansas.	Parts of agricultural implements, wheel hubs, chair frames, ornamental planting.	1875

[1]Prof. M. L. Fernald has recently indicated that the name *Ulmus rubra* Muhl. (1793), which has 10 years' priority over *U. fulva* Michx., will probably have to be accepted for slippery elm.
[2]Long cultivated.

270) ripens a few weeks after flowering. Dispersal is by wind as the seed ripens. Commercial "seed" consists of the ripened samara and varies greatly in size among species. Comparative seed developments are given in table 218.

COLLECTION, EXTRACTION, AND STORAGE. — Elm seed can be collected by sweeping it up from the ground soon after it falls, or by beating or stripping it from the branches at maturity.[69] Freshly collected fruits should be air-dried a few days previous to sowing or storage; excessive drying, however, is injurious. Although wings are not usually removed, this is occasionally practiced when the seed is to be sown by seeders. Such dewinging is accomplished by putting the fruit in bags and beating with flails. The available data on size, purity, soundness, and cost for the six species discussed here are given in table 219. Elm seed keeps best if stored at a low moisture content in sealed containers at temperatures not far from freezing. For example, seed of *Ulmus americana* has been stored for over 2 years in sealed jars at 40° F. with no loss in germinability. Seed of *U. pumila* stored in sealed drums buried in a cool place for 1 year and then stored at 40° for 4 years showed a germination of 50 percent. On the other hand seed of *U. thomasi,* which had been stored in sealed containers at temperatures fluctuating from near freezing to about 70°, dropped in germination from 64 to 13 percent in 1 year. Viability in most species decreases rapidly if the seed is stored at air temperatures.

[69] The large seed of *Ulmus thomasi,* with a flavor much like filberts, is greatly relished by rodents and should be picked from the trees just before it falls.

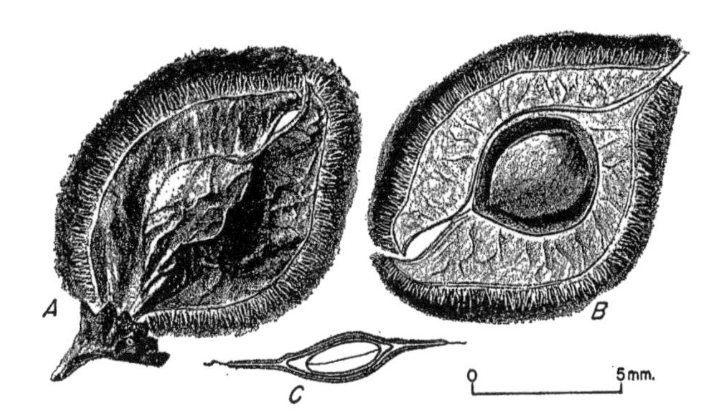

FIGURE 269.—Seed views of *Ulmus americana: A,* Exterior view of fruit with attached calyx; *B,* view with upper section of wing removed; *C,* cross section.

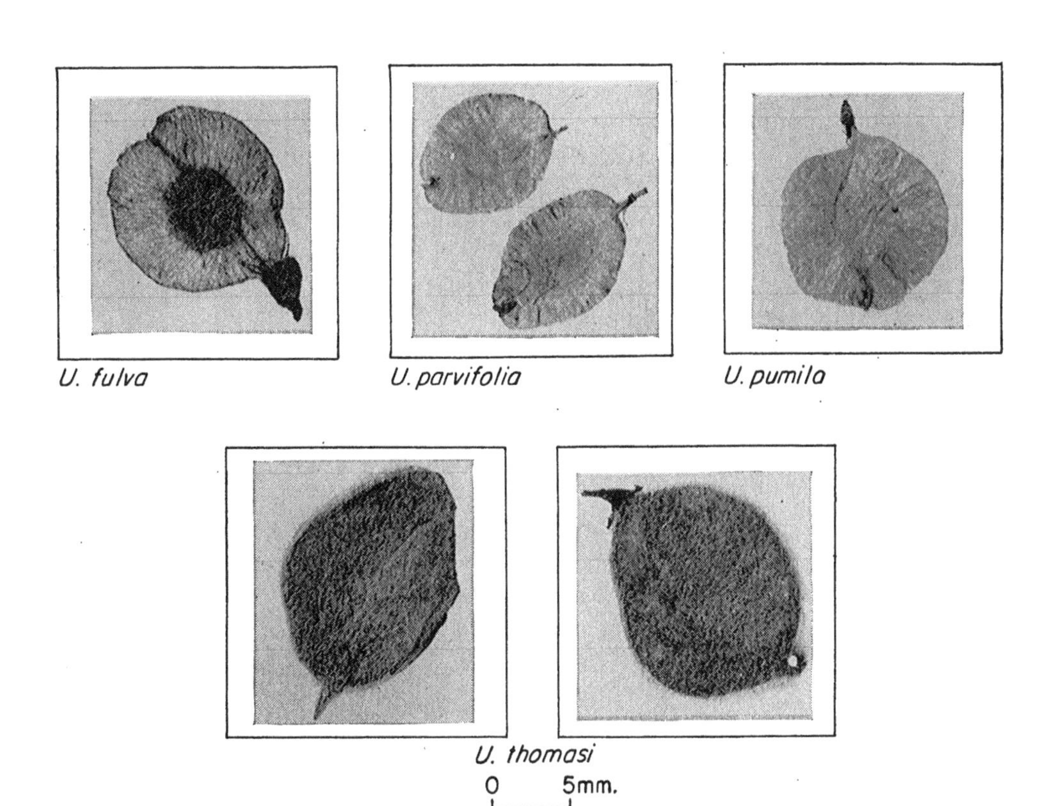

FIGURE 270.—Seed of four species of *Ulmus.*

ULMUS

Table 218.—*Ulmus: Time of flowering, fruit ripening, and seed dispersal, and frequency of seed crops*

Species	Time of—			Commercial seed-bearing age			Seed year frequency	
	Flowering	Fruit ripening	Seed dispersal	Minimum	Optimum	Maximum	Good crops	Light crops
				Years	*Years*	*Years*	*Years*	*Years*
U. americana	February–April	March–June	March–June	15	40–150	300	Most	Intervening.
U. fulva	Late February–April	April–June	April–June	15	25–125	200	2–4	do.
U. laevis	April–May	May–June	May–June	----------	----------	----------	Most	do.
U. parvifolia	August–September[1]	September–October	September–October	----------	----------	----------	----------	----------
U. pumila	March–April	April–May	April–May	8	15–75	----------	Most	Intervening.
U. thomasi	March–May	May–early July	May–early July	20	45–125	250	3–4	do.

[1] 2 other species, *U. crassifolia* Nutt. and *U. serotina* Sarg., both of which are native to the United States, flower and ripen their fruit in autumn.

Germination. — Elm seeds which ripen in the spring usually germinate the same season, whereas those ripening in the fall germinate the following spring under natural conditions. Sometimes practically all of the seeds in some lots of *Ulmus americana* will lie dormant until the second spring. Such dormancy is apparently due to conditions within the embryo and can be overcome by stratification in sand for 60 days at 41° F. Many lots of *U. fulva* seed, particularly from northern sources, also show dormancy. For these no satisfactory pretreatment has yet been worked out although stratification for 60 to 90 days at 41° was of benefit. Germination is epigeous (fig. 271). Adequate tests may be made on 200 to 400 seeds in sand flats, peat mats, or "rag dolls." Further details of methods used and average results of available tests are given in table 220.

Nursery and Field Practice.—Seed of elm species ripening in the spring is usually sown immediately after collection, while that of fall ripening species is usually dry-stored over winter and spring sown.[70] Approximately 15 seeds are sown per linear foot in drills 8 to 10 inches apart, and covered with about one-fourth inch firmed soil. The beds should be kept moist until germination is complete; shade is not usually necessary. A few diseases often interfere with production of elm seedlings, affecting some species more than others. Damping-off, a common cause of early loss, may be controlled with formaldehyde treatment of the soil previous to seeding. American elm is subject to attack by a leaf-disease fungus, *Gnomonia ulmea*, which may be controlled by early spraying with 4–4–50 bordeaux mixture or with sulfur dust and sprays. Ordinarily about 12 percent of the viable seed sown may be expected to produce plantable stock. One-year-old seedlings are generally sufficiently large for field planting.

[70] Fall sowing is probably preferable.

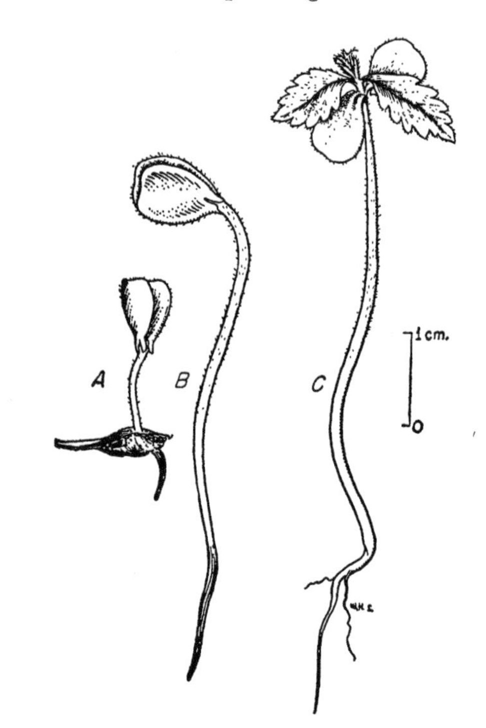

Figure 271.—Seedling views of *Ulmus americana: A,* At 1 day; *B,* at 3 days; *C,* at 21 days.

TABLE 219.—*Ulmus: Yield of cleaned seed, and purity, soundness, and cost of commercial seed*

Species	Cleaned seed					Commercial seed		
	Yield per 100 lbs. of fruit	Number per pound[1]			Basis, Samples	Purity[2]	Sound-ness	Cost per pound
		Low	Average	High				
	Pounds	*Number*	*Number*	*Number*	*Number*	*Percent*	*Percent*	*Dollars*
U. americana	[3]50	48,000	68,000	[4]95,000	13	92 (79–99)	96	0.50–1.00
U. fulva	----------	35,000	41,000	54,000	10	94	96	.50–1.00
U. laevis	60	53,000	71,000	93,000	3+	85	55	----------
U. parvifolia	50	151,000	----------	169,000	2	64	----------	6.00–12.00
U. pumila	50	40,000	65,000	112,000	24+	90	85	2.15–5.70
U. thomasi	(3)	5,000	7,000	9,000	5	95	97	.50–1.00

[1]Since dewinging of seed is seldom practiced, figures in these columns refer to seed with wings.
[2]First figure is average percent; figures in parentheses indicate range of lowest to highest.
[3]Weight per bushel of fruit; *Ulmus americana*, 4.5 pounds; *U. thomasi*, 7.8 pounds.
[4]Dewinged seed, 164,000 per pound.

TABLE 220.—*Ulmus: Recommended conditions for germination tests, and summary of germination data*

Species	Test conditions recommended			Germination data from various sources					
	Temperature		Duration	Germinative energy		Germinative capacity			
	Night	Day		Amount	Period	Low	Average	High	Basis, tests
	°F.	*°F.*	*Days*	*Percent*	*Days*	*Percent*	*Percent*	*Percent*	*Number*
U. americana	[1]68	[1]86	[2]13–60	35–90	6–12	3	63	94	14
U. fulva	68	86	90–120	13–32	8–20	4	17	34	6
U. laevis	----------	----------	----------	----------	----------	45	65	85	3+
U. parvifolia	70	85	60	----------	----------	40	----------	70	2+
U. pumila	68	86	14	2–48	9	20	60	90	26+
U. thomasi	68	86	4–10	61–90	4–10	64	85	100	4

[1]Alternations of 50° to 70° F. and 65° to 75° F. are almost as effective.
[2]Dormant lots require at least 90 days and then show only a small fraction of the germination obtained after stratification. Stratified seed usually completes its germination in 10 to 30 days.

UMBELLULARIA CALIFORNICA (Hook. & Arn.) Nutt. California-laurel

(Laurel family—Lauraceae)

Also called Oregon-myrtle, bay-tree.

Native to the lower mountain slopes and canyons of southern Oregon and California, this evergreen tree has hard, fine-grained wood that is valuable for cabinet work. Its leaves contain an appreciable quantity of menthol. Introduced into cultivation in 1829, it is occasionally planted in parks, but at present, it is grown mostly in pots for horticultural purposes.

The seed is large and thin-shelled, and is enclosed in a yellow-green—sometimes purplish—leathery, fleshy covering (fig. 272). Flowers open from January to May, and the fruits ripen from September to October, at which time they fall and are dispersed by gravity or, later in the season, by water. Seed crops are abundant almost every year.

Fruits are collected from the ground in October and November, and the flesh is removed by hand. One pound of fruit contains about 300 cleaned seed (1 sample). Commercial seed costs $2 to $2.50 per pound. Seed of this species cannot be stored long, even at low temperatures. It is advisable to collect fresh seed every year.

Under natural conditions, seed germinate probably late in the winter or early spring. Germination tests can be made in deep, wooden boxes filled with light soil. Untreated seed germinate fairly well (20 to 25 percent); stratification doubles the germination percent (41 percent, one test).

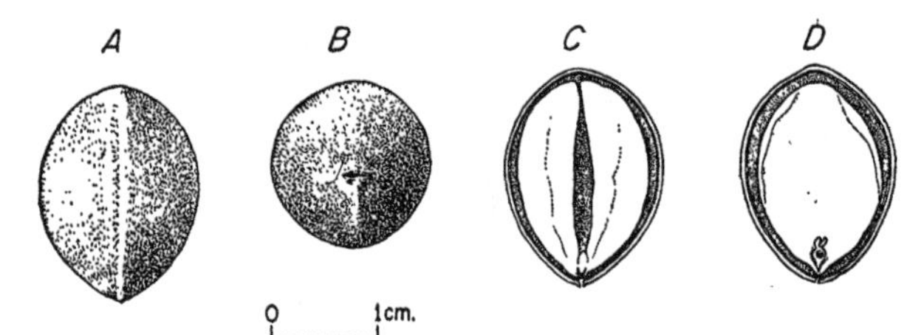

FIGURE 272.—Seed views of *Umbellularia californica*: *A* and *B*, Exterior views in two planes; *C* and *D*, longitudinal sections in two planes, showing embryo with large cotyledons.

VIBURNUM L. Viburnum

(Honeysuckle family—Caprifoliaceae)

Distribution and Use.—The viburnums consist of about 120 species of deciduous, or sometimes evergreen, large shrubs or small trees which are widely distributed in North and Central America, Europe, North Africa, and in Asia south to Java. Their attractive flowers, foliage, and fruit make them valuable ornamentals. The fruit is used by birds and mammals, and that of some species by man. The twigs of some species are used as browse by some of the larger game animals; the bark of others is used in medicine. Eight species, which are suitable for planting in the United States, are described in table 221.

Seeding Habits.—The white, or sometimes pinkish, perfect flowers are borne in flat clusters (cymes) in the spring. The fruit, a red, dark blue, or black fleshy drupe containing one stone, usually flattened, ripens in the late summer or fall. Dis-

Table 221.—*Viburnum: Growth habit, distribution, and uses*

Accepted name	Synonyms	Growth habit	Natural range	Chief uses	Date of earliest cultivation
V. acerifolium L. (mapleleaf viburnum).	dockmackie	Small, sometimes thicket-forming, erect shrub.	On drier soils from New Brunswick to Wisconsin, south to Georgia and Alabama.	Ornamental	1736
V. alnifolium Marsh. (hobblebush viburnum).	American wayfaring tree.	Medium-tall shrub, sometimes procumbent.	New Brunswick to Michigan, south to the mountains of North Carolina.	Wildlife food and cover, ornamental.	1820
V. cassinoides L. (witherod viburnum).	*V. squamatum* Muhl. ex Willd. (Appalachian tea).	Upright, sometimes treelike shrub.	In swamps and wet thickets from Newfoundland to Wisconsin, south to Georgia and Alabama.	Wildlife food (berries and twigs), ornametal, leaves used for tea in Northeast.	1761
V. dentatum L. (arrowwood viburnum).	arrowwood	Upright, bushy, tall shrub.	New Brunswick to Minnesota, south to Georgia.	Wildlife food, possible ornamental.	1736
V. lentago L. (nannyberry).	nannyberry viburnum, SPN; sheepberry, sweet viburnum, black haw.	Large shrub or small tree.	Rich, moist soils from Quebec to Saskatchewan, south to Georgia and Colorado.	Wildlife food, human food; ornamental, shelter belt.	1761
V. opulus L. (European cranberrybush viburnum).	European cranberrybush.	Medium-tall shrub.	Europe, northern Africa, and northern Asia.	Human food, wildlife food, ornamental (especially var. *roseum*).	[1]
V. prunifolium L. (blackhaw).	blackhaw viburnum, SPN; stagbush.	Large shrub or small tree.	Connecticut to Michigan, south to Georgia and Texas.	Ornamental, human food, bird food.	1727
V. trilobum Marsh. (American cranberrybush viburnum).	*V. americanum* auth., not Mill.; *V. opulus* var. *americanum* Ait. (American cranberrybush, pembina cranberry).	Tall shrub	Swamps and low ground from Newfoundland to British Columbia, south to New York and Oregon.	Wildlife food, human food, ornamental.	1812

[1]Long cultivated.

persal is chiefly by animals or gravity. Commercial seed consists of the dried fruit or clean stones (figs. 273 and 274). The comparative seeding habits of eight species discussed here are given in table 222. In *Viburnum prunifolium, V. cassinoides* and *V. opulus,* good seed crops are borne every 1 or 2 years. Information for other species is lacking. No information as to the development of climatic races in *Viburnum* species is available. Several horticultural varieties of *V. opulus* are known, and it is likely that most of the species discussed here have developed races or strains in view of their wide distribution.

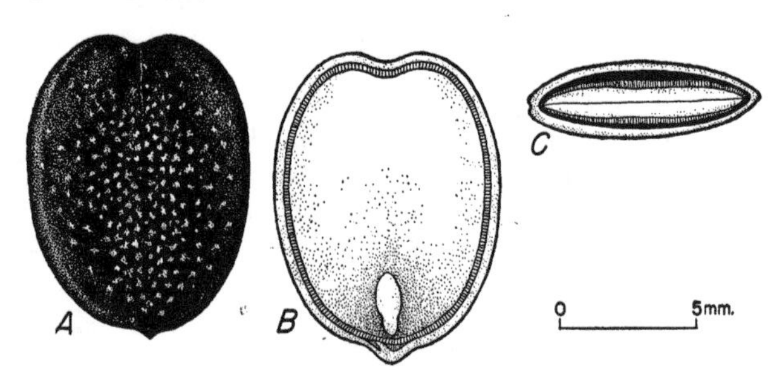

FIGURE 273.—Seed views of *Viburnum lentago: A,* Exterior view; *B,* longitudinal section; *C,* cross section.

COLLECTION, EXTRACTION, AND STORAGE. — The fruits may be picked from the branches as soon as they are ripe, and care should be taken to prevent heating. If they are to be used as whole fruits, they should be spread out to dry. Otherwise they should be run through a macerator with water and the pulp floated off. The seeds should be dried for storage or use. The size, form, purity, and soundness of commercial cleaned seed vary considerably by species, as shown in table 223.

The seed of some viburnums can be stored dry in sealed containers at temperatures not far above freezing for 1 or 2 years with little loss of viability. Although the optimum storage conditions are not yet known for any viburnums, the following results are known: Seed of *V. acerifolium* stored in the dried fruit in a sealed jar at 41° F. showed no loss in germinability in over 2 years. That of *V. cassinoides* stored for 14 months—about half of

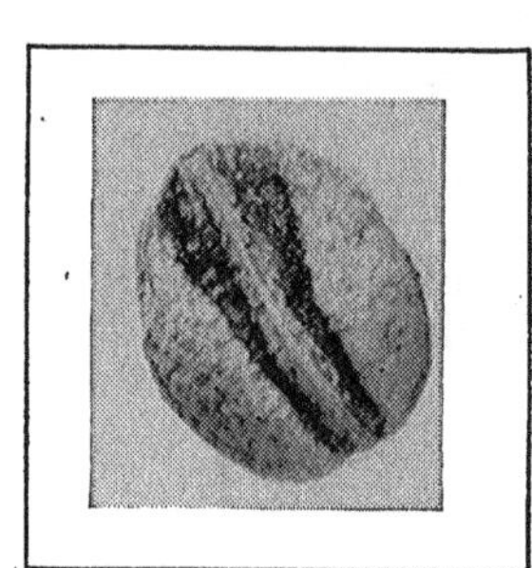

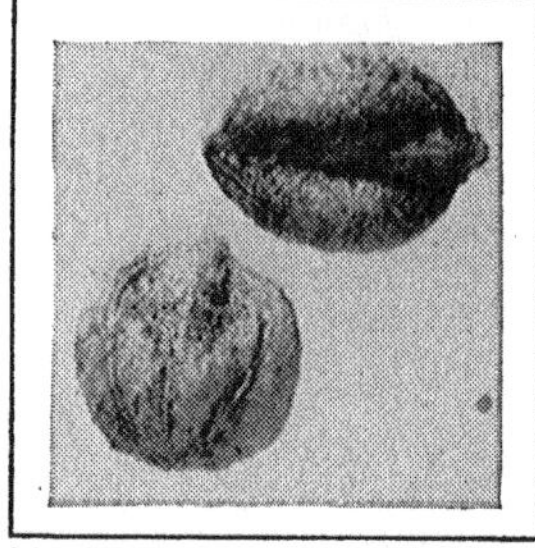

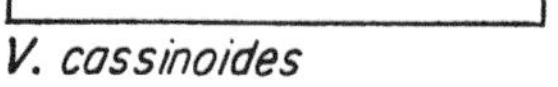

FIGURE 274.—Seed of four species of *Viburnum.*

TABLE 222.—*Viburnum: Time of flowering, fruit ripening, and seed dispersal*

Species	Time of—			Color of ripe fruit	Size of fruit
	Flowering	Fruit ripening	Seed dispersal		
V. acerifolium	May-June	September-October	Fall	Black	3/8 inch in diameter.
V. alnifolium	do	September	do	Purple black	1/3 inch long.
V. cassinoides	June-July	September-October	do	Blue to black	1/4 inch long.
V. dentatum	May-June	do	do	Blue black	Do.
V. lentago	April-June	August-October	Fall-spring	do	1/2 to 3/4 inch long.
V. opulus	May-June	August-September	Fall	Red	1/3 inch in diameter.
V. prunifolium	do	September-October	do	Blue black	1/3 to 1/2 inch long.
V. trilobum	May-July	August-September	do	Red	3/8 to 1/2 inch long.

Table 223.—*Viburnum: Yield of cleaned seed, and purity, soundness, and cost of commercial seed*

Species	Dried berries per pound	Cleaned seed					Commercial seed		
		Yield per 100 pounds of whole fruit	Per pound			Basis, samples	Purity	Soundness	Cost per pound dried fruit
			Low	Average	High				
	Number	*Pounds*	*Number*	*Number*	*Number*	*Number*	*Percent*	*Percent*	*Dollars*
V. acerifolium	4,800	30	10,900	13,200	16,600	4	--------	96	1.00–1.25
V. alnifolium	----------	[1]43	----------	11,800	----------	1	--------	94	1.75–3.00
V. cassinoides	----------	----------	25,000	----------	29,000	2	--------	94	1.25–2.00
V. dentatum	----------	[1]34	24,500	----------	32,600	2	97	98	.75–2.50
V. lentago	2,200	[2]10–25	3,300	5,700	12,400	12	98	90	.50–1.25
V. opulus	5,500	10–30	12,000	14,000	16,000	5	--------	90	(3)
V. prunifolium	----------	20–50	4,000	----------	6,000	2	--------	95	(3)
V. trilobum	----------	7–9	9,400	13,200	17,800	4	--------	94	[3]1.25–2.00

[1] Yield per 100 pounds of dried fruit.
[2] 100 pounds of fruit yields 50 pounds of dried berries, or 10 to 25 pounds of cleaned seed.
[3] Cost of clean seed per pound: *V. opulus*, \$1.25 to \$2; *V. prunifolium*, \$1 to \$2.20; *V. trilobum*, \$3.50.

that time in the form of dried fruit—in a sealed container at 41° showed practically no loss in viability. Seed of *V. lentago* stored at about 40° in moist sand showed 23 percent germination after 3½ years, and stored dry in a sealed container at 41°, 88 percent germination after 2½ years. That of *V. opulus* remains viable for 2 years under ordinary storage. Seed of *V. trilobum* stratified in moist sand showed practically no loss in viability after 1 year and stored dry in a sealed jar at 41°, no loss after 1½ years.

Germination.—Viburnum seed ordinarily is slow to germinate. For *Viburnum prunifolium* the best natural seedbed is a fertile, well-drained soil protected by partial vegetation; natural germination takes place the second spring after seed fall. Seed of some species appear to germinate quite readily following stratification at low temperatures, while that of others requires stratification at both high and low temperatures, the former to induce root growth, the latter shoot production. Most species have embryo dormancy (determined to be localized in the plumule for several species), and in some of them germination may also be impeded by an impermeable seed coat. Germination is epigeous (fig. 275). Further information on seed dormancy and pretreatment is given in table 224. As is evident, the best practices have not yet been worked out for several species and only suggested treatments can be given.

Germination tests should be made in sand or soil flats, using 400 to 1,000 properly pretreated seeds per test; for most species temperatures alternating from 68° F. (night) to 86° (day) appear to be satisfactory. Methods recommended and average results for the 8 species discussed here are given in table 225.

Nursery and Field Practices.—Viburnum seed should be sown in drills 8 to 12 inches apart in the spring or early enough in the late summer or fall so that about 60 warm days will elapse before winter.[71] They should be covered with one-half inch of soil. The seedbeds should be mulched with straw, and the mulch removed after germination begins. Also, viburnum seed are sometimes stratified a full year and than fall sown. In any case, germination does not occur until the following spring. The beds should be protected against birds and rodents by screens. For *V. lentago* about 25 percent of the viable seed sown produce usable seedlings. Viburnums can also be propagated from greenwood or hardwood cuttings, or from layers.

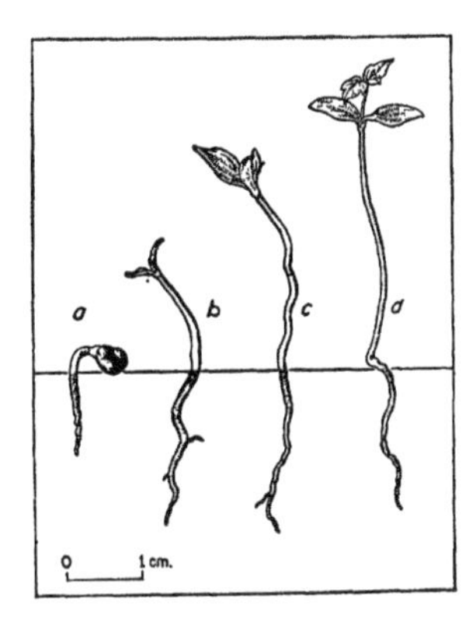

V. dentatum
a, At 1 day.
b, At 2 days.
c, At 11 days.
d, At 29 days.

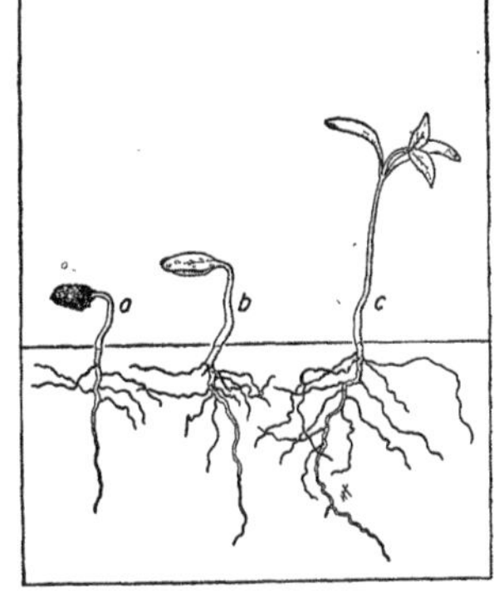

V. lentago
a, Development during prolonged, warm stratification.
b, Development during ensuing cold stratification.
c, Subsequent development at germination temperature.

Figure 275.—Seedling views of *Viburnum dentatum* and *V. lentago*. Note differences in root development of two species. In *V. lentago* considerable root development takes place before shoot growth starts; in *V. dentatum* both root and shoot growth begin together.

[71] In one case seed of *V. dentatum* collected in August in Iowa, and sowed immediately, gave 90 percent germination the following spring.

VIBURNUM

TABLE 224.—*Viburnum: Methods of seed pretreatment for dormancy*

Species	Stratification: Medium	Stratification: Duration and temperature[1]	Other methods used	Remarks
		Days °*F.*		
V. acerifolium	Moist soil, or peat	180–510 at 68–86 + 60–120 at 41	150 days at 32°–50 F. for emergence tests.	Roots produced by warm temperatures, shoots following cold temperatures; with cold stratification alone, germination will not occur.
V. alnifolium[2]	Moist soil	90+ at 68–86 + 60 at 41	----------	Dormancy known but satisfactory pretreatment not yet worked out.
V. cassinoides	Moist sand	60 at 68–86 + 90 at 50	120 days at 41° F. and 60 days at 68°–86° F. plus 120 days at 41° F. almost as effective.	Roots and shoots produced simultaneously.
V. dentatum	Moist peat	180–510 at 68–86 + 15– 60 at 41–50	----------	Although warm plus cold stratification seems to be the only treatment which will induce germination, development of roots and shoots is simultaneous.
V. lentago[2]	Moist sand	150–270 at 68–86 + 60–120 at 41	----------	See remarks for *V. acerifolium.*
V. opulus	Moist sand or peat	60– 90 at 68–86 + 30– 60 at 41	----------	----------
V. prunifolium	Moist peat or sand plus peat	150–200 at 68–86 + 30– 45 at 41	----------	----------
V. trilobum[2]	Moist sand	90–150 at 68–86 + 60 at 41	----------	See remarks for *V. acerifolium.*

[1] Temperatures in this column to be alternated diurnally.
[2] Treatment only suggested for this species; experimental data not complete.

TABLE 225.—*Viburnum: Recommended germination test duration, and summary of germination data*

Species	Test duration recommended: Untreated seed	Test duration recommended: Stratified seed	Germinative capacity: Low	Germinative capacity: Average	Germinative capacity: High	Basis, tests	Potential germination
	Days	*Days*	*Percent*	*Percent*	*Percent*	*Number*	*Percent*
V. acerifolium	240+	[1]60	0	32	81	5	82–100
V. alnifolium	----------	[1]100	26	----------	60	2	----------
V. cassinoides[2]	90+	90–120	60	----------	74	2	74– 80
V. dentatum	500	[1]60	53	----------	85	2	----------
V. lentago	270+	60–120	25	51	77	3	80– 85
V. opulus	----------	60	1	60	87	3+	----------
V. prunifolium	300	[1]60	55	----------	95	2	----------
V. trilobum	210+	60– 90	0	33	85	5	17–100

[1] Suggested treatment; experimental data not complete.
[2] Germinative energy 67 percent in 50 days.

VITEX AGNUS-CASTUS L. Lilac chastetree

(Verbena family—Verbenaceae)

Also called chaste-tree, hemp-tree, monks' pepper-tree. Includes var. *alba* West., var. *latifolia* (Mill.) Loud., and f. *rosea* Rehd.

Distribution and Use. — Native to southern Europe and western Asia, the lilac chastetree is a shrub useful for ornamental, wildlife food, and shelter-belt purposes. It is one of the few out of about 60 species in the genus which is native to the temperate regions; most of them grow in the tropical and subtropical regions of both hemispheres. It was introduced into cultivation in 1570.

Seeding Habits.—The seeds are small four-celled, roundish stones, about one-eighth inch long, brownish to purplish brown in color, frequently one-half to two-thirds covered with a lighter colored membranous cap (fig. 276). The seed is without endosperm. The fragrant lilac or pale violet flowers (white in var. *alba* and pink in f. *rosea*), occurring in dense spikes, bloom from June to September; the pungent-flavored fruits, small drupes about one-eighth to three-sixteenth inch across, ripen in late summer or early fall. Seed crops are generally abundant.

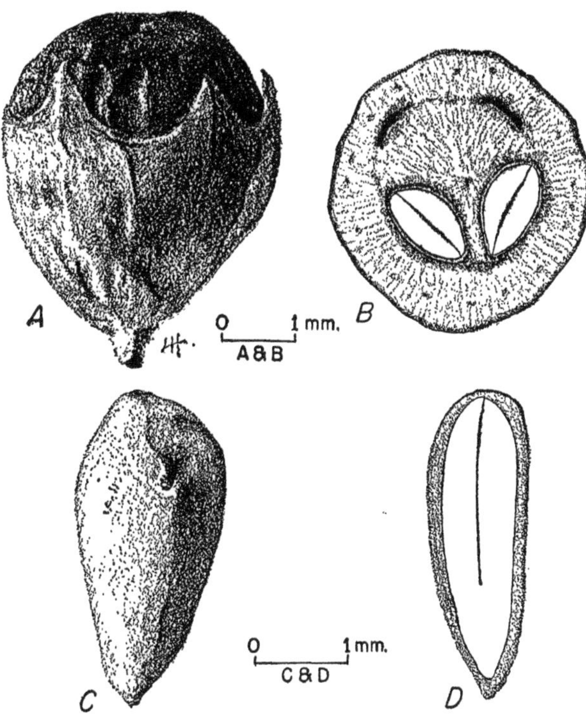

Figure 276.—Seed views of *Vitex agnus-castus: A,* Exterior view of fruit; *B,* cross section of fruit showing two seeds; *C,* exterior view of seed; *D,* longitudinal section of seed without endosperm.

Collection, Extraction, and Storage. — The fruits may be gathered in late summer or early fall by picking them from the shrubs by hand, or by flailing or stripping them onto canvas. Seed can be removed by running the fruits dry through a macerator and fanning to remove impurities. Yield per 100 pounds of fruit is about 75 pounds of cleaned seed. Number of cleaned seeds per pound ran from 34,000 to 59,000 in 4 samples. Commercial seed averages about 80 percent in purity (2 samples), 55 percent in soundness (4 samples), and costs $1.50 to $2.50 per pound. In one test seed was stored in moist sand and peat at 41° F. for 1 year with no loss of viability.

Germination. — The seeds exhibit dormancy which may be overcome by stratification in moist sand and peat for 90 days at about 41° F. Germination tests should be made in sand flats for 40 days at 70° (night) to 85° (day) using 1,000 stratified seeds per test. Germinative energy with stratified seed, 18 to 60 percent in 10 to 22 days (3 tests); germinative capacity with untreated seed, 0.4 percent in 71 days (1 test), with stratified seed, 20 to 72 percent (3 tests).

Nursery and Field Practice. — Seed of lilac chastetree should be sown in the spring, and covered with one-fourth inch of soil. On the average, about 16 percent of the viable seed sown produce usable 1–0 seedlings. This species is hardy as far north as New York in sheltered positions. It is quite drought-resistant and will grow on almost any kind of soil but prefers rather dry, sunny situations. It frequently is propagated by greenwood cuttings under glass and by layers.

VITIS RIPARIA Michx. Riverbank grape

(Grape family—Vitaceae)

Botanical syns.: ***V. vulpina*** **auth., not L.,** ***V. cordifolia*** **var.** ***riparia*** **(Michx.) A. Gray**

Distribution and Use.—Riverbank grape occurs on alluvial soils along streams and in other moist places from Nova Scotia to Manitoba, south to West Virginia, Arkansas, Colorado, and Texas. It is a deciduous, climbing or trailing vine, often ascending high into tall trees, and is valuable chiefly for the food and cover it affords game. It is a useful honey plant. The fruit is also used by man in making wine and preserves. Introduced into cultivation in 1656, it is often grown in this country and in Europe for its attractive foliage and fragrant staminate flowers.

Seeding Habits. — The small flowers which are polygamo-dioecious, greenish in color and borne in clusters, open from May to June; the fruit, a cluster of blue-black to purplish-black, juicy and sour berries with heavy, blue bloom, ripens from September to November, even after frost, and may persist into winter. Individual berries are one-fourth to three-eighths inch in diameter and contain one to four seeds (fig. 277). Dispersal is usually by birds or mammals.

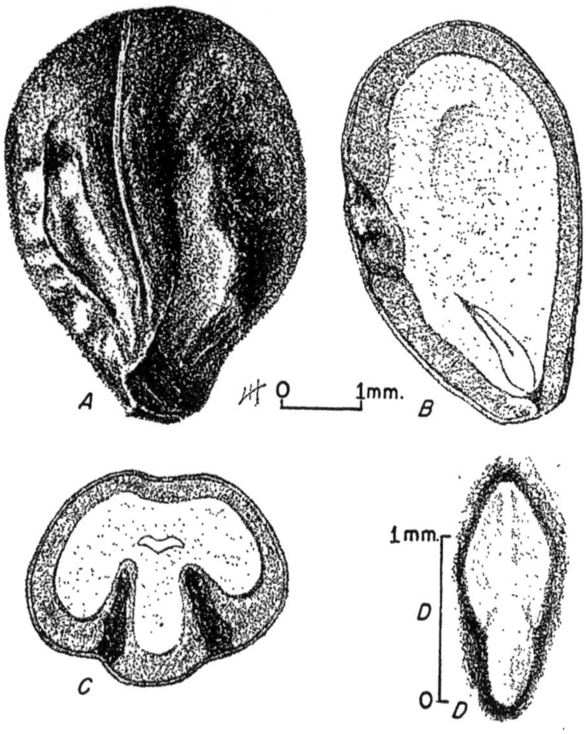

Figure 277.—Seed views of *Vitis riparia: A,* Exterior view; *B,* longitudinal section; *C,* cross section; *D,* embryo.

Collection, Extraction, and Storage. — The fruit may be collected as soon as it is ripe by stripping the clusters from the vines or, later in the season, by shaking the vines and collecting the fruit on canvas. Unless the seed is to be extracted soon, the fruit should be piled in rather shallow layers to prevent heating. The seed is extracted by running the fruit with water through a macerator or hammer mill and allowing the pulp and the empty and wormy seeds to float away. One hundred pounds of fruit will yield about 10 to 12 pounds of clean seed. Clean seeds per pound (6 samples): low, 11,300; average, 15,200; high,

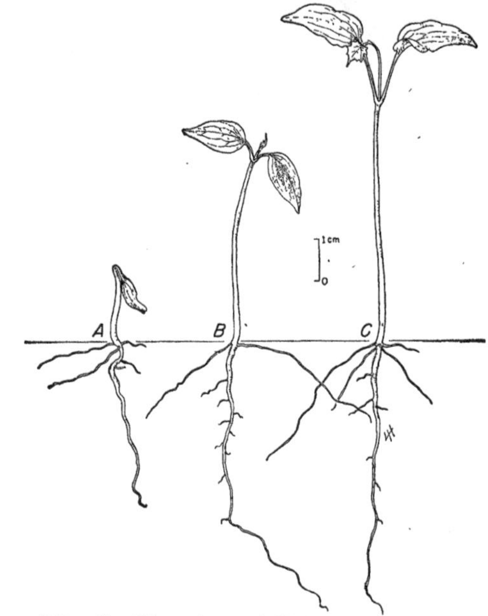

Figure 278.—Seedling views of *Vitis riparia: A,* At 2 days; *B,* at 4 days; *C,* at 17 days.

17,200. Purity, 99 percent; soundness, 91 percent (6 samples). Commercial seed costs $2.50 per pound. Seed stored in a sealed container, about 7 months in the form of dried berries, showed no loss in viability after 26 months at 41° F.

GERMINATION.—Seed of riverbank grape has a dormant embryo and will germinate very slowly unless stratified. Suggested treatment, stratification for 60 to 120 days at 41° F. Germination is epigeous (fig. 278). Germination tests may be run in sand flats at temperatures alternating diurnally from 68° to 86°. Germinative energy: 41 to 96 percent in 13 to 16 days; germinative capacity (6 samples): low, 43 percent; average, 82 percent; high, 96 percent.

NURSERY PRACTICE.—Seed of this species should be fall sown or stratified for about 60 days prior to spring sowing. Nursery germination, 90 percent. This species can also be propagated from cuttings or by layering.

ZANTHOXYLUM AMERICANUM Mill. Common prickly-ash

(Rue family—Rutaceae)

Also called toothache-tree.

Distribution and Use.—Occurring along stream banks and the edges of woodlands from Quebec to North Dakota, south to Virginia and Oklahoma, the common prickly-ash is an erect, prickly shrub or small tree. It has aromatic foliage and fruit and often forms dense thickets which provide food and cover for game. It is also useful as a honey plant. The fruit and bark are sometimes used in the drug trade. Introduced into cultivation about 1740, it is occasionally used for ornamental purposes.

Seeding Habits. — The rather inconspicuous, greenish, dioecious flowers are borne in small axillary clusters from April to May; the fruit, a reddish, dehiscent capsule containing one or two black, shining seeds (fig. 279), about one-eighth inch long, ripens from June to August.

Collection, Extraction, and Storage.—The capsules should be stripped from the branches as soon as they begin to open and release the seeds. Leather gloves should be worn as protection against the prickles. After collection, the fruit should be spread out to dry for several days in a warm room during which period practically all the capsules with good seed will open and the seed fall out. The seed can be separated from the empty capsules by screening or fanning. Clean seeds per pound (2 samples) : 22,100 to 32,900; soundness, 85 percent (3 samples). Commercial seed costs $3.40 per pound. Optimum storage conditions are not known; however, seed stored in sealed containers at 41° F. showed practically no loss in germinability in 25 months.

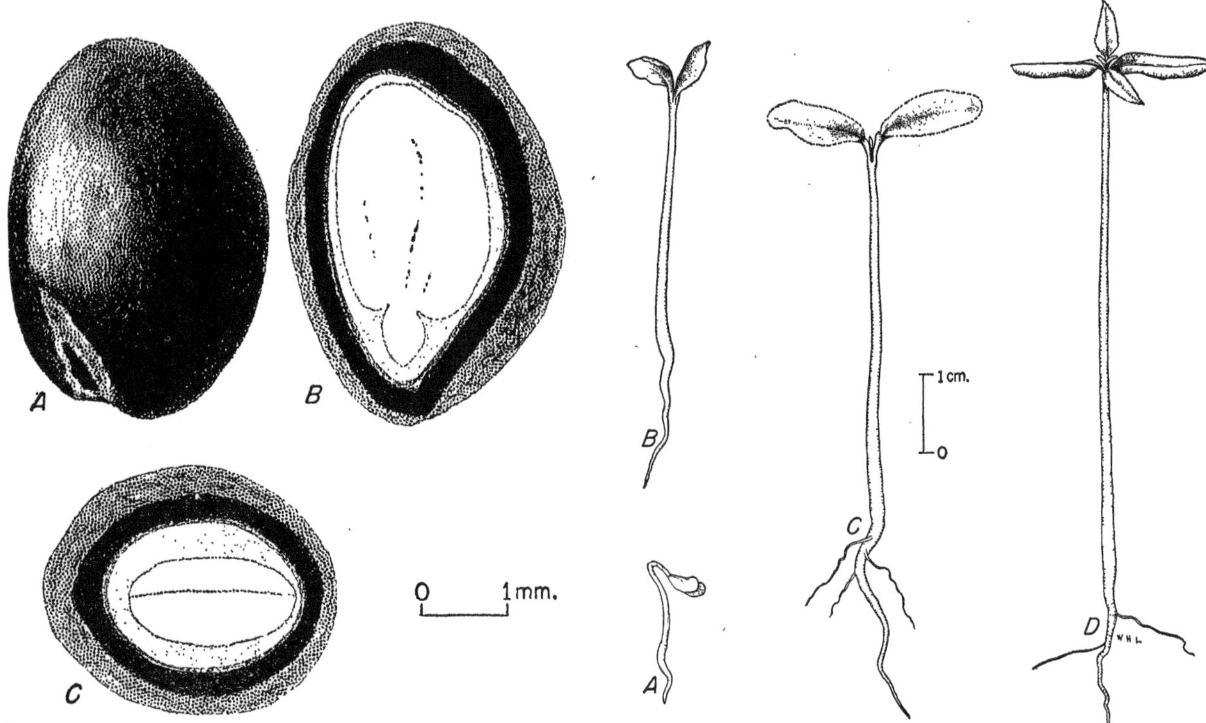

Figure 279.—Seed views of *Zanthoxylum americanum: A*, Exterior view; *B*, longitudinal section, showing large embryo; *C*, cross section.

Figure 280.—Seedling views of *Zanthoxylum americanum: A*, At 1 day; *B*, at 3 days; *C*, at 13 days; *D*, at 18 days.

GERMINATION.—Prickly-ash seed is slow to germinate, apparently because of a dormant embryo. Stratification in sand for 120 days at 41° F. has given a fair amount of germination. There is also some indication that the seed coat may have an inhibiting effect. Treatment suggested: stratification for 30 to 60 days at 68° to 86° F. (alternated diurnally), plus 120 days at 41°. Germination is epigeous (fig. 280). Germination tests may be run in sand flats; alternating diurnal temperatures of 68° to 86° and of 50° to 77° appear to be equally successful. Constant temperatures of 50° and 41° gave no results. Germinative energy, 20 percent in 20 days. Germinative capacity, 24 percent; potential germination, 95 percent (1 sample). Seed kept in soil at 68° to 86° will spoil and decay within a few months.

NURSERY PRACTICE.—Sowing the seed as soon as possible after collection in the late summer is suggested, with most of the germination to be looked for the next spring. Prickly-ash can also be propagated from root cuttings and suckers.

ZIZIPHUS JUJUBA Mill. Common jujube

(Buckthorn family—Rhamnaceae)

Botanical syns.: *Z. vulgaris* Lam., *Z. sativa Gaertn.* Includes var. *inermis* (Bunge) Rehd.

Distribution and Use.—Native from southeastern Europe to southern and eastern Asia, common jujube is a small to moderate-sized deciduous tree valued for its edible fruit, for wildlife food, and for shelter-belt purposes. In India it is used for fuel, small timber, and cattle fodder, and provides food for the tasar silkworm and the lac insect. It has some medicinal value. Common jujube is one of some 40 species in the genus found chiefly in the tropical or subtropical regions of both hemispheres. It was introduced into cultivation about 1640, and has become naturalized in Alabama.

Seeding Habits.—The seed is a deeply furrowed, two-celled and two-seeded oblong stone, light reddish brown to deep gray in color, three-fourths to 1 inch long, pointed at both ends, and has a long spurlike point at one end (fig. 281). The yellow flowers bloom from April to May (to October in India); the reddish-brown drupe, about three-fourths to 1 inch long one-half to three-fourths inch in wild forms), ripens from September to November (to March in India) and becomes black if allowed to stay on the trees. Dispersal is chiefly by birds and mammals which eat the fruit. Good fruit crops are borne annually. Since it grows over a wide range and is known to vary considerably in size and form, it is likely that geographic strains have developed.

Collection, Extraction, and Storage.—The fruits should be picked by hand from the trees in the fall, run wet through a macerator with water, and the pulp floated off or screened out. The seed should then be dried. Yield of cleaned seed per 100 pounds of fruit averages 25 to 35 pounds. Cleaned seed per pound (13 samples): low, 600; average, 750; high, 1,150. One sample of commercial seed averaged 99 percent in purity and 64 percent in soundness. Commercial seed costs $1.75 to $2.50 per pound. It is suggested that the seed be stored dry in sealed containers at 41° F. How long viability will be retained in storage is not known.

Germination.—Natural germination is best on exposed mineral soil where the seed has been partly covered. Germination takes place in the first, or sometimes the second, spring and is epigeous. There are usually 2 seedlings per stone. The seed exhibit dormancy due probably to both a hard seed coat and internal conditions of the embryo. Both stratification in moist sand for 60 to 90 days at 41° F. and scarification for 6 hours have succeeded fairly well in breaking dormancy. Germination tests can be run in sand flats for 50 days at 70° (night) to 85° (day), using 400 pretreated seeds per test. Germinative capacity averaged 31 percent for stratified seed (two tests) and 39 percent for scarified seed (one test); it was 70 percent when the seed was removed from the shell and stratified (one test).

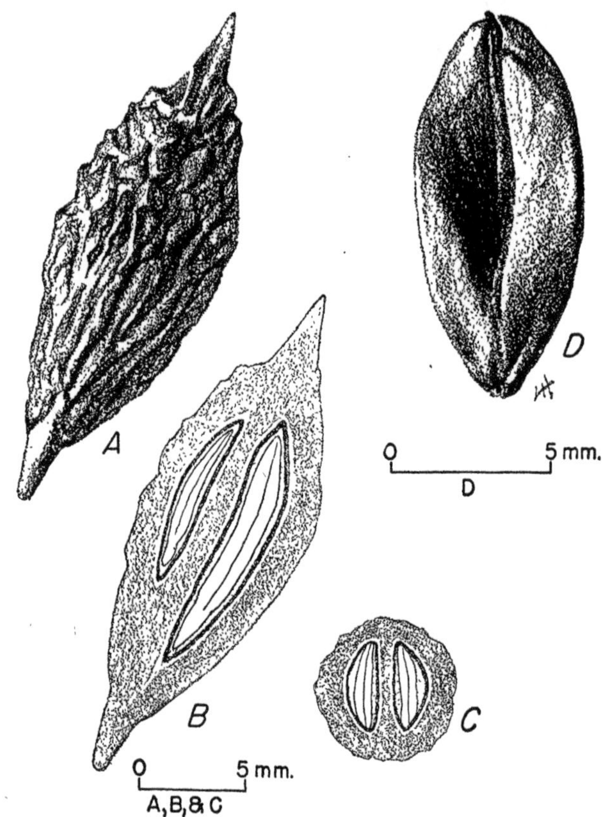

Figure 281.—Seed views of *Ziziphus jujuba: A,* Exterior view of stone; *B,* longitudinal section of stone; *C,* cross section of stone; *D,* exterior view of seed.

Nursery and Field Practice.—Jujube seed should be sown in drills in the fall, or seed stratified 90 days may be sown in the spring. They should be covered with 1 inch of soil. About one-third of the viable seed sown produce usable 1–0 seedlings. The jujube is suited only for the Southern and Southwestern States. It thrives on most kinds of soil except clays and in wet locations. It is also reproduced by offsets. One-year-old or pruned two-year-old seedling stock is suitable for field planting. In its natural habitat, rainfall varies from 5 to 90 inches per year and temperatures range between 20° and 120° F.

BIBLIOGRAPHY

ABIES Mill.

(1) ALLEN, G. S.
1941. LIGHT AND TEMPERATURE AS FACTORS IN THE GERMINATION OF THE SEED OF DOUGLAS FIR [PSEUDOTSUGA TAXIFOLIA (LAMB.) BRITT.]. Forestry Chron. 17(2): 99-109, illus.

(2) BARTON, L. V.
1930. HASTENING THE GERMINATION OF SOME CONIFEROUS SEEDS. Amer. Jour. Bot. 17: 88-115, illus.

(3) CHITTENDEN, A. K.
1927. FOREST PLANTING IN MICHIGAN. Mich. Agr. Expt. Sta. Spec. Bul. 163, 24 pp., illus.

(4) DALLIMORE, W., AND JACKSON, A. B.
1923. A HANDBOOK OF CONIFERAE, INCLUDING GINKGOACEAE. 570 pp., illus. New York and London.

(5) DAMBERG, E. F.
1915. LESOVODI-LÏUBITELI. RUKOVODSTVO KI SBORY DREVESNYKH SEMÏAN, POSEVY I POSADKE LESNYKH POROD. [FRIENDS OF FORESTRY. GUIDE FOR SEED COLLECTION AND PLANTING AND SEEDING OF FOREST SPECIES.] 64 pp., illus. Petrograd.

(6) DAVIS, W. C., WRIGHT, E., AND HARTLEY, C.
1942. DISEASES OF FOREST-TREE NURSERY STOCK. Fed. Security Agency Civilian Conserv. Corps. Forestry Pub. 9, 79 pp., illus.

(7) FISHER, G. M.
1935. COMPARATIVE GERMINATION OF TREE SPECIES ON VARIOUS KINDS OF SURFACE-SOIL MATERIAL IN THE WESTERN WHITE PINE TYPE. Ecology 16: 606-611.

(8) HAGEM, O.
1931. FORSØK MED VESTAMERIKANSKE TRAESLAG. Vestland. Forstl. Forsøkssta. Meddel. 12(4): 137-142.

(9) HEIT, C. E., AND ELIASON, E. J.
1940. CONIFEROUS TREE SEED TESTING AND FACTORS AFFECTING GERMINATION AND SEED QUALITY. N. Y. State Agr. Expt. Sta. Tech. Bul. 255, 45 pp., illus.

(10) HOFMANN, J. V.
1925. BEST TIME FOR SOWING SILVER FIR IN THE NURSERY. Jour. Agr. Res. 31: 261-266, illus.

(11) ISAAC, L. A.
1934. COLD STORAGE PROLONGS THE LIFE OF NOBLE FIR SEED AND APPARENTLY INCREASES GERMINATIVE POWER. Ecology 15: 216-217.

(12) LARSEN, J. A.
1922. SOME CHARACTERISTICS OF SEEDS OF CONIFEROUS TREES FROM THE PACIFIC NORTHWEST. Natl. Nurseryman 30: 246-249.

(13) MÜLLER, K. M.
1935. ABIES GRANDIS UND IHRE KLIMARASSEN. Teil I. Deut. Dendrol. Gessell. Mitt. 47: 54-123, illus.

(14) ———
1936. ABIES GRANDIS UND IHRE KLIMARASSEN. Teil II. Deut. Dendrol. Gesell. Mitt. 48: 82-127, illus.

(15) OLSON, D. S.
1930. GROWING TREES FOR FOREST PLANTING IN MONTANA AND IDAHO. U. S. Dept. Agr. Cir. 120, 92 pp., illus.

(16) PEARSON, G. A.
1931. FOREST TYPES IN THE SOUTHWEST AS DETERMINED BY CLIMATE AND SOIL. U. S. Dept. Agr. Tech. Bul. 247, 144 pp., illus.

(17) PETTIS, C. R.
1909. HOW TO GROW AND PLANT CONIFERS IN THE NORTHEASTERN STATES. U. S. Dept. Agr., Forest Serv. Bul. 76, 36 pp., illus.

(18) RAFN, J.
1915. THE TESTING OF FOREST SEEDS DURING 25 YEARS, 1887-1912. 91 pp., illus. Copenhagen.

(19) ——— & SON.
[n.d.] SKOVFRÖKONTORET'S FRÖANALYSER GENNEM 40 AAR, 1887-1927, UDFÖRT PAA STATSFRÖKONTROLLEN I KÖBENAVN. 5 pp. Copenhagen.

(20) SHOW, S. B.
1918. THE RELATION OF GERMINATION IN THE GREENHOUSE AND NURSERY. Jour. Forestry 16: 319-328.

(21) ———
1930. FOREST NURSERY AND PLANTING PRACTICE IN THE CALIFORNIA PINE REGION. U. S. Dept. Agr. Cir. 92, 75 pp., illus.

(22) SUDWORTH, G. B.
1900. THE FOREST NURSERY: COLLECTION OF TREE SEEDS AND PROPAGATION OF SEEDLINGS. U. S. Dept. Agr., Div. Forestry Bul. 29, 63 pp., illus.

(23) ———
1908. FOREST TREES OF THE PACIFIC SLOPE. 441 pp., illus. Washington, D. C.

(24) ———
1916. THE SPRUCE AND BALSAM FIR TREES OF THE ROCKY MOUNTAIN REGION. U. S. Dept. Agr. Bul. 327, 43 pp., illus.

(25) TILLOTSON, C. R.
1917. REFORESTATION ON THE NATIONAL FORESTS. U. S. Dept. Agr. Bul. 475, 63 pp., illus.

(26) ———
1925. GROWING AND PLANTING CONIFEROUS TREES ON THE FARM. U. S. Dept. Agr. Farmers' Bul. 1453, 38 pp., illus.

(27) TOUMEY, J. W., AND KORSTIAN, C. F.
1942. SEEDING AND PLANTING IN THE PRACTICE OF FORESTRY. Ed. 3, 520 pp., illus. New York.

(28) ——— AND STEVENS, C. L.
1928. THE TESTING OF CONIFEROUS TREE SEEDS AT THE SCHOOL OF FORESTY, YALE UNIVERSITY, 1906-1926. Yale Univ., School Forestry Bul. 21, 46 pp., illus.

(29) WOOLSEY, T. S., JR.
1920. STUDIES IN FRENCH FORESTRY. 550 pp., illus. New York.

ACACIA Mill.

(30) DREES, E. M.
1941. KIEMPROEVEN EN KIEMPLANTEN. I. ACACIA. Tectona 34: 1-45, illus. [In Dutch. English summary, p. 29.]

(31) EWART, A. J.
1908. ON THE LONGEVITY OF SEEDS. Roy. Soc. Victoria, Proc. 21, 210 pp.

(32) OSBORN, J. B., AND OSBORN, E.
1931. STUDIES ON THE GERMINATION OF SEED OF THE BLACK WATTLE (ACACIA MOLLISSIMA) AND THE GREEN WATTLE (ACACIA DECURRENS NORMALIS). So. African Jour. Sci. 28: 222-237.

(33) PHILLIPS, J. F. V.
1928. THE BEHAVIOR OF ACACIA MELANOXYLON R. BR. ("TASMANIAN BLACKWOOD") IN THE KNYSNA FOREST: AN ECOLOGICAL STUDY. Roy. Soc. So. Africa, Cape Town, Trans. 16: 31-41.

(34) SHINN, C. H.
1913. AN ECONOMIC STUDY OF ACACIAS. U. S. Dept. Agr. Bul. 9, 38 pp., illus.

ACER L.

(35) ANONYMOUS.
1907. SUGAR MAPLE (ACER SACCHARUM). U. S. Dept. Agr., Forest Serv. Cir. 95, 4 pp.

(36) CROCKER, W.
1925. SEEDS: THEIR TRICKS AND TRAITS. N. Y. Bot. Gard. Jour. 26: 178-187.

891183°—50——25

(37) JONES, H. A.
1920. PHYSIOLOGICAL STUDY OF MAPLE SEEDS. Bot. Gaz. 69: 127-152, illus.
(38) MIROV, N. T., AND KRAEBEL, C. J.
1939. COLLECTING AND HANDLING SEEDS OF WILD PLANTS. Civilian Conserv. Corps Forestry Pub. 5, 42 pp., illus.
(39) ROE, E. I.
1941. EFFECT OF TEMPERATURE ON SEED GERMINATION. Jour. Forestry 39: 413-414.
(40) SCHAAF, M.
1937-38. DIVISION OF FORESTRY. Mich. Conserv., Ninth Bien. Rpt. 1937-38: 267-290, illus.
(41) SUS, N. I.
1925. PITOMNIK. [THE FOREST NURSERY.] 227 pp., illus. Moscow.
(42) TESOV, N. E., AND TOLSKI, A. P.
1935. TREE SEEDS OF THE SOVIET UNION.
(43) TILLOTSON, C. R.
1921. GROWING AND PLANTING HARDWOOD SEEDLINGS ON THE FARM. U. S. Dept. Agr. Farmers' Bul. 1123, 29 pp., illus.
See also (3), (6), (18), (19), (22), (23), and (27).

AESCULUS L.

(44) JENKINS, E. M.
1936. SEED PRACTICES IN NURSERY. Amer. Nurseryman 63(11): 9-11.
(45) ZEDERBAUER, E.
1910. VERSUCHE ÜBER AUFBEWAHRUNG VON WALDSÄMEREIEN. Centbl. f. das Gesam. Forstw. 36: 116-121.
See also (23), (27), and (38).

AILANTHUS ALTISSIMA (Mill.) Swingle

(46) BAILEY, L. H.
1939. THE STANDARD CYCLOPEDIA OF HORTICULTURE. 3 v. 3639 pp., illus. New York.
(47) BARTON, L. V.
1939. EXPERIMENTS AT BOYCE THOMPSON INSTITUTE ON GERMINATION AND DORMANCY IN SEEDS. Sci. Hort. [Wye, Kent] 7: 186-193, illus.
(48) ENGSTROM, H. E., AND STOECKELER, J. H.
1941. NURSERY PRACTICE FOR TREES AND SHRUBS SUITABLE FOR PLANTING ON THE PRAIRIE-PLAINS. U. S. Dept. Agr. Misc. Pub. 434, 159 pp., illus.
(49) THORNBER, J. J.
1904. SEED GERMINATION. Ariz. Agr. Expt. Sta. Ann. Rpt. 1904: 491-492.
See also (41).

ALBIZIA JULIBRISSIN Durazz.

(50) TROUP, R. S.
1921. THE SILVICULTURE OF INDIAN TREES. 3 v. 1195 pp., illus. Oxford.
See also (31).

ALNUS Mill.

(51) MACDONALD, J.
1935. NURSERY INVESTIGATIONS. Forestry 9: 24-41, illus. London.
(52) NICHOLS, G. E.
1934. THE INFLUENCE OF EXPOSURE TO WINTER TEMPERATURES UPON SEED GERMINATION IN VARIOUS NATIVE AMERICAN PLANTS. Ecology 15: 364-373.
(53) STEBLER, F. G., ET AL.
1908. DREISSIGSTER JAHRESBERICHT, DER SCHWEIZ. SAMENUNTERSUCHUNGS- UND VERSUCHSANSTALT IN ZÜRICH. 35 pp., illus.
See also (5), (18), (19), (38), and (48).

AMELANCHIER Medic.

(54) CROCKER, W., AND BARTON, L. V.
1931. AFTER-RIPENING, GERMINATION, AND STORAGE OF CERTAIN ROSACEOUS SEEDS. Boyce Thompson Inst. Contrib. 3: 385-404, illus.
(55) HARGRAVE, P. D.
1937. SEED GERMINATION OF THE SASKATOON AND PIN-CHERRY. Sci. Agr. 17: 736-739.
See also (48).

AMORPHA L.

(56) GORSHENIN, N. M.
1941. AGROLESOMELIORATSIIA. [AGRO-FOREST MELIORATION.] 392 pp., illus. Moscow.
(57) HUTTON, M. E., AND PORTER, R. H.
1937. SEED IMPERMEABILITY AND VIABILITY OF NATIVE AND INTRODUCED SPECIES OF LEGUMINOSAE. Iowa State Col. Jour. Sci. 12: 5-24, illus.
(58) ROGERS, M. S.
1931. SEED GERMINATION EXPERIMENTS WITH THE FRAGRANT DWARF INDIGO. Colorado Univ., Studies 18: 205-213.
See also (44).

APLOPAPPUS PARISHII (Greene) Blake

(59) SCIENCE SERVICE.
1942. WILD RUBBER ON WASTELANDS IN THE WEST. Sci., Sup. 95(2457): 6.

ARALIA L.

(60) REHDER, A.
1940. MANUAL OF CULTIVATED TREES AND SHRUBS. Ed. 2, 996 pp., illus. New York.
See also (46).

ARBUTUS MENZIESII Pursh

See (46).

ARCTOSTAPHYLOS Adans.

(61) GIERSBACH, J.
1937. GERMINATION AND SEEDLING PRODUCTION OF ARCTOSTAPHYLOS UVA-URSI. Boyce Thompson Inst. Contrib. 9: 71-78, illus.

ARONIA Med.

See (46) and (54).

ASIMINA TRILOBA (L.) Dunal

(62) GOULD, H. P.
1939. THE NATIVE PAPAW. U. S. Dept. Agr. Leaflet 179, 6 pp.
(63) ZIMMERMAN, G. A.
1941. HYBRIDS OF THE AMERICAN PAPAW. Jour. Hered. 32: 83-91, illus.
See also (46) and (47).

ATRIPLEX CANESCENS James

(64) WILSON, C. P.
1928. FACTORS AFFECTING THE GERMINATION AND GROWTH OF CHAMIZA (ATRIPLEX CANESCENS). N. Mex. Agr. Expt. Sta. Bul. 169, 29 pp., illus.

BERBERIS L.

(65) CHADWICK, L. C.
1936. IMPROVED PRACTICES IN PROPAGATION BY SEED. Amer. Nurseryman 62(8): [3]-4, (9): 5-6, (10): 7-8, (12): [3]-9.

(66) Davis, O. H.
1927. Germination and early growth of Cornus florida, Sambucus canadensis, and Berberis thunbergii. Bot. Gaz. 84: 225-263, illus.
(67) Kern, F. D.
1921. Observations of the dissemination of the barberry. Ecology 2: 211-214.
(68) Morinaga, T.
1926. Effect of alternating temperatures upon the germination of seeds. Amer. Jour. Bot. 13: 141-158, illus.

BETULA L.

(69) Dana, S. T.
1909. Paper birch in the northeast. U. S. Dept. Agr., Forest Serv. Cir. 163, 37 pp., illus.
(70) Joseph, H. C.
1929. Germination and vitality of birch seeds. Bot. Gaz. 87: 127-151, illus.
(71) Patterson, C. F., and Bunce, A. C.
1931. Rapid methods of determining the percentages of fertility and sterility in seeds of the genus Betula. Sci. Agr. 11: 704-708, illus.
(72) Weiss, F.
1926. Seed germination in the gray birch (Betula populifolia). Amer. Jour. Bot. 13: 737-742.
See also (3), (5), (18), (19), (27), and (52).

BUMELIA LANUGINOSA (Michx.) Pers.

(73) Clark, R.
1940. A hardy woody plant new to horticulture. Mo. Bot. Gard. Bul. 28: 216-220, illus.

CAMPSIS RADICANS (L.) Seem.

(74) Howard, W. L.
1915. An experimental study of the rest period in plants. Seeds. Mo. Agr. Expt. Sta. Res. Bul. 17, 58 pp.

CARAGANA ARBORESCENS Lam.

See (19), (41), (48), (56), and (74).

CARPENTERIA CALIFORNICA Torr.

See (46).

CARPINUS CAROLINIANA Walt.

(75) Sandahl, P. L.
1941. Seed germination. Parks & Recreation 24: 508.
See also (19) and (60).

CARYA Nutt.

(76) Anonymous.
1909. Big shellbark: king-nut hickory. U. S. Dept. Agr., Forest Serv. Silvic. Leaflet 50, 4 pp.
(77) ———
1909. Pignut hickory. U. S. Dept. Agr., Forest Serv. Silvic. Leaflet 48, 4 pp.
(78) ———
1909. Shagbark hickory. U. S. Dept. Agr., Forest Serv. Silvic. Leaflet 49, 4 pp.
(79) Barton, L. V.
1936. Seedling production in Carya ovata (Mill.) K. Koch, Juglans cinerea L., and Juglans nigra L. Boyce Thompson Inst., Contrib. 8: 1-5, illus.
(80) Brown, H. P.
1921. Trees of New York State, native and naturalized. N. Y. State Col. Forestry, Syracuse Univ., Tech. Pub. 15, 401 pp., illus.
(81) Hough, R. B.
1924. Handbook of the trees of the northern states and Canada. Ed. 3, 470 pp., illus. New York.
(82) Sargent, C. S.
1933. Manual of the trees of North America. Ed. 2, 910 pp., illus. New York.
See also (27), (48), and (65).

CASTANEA DENTATA (Marsh.) Borkh.

(83) Anonymous.
1907. Chestnut (Castanea dentata). U. S. Dept. Agr., Forest Serv. Cir. 71, 4 pp.
(84) Clapper, R. B.
1943. New chestnuts for our forests? Amer. Forests 49: 331-333, 365, illus.
(85) Kains, M. G., and McQuesten, L. M.
1938. Propagation of plants. A complete guide for professional and amateur growers of plants by seeds, layers, grafting, and budding, with chapters on nursery and greenhouse management. 555 pp., illus. New York.
See also (27).

CASTANOPSIS (D. Don) Spach

See (46).

CATALPA Scop.

(86) Anonymous.
1907. Hardy catalpa (Catalpa speciosa). U. S. Dept. Agr., Forest Serv. Cir. 82, 8 pp.
(87) Lamb, G. N.
1915. A calendar of the leafing, flowering, and seeding of the common trees of the eastern United States. U. S. Monthly Weather Rev., Sup. 2, pp. 3-19, illus.
(88) Scott, C. A.
1911. The hardy catalpa in Iowa. Iowa Agr. Expt. Sta. Bul. 120: 310-325, illus.
See also (6), (18), (48), (60), and (82).

CEANOTHUS L.

(89) Quick, C. R.
1935. Notes on the germination of Ceanothus seeds. Madroño 3: 135-140.
(90) U. S. Forest Service.
1937. Range plant handbook. Washington, D. C.
See also (82).

CEDRUS Trew.

(91) Martin, E.
1934. Note sur le semis du cèdre en pépinière. Rev. des Eaux et Forêts 72: 777-779.
(92) Sen Gupta, J. N.
1936. Seed weights, plant percents, etc., for forest plants in India. Indian Forest Rec. (n.s.) 2: 175-221.
See also (4), (18), (19), (46), and (50).

CELASTRUS SCANDENS L.

(93) Joseph, H. C.
1928. Growing bittersweet (Celastrus scandens) from seed. Florists' Exch. 68: 499, illus.

CELTIS L.

(94) Steavenson, H. A.
1940. The hammer mill as an important nursery implement. Jour. Forestry 38: 356-361, illus.
See also (48).

CERCIS CANADENSIS L.

See (75) and (82).

CERCOCARPUS H. B. K.

See (60).

CHAMAECYPARIS Spach

(95) KORSTIAN, C. F., AND BRUSH, W. D.
1931. SOUTHERN WHITE CEDAR. U. S. Dept. Agr. Tech. Bul. 251, 76 pp., illus.

(96) LARSEN, C. S.
1937. THE EMPLOYMENT OF SPECIES, TYPES, AND INDIVIDUALS IN FORESTRY. [Denmark] K. Vet. og Landbohøjskole, Aarsskr. 1937: [69]-222, illus.

(97) SUDWORTH, G. B.
1913. BIENNIAL FRUCTIFICATION OF ALASKA CYPRESS. Rev. Forest Serv. Invest. 2: 7-8, illus.

See also (19) and (28).

CHILOPSIS LINEARIS (Cav.) Sweet

See (48).

CHIONANTHUS VIRGINICUS L.

(98) FLEMION, F.
1941. FURTHER STUDIES ON THE RAPID DETERMINATION OF THE GERMINATIVE CAPACITY OF SEEDS. Boyce Thompson Inst. Contrib. 11: 455-464.

(99) YERKES, G. E.
1932. PROPAGATION OF TREES AND SHRUBS. U. S. Dept. Agr. Farmers' Bul. 1567, 52 pp., illus. (Revised.)

See also (19).

CHRYSOTHAMNUS NAUSEOSUS (Pursh) Britton

See (59).

CLEMATIS L.

See (18) and (60).

CORNUS L.

(100) ADAMS, J.
1927. THE GERMINATION OF THE SEEDS OF SOME PLANTS WITH FLESHY FRUITS. Amer. Jour. Bot. 14: 415-428.

(101) DAVIS, O. H.
1926. GERMINATION OF SEEDS OF CERTAIN HORTICULTURAL PLANTS. Florists' Exch. 63: 917, 922.

See also (6), (27), (38), (40), (47), (52), (65), (66), (75), and (94).

CORYLUS L.

(102) SHOEMAKER, J. S., AND HARGRAVE, P. D.
1936. PROPAGATING TREES AND SHRUBS FROM SEED. Alberta Univ., Col. Agr. Cir. 21, 22 pp.

See also (19), (40), and (56).

COTONEASTER B. Ehrh.

(103) GIERSBACH, J.
1934. AFTER-RIPENING AND GERMINATION OF COTONEASTER SEEDS. Boyce Thompson Inst. Contrib. 6: 323-338, illus.

See also (60), (65), and (75).

CRATAEGUS MOLLIS Scheele

(104) DAVIS, W. E., AND ROSE, R. C.
1912. THE EFFECT OF EXTERNAL CONDITIONS ON THE AFTER-RIPENING OF THE SEEDS OF CRATAEGUS MOLLIS. Bot. Gaz. 54: 49-62.

(105) FLEMION, F.
1938. BREAKING THE DORMANCY OF SEEDS OF CRATAEGUS SPECIES. Boyce Thompson Inst. Contrib. 9: 409-423, illus.

See also (100).

CUPRESSUS L.

(106) BHOLA, M. P.
1918. GERMINATION OF CUPRESSUS TORULOSA SEED. Indian Forester 44: 175-176.

(107) SUDWORTH, G. B.
1915. THE CYPRESS AND JUNIPER TREES OF THE ROCKY MOUNTAIN REGION. U. S. Dept. Agr. Bul. 207, 36 pp., illus.

(108) TOUMEY, J. W.
1921. ON THE LIABILITY [VIABILITY] OF TREE SEEDS AFTER STORAGE FOR TEN YEARS. Jour. Forestry 19: 814.

See also (2), (4), (19), (28), (50), and (92).

DENDROMECON RIGIDA Benth.

See (46).

DIOSPYROS VIRGINIANA L.

(109) BLOMQUIST, H. L.
1922. DORMANCY IN SEED OF PERSIMMON (DIOSPYROS VIRGINIANA). Elisha Mitchell Sci. Soc. Jour. 38: 14.

(110) FLETCHER, W. F.
1935. THE NATIVE PERSIMMON. U. S. Dept. Agr. Farmers' Bul. 685, 22 pp., illus. (Revised.)

See also (47), (48), and (74).

ELAEAGNUS L.

See (41) and (56).

EPIGAEA REPENS L.

(111) BARROWS, F. L.
1936. PROPAGATION OF EPIGAEA REPENS L. I. CUTTINGS AND SEEDS. Boyce Thompson Inst. Contrib. 8: 81-97, illus.

(112) COVILLE, F. V.
1911. THE USE OF ACID SOIL FOR RAISING SEEDLINGS OF THE MAYFLOWER, EPIGAEA REPENS. Science 33: 711-712.

(113) ———
1915. THE CULTIVATION OF THE MAYFLOWER. Natl. Geog. Mag. 27: 518-519, illus.

(114) LEMMON, R. S.
1935. THE TRAILING ARBUTUS IN HOME GARDENS. Horticulture 13: 101-102, illus. Boston.

EUCALYPTUS GLOBULUS Labill.

(115) METCALF, W.
1924. GROWTH OF EUCALYPTUS IN CALIFORNIA PLANTATIONS. Calif. Agr. Expt. Sta. Bul. 380, 61 pp., illus.

See also (31), (50), and (92).

EUONYMUS L.

See (41), (46), (56), and (60).

EUROTIA LANATA (Pursh) Moq.

(116) HILTON, J. W.
1941. EFFECTS OF CERTAIN MICRO-ECOLOGICAL FACTORS ON THE GERMINABILITY AND EARLY DEVELOPMENT OF EUROTIA LANATA. Northwest Sci. 15: 86-92, illus.

(117) WILSON, C. P.
1931. THE ARTIFICIAL RESEEDING OF NEW MEXICO RANGES. N. Mex. Agr. Expt. Sta. Bul. 189, 37 pp., illus.

FAGUS L.

(118) JOHANNSEN, W.
1921. ORIENTERENDE FORSØG MED OPBEVARING AF AGERN OG BØGEOLDEN. Forstl. Forsøgsv. i Danmark 5: 372-390. [In Danish. English summary, p. 390.]

(119) KAIGORODOV, D.
1907. DREVESNII KALENDAR EVROPEISKOI ROSSII. [TREE CALENDAR FOR EUROPEAN RUSSIA.] 2 pp., illus. St. Petersburg.

See also (3), (18), (27), and (29).

FALLUGIA PARADOXA (D. Don) Endl.

See (38) and (117).

FRAXINUS L.

(120) KALELA, A.
1937. ZUR SYNTHESE DER EXPERIMENTELLEN UNTERSUCHUNGEN ÜBER KLIMARASSEN DER HOLZARTEN. Inst. Forest Fenniae Commun. 26, 445 pp. Helsinki.

(121) LAKON, G.
1911. BEITRÄGE ZUR FORSTLICHEN SAMENKUNDE. II. ZUR ANATOMIE UND KEIMUNGSPHYSIOLOGIE DER ESCHENSAMEN. Naturw. Ztschr. Forst u. Landw. 9: 285-298.

(122) LESUEUR, A. D. C.
1924. THE COMMON ASH. Quart. Jour. Forestry 18: 217-227.

(123) MEULI, L. J., AND SHIRLEY, H. L.
1937. THE EFFECT OF SEED ORIGIN ON DROUGHT RESISTANCE OF GREEN ASH IN THE PRAIRIE-PLAINS STATES. Jour. Forestry 35: 1060-1062, illus.

(124) PERRY, G. S., and Coover, C. A.
1933. SEED SOURCE AND QUALITY. Jour. Forestry 31: 19-25.

(125) SOBOLEV, A.
1908. SEMENA LESNIÎA. [FOREST TREE SEED.] Entsiklopediâ Russkago Lesnogo Khoziâistva [Encyclopedia for Russian Forestry] 2: 1150.

(126) STEINBAUER, G. P.
1937. DORMANCY AND GERMINATION OF FRAXINUS SEEDS. Plant Physiol. 12: 813-824, illus.

(127) STERRETT, W. D.
1915. THE ASHES: THEIR CHARACTERISTICS AND MANAGEMENT. U. S. Dept. Agr. Bul. 299, 88 pp., illus.

See also (6), (19), (22), (27), (38), (40), (41), (43), (48), (56), and (92).

FREMONTODENDRON Cov.

(128) MCMINN, H. E., AND MAINO, E.
1937. AN ILLUSTRATED MANUAL OF PACIFIC COAST TREES, WITH LISTS OF TREES RECOMMENDED FOR VARIOUS USES ON THE PACIFIC COAST. Ed. 2, 409 pp., illus. Berkeley, Calif.

See also (23) and (46).

GARRYA FREMONTII Torr.

(129) MCMINN, H. E.
1936. THE GARRYAS OR SILKTASSEL BUSHES. Gard. Quart. 4(2): 20-22, illus.

See also (46).

GAULTHERIA PROCUMBENS L.

See (46) and (47).

GAYLUSSACIA BACCATA (Wangenh.) K. Koch

(130) DARROW, G. M.
1941. SEED SIZE IN BLUEBERRY AND RELATED SPECIES. Amer. Soc. Hort. Sci. Proc. 38: 438-440.

GLEDITSIA L.

(131) PAMMEL, L. H., AND KING, C. M.
1922. GERMINATION STUDIES OF SOME SHRUBS AND TREES. Iowa Acad. Sci. Proc. 29: 257-266, illus.

See also (27), (48), (60), and (82).

GYMNOCLADUS DIOICUS (L.) K. Koch

(132) ANONYMOUS.
1907. COFFEETREE (GYMNOCLADUS DIOICUS). U. S. Dept. Agr., Forest Serv. Cir. 91, 4 pp.

(133) WIESEHUEGEL, E. G.
1935. GERMINATING KENTUCKY COFFEE TREE. Jour. Forestry 33: 533-534.

See also (48).

HALESIA CAROLINA Ellis

(134) GIERSBACH, J., AND BARTON, L. V.
1932. GERMINATION OF SEEDS OF THE SILVER BELL, (HALESIA CAROLINA). Boyce Thompson Inst. Contrib. 4: 27-38, illus.

See also (46).

HAMAMELIS VIRGINIANA L.

See (65) and (75).

HIPPOPHAE RHAMNOIDES L.

(135) SHARROV, N.
1908. OVLYEPIKHA [SEA-BUCKTHORN.] Entsiklopediâ Russkago Lesnogo Khoziâistva 2: 83.

See also (19), (56), and (60).

ILEX L.

(136) ANONYMOUS.
1936. HOLLY (ILEX OPACA) AITON. Amer. Forests 42: 30-31, illus.

(137) COVILLE, P.
1932. GROWING CHRISTMAS HOLLY ON THE FARM. U. S. Dept. Agr. Farmers' Bul. 1693, 22 pp., illus.

(138) CROCKER, W.
1930. HARVESTING, STORAGE AND STRATIFICATION OF SEEDS IN RELATION TO NURSERY PRACTICE. Florists' Rev. 65: 43-46.

(139) GIERSBACH, J., AND CROCKER, W.
1929. GERMINATION OF ILEX SEEDS. Amer. Jour. Bot. 16: 854-855.

(140) IVES, S. A.
1923. MATURATION AND GERMINATION OF SEEDS OF ILEX OPACA. Bot. Gaz. 76: 60-77, illus.

(141) STODDARD, H. L.
1931. THE BOBWHITE QUAIL. 559 pp., illus. New York.

(142) ZIMMERMAN, P. W., AND HITCHCOCK, A. E.
1933. SELECTION, PROPAGATION, AND GROWTH OF HOLLY. Boyce Thompson Inst. Prof. Paper 1: 252-260, illus.

See also (52) and (60).

JUGLANS L.

(143) KAYLOR, J. F., AND RANDALL, L. R.
1931. METHODS OF COLLECTING, STRATIFYING, AND PLANTING BLACK WALNUTS IN INDIANA. Ind. Dept. Conserv., Forestry Bul. 12, 8 pp., illus.
(144) SUDWORTH, G. B.
1934. POPLARS, PRINCIPAL TREE WILLOWS AND WALNUTS OF THE ROCKY MOUNTAIN REGION. U. S. Dept. Agr. Tech. Bul. 420, 112 pp., illus.
(145) WITT, A. W.
1930. FURTHER OBSERVATIONS ON WALNUT GROWING IN ENGLAND. Royal Hort. Soc. Jour. 55: 257-265, illus.

See also (46), (48), (50), (56), (79), and (92).

JUNIPERUS L.

(146) EASTMAN, R. E.
1911. NEWS AND NOTES. Forestry Quart. 9: 173-174.
(147) JELLEY, M. E.
1937. EASTERN RED CEDAR. Jour. Forestry 35: 865-867.
(148) PACK, D. A.
1921. AFTER-RIPENING AND GERMINATION OF JUNIPERUS SEEDS. Bot. Gaz. 71: 32-60, illus.
(149) ———
1921. CHEMISTRY OF AFTER-RIPENING, GERMINATION, AND SEEDLING DEVELOPMENT OF JUNIPER SEEDS. Bot. Gaz. 72: 139-150.
(150) PHILLIPS, F. J., AND MULFORD, W.
1912. UTAH JUNIPER IN CENTRAL ARIZONA. U. S. Dept. Agr., Forest Serv. Cir. 197, 19 pp., illus.
(151) STEAVENSON, H. A., AND LANGFORD, J. K. R.
1940. SEED MATURITY IN WILD PLUM AND RED CEDAR. Soil Conserv. Serv., 13 pp., illus.
(152) SWINGLE, C. F.
1937. A PROMISING NEW CEDAR FOR EROSION CONTROL. Soil Conserv. 3: 75-78, illus.
(153) TILLOTSON, C. R.
1917. NURSERY PRACTICE ON THE NATIONAL FORESTS. U. S. Dept. Agr. Bul. 479, 86 pp., illus.
(154) TURNER, H. C.
1913. THE GERMINATION OF ALLIGATOR JUNIPER SEED. Rev. Forest Serv. Invest. 2: 49-52.

See also (22), (27), (92), (94), and (107).

KALMIA LATIFOLIA L.

See (46) and (52).

KOELREUTERIA PANICULATA Laxm.

See (46).

LARIX L.

(155) HUNT, S. S.
1932. EUROPEAN LARCH IN THE NORTHEASTERN UNITED STATES. A STUDY OF EXISTING PLANTATIONS. Harvard Forest Bul. 16, 45 pp., illus.
(156) LARSEN, J. A.
1916. SILVICAL NOTES ON WESTERN LARCH. Soc. Amer. Foresters Proc. 11: 434-440.
(157) MILLER, F. G., ET AL.
1927. THE IDAHO FOREST AND TIMBER HANDBOOK. Idaho Univ. School Forestry Bul. 22(22): 155 pp., illus.
(158) TKACHENKO, M. E., ASOSKOV, A. I., AND SINEV, V. N.
1939. OBSHEE LESOVODSTVO. [GENERAL FORESTRY.] 745 pp., illus. Leningrad.

See also (4), (9), (12), (15), (19), (28), (41), (53), (56), (82), (92), (119), and (120).

LIBOCEDRUS DECURRENS Torr.

(159) MITCHELL, J. A.
1918. INCENSE CEDAR. U. S. Dept. Agr. Bul. 604, 40 pp., illus.

See also (20) and (21).

LIGUSTRUM VULGARE L.

See (46), (56), and (65).

LINDĚRA BENZOIN (L.) Blume

(160) SCHROEDER, E. M.
1935. DORMANCY IN SEEDS OF BENZOIN AESTIVALE L. Boyce Thompson Inst. Contrib. 7: 411-419, illus.

See also (46).

LIQUIDAMBAR STYRACIFLUA L.

(161) CHITTENDEN, A. K.
1906. THE RED GUM. U. S. Dept. Agr., Forest Serv. Bul. 58, 56 pp., illus. (Revised.)

See also (87).

LIRIODENDRON TULIPIFERA L.

(162) GIERSBACH, J.
1929. THE EFFECT OF STRATIFICATION ON SEEDS OF LIRIODENDRON TULIPIFERA. Amer. Jour. Bot. 16: 855.
(163) GUARD, A. T., AND WEAN, R. E.
1941. SEED PRODUCTION IN TULIP POPLAR. Jour. Forestry 39: 1032-1033.
(164) HINSON, E.
1935. COLLECTION OF YELLOW POPLAR SEED. Jour. Forestry 33: 1007-1008.
(165) MCCARTHY, E. F.
1933. YELLOW POPLAR CHARACTERISTICS, GROWTH, AND MANAGEMENT. U. S. Dept. Agr. Tech. Bul. 356, 58 pp., illus.
(166) WEAN, R. E., AND GUARD, A. T.
1940. THE VIABILITY AND COLLECTION OF SEED OF LIRIODENDRON TULIPIFERA L. Jour. Forestry 38: 815-817, illus.
(167) YOUNG, L. J.
1919. GERMINATION OF YELLOW-POPLAR SEED. Jour. Forestry 17: 101.

See also (94) and (124).

LITHOCARPUS DENSIFLORUS (Hook & Arn.) Rehd.

(168) JEPSON, W. L.
1911. CALIFORNIA TANBARK OAK. PART I. TANBARK OAK AND THE TANNING INDUSTRY. U. S. Dept. Agr., Forest Serv. Bul. 75: 5-23, illus.

LONICERA L.

See (48), (52), (56), and (65).

MACLURA POMIFERA (Raf.) Schneid.

(169) ANONYMOUS.
1907. OSAGE ORANGE (TOXYLON POMIFERUM). U. S. Dept. Agr., Forest Serv. Cir. 90, 3 pp.
(170) BEILMANN, A. P.
1938. COMMON NATIVE TREES OF MISSOURI. III. OSAGE ORANGE [MACLURA POMIFERA (RAF.) SCHNEIDER]. Mo. Bot. Gard. Bul. 26: 75-78, illus.

MAGNOLIA L.

(171) AFANASIEV, M.
1937. A PHYSIOLOGICAL STUDY OF DORMANCY IN SEED OF MAGNOLIA ACUMINATA. N. Y. (Cornell) Agr. Expt. Sta. Mem. 208, 37 pp., illus.

(172) EVANS, C. R.
1933. GERMINATION BEHAVIOR OF MAGNOLIA GRANDIFLORA. Bot. Gaz. 94: 729-754.

See also (19), (27), (43), and (65).

MALUS Mill.

(173) CROCKER, W.
1928. STORAGE, AFTER-RIPENING, AND GERMINATION OF APPLE SEEDS. Amer. Jour. Bot. 15: 625-626.

See also (41), (46), (48), (56), and (75).

MELIA AZEDARACH L.

(174) PAMMEL, L. H., AND KING, C. M.
1924. FURTHER STUDIES OF THE GERMINATION OF WOODY PLANTS. Iowa Acad. Sci. Proc. 31: 157-167, illus.

See also (50) and (92).

MORUS L.

(175) ANONYMOUS.
1907. RUSSIAN MULBERRY (MORUS ALBA TATARICA). U. S. Dept. Agr., Forest Serv. Cir. 83, 3 pp.

See also (19), (41), (48), (50), (99), and (153).

MYRICA CERIFERA L.

(176) BARTON, L. V.
1932. GERMINATION OF BAYBERRY SEEDS. Boyce Thompson Inst. Contrib. 4: 19-26, illus.

NEMOPANTHUS MUCRONATUS (L.) Trel.

See (46), (52), and (100).

NYSSA L.

(177) SHUNK, I. V.
1939. OXYGEN REQUIREMENTS FOR GERMINATION OF SEEDS OF NYSSA AQUATICA—TUPELO GUM. Science 90: 565-566.

See also (47), (52), and (82).

OLNEYA TESOTA A. Gray

(178) VAN DERSAL, W. R.
1938. NATIVE WOODY PLANTS OF THE UNITED STATES: THEIR EROSION-CONTROL AND WILDLIFE VALUES. U. S. Dept. Agr. Misc. Pub. 303, 362 pp., illus.

OSTRYA VIRGINIANA (Mill.) K. Koch

See (75) and (82).

PARTHENOCISSUS Planch.

See (74) and (100).

PHOTINIA ARBUTIFOLIA (Ait. f.) Lindl.

See (46).

PICEA A. Dietr.

(179) ANONYMOUS.
1903. SEED COLLECTING. Forestry Quart. 1: 79-80.

(180) BALDWIN, H. I.
1935. SEASONAL VARIATIONS IN THE GERMINATION OF RED SPRUCE. Amer. Jour. Bot. 22: 392-394, illus.

(181) CARY, N. L.
1922. SITKA SPRUCE: ITS USES, GROWTH, AND MANAGEMENT. U. S. Dept. Agr. Bul. 1060, 38 pp., illus.

(182) CIESLAR, A.
1895. DIE ERBLICHKEIT DES ZUWACHSVERMÖGENS BEI DEN WALDBÄUMEN. Centbl. f. das. Gesam. Forstw. 21: 7-29.

(183) COLLINGWOOD, G. H.
1937. KNOWING YOUR TREES. 109 pp., illus. Washington, D. C.

(184) HILF, R. B.
1927. WIE WIRKEN ERNTEZEIT, ALTER DES MUTTERBAUMES, UND HOHENLAGE AUF DIE GÜTE DES FICHTEN SAATGUTS. Ztschr. Forst u. Jagdw. 59: 65-87, illus.

(185) HOFFMAN, B. E.
1912. SITKA SPRUCE OF ALASKA. Soc. Amer. Foresters Proc. 7: 226-238.

(186) HOPKINSON, A. D.
1931. NOTES ON THE SITKA SPRUCE AND OTHER CONIFERS ON THE QUEEN CHARLOTTE ISLANDS. Forestry 5: 9-13.

(187) ILLICK, J. S.
1923. PENNSYLVANIA TREES. Pa. Dept. Forestry Bul. 11, Ed. 4, 237 pp., illus.

(188) KORSTIAN, C. F. AND BAKER, F. S.
1925. FOREST PLANTING IN THE INTERMOUNTAIN REGION. U. S. Dept. Agr. Bul. 1264, 57 pp., illus.

(189) LEBARRON, R. K.
1939. THE ROLE OF FOREST FIRES IN THE REPRODUCTION OF BLACK SPRUCE. Minn. Acad. Sci. Proc. 7: 10-14, illus.

(190) MURPHY, L. S.
1917. THE RED SPRUCE: ITS GROWTH AND MANAGEMENT. U. S. Dept. Agr. Bul. 544, 100 pp., illus.

(191) REUSS.
1916. AUS DEN REUSS'SCHEN FICHTEN-REINZUCHTVERSUCHEN VOM JAHRE 1878 AUF DER FÜRSTLICH COLLOREDOMANNSFELD'SCHEN DOMÄNE DOBRISCH IN BÖHMEN. Centbl. f. das. Gesam. Forstw. 42: 383-417, illus.

(192) ROHMEDER, E.
1936. FORSTLICHE VERSUCHE. 19. ZUSAMMENHANG ZWISCHEN ZAPFENFARBE UND KNOSPENENTFALTUNG DER FICHTE. Forstwiss. Centbl. 58: 253-261, illus.

(193) RUBNER, K.
1927. BEDECKUNGSTIEFE UND KEIMUNG DES FICHTENSAMENS. Forstwiss. Centbl. 49: 168-183.

(194) ———
1936. BEITRAG ZUR KENNTNIS DER FICHTENFORMEN UND FICHTENRASSEN. Tharandter Forstl. Jahrb. 87: 101-176.

(195) SCHMIDT, W.
1923. BESTANDESALTER UND SAMENGÜTE IM FICHTENSAMENJAHR 1921-1922 (IN DER OBERFÖRSTEREI EBERSBACH, BEZ. WEISBADEN). Ztschr. f. Forst u. Jagdw. 55: 490-495.

(196) SUKACHEV, V. N.
1928. LESNYE PORODY. SISTEMATIKA, GEOGRAFIYA I. FITOSOTIOLOGIYA IKH. KHVOINYE VIP. I. [FOREST TREES: TAXONOMY, RANGE, AND ECOLOGY, PT. I. CONIFERAE.]. 80 pp. Moscow.

(197) TILLOTSON, C. R.
1921. STORAGE OF CONIFEROUS TREE SEED. Jour. Agr. Res. 22: 479-510, illus.

See also (2), (4), (9), (16), (18), (19), (24), (50), (53,) (82), (87), and (92).

PINUS L.

(198) ANONYMOUS.
1939. SLASH PINE IN TEXAS. Texas Forest News 19(5-6): 3-4, illus.

(199) ADAMS, W. R.
1934. STUDIES IN TOLERANCE OF NEW ENGLAND FOREST TREES. XI. THE INFLUENCE OF SOIL TEMPERATURE ON THE GERMINATION AND DEVELOPMENT OF WHITE PINE SEEDLINGS. Vt. Agr. Expt. Sta. Bul. 379, 18 pp., illus.

(200) ASHE, W. W.
1894. THE FORESTS, FOREST LANDS, AND FOREST PRODUCTS OF EASTERN NORTH CAROLINA. N. C. Geol. Survey Bul. 5, 128 pp., illus.

(201) ———
1913. SHORTLEAF PINE IN VIRGINIA—THE INCREASE IN ITS YIELD BY THINNING. Va. Dept. Agr. and Immgr., 44 pp., illus.

(202) ———
1915. LOBLOLLY OR NORTH CAROLINA PINE. N. C. Geol. and Econ. Survey Bul. 24, 176 pp., illus.

(203) BALDWIN, H. I.
1934. EFFECT OF AFTER-RIPENING TREATMENT ON GERMINATION OF WHITE PINE SEEDS OF DIFFERENT AGES. Bot. Gaz. 96: 372-376, illus.

(204) BARTON, L. V.
1928. HASTENING THE GERMINATION OF SOUTHERN PINE SEEDS. Jour. Forestry 26: 774-785, illus.

(205) ———
1935. STORAGE OF SOME CONIFEROUS SEEDS. Boyce Thompson Inst. Contrib. 7: 379-404, illus.

(206) BATES, C. G.
1930. ONE-YEAR STORAGE OF WHITE PINE SEED. Jour. Forestry 28: 571-572.

(207) ———
1930. THE PRODUCTION, EXTRACTION, AND GERMINATION OF LODGEPOLE PINE SEED. U. S. Dept. Agr. Tech. Bul. 191, 92 pp., illus.

(208) ———
1931. A NEW PRINCIPLE IN SEED COLLECTING FOR NORWAY PINE. Jour. Forestry 29: 661-678, illus.

(209) CHAMPION, H. G., AND PANT, B. D.
1932. NOTES ON PINUS LONGIFOLIA ROXB. Indian Forest Rec. 16, 25 pp., illus.

(210) CHAPMAN, H. H.
1922. A NEW HYBRID PINE (PINUS PALUSTRIS X PINUS TAEDA). Jour. Forestry 20: 729-734, illus.

(211) CIESLAR, A.
1893. APHORISMEN AUS DEM GEBIETE DER FORSTLICHEN SAMENKUNDE. Centbl. f. das Gesam. Forstw. 19: [145]-158, illus.

(212) CLEMENTS, F. E.
1910. THE LIFE HISTORY OF LODGEPOLE BURN FORESTS. U. S. Dept. Agr., Forest Serv. Bul. 79, 56 pp., illus.

(213) COKER, W. C.
1909. VITALITY OF PINE SEEDS AND THE DELAYED OPENING OF CONES. Amer. Nat. 43: 677-681.

(214) CONNER, J. F.
1931. WESTERN YELLOW PINE SEED EXTRACTION IN THE BLACK HILLS. Jour. Forestry 29: 1165-1167.

(215) DELEVOY, G.
1935. NOTE PRÉLIMINAIRE SUR L'INFLUENCE D L'ORIGINE DES GRAINES CHEZ LE PIN MARITIME. Soc. Cent. Forest. de Belg. Bul. 42(3): 97-105.

(216) DIMITROFF, T.
1926. STUDY OF THE SEED MATERIAL OF PINUS PEUCE. Sofisk. Univ., Agron. Fakult. God. (Sofia Univ., Faculté Agron. Ann.) 4: 259-306. [In Bulgarian. French summary, pp. 303-306.]

(217) DUFF, C. E.
1928. THE VARIETIES AND GEOGRAPHICAL FORMS OF PINUS PINASTER, SOLAND, IN EUROPE AND SOUTH AFRICA. Brit. Empire Forestry Conf. Australasia, 55 pp., illus.

(218) ECKBO, N. B.
1916. IMPORTANCE OF SOURCE OF SEED IN FORESTATION. Soc. Amer. Foresters Proc. 11: 240-243.

(219) ELIASON, E. J., AND HEIT, C. E.
1940. THE RESULTS OF LABORATORY TESTS AS APPLIED TO LARGE SCALE EXTRACTION OF RED PINE SEED. Jour. Forestry 38: 426-429, illus.

(220) ENTRICAN, A. R.
1939. ANNUAL REPORT OF THE DIRECTOR OF FORESTRY, STATE FOREST SERVICE, NEW ZEALAND, FOR THE YEAR ENDED 31ST MARCH, 1939. 37 pp., illus. Wellington.

(221) FIELD, J. F.
1934. EXPERIMENTAL GROWING OF INSIGNIS PINE FROM SLIPS. Te Kura Ngahere 3: 185-186.

(222) FORBES, R. D.
1930. TIMBER GROWING AND LOGGING AND TURPENTINING PRACTICES IN THE SOUTHERN PINE REGION. U. S. Dept. Agr. Tech. Bul. 204, 114 pp., illus.

(223) GOMILEVSKI, V. T.
1893. AVSTRIISKAIA CHERNAIA SOSNA (PINUS LARICIO AUSTRIACA ANT.): SVOISTVA I LESORAZVEDENIE ETOI PORODY. [AUSTRIAN BLACK PINE: CHARACTERISTICS AND PROPAGATION.] 73 pp. St. Petersburg.

(224) [GRENNING, V.]
1936-37. REPORT OF THE DIRECTOR OF FORESTS FOR THE YEAR ENDED 30TH JUNE, 1937. Queensland Dept. Pub. Lands, Dir. Forests Ann. Rpt. 59 pp.

(225) ———
1937-38. REPORT OF THE DIRECTOR OF FORESTS FOR THE YEAR ENDED 30TH JUNE, 1938. Queensland Dept. of Pub. Lands, Dir. Forests Ann. Rpt. 48 pp., illus.

(226) HAACK.
1905. UNTERSUCHUNGEN ÜBER DEN EINFLUSS VERSCHIEDEN HÖHER DARRHITZE AUF DES KIEMPROZENT DES KIEFERNSAMENS. Ztschr. Forst u. Jagdw. 37: 296-312, illus.

(227) HAASIS, F. W.
1928. GERMINATIVE ENERGY OF LOTS OF CONIFEROUS-TREE SEED, AS RELATED TO INCUBATION TEMPERATURE AND TO DURATION OF INCUBATION. Plant Physiol. 3: 365-412, illus.

(228) HAIG, I. T.
1932. PREMATURE GERMINATION OF FOREST TREE SEED DURING NATURAL STORAGE IN DUFF. Ecology 13: 311-312, illus.

(229) ———
1936. FACTORS CONTROLLING INITIAL ESTABLISHMENT OF WESTERN WHITE PINE AND ASSOCIATED SPECIES. Yale Univ., School Forestry Bul. 41, 149 pp., illus.

(230) HEDGCOCK, G. G., HAHN, G. G., AND HUNT, N. R.
1922. TWO IMPORTANT PINE CONE RUSTS AND THEIR NEW CRONARTIAL STAGES. Phytopathology 12: 109-122, illus.

(231) HUBERMAN, M. A.
1938. GROWING NURSERY STOCK OF SOUTHERN PINES. U. S. Dept. Agr. Leaflet 155, 8 pp., illus.

(232) ———
1940. STUDIES IN RAISING SOUTHERN PINE NURSERY SEEDLINGS. Jour. Forestry 38: 341-345.

(233) ———
1940. NORMAL GROWTH AND DEVELOPMENT OF SOUTHERN PINE SEEDLINGS IN THE NURSERY. Ecology 21: 323-334, illus.

(234) HUNT, L. O.
1939. TORREY PINE. Jour. Forestry 37: 267-268.

(235) ILLICK, J. S., AND AUGHANBAUGH, J. E.
1930. PITCH PINE IN PENNSYLVANIA. Pa. Dept. Forests and Waters Res. Bul. 2, 108 pp., illus.

(236) JACOBS, A. W.
1925. HASTENING THE GERMINATION OF SUGAR PINE SEED. Jour. Forestry 23: 919-931.

(237) JOHNSTONE, G. R. AND CLARE, T. S.
1931. HASTENING THE GERMINATION OF WESTERN PINE SEEDS. Jour. Forestry 29: 895-906, illus.
(238) KESSELL, S. L.
1934. FORESTRY IN NEW SOUTH WALES. A REPORT ON THE INDIGENOUS FORESTS, PLANTATIONS, AND FOREST INDUSTRIES OF NEW SOUTH WALES, WITH RECOMMENDATIONS FOR FUTURE ADMINISTRATION AND PRACTICE. 70 pp., illus. Sydney.
(239) KOBLET, R.
1932. UEBER DIE KEIMUNG VON PINUS STROBUS UNTER BESONDERER BERÜCKSICHTIGUNG DER HERKUNFT DES SAMENS. Schweiz. Bot. Gesell. Ber. 41(2): 199-283, illus.
(240) LARSEN, J. A.
1925. METHODS OF STIMULATING GERMINATION OF WESTERN WHITE PINE SEED. Jour. Agr. Res. 31: 889-899, illus.
(241) LARSEN, L. T., AND WOODBURY, T. D.
1916. SUGAR PINE. U. S. Dept. Agr. Bul. 426, 40 pp., illus.
(242) LINDFORS, J.
1928. MURRAYANA-TALLEN (LODGEPOLE PINE). Skogen 15: 367-368, illus.
(243) MACKINNEY, A. L., AND KORSTIAN, C. F.
1938. LOBLOLLY PINE SEED DISPERSAL. Jour. Forestry 36: 465-468, illus.
(244) ——— AND MCQUILKIN, W. E.
1938. METHODS OF STRATIFICATION FOR LOBLOLLY PINE SEEDS. Jour. Forestry 36: 1123-1127.
(245) MCGAVOCK, A. D.
1937. ANNUAL REPORT OF THE DIRECTOR OF FORESTRY FOR THE YEAR ENDED 31ST MARCH, 1937. New Zeal. State Forest Serv. Rpt. 21 pp.
(246) MCINTYRE, A. C.
1929. A CONE AND SEED STUDY OF THE MOUNTAIN PINE (PINUS PUNGENS LAMBERT). Amer. Jour. Bot. 16: 402-406.
(247) MCQUILKIN, W. E.
1940. THE NATURAL ESTABLISHMENT OF PINE IN ABANDONED FIELDS IN THE PIEDMONT PLATEAU REGION. Ecology 21: 135-147, illus.
(248) MAKI, T. E.
1940. SIGNIFICANCE AND APPLICABILITY OF SEED MATURITY INDICES FOR PONDEROSA PINE. Jour. Forestry 38: 55-60, illus.
(249) MASON, D. T.
1915. THE LIFE HISTORY OF LODGEPOLE PINE IN THE ROCKY MOUNTAINS. U. S. Dept. Agr. Bul. 154, 35 pp., illus.
(250) MATHEWS, A. C.
1932. SEED DEVELOPMENT IN PINUS PALUSTRIS. Elisha Mitchell Sci. Soc. Jour. 48: 101-118.
(251) METCALF, W.
1921. NOTES ON THE BISHOP PINE (PINUS MURICATA). Jour. Forestry 19: 886-902, illus.
(252) MILLS, E. A.
1915. THE ROCKY MOUNTAIN WONDERLAND. 363 pp., illus. Boston and New York.
(253) MIROV, N. T.
1936. A NOTE ON GERMINATION METHODS FOR CONIFEROUS SPECIES. Jour. Forestry 34: 719-723.
(254) MUNGER, T. T.
1917. WESTERN YELLOW PINE IN OREGON. U. S. Dept. Agr. Bul. 418, 48 pp., illus.
(255) MUNNS, E. N.
1921. COULTER PINE. Jour. Forestry 19: 903-906.
(256) NELSON, M. L.
1940. SUCCESSFUL STORAGE OF SOUTHERN PINE SEED FOR SEVEN YEARS. Jour. Forestry 38: 443-444.
(257) NESS, H.
1927. THE DISTRIBUTION LIMITS OF THE LONGLEAF PINE AND THEIR POSSIBLE EXTENSION. Jour. Forestry 25: 852-857.
(258) OLSON, D. S.
1932. SEED RELEASE FROM WESTERN WHITE PINE AND PONDEROSA PINE CONES. Jour. Forestry 30: 748-749.
(259) PHILLIPS, F. J.
1909. A STUDY OF PIÑON PINE. Bot. Gaz. 48: 216-223.
(260) READ, A. D.
1932. NOTES ON ARIZONA PINE AND APACHE PINE. Jour. Forestry 30: 1013-1014.
(261) RICHARDSON, A. H.
1925. GATHERING AND EXTRACTING RED PINE SEED. Jour. Forestry 23: 304-310.
(262) RIETZ, R. C.
1939. INFLUENCE OF KILN TEMPERATURES ON FIELD GERMINATION AND TREE PERCENT IN NORTHERN WHITE PINE. Jour. Forestry 37: 343-344.
(263) ———
1939. EFFECT OF FIVE KILN TEMPERATURES ON THE GERMINATIVE CAPACITY OF LONGLEAF PINE SEED. Jour. Forestry 37: 960-963, illus.
(264) ———
1939. KILN DESIGN AND DEVELOPMENT OF SCHEDULES FOR EXTRACTING SEED FROM CONES. U. S. Dept. Agr. Tech. Bul. 773, 70 pp., illus.
(265) ——— AND KIMBALL, K. E.
1940. KILN SCHEDULE FOR EXTRACTING RED PINE SEED FROM FRESH AND STORED CONES. Jour. Forestry 38: 430-434, illus.
(266) ——— AND TORGESON, O. W.
1937. KILN TEMPERATURES FOR NORTHERN WHITE PINE CONES. Jour. Forestry 35: 836-839, illus.
(267) ROE, E. I.
1940. LONGEVITY OF RED PINE SEED. Minn. Acad. Sci. Proc. 8: 28-30, illus.
(268) ROHMEDER, E., AND LOEBEL, M.
1940. KEIMVERSUCHE MIT ZIRBELKIEFER. Forstwiss. Centbl. 62: 25-36, illus.
(269) ROSSI, E.
1929. SULLA GERMINABILITÀ DEL SEME DI PINUS MARITIMA LAM. IN RAPPORTO ALLA TEMPERATURA. Ist Bot. della R. Univ. Pavia Atti., Ser. IV, 1: 107-115.
(270) RUDOLF, P. O.
1940. WHEN ARE PINE CONES RIPE? Minn. Acad. Sci. Proc. 8: 31-38, illus.
(271) ST. GEORGE, R. A. AND BEAL, J. A.
1929. THE SOUTHERN PINE BEETLE: A SERIOUS ENEMY OF PINES IN THE SOUTH. U. S. Dept. Agr. Farmers' Bul. 1586, 18 pp., illus.
(272) SAWYER, L. E.
1929. NURSERY PRACTICE IN GEORGIA. Jour. Forestry 27: 428-429.
(273) SCHANTZ-HANSEN, T.
1941. A STUDY OF JACK PINE SEED. Jour. Forestry 39: 980-990.
(274) SCHMIDT, W.
1930. UNSERE KENNTNIS VOM FORSTSAATGUT. 256 pp., illus. Berlin.
(275) SHOW, S. B.
1917. METHODS OF HASTENING GERMINATION. Jour. Forestry 15: 1003-1006.
(276) SIGGERS, P. V.
1944. THE BROWN SPOT NEEDLE BLIGHT OF PINE SEEDLINGS. U. S. Dept. Agr. Tech. Bul. 870, 36 pp., illus.
(277) SIM, T. R.
1927. TREE PLANTING IN SOUTH AFRICA. Natal Witness, Ltd., 452 pp., illus.
(278) STERRETT, W. D.
1911. SCRUB PINE (PINUS VIRGINIANA). U. S. Dept. Agr., Forest Serv. Bul. 94, 27 pp., illus.
(279) SUDWORTH, G. B.
1917. THE PINE TREES OF THE ROCKY MOUNTAIN REGION. U. S. Dept. Agr. Bul. 460, 47 pp., illus.

792026°—49—26

(280) TOUMEY, J. W., AND DURLAND, W. D.
1923. THE EFFECT OF SOAKING CERTAIN TREE SEEDS IN WATER AT GREENHOUSE TEMPERATURES ON VIABILITY AND THE TIME REQUIRED FOR GERMINATION. Jour. Forestry 21: 369-375.
(281) TOZAWA, M.
1926. METHODS OF HASTENING GERMINATION OF TREE SEEDS. Chosen Govt.-Gen., Forest Expt. Sta. Bul. 5, 33 pp.
(282) TROUP, R. S.
1932. EXOTIC FOREST TREES IN THE BRITISH EMPIRE. 259 pp., illus. Oxford.
(283) UNITED STATES FOREST SERVICE.
1907. GERMINATION OF PINE SEED. 12 pp.
(284) VILJOEN, P. R.
1935. DIVISION OF FORESTRY ANNUAL REPORT FOR THE YEAR ENDED 31ST MARCH, 1935. So. Africa Dept. Agr. and Forestry, Div. Forestry Ann. Rpt. 43 pp., illus.
(285) ———
1937. DIVISION OF FORESTRY ANNUAL REPORT FOR THE YEAR ENDED 31ST MARCH, 1937. So. Africa Dept. Agr. and Forestry, Div. Forestry Ann. Rpt. 61 pp., illus.
(286) WAHLENBERG, W. G.
1924. CIRCUMVENTING DELAYED GERMINATION IN THE NURSERY. Jour. Forestry 22: 574-575.
(287) ———
1924. FALL SOWING AND DELAYED GERMINATION OF WESTERN WHITE PINE SEED. Jour. Agr. Res. 28: 1127-1131, illus.
(288) WAKELEY, P. C.
1930. SEED YIELD DATA FOR SOUTHERN PINES. Jour. Forestry 28: 391-394.
(289) ———
1931. SOME OBSERVATIONS ON SOUTHERN PINE SEED. Jour. Forestry 29: 1150-1164, illus.
(290) ———
1932. PEAT MATS FOR GERMINATION TESTS OF FOREST TREE SEEDS. Science 76: 627-628, illus.
(291) ———
1935. ARTIFICIAL REFORESTATION IN THE SOUTHERN PINE REGION. U. S. Dept. Agr. Tech. Bul. 492, 115 pp., illus.
(292) ———
1938. HARVESTING AND SELLING SEED OF SOUTHERN PINES. U. S. Dept. Agr. Leaflet 156, 8 pp., illus.
(293) ———
1938. PLANTING SOUTHERN PINES. U. S. Dept. Agr. Leaflet 159, 8 pp., illus.
(294) ———
1939. STORING SOUTHERN PINE SEED ON A COMMERCIAL SCALE. South. Lumberman 159: 114.
(295) WEBSTER, C. B.
1931. NOTES ON GROWTH OF SLASH PINE IN TEXAS. Jour. Forestry 29: 425-426.
(296) WEDDELL, D. J.
1935. VIABLE SEED FROM NINE-YEAR-OLD SOUTHERN PINE. Jour. Forestry 33: 902.
(297) ———
1939. EXTENDING THE NATURAL RANGE OF SLASH PINE IN ALABAMA. Jour. Forestry 37: 342-343, illus.
(298) WEIDMAN, R. H.
1939. EVIDENCES OF RACIAL INFLUENCE IN A 25-YEAR TEST OF PONDEROSA PINE. Jour. Agr. Res. 59: 855-887, illus.
(299) WIBECK, E.
1928. DET NORRLANDSKA TALLFROETS GROBARHET OCH ANATOMISKA BESKAFFENHET. Norrlands Skogsvärdsförbunds Tidskr. 1: 4-35.
See also (1), (2), (4), (7), (9), (12), (15), (16), (17), (18), (19), (20), (21), (23), (25), (27), (28), (45), (50), (53), (74), (82), (92), (108), (120), (124), (141), (158).

PLATANUS L.

See (18), (19), (38), and (48).

POPULUS L.

(300) BULL, H., AND MUNTZ, H. H.
1943. PLANTING COTTONWOOD ON BOTTOMLANDS. Miss. Agr. Expt. Sta. Bul. 391, 18 pp., illus.
(301) BUSSE, I.
1935. SAMENAUFBEWAHRUNG IM VAKUUM. Ztschr. f. Forst u. Jagdw. 67: 321-326.
(302) FAUST, M. E.
1936. GERMINATION OF POPULUS GRANDIDENTATA AND P. TREMULOIDES, WITH PARTICULAR REFERENCE TO OXYGEN CONSUMPTION. Bot. Gaz. 97: 808-821, illus.
(303) MOSS, E. H.
1938. LONGEVITY OF SEED AND ESTABLISHMENT OF SEEDLINGS IN SPECIES OF POPULUS. Bot. Gaz. 99: 529-542.
(304) REIM, P.
1929. DIE VERMEHRUNGSBIOLOGIE DER ASPE AUF GRUNDLAGE DES IN ESTLAND UND FINNLAND GESAMMELTEN UNTERSUCHUNGSMATERIALS. 60 pp., illus. Tartu.
(305) ROHMEDER, E.
1941. DIE VERMEHRUNG DER PAPPELN DURCH SAMEN. Forstarchiv 17: 73-80, illus.
(306) ROSENDAHL, C. O., AND BUTTERS, F. K.
1928. TREES AND SHRUBS OF MINNESOTA. 385 pp., illus. Minneapolis, Minn.
(307) STEVENS, R. D.
1940. SEED STUDIES. Ark. Agr. Expt. Sta. Ann. Rpt. (1939), Bul. 386: 87-88.
(308) SUDWORTH, G. B.
1927. CHECK LIST OF THE FOREST TREES OF THE UNITED STATES: THEIR NAMES AND RANGES. U. S. Dept. Agr. Misc. Cir. 92, 295 pp.
(309) WILLIAMSON, A. W.
1913. COTTONWOOD IN THE MISSISSIPPI VALLEY. U. S. Dept. Agr. Bul. 24, 62 pp., illus.
See also (82).

PROSOPIS JULIFLORA (Sw.) DC.

(310) CROCKER, W.
1909. LONGEVITY OF SEEDS. Bot. Gaz. 47: 69-72.
See also (92).

PRUNUS L.

(311) ANONYMOUS.
1936. METHOD FOR TESTING PEACH SEEDS IS DEVISED. Fert. Rev. 11(4): 10.
(312) GIERSBACH, J.
1932. GERMINATION AND STORAGE OF WILD PLUM SEEDS. Boyce Thompson Inst. Contrib. 4: 39-52, illus.
(313) HAUT, I. C.
1938. PHYSIOLOGICAL STUDIES ON AFTER-RIPENING AND GERMINATION OF FRUIT-TREE SEEDS. Md. Agr. Expt. Sta. Bul. 420, 52 pp., illus.
See also (19), (38), (39), (40), (41), (46), (48), (50), (52), (56), (60), (65), (82), (94), and (131).

PSEUDOTSUGA Carr.

(314) ALLEN, G. S.
1941. A STANDARD GERMINATION TEST FOR DOUGLAS FIR SEED. Forestry Chron. 17 (2): 75-78, illus.
(315) MUNGER, T. T., AND MORRIS, W. G.
1936. GROWTH OF DOUGLAS FIR TREES OF KNOWN SEED SOURCE. U. S. Dept. Agr. Tech. Bul. 537, 40 pp., illus.
See also (1), (6), (15), (19), (21), (23), (28), (51), (53), (120), and (188).

PTELEA TRIFOLIATA L.

(316) SCHROEDER, E. M.
1937. GERMINATION OF FRUITS OF PTELEA SPECIES. Boyce Thompson Inst. Contrib. 8: 355-359.

PURSHIA TRIDENTATA DC.

See (46).

PYRUS COMMUNIS L.

See (19), (41), (46), (56), and (313).

QUERCUS L.

(317) ANONYMOUS.
1935. WILLOW OAK (QUERCUS PHELLOS) LINNAEUS. Amer. Forests 41: 332-333, illus.

(318) BAGNERIS, G.
1882. ELEMENTS OF SYLVICULTURE: A SHORT TREATISE ON THE SCIENTIFIC CULTIVATION OF THE OAK AND OTHER HARDWOOD TREES. 283 pp. London.

(319) BRITTON, N. L., AND SHAFER, J. A.
1908. NORTH AMERICAN TREES—BEING DESCRIPTIONS AND ILLUSTRATIONS OF THE TREES GROWING INDEPENDENTLY OF CULTIVATION IN NORTH AMERICA, NORTH OF MEXICO, AND THE WEST INDIES. 894 pp., illus. New York.

(320) DEUBER, C. G.
1932. CHEMICAL TREATMENTS TO SHORTEN THE REST PERIOD OF RED AND BLACK OAK ACORNS. Jour. Forestry 30: 674-679, illus.

(321) GARDNER, R. C. B.
1937. STORAGE OF ACORNS. Quart. Jour. Forestry 31: 32-33.

(322) GREELEY, W. B., AND ASHE, W. W.
1907. WHITE OAK IN THE SOUTHERN APPALACHIANS. U. S. Dept. Agr., Forest Serv. Cir. 105, 27 pp.

(323) KORSTIAN, C. F.
1927. FACTORS CONTROLLING GERMINATION AND EARLY SURVIVAL IN OAKS. Yale Forest School Bul. 19, 115 pp., illus.

(324) ———
1930. ACORN STORAGE IN THE SOUTHERN STATES. Jour. Forestry 28: 858-863.

(325) WINTERS, R. K.
1933. WEIGHT OF FRUIT OF NUTTALL'S OAK. Jour. Forestry 31: 340.

See also (3), (18), (19), (25), (27), (29), (40), (48), (60), (118), and (120).

RHAMNUS L.

(326) STARKER, T. J., AND WILCOX, A. R.
1931. CASCARA. Amer. Jour. Pharm. 103: 53 pp., illus.

See also (23), (46), (48), (52), (56), and (60).

RHODODENDRON L.

(327) MORRISON, B. Y.
1929. AZALEAS AND RHODODENDRONS FROM SEED. U. S. Dept. Agr. Cir. 68, 8 pp., illus.

See also (18), (46), and (60).

RHODOTYPOS SCANDENS (Thunb.) Mak.

(328) FLEMION, F.
1933. PHYSIOLOGICAL AND CHEMICAL STUDIES OF AFTER-RIPENING OF RHODOTYPOS KERRIOIDES SEEDS. Boyce Thompson Inst. Contrib. 5: 143-159, illus.

See also (60) and (75).

RHUS L.

(329) WRIGHT, E.
1931. THE EFFECT OF HIGH TEMPERATURES ON SEED GERMINATION. Jour. Forestry 29: 679-687.

See also (48).

RIBES L.

(330) FIVAZ, A. E.
1931. LONGEVITY AND GERMINATION OF SEEDS OF RIBES, PARTICULARLY R. ROTUNDIFOLIUM, UNDER LABORATORY AND NATURAL CONDITIONS. U. S. Dept. Agr. Tech. Bul. 261, 40 pp., illus.

See also (48), (52), and (100).

ROBINIA L.

(331) BURTON, C. L.
1932. VARIATION IN CHARACTERISTICS OF BLACK LOCUST SEEDS FROM TWO REGIONS. Jour. Forestry 30: 29-33.

(332) CHAPMAN, A. G.
1936. SCARIFICATION OF BLACK LOCUST SEED TO INCREASE AND HASTEN GERMINATION. Jour. Forestry 34: 66-74, illus.

(333) FLEISCHMANN, R.
1933. BEITRÀGE ZÜR ROBINIENZUCHTUNG. Züchter 5: 85-88, illus.

(334) GOSS, W. L.
1924. THE VITALITY OF BURIED SEED. Jour. Agr. Res. 29: 349-362, illus.

(335) MCKEEVER, D. G.
1937. A NEW BLACK LOCUST SEED TREATMENT. Jour. Forestry 35: 500-501.

(336) MADDOX, R. S.
1922. DIRECTIONS FOR PLANTING BLACK LOCUST SEED, AND BLACK LOCUST SEEDLINGS AND SPROUTS. Tenn. Forestry Bur. Cir. 3, 7 pp., illus. Nashville.

(337) MEGINNIS, H. G.
1937. SULPHURIC ACID TREATMENT TO INCREASE GERMINATION OF BLACK LOCUST SEED. U. S. Dept. Agr. Cir. 453, 35 pp., illus.

(338) RABER, O.
1936. SHIPMAST LOCUST, A VALUABLE UNDESCRIBED VARIETY OF ROBINIA PSEUDOACACIA. U. S. Dept. Agr. Cir. 379, 8 pp., illus.

See also (48) and (90).

ROSA BLANDA Ait.

See (54).

RUBUS L.

(339) ROSE, R. C.
1919. AFTER-RIPENING, AND GERMINATION OF SEEDS OF TILIA, SAMBUCUS, AND RUBUS. Bot. Gaz. 67: 281-308.

SALIX L.

(340) GRAMS.
1910. VERMEHRUNG DER WEIDEN DURCH AUSSAAT. Allg. Forst u. Jagd Ztg. 86: 265-266.

See also (6), (144), and (303).

SAMBUCUS L.

See (40), (48), (52), (60), (66), (100), and (339).

SAPINDUS DRUMMONDI Hook. & Arn.

See (48).

SEQUOIA Endl.

(341) BUCHHOLZ, J. T.
1938. CONE FORMATION IN SEQUOIA GIGANTEA. I. THE RELATION OF STEM SIZE AND TISSUE DEVELOPMENT TO CONE FORMATION. II. THE HISTORY OF THE SEED CONE. Amer. Jour. Bot. 25: 296-305, illus.

(342) DETWILER, S. B.
1916. THE REDWOODS—IDENTIFICATION AND CHARACTERISTICS. Amer. Forestry 22: 323-332, illus.

(343) FRY, W., AND WHITE, J. R.
1930. BIG TREES. 114 pp., illus. Stanford University, Calif., and London.
(344) METCALF, W.
1924. ARTIFICIAL REPRODUCTION OF REDWOOD (SEQUOIA SEMPERVIRENS). Jour. Forestry 22: 873-893.
See also (18), (19), (20), and (108).

SHEPHERDIA Nutt.

See (46), (48), (52), and (90).

SIMMONDSIA CHINENSIS (Link) Schneid.

(345) GREEN, T. G., HILDITCH, T. P., AND STAINSBY, W. J.
1936. THE SEED WAX OF SIMMONDSIA CALIFORNICA. [London] Jour. Chem. Soc. 2: 1750-1755.

SOLANUM DULCAMARA L.

See (60) and (74).

SORBUS L.

(346) FABRICIUS, L.
1931. DIE SAMENKEIMUNG VON SORBUS AUCUPARIA L. Forstwiss. Centbl. 53: 413-418.
(347) FLEMION, F.
1931. AFTER-RIPENING, GERMINATION, AND VITALITY OF SEEDS OF SORBUS AUCUPARIA L. Boyce Thompson Inst. Contrib. 3: 413-440, illus.
See also (56) and (102).

SYMPHORICARPOS Duham.

(348) FLEMION, F.
1934. PHYSIOLOGICAL AND CHEMICAL CHANGES PRECEDING AND DURING THE AFTER-RIPENING OF SYMPHORICARPOS RACEMOSUS SEEDS. Boyce Thompson Inst. Contrib. 6: 91-102, illus.
(349) ——— AND PARKER, E.
1942. GERMINATION STUDIES OF SEEDS OF SYMPHORICARPOS ORBICULATUS. Boyce Thompson Inst. Contrib. 12: 301-307, illus.
See also (38), (98), and (100).

SYRINGA VULGARIS L.

See (19) and (48).

TAXODIUM DISTICHUM (L.) Rich.

(350) DEMAREE, D.
1932. SUBMERGING EXPERIMENTS WITH TAXODIUM. Ecology 13: 258-262.
(351) DETWILER, S. B.
1916. THE BALD CYPRESS (TAXODIUM DISTICHUM)—IDENTIFICATION AND CHARACTERISTICS. Amer. Forestry 22: 577-585, illus.
(352) TOUMEY, J. W.
1930. SOME NOTES ON SEED STORAGE. Jour. Forestry 28: 394-395.
See also (2), (18), (28), and (280).

TAXUS L.

See (4), (25), (40), (50), (52), and (60).

THUJA L.

See (1), (2), (4), (8), (9), (12), (15), (19), (27), (28), and (82).

TILIA L.

(353) BARTON, L. V.
1934. DORMANCY IN TILIA SEEDS. Boyce Thompson Inst. Contrib. 6: 69-89, illus.
(354) GOLOSOV, N. A.
1938. SBOR I KHRANENIE SEMÎAN LIPY. [COLLECTION AND STORAGE OF LINDEN SEED.] Lesnoe Khoz. 1: 55-60.
(355) PUCHNER, H.
1922. DIE VERZÖGERTE KEIMUNG VON BAUMSÄMEREIEN. Forstwiss. Centbl. 44: 445-455.
(356) SPAETH, J. N.
1934. A PHYSIOLOGICAL STUDY OF DORMANCY IN TILIA SEED. N. Y. (Cornell) Agr. Expt. Sta. Mem. 169, 78 pp., illus.
See also (5), (41), (56), and (339).

TSUGA L.

(357) BALDWIN, H. I.
1930. THE EFFECT OF AFTER-RIPENING TREATMENT ON THE GERMINATION OF EASTERN HEMLOCK SEED. Jour. Forestry 28: 853-857, illus.
(358) ———
1934. FURTHER NOTES ON THE GERMINATION OF HEMLOCK SEED. Jour. Forestry 32: 99-100.
(359) HOFMANN, J. V.
1918. THE IMPORTANCE OF SEED CHARACTERISTICS IN THE NATURAL REPRODUCTION OF CONIFEROUS FORESTS. Minn. Univ., Studies Biol. Sci. 2, 25 pp., illus.
(360) LARSEN, J. A.
1918. COMPARISON OF SEED TESTING IN SAND AND IN THE JACOBSEN GERMINATOR. Jour. Forestry 16: 690-695.
See also (1), (9), (12), (15), (18), and (19).

ULMUS L.

(361) ANONYMOUS.
1907. WHITE ELM (ULMUS AMERICANA). U. S. Dept. Agr., Forest Serv. Cir. 66, 3 pp.
(362) ———
1907. SLIPPERY ELM (ULMUS PUBESCENS). U. S. Dept. Agr., Forest Serv. Cir. 85, 4 pp.
(363) BARTON, L. V.
1939. STORAGE OF ELM SEEDS. Boyce Thompson Inst. Contrib. 10: 221-233, illus.
(364) CROIZAT, L.
1937. CHINESE AND SIBERIAN ELMS. Amer. Nurseryman 65 (11): 3-4, 12, illus.
(365) GEORGE, E. J.
1937. STORAGE AND DEWINGING OF AMERICAN ELM SEED. Jour. Forestry 35: 769-772.
(366) RUDOLF, P. O.
1937. DELAYED GERMINATION IN AMERICAN ELM. Jour. Forestry 35: 876-877.
(367) STEINBAUER, C. E., AND STEINBAUER, G. P.
1931. EFFECTS OF TEMPERATURE AND DESICCATION DURING STORAGE ON GERMINATION OF SEEDS OF THE AMERICAN ELM (ULMUS AMERICANA L.). Amer. Soc. Hort. Sci. Proc. 28: 441-443.
See also (19), (27), (41), (43), (48), (56), (60), and (87).

VIBURNUM L.

(368) GIERSBACH, J.
1937. GERMINATION AND SEEDLING PRODUCTION OF SPECIES OF VIBURNUM. Boyce Thompson Inst. Contrib. 9: 79-90, illus.
(369) MITCHELL, E.
1926. GERMINATION OF SEEDS OF PLANTS NATIVE TO DUTCHESS COUNTY, NEW YORK. Bot. Gaz. 81: 108-112.
See also (41), (47), (52), (56), (60), (65), (75), and (101).

VITEX AGNUS-CASTUS L.

See (46), (48), and (60).

VITIS RIPARIA Michx.

See (74), (94), and (100).

ZIZIPHUS JUJUBA Mill.

See (46), (48), and (50).

USE LIST[1]

WOOD PRODUCTION[2]

SPECIES INCLUDED

Abies alba
A. amabilis
A. balsamea
A. concolor
A. fraseri
A. grandis
A. lasiocarpa
A. lasiocarpa var. *arizonica*[3]
A. magnifica
A. procera
Acacia decurrens var. *mollis*
A. decurrens var. *normalis*
A. melanoxylon
Acer macrophyllum
A. rubrum
A. saccharinum
A. saccharum
Aesculus glabra
A. hippocastanum
A. octandra
Alnus glutinosa
A. rubra
Arbutus menziesii

Betula lenta
B. lutea
B. nigra
B. papyrifera
B. pendula
B. populifolia
B. pubescens

Carpinus caroliniana
Carya spp.
Castanea dentata[4]
Castanopsis chrysophylla
Catalpa spp.
Cedrus spp.
Chamaecyparis spp.
Cladrastis lutea
Cornus florida
C. nuttalli
Cupressus arizonica
C. torulosa

Diospyros virginiana

Eucalyptus globulus
Euonymus europaeus
E. verrucosus[5]

Fagus spp.
Fraxinus (except *F. dipetala*)

Gleditsia spp.
Gymnocladus dioicus

Ilex opaca

Juglans (except *J. hindsii*)
Juniperus ashei
J. monosperma
J. occidentalis
J. pachyphloea
J. scopulorum
J. utahensis
J. virginiana

Larix (except *L. lyallii*)
Liquidambar styraciflua[6]
Liriodendron tulipifera

Maclura pomifera
Magnolia spp.
Morus spp.

Nyssa spp.

Olneya tesota
Ostrya virginiana
Oxydendrum arboreum

Picea (except *P. breweriana*)

Pinus banksiana
P. canariensis
P. caribaea[7]
P. cembra
P. clausa
P. contorta
P. contorta var. *latifolia*
P. densiflora
P. echinata
P. flexilis
P. flexilis var. *reflexa*
P. griffithii[7]
P. halepensis[7]
P. jeffreyi[8]
P. lambertiana
P. latifolia
P. monticola
P. muricata
P. nigra[7]
P. palustris[7]
P. peuce
P. pinaster[7]
P. ponderosa
P. ponderosa var. *arizonica*
P. radiata
P. resinosa
P. rigida
P. roxburghii[7]
P. strobus
P. sylvestris[7]
P. taeda
P. thunbergii[7]
Platanus occidentalis
Populus deltoides
P. grandidentata
P. sargentii
P. tremula
P. tremuloides
Prosopis juliflora
Prunus serotina
Pseudotsuga spp.

Quercus alba
Q. bicolor
Quercus borealis
Q. borealis var. *maxima*
Q. chrysolepis
Q. coccinea
Q. ellipsoidalis
Q. falcata
Q. garryana
Q. imbricaria
Q. lyrata
Q. macrocarpa
Q. montana
Q. nigra
Q. nuttallii
Q. palustris
Q. petraea
Q. phellos
Q. prinus
Q. robur
Q. stellata
Q. suber[3]
Q. velutina
Q. virginiana
Q. wislizeni

Robinia spp.

Sassafras albidum
Sequoia spp.
Sorbus aucuparia

Taxodium distichum
Taxus baccata
Thuja spp.
Tilia spp.
Tsuga canadensis
T. heterophylla

Ulmus americana
U. fulva
U. laevis
U. thomasi
Umbellularia californica

OTHER SPECIES

Abies nordmanniana
A. veitchii
Acer nigrum
Albizia kalkora

Betula fontinalis
B. pumila

Carya ovalis
Castanea sativa

Eucalyptus camaldulensis
E. rudis
E. viminalis

Fraxinus berlandieriana
F. caroliniana
F. lowellii

Magnolia tripetala

Nyssa ogeche

Platanus wrightii
Populus acuminata
P. alba
P. angustifolia
P. arizonica
P. fremontii
Populus heterophylla
P. macdougali
P. trichocarpa
P. wislizeni

Quercus durandii
Q. emoryi
Q. gambelii
Q. hypoleucoides
Q. laurifolia
Q. marilandica
Q. muehlenbergii
Q. oblongifolia
Q. obtusa
Q. shumardii

Salix alba
S. alba var. *vitellina*
S. exigua
S. fragilis
S. laevigata
S. lasiandra
S. missouriensis
S. petiolaris
Swietenia mahagoni

Taxodium ascendens
Tilia heterophylla
T. heterophylla var. *michauxii*
T. neglecta

Ulmus alata

Footnotes on page 395.

EROSION CONTROL[9]

SPECIES INCLUDED

Abies fraseri
A. lasiocarpa
A. lasiocarpa var. *arizonica*
A. venusta
Acacia decurrens var. *mollis*
A. decurrens var. *normalis*
Acer ginnala
A. glabrum
A. saccharinum
Alnus incana
A. tenuifolia
Amelanchier alnifolia
Amorpha spp.
Aplopappus parishii[5]
Arctostaphylos spp.
Atriplex canescens

Baccharis spp.
Betula nigra

Campsis radicans
Caragana arborescens
Castanopsis sempervirens
Ceanothus crassifolius
C. cuneatus
C. integerrimus
C. oliganthus
C. sanguineus
C. thyrsiflorus
C. velutinus
Cercocarpus spp.
Chamaecyparis nootkatensis
Chilopsis linearis
Chrysothamnus nauseosus[5]
Clematis spp.
Corylus californica
Cupressus arizonica
C. macnabiana

Elaeagnus commutata

Fallugia paradoxa
Fraxinus dipetala
F. excelsior
F. pennsylvanica var. *lanceolata*
F. velutina
Fremontodendron spp.

Garrya fremontii

Hippophae rhamnoides

Juglans californica
J. nigra
J. rupestris
Juniperus communis var. *depressa*
J. monosperma
J. scopulorum

Larix leptolepis
L. lyallii
Lithocarpus densiflorus[4]
Lonicera maackii
L. tatarica
Lupinus longifolius

Maclura pomifera

Parthenocissus inserta
Photinia arbutifolia
Picea breweriana
P. pungens
Pinus albicaulis
P. aristata
P. attenuata
P. banksiana
P. caribaea
P. cembra
P. contorta var. *latifolia*
P. coulteri
P. echinata
P. flexilis
P. halepensis
P. heldreichii var. *leucodermis*
P. latifolia
P. mugo
P. nigra
P. pinaster
P. ponderosa var. *arizonica*
P. pungens
P. sabiniana[8]
P. taeda
P. thunbergii
P. virginiana
Platanus racemosa
Prosopis juliflora
Prunus angustifolia
P. besseyi
Prunus pumila
P. pumila var. *susquehanae*
P. virginiana
Pseudotsuga spp.
Purshia tridentata

Quercus agrifolia
Q. chrysolepis
Q. douglasii
Q. dumosa
Q. wislizeni

Rhus glabra
R. integrifolia
R. laurina
R. trilobata
R. typhina
Ribes aureum
Robinia spp.
Rosa blanda
Rubus spp.

Salix spp.
Sambucus canadensis
Sequoia gigantea
Shepherdia spp.
Symphoricarpos spp.

Taxus brevifolia
Tsuga mertensiana

OTHER SPECIES

Alnus sinuata
Aplopappus bloomeri
Artemisia tridentata
Atriplex lentiformis var. *breweri*

Baccharis glutinosa
B. halimifolia

Carya ovalis
Cercis occidentalis
Crataegus rivularis
Cytisus scoparius

Encelia californica
Eucalyptus camaldulensis
E. rudis
E. viminalis

Fraxinus cuspidata

Hedera helix
Hibiscus syriacus

Juniperus communis
J. horizontalis

Lonicera arizonica
X *L. bella*
L. chrysantha
L. coerulea
L. involucrata
L. korolkowii
L. morrowii
L. utahensis
L. xylosteum

Mahonia repens
Myrica pensylvanica

Nicotiana glauca

Philadelphus lewisii
Physocarpus malvaceus
Populus fremontii
X *P. petrowskyana*
P. wislizeni
Potentilla fruticosa
Prunus emarginata
P. ilicifolia
P. lyonii
P. maritima
P. mexicana
P. nigra
P. spinosa
P. virens
Pueraria thunbergiana

Rhamnus crocea
Rhus aromatica
R. copallina
Robinia hispida
Rosa arkansana
R. canina
R. laevigata
R. setigera
Rubus flagellaris
R. leucodermis
R. ursinus

Salix alba
S. exigua
S. fragilis
S. laevigata
S. lasiandra
Spiraea trichocarpa
Symphoricarpos oreophilus

Tamarix aphylla
T. gallica
Toxicodendron vernix

Ulex europaeus

Vitis arizonica

Zanthoxylum clava-herculis

SHELTER BELTS

SPECIES INCLUDED

Acer negundo
A. negundo var. *californicum*
A. saccharinum
Ailanthus altissima
Bumelia lanuginosa

Caragana arborescens
Catalpa spp.
Chamaecyparis lawsoniana
Chilopsis linearis
Cornus asperifolia
Corylus americana
C. avellana
Cupressus arizonica
C. macrocarpa

Elaeagnus spp.
Eucalyptus globulus
Euonymus europaeus

Fraxinus excelsior
F. pennsylvanica var. *lanceolata*
F. velutina

Hippophae rhamnoides

Footnotes on page 395.

Juglans nigra
J. rupestris
Juniperus scopulorum
J. virginiana

Lonicera maackii
L. tatarica

Maclura pomifera
Malus baccata
Morus alba var. *tatarica*

Picea abies
P. glauca

Picea glauca var. *albertiana*
P. pungens
Pinus attenuata
P. banksiana
P. halepensis
P. muricata
P. nigra
P. ponderosa
P. sabiniana
Populus deltoides
P. nigra
P. sargentii
Prunus americana
P. angustifolia

Prunus angustifolia var. *watsoni*
P. armeniaca
P. virginiana
Pseudotsuga taxifolia
Ptelea trifoliata
Pyrus communis

Quercus macrocarpa

Rhamnus cathartica
R. davurica
Rhus glabra
R. trilobata
Ribes aureum

Ribes odoratum
Robinia pseudoacacia

Sambucus canadensis
Sapindus drummondi
Shepherdia argentea
Syringa vulgaris

Ulmus parvifolia
U. pumila

Viburnum lentago
Vitex agnus-castus

Ziziphus jujuba

OTHER SPECIES

Celtis reticulata
Colutea arborescens

Juniperus horizontalis

Lonicera arizonica
X *L. bella*

Lonicera chrysantha
L. coerulea
L. involucrata
L. korolkowii
L. morrowii
L. utahensis
L. villosa

Lonicera xylosteum

X *Populus petrowskyana*
Prunus virens
P. virginiana var. *melanocarpa*

Salix alba
S. alba var. *vitellina*
S. missouriensis
Syringa persica

Tamarix gallica
T. pentandra

WILDLIFE PURPOSES[10]

SPECIES INCLUDED

Abies balsamea
A. lasiocarpa
Acer glabrum
A. negundo
A. negundo var. *californicum*
A. pensylvanicum
A. rubrum
A. spicatum
Aesculus glabra
A. octandra
Ailanthus altissima
Albizia julibrissin
Alnus crispa
A. incana
A. rubra
A. tenuifolia
Amelanchier alnifolia
A. arborea
A. sanguinea
Amorpha spp.
Aralia spp.
Arbutus menziesii
Arctostaphylos spp.
Aronia spp.
Asimina triloba
Atriplex canescens

Baccharis pilularis
Berberis spp.
Betula glandulifera
B. glandulosa
B. pendula
B. pubescens
Bumelia lanuginosa

Campsis radicans
Caragana arborescens
Carpinus caroliniana
Carya cordiformis
C. illinoensis
C. laciniosa
C. ovata
C. tomentosa
Castanea dentata
Castanopsis chrysophylla
Ceanothus americanus
C. crassifolius
C. cuneatus
C. integerrimus
C. oliganthus
C. prostratus
C. sanguineus
C. velutinus
Celastrus scandens
Celtis spp.
Cephalanthus occidentalis
Cercis canadensis
Cercocarpus ledifolius
Chamaecyparis lawsoniana
C. nootkatensis
Chilopsis linearis
Clematis spp.
Comptonia peregrina
Cornus spp.
Corylus spp.
Cotoneaster acutifolia
C. horizontalis
Cowania stansburiana
Crataegus mollis
Cupressus arizonica

Dendromecon rigida
Diospyros virginiana

Elaeagnus spp.
Epigaea repens
Eriogonum fasciculatum
Euonymus spp.
Eurotia lanata

Fagus spp.
Fallugia paradoxa
Fraxinus dipetala
F. excelsior
F. velutina
Fremontodendron spp.

Gaultheria procumbens
Gaylussacia baccata
Gleditsia spp.
Gymnocladus dioicus

Ilex spp.

Juglans spp.
Juniperus spp.

Kalmia latifolia

Larix decidua
L. lyallii
L. occidentalis
Ligustrum vulgare
Lindera benzoin
Lonicera spp.

Magnolia spp.
Mahonia haematocarpa
Malus spp.
Melia azedarach
Menodora scabra
Morus spp.

Nemopanthus mucronatus
Nyssa spp.

Olneya tesota
Ostrya virginiana

Parthenocissus spp.
Photinia arbutifolia
Picea engelmanni
P. pungens
Pinus albicaulis
P. caribaea
P. cembra
P. clausa
P. echinata
P. edulis
P. glabra
P. monophylla
P. palustris
P. quadrifolia
P. rigida
P. sabiniana
Pinus taeda
P. virginiana
Platanus occidentalis
Prunus spp.
Pseudotsuga taxifolia
Ptelea trifoliata
Purshia tridentata
Pyrus communis

Quercus spp.[4]

Rhamnus spp.
Rhododendron maximum
Rhus spp.[4]
Ribes spp.
Robinia neo-mexicana
Rosa blanda
Rubus spp.

Salix bebbiana
S. discolor
Sambucus spp.
Shepherdia spp.
Simmondsia chinensis
Solanum dulcamara
Sorbus spp.
Symphoricarpos spp.

Taxodium distichum
Taxus spp.
Thuja spp.
Tilia americana
Tsuga spp.

Ulmus laevis
U. parvifolia
U. pumila

Viburnum spp.
Vitex agnus-castus
Vitis riparia

Zanthoxylum americanum
Ziziphus jujuba

Footnotes on page 395.

OTHER SPECIES

Acer buergerianum
A. circinatum
A. floridanum
A. grandidentatum
A. leucoderme
A. nigrum
A. tataricum
Aesculus neglecta var. *georgiana*
A. pavia
Alnus oblongifolia
A. rhombifolia
A. sinuata
Amelanchier florida
A. laevis
A. stolonifera
Amorpha californica
Ampelopsis cordata
Aplopappus bloomeri
Arctostaphylos glauca
Artemisia tridentata
Atriplex semibaccata

Baccharis glutinosa
Betula fontinalis
B. pumila
Bumelia texana

Callicarpa americana
Calycanthus occidentalis
Castanea ozarkensis
C. pumila
C. sativa
Celtis reticulata
C. sinensis
Chaenomeles lagenaria
Chiogenes hispidula
Cissus incisa
Cliftonia monophylla
Colutea arborescens
Cornus alba
C. californica
C. foemina
C. rugosa
Crataegus chrysocarpa
C. crus-galli
C. douglasii
C. intricata
C. macracantha
C. phaenopyrum
C. pinnatifida
C. punctata

Cornus rivularis
C. sanguinea
Cytisus scoparius

Empetrum nigrum
Euonymus bungeanus

Fraxinus caroliniana
F. cuspidata
F. lowellii

Gordonia lasianthus

Ilex cassine
I. decidua
I. vomitoria
Indigofera spp.

Juglans major
Juniperus californica
J. chinensis
J. communis
J. communis var. *saxatilis*
J. horizontalis
J. phoenicea
J. pinchoti

Kalmia polifolia

Lagerstroemia indica
Ligustrum amurense
L. ovalifolium
L. quihoui
Lonicera arizonica
X *L. bella*
L. chrysantha
L. coerulea
L. involucrata
L. korolkowii
L. morrowii
L. utahensis
L. villosa
L. xylosteum
Lycium chinense
L. halimifolium

Mahonia aquifolium
M. repens
M. swaseyi
M. trifoliolata
Malus bracteata
M. floribunda
M. glaucescens

X *Malus robusta*
X *M. soulardi*
Mitchella repens
Myrica pensylvanica

Osmaronia cerasiformis

Parthenocissus tricuspidata
Persea borbonia
Physocarpus opulifolius
Planera aquatica
Poinciana gilliesii
Populus acuminata
P. alba
P. angustifolia
P. laurifolia
P. trichocarpa
Prunus emarginata
P. gracilis
P. maacki
P. maritima
P. maximowiczii
P. mexicana
P. munsoniana
P. nigra
P. spinosa
P. subcordata
P. virens

Quercus durandii
Q. emoryi
Q. gambelii
Q. hypoleucoides
Q. ilicifolia
Q. laceyi
Q. laurifolia
Q. marilandica
Q. mohriana
Q. muehlenbergii
Q. obtusa
Q. shumardii
Q. shumardii var. *texana*
Q. undulata
Q. undulata var. *vaseyana*

Rhamnus caroliniana
R. crocea
R. crocea var. *ilicifolia*
R. lanceolata
R. tomentella
Rhus aromatica

Rhus copallina
Ribes glandulosum
R. triste
Rosa arkansana
R. canina
R. eglanteria
R. laevigata
R. multiflora
R. setigera
R. woodsii var. *fendleri*
Rubus flagellaris
R. leucodermis
R. ursinus

Salix alba
S. babylonica
S. candida
S. exigua
S. fragilis
S. laevigata
S. lasiandra
S. petiolaris
S. planifolia
Sambucus callicarpa
S. melanocarpa
S. simpsonii
Smilax hispida
S. rotundifolia
Sophora japonica
Sorbus alnifolia
S. occidentalis
Spiraea trichocarpa
Symphoricarpos oreophilus

Tamarix gallica
T. pentandra
Taxodium ascendens
Thuja orientalis
Toxicodendron vernix

Ulmus alata

Vaccinium angustifolium
V. arboreum
V. cespitosum
V. membranaceum
V. scoparium
Viburnum nudum
V. pauciflorum
Vitis amurensis
V. arizonica
V. labrusca

ORNAMENTAL PLANTING[11]

SPECIES INCLUDED

Abies amabilis
A. concolor
A. grandis
A. procera
A. venusta
Acacia spp.
Acer ginnala
A. negundo var. *californicum*
A. pensylvanicum
A. platanoides
A. pseudoplatanus
A. rubrum
A. saccharinum

Acer saccharum
Aesculus spp.
Ailanthus altissima
Albizia julibrissin
Alnus crispa
Amorpha canescens
A. fruticosa
Aralia spp.
Arbutus menziesii
Aronia spp.
Asimina triloba

Berberis spp.
Betula nigra

Betula papyrifera
B. pendula
B. pubescens
Bumelia lanuginosa

Caragana arborescens
Carpenteria californica
Catalpa spp.
Ceanothus americanus
C. impressus
C. rigidus
C. thyrsiflorus
Cedrus spp.
Celastrus scandens
Celtis spp.

Cephalanthus occidentalis
Cercis canadensis
Cercocarpus montanus
Chamaecyparis spp.
Chionanthus virginicus
Cladrastis lutea
Clematis spp.
Comptonia peregrina
Cornus spp.
Corylus americana
Cotoneaster spp.
Cowania stansburiana
Crataegus mollis
Cupressus arizonica
C. macrocarpa
C. torulosa

Footnotes on page 395.

Diospyros virginiana

Epigaea repens
Eucalyptus globulus
Euonymus spp.
Eurotia lanata

Fagus spp.
Fallugia paradoxa
Fraxinus americana
F. excelsior
F. oregona
F. pennsylvanica
F. pennsylvanica var. *lanceolata*
F. velutina
Fremontodendron spp.

Garrya fremontii
Gleditsia triacanthos
Gymnocladus dioicus

Halesia carolina
Hamamelis virginiana
Hippophae rhamnoides

Ilex aquifolium
I. opaca
I. verticillata

Juglans hindsii
J. nigra
J. rupestris
Juniperus communis var. *depressa*
J. virginiana

Kalmia latifolia
Koelreuteria paniculata

Larix decidua
L. leptolepis
L. occidentalis
Libocedrus decurrens
Ligustrum vulgare
Lindera benzoin
Liquidambar styraciflua
Liriodendron tulipifera
Lithocarpus densiflorus
Lonicera hirsuta
L. maackii
L. oblongifolia
L. tatarica
Lupinus longifolius

Maclura pomifera
Magnolia spp.
Malus spp.
Melia azedarach

Nyssa sylvatica

Ostrya virginiana
Oxydendrum arboreum

Parthenocissus spp.
Photinia arbutifolia
Picea abies
P. glauca
P. glauca var. *albertiana*
P. pungens
P. rubens
P. sitchensis
P. smithiana
Pinus aristata
P. coulteri
P. heldreichii var. *leucodermis*

Pinus mugo
P. nigra
P. peuce
P. roxburghii
P. thunbergii
P. torreyana
Platanus spp.
Populus deltoides
P. nigra
P. tacamahaca
Prunus amygdalus
P. avium
P. caroliniana
P. cerasus
P. padus
P. pensylvanica
P. pumila
P. virginiana
Pseudotsuga taxifolia
Ptelea trifoliata
Pyrus communis

Quercus agrifolia
Q. alba
Q. borealis
Q. chrysolepis
Q. coccinea
Q. nigra
Q. palustris
Q. phellos
Q. suber
Q. virginiana
Q. wislizeni

Rhamnus frangula
R. purshiana
Rhododendron spp.

Rhodotypos scandens
Rhus glabra
R. ovata
R. typhina
Ribes americanum
R. aureum
R. odoratum
Robinia spp.

Salix bebbiana
S. discolor
Sambucus spp.
Sapindus drummondi
Sequoia spp.
Shepherdia argentea
Solanum dulcamara
Sorbus spp.
Symphoricarpos spp.
Syringa vulgaris

Taxodium distichum
Taxus spp.
Thuja occidentalis
Tilia spp.
Torreya californica
Tsuga spp.

Ulmus americana
U. laevis
U. parvifolia
U. pumila
U. thomasi
Umbellularia californica

Viburnum spp.
Vitex agnus-castus
Vitis riparia

Zanthoxylum americanum

OTHER SPECIES

Abies nordmanniana
A. veitchii
Acer circinatum
A. floridanum
A. grandidentatum
A. leucoderme
A. tataricum
Aesculus pavia
Amelanchier florida
A. laevis
A. stolonifera
Ampelopsis cordata

Baccharis halimifolia
Betula fontinalis

Callicarpa americana
Calycanthus occidentalis
Chiogenes hispidula
Cliftonia monophylla
Cornus alba
C. foemina
C. rugosa
Crataegus phaenopyrum
C. succulenta
Cytisus scoparius

Empetrum nigrum
Encelia californica

Eucalyptus rudis
Euonymus bungeanus

Ginkgo biloba
Gordonia lasianthus

Hedera helix
Hibiscus syriacus

Indigofera spp.

Juniperus chinensis
J. horizontalis

Lagerstroemia indica
Ligustrum amurense
L. quihoui
Lonicera arizonica
X *L. bella*
L. chrysantha
L. coerulea
L. involucrata
L. korolkowii
L. morrowii
L. utahensis
L. villosa
L. xylosteum
Lycium chinense
L. halimifolium

Magnolia tripetala
Mahonia aquifolium
Malus bracteata
M. floribunda
M. glaucescens
X *M. robusta*
X *M. soulardi*
Mitchella repens

Nicotiana glauca

Osmaronia cerasiformis

Persea borbonia
Philadelphus lewisii
Physocarpus malvaceus
P. opulifolius
Platanus wrightii
Poinciana gilliesii
Populus alba
Potentilla fruticosa
Prunus lyonii
P. maacki
P. maritima
P. maximowiczii
P. munsoniana
P. spinosa
Pueraria thunbergiana

Quercus laurifolia

Rhamnus caroliniana
R. crocea
R. crocea var. *ilicifolia*
R. lanceolata
R. tomentella
Rhus copallina
Robinia hispida
Rosa eglanteria

Salix alba
S. babylonica
S. candida
S. fragilis
Sambucus callicarpa
S. melanocarpa
Sophora japonica
Sorbus alnifolia
Spiraea trichocarpa
Symphoricarpos oreophilus

Tamarix aphylla
T. gallica
T. pentandra
Thuja orientalis

Ulex europaeus

Vitis labrusca

---

[1] "Other species" in this list includes species not discussed in part 2 because of lack of authentic information, but which can be grown in the United States.

[2] Includes all forms of wood products—lumber, pulpwood, ties, etc.; species used occasionally for fuel only are not listed.

[3] Also used for cork.

[4] Also used for tanning.

[5] Possible source of rubber.

[6] Also used for storax.

[7] Also used for naval stores.

[8] Also used for heptane.

[9] Includes wind erosion as well as water erosion.

[10] Includes use for food, cover, or both.

[11] Includes planting for landscaping, shade, hedges, etc.

891183°—50——26

GLOSSARY[1]

Abortive. Imperfectly or not developed; hence sterile.

Achene. A small, dry, nonsplitting, one-celled, one-seeded fruit (as in *Baccharis, Cowania,* and *Eriogonum*).

After-ripening. Complex biochemical or physical changes occuring in seeds, bulbs, tubers, and fruits after harvesting, when ripe in the ordinary sense, and often necessary for subsequent germination.

Ament. *See* Catkin.

Angiosperm. A flowering plant, or member of the angiosperms, one of the two groups of seed plants and characterized by the seeds enclosed in a fruit. *See* Gymnosperm.

Apetalous. Without petals.

Aril. An exterior appendage growing out from the hilum (scar or point of attachment of the seed) and covering the seed partly or wholly (as in *Celastrus, Euonymus,* and *Taxus*).

Asexual. Reproduction without union of male and female elements, such as in vegetative propagation.

Axil. The upper angle formed by a leaf or branch with the stem.

Berry. A fleshy or pulpy, usually many-seeded fruit (as in *Arctostaphylos, Lonicera,* and *Ribes*).

Bract. A modified reduced leaf from the axil of which a flower or flower stem arises.

Broadcast sowing. Sowing seed by scattering as uniformly as possible over an area.

Browse. Twigs and shoots, with their leaves, cropped by livestock and wild animals from shrubs, trees, and woody vines.

Bur. A prickly or spiny fruit envelope (as in *Castanea* and *Fagus*).

Calyx. The outer series of the floral envelope, or perianth; the sepals as a unit.

Capsule. A dry, splitting, usually many-seeded fruit of more than one carpel (as in *Ceanothus* and *Kalmia*).

Carpel. A modified floral leaf, one or more of which form a pistil.

Catkin. A scaly-bracted spike of usually unisexual flowers (as in *Alnus* and *Betula*). Syn. Ament.

Ceresan. An organic mercury dust used for treating seeds. *See* Uspulun.

Clon (or clone). A group of plants propagated only by vegetative and asexual means, all members of which have been derived by repeated propagation from a single individual.

Coalescent. Two or more similar parts united or growing together.

Cone. The seed-bearing (female) or pollen-bearing (male) structure of gymnosperms, including conifers, consisting of an axis with many overlapping scales (as in *Abies, Larix,* and *Pinus*).

Conelet. A small, immature female cone.

Coppice. To cut back so as to produce shoots from old stumps.

Corolla. The inner series of the floral envelope; the petals as a unit.

Cotyledon. The seed leaf or primary leaf (or leaves) in the embryo, containing stored food for the developing seedling and often resembling true leaves and manufacturing food.

[1] In preparing the definitions listed here, the following references have been consulted and freely drawn upon: Gray's Lessons in Botany, by A. Gray, 1887 ed.; Manual of Cultivated Trees and Shrubs, by A. Rehder, 1940 ed.; Manual of the Trees of North America, by C. S. Sargent, 1922 ed.; Webster's International Dictionary; A Glossary of Tree Seed Terms, by H. I. Baldwin, et al., 1936 (mimeographed); Standardized Plant Names, by H. P. Kelsey and W. A. Dayton, 1942 ed.

Crustaceous. Hard and brittle in texture; crustlike.

Cutting. A severed vegetative or asexual part of a plant used in propagation.

Cutting test. Cutting or otherwise opening seed for the purpose of determining their soundness or viability.

Deciduous. Falling at maturity, not persistent.

Dehiscent Splitting open to discharge the contents, as a capsule or anther.

Dewing. To remove wings from seed.

Dioecious. Having staminate (male) and pistillate (female) flowers borne on different plants (as in *Acer, Fraxinus,* and *Ilex*).

Dormancy. Continued suspension of growth or development in the presence of external conditions favorable for germination.

Dormancy, double. A combination of seed coat and internal dormancy.

Dormancy, embryo. Dormancy due to internal conditions of the embryo.

Dormancy, internal. Dormancy due to internal conditions of the stored food or embryo.

Dormancy, seed coat. Dormancy due to a seed coat impermeable to water or oxygen.

Dormancy, secondary. Suspension of growth or development after original dormancy has been broken.

Drupe. A stone fruit, or fleshy nonsplitting fruit with a bony inner layer (endocarp) and usually one-seeded (as in *Chionanthus, Cornus,* and *Prunus*).

Ecotype. A race or subspecies naturally selected on the basis of a certain local habitat and climate.

Ellipsoid. A solid body elliptic in the longitudinal section.

Embryo. The rudimentary plant within the seed; sometimes called germ.

Embryo, immature. An embryo which is not fully developed or capable of germination when the seed is harvested.

Endocarp. The inner layer of the pericarp (for example, the bony part or stone of the fruit in *Prunus*).

Endosperm. The nutritive tissue of seeds, in which the embryo is imbedded.

Epicotyl. *See* Plumule.

Epigeous. *See* Germination.

Extraction factor. The weight of cleaned seed per given weight of fresh fruits, usually expressed in percent.

Fertilization. The union of a sperm or male nucleus and an egg or female nucleus within an ovule, resulting in the development of an embryo plant within a seed.

Flower, bisexual. A flower having both stamens (male element) and pistil (female element).

Flower, imperfect. A flower having either stamens or pistil but not both.

Flower, perfect. A flower having both stamens and pistil; bisexual.

Flower, pistillate. A flower having a pistil but no stamens; a female flower.

Flower, staminate. A flower having stamens but no pistil; male or pollen-producing flower.

Flower, unisexual. A flower having organs of a single sex, either stamens or pistil.

Follicle. A dry-one-celled fruit which opens along one side only (as in *Magnolia*).

Form. A subdivision of a botanical species or variety distinguished by some minor character and designated by the abbreviation "f." (forma) preceding the Latin name. *See* Variety.

Fruit. The ripened ovary of a flower, containing the seeds and composed of the usually thickened ovary wall (pericarp) and any other closely associated parts.

Genuineness. True to species or variety.

Germination. The development of the seedling from the seed; sprouting.

Germination, epigeous. The usual type of germination in which the cotyledons are brought above the ground.

Germination, hypogeous. A type of germination in which the cotyledons remain below ground (as in *Juglans, Quercus,* and *Torreya*).

Germination percent. Percent of a given number of seeds germinating (usually in a given time).

Germination, potential. Number of seeds germinating plus the number of sound seeds ungerminated at the close of the test, expressed in percent of the total number of seeds tested.

Germination, real. Percent of sound seed germinating.

Germination tests, indirect. Tests to determine the viability of seeds without actually germinating them, such as cutting tests, staining with chemicals, excising embryos, and the like.

Germinative capacity. The percent of seed actually germinating, regardless of time.

Germinative energy. The percent of seed germinated at the time the trend of germination reaches its peak.

Germinator. An apparatus in which seeds are tested for germination.

Germinator, standard. An apparatus for testing seed germination. It consists of a porous surface (porous clay, plaster of paris, blotting paper, or cloth) on which the seed is sown; the porous surface rests on an impervious surface, is kept continuously moist by wicks or similar arrangements, and usually is covered by an impervious transparent object such as a bell jar or inverted glass funnel. (In general use are Geneva, Jacobsen, Rodewald, Stainer, Toumey, and some other germinators.)

Glabrous. Smooth; not hairy.

Glaucous. Covered with a bluish or whitish bloom.

Globose. Spherical in form; globe-shaped.

Gymnosperm. A member of the gymnosperms, one of the two groups of seed plants and characterized by naked seeds borne usually on scales of a cone, or sometimes singly. *See* Angiosperm.

Head (flower). A dense, rounded flower cluster with the individual flowers stalkless or nearly so (as in *Baccharis* and *Cephalanthus*).

Hilum. The scar or point of attachment of the seed.

Hull. The outer covering of a fruit, especially if smooth or relatively so, as distinguished from the shell (as in *Carya* or *Juglans*).

Husk. An outside envelope of a fruit, especially if coarse, harsh, or rough (as in *Corylus*).

Hybrid. A plant resulting from a cross between two or more parents that are more or less unlike.

Hypocotyl. The stalk or stem of the embryo or seedling between the attachment of the cotyledons and the radicle or root.

Hypogeous. *See* Germination.

Indehiscent. Not splitting open. *See* Dehiscent.

Integument. *See* Seed coat; Testa.

Involucre. A whorl of bracts surrounding a flower cluster or a single flower or the fruits developed therefrom.

Kernel. The inner, often edible portion of a nut. Also, the portion of the seed which contains the embryo.

Layer. A stem or branch which takes root while still attached to the parent plant and tends eventually to become a separate individual plant.

Macerator. A special apparatus for separating seeds from fruits, either fleshy or dry.

Membranous. Thin, pliable, and rather soft; of the texture of a membrane. Syn. Membranaceous.

Micropyle. The pore of the ovule through which the pollen tube enters, and the corresponding scar in the seed.

Monoecious. Having staminate (male) and pistillate (female) flowers borne separately on the same plant (as in *Betula, Pinus,* and *Quercus*).

Nucellus. A tissue composing the central part of the young ovule.

Nucleus. The central denser structure of a cell. Also the kernel of a seed.

Nut. A nonsplitting one-seeded fruit, with hard woody shell (as in *Carya* and *Corylus*). In common usage, a hard-shelled fruit or seed containing an edible kernel.

Nutlet. A small nut or nutlike fruit or seed (as in *Alnus, Carpinus,* and *Ostraya*).

Obovoid. Inversely egg-shaped, with the broader end uppermost.

Ovoid. Oval or egg-shaped.

Ovule. The body which after fertilization becomes the seed.

Panicle. A compound raceme, or an open and branched flower cluster (as in *Aesculus*).

Pappus. The modified calyx-limb in the composite family (Compositae), forming a crown of bristles, awns, or scales (as in *Baccharis* and *Chrysothamnus*).

Parthenocarpy. The development of fruit without fertilization.

Pedicel. The stalk of each individual flower in a cluster.

Peduncle. The stalk of a flower cluster; also used for the stalk of a solitary flower.

Pendulous. Hanging, or drooping.

Pericarp. The wall of the ripened ovary, or fruit.

Persistent. Remaining attached; not falling off.

Phenology. The science of the relations between climate and periodic biological phenomena, as the flowering and fruiting of plants.

Pistil. The female or seed-producing organ of a flower, composed of stigma, style, and ovary and derived from one or more modified floral leaves (carpels).

Pistillate. Female; bearing pistils or seed-producing organs but no stamens.

Plant percent. *See* Tree percent.

Plumule. The bud or growing point of the embryo, which develops into the stem and leaves. Syn. Epicotyl.

Polyembryony. The production of more than one embryo in a single seed.

Polygamous. Having both perfect flowers and unisexual, or imperfect (i.e., staminate and pistillate) flowers.

Polygamo-dioecious. Having both perfect and unisexual flowers on the same plant, the staminate flowers and pistillate flowers on different plants.

Polygamo-monoecious. Having perfect flowers and the two kinds of unisexual flowers (staminate and pistillate) all on the same plant.

Pome. A fleshy fruit derived from several carpels and formed in part by the fleshy receptacle (as in the apple, pear, and some others of the rose family).

Prechilling. A method of pretreatment to overcome dormancy in which seed are held at near-freezing temperatures and on a moist medium.

Pregermination. *See* Stratification.

Provenience (or provenance). The geographical source or place of origin of a lot of seed.

Purity. Percent by weight of clean, whole seed, true to species, in a sample of mixed impurities and seed.

Pyriform. Pear-shaped.

Race. A group of plants that possess certain well-marked differentiating characters, and which propagate true from seed.

Race, altitudinal. A race adapted by inheritance to a certain altitudinal belt.

Race, climatic. A race adapted by heredity to specific climatic conditions.

Race, geographic. A race peculiar to a definite geographic region.

Race, local. An aggregation of individuals of a given species by inheritance adapted to a given environment more perfectly than other groups of the same species.

Race, soil. A race peculiarly adapted by inheritance to a specific kind of soil.

Raceme. A flower cluster, with stalked flowers arranged along the sides of an elongated axis (as in *Prunus serotina*).

Radicle. The portion of the embryo from which the root develops.

"Rag-doll." A cloth cushion which is kept moist and used as a crude germination apparatus.

Rugose. Wrinkled or creased.

Samara. A nonsplitting winged or key fruit (as in *Acer, Fraxinus,* and *Ulmus*).

Scarification. The wearing down by abrasion of an outer more or less impervious seed coat, to facilitate water absorption and to hasten germination.

Seed. The mature ovule, resulting from fertilization and consisting of the embryo, usually endosperm, and seed coats.

Seed certification. Guarantee of seed character and quality by an officially recognized organization, usually evidenced by a certificate including such information as genuineness of species and variety, origin, purity, soundness, and germinative capacity.

Seed coat. The covering of the seed. Syn. Integument. *See* Testa.

Seed coat, impermeable. A seed coat which is resistant to the absorption of water, oxygen, or both.

Seed extraction. Separation of seed from fruit by artificial means.

Seed origin. The locality in which the seed was collected. *See* Provenience.

Seed pretreatment. Any process, such as soaking, stratification, scarification or acid treatment, to which seeds are subjected to improve and hasten germination.

Seed setting. Formation of a seed crop on plants.

Seedling. A young plant grown from seed. In a nursery, a young tree which has not been transplanted.

Semesan. An organic mercury compound for treating seeds. *See* Uspulun.

Serotinous. Flowering or fruiting late in the season, as in autumn. Also, bearing persistent cones or fruits.

Sessile. Without a stem or stalk.

Shell. The hard layer of a nut as distinguished from its hull and kernel.

Soundness. Percent of seeds which are fully developed, or sound.

Species. A group of individuals with so many characteristics in common as to indicate a very high degree of relationship and a common descent; the unit of plant (or animal) classification and designated by a binomial Latin name.

Spike. An elongated cluster of stalkless flowers (as in *Amorpha*).

Stamen. The pollen-bearing or male organ of a flower.

Staminate. Male; bearing stamens or pollen-producing organs but no pistils.

Stigma. The part (usually the tip and mostly sticky or hairy) of the pistil which receives the pollen.

Stock, class of. Age of nursery stock denoted by two or more figures (as 2-0, 2-1, or 1-1-1), the first figure indicating years in the seedbed and the succeeding figure or figures the years in the transplant bed or beds.

Stolon. A trailing or reclining stem above ground, which strikes root where it touches the soil, there sending up new shoots which later may become separate plants.

Stoloniferous. Bearing stolons.

Stone. A hard, bony fruit part containing the seed (as in *Cornus* and *Prunus*).

Strain. A group of cultivated plants differing from the race to which it belongs by no apparent morphological characters, but by some enhanced, or improved physiological tendency propagated from seed.

Stratification. The operation or method of burying seeds, often in alternate layers, in a moist medium, such as sand or peat, to overcome dormancy.

Stratification, cold. Stratification at low temperatures, generally just above freezing, usually to overcome embryo dormancy.

Stratification, warm. Stratification at higher temperatures, usually about room temperature, chiefly to overcome seed-coat dormancy.

Striate. Marked with fine, longitudinal lines or ridges.

Strobile. A multiple fruit in the form of a small cone or head (as in *Alnus* and *Betula*). Also, a cone.

Style. The stalk of the pistil between the ovary and stigma.

Sucker. A branch or shoot from a creeping underground stem or root which ascends above ground and tends eventually to become a separate individual plant.

Suture. A line of splitting.

Syncarp. A fleshy multiple fruit, consisting of the enlarged ovaries of several flowers more or less united (as in *Morus*).

Temperature, room. In laboratory work, 68° to 70° F.

The outer, and usually the harder, seed coat. *See* Seed coat.

Tree percent. The percent of viable seed sown which produce usable (ordinarily 1–0) seedlings.

Umbel. An umbrellalike flower cluster with the pedicels arising from the same point (as in *Aralia nudicaulis*).

Uspulun. An organic mercury compound (hydroxymercurichlorophenol sulfate and similar compounds) used for disinfecting seeds and seedbeds. Ceresan and Semesan are somewhat similar preparations.

Variety. A subdivision of a botanical species having some characters different from the typical and designated by the abbreviation "var." (varietas) preceding the Latin name. *See* Form.

Veneer grafting. Grafting by beveling the scion and fastening it to a groove on the stock.

Verticillate. Arranged in a whorl.

Viability. The potential capacity to germinate.

Whorl. An arrangement of three or more organs (such as leaves or branches) in a circle around the axis.

Wing. A membranous or thin and dry expansion or appendage of a seed or fruit.

Winnow. To separate and drive off chaff and debris as by fanning.

INDEX OF PLANT NAMES USED IN PART 2

This index includes all genera, species, and varieties mentioned in part 2. Accepted plant names, both English and Latin, appear in heavy type. Others are shown in ordinary type.

891183°—50——27

— U. S. GOVERNMENT PRINTING OFFICE: 1950

Printed in the USA
CPSIA information can be obtained
at www.ICGtesting.com
LVHW080908210224
772391LV00003B/14